Dr. Dennis S. Hill
KT-551-467

Grain Protectants

INKATA PRESS PTY LTD
BOOK AND JOURNAL PUBLISHERS
Sole distributors of ACIAR publications

13/170 FORSTER ROAD, MOUNT WAVERLEY, VICTORIA 3149, AUSTRALIA
TEL: (03) 562 9022 TELEX: 30625 ATT ME 3733 FAX: (03) 562 8620

20.3.1990

With the compliments of

Australian Centre for International Agricultural Research

The Australian Centre for International Agricultural Research (ACIAR) was established in June 1982 by an Act of the Australian Parliament. Its mandate is to help identify agricultural problems in developing countries and to commission collaborative research between Australian and developing country researchers in fields where Australia has a special research competence.

Where trade names are used this does not constitute endorsement of nor discrimination against any product by the Centre.

ACIAR MONOGRAPH SERIES

This peer-reviewed series contains the results of original research supported by ACIAR, or material deemed relevant to ACIAR's research objectives. The series is distributed internationally, with an emphasis on the Third World.

Snelson, J.T. 1987. Grain protectants. ACIAR Monograph No. 3, x + 448 p.

ISBN 0949511 34 X

Typeset and laid out by Canberra Publishing and Printing Co., Fyshwick, ACT.
Printed by Ruskin Press, Melbourne
Cover design: Ingrid Slamer

Grain Protectants

J.T. Snelson

Published in collaboration with the
Department of Primary Industry
by the
Australian Centre for International Agricultural Research
Canberra 1987

About the author

J.T. Snelson's career in agricultural chemistry has spanned the 40 years during which the various grain protectants have progressed from the status of experimental compounds showing some promise to a suite of extremely effective chemicals in widespread use.

Mr Snelson has been a key player in international activity to develop a rigorous code of practice for the safe and effective use of grain protectants, stimulated by a commitment to help in reducing the devastation that insect pests can cause to stored foodstuffs, and by the serious impact that these losses have on people in the developing countries of the world.

Mr Snelson was, from 1969 to 1986, a member of both the FAO Panel of Experts on Pesticide Residues in Food and the Environment and the FAO/WHO Joint Meeting of Experts on Pesticide Residues, and has held office on numerous other national and international committees and working parties concerned with maximising the benefits of pesticide use. He was for many years chairman of the Australian Wheat Board Working Party on Grain Protectants and, from 1967 to 1983, was Australian delegate to the UN Codex Committee on Pesticide Residues.

Although Mr Snelson retired as Pesticides Coordinator in the Australian Department of Primary Industry in 1986, he is still actively involved in pesticide matters as a consultant. In recognition of his service to agriculture, he has been appointed a Member of the Order of Australia, one of the highest honours that can be bestowed on a citizen of Australia.

Contents

FOREWORD

The Department of Primary Industry and the Australian Centre for International Agricultural Research (ACIAR) are delighted to be associated in the publication of Mr J.T. Snelson's monograph 'Grain Protectants'.

The Department's responsibility for administering the Export (Grain) Regulations has brought it into a close working relationship with those people and organisations responsible for the production, storage, transport, inspection and marketing of all types of grain, and thus into an involvement with the control of stored-product pests. The importance of the wealth of scientific information on effective, efficient and acceptable procedures for controlling stored-product pests is therefore well understood.

The problems involved in providing Australia's international trading partners with high quality grain, free from insect pests, their damage, excreta and fragments have received high priority at all levels of government, industry and private enterprise. Innovative measures that are now regarded as mileposts in grain storage technology gradually evolved until they reduced these problems to manageable proportions. There is no doubt that the dedication, determination and co-operation of a relative few key personnel in the grain trade, grain storage industry, research organisations, government and chemical industry have been mainly responsible for the high levels of freedom from stored-product pests for which Australian grain is now recognised.

During the period 1967–1986 when Jack Snelson served as Commonwealth Pesticides Co-ordinator in the Department of Primary Industry he was initimately involved in all aspects of the efforts to eliminate stored-product pests from commercial grain handled in and exported from Australia. For 13 years he served as Chairman of the Australian Wheat Board Working Party on Grain Protectants where he made a substantial contribution in co-operation with all who served to ensure the success of its research and development.

Mr Snelson was a member of the FAO Panel of Experts on Pesticide Residues in Food and the Environment and the Joint FAO/WHO Meeting of Experts on Pesticide Residues for 17 years. During this period he was responsible for evaluating much of the scientific data on residues of grain protectant insecticides and, with his colleagues, for establishing recommendations for international maximum residue limits for these insecticides in food. He served in a similar capacity on the corresponding Australian body, the Pesticides and Agricultural Chemicals Committee of the National Health and Medical Research Council.

As Australian delegate to the Codex Committee on Pesticide Residues between 1967 and 1985, Mr Snelson found that many officials concerned with food inspection and food laws in both industrialised and developing countries had not had the opportunity to consider the need to use grain protectant insecticides to protect world food supplies or to gain an insight into the nutritional, social and economic consequences of stored-product pests. He recognised the difficulty which administrators and scientists faced because the extensive scientific literature was fragmented, and realised that few had had the chance of considering the many data relating to the biological activity of grain protectants, their properties and fate, and the safety of residues remaining in prepared foods.

In 1978 he began the task of reviewing and summarising the literature in his spare time. However, the volume of information to be handled and work pressure saw this activity shelved until just before his retirement in February 1986. Even so, it has taken almost 12 months of meticulous work to collect, collate and review the more than 2300 publications which form the basis of this monograph.

The Department was pleased to encourage and assist by providing word processing and library facilities, and it is a pleasure to see how these have been put to such good effect.

ACIAR's decision to publish 'Grain Protectants' is in accord with the Australian Government's support for the greater and more effective utilisation of Australia's scientific research and technology capability to assist the economic and social advancement of our Asian-Pacific neighbours.

In important areas of science and technology, particularly in aspects of agriculture, Australia has developed unique capabilities. In particular, we have succeeded in applying many of the principles developed in relation to temperate agriculture to the solution of problems of the tropics.

The subject matter of this monograph is a good example of this. It should enable developing countries to apply the scientific information and practical experience gained in Australia and elsewhere to resolve many of their own problems in the confident knowledge that the safety and acceptability of the control measures have been carefully and independently evaluated and approved. It also draws attention to the fine work of scientists in many countries who have devoted their talents to finding ways of safely eliminating losses and difficulties due to stored-product pests and provides encouragement to apply modern scientific discoveries in solving age-old problems.

The book is published as part of the activities of the ACIAR Grain Storage Research Program, which aims to enhance the capacity of developing countries to identify and overcome constraints to the safe storage of grain, thereby decreasing losses through spoilage. Ensuring correct use of suitable grain protectants is central to this aim.

We are sure that this monograph will serve the needs of scientists, grain storage technologists, the grain trade and administrators concerned with grain handling, food standards and consumer safety now and for many years to come.

It should also bring about a better understanding of the problems which face peoples living in those regions where temperatures and/or humidities are high and where insect pests present serious problems to those producing, storing and handling grain. The realisation that these problems can be brought under control by simple measures which have proven effective, convenient, economical and safe should bring forward the day when the world no longer has to produce food merely to replace what has been damaged or devoured by hungry pests.

If this can happen the work of the author and all those scientists whose labours made this monograph possible will not have been in vain.

Geoff L. Miller
Secretary
Department of Primary Industry
Canberra

J.R. McWilliam
Director
ACIAR
Canberra
March 1987

Preface

Grains, such as wheat, rice, corn, millet etc., and legumes, such as beans, lentils and peas, form a large part of the diet of the world's population. These products are stored as dry seeds and form the only real reserve of food. However, all of them are subject to attack by a variety of insects that cause great damage and loss of nutritious foods, and thus give rise to one of the causes of malnutrition in many lands. **Subrahmanyan (1962)** indicated that the world's total food could be increased by 25–30% if we could avoid post-harvest losses of our food. **Bakal (1963),** in his *Mathematics of Hunger,* estimated the annual food losses due to rats, insects and fungi at 33 million tonnes, enough seed foods to feed the population of the United States of America for 1 year. Why does the world continue to tolerate the feeding of enormous populations of insects at the expense of hungry and starving people?

The Food and Agriculture Organization of the United Nations (FAO) pointed out in 1973 that world food supplies no longer provided an adequate buffer against crop failure and starvation; food supplies for millions of people depend almost entirely on each year's harvest. At least 10% of harvested food crops are destroyed by pests in storage, and losses of 30% are currently common in large areas of the world **(Hall 1970)**, especially in some of the tropical and sub-tropical areas where the need for increased food is greatest. In some of these regions, a certain amount of insect infestation in grains and other food products is accepted as normal but in many developed countries strict control of stored-product insects is dictated not only by economic necessity but also by strict government standards concerning permissible contamination.

The inadequate supply of food in developing countries has long been recognised as one of the major problems that needs attention because a population that is not well fed is prone to deficiency diseases, more susceptible to infectious diseases and does not have the energy to work hard. Both multi-lateral and bi-lateral organisations have recognised this problem and have devoted a large proportion of their aid efforts to increasing the production of food. However, increasing the production of food is not the real goal of these activities. The real goal is to put more food regularly into the mouths of the people and this requires not only the production of more food but also moving that food through the delivery system all the way to the point of consumption. Most attempts to increase the world food supply have failed to recognise that the food must be delivered to the ultimate consumer through the post-harvest system without significant loss. **Bourne (1977)** in his excellent review *Post-Harvest Food Losses — the Neglected Dimension in Increasing the World Food Supply* drew attention to the critical position and proposed a series of practical measures for bringing about the necessary improvement.

The attention of the world was drawn to the problem of post-harvest food losses by the US Secretary of State, Dr Henry Kissinger, at the World Food Conference that was held in November 1974 in Rome. In his keynote address at the opening session of the World Food Conference, Secretary Kissinger said:

> Another major priority must be to reduce losses from inadequate storage, transport and pest control. Tragically, as much as 10% of the country's food production is often lost after harvesting because of pests that attack grain in sub-standard storage facilities. Better methods of safe storage must be taught and spread as widely as possible. Existing pesticides must be made more generally available. Many of these techniques are simple and inexpensive; investment in these areas could have a rapid and substantial impact on the world's food supply.

The attention of the political world was again drawn to the problem of post-harvest food losses by Dr Kissinger in an address presented at the Seventh Special Session of the UN General Assembly on 1 September 1975 in New York. In this address, Dr Kissinger said:

> Another priority in the poorest countries must be to reduce the tragic waste of losses after harvest from inadequate storage, transportation and pest control. There are often simple and inexpensive techniques to resolve these problems. Investment in such areas as better storage and pesticides can have a rapid and substantial impact on the world's food supply; indeed, the savings could match the total of the food aid being given around the world. Therefore, we urge that the Food and Agriculture Organization, in conjunction with the UN Development Program and the World Bank, set a goal of cutting in half these post-harvest losses by 1985, and develop a comprehensive program to this end.

This challenge from Dr Kissinger was taken up by the UN General Assembly in that Session and on 19 September 1975 one of the resolutions of the UN General Assembly was:

> The further reduction of the post-harvest food losses in developing countries should be undertaken as a matter of priority, with a view to reaching at least a

50% reduction by 1985. All countries and competent international organisations should co-operate financially and technically in the effort to achieve this objective.

This resolution has drawn the attention of the highest levels of government around the world to the problem of post-harvest food losses.

A recommendation to use or not to use pesticides at the small farm level is an important one, as it affects many people. It is estimated that over 100 million farms are less than 5 ha in size and of these over 50 million are less than 1 ha (the President of the World Bank quoted in **Brader 1979**).

As long as the farming practices and varieties grown remain 'traditional', food losses on the small farm are probably of little importance. There is at present ample evidence that storage losses may increase considerably when traditional farming is changed, so a modernisation of agricultural production makes changes in storage practices unavoidable. The financial means of each individual are very limited. Sophisticated storage methods are mostly out of reach. With insecticides, storage losses can be prevented at relatively low cost. The distribution of insecticides and the organisation of the extension of their use poses, however, seemingly insoluble problems.

De Las Casas (1984) described FAO's Program on Prevention of Food Losses which began with a request of the Director-General of FAO in 1975 for a survey of the position in member countries. Responses were received from 51 countries. Most of the reports concerned losses of durables and the need to reduce them, but surprisingly enough, despite the high losses reported from many countries, there was no indication of steps being taken to bring about their reduction. Insects were clearly identified as the most important cause of loss in all types of storage, followed by fungi and rodents. The report regarding the various countries reflected a general unawareness of the cause/effect relationship of losses. The magnitude of such losses was generally not well known.

A call for international action was made in 1975 at the 7th Special Session of the United Nations General Assembly, where a resolution was passed calling upon member states to reduce post-harvest food losses by 50% by 1985. It was clearly stated that countries should commit themselves fully to food loss reduction by establishing a national policy and implementing an effective plan of action. FAO has been the leader within the UN system. Its Action Program for the Prevention of Food Losses (PFL) was approved by the 19th Session of the FAO Conference, in November 1977. This action program, as formulated in 1976, identified three major constraints to post-harvest loss prevention in the developing world:

1. lack of information concerning the magnitude of the losses, their nature, causes and the most effective techniques for reducing or preventing them;
2. lack of infrastructure for the implementation of loss prevention measures;
3. lack of investment in food loss prevention.

In formulating the FAO Action Program for the Prevention of Food Losses certain priorities have been set in order to achieve a significant impact within a relatively short time. Thus, in its initial phase, the program has concentrated on reducing post-harvest losses in food grains, roots and tubers, since these foods constitute the staple foods of developing countries. The basic purpose of the program is to assist the countries in planning and implementing national food loss reduction programs. Direct action projects have been implemented to improve storage structures and design, construct and manage pilot warehouses, provide small-scale grain dryers, improve processing facilities, organise activities on quality control and pest management, improve rodent control and carry out training at all levels and in all aspects of the post-production system.

PFL direct action projects carry out activities in five main areas:

1. food loss assessment surveys, which are considered as the basis for action programs for food loss reduction;
2. provision of practical assistance to governments to combat losses at various stages in the post-harvest food system;
3. training and building up national capabilities;
4. establishing national focal points for food loss reduction programs;
5. strengthening research, development and information programs.

At the end of December 1982, the number of projects had reached 108, with a total allocation of $US34.2 million. Of these projects, 88 were either completed or had reached the state of generating some outputs. The regional distribution of PFL activities shows 30% in Africa, 29% in Asia and the Pacific, 19% in Latin America and the Caribbean, and 14% in the Near East and North Africa.

PFL activities are restricted to staple foods: cereals (rice, maize, wheat, sorghum and millet), pulses and roots and tubers. Pest control activities were carried out by 46 of the 88 completed projects. In most of them, pest control activities have proved to be acceptable, but it has been concluded that this particular component needs to be reviewed. A review of the pest management practices and an evaluation of their adequacy, an assessment of the validity of experimental trials and interpretation of resulting data are now being carried out. The aim is to obtain guidelines for more effective preventative sanitation practices, more economical and safer pest management action and more efficient training instruction, which will be used by the PFL projects.

In the light of the experience gained, many remedial measures have already been incorporated in subsequent projects. For example, more attention has been devoted to the socio-economic analysis of post-harvest practices, pest management and quality control, the right approach to technology transfer, as well as the economic viability and social acceptance of the proposed new technology. Experience to date indicates the need to broaden the scope of overall activities, to include expertise on farm management and economics, food quality, home economics, training and extension.

Acknowledgments

The realisation that there was such a wealth of valuable information about simple, safe, effective and economical measures for preventing the loss of enormous quantities of valuable foodstuffs and that this was not known to or accessible by those needing to put such information to good use prompted me to attempt to make it more readily available. I commenced the task for presentation at the Second International Working Conference on Stored-Product Entomology held in Ibadan, Nigeria in September 1978. I did manage to prepare about 300 pages of manuscript at that time but this did not cover more than a small fraction of the available literature and I therefore resigned myself to a much more extensive exercise.

The large volume of information becoming available since then and the need to explore many issues in greater depth has greatly extended the task and I am sorry that my many friends who have been awaiting this monograph for several years have had to get along without it. However, in spite of the long development period, I trust that the final outcome will be even more useful than it would have been a few years earlier.

I am grateful to my wife Eva for the sacrifice she made so that I could continue the work rather than attend to urgent tasks around the home and for our family. I appreciate the encouragement of colleagues, friends and professional associates in Australia and overseas. The support of the Executive of the Department of Primary Industry in allowing me to devote myself full time to the work for three months before my retirement is gratefully acknowledged.

Most especially I want to acknowledge the dedication of several ladies of the Word Processing Centre of the Department of Primary Industry who skilfully transcribed the whole of the text from my taped dictation. Without such help there would have been no way for me to complete the task. Their patience and skill has made a wonderful contribution.

Dr Jim Desmarchelier reviewed the manuscript and I acknowledge his comments and encouragement. The scientists in industry who reviewed the chapters on the insecticides developed by their companies have helped to keep the facts and my interpretation in line with their considerable experience. I appreciate their help and also permission to quote many reports which have not been published. I trust that they will feel that the final presentation does justice to the extensive database available on these insecticides.

To my long-standing associates on the Codex Committee on Pesticide Residues, the FAO Panel of Experts on Pesticide Residues in Food and the Environment and the Joint FAO/WHO Meeting on Pesticide Residues I extend thanks for sharing knowledge, experience and concerns. I trust you will find the answers you need to all your questions on grain protectant insecticides.

And to those who are devoting their skills, knowledge and experience to the task of bringing the benefits of scientific development to improve the health, nutrition and economic well-being of those whose agriculture and industry is still in great need of development I express the hope that this monograph will save you time, effort and money and that you will be able to adapt much of the experience that it represents to resolve problems of your part of the world.

Finally, I wish to acknowledge the support of ACIAR, which, by agreeing to publish this monograph, has brought the enterprise to fruition.

J.T. Snelson

1. Importance of Stored-Product Pests

That insects are a major cause of loss in stored grains and seeds, as well as in many other kinds of stored food products, should be too well recognised to need emphasis here. They not only consume these materials but also contaminate them with insect fragments, feces, webbing, ill-smelling metabolic products and with a variety of microflora; they therefore constitute a major sanitation and quality control problem.

Many of these most important insects and mites found in grain throughout the world have originated from the Old World. Ships, trucks, trains and planes have facilitated their distribution throughout the grain growing and grain handling regions of the world. Once established in a granary, insects can damage stored grain by feeding on it, lowering germination, spoiling its appearance, lowering its milling quality, producing off-odours, causing it to heat and by creating conditions favourable to the growth of moulds.

The insects and mites that attack stored grain have features which distinguish them from those found on growing crops. Most are small, light-avoiding pests capable of hiding in cracks and crevices, and of producing several generations a year. They have become specialised for feeding on dried animal and vegetable matter. Some may survive on foods that contain as little as 1% water, although the species that live on grain usually require not less than 8% water. Some can also tolerate extremes of temperature; indeed, these are the insect pests with which we are chiefly concerned since those that cannot survive the conditions under which grain is stored are not likely to give us trouble **(Watters 1959b).**

Before crops were cultivated by man, the weevils and Angoumois grain moth (*Sitotroga cerealella*) occurred naturally in seeds on wild plants or in seeds gathered by rodents, ants and other seed harvesting animals **(Linsley 1944)**. Certain species were fungus feeders or scavengers; several were wood borers or lived under loose, dead bark; others inhabitated nests of social insects and birds. When man began to store grain, these insects readily moved from their original native habitat to man-made stores. These pests have been carried by commerce to all parts of the world. In the days of the sailing vessels, grain and other dry foods swarmed with insects at the end of long voyages.

Today, several hundred species of insects and mites are associated in one way or another with stored grains and their products. Fortunately, only a few cause serious damage to seeds or cereal products in good condition. Some feed on the fungi growing on spoiled grain or on other dried vegetable materials. Others are predators or parasites that attack the true grain pests. Environmental conditions in various parts of the world are not equally favourable for the development of all insect species, and those species that are injurious in some areas barely exist in others.

Cotton and Good (1937) listed the species found in stored grain and processed cereal products in four categories.

1. *Major pests* comprise those species responsible for most of the insect damage to stored grain and cereal products. They are particularly well adapted to life in the stringent environment imposed by a bin of grain.
2. *Minor pests* include an appreciably larger group that may become damaging locally and occasionally may approach the status of major pests. Frequently, large populations develop in grain or in cereal products going out of condition because of high moisture and poor sanitation. Certain species *Ptinus villiger* and *Ahasverus advena*, are indicators of high-moisture grain; when they occur in abundance in a grain mass, spoilage from moulds probably has occurred.
3. *Incidental pests* are those insects that stray into an unscreened or open-doored processing plant or into open-doored box cars being loaded or standing in the railroad yards. They include houseflies, roaches, moths, or other insects that might have been attracted to lights, odours or shelter — in fact any insect that happens to alight on or crawl into grain or grain products. Usually these insects are ignored by grain handlers and food processors, but when a moth or beetle accidentally gets into a package of cake mix or breakfast food, it suddenly becomes a major pest to the householder who opens the package.
4. *Parasites and predators* of grain-infesting insects frequently are found in bins of infested grain, where they are as unwanted by grain handlers as are the major pests.

Lists of stored-product pests, categorised by **Cotton and Wilbur (1974)**, are given in **Tables 1.1 and 1.2**.

Bauer (1984a) has provided useful data on important behavioural aspects of the more important stored-product pests in the form of a table which makes it easy to compare vital features of different species and thus deduce why some are much more pestiferous than others.

Singh *et al.* (1978) have edited a universal review of

Table 1.1. Alphabetical list of the major insect pests of stored grain **(Cotton and Wilbur 1974)**

Scientific Name	Common Name	Family
Acarus siro L.	Grain mite	Acaridae
Anagasta kuehniella (Zeller)	Mediterranean flour moth	Phycitidae
Cadra cautella (Walker)	Almond moth	Phycitidae
Cryptolestes ferrugineus (Stephens)	Rusty grain beetle	Cucujidae
Cryptolestes pusillus (Schönherr)	Flat grain beetle	Cucujidae
Cryptolestes turcicus (Grouv.)	Flour-mill beetle	Cucujidae
Ephestia elutella (Hübner)	Tobacco moth	Phycitidae
Oryzaephilus surinamensis (L.)	Saw-toothed grain beetle	Cucujidae
Oryzaephilus mercator (Fauv.)	Merchant grain beetle	Cucujidae
Plodia interpunctella (Hübner)	Indian-meal moth	Phycitidae
Rhyzopertha dominica (F.)	Lesser grain borer	Bostrichidae
Sitophilus granarius (L.)	Granary weevil	Curculionidae
Sitophilus oryzae (L.)	Rice weevil	Curculionidae
Sitophilus zeamais Motschulsky	Maize weevil	Curculionidae
Sitotroga cerealella (Olivier)	Angoumois grain moth	Gelechiidae
Tenebroides mauritanicus (L.)	Cadelle	Ostomatidae
Tribolium castaneum (Herbst)	Red flour beetle	Tenebrionidae
Tribolium confusum Duval	Confused flour beetle	Tenebrionidae
Trogoderma granarium Everts	Khapra beetle	Dermestidae

grain legume entomology involving contributions from over 20 individual countries concerning grain legume pest complexes.

Newly harvested, uninfested grain in farm storage may become infested from many sources, including hold-over infested grain, cracks, crevices, cadelle tunnels and

Table 1.2. Alphabetical list of the minor pests most frequently encountered in stored grain **(Cotton and Wilbur 1974)**

Scientific Name	Common Name	Family
Ahasverus advena (Waltl.)	Foreign grain beetle	Cucujidae
Alphitobius diaperinus (Panzer)	Lesser mealworm	Tenebrionidae
Araecerus fasciculatus (DeGeer)	Coffee-bean weevil	Platystomidae
Attagenus piceus (Olivier)	Black carpet beetle	Dermestidae
Carpophilus dimidiatus (F.)	Corn sap beetle	Nitidulidae
Carpophilus hemipterus (L.)	Dried fruit beetle	Nitidulidae
Caulophilus oryzae (Gyllenhal)	Broad-nosed grain beetle	Curculionidae
Corcyra cephalonica (Staint.)	Rice moth	Galleriidae
Cynaeus augustus (LeConte)	Larger black flour beetle	Tenebrionidae
Gnathocerus cornutus (F.)	Broad-horned flour beetle	Tenebrionidae
Lasioderma serricorne (F.)	Cigarette beetle	Anobiidae
Latheticus oryzae (Waterhouse)	Long-headed flour beetle	Tenebrionidae
Liposcelis spp.	Psocids	Psocoptera
Palorus ratzeburgi (Wissmann)	Small-eyed flour beetle	Tenebrionidae
Palorus subdepressus (Wollaston)	Depressed flour beetle	Tenebrionidae
Ptinus claviceps (Panzer)	Brown spider beetle	Ptinidae
Ptinus villiger (Reitter)	Hairy spider beetle	Ptinidae
Prostephanus truncatus (Horn)	Larger grain borer	Bostrichidae
Stegobium paniceum (L.)	Drugstore beetle	Anobiidae
Tenebrio molitor (L.)	Yellow mealworm	Tenebrionidae
Tenebrio obscurus (F.)	Dark mealworm	Tenebrionidae
Tribolium audax (Halstead)	Black flour beetle	Tenebrionidae
Trogoderma spp.	Grain-feeding dermestids	Dermestidae
Typhaea stercorea (L.)	Hairy fungus beetle	Mycetophagidae

double-wall construction in bins or other parts of the granary, whether empty or occupied; accumulation of waste or spilled grain in or under the granary or machinery and implements, or in empty grain and feed sacks; imported feeds or seeds from infested sources, especially elevator sweepings; migration by flight from nearby infested sources; and, in many regions, field-infested grain.

Sinclair and Haddrell (1985) conducted 3 surveys to determine the extent of migration in a grain farming area by 4 stored-products beetles: *Tribolium castaneum* and *Rhyzopertha dominica* were found to be the predominant migrants being active in the field from early spring to late autumn. There was no clear relationship of catch with nearby infestations, except over short distances. Average density of beetles during flight peaks was estimated to be about 52 per 100 m^3. These insects represent a serious potential for reinfestation of stored grain.

In commercial and government storage, uninfested grain becomes contaminated from infested grain or wastes on the premises unless exceptionally favourable sanitation procedures have been followed. Many of the older elevators, especially those of the wooden crib type, are difficult to disinfest. Much commercial grain received from farm storage is infested. Grains shipped in wooden box cars is almost certain to become infested during shipment. Insects capable of flight migrate in all directions from feed mills that process infested grain.

Bauer (1984) edited a valuable monograph on insect management for food storage and processing which starts off by demonstrating that losses due to insects in food storage, transportation and processing in the United States run 'in the low billions of dollars' annually. It quotes the report of the Council of Environmental Quality on Integrated Pest Management by Dale R. Bottrill of December 1979 as an authority for the statement that the post-harvest loss due to insects in the US is 6% of the value of all foods marketed.

Insects are attracted to damp grain. If leaks occur and portions of a bin of grain become damp, insects will be found there in greater abundance than in dry areas. Surface grain that is cooler than the bulk of grain may become damp as a result of translocation of water from the interior of the grain mass. Whenever a bulk of grain has parts with different temperatures, moisture moves from the warmest to the cooler parts. When heating by insects occurs beneath the surface grain and outside temperatures are low, the temperature gradient is very steep. Water movement may be so excessive that considerable rotting of the surface grain will result. Damp surface grain may attract scavenger and fungus-feeding insects and their presence serves as a certain indicator of high-moisture grain.

In most developing countries 70–80% of all the staple produce never leaves the farm on which it was grown and many factors affect farm-stored produce. Of these, damage due to insects, rodents and fungi is probably the main cause of loss in quality and quantity of the food stored.

Insect infestations result in contamination by entire insects or parts thereof. Infestations may include eggs, egg shells and egg cases, larvae and cast-larval exoskeletons, pupae and pupal cases and cocoons, and mature insects. Internal-feeding weevils and *Sitotroga* larvae deposit their excreta or fecal pellets inside the kernels; *Rhyzopertha* push excreta along with some starch particles out of the kernel through the entrance hole and leave excreta/starch accumulations as dockage in the infested area **(Cotton and Wilbur 1974)**. In addition to the flour consumed, the flour beetles secrete a pungent, irritating liquid consisting of ethylquinone, toluquinone and methylquinone from certain abdominal and thoracic odoriferous glands. Contamination from large populations of flour beetles turns the flour pink.

Freeman and Turtle (1947) stated that flour infested with mites has a particularly objectionable smell. When such flour is used for bread making, the bread has a sour taste, poor colour and may not rise adequately.

Larvae of *Plodia interpunctella* and certain other moths spin silken webs over their foods, so that their presence in cereal products ruins much more than the amount eaten. Frequently a thick carpet of webbing may cover the surface of stored grain infested by such moths. Older millers recall that, when *Plodia* infested their mills, the webbing would tie moving mill parts so tightly during the weekend shutdown that it had to be removed to get the machinery into operation again **(Cotton and Wilbur 1974)**.

When insects feed on grain, the breakdown of the grain starches during insect metabolism results in heat, moisture and energy as end products. Lowering of the grade of grain as a result of damaged kernels and of off-colour and off-odours may cause a substantial price decrease **(Cotton and Wilbur 1974)**.

Storage fungi are recognised as a major cause of spoilage in stored grain and seeds; insects and mites are frequently responsible for their distribution and development. Grain-infesting insects carry fungus spores into the grain and may increase the moisture content of the grain sufficiently so that the spores can develop. **Christensen and Kaufmann (1969)** summed up the situation by stating that at least some of the common grain-infesting insects regularly carry into the grain they infest a large load of inoculum of storage fungi, and, as they develop in the grain, provide conditions favourable to the development of these fungi. These authors stated that what appears to be an insect problem may be an insect-plus-storage-mould problem.

Apart from losses caused by moulds, storage fungi produce an unknown number of toxic chemicals, some of which are carcinogenic, such as aflatoxins and okratoxin and many of which are suspected as causing serious diseases in livestock.

Of great significance is the possible distribution of such

harmful bacteria as *Salmonella, Streptococcus*, and *Escherichia coli* by grain-infesting insects. **Husted *et al.* (1969)**, found that *Sitophilus oryzae* retained *Salmonella montevideo* internally and externally after being in contaminated wheat for 7 days, and for at least 5 weeks after being in contaminated wheat for 14 and 21 days.

The role of insects in promoting, spreading and intensifying the development of mycotoxins in stored grain has not been adequately researched. **Ragunathan *et al.* (1974)** showed that there was a close association between storage fungi and *Sitophilus oryzae*. The fungi were mainly those responsible for producing mycotoxins. The incidence of infection of the insects ranged from 20% to 100%, the fungi being carried in the alimentary canal of the insect along with the food and excreta. Weevils collected in the field at the time of corn harvest carried *Aspergillus flavus*.

Senappa *et al.* (1979) reported that dried Indian peppers (*Capsicum annuum*) containing *Aspergillus flavus* and aflatoxin had been subjected to insect infestation before storage and the small number of infested pods had become completely filled with aspergilli. These pods showed no surface growth of mould, but during storage the incidence of aspergillus-loaded pods increased, depending on insect infestation and relative humidity.

Phillips and Burkholder (1984) have reviewed the literature on health hazards of insects and mites in food. The fact that stored-product insects can and do cause illness in humans is well known to entomologists but less so to the food industry. These illnesses range from mechanical irritations of skin, eyes and upper respiratory tract due to the abundance of insect fragments following on even modest levels of infestation to serious allergenic reactions induced by the foreign protein of the anatomical structures of pests and the poison glands of insect bodies.

Tapeworm diseases, including dwarf tapeworm and rat tapeworm, are transmitted from rodents to man by cockroaches and meal worms as well as by fleas **(Scott 1962)**. Such enteric bacterial and protozoal diseases as typhoid, diarrhoea and dysentery are transmitted mechanically from man to man by cockroaches and other stored-food pests. Various intestinal infestations such as intestinal acariasis, intestinal myiasis and canthariasis result from the ingestion of live stored-food mites, from stored-food flies, and from yellow meal worms and other stored-food beetles. A severe dermatitis known as grain itch, grocer's itch, vanillism, or other names, is caused by bites of stored-food mites. In addition, some persons are allergic to mill dust containing fragments and fecal pellets of stored-food insects **(Cotton and Wilbur 1974)**.

Insects render unfit for human consumption much more grain and grain products than they eat, because every infestation contaminates its host product with insect fragments and excreta. Excreta from flour beetles and most other major insect pests of cereal products are deposited directly into the product and cannot be removed.

Most persons have a natural revulsion to all forms of filth in their food. The thought that they may be eating insect fragments and excreta affects the aesthetic sensitivities of most people, but when such filth is actually observed it may have serious effects. The public wants to believe that the cereal products it is using are as free of insect filth as possible. If there were to be a statement on a package of flour to the affect that 'this flour contains fewer than 10 insect fragments per 100 g', it would have an adverse effect on the consumer, even though such a low count represents a major achievement over the past 20 years.

Flour infested with insects is generally rejected for aesthetic and assumed health reasons, without consideration of any undesirable effects, which may be produced by secretions and excretions of the insects, and which may affect the baking qualities and taste of the product. **Smith *et al.* (1971)** reviewed the work of many authors and showed that cereals infested with *Tribolium* spp. give rise to persistent and disagreeable odours in bread. They have pointed out that the presence of *Tribolium* spp. in flour affected the odour, taste and colour of bread, considerably reducing the volume of the loaf and destroying the texture.

Smith *et al.* (1971) referred to a number of studies of the nature and properties of these insect secretions, which are predominantly quinones. Numerous authors have pointed out that these quinone secretions of *Tribolium* spp. may be important carcinogens since they produce carcinoma in mice. They also showed that whilst bread prepared from flour infested with *Tenebrio, Trogoderma* and *Oryzaephilus* spp. showed no changes from the control bread with regard to most quality criteria, the presence of a characteristic, low intensity, taste and/or odour was identified. Bread prepared from flour previously infested with *Tribolium* disclosed a variety of property changes and a distinct offensive taste and odour. These effects were observed notwithstanding the fact that the insects and their body fragments were separated from the flour by sifting, prior to making the bread.

Storey *et al.* (1984b), reporting the outcome of a survey of over 8000 farm storages across 27 states of the USA indicated that 1 or more live, stored-product insect species were found in 25.1% of the wheat, 56.4% of the oats and 79.7% of the corn samples obtained. The average number of insects per 1000 g in the infested samples was 26 for corn, 39 for oats and 105 for wheat. Insect species that prefer high moisture and which feed on moulds were found in 28.8% of the corn samples. The incidence of insects generally increased with increased grain moisture and average test weights per bushel were lower in infested grain than in uninfested grain. The incidence of insects and the composition of insect populations within each commodity did not vary significantly with the length of time the grain had been in storage.

Sinclair (1982) conducted 3 surveys between 1977 and

1979 on about 60 farms on the Darling Downs, Queensland. Farms in this area would be considered to be well above average in the application of management practices. Although predominantly grain producing farms, no commercial grain would be held on the farms. The mean amounts of grain per farm, estimated visually, were 950 kg of bulk and bagged stocks, 270 kg of grain spills, 165 kg in farm machinery and 75 kg in other residues. The mean farm populations of the 4 major pest species combined was estimated at 360 000 adults, of which about 88% were in bulk and bagged stocks. The insect populations peaked in late autumn and reached a minimum in early summer.

During the past decade concerted efforts have been made to expand the production capability of the major staple food and feed grains, to the point of self-sufficiency or even export, both of which would markedly reduce the burgeoning foreign exchange deficits that are presently being experienced in many developing countries. Increasing areas are being brought under irrigation, and existing and new irrigation areas are being more intensively managed, through the utilisation of high-yielding varieties, in a multi-cropping system **(Russell 1980)**. **Pomeranz (1982)** stated that 'increased crop productivity is the key to feeding the world's expanding population', and this becomes more evident with the realisation of the disproportionate increase in population that is occurring in the developing world, compared with the industrialised nations. Multi-cropping, simply translated, means that at least one crop is grown during the monsoon season, creating problems and difficulties in harvesting, threshing, drying and storing, resulting in rapid biological deterioration. In addition, the high-yielding varieties are easily shattered, generally softer and, when harvested, tend to have a wider range of kernel maturity than the traditional non-improved varieties. The dichotomy that exists in enhancing productivity utilising high yielding varieties is in the associated problems that are now so prevalent in the post-production system. Traditional systems of storage and handling are now inappropriate, and their shortcomings are accentuated more with the high-yielding varieties than they are with the traditional varieties. One of the most challenging problems of the 1980s will be to reduce losses caused by pests, especially insects, during the food production, storage and processing operations. Without exception, greater benefits would be derived if more intensive efforts were directed towards conservation and quality maintenance of what has already been produced, rather than on energy-expensive methods to produce more. The challenge facing the Association of South East Asian Nations (ASEAN) has been clearly documented by **Semple (1985)**.

Arthur N. Hibbs (1968), a former president of the Association of Operative Millers, in an address to a conference sponsored by his organisation, sharply defined the sanitation problems of the grain-storage and cereal-processing industries:

> We all have our common goal, regardless of our affiliation, to do the best possible job in the field of sanitation. Today, as never before, the citizens of this nation, as well as foreign countries, are keenly aware of proper sanitation and pest control. Not only are they aware of it, but they demand it. And this is true no matter what industry we serve or service. Because of the enormous publicity given to sanitation standards and the use of certain chemicals to control pests, we are faced with a far more sophisticated and educated customer for our end products. And, I might add a much more critical customer, too.

If the information now available were utilised, most of the damage and loss due to insects could be prevented.

Golob (1984) pointed out that the average small farmer has very little cash. Consequently, improvements in his agricultural practices must be cheap if they are to be acceptable. There are 3 main ways in which improvements could be made to the existing small-holder storage system. These are:

1. improve the storage qualities of the varieties of produce grown;
2. improve the structure of the food store;
3. use protectants to control insect pest damage.

Plant breeders have begun to introduce into their programs factors which will enhance resistance to attack by storage insect pests. Losses can be reduced by improving the structure that the smallholder uses. In several countries the need to do this has led to the development of highly durable but complicated structures which have commonly proved too expensive for the farmer to purchase or build. Even small-farm stores can be made rodent-proof and it is generally easier to do this than to kill rodents once they have entered. Drying rates after harvest can be improved by storing freshly harvested produce in a structure which takes full advantage of prevailing wind. Improved means of access to the grain which avoids damage to the storage structure or its weather-proofness contributes greatly to improving the quality of stored grains. These modifications are simple to implement and of little or no cost. They enable the farmer to keep the produce in good condition during storage. In order to produce a recommendation acceptable to the farmer for the application of grain protectants, the cost proposed must be low to be seen as economical. The costs and benefits, however, change with the duration of the storage period and with the variety of grain stored. Since the development of the insect population only becomes significant during the final stages of storage, the damage to the farmer's grain becomes significant at the end of the storage period.

Greening (1979a) produced evidence of the dispersal of grain insects in commerce over quite large distances and from different climatic regions. This was particularly noticeable in the case of dispersal of resistant strains from bulk grain storages to stock-feed mills and produce stores and later to farms. These examples should serve as a warning of the ease with which insect pests may enter regions were they have not previously been found.

Joffe (1958) described additional problems of protecting stored grain from insects under tropical conditions. Loss of quality takes priority over loss of quantity in temperate areas. Loss of quantity is of primary concern in tropical countries where insects have a near optimum environment most of the year.

In the main, storage practices in the tropics have evolved through traditional methods which depend on physical factors such as natural drying, storage and preservation in simple structures. There is little or only ineffective provision for the prevention of reinfestation, and the preservation of quality. Traditional methods include little or no use of insecticides. Hence, in many parts of the developing world, storage practices involve traditional structures such as simple wooden platforms, storage baskets, timber tripods, vertical or horizontal poles or racks, varieties of cribs made from wooden or grass materials, pots or other earthen containers, gourds and other types of granaries. The structures and practices are largely ineffective in the prevention of deterioration of stored agricultural produce. Under such conditions, **Taylor (1974)** indicated that losses due to insect infestation, biodeterioration, microorganisms and rodent damage are considerable. Until recently very limited inputs of pesticides for safe disinfestation of produce were used but recent developments which combined improvements in the design of cribs and storage bins with the wise use of pesticides such as lindane, malathion, pirimiphos-methyl and synthetic pyrethroids show great promise for the future. In particular, successes that have been achieved in incorporating these grain disinfestation and protection methods into traditional storage practices in Nigeria, Kenya, Zambia and many other countries in tropical Africa are now beginning to have a major impact on the reduction of losses at the farm level and therefore on the economy and nutrition of the small farmer.

The present trends in agricultural production patterns and systems and a displacement of populations from rural to urban areas seems to indicate that in the remaining years of this century decreasing percentages of human populations are likely to be directly involved in farming; small farms are likely to give way to medium scale and large farms; and the storage of agricultural produce is likely to evolve along the lines of greater co-ordination and co-operation in the use of modern techniques of disinfestation and storage. Already in many developing countries, the percentages of agricultural workers are going down from 60–70% of the total working force to 40–50% and greater efforts are going into the organisation of co-ordinated and centralised storage systems which would receive surplus agricultural produce and utilise modern techniques of storage to ensure their preservation and release into the marketing channels as and when necessary. Although this trend is likely to continue and be strengthened in the immediate future, it must be realised that it would depend on a fundamental understanding of the role of the farmer, small or large scale, in the first stage of handling and disinfestation necessary to ensure that produce delivered to medium and long-term storage depots does not constitute foci of infestation and deterioration of the bulk of stored items. In other words, clear guidelines which now form the basis of on-farm disinfestation and handling of agricultural produce should be retained and strengthened and the inputs necessary for their observance ensured at the farm level.

The rest of this century will be crucial in the efforts of man to expand and improve agricultural production and protection in order to feed the population at a level compatible with modern and acceptable concepts of social and economic welfare. The intensification of agricultural production must be matched by the pursuit and application of modern storage technologies if the gains to be realised by such intensification are to be maintained so that they reach the consumer. Tropical storage entomology has many of the answers in terms of developed technology, organisation and relevant research. If this is put to work we may yet avert a possible food crisis before the end of this century and indeed close the food and nutrition gap and strengthen the economies of the developing countries of the tropical world.

2. Loss of Grain in Storage

2.1 Introduction

In many countries the presence of insects and other contaminants in stored food has become an accepted phenomenon. According to one belief, products in store for a few weeks 'spontaneously generate' insect life. Infested stored produce is also believed to have 'matured' and therefore be better than freshly harvested produce, so that there are many merchants and consumers who are accustomed to, and therefore accept, products of inferior or poor quality **(Hall 1970)**.

As FAO and international conferences are now pointing out, in many countries the extent and level of losses after harvest have not been fully assessed, but those assessments which have been made of the quantities of food damaged and lost indicate that there is a serious wastage.

An FAO estimate of world-wide annual losses in store has been given as 10% of all stored grain, i.e. 13 million tonnes of grain lost due to insects or 100 million tonnes to failure to store properly **(Wolpert 1967)**, but losses in the tropical and sub-tropical regions of the world (where food is in short supply) can be expected to be higher than in the temperate climate zones. An FAO publication **(Anon 1979e)** referred to a US study that calculated that in 1976 post-harvest cereal losses in developing countries amounted to 42 million tonnes or 95% of the total normal Canadian grain crop. The economic impact of this may have been diffused, of course, but a calorie is a calorie and those that went missing in 1976 were nearer to the real need. In the United States, grain storage losses each year have been stated **(Powley 1963)** to be between 15 and 23 million tonnes (some 7 million tonnes due to rats and between 8 and 16 million due to insects) and a breakdown of the quantity and quality losses together with the dollar value has been given by the **US Department of Agriculture (1965)** to be $465 million. In Latin America it has been estimated that there is a loss of 25–50% of harvested cereals and pulses; in certain African countries about 30% of the total subsistence agricultural production is lost annually, and in areas in South-East Asia some crops suffer losses of up to 50% **(Hall 1970)**.

Although these figures must be considered in relation to many millions of tonnes of food production in each country, it must be emphasised that such losses do not involve every commodity in all countries in the tropics and sub-tropics. Current losses of about 30% are apparently occurring throughout large areas of the world. Prevention of these losses would result in:

1. more food for consumption by the farmers;
2. more food available for farmers to sell;
3. higher living standards for farmers;
4. more food available for non-farming populations;
5. higher quality and competitiveness of export commodities in rural trade;
6. sounder economy for the country and improvement of its international standing.

Losses are manifested in several ways:

1. weight loss;
2. food loss;
3. quality loss;
4. loss due to contamination with metabolites of insects and moulds;
5. monetary loss;
6. loss of goodwill (i.e. reputation);
7. seed loss.

Howe (1965a) reviewed the world literature concerning the losses caused by insects and mites in stored foods and feeding stuffs. **Adams (1977b)** presented a comprehensive review of the information published between 1965 and 1976 on the extent of losses occurring during the transportation, handling and storage of cereals and pulses. The bibliography contains over 100 references.

In introducing his review of the literature concerning losses in stored cereals and pulses, **Adams (1977b)** stated '**Howe (1965a)** reviewed the losses caused by insects and mites in stored foods and feeding stuffs. It is significant that 10 years after this excellent paper appeared the Seventh Session of the United Nations General Assembly passed a resolution calling for the reduction of post harvest losses to be given high priority, with a target of a 50% reduction in these losses by 1985. This has resulted in an upsurge of interest in the post-harvest sector since it was necessary to know the present extent of these losses.' It might be pertinent to observe that now, in 1986, we are still not sure of the nature and extent of these losses or whether the measures taken over the past 10 years have effectively reduced losses which would otherwise have occurred and, if so, to what extent.

Singh and Benazet (1974) considered that an overall loss of 30%, as estimated by some workers, to be conservative rather than high.

Tyler and Boxall (1984) have reviewed the activities of post-harvest loss reduction programs over the previous 10 years and have made observations about the consequences. They have described the initiatives taken under the main multilateral and bilateral aid programs and have provided examples of loss assessment and reduction projects in a number of countries. They concluded that the prescribed methodology has worked adequately for the assessment of on-farm storage and processing losses particularly in relation to weight loss. There is a lack of proven guidelines for the commercial sector including procurement, marketing, storage and distribution. Such guidelines are needed and should embrace a wider approach which is more concerned with reducing insufficiencies in the total system. Selected examples of results from studies of losses at farm level are given in **Table 2.1.**

Table 2.1. Examples of comprehensive studies to measure post-harvest weight loss in farm level storage

Country	Crop	Period of storage (months)	Cause of loss	Estimated % loss of weight and range	Reference
Zambia	maize	7	insects	1.7 to 5.6	**Adams and Harman 1977**
India	paddy	7	insects, rodents mould	4.26±1.33	**Boxall *et al.* 1978**
Kenya	maize	up to 9	insects, rodents	3.53±0.25	**De Lima 1979**
Malawi	maize	up to 9	insects	3.2±3.4 1.8±3.5	**Golob 1981**
	sorghum	up to 9	insects	1.7±0.5	
Nepal	maize	6	insects, rodents	5.7±3.2	**Boxall and**
	wheat	3	mould	2.4±1.9	**Gillett 1982**
Turkey	wheat	8	insects, mould	3.7±1.9	**Boxall, pers. comm.**
Tanzania	maize	3–6.5	insects	8.7	**Hodges *et al.* 1983**
Swaziland	maize	unspecified	insects moulds rodents	3.66 0.53 0.16	**De Lima 1982**
Bangladesh	raw and parboiled paddy	3–4 rodents	insects	2.4 (rice equivalent) (average for 3 seasons)	**Huq 1980**
Honduras	maize	7	insects	5.5	**de Breve *et al.* 1982**

The World Food Conference held in Rome in 1974 proposed that a world food reserve of approximately 30 million tonnes of grain be established to carry the developing countries of the world through critical times of food shortage in future years. At least some of these long-term storages will be held in the developing countries. One technical aspect that appears to have been given insufficient attention is the ability to store cereals for periods of many years under tropical conditions. We know that cereals can be successfully stored for long periods of time in a temperate climate but there has been little experience with long-term storage of cereals under tropical conditions, particularly in the humid tropics.

2.2 Nature and Origin of Losses

One of the basic problems in reviewing loss estimates is the definition of the term 'loss'. In many cases, authors have not adequately defined what they mean by loss and it is therefore difficult, particularly with field estimates, to compare the estimates provided by different authors even within the same locality.

'Weight loss' may be defined as the loss in weight of the commodity over the period under investigation. It is essential to distinguish between the 2 types of weight loss, 'apparent' and 'real'.

The apparent weight loss is defined as the difference in weight of the commodity before and after the particular activity under study. In commercial practice, a weight loss may, to some extent, be hidden by an increase in moisture content and by the transformation of the usable material into dust and frass which, since it remains within the bag, may still be recorded as usable commodity in transactions. Since this hidden loss is being passed from seller to buyer, it is important to know its extent.

Any loss in weight of the edible matter will inevitably lead to a proportionally higher nutritional loss because many insects consume, preferentially, the highly nutritious portion of the grain. Thus, weight loss should not be used to estimate gross food losses, although in the majority of situations it is used to estimate nutritional loss because the analysis of the various nutrients requires elaborate equipment and is expensive and time consuming.

The presence of damaged grains, dust contaminants, such as insect fragments and rodent hairs and excreta within the commodity contributes to a loss of quality which, in many cases, will lead to a loss in monetary value of the commodity. Similarly, changes in the biochemical composition may also rank as losses in quality, e.g. an increase in free fatty acid content.

Loss in seed viability is one of the easiest losses to estimate but such losses may not be immediately apparent in situations where there is replacement seed grain available. In a situation where seed is a prime consideration for rebuilding the agrarian structure, loss in viability is a potential disaster.

Hurlock (1965) reported an investigation into the weight of grain eaten by a small population of *Sitophilus* larvae at 3 different temperatures. The damage caused was assessed by weighing and radiography. There was no significant difference between the amounts of damage caused by each insect at the 3 temperatures (25°, 27.8°, 31.1°C); the average amount of grain consumed by each larvae during development from egg to adult being 28.7 mg. The mean weight of each adult produced was 2.7 mg.

A similar study with *Oryzaephilus surinamensis* produced data which indicated a rate of growth such than 100 insects would become over 12 million in 3 months and would consume grain at the rate of 54 kg per month **(Hurlock 1967)**.

Moore *et al.* (1966) carried out a study to determine the losses caused by *Sitotroga cerealella* in dent corn. The weight, moisture and feed value of kernels damaged by *Sitotroga cerealella* and undamaged kernels taken from the butt, mid and tip section of ears of dent corn were compared. A single moth consumed an average of 32.9 mg during its development in a kernel of dent corn, which

Table 2.2. Data recorded for weight losses to products during storage in a number of countries

Country	Commodity	Loss		Level of storage	Period of storage	Cause	Source
		(percent)	(million tonnes)		(months)		
Ghana	Legumes	9.3	0.006	FTC	12	I	**Hayward, L.A.W., 1964. Personal communication to FAO Plant Production and Protection Division.**
Upper Volta	Legumes	50–100		FTC	12	I	**Hayward, L.A.W., 1964.**
Nigeria	Legumes	10		T	6	I	**Hayward, L.A.W., 1964.**
	Wheat	34		F	24	I	**Ahmadu Bello University, Institute for Agricultural Research, 1963.**
Somalia	Grain	20–50		FTC	12	RI	**Bethke, S. 1963. FAO report.**
South Africa	Legumes	50		FTC	12	I	**Hayward, L.A.W., 1964.**
Tanzania	Legumes	50		FTC	12	I	**Hayward, L.A.W., 1964.**
Malaysia	Rice	17		C	8–9	A	**Wright, F.N., 1963.**
India	Rice		11.0	FTC	12		**Central Food Technological Research Institute, 1965.**
	Food grains	20.0		FTC	12		**Johnston, J., 1966, personal communication.**
	Food grains	5.0	2.8	F	12		**Wolpert, V., 1967.**
	Grain	4.0	0.2	C	3–4		**Pingale, S.V., 1964.**
	Wheat	8.3		C	12		**Pingale, S.V., 1964.**
	Food grains	9.3		FTC	12	A	
Japan	Rice	5		FTC	12	I	**Kiritani, K., 1964.**
United Arab Republic	Rice	0.5			12	IR	**Kamel, A.H., 1951, personal communication.**
United States of America	Maize	0.5	5.9		12	I	**U.S. Department of Agriculture, 1965.**
	Rice	1.5	0.04		12	I	**U.S. Department of Agriculture, 1965.**
	Sorghum	3.4	0.41		12	I	**U.S. Department of Agriculture, 1965.**
	Wheat	3.0	1.6		12	I	**U.S. Department of Agriculture, 1965.**
Thailand	Paddy } Maize }	10		F		I	**Boon-Long, S., 1965, personal communication.**

F = farmer storage; T = trader storage; C = central storage depots; R = rodents; I = insects; A = all causes.

Table 2.3. Figures of losses of products due to insect pests at different levels of storage (nonexperimental) in Africa

	Commodity	Apparent loss: Damaged grains/ kernels (percent)	Apparent loss: Weight loss (percent)	Storage period (months)	Major pest	Country
Producer storage	Beans	38–69	6	6	Pulse beetles	Uganda
	Beans	3.6		4	Pulse beetles	Zambia
		80.7		12	Pulse beetles	Zambia
	Cowpeas	13.0		4	Pulse beetles	Zambia
		81.6		12	Pulse beetles	Zambia
	Maize		20+	8	Weevils	Ghana
	Maize	30–50		5	Weevils	Dahomey
	Maize	30		5	Weevils	Togo
	Maize	45–75	20+	7	Weevils	Uganda
	Maize	90–100		12	Weevils and moths	Zambia
	Maize cobs	5–10		12	Weevils	Ivory Coast
	Sorghum, unthreshed	3–78	1–26	9	Weevils	N. Nigeria
	Sorghum, unthreshed	2–33	0–13	6	Weevils	N. Nigeria
	Sorghum, unthreshed	2–62	3–13	14	Weevils	N. Nigeria
	Sorghum	11–88	6–37	26	Weevils	N. Nigeria
Trader storage	Beans	35–44	6	12	Pulse beetles	Uganda
	Maize	20			Beetles	Togo
	Maize		5–10	6	Beetles	Uganda
	Maize	16.7	4	3	Beetles	Uganda
	General produce		10–15	12	Beetles and moths	Tanzania (Zanzibar)
Central depot	Maize		5–6	11	Weevils and moths	Rhodesia
Storage	Maize		12–19	24	Weevils and moths	Rhodesia
	Maize	35–38	10	9		Uganda
	Wheat	5 8	2	6	Weevils	Kenya

Source: Data based on personal communications by specialists to FAO.

Table 2.4. Laboratory estimates of weight loss caused by certain insect species

Insect species	Commodity	Conditions °C	Conditions % rh	Weight loss per insect	Reference
Sitophilus granarius (L.)	Wheat	28.0	75	30 mg egg: adult +5.70 mg frass	**Golebiowska (1969)**
Sitophilus granarius (L.)	Wheat	27.8	70	29.6 mg egg: adult	**Hurlock (1965)**
Sitophilus granarius (L.)	Wheat	30.0	70	18.8 mg egg: pupa +4.4 mg frass	**Campbell and Sinha (1976)**
Sitophilus zeamais Motsch.	Maize	27.0	70	25.8 mg egg: adult +9.3 mg frass	**Adams (1976)**
Sitophilus oryzae (L.)	Wheat	28.0	75	10 mg egg: adult +0.690 mg frass	**Golebiowska (1969)**
Rhyzopertha dominica (Fab.)	Wheat	28.0	75	12 mg egg: adult +7 mg frass	**Golebiowska (1969)**
Cryptolestes ferrugineus (Steph.)	Wheat	30.0	70	1.1 mg egg: pupa +0.46 mg frass	**Sinha and Campbell (1975)**
Oryzaephilus surinamensis (L.)	Wheat	27.7	70	4.3 mg over four weeks	**Hurlock (1967)**
Sitotroga cerealella (Oliv.)	Maize	13.0	–	32.9 mg egg: adult	**Moore *et al.* (1966)**

amounts to 10.35% weight loss per kernel per insect.

Hall (1970) in a 350 page monograph published by FAO on *Handling and Storage of Foodgrain in Tropical and Sub-tropical Areas* discusses the many manifestations of losses due to insects under tropical storage conditions. **Tables 2.2 and 2.3** summarise a few of the important data.

Various investigators have estimated the weight loss caused by certain insect species and these results were summarised by **Adams (1977b)** in **Table 2.4**.

The effect of the initial population density on the loss of weight in samples of wheat was investigated by **Stojanovic (1966)**. He found that after 200 days at 20°C the weight loss caused by *Sitophilus granarius* varied from 59 to 78%, depending upon whether the initial population was 1, 2 or 3 pairs of adults per 500 g of grain.

Weight loss of stored grain caused by the feeding of insects depends largely on the pest species and its feeding behaviour. **Campbell and Sinha (1976)** determined the damage to stored wheat caused by larval and adult feeding of *Cryptolestes ferrugineus, Rhyzopertha dominica* and *Sitophilus granarius* by feeding individual insects on single wheat kernels at 30°C and 70% relative humidity. Weight loss caused by the feeding activities of each species was measured and related to kernel site damage and frass production. When insects were allowed to develop from egg to pupae, *Sitophilus granarius* caused 60%, *Rhyzopertha dominica* 17% and *Cryptolestes ferrugineus* 4% weight loss in single kernels weighing a mean of 29.5 mg. Weight loss of kernels and frass production caused by the feeding of *R. dominica* adults was greatest, whilst *S. granarius* and *C. ferrugineus* adults caused smaller losses. *C. ferrugineus* fed exclusively on wheat germ whereas the other two species fed on both germ and endosperm.

Adams (1976b) reported an experiment in which 70 single maize kernels were exposed to oviposition by *Sitophilus zeamais*. Weevils developed in 50 kernels, 16 yielding one adult, 24 two, 7 three and 2 four and five adults. Weight loss of the kernels was observed throughout the developmental period at 27°C and 70% r.h. Development took a mean of 37 days resulting in adults with a mean weight of 3.1 mg. The overall mean loss in weight per grain was 18.3%. The total consumption by an insect was 35.1 mg in grains with 1 emerging adult and approximately 25 mg for those with more than 1. Half the total loss occurred between days 13 and 24.

Bitran *et al.* (1981, 1982a) reported a significant correlation between the loss of weight of stored maize and the level of infestation by *Sitophilus zeamais* and *Sitotroga cerealella*.

Al-Saffar and Kansouh (1979) reporting a survey of the insects infesting stored grain in Iraq in 1977–78 showed that the most abundant species in wheat was *Trogoderma granarium* which was present in more than 50% of the samples. Infestation levels ranged up to 685 insects/kg grain; the mean percentage of infested grains ranged from 2.5 to 5.7 according to the origin of the wheat, and the percentage weight loss ranged from 3.1 to 6.6%.

Minor pests can cause enormous damage when they are introduced into a region where they have previously not been found. FAO reported **(Anon. 1982)** that the larger grain borer *Prostephanus truncatus,* an established but minor pest of stored produce in central America, Brazil, Colombia and southern USA, had recently been reported for the first time in Africa where it had devastated stored maize in Tanzania. Official investigators recorded weight losses of up to 34% in stored maize, implying that at least 70% of grains were heavily damaged and useless as either seed or food. Losses were high even on indigenous sheathed varieties. The beetle poses a very real threat throughout east and central Africa. The consequences would be enforced changes in the food habits of the people through increased growing of sorghum millet and rice.

A major concern is that stored-product insects may be an important factor in the development and spread of mycotoxins in stored grains and related commodities. Mycotoxins are of great public health concern since several have been identified as the cause of a variety of human and livestock diseases including cancer (aflatoxins). It is recognised that stored-product insects can absorb, carry and spread the microorganisms responsible for the production of these toxins. **Ragunathan *et al.* 1974** studied the association of storage fungi with *Sitophilus oryzae*. When eggs, lavae, pupae, and adults were surface-steralised and plated on three media, only the eggs were free of fungi. *Aspergillus ochraceus, A. flavus*, and a number of other fungi were found in the larvae, pupae and adults. In weevils from sorghum, the most common species was *A. restrictus*, but in weevils from wheat and rice, *A. flavus* predominated. The incidence of infected weevils ranged from 20 to 100%.

In addition to spreading the inoculum, stored-product insects create an environment that favours the multiplication of the fungus. The increase in temperature due to insect metabolism which results in 'hot spots' triggers the growth of fungi which is promoted by the increased humidity generated by insect activity and the metabolism of grain components.

Bulk grain with a moisture content of 11–14%, and in apparently good condition except for the presence of insects, often becomes hot. Heat resulting from the metabolism of grain and microoganisms under such conditions is not sufficient to account for the pronounced rise in temperature. According to numerous investigators, the source of heat is the metabolism of the insects themselves. Temperatures may rise to as high as 60°C.

An 'insect caused' hot spot begins when insects produce metabolic heat faster than the heat can escape. **Howe (1962)** pointed out that a small local rise of temperature accelerates the metabolism of each insect and speeds the rate of population increase; thus, the amount of metabolic heat continuously increases until it reaches a level that is unfavourable to insects. At this point the insects move to the periphery of the hot spot, thereby increasing its size and the spread of microorganisms.

Ciegler and Lillhoj (1968) list 15 reviews and symposia proceedings dealing with mycotoxins in the 6-year period from 1962 through 1967. Additional reviews have been published since by **Wogan (1965), Moreau (1968), Hesseltine (1969), Goldblatt (1969), Herzberg (1970), Meyer (1970), Christensen (1971), Purchase (1974), Rodricks (1974),** and **Rodricks *et al.* (1977).**

2.3 Methods of Assessing Losses

There has been concern in recent years about the scarcity, unreliability and lack of standardisation of observations on post-harvest losses, particularly in tropical countries. For many years the estimation of such losses has been based on extrapolation of comparatively few unstandardised studies, leavened by subjective assessment **(Haines 1982a)**. Recent efforts have been made to design standard methodologies for post-harvest loss assessment, e.g. the work of **Adams and Harman (1977)**. The current state of knowledge and recommendations for future studies are given in 2 books arising from the collaboration of various national and international organisations concerned with the conservation of natural resources **(Board on Science and Technology for International Development 1978; Harris and Lindbad 1978)**. A number of authors have made valuable observations and recommendations to improve the objectivity and reliability of estimates of food losses in small-scale farm storage and in warehouses **(Adams 1976a; Boxall *et al.* 1978; DeLima 1979a; Golob 1981a, b, c; Haines 1982a). Proctor and Rowley (1983)** have proposed a new method developed in the UK for assessing weight losses caused by arthropod pests in stored grain. The method, called the Thousand Grain Mass method, is considered to be based on sounder principles than any method previously described.

Adams (1977a) reported on research into quantitative and qualitative losses incurred during maize storage on selected small farms in Zambia. Various methods of analysing samples for loss were tested. The best estimate of loss within a sample was obtained by comparing the weight of a standard volume of grain from the sample with that of a reference sample obtained at the time of storage. Estimates of loss over a storage season were obtained by integrating losses within samples with the pattern of consumption over the storage period. Monetary values were placed on these losses. Changes in storage practice were evaluated with the aid of these methods which enabled the costs and benefits of a simple improved storage technique to be calculated. It was not intended that the loss estimates produced by the project should be interpreted as representative of Zambia as a whole or of any particular region but they provided a practical technique for carrying out studies under typical field conditions.

Bourne (1977) drew attention to another precaution that needs to be considered in assessing overall losses to see that the arithmetic of loss figures is correctly calculated. In quoting loss figures at various steps along the post-harvest chain there is a common error of adding the percentage of loss at each stage to obtain the total loss. This leads to overall loss figures that are too large because it assesses that each loss figure is a percentage of the original weight of commodity. In fact, each percentage loss is expressed as a percentage of the amount in the previous step, the percentages being applied to a diminishing base.

Adams (1977c) published the results of a questionnaire circulated widely to authorities concerned with post-harvest losses in cereals and pulses. As a result of the replies received, the author concluded that it would continue to be difficult to interpret findings of individual workers and to compare their results until agreement had been reached on a suitable method for the measurement of post-harvest losses in all stages of the system and the method had been published and adopted. Nevertheless, many of the estimates of loss that have been provided are of sufficient magnitude to indicate that remedial action is urgently required. The estimates provided in this survey range from below 5% to above 40% and the author expressed the hope that the information would stimulate a concerted effort to achieve the maximum possible reduction in these losses.

Boxall *et al.* (1978), commenting on the need for objectiveness in loss assessment, drew attention to the fact that of 126 field estimates of loss in the bibliography of post-harvest losses in cereals **(Adams 1977a)** only 11 were complete estimates providing the reader with sufficient information on the methods employed to enable a decision to be made about the reliability of the estimates. Most of these were also limited to a study of one aspect of the post-harvest system and data on which to base loss reduction imputs were frequently lacking.

DeLima (1979a) described appropriate techniques for use in the assessment of loss in stored produce in the tropics, giving examples of many of the difficulties involved. In a separate paper **DeLima (1979b)** reported that several of the techniques described in the literature were found to be subject to variation from several sources and therefore not suitable in surveys covering widely varying conditions. Alternative methods were developed and by the use of these techniques the annual loss in subsistence maize due to insects in Kenya was found to be 4.5%.

Adams and Harman (1977) carried out an evaluation of losses in maize stored in a selection of small farms in Zambia paying particular attention to evaluation methodology. **Schulten (1981)**, commenting on this study, stated that the method of cost/benefit analysis for insecticide use at the small-farm level seemed most appropriate. The method compared the extra costs arising from introducing small changes with the expected benefits. It was shown that under Zambian conditions it pays to follow the government's recommendations of admixing malathion dusts with shelled maize. A farmer who presently stores his maize as ears with husks will, in general, also benefit when changing to storing shelled maize treated with malathion.

Calderon (1981) presented a new approach for apprehending the extent of post-harvest grain losses. The author discussed the grain bulk as an ecosystem, comprising biotic elements (grain, insects, microflora) and abiotic elements (dockage, intergranular air, water vapours, storage structure). The effect of the stored grain bulk environment and micro-environment on each of the system components is analysed and the possible implications of these effects in the cause or prevention of grain storage losses are discussed. It is suggested that the ecosystem approach for evaluating the potential loss of stored grain at a given storage site can be used not only to assess the present status of damage and the storability of grain but also to indicate the steps needed to prevent losses.

In order to improve the accuracy and comparability of loss assessments, an accounting and inventory system **(Caliboso 1982; Caliboso and Teter 1983)** is being field evaluated in the Philippines for paddy and for maize.

2.4 Effect of Cereal Type and Variety

Howe (1965a) suggested that rice is probably the most resistant of cereals, because each kernel is protected by a hull. Nevertheless, the extent of protection afforded by the hull varies both with the variety of rice attacked and the insect species attacking it. Rice varieties differ in their attractiveness to stored-product insects and possibly in their infestability.

Cogburn (1975) showed that rice in hull is much less subject to damage by insects than is hulled rice. Damage to hulled rice varies according to insect species and rice variety. The maximum loss observed to date equals 30% of the value of the rice. A starting population equivalent to 1 gravid female *Sitotroga cerealella* in 500 g will totally destroy hulled rice within 3 generations. Approximately 1000 varieties of rice were screened for susceptibility to attack by *Sitotroga cerealella*. About 10% appeared to be resistant.

Cogburn (1977) showed that 6 commercial varieties of rough rice in regular production in southern United States differed greatly in their susceptibility to 3 insect species and that the weight loss of rough rice, loss of milling yield and loss of monetary value differed greatly, depending upon the variety.

Golob (1981a, 1981b) reported 2 surveys of farm-level storage losses in southern and central Malawi respectively. He described the practical problems of such surveys and the limitations of 2 different loss-assessment methods but concluded that losses in up to 10 months storage were 3% or less for maize in southern Malawi and less than 1% for maize in central Malawi. Such losses demonstrate the suitability of local crop varieties and methods of storage to conservation in the area. The likelihood of increased losses of high-yielding varieties of maize, if these were introduced, and the consequent possible need for insecticides, which are not needed at present, were noted.

Golob (1981c) in reviewing the use of pesticides at farmer and village level stated that surveys of on-farm storage of food for consumption have demonstrated loss due to insect infestation of less than 3% for untreated maize, sorghum, paddy and ground nuts. There are several reasons for these low estimates:

1. generally the storage of food for consumption is short, mostly not longer than 6 months, so that insect populations do not have sufficient time to develop before most of the food is consumed;
2. farmers grow and store commodities that naturally resist infestation; they tend to store maize cobs that have well developed husks; they store paddy rather than milled rice, and undecorticated ground nuts rather than kernels;
3. the indigenous variety of grain is relatively resistant to insect attack compared with many hybridised or composite varieties.

Pulses, particularly, are usually very susceptible to insect damage and are subject to large losses. Likewise the high-yielding varieties of other grains. As governments encourage the production of high-yielding varieties farmers will gradually store food which is more susceptible to infestation than that which they are storing at present. It has been found in trials in Malawi that in 6 months, losses of hybrid maize can exceed 20% by weight but that the loss can be reduced to the levels occurring in the indigenous varieties, i.e. 3% or less, if pirimiphos-methyl is applied to the grain just before storage.

Reports by **Reader (1971)** and **Schulten (1975)** point to the very high susceptibility of hybrid maize to insect attack. They quoted weight losses caused by insect infestation in local varieties of maize as 1–2%, compared with losses in improved varieties of 5% and hybrid varieties of 10%. The reason for the low level of loss due to insect attack in local maize varieties is that the grain is inherently less susceptible to attack and that the sheath covering the grain offers good protection against field infestation by storage insects. In comparison, the most commonly used hybrid has grains of high susceptibility to insect attack and sheaths which do not completely cover the grain on the cob.

2.5 Losses Reported for Various Regions

Adams (1977b) found that most of the references dealt with situations in central and east Africa, west Africa and the Indian sub-continent with very few reports from South America, Sahel and the Far East. By far the majority of these studies dealt with the effect on quantity with considerably lesser attention being given to quality. Questions involving nutritional consideration or viability of seed received scant attention. Most of the studies were concerned with losses at the farm level, relatively little attention being given to losses in large-scale storage or in the hands of traders.

Serious losses may also occur in large stores in which pest control techniques and hygiene are poor or lacking. **Aboul-Nasr *et al.* (1973)** investigated loss of wheat stored in bags in Egypt. Their results showed a loss of from 37 to 48% depending upon the location within the stack.

Dichter (1976) estimated that in the sub-Saharan regions, losses of food grains during storage at farm or village level can amount to 25–40%. About 80% of grain produced is kept at village level for seed or consumption. The remaining 20% is sold as the farmer's only cash income. The author pointed out that if this grain could be kept in good storage conditions, farmers would benefit from higher prices during times of shortage.

Adams *et al.* (1975), reporting an intensive study of a group of farmers in Zambia, stated that they used a system to reduce losses of maize during storage. Originally the farmers grew a local low-yielding variety, resistant to storage pests because of its tight husk and hard grain. When the high-yielding hybrid SR52 was introduced they soon discovered its poor storage qualities. Some reverted to the local variety for their food store, growing SR52 for sale alone. Others continued growing SR52 for sale and food but carefully selected cobs for storage. They chose small tight-husked undamaged cobs and stacked these carefully in the store. These often remained untouched for 2 to 3 months after storage. The largest cobs were shelled and sold as soon as possible, leaving some less suceptible cobs which remained on the drying platform and were removed when needed for food. Only when these are finished is the stored maize utilised.

Bindra (1974) has pointed out that India has a vast rural population most of which is engaged in agriculture. Consequently, the requirements for food grains in the rural areas for food, feed and seed is immense. Rural dwellers and many urban dwellers store their seasonal requirement of food grains themselves because of a continuing food deficit. Thus, of a total of about 105–110 million tonnes of food grains, about 85–90 million tonnes are stored in rural and urban homes in small quantities of half to 5 tonnes in heaps in corners of houses, in gunny sacks, in bins made of mud, bricks, metal or wood, and in a variety of other small receptacles. Most of the storage is unsuitable and cross-infestation is high. Hence the losses are high. Since the grain is generally stored inside the house, fumigation is hazardous and inadvisable. Even if it is fumigated, despite the risk involved, the grain frequently gets reinfested because the infestation in the houses is not eliminated and in practice the storage structures are rarely insect-proof.

Zutzhi (1966) investigating a selection of farm wheat stores in Delhi observed a loss of 15% for wheat stored in bulk in a room. Likewise, **Willson *et al.* (1970)** recorded losses of 3.1% for bulk grain heaped on the floor, 2.3% in sacked grain, 1.1% in grain stored in a metal bin and 0.5% in malathion-treated grain in sacks.

Khare (1972), reporting investigations carried out in several climatic zones of India, observed maximum and minimum weight losses of 4.5% and 1.7%, depending upon the store type and climate.

Reports by **Khare (1972, 1973)** demonstrated the importance of insect attack in India for which are quoted weight losses of 1.5–4.3% in stored wheat per annum, 0.7–4.0% in paddy after 6 months and 2–11% for maize after 9 months. Protein losses of 17–31 mg/g in wheat and 9–51 mg/g in maize were also recorded.

Krishnamurthy (1975) stated that the post-harvest losses in food grains in India were of the order of 9%, whereas **Khare (1973)**, reported that losses in storage could be as high as 30% in Uttar Pradesh.

Boxall *et al.* (1979) reported a study of farm level food grain losses in Andhra Pradesh, India, from 1976 to 1978. The object of the project was to provide a social cost/benefit analysis of farm-level storage improvements. The study concentrated on the storage of paddy rice as the staple crop. Eighteen villages provided approximately 20 sample stores each during the 2-year period of the project. Samples were subjected to laboratory analysis with emphasis on physical dry weight loss. The final estimate of loss was 4.26%. In the second year various improved storage practices were tested, including improvements to traditional stores, use of metal bins and pesticides. Analysis of the data suggested a positive cost/benefit from these measures and support for the extension of India's storage extension service, the 'Save Grain Campaign'. Because of the physical characteristics of paddy, the loss estimates are likely to be lower than for other food grains which are more susceptible to insect infestation.

Thomas (1974), reporting on the development of the Bangladesh Warehousing Corporation covering 820 central and state warehouses with a total storage capacity of 3.6 million tonnes, estimated that between 5 and 10% of the stored commodities was lost through insects, rodents and other pests.

Fazlul Huq (1980) presented results of a study of farm-level storage of winter season unhusked rice and milled rice in Bangladesh in 1978–79. The results were obtained from a study in 1 district over a period of 3 to 5 months. The project was confined to physical weight loss which was regarded as the most significant type of storage loss because it can be readily measured. Qualitative losses which involve changes to taste or texture are more difficult to measure and evaluate particularly when they involve consumer preferences. Nutritional losses and chemical changes occurring during storage were not considered because they are too difficult to measure and interpret. The authors emphasised that there is a need for further research on quality deterioration in cereals and cereal products and the effect on human health of consumption of deteriorated and mould-infected grains.

The study revealed a loss, due to insects, of from 1 to 2.5%, being about half that due to rodents.

Sze-Peng (1979) in a post-graduate thesis on *Sitophilus oryzae* stated that up to 10% of the total rice harvest of Malaysia has been lost due to storage pests.

Caliboso *et al.* (1985) reported that maize loses about 35% of its weight when it is stored for 8 months in the Philippines without protection from insects. Based on the 1983 procurement of the National Food Authority, the Philippines could expect to lose about 44.8 million kg of maize valued at $US8.8 million if appropriate pest control measures were not adequately applied. This volume represents about 22.5% of the country's import requirements for maize. In milled rice, where the government must stockpile 260 million kg to constitute 45 days' consumption requirements, insect infestation that is left unchecked can result in a loss of about 18.5 million kg valued at $US6.2 million. A recent survey of government storages conducted by the National Post-Harvest Institute for Research and Extension revealed that, although under present conditions warehouse designs have improved and chemicals are used to a certain extent to control insect infestations, significant losses to insects still occur. Paddy rice stored for 7 months lost 5% of its weight, equivalent to 24.5 million kg valued at $US2.46 million.

Prevett (1974) discussed the problem of food storage losses in general and gave an estimate for Brazil, pointing out the importance of pre-storage factors. The use of a time-controlled space sprayer using dichlorvos for control of *Ephestia cautella* is mentioned. For bulk storage, modern vertical silos offer few problems of insect control, but horizontal silos are less satisfactory, and direct admixture of insecticides with the grain seems to offer the most satisfactory solution.

Murray (1979) reviewed the infestation pattern in the Australian wheat industry during the past 2 decades and described the immediate short-term solutions and a prospective for the future. In the late 1950s and early 1960s, insect infestation became recognised in Australia as a significant factor in the marketing of wheat by the Australian Wheat Board. The introduction in 1961–64 of malathion as a grain protectant almost completely suppressed insect populations in stored wheat. Very few insects were detected in stored wheat in Australia or at overseas ports during the period 1968–73.

Malathion resistance in *Tribolium castaneum* was detected in Australia in 1968 and inspection of wheat held in the country showed a gradual increase in the incidence of infestation during the early 1970s. In 1971, an integrated pest control plan was formulated which stated that, in the short-term, new protectants should be sought to replace malathion. These protectants were found, evaluated and put into use causing a significant reduction in insect populations. The long-term aims of the plan require the increased use of fumigation as a control measure, and the utilisation of various physical means of control in association with chemical protectants.

Murray (1981) reported that losses within the Australian grain storage system are absolutely minimal. Grain is harvested in very clean condition and is not graded or mixed after it has been received from the grower. Dust extraction systems remove dust at sub-terminals and at seaboard terminals. The purpose of removing dust is not to clean the grain, but to provide a safe working environment and reduce the risk of dust explosions. Unlike in many countries the dust is not remixed with the grain but is treated as wastage. In the average year, the amount of dust removed is approximately 0.1% of the total tonnage handled. In addition to the dust loss, the average losses during storage over the previous 5 years had been of the order of 0.2% by weight. The average loss is considered to be acceptable. The loss may be actual and partly caused by many factors including spillage and loss of moisture during the use of aeration systems. Alternatively, it may be only an apparent loss due to inaccuracies in stock recording systems. Murray stated emphatically that there is little or no loss due to the activity of either insect or rodent pests.

2.6 Social, Political and Economic Considerations

Losses which involve commercial relationships may not be easily quantifiable. Examples include: loss of goodwill, which may affect potential markets; social losses, such as the effect of chronic disease arising from the ingestion of mycotoxins or allergies arising from the inhalation of spores or attacks by mites.

If, as appears likely, stored-product pests are to any extent implicated in the promotion of mycotoxins and the human and animal diseases that result from the ingestion of such highly bio-active materials, the social consequences are indeed great. The knowledge that those peoples on the poorest level of nutrition, and therefore the lowest ability to remain in prime condition are the ones whose food sources are most likely to be infested by stored-product pests and contaminated by mycotoxins should serve to promote appropriate action by members of the more fortunate segment of society. In-depth research into the nature of food contaminants including mycotoxins and how they interact with nutrients requires the highest order of priority in the scientific community.

Van Rensburg (1977) and **Pier *et al.* (1977)** have provided a brief overview of the human and animal health risks associated with mycotoxins. Twenty separate papers each dealing with a separate mycotoxin and its attendant human and animal diseases have been published (**Purchase 1974**). A further 15 major papers dealing with various facets of aflatoxins and aflatoxicosis leave no doubts about the social, political and economic implications of aflatoxins (**Goldblatt 1969**).

The paper by Kraybill and **Shapiro (1969)** deals with the implications of fungal toxicity to human health and leaves one in no doubt of the far-reaching consequences that flow from the developments of moulds in grain and other stored commodities.

Goresline (1973), reviewing the need for action on the

disinfestation of grain and grain products, indicated that the estimated post-harvest losses are equivalent to the agricultural product on over 12 million acres of land. The agricultural production making up these enormous losses is equivalent to a great industry occupying large areas of land and representing loss of the fertility of the soil and wastage thousands of man-years of effort. It constitutes a large drain on the economy of the countries involved, yet there seems to be but little effort expended in saving this food material that has already been produced, harvested and is in our hands. He pointed out that the 'green revolution' increased food production in many areas by 15% or more but the food losses were also increased because little attention had been given to providing means for protecting that food from infestation. From a purely monetary standpoint it would be a sound investment to provide adequate storage facilities and sanitary methods of handling. However, this is not enough, for the means of eradication and control of the agents responsible for these losses must also be provided.

Shuyler *et al.* (1976) commented that the lack of knowledge of losses in stored products retards approval of programs to reduce post-harvest food losses in developing countries. FAO, using information recently assembled by the Senior Agricultural Advisor/FAO Country Representatives and some other sources, proposed continuing to gather loss-assessment data during the execution of projects designed to reduce the losses. For example, assistance in training (widely needed), during which losses are assessed, can lead to justification for additional programs to further reduce losses. Projects to assist in reducing post-harvest losses have seldom had sufficient resources of time, personnel, equipment, materials and funds to allow assessment of losses, let alone the development of methodology for this task. This has occurred for pragmatic reasons. For these same reasons, reports to governments summarising the results of projects have not stated the quantities or proportional reduction in post-harvest losses achievable through the use of project recommendations, with the exception of losses reduced through improvement of milling/processing equipment where reasonably accurate judgments can be made. Similarly, data are not available to assist in evaluating the results of such projects. Despite these facts, it is deemed that essentially all projects to assist in reducing post-harvest losses have had laudable results, though not measurable. For all practical purposes, though loss reduction has assuredly occurred, it is not presently appropriate to state that post-harvest losses have been reduced by a specific proportion as a result of development assistance.

Bourne (1977) in a 50 page monograph on post-harvest food losses described this as the neglected dimension in increasing the world food supply. He complained that much of the data are unreliable because the amount of loss has been estimated and has not been obtained by actual measurement. There is often the temptation to cite 'worst case' figures to dramatise the problem. Care must be taken when looking at one lot of food and seeing, for example, extensive infestation with insects not to assume that all the food in the country is similarly heavily infested with insects. Extrapolation of loss from a limited sample to the entire food of a country is an unsound procedure. He believes that another problem is that even some of the figures that have been obtained by careful measurements are manipulated for various reasons. In some cases there is the temptation to exaggerate the figures of loss particularly if there is a prospect that high figures of loss will prompt aid or grants from some donor. In other cases there is a temptation to minimise the actual loss figures in order to prevent the embarrassment of acknowledging the magnitude of losses, or for political, financial and trading reasons.

Adams and Harman (1977) discussed the social costs and benefits. These are the costs a country has to make to improve the storage system and the benefits it will obtain. Among the social costs are counted subsidies for equipment or pesticides and the costs of the extension effort to induce farmers to modify their traditional systems. Possible benefits could be an increase in available maize, and increase in the income of the farming community etc.

Based on loss estimates of **Caliboso (1977)** for maize in the Philippines, **Morallo-Rejesus (1982b)** calculated that 71.3% of the increased production of maize in 1977 over 1976 was lost to insects in storage. This clearly demonstrates that the gains in food production are less impressive if total food availability reflects the magnitude of losses sustained after harvest, particularly in storage.

Morallo-Rejesus (1982a, 1982b) and **Haines (1982b)** drew attention to the major reasons why the loss of stored grain is unnecessarily high in the humid tropics. These are:

1. lack of information and understanding of the pest problem in storage;
2. lack of adequate storage facilities;
3. lack of information on adequate methods of control;
4. lack of adequate store management.

Schulten (1982) listed several constraints on the effective implementation of post-harvest loss reduction techniques, and in the view of **Semple (1985)** these appear quite valid for ASEAN member countries as well. These include:

1. lack of coordination among the various national institutes/agencies involved in loss prevention;
2. lack of trained personnel in research, warehouse management, quality control and extension;
3. lack of information on post-harvest technologies that have proved effective elsewhere;
4. lack of accurate information on the magnitude of losses in different operations within post-harvest systems;
5. lack of appropriate loss assessment methods;
6. lack of storage capacity;

7. lack of an effective transport and distribution system;
8. lack of grades and standards that can be applied in food for assessing quality;
9. lack of differential pricing of the various grades to create incentives for farmers to deliver better quality grain, and for the investment in improved facilities such as dryers and appropriate storage systems that facilitate the implementation of suitable pest control strategies.

The post-harvest losses in farm level storage utilising traditional methods appear quite low, but the introduction of the high-yielding varieties have taxed the ability of traditional handling, drying and storage systems to cope with the larger quantities being produced, especially from the wet season harvest. The low bench-mark of losses encountered in traditional, improved farm-level storage systems should be recognised as the acceptable level of losses attainable under all conditions. Low-cost control methods become more relevant in this form of storage and care should be exercised that improved technologies often advocated do not put the farmer at a disadvantage (**Tyler 1982**). At the national and commercial storage level, capital intensive but cost-effective control measures assume greater importance, since reserves or carry-over buffer stock are often stored for more than 12 months, and losses in this type of storage can be extremely high. As reiterated by **Tyler (1982)**, there must exist a national commitment toward identifying the major causes of loss, the extent and where they occur, and developing a co-ordinated national plan of action to reduce these losses, in line with ongoing productivity programs.

The Southern African Development Co-ordination Conference (SADCC) (**Anon. 1985**) concluded that the use of chemical pesticides to control post-harvest losses is probably unavoidable, but improved storage facilities continue to be important. The first step must be taken: cereals, pulses and tubers must be dried before they are stored. Substantial improvements in farm-level storage can be made by a greater extension effort and some small-scale drying facilities. Large-scale, central storage is a different problem. Many SADCC countries have built improved silos and warehouses in recent years but many more will be needed in the future, involving high capital costs. FAO's projections indicate the need for a doubling of annual gross investment in dry storage between 1980 and 2000. It was pointed out that much can be achieved in the short term to reduce post-harvest losses since there is a backlog of research results awaiting broader dissemination in the field. Similarly, in the longer term there are many research opportunities that could fill gaps in the present spectrum of control methods. However, both a wider use of known methods and the development of new methods are severely constrained by a lack of manpower for research and extension, and high recurrent costs for such services.

Morley (1979) pointed out that whilst it is relatively easy to establish the magnitude of losses due to insects, it is quite a different thing to convey the urgency of the situation to the farmer. The farmer may be conditioned over many years to accept such losses as inevitable and whilst some farmers always eat or sell the same amount of produce each year, they do not consider they have lost anything. The farmer is unlikely to recognise the cause of loss, particularly nutritional loss, unless it is visible. In a recent study in Nigeria, 62% of farmers questioned were not concerned about the loss in quantity or quality of the stored-product produce, but they were concerned about the problems of termites eating the storage stucture.

There are various benefits to be obtained from reducing the wastage of food in the post-harvest chain. The first of these is nutrition. Since less food will be lost from whatever cause, there will be more nutrients available for the population. The loss of food also represents an economic loss, which increases as the food moves through the distribution system because to the cost of food that is lost at each step must be added the cost of storing and handling the food in all the previous steps in the chain. Another important economic aspect is that those developing countries that need to import substantial quantities of food find that the cost of paying for this food places a great burden upon their overseas balance of payments. The reduction of post-harvest losses within the country should reduce dependence upon imported foods and reduce the burden that purchasing this food places upon overseas currency reserves.

There is yet another important aspect of post-harvest food losses and this is 'the feed-back incentive'. In some countries farmers could well increase their production but at present they are unable to store food for any lengthy period of time. There is no incentive to increase production when they know that the extra production will spoil before it can be utilised or sold. If the post-harvest losses that they suffer at present could be reduced or eliminated there would be more incentive for them to increase production.

Webley (1984), in presenting a series of lectures on loss assessment to officials from developing countries during a course on preservation of stored cereals, introduced an interesting philosophy as follows:

> It is also necessary to be clear about the relationship of loss assessment to loss reduction. It is, of course, the general intention that the measurement of losses will lead to steps being taken to reduce them. However, unless one is careful, it is easier to suggest some loss reduction measures and in so doing, prevent loss assessments being made. Thus, it is easier to tell a marketing board that it has an insect infestation and ought to fumigate and perhaps carry out a demonstration of fumigation, than it is to tell the board that it has an insect infestation problem and that there must be an exercise to measure how much commodity, quality or money the board is losing by its current inefficient storage program. Marketing

boards often do not welcome that sort of information, whereas they would generally listen to advice about cleaning the store.

2.7 Conclusions

The spectre of malnutrition, starvation and famine should bring a shudder to those fortunate enough to live in an affluent society but let us not fail to recognise that this spectre in each of its phases is largely related to economics. Whilst it is affected by the wastage of food to insects it is not so much related to average loss but to the price of grain at its peak — just prior to the next harvest. This can be relieved by the ability and capacity to store grain for long periods. Such long-term storage means (a) a better price for farmers at harvest time and (b) a lower price for consumers just before harvest. These objectives are addressed by some national and some international government support programs but need greater effort and greater resources.

Semple (1985) reviewed the current situation in ASEAN countries and made recommendations for future requirements. He pointed out that the constraints to safe storage of cereals and secondary food and feed crops in ASEAN countries have been amplified by the introduction of high-yielding varieties and multiple cropping systems resulting in harvesting during the wet season. Traditional systems of drying and storing this increased production have proved to be inadequate, resulting in grain that is subsequently of poor quality. Improvements in storage facilities, warehouse management and the utilisation of pest control techniques to reduce losses will only be initiated in the private sector, and at farm or rural village level, if financial incentives for higher quality and workable grades and standards are devised. He provided extensive information on the species of pests most commonly encountered, the commodities they infest, and the estimated storage losses based on limited field evaluations and laboratory trials in order to give an indication of the severity of the problem with respect to storage type and duration.

Snelson (1979d) in a paper to the Second International Working Conference on Stored Product Entomology suggested that everyone should accept a share of the responsibility for preventing avoidable losses of stored products. Suitable technologies exist and have been evaluated under a wide variety of conditions. Chemicals, including grain protectants and fumigants, currently offer the major line of attack against insects but must be recognised as adjuncts to good management, to reinforce hygiene and sanitation and to enhance the effectiveness of suitable storage facilities.

Webley (1981b) strongly recommended to the GASGA Seminar on Appropriate Use of Pesticides for the Control of Stored Product Pests in Developing Countries the acceptance of insecticides approved through the FAO/WHO Joint Meeting on Pesticide Residues as the best remedy available for reducing insect attack on susceptible commodities even when orderly and hygienic storage systems are available. He indicated that such treatment is necessary to avoid losses and, notwithstanding the theoretical risks involved, the use of approved insecticides is certainly safer than many traditional methods of insect control.

As if the loss of badly needed food and the increased cost of food staples were not sufficiently serious, particulary in the poorest countries, the realisation that stored-product pests cause selective loss of vitamins and vital amino acids and an increase in dangerous mycotoxins should spur the world community to develop the capabilities and capacities necessary to reduce insect infestation to the smallest level possible.

The short- and long-term effects on human health and the reduction in animal production caused by mycotoxins and related diseases warrant additional research effort and an educational campaign to bring home to those responsible for grain storage the importance of correct and adequate storage management.

Most of the developing countries of the world lie within the tropics where the prevailing high ambient temperatures make the preserving of stored foods considerably more difficult than it is in temperate climates. Naturally occurring deteriorative chemical reactions are accelerated in the tropics. In addition, the year-round high temperatures of tropical countries allow insect pests to feed and multiply throughout the year, whereas the cold winters of the temperate climate countries stop reproduction and reduce the feeding activities of these pests. In very cold climates the low temperatures kill many of the insect pests. Economic constraints prevent tropical countries from adopting measures that would enable grain to be cooled to the point where insect reproduction slows down or stops.

In simple terms, post-harvest losses mean that a large sector of the world's agricultural resources is being wasted. This sector producing food destined never to reach the mouths of needy consumers occupies large land areas, represents soil fertility loss, is wasteful of human labour, and is a drain on the economy. It is therefore surprising to discover what little effort is being made to prevent wastage by saving food that has already been produced and harvested and is actually in our hands. How much longer can we ignore the spectre of millions of tonnes of grain going to waste each year due to the ravage of insects, when it could have gone into the stomachs of hungry people? At present there is no hope of recovering or preventing all of the losses but the application of available knowledge and relatively inexpensive techniques could quickly reduce the worst of them.

Any such program to control insects in grain products must be an organised, co-ordinated system of approach, regardless of the method of controlling the insects. At each stage of the progress of grain from the harvest field to the consumer there must be contributions to the objective of preventing loss of food. Constant surveillance is a vital factor in the establishment and maintenance of an effective program to control insect pests of stored products.

3. International Grain Trade

The FAO Production and Trade Year-Books provide a continuing record of grain production and export and import movements from individual countries. The pattern of these movements is continually changing as crop prospects and harvests vary — thus countries which normally export grain may become nett importers in adverse seasons. Only 10% of the world's grain production enters international trade and though countries such as Australia may be significant exporters of grain they are minor producers when compared with more populous countries such as China, India and the USSR where most grain produced is consumed locally. The insect infestation problem is of major concern for those responsible for the preservation of grain within the storage and distribution systems of the countries where it is produced for export because grain-importing countries are no longer prepared to accept grain with significant infestations and many have established strict legal, administrative or contractual requirements which have to be met.

Most of the insects and mites found in grain have originated in the Old World. Ships, trucks, trains, planes and every other conceivable type of conveyance have facilitated their distribution throughout the grain growing and grain handling regions of the world. Once established in a grain store, insects can damage stored grain by feeding on it, lowering germination, spoiling its appearance, lowering its milling quality, producing off-odours, causing it to heat and by creating conditions favourable for the growth of moulds, greatly increasing indirect losses of valuable food and feed.

Quantitative data on the importance of the major stored-grain pests and the losses they cause are not available from most countries. As the situation can vary markedly from place to place within a country, and from year to year because of differences in weather, changes in varieties grown, management of grain stocks, and pest control procedures and their effectiveness, precise figures are unobtainable.

Apart from the direct losses, indirect financial loss may result if the presence of insects and damage causes the commodity to be placed in a lower grade or be rejected entirely — this is particularly so when official grading systems are operated in national and international commerce. Infested grain may yield less flour and may contain excessive amounts of insect fragments. In some countries, grain products so contaminated may be condemned by public health authorities, but may still be used as feed for livestock.

The presence of insects may lead to grain or grain products being rejected if phytosanitary standards are applied on import into certain countries. Some commercial contracts may give the buyer the right to reject grain if insects are found, particulary in grain for malting. In other instances buyers may impose penalties operable under threat of not negotiating further sales. The cost of preventative and curative measures must also be taken into account. Disinfestation at port of entry often entails expensive demurrage charges.

The presence of insects or mites is recognised in national and international trade by clauses in commercial contracts and in the various official grading regulations of exporting countries. These place limits on the amount of insect damage which will be accepted, and may or may not tolerate the presence of living insects or mites. Thus, the Canadian and Australian regulations stipulate that wheat presented for export must contain no living insects or mites (the requirement being referred to as 'nil tolerance'); while the USA allows only one live weevil or a larger number of other insects per kilogram of wheat or maize. Similar official regulations apply to cocoa beans exported from Nigeria and Ghana and to groundnuts from South Africa and the USA. The mere presence of insects can result in severe financial loss if an importing country imposes strict quality standards. **Freeman (1976)** stated that this is so for China, where the discovery of one living insect in a cargo of wheat can result in the shipment being either rejected or accepted only after expensive fumigation.

The movement of infested parcels of grain and grain products in international trade has provided an effective means of ensuring the widest possible geographic spread of storage pests since cereals were first harvested. The continued introduction of the major pest species into climates outside their normal distribution ensures also that all these species may be found in all parts of the world notwithstanding that infestations of some species may be somewhat transient under local conditions, for example, *Sitophilus granarius* in the tropics **(Champ 1984)**. Resistant strains may also be transported **(Champ and Dyte 1976)**.

Because of the bulk and relatively low value of cereals, international transport is almost exclusively by sea, although rail and road transport may become significant in limited areas. The infestations which occur in ships originate either in the commodity before loading or from cross-infestation from residues of previous cargoes or other infested cargoes being carried. It is often difficult, however, to determine the precise origin of particular infestations.

It is interesting and useful to look back to the situation which prevailed 30 years ago and to recognise that enormous progress has been made in reducing infestation of grain in international trade. In the light of this success we must accept that further progress is possible and worth working for. John A. Freeman, who spent a lifetime in research, administration and teaching on the subject of pest infestation, made a review of problems and control methods **(Freeman 1957)** in which he placed great emphasis on the problems of controlling insect infestation in grain moving in course of trade. He pointed out that even if grain is not stored in the producing country, it may be in the consuming country. There is always the risk that grain may be damaged by development of infestation during carriage by sea. For these and other reasons, he felt that the prevention and control of infestation should be regarded as essential through the whole process from production to consumption.

In Freeman's opinion no country, whatever its geographical situation, can safely dispense with measures for prevention and control of infestation if it is to conserve its own reserve stocks and export grain in good condition. The need to protect grain from attack by insects as distinct from control of outbreaks, is particularly important in countries like Canada and the United States, where grain is marketed according to a uniform official grading system in which living insects and the damage done by them are grading factors. He referred to similar policies in Argentina, Morocco and Algeria. The existence of grading and quality control schemes which include tolerances for living insects has the effect of forcing those responsible for handling grain to carry out prevention and control of infestation.

Freeman spent several pages in reviewing methods of control which, in the absence of grain protectant insecticides other than DDT and BHC dusts, depended almost entirely upon fumigants, mainly the now superseded halogenated liquid grain fumigants together with methyl bromide and hydrogen cyanide.

Having recently made a survey of methods of infestation control used in grain exporting countries, **Freeman (1957)** was able to lay stress on the fact that a major factor in causing control measures to be carried out on a wide scale has been the existence of official quality standards for grain. Official controls had reduced the extent of the problem for the importing countries, but these measures appeared to have been taken primarily on the initiative of the exporters rather than as a result of any direct action by importing countries. Importing countries had for many years accepted insects and mites as something inseparable from the commodity and therefore not a matter which could be reasonably objected to. During recent years, however, this attitude has been disappearing and more pressure is being put on exporters to send goods free from infestation. Owing to this previous attitude to infestation, the United Kingdom, in particular, had at that time very few fixed installations in the ports and elsewhere for the treatment of grain and other commodities. There were no port silos equipped with built-in circulatory fumigation although these had been installed in port silos on the Continent, e.g. Antwerp and Rotterdam, for many years. Whereas the major French ports had modern vacuum fumigation installations used mainly for the treatment of pulses, there were no public fumigation chambers — atmospheric or vacuum — in British ports. This lack of fixed facilities, combined with the demand for treatment called for by official action or by enlightened importers, had caused the development of substitute methods such as fumigation in hatched store barges and under gas-proof sheets on quays and in dock sheds.

It is of interest to examine the report of the Infestation Control Laboratory of the U.K. Ministry of Agriculture, Fisheries and Food which is published every three years and which deals with many scientific aspects of pest infestation control. The plaintive statement which introduced the 1962–64 report on infestation of imported produce — 'Problems of control of infestation within the country would be simpler if imported food, feeding stuffs and raw materials were not infested. It is the policy of the Laboratory to encourage, by all possible means, the proper care of product in countries of origin and during transport so that imports are free, or nearly free from infestation, represent better value for money and do not give rise to all kinds of secondary losses' — is as scientifically true as it is politically obvious.

The report for that period went on to state that — 'during the period there was a general decline in overall infestation and the rate of occurrence of certain species, particularly those associated with cereals. This was mainly due to the marked reduction in infestation of wheat from Australia, and to a lesser extent in that from Argentina.' It should be noted that 1961–62 saw the first use of malathion for grain protection on a commercial scale. In the case of Australia this period not only involved the extensive use of malathion but also fumigation with hydrogen phosphide, together with inspection of grain at time of shipment and preloading inspection of ships.

Turtle (1965) reported results of a survey of residues in grain arriving in the United Kingdom from overseas. This revealed that wheat from Australia and Argentina and maize from South Africa contained low levels of malathion, indicating that the use of this insecticide as a grain protectant was becoming widespread.

The report of the Infestation Control Laboratory for 1968–70 contained the observation that recent legislative action overseas had resulted in the quarantine of the port of Mombasa, Kenya against storage pests (1968), the compulsory certification of Tanzanian export products as free from pests (1968) and a compulsory preloading inspection of the holds of ships carrying Nigerian produce (1969). The insistence of food manufacturers that many imported food products be completely free from infestation had brought about an extension of a practice of routine fumigation in the producing country to include Australian and U.S. rice and some Australian oats. The most notable change affecting the inspection of imports had been the rapid development of containerisation and development work in Australian which had shown that commodities can be satisfactorily fumigated after being loaded into freight containers.

The report for 1971–73 started off: 'The incidence of infested cargos of wheat has generally remained low with the virtual absence of infestation on Australian wheat being maintained. An exception has been that on French wheat which, with a considerable increase in the number of shipments, has shown an upward trend. The fumigation of commodities in containers before shipment, using either methyl bromide or phosphine, has now become routine practice for consignments from Australia, USA and Italy'.

The report for 1974–76 commented that 'the proportion of infested cargos found changed little compared with the previous 3 years, though this was partly due to an increased tendency to select for examination cargos that were more likely to be infested. The figure of 22% infested, however, maintained the general decline in the proportion of infested cargos intercepted over the last 10 years. It is apparent that the decline in infestation on certain foods and animal feeding stuffs reflects improved pest control. This has been most marked where the authorities in the producing country have been able to assess the efficacy of their control measures from regular intellegence about the condition of the commodity on arrival in the U.K. supplied by this Laboratory.'

Freeman (1974b) indicated that cargos imported into Great Britain were frequently infested with storage insects. The incidence ranged from 8% for cargos coming from Europe to 33% from Africa and the Far East and 38% from the Middle East **(see Table 3.1).**

Freeman (1974a) drew attention to the spectacular improvement which had been effected since modern pest control procedures were introduced in 1961. As indicated in **Figure 3.1**, from 80 to 100% of wheat shipments from Australia prior to 1962 were found to have insects in them on arrival in the U.K. This had declined to less than 5% by 1970 and by 1973 no infestations were found. This extremely satisfactory position has been maintained over the past eight years. The number of shipments inspected in 1970 was 166. **Banks and Desmarchelier (1978)** pointed out that Australian grain, once notorious for its high insect levels, now enjoys a reputation as one of the cleanest on the international market, a reputation achieved largely through the use of chemical pesticides, particularly malathion.

Table 3.1. Frequency of occurrence of storage insects intercepted in cargoes of plant and animal products imported into Great Britain in 1972, from certain representative areas of origin (total numbers of cargoes seen from each shown in brackets) **(Freeman 1974b)**

	Percentage of cargoes infested				
Principal insects	Europe (819)	Mediterranean (706)	Middle East (721)	Far East (603)	Africa (1,622)
Ephestia cautella	1	4	15	15	22
Tribolium castaneum	<1	1	25	24	16
Dermestes maculatus	–	1	7	–	2
Necrobia rufipes	–	1	4	5	2
Oryzaephilus mercator	–	–	4	10	1
Corcyra cephalonica	–	–	1	7	3
Lasiodernia serricorne	–	–	5	–	2
Trogoderma granarium	–	–	–	8	1
Alphitobias diaperinus	–	–	3	4	–
Sitophilus granarius	5	–	–	–	–
Ahasverus advena	–	–	–	3	–
Tenebroides mauritanicus	–	–	1	4	–
Sitophilus oryzae	2	–	–	2	–
Oryzaephilus surinamensis	<1	1	1	4	–
Dermestes frischii	–	1	4	–	–
Plodia interpunctella	–	1	–	–	–
Sitophilus zeamais	–	–	–	2	–
Infested with any insect	8	9	38	33	32

Based on inspections carried out by the Ministry of Agriculture, Fisheries & Food and the Department of Agriculture and Fisheries for Scotland.

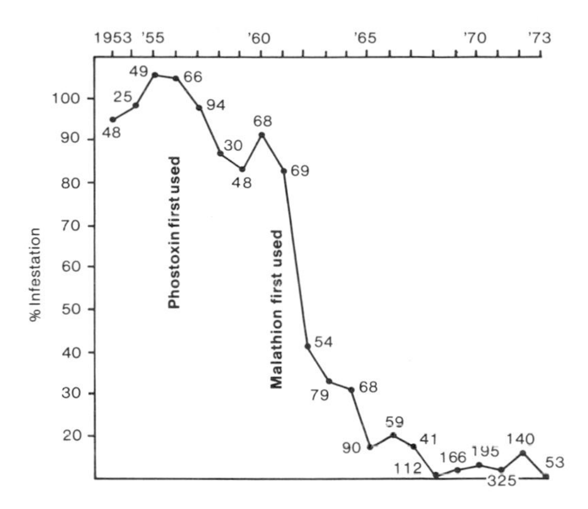

Figure 3.1. Level of infestation of Australian wheat imported into Great Britain. Numbers indicate cargoes inspected **(Freeman 1974a).**

Freeman (1974b) discussed the various insects and mites attacking stored grain, flour and malt in temperate countries with special reference to Great Britain. He pointed out that in most industrialised countries with temperate climates direct losses of food due to attack by insects and mites are relatively small and are of much less economic importance than the consequence of the discovery of small numbers of living or dead insects or mites by the purchaser. This means that in addition to the development of measures for control of existing infestations, considerable effort must be devoted to the prevention of infestation, especially in manufactured foods and those which pass to the shop with little or no processing, such as nuts and dried fruit.

Losses in temperate countries occur not only because of the development of insect and mite populations already established in unheated and heated buildings but also because of the introduction of insects in produce imported from other countries, both temperate and tropical, although the worst infested commodities are mainly of tropical origin. Uncontrolled import of infested commodities not only offers risks of spread of storage pests to new areas where they are not already established, including heated premises, but also of continued damage to the imported commodities during storage, and of spread of insects to clean commodities stored nearby **(Freeman 1974b)**.

Champ and Dyte (1976), in the report of the FAO Global Survey of Pesticide Susceptibility on Stored Grain Pests provide extensive information on grain production and trade, and the importance of major storage pests in each of 30 contries. The information is segregated for each of the major cereal varieties — wheat, barley, millet, sorghum, rice, and maize. The 300 page report recognises two distinct levels of pest controls. The first is at the local level and is essentially concerned with preventing physical damage or loss of commodities as for example in subsistence agriculture. The second type of control is concerned with contamination by insects of commodities in commerce, particularly in international trade. There is usually negligible physical loss of foodstuffs but partly for aesthetic reasons and particularly because of a growing appreciation of the potential damage and economic losses that insects represent, such contamination has become unacceptable in international trade. It is becoming increasingly apparent that technological advances in agriculture have created surpluses for marketing in both large- and small-scale farm production and standards of insect freedom applicable to international trade are now becoming accepted as a general requirement throughout the industry — both in local and export trade.

Champ (1976) pointed out that, although there appears to be a steady decline in the severity of the problem resulting from consumer demand and the associated introduction of compulsory inspection and fumigation of export commodities in the countries of origin, the extensive and spectacular control exercised by grain protectant insecticides, and less importantly containerisation, the problem remains serious. Of particular concern are the residual infestations in ships as these provide a continuing source of infestation by populations containing the various species and strains which have been carried previously in infested cargos. Less important but of significance also are the resident populations in the food storage and handling areas of the ship. The extent of these residual infestations has been highlighted by the introduction of compulsory inspection and treatment of ships' holds in exporting countries to supplement pre-loading control measures, as for example in Australia, Canada, Nigeria and the USA, where export grain and other commodities can be loaded only into ships that have been certified by inspection to be free of infestation.

Jeffries (1979) described the legislative, administrative and technical requirements established in Australia to maintain the quality of grains and allied food and feedstuffs held and handled for export and domestic markets. Because of considerable problems being encountered with insect infestation in export wheat in the late 1950s there were many complaints from overseas buyers leading to the promulgation of the Exports (Grain) Regulations in 1963. These regulations imposed obligations on the government, exporters and shipping companies. It is mandatory for inspections to be carried out prior to export to ensure that:

- the grain is not infested with insect pests;
- the grain is not otherwise in a condition that renders it unfit for export;
- the cargo spaces and dry stores of a ship presented to carry grain are not likely to cause the grain to become infested or contaminated in transit;
- other cargo likely to cause infestation or contamination is not loaded into the same cargo spaces;
- places in which grain for export is stored or vehicles

or equipment in which it is moved are not infested by insects or in a condition which may cause it to become unfit for export.

The acceptance of these regulations by grain handling authorities and exporters has enabled the Australian Government to establish a nil tolerance for insects in grain presented for export. This means that grain which is continuously under inspection as it is being loaded on the ship will be rejected if any living insect is detected. The rejected grain has to be fumigated or otherwise treated in an approved manner to ensure that all insects are destroyed before it can be re-presented for inspection prior to export. To achieve this high standard it has been necessary to ensure that storage and handling facilities are cleaned meticulously, inspected regularly and subjected to regular and systematic sanitation. Since grain used for domestic markets is not segregated from that destined for export the Australian consumer benefits from these strict export standards. The stimulus to devise techniques for handling, storing and safeguarding grain may have applications in many other countries.

The Council of Europe (1973) passed a resolution recommending that the governments of the 7 countries who are parties to the Partial Agreement, together with Denmark, Ireland and Switzerland should endeavour to persuade the countries from which they import cereals to adopt the following measures:

1. treatment with pesticides to be under responsible and expert supervision;
2. use of persistant organochlorine insecticides such as DDT, aldrin, dieldrin, heptachlor and chlordane to be prevented and the reason for this explained locally;
3. specified pesticides to be used only in accordance with prescribed procedures;
4. exporters to consult with authorities in importing states regarding the acceptability of cereals treated with pesticides.

The regulations list pesticides generally acceptable on cereals in importing countries and give notes on their application. They also describe acceptable methods of applying insecticides to cereals and storage premises and indicate effective dosage rates.

Snelson (1979d), in reviewing the prospects for the use of chemicals in protecting world food supplies, drew attention to the significant increase in the standards demanded by purchasers of grain and other stored products and how this has been reflected in national and international trade by clauses in commercial contracts. These place limits on the amount of insect damage which will be accepted, and may or may not tolerate the presence of living insects or mites. The presence of insects can result in severe financial loss if an importing country imposes phytosanitary or quality standards. He pointed to efforts made by FAO and other international agencies to bring about a better understanding and acceptance of the need to use chemicals in protecting world food supplies.

Snelson (1979b), in discussing the significance and safety of pesticide residues in stored grain, pointed out that they have become one of the quality standards which are important in the marketing of grain. He outlined the efforts that are being made to obtain international agreement on legal limits for residues. Whether by accident or by design, chemical residues, including residues of a wide range of agricultural chemicals, have become a hazard to international trade in food commodities and in a number of instances have become barriers to free trade in important foodstuffs. Concern over the development of such trade barriers and the need to have assurance on matters of public health concern, prompted the Food and Agriculture Organization (FAO) and the World Health Organization (WHO) of the United Nations to sponsor meetings of member governments and provide a forum for discussion and agreement on international standards for residues in food commodities. An ability to meet these standards is not only advantageous but essential to secure a place in the international grain market.

In a more comprehensive review of pesticide residues in international trade, **Snelson (1980)** stressed the importance of the efforts being made to bring about a better understanding of the many aspects of the evaluation and regulation of pesticide residues and the harmonisation of legal limits in the interests of public health and international trade in foodstuffs.

Desmarchelier (1981a), in discussing the use of insecticides for the control of infestation on exported grain and pulses, pointed out that acceptance by importers is the limiting factor for the use of insecticides to control insects on grains and pulses. A major rationale behind the philosophy of the Codex Alimentarius Commission is acceptance by all countries of recommended maximum residue limits, so that trade can be free and exporters can plan strategies for pest control. An exporting country can only segregate grain according to pesticide preference of importing countries at great costs, and only if the preferences of such importing countries are effective in the exporting country. The problems associated with the control of insect pests in hotter climates are not always recognised by those living in colder climates.

Desmarchelier (1981a) believed that we face an enormous task over the next 20 years to prevent the food shortages foreseen by President Carter's Global Survey. Protectants are an indespensible part of an overall strategy to alleviate famine. This is because they can be used in all types of storages and because they provide protection during transport of grain. This point is important when it is considered that imports in 20 years' time in Asia are predicted to nearly equal current world imports. This task may well be impossible, given the best circumstances, but non-acceptance now by importers of the international standards proposed by the Codex Alimentarius Commission will make the difficult task even harder. We need to develop new chemicals that will not be used until 1990, but this will not be done if non-acceptance of insecticides is a barrier to trade.

Snelson (1981c), describing the regulation of chemicals and chemical residues, pointed out that the buyers of agricultural produce dictate not only the demand and price, but standard of quality as well. In order to meet the high standards demanded by overseas markets or set by foreign competitors, producers must employ modern technology to prevent blemish from pests and diseases. In so doing, it is absolutely necessary to avoid odours and visible residues and to control invisible residues to ensure that these never exceed the limits fixed by legislation or convention in the market place.

Wilson and Desmarchelier (1981), having encountered great difficulties in sampling bulk grain held in vertical and horizontal silos, developed a technique in which the railway wagons, and presumably also the motor trucks, used to transport the grain to terminals and seaports could be used to obtain representative samples of the grain for the determination of chemical residues.

Story *et al.* (1982b) advised that insect infestation is a source of foreign complaints about the quality of US grains. To investigate what was considered to be the most significant and prevalent quality problem, the incidence of insects in wheat and corn for export from the United States was determined in grain loaded at 79 port terminals in the United States and Canada during a 2-year period from January 1977 to December 1978. One or more live, stored-product insect species were found in 17.9% of the wheat samples and 22.4% of the corn samples. Insects of the genus *Sitophilus* were the predominant ones in all samples, with 7.7% of the wheat samples containing an average of 4.2 weevils per 1000 g and 14.4% of the corn samples containing an average of 5.8 weevils per 1000 g.

Story *et al.* (1982a) measured the incidence, concentration and effectiveness of malathion treatments in wheat and maize samples obtained from 79 port terminals in the United States. The presence of biologically active levels of malathion was inferred by bioassay assessment in 28% of 2058 wheat samples and in 8.4% of 2383 maize samples examined during a 2-year period from January 1977 to December 1978. At terminals where the grain was not treated immediately prior to sampling, the incidence of malathion on grain arriving at the ports was only 11.6% for wheat and 7.7% for maize. Among all grain samples in which the presence of malathion was inferred from the bioassay data, only 2.7% of the wheat samples and 0.7% of the maize samples contained live insects. In contrast, the overall incidence of live insects in all exports samples examined during the 2-year period was 17.9% for wheat and 22.4% for maize. These data suggest that malathion is still sufficiently effective to limit insect infestations in grain. The investigators deduced that the overall use of malathion in stored-grain insects control programs from farms to export was minimal.

Wilkin *et al.* (1983) carried out a survey of pesticide use on grain and commercial grain stores in the United Kingdom during 1982. They found that grain protectants were used in 96% of commercial stores including 28% in which they were used to treat all the grain received. The authors considered that this was in response to the more stringent marketing requirements which have affected the UK grain trade as a result of considerably increased production, the growth of the export trade and membership of the EEC. Approximately 25% (5 million tonnes) of the UK cereal production was exported in 1981.

Horrigan (1984), in discussing the export inspection of grain in the future, considered that the ability to export grain to the current standard of freedom from insects is dependent on the availability of grain protectants. The desire to reduce the dependence on pesticides has led to the investigation of alternative methods of pest control. The alternative methods now available have been developed to the point where the economics of some approach that of grain protectants. These methods, e.g. ambient aeration, use of carbon dioxide and fumigants in sealed storages and thermal disinfestation, will, if used carefully, still allow the industry to export grain that meets the phytosanitary requirements of overseas countries. However, these methods also present practical problems in that they provide no residual protection during transport of grain overseas. It has been demonstrated that storage insects are very mobile, with the ability to seek out small amounts of grain over considerable distances. Observations in export grain terminals show that endemic populations of grain insects are usually present and extremely difficult to eradicate. Reinfestation from grain brought from country storages occurs regularly. The inference is that unprotected grain will, during transport and handling, become reinfested prior to export unless greater efforts are made to provide a pest-free grain pipeline.

Snelson (1984) in an extensive presentation concerning pesticide residues and their significance mentioned that the presence of chemical residues in food and agricultural commodities has resulted in many 'unofficial' trade restrictions. When the importing country has established very low legal limits for such residues, it is possible to reject or refuse imports on the grounds that the rejection is legally justified. The defence that legal residue limits are designed to protect the health of the consumers is hard to challenge. Many countries have attempted to seek protection for their local agricultural interests by designing legislation refusing importation of food commodities containing even insignificant quantities of residues. Such moves have been attempted in countries with heavily subsidised agriculture where efforts to invoke low maximum residue limits have been proposed in an effort to achieve protection for an uneconomic local agriculture.

Semple (1985), in reviewing problems related to pest control and use of pesticides in grain storage, drew attention to the current situation in ASEAN countries where concerted efforts are being made to expand the production capabilities of the major staple food and feed grains to the point of self-sufficiency or even export. Such a policy is necessary to reduce burgeoning foreign exchange deficits. To meet this objective, considerable attention is being given to improve storage facilities, warehouse management and the utilisation of pest control techniques. These will have to be applied at the farm or rural village level as well as in the commercial sector and for long-term storage of national reserves. The rapidly expanding population throughout South East Asia means that considerable quantities of grains, far above those possible under traditional methods, must be accommodated.

Snelson (1985a), in dealing with safety considerations for use with pesticides in humid tropical grain storage systems, stressed the need to ensure that at no time do residues of grain protectant insecticides and fumigants exceed the limits fixed by legislation or convention in the market place. When a consignment is rejected in a foreign market the value of that consignment is lost and the resulting publicity can often damage the future prospects for the exporter and for others engaged in similar trade. If the loss of the consignment means that the shipper can no longer meet contractual obligations, substantial damages may be claimed by the importer and the exporter is likely to face a considerable increase in insurance premiums on future shipments. It is not unlikely that such an incident could result in protracted negotiations on a government to government basis. More often than not the incident arises merely because in the importing country:

1. there is not yet a legal limit for the particular residue in that food commodity;
2. the limit is set at a slightly lower figure than is accepted internationally;
3. the definition of the residue is different to some minor extent;
4. sampling and analytical problems have resulted in apparent violation of the laws.

Snelson and Desmarchelier (1975) presented results of an international collaborative study on the analysis of grain for residues of grain protectant insecticides with reference to the following criteria: half-lives, representative sampling, analytical variability and evaluation of extraction procedures. They pointed to the need for international agreement on many of these issues and particularly the desirability of incorporating confidence limits into methods of regulatory analysis.

Although there appears to be a steady decline in the severity of the problem of insect infestation resulting from consumer pressure and the associated introduction of compulsory inspection and fumigation of export commodities in the countries of origin, the extensive and spectacular control exercised by malathion and the newer grain protectant insecticides, and less importantly, containerisation, has had a marked influence on the incidence and level of insect infestation in infestable commodities moving in trade. Unncessary concern over residues has, to some extent, taken the place of the real problems produced by insect pests.

4. Control Measures

In order to appreciate the place of and prospect for grain protectant insecticides it is important to understand how they compare with and relate to alternative control methods.

Each of the many methods of insect control has its place and its advantages. There are many situations where two or more methods can be used in an integrated program. Not all methods, however, are suitable for use in every situation. Without effective and well-planned storehouse management all the available control techniques will fail.

Certain non-chemical methods of pest control appear attractive since they do not suffer from the disadvantage of residues and the development of resistance. However, they require specialised storage structures and equipment involving high capital investment, sophisticated management and facilities. Their introduction to industry on a significant scale is therefore likely to take many years, probably many decades, in much of the world.

Chemical control methods, on the other hand, have the advantage of effectiveness, simplicity, versatility, low cost and immediate availability and are therefore the ones most likely to be used during the next several decades. Before looking at grain protectant insecticides let us first review the non-chemical and alternative chemical methods of insect control.

Chemical control must be placed in its correct perspective. It is necessary to reiterate the framework on which practical infestation control programs are based. Practical infestation control is an integration of the following:

- adequate drying of the commodity to be stored;
- use of suitable storage facilities and, if necessary, their improvement to an acceptable standard;
- use of aeration and other physical control methods if practicable;
- good warehouse keeping including sanitation;
- regular inspection for infestation or other causes of deterioration;
- use of commodity protectants;
- use of residual insecticides on storage structures;
- fumigation, where infestations become established;
- if appropriate, changes in varieties of grain produced and changes in harvesting techniques.

4.1 Biological Control of Stored Grain Pests

Biological control may be defined as a method of reducing damage done by a pest by means of a biological agent — traditionally a parasite, predator or disease. In a broader sense, biological control may be considered to include the manipulation of other biotic facets of the pest's life system, such as its reproductive processes, its behaviour, the quality of its food, and so forth. Here, the broader view will be taken so that some relatively new approaches, as well as the classical ones, may be considered.

The stored grain ecosystem places several constraints on the potential use and success of biological control. The level of pest infestation tolerable to grain exporters and consumers makes it very difficult for pest populations to be sufficiently abundant and persistent for biological agents to establish themselves and become effective. Predators, parasites and genetically manipulated pest populations found amongst the grain would themselves be regarded as contaminants. The physical conditions of well-stored grain are not those which favour the spread of pathogens.

All in all, the potential for biological control of stored-grain pests appears limited.

Arbogast (1984) reviewed the status and prospects for biological control of stored-product insects and concluded that the potential value of biological control has been clearly demonstrated, but technologically it is still rudimentary. Further development will require field trials to answer questions concerning: (1) efficacy under actual conditions of commercial storage; (2) compatibility with other storage procedures; (3) cost effectiveness; and (4) undesirable side effects.

Evans (1984b) has given a comprehensive overview of the subject and much of the information in this section is taken directly from his work.

4.1.1 Parasites, Parasitoids and Predators

The parasite is an organism that benefits by feeding upon, or securing shelter or transportation from, one or more other organisms, its hosts, which may not necessarily be killed. By contrast, a parasitoid invariably kills its host and requires only one host for development. A predator is an organism that feeds upon 1 or more hosts but seldom obtains shelter or transport from them. Predators always kill their hosts.

The abundance or density of a parasite or predator generally depends upon the abundance of its host. Characteristically, the relative number of the host and parasite oscillate somewhat out of phase i.e. an increase in the host population is followed by a corresponding, but delayed, increase in the parasite or predator population. The reduced host population cannot support the increased parasite population, which then gradually wanes and allows the pest population to recover and a another oscillation to start.

A successful parasite or predator **(Stehr 1975)** has attributes such as:

- a general adaptation to the environment and the host;
- a high searching capacity;
- a high rate of increase;
- a mobility adequate for dispersal;
- a small time lag in responding to a change in the abundance of the host.

Evans (1984b) discussed each of the parasites and parasitoids known to attack stored-grain pests. Even the most aggressive failed to reduce pest populations either within an acceptable period of time or to tolerable levels.

Insects of the order Coleoptera (beetles), Heptoptera (bugs) and Hymenoptera (wasps) are capable of attacking stored-product pests and several have received much attention as possible biological control agents. Some of these species will attack humans and none appears to have characteristics that make it suitable for controlling stored product pests under practical conditions.

Mites are frequent predators of stored-products insects and other mites **(Bare 1942)**. None is sufficiently predaceous or adaptable to typical storage conditions to exert an acceptable degree of control on their hosts.

Haines (1984), in an evaluation of the use of predators and parasites in biological methods for integrated control of insects and mites in tropical stored products, concluded that there have not yet been any large-scale attempts to use biological control by natural enemies in an integrated program of storage pest management. Many species of naturally occurring predators and parasites are commonly and widely distributed in tropical storage. There is thus little potential for the classical concept of widespread artificial introduction of natural enemies of storage pests to a geographical area. However, considerable evidence exists to suggest that biological control, by the existing natural enemies in stores, could be achieved by the following: modification of conventional control procedures to reduce their effects on beneficial organisms; modification of aspects of the storage environment to favour natural enemies; and augmentation or re-introduction of predators and parasites in particular situations. Nevertheless, there have as yet been very few field trials to determine what levels of pest control can be achieved by these techniques and to assess their usefulness in practice.

Haines (1984) concluded that before any suitable control strategies can be developed and implemented, many more observations and studies, basic and applied, need to be made. The major priorities for future research are:

1. observations on the prey/host range of many of the natural enemies for which such basic data are lacking, and surveys of their distribution and occurrence;
2. ecological and behavioural studies, under relevant environmental conditions, on species that appear to have potential as control agents;
3. increased emphasis on the use of field trials to evaluate the efficiency of potential biological control agents. Applied research has been concentrated, in the past, on only a few groups of natural enemies. There has been a lack of relevant research on the parasitic Hymenoptera, which deserve more intensive study. **Kent-Jones and Amos (1957)** reported that such wasps serve as a useful indicator of the presence of *Tribolium* species, but do not effectively control such species. Wasps are unable to penetrate bulk grain.

4.1.2 Diseases

The successful use of diseases for biological control depends on the biology of both host and micro-organism **(Maddocks 1975)**. The host insects must occupy habitats suitable for the introduction of the pathogen and have habits that facilitate its transmission and spread. As diseases act in a density-dependent manner, insects that aggregate or are abundant are more likely to be suitable subjects for this approach than those that are well dispersed and seldom attain high density.

The principal attraction of control by pathogens is that they are rather specific and thus harmless to non-target organisms; they are compatible with and even, at times, synergistic with insecticides; they are relatively easy and cheap to culture; and they are not associated with rapidly developing mechanisms of host resistance. The main disadvantages are their requirements for careful timing of application relative to incubation periods; their specificity, which may limit their effectiveness in situations where a complex of species is involved; the maintenance of their virulence which may be difficult; and, finally, their infectivity which may depend greatly on favourable climatic conditions.

To store grain successfully the humidity must be controlled preferably below 65% to prevent the proliferation of moulds which degrade the grain, destroy its food and feed value and produce mycotoxins which have far-reaching consequences for human and animal health **(Goldblatt 1969; Purchase 1974; Rodricks *et al.* 1974)**. Such controlled humidities are less than ideal for the survival and proliferation of entomophagous fungi and bacteria. Whilst the spread of fungi and bacteria may be promoted by physical contact and cannibalism between insects, such processes are density-dependent and therefore run counter to the objectives of good grain storage management.

Much effort has gone into the commercial production and use of *Bacillus thuringiensis*, which was first isolated from diseased larvae of *Ephestia kuehniella*. The bacteria can infect *Ephestia cautella, Ephestia kuehniella* and *Plodia interpunctella*. Treatment of the top 10 cm layer of wheat in small-capacity bins with *Bacillus thuringiensis*, reduced *P. interpunctella* and *E. cautella* populations by about 81% and reduced feeding by more than 92% **(McGaughey 1980a)**.

Subramanyan and Cutkomp (1985) published an extensive review of moth control in stored grain and the role *Bacillus thuringiensis* might play in this. They pointed out that very little has been published on the efficacy of *Bacillus thuringiensis* in controlling moth pests of stored grain. Most of the research has been done in the United States by McGaughey and co-workers. The species studied were *Plodia interpunctella, Cadra cautella*, and *Sitotroga cerealella*. Although doses as low as 10 mg/kg have been reported to be effective against some moths in some cereal grains, doses as high as 250 mg/kg did not provide effective control of *Sitotroga cerealella* in maize.

Even if *Bacillus thuringiensis* were to be regarded as a 'disease' rather than a biologically-produced insecticide it would have to be used in a manner analogous to other insecticides e.g. pyrethrum, where its lack of effect on the major pest species (beetles) would limit its usefulness.

Other biologically produced insecticides such as avermectin may have potential though this needs to be further explored. Once again, such materials are insecticides rather than insect-pathogens.

The use of viruses for the control of pests of field crops and forests has met, in some instances, with considerable success. The granulosis virus and nuclear-polyhedrosis virus have been found in the larvae of *Ephestia cautella*. **McGaughey (1980b)** used surface applications of a granulosis virus to control this pest in small bins of stored maize and wheat.

A variety of fungi are spectacular, if somewhat erratic pathogens of insects. The performance of entomogenous fungi is often erratic due to the necessity for suitable temperature and humidity conditions to prevail before the spores can germinate and infect their hosts. In addition, the density-dependence of these pathogens frequently results in their achieving adequate levels of control only when the pest has already done considerable damage.

The protozoa have received relatively little attention to date because they cause chronic rather than acute infections. However, the reproduction of the hosts is often curtailed with resultant long-term reduction of the host population rather than high initial mortality.

Hodges (1984b), in reviewing the use of insect diseases for integrated control of insects and mites in tropical stored products, concluded that insect pathogens offer distinct possibilities for the control of stored-product pests. They may be admixed with bulk grain or disseminated by insects lured to a source of pathogen by an attractant. One insect pathogen, *Bacillus thuringiensis*, is already commercially available for use against moths. It may be especially useful in situations where the moths are particularly tolerant to insecticides or where insecticide application is either not permitted or cannot be carried out effectively because of unsuitable storage buildings. Before *Bacillus thuringiensis* is brought into more widespread use, thorough testing of its efficacy under tropical conditions is necessary.

Hodges (1984b) also believed that considerable research is needed on the potential use of protozoa and viruses. For example, the use of pheromones and other attractants in pathogen dissemination would seem a promising, yet almost unexplored field. It would also be valuable to discover a single virus that could control all storage moths. Owing to the rather specific nature of pathogens it is almost inevitable that they will have to be integrated with other control procedures for the protection of stored food against a wide range of pests. Procedures for the integration of pathogens with other control measures offer a further area of necessary research.

In recent years, much thought has been given to the possibility of controlling pests by means of manipulating their genetic make-up. Several distinct types of genetic manipulations or combinations of 2 or more types are now proposed, usually for control of medically important insects or field-crops pests. However, successes have been few because the biological and logistical problems of controlling large, dispersed and often mobile populations in nature are formidable. In contrast, stored-product insects, which might be more easily handled, have been largely ignored by those interested in genetic manipulation for insect control. **Brower (1974, 1975)** has made a comprehensive evaluation of the potential for genetic control of stored-product insect populations and his paper contains a bibliography of 77 references and provides a comprehensive appraisal of this form of approach to pest control. However, genetic control is not a panacea. Because it is species-specific, it would be more difficult to apply if many different pest species were present in the target area. However, some commodities are normally attacked by only 1 or 2 species, and some species presently have developed resistance to some insecticides. In such instances, control by genetic manipulation may well have an important role in an integrated pest control approach. However, on present indications there appears to be little likelihood that genetic control could be adapted for the control of pests in grain destined for international trade or for the management of pest populations at farm or village level.

4.1.3 Pheromones

Pheromones are chemical 'messages' released by organisms to influence the behaviour of other organisms of the same species. In the case of non-social insects, pheromones seem to be employed mainly in sexual situations, for example, to advertise sexual receptivity and/or to attract potential mating partners **(Burkholder 1979; Levison and Levison 1979; Hodges 1984a)**.

Stored-grain Lepidoptera and Coleoptera are both known to use pheromones in sexual situations. The pheromones have usually been shown to be blends of several different chemicals. Certain of the more volatile types of identified pheromones have been synthesised and made available commercially. In practice, however, the majority of pheromones identified in recent years seem to be relatively non-volatile, acting more as 'arrestants' (i.e. insects tending to remain in the vicinity of a source of such chemicals once they have 'fortuitously' arrived in that vicinity) rather than as 'attractants'. The potential for control strategies based on the use of 'arrestants' seems to be very much less than that of 'attractants'.

Another obvious way in which 'attractant' pheromones might be employed is to prevent mating and thereby eliminate the next generation before it is conceived. Attempts to achieve this end by trapping out the males of a population have never proved of much value, mainly because the non-trapped males are able to mate a sufficient number of times to minimise the effect.

However successful such methods eventually prove, it must be recognised that on their own they will almost never provide complete control of field populations. As attractants for traps that might enable population build-ups to be detected and monitored, however, they may yet prove completely successful when combined with other control measures such as appropriately timed insecticide treatments **(Hodges 1984a)**.

There has been success from pheromone traps containing a moth-attracting pheromone and dichlorvos. These have proved effective in European flour mills and stores where the standard of hygiene is high and moths are the only species that invade premises and then usually only in low numbers by tropical standards. This specialised application was proven worthwhile but it is unlikely to meet the requirements where massive infestations of mixed populations are encountered.

Burkholder (1984) reviewed the use of pheromones and food attractants for monitoring and trapping stored-product pests and concluded that they are powerful tools and, when handled properly, can aid in the effective management of pest insects. He forecast that pheromones for other stored-product pests such as *Sitophilus* spp. will shortly be available and these should offer a bright future for detecting and monitoring these insects.

4.1.4 Host Resistance

Host resistance to insect attack is an important aspect of biological control. Some of the factors involved in stored grains are as follows:

1. *Husk cover of maize*. Long tight husks reduce attack by weevils and grain moths.
2. *Integrity of the husks of rice, barley, and oats*. Seeds with gaping or damaged husks are more susceptible to attack.
3. *Antibiosis*. Seeds of some plants contain chemical substances unacceptable to certain insects.
4. *Physical and nutritional properties of seeds*. A hard seed coat and/or a vitreous endosperm contribute to resistance. Maize seeds lower in certain carbohydrates are less susceptible to weevil attack than normal maize seeds.
5. *Definite oviposition preferences* exist amongst seed species and varieties. Prospects of breeding or selecting plants with less susceptible seeds are promising.

The resistance to pest attack of many varieties and hybrids of cereals has been repeatedly demonstrated in laboratory experiments. At more subtle levels, several varieties of shelled maize may be equally susceptible to *Sitophilus zeamais* in terms of the number of eggs laid but resistant in that they may share marked differences in the survival of the immature stages **(Dobie 1975)**. Thus, Kenya maize hybrid H622 is less resistant than ordinary Kenya Yellow maize in that almost 3 times more weevils emerged about 14% more rapidly from H622 than from Kenya Yellow, following equal oviposition by attacking weevils. The differences in survival appear to be related to the flouriness of the endosperm **(Dobie 1975)**.

Dobie (1984a) reviewed the use of resistant varieties for integrated control of insects and mites in tropical stored products and concluded that the use of resistant varieties can contribute effectively to the safe storage of food crops. Most crop species that have been investigated have shown a certain amount of variation in resistance to major storage pests and plant breeding programs could be designed to exploit such variation and to produce resistant varieties. Farmers who store their own produce in tropical and sub-tropical areas are usually very aware of the need for crop varieties to be resistant to storage pests. It remains necessary for storage entomologists and plant breeders to collaborate increasingly in the production of varieties that have other desirable agronomic characteristics together with resistance to storage pests.

In considering the question of resistance of stored commodities to attack by traditional stored-product pests, it is important not to jump to the wrong conclusion as a result of limited experimentation. Many species of stored-product pests have difficulty in adapting quickly to an unfamiliar cultivar of the same type of grain that they and their antecedents have been accustomed to. To take insects that have been reproducing successfully on one cultivar and transfer them to another almost always results in a sudden lowering of their productive (and reproductive) capacity — but only for 1 or 2 generations. Workers at the Stored Grain Research Laboratory, CSIRO, Canberra always breed insect cultures on a new cultivar/variety of wheat for 3 generations before initiating detailed experiments using the new insect strain or new grain variety.

4.1.5 Use of Sterile Insects

Hodges (1984c) reviewed the use of sterile insects for integrated control of insects and mites in tropical stored products and concluded that at present no sterile insects method is used for the control of pests in tropical stored products. There has been considerable research into certain methods of producing sterile insects but this has resulted in only 1 trial of 1 technique in a real storage system. This is possibly because researchers are deterred by three problems:

1. The difficulties in supplying sterile insects. This may be due to problems of organisation and expense. Such problems are increased if control is attempted where there is a large population of infesting insects. This is particularly true in species where effective control can only be achieved when there is a high ratio of sterile to normal insects. Examples are 24:1 for *Ephestia cautella* (**Brower 1980b**) and 10:1 for *Tribolium castaneum* (**Pradham *et al.* 1971**).
2. The resistance of store managers to the release of large numbers of sterile insects into stores. In many cases this is quite justified as the insects would cause significant contamination and food loss.
3. The need to release several and often many species of sterile insects.

It is clear that the uses of sterile-insect control are limited. However, there are some, as yet, untried possibilities for this technique. For example, small numbers of sterile insects, males and perhaps also females, could be released at an appropriate time after routine chemical control procedures. The results of this may be a reduction in the reinfestation rate and consequently less frequent use of expensive fumigants and insecticides. The size and frequency of insect release for such a method would have to be determined either by a study of the normal reinfestation rate of treated produce or by trial and error. As an alternative, insect release could be avoided altogether if pest populations were sterilised by contact with suitable chemosterilants. It remains to be seen whether or not the sterile-insect method can eventually be used as an alternative method for control of tropical stored-product pests. However, before any judgment can be made on the value of this technique it is important that the more feasible possibilities should be thoroughly tested.

Brower (1975) pointed out that although genetic manipulation of natural insect populations is being used to control at least 3 pest species and research with others is promising, little effort has been directed against post-harvest insect pests. Nevertheless, stored-product insects offer unique advantages for the successful application of these new control techniques and could well serve as models for genetic control of other pest species. They occur as discrete populations in storage structures that can be controlled individually, thus greatly simplifying logistical problems. Also, populations of stored-product insects are easy to sample, are easy and relatively inexpensive to rear, have short generation times, can be reduced by conventional means, are compatible with an integrated control approach, have insecticide-susceptible strains that are still available, and have known biology and genetics, at least in the more important species. In the final analysis the technique must be evaluated as a commercial proposition.

4.1.6 Constraints on Biological Control

Dobie (1984b) summarised the role of biological methods for integrated control of insects and mites in tropical stored products and concluded that the integration of biological, physical and chemical control measures is a young science. A number of good laboratory-based studies have provided a sound basis for full-scale field trials which would identify effective integrated control strategies for use in stores in the tropics. The relative cost of integrated methods cannot easily be predicted as the biological agents necessary are often not yet available commercially. The true cost of novel methods will only be determined when the demand has been created for the supply of biological agents and their commercial production has been stimulated.

The stored-grain ecosystem places several constraints on the potential use and success of biological control. Some of these are as follows:

- The level of pest infestation tolerable to grain exporters and consumers makes it very difficult for pest populations to be sufficiently abundant and persistent for biological agents to establish themselves and become effective.
- Predators, parasites and genetically manipulated pest populations found among the grain would themselves be regarded as contaminants.
- Grain in storage is frequently treated with insecticides that are toxic to both pest and beneficial species.
- The large bulks in which grain is frequently stored are inimical to these searching predators and parasites whose activities are restricted to superficial layers.
- The physical conditions, especially moisture, of well-stored grain are not those which favour the spread of pathogens.
- There is generally considerable emotional, yet understandable, reaction on the part of consumers to eating food 'treated' or, in their minds, 'contaminated' with pathogens.

All in all, the potential for the biological control of stored grain pests appears limited. However, some approaches, particularly the use of pheromones and of resistant varieties, do seem feasible and should be further investigated and evaluated particularly with a view to integrating their use with existing physical and chemical control methods.

4.2 Physical Methods

Whitney (1974) has provided a bibliography of over 300 references to physical control methods and the following arbitrary groupings:

- management and sanitation;
- ecological control;
- radiation;
- mechanical forces, physical barriers and inert dusts;

- interactions of physical control with insecticidal protectants.

In considering the information on these topics **Whitney (1974)** asks us to keep in mind that control in any form depends primarily on making the environment *un*favourable for the development of pest species.

Banks (1976, 1981a) reviewed recent developments in physical control of insects covering quite a number of topics that are not specifically mentioned by Whitney. He described physical control as 'alteration of the environment by physical means to make it hostile or inaccessible to the pest insects'. A bibliography of some 170 papers was provided.

4.2.1 Hygiene and Physical Removal

No review of physical control is complete without mention of the necessity for hygiene, particularly the physical removal of infestation foci such as commodity residues, secondary or unproductive primary hosts for field pests. Little research attention is paid to this aspect of physical control. For instance, while residual infestation in commodity storages or handling equipment is acknowledged as a source of infestation the quantitative influence of these sources on subsequent insect population is not known.

Ripp (1981) summarised extensive knowledge gained in the management of a large, diverse grain storage system. He pointed out that general tidiness and an absence of visible grain residues had been the standard of acceptable cleanliness until the development of insecticide resistance in insects had shown that minute cracks and crevices can hold grain residues and provide protection for insects. Hygiene implies a change in attention towards acceptable standards of cleanliness.

Effective hygiene is necessary in any pest control program to:

- reduce residual populations of insects in grain stores, surrounds and machinery;
- avoid seeding of clean grain with insects;
- avoid cross infestation of adjacent stores;
- reveal structural weaknesses and possible sources of water ingress.

All areas, both within and surrounding the storage, require meticulous inspection and cleaning. Cleaning should involve brushing and washing, and disposal of all residues that may contain or support live insects. Non-chemical control is not a possibility without strict attention to hygiene. Poor hygiene practices can negate the value of grain protectant insecticides **(Ripp 1981)**.

Evans (1981) regards hygiene as an essential precursor to the successful application of all other (biological, chemical and physical) methods of pest control.

Surroundings of grain stores should be kept permanently free of weeds and other vegetation to make access by crawling insects and rodents more difficult. The use of residual herbicides, good drainage and good housekeeping is essential.

4.2.2 Physical Exclusion

The packaging of foodstuffs to exclude insects has a long history. Recent developments have included the extensive use of plastics **(e.g. Wilkin and Green 1970; Mallikarjuna Rao *et al.* 1972)** and sealable tins in the tropics **(McFarlane 1970)** for small-scale storage of products. The physical barrier used to contain grain under semi-hermetic conditions (e.g. butyl rubber silos) helps to prevent infestation from external sources. Underwater storage, combining an excellent physical barrier to storage pests with temperature and environmental gas control, is under evaluation for bulk rice storage **(Mitsuda *et al.* 1971; Tani *et al.* 1972)**.

Both traditional and modern forms of hermetic and controlled atmosphere storage require a high degree of gas-tightness to be effective **(Banks 1976)**. The sealing that provides such gas-tightness hinders insects attempting to invade the grain within the storage and is an important adjunct to the long-term effectiveness of control measures, such as fumigation and single-treatment controlled atmosphere and heat disinfestation, that, of themselves, provide no residual protection of the grain after treatment.

4.2.3 Grain Drying

One of the penalties for delaying harvesting until the grain is dry enough to store is that shedding and other losses will occur. In humid climates there is little choice: grain has to be harvested damp and then, usually, dried. In its early states of development, grain has a moisture content of some 70–80%, but as the grain swells this begins to fall towards 35–40%, the level at which the accumulation of dry matter ceases. Periods of wet or dry weather during this time have little effect on grain moisture content: the overall trend being downwards. After the grain has stopped 'growing' it continues to lose water, but changes in moisture content, in response to changing weather, become more marked. For example, rain falling on a ripe crop may cause grain moisture content to rise by 10% and, in a cool climate, it can require 2 fine days to bring it back to its original level.

In countries with a maritime climate, most grain is harvested at moisture contents of between 17–22%: in exceptionally dry conditions it may be 14% or 15%, although by this time, shedding losses can be substantial in some varieties. In countries with high summer temperatures and low relative humidities, the moisture content range will tend to be smaller than in countries with a maritime climate and the overall moisture content level will be lower. Nevertheless, early harvesting of relatively high moisture content grain is sometimes advisable in these areas, because shedding losses can become unacceptably high, and losses caused by high populations of grain-eating birds can be a major factor in determining final yield.

Several factors have a bearing on whether or not grain will store safely and they all tend to interact with one another. Moisture content, temperature, the volume of grain and the presence of pests, dusts and other contaminants all influence the rate at which changes in stored grain take place. Deterioration will normally involve the development of moulds and infestation with mites and insects, all of which are accompanied by heating. Changes may not be easy to detect — and they can take place quickly and be difficult to arrest. Given these facts, it is not surprising that recommendations relating to storage tend to err on the side of safety. The storage of grain with an equilibrum relative humidity of less than 70% should certainly be safe, but for the bulk of the world's grain-producing countries it is generally not practicable to achieve this level without artificial drying.

Modest variations in diurnal temperature set up convection currents in bulk grain. These carry water vapour from a slightly warm zone within the store to a cooler one where it condenses, raising the grain moisture content to a level at which mite and fungal activity can begin. Sun shining on one side of a store can cause moisture migration to the cooler, shaded side where it can initiate the spoilage process. Similar problems can occur in grain while it is being shipped, with dire financial consequences. Steps can be taken to prevent the development of this type of problem; these vary in cost, flexibility and ease of management but frequently involve forced ventilation at regular intervals.

There are so many ways in which man has contrived to dry grain that it is almost impossible to review them, but in the last 20 years there has been a general trend towards some form of storage dryer. Currently, there is some renewed interest in high capacity, continuous dryers, particularly by farmers who have formed co-operative handling and marketing groups.

Until recently, the energy costs of grain drying have been regarded as being relatively unimportant, as it has nearly always cost less than 5% of the value of the crop. Greater emphasis has tended to be given to ease of management, the ability to handle wet crops, the ease with which batches can be kept separate and the avoidance of bottlenecks that interfere with harvesting. Only in the last few years have energy cost aspects come to the fore **(Wilton 1980)**.

Continuous machines which use direct-fired oil or gas heaters to provide drying temperatures between 40° and 100°C have many attractive features but **Nellist (1979)** stated that the energy requirements range from 3.5 to 9 MJ per kg of water evaporated.

Storage dryers which use fans to blow large volumes of either slightly warmed air or air at ambient temperatures through the grain are more economical but their capacity is related to fan output and the relative humidity of the air and air temperature. **McLean (1979)** calculated the energy requirements to range from 1.5 to 1.8 MJ per kg of water evaporated. McLean pointed out that factors such as grain species, depth of storage, the presence of impurities, the method of filling and the design of ducting, all influence the system's energy requirement. In high humidity areas, and particularly with wet grain, drying may be slow, and hence expensive.

Drying renders stored grain less liable to attack by most insects, mites and fungi. For example, *Sitophilus oryzae* does not oviposit in wheat of less than about 10.5% moisture content even when temperature is favourable. Both period of development and level of mortality of the immature stages are increased in grain of less than about 13% moisture content. The influence of aridity often interacts with that of temperature. Thus, at 15–21°C, the longevity and fecundity of the adult, and the development and mortality rates of the immature stages are strongly influenced by interaction between the effects of the 2 factors. Although *Rhyzopertha dominica* oviposits in wheat whose moisture content is as low as 8%, mortality of the larvae is increased in grain of less than about 11% moisture content. Mites are very sensitive to aridity and do not increase in grain of less than about 13% moisture content (equilibrium relative humidity about 65%) **(Evans 1981)**. Information on the influence of relative humidity on the increase in numbers of grain pests is summarised by **Howe (1965b)**.

Hyde (1969), in evaluating the hazards of storing high-moisture grain in airtight silos in tropical countries pointed out that moulds and mites will only develop at high humidities, usually above 70–75%, and if the product is dried so that its equilibrium relative humidity is less than this, it will be safe from fungal attack and mite infestation. For cereal grains such as maize, a 'safe' moisture content will be 12–13%; for oily products it is somewhat less, e.g. 8% for gound-nuts and 6% for palm kernels **(Davey and Elcoate 1966)**.

Delouche (1975) stressed the importance of drying in the storage of seeds and **Desmarchelier and Bengston (1979a)** pointed to the importance of partial drying to improve the efficacy of grain protectant insecticides. The deleterious effect of moisture on the stability and biological efficacy of all grain protectant insecticides has been the subject of considerable research and pronouncement by many investigators. Attention is given to this aspect in connection with the review of individual grain protectants.

Desmarchelier *et al.* (1979b) pointed out that combining cooling and/or drying of grain with appropriate use of insecticides provides flexible procedures of pest control that are generally superior to, and cheaper than, the control given by cooling alone, or drying alone, or insecticides alone. They discussed the effect of drying and cooling on both pest biology and the chemistry of protectants. It is argued that control of mites and microflora by grain drying is by far the most useful and, perhaps, the only completely reliable manipulation suitable for an exporting country.

Quinlan (1980) published results of a preliminary study with a thermally-generated malathion aerosol applied with a corn-drying system for the control of insects.

Using approximately 100 tonnes of shelled corn in a circular metal drying bin, malathion aerosol was introduced in the hot air stream. A high proportion of the test insects placed in cages through the grain mass were killed and reproduction was reduced by about 99%.

4.2.4 Cold/Refrigeration

Low ambient temperatures are of great value in insect control in stored grains. It is very likely that the high reputation of Canadian grain for freedom from insects is due to the low temperatures that prevail in that country after harvest. Under such conditions insect populations can only build up in very exceptional circumstances. Apart from such situations where low ambient temperatures prevail, the temperature of grain in storage is sometimes lowered by manipulative procedures. The moderately low temperatures achieved by controlled aeration (Section 4.2.5) are effective in preventing the build-up of insect populations. Providing the grain can be cooled to 12°C breeding of many important pests of stored grain can be reduced **(Burrell 1967, 1974; Elder *et al.* 1975; Howe 1965b; Sutherland *et al.* 1970; Evans 1977a, 1977b, 1977c, 1979a, 1979b; Desmarchelier and Bengston 1979a; Desmarchelier *et al.* 1979b)**.

Forced aeration with ambient air is used fairly extensively in Australia — some 25% of the storage capacity of 20 million tonnes currently has aeration facilities. Wherever climatic conditions are suitable, aeration is effective provided the temperature of bulk grain can be lowered from the high levels (30–40°C), which occur at harvest, to about 12°C. However, such reductions in temperature may take up to 3 months when ambient temperatures remain comparatively high during the whole of the day and most of the night, falling low enough to cool stored grain only during limited periods in the early morning.

Forced aeration is beneficial, especially in winter, but in many regions of the world it cannot produce grain temperatures low enough to eliminate insect populations. Extensive work has been carried out to develop techniques and facilities employing artificially chilled air to cool bulk grain in storage. This work has been reported by **Hunter and Taylor (1980)**. The major requirement is to cool grain down over a period of several weeks to temperatures below the reproduction threshold of the common grain insect pests **(Howe 1965b)**. This must be done without changing the grain condition to a state where moulds or other pests are viable.

To be economic in warm to hot climates, refrigerated aeration requires both recirculation of the cooled air and thermal insulation of the storage, whatever the storage type.

Desmarchelier (1985) discussed the manipulation of such factors as cooling and drying in order to alter the chemical behaviour of insecticides deposited on stored products.

Mills (1978), in a discussion of the potential and limitations of the use of low temperatures to prevent insect damage to stored grain, regretted that the use of low temperatures to control stored-grain insects had not been realised, even in geographical locations where low temperatures are available and are relatively inexpensive to use in cooling grain sufficiently to inactivate or kill stored-grain insects and mites. He then reviewed the susceptibility of insects to low temperatures as did **Howe (1965a)** and pointed out that temperature just low enough to effectively retard insect development (17°C) will be quite suitable for build-up of certain mite populations. He also discussed the cost of cooling by aeration in those regions were the ambient air temperature is sufficiently low for the purpose.

Sitophilus granarius is one of the most resistant pests and will survive for over 2 months at 0°C **(Solomon and Adamson 1955)**. In tests with this species, **Burrell (1967)** found that 97% of the adults move from a cold bulk during storage, and only 0.5% remained and survived the storage period. However, larvae and pupae of *Sitophilus granarius* live within the kernels during development, and so cannot migrate.

One of the most common grain insects, *Oryzaephilus surinamensis*, survived for over 10 weeks at 5°C to 10°C in a heavily infested, 100 tonne bin of dry barley **(Burrell 1967)**. Although the insects were not killed by the treatment, grain below 10°C showed little tendency to reheat because this species does not breed at this temperature and their activity and heat production were greatly reduced.

Mathlein (1968) suggested that refrigeration of infested goods is cheaper than fumigation for killing insect pests in rice and other commodities, and also eliminates the problem of residues in the commodity. He gives a temperature of minus 20°C as being sufficient to kill most pests after exposure for 1 to 24 hours, but suggests that a temperature of minus 30°C is preferable for complete eradication.

Hunter and Taylor (1980) described an installation in a 1700 tonne vertical steel silo that was insulated with polyurethane foam and equipped with a refrigeration unit and air recirculation ducting. The cost of insulating the storage was of the order of US$15.00–US$17.00 per tonne of storage. This is in addition to the current construction costs of silos which range from $A85.00 to $A130.00 per tonne. The energy cost required to cool and maintain the grain at the required temperature was 8 cents per month per tonne of storage capacity when using energy costing 3 cents per kWh over 8 months.

Thorpe and Elder (1980) showed that the use of mechanical refrigeration improved the storage of insecticide-treated grain. Cooling a bulk of stored grain greatly increases the persistence of insecticides applied to it. It was demonstrated that blowing mechanically-refrigerated air through grain can be accomplished by standard commercial air conditioners fitted with appropriate high capacity fans. Results from their theoretical studies showed that the loss of the insecticide methacrifos on treated wheat, initially at 30°C and 11% moisture content, could be reduced during 6 months storage from 90% to about 30% by prompt cooling to 20°C. The energy requirement is 1.5 kWh/tonne of grain. Since cooling will also retard the growth of insect populations, it should retard the development of insect resistance to the insecticide.

Smith (1974) reviewed the role of low temperature to control stored food pests and concluded that the major consideration in the use of refrigeration to cool grain will be the cost of power. The method used to generate power in the country will determine the economics of this methods. The many other problems which have to be overcome are discussed.

Connell and Johnston (1981) published an extensive survey of the costs of introducing refrigerated aeration as an alternative method of grain insect control. They provided a summary of the cost comparisons between the current and proposed control measures and provided details of the derivation of their data.

Mullen and Arbogast (1984) reviewed the use of low temperature to control stored-product insects and pointed out that more basic information is needed on the effect of cold on the pests and on the commodities. Whilst pointing to the susceptibility of 17 important species to low temperatures, these authors appear to have ignored the constraints of cost, both capital and operating, and the problems of insulation and condensation which operate in the tropics.

4.2.5 Aeration

Aeration is the process of passing cold outside air through a grain bulk in order to cool the grain and thus assist in preserving it from damage. The common grain infesting insects reproduce rapidly at temperatures of 25°C to 35°C which can readily occur in stored grain, even in cold climates, because of the self-insulating properties of the grain. Reduction of temperatures only to as low as 15°C to 18°C, however, is sufficient to prevent rapid increase in the insect population, though the degree of control depends on species and grain moisture content, as well as on temperature. In temperate and cool climates, aeration can eliminate significant insect damage, though it will not completely eliminate all insects. The system has little chance of success in semi-tropical and tropical climates **(Hyde 1969)**, or where grain moisture and atmospheric humidity levels are consistently high but could be effective in preventing local hot or wet spots.

Burrell (1974) prepared a comprehensive assessment of aeration which should be consulted by anyone wishing to obtain an overview of the subject.

The use of aeration to maintain high quality in stored grains is not new. **DuMonceau (1753, 1765)** described a ventilated bin with a capacity of 90 tonnes which was aerated with an airflow rate of up to 5 cubic metres per minute by a pair of bellows operated by a windmill. However, it was not until the early 1950s that aeration came into general use with mass production of the necessary equipment such as ducts and fans.

Aeration of grain consists of blowing or drawing ambient air through the grain mass, usually by means of fans attached to perforated ducts. Its major function is to establish and maintain a moderately low and uniform temperature throughout the bulk. The rate of airflow is variable and usually low being only one-tenth to one-twentieth of the rate used in grain drying. Maintenance of a uniform temperature throughout the grain mass reduces moisture transfer, which otherwise almost invariably occurs, and this prevents the development of local and often hidden or unexpected hot spots with their accompanying deterioration.

Burges and Burrell (1964) showed that cooling grain to 20°C greatly reduced the hazard from insects. If the highest temperature in a grain bulk was 17°C, the grain was secure against all the major insects unless the infestation was so numerous that it could raise the temperature. If the cooling period was protracted the period of risk was lengthened. With aeration the grain temperature in temperate or cold climates can be brought to a level at which security is given even against heavy infestations. Cooling to 5° to 10°C will not kill some insect pests and will kill some stages of others only very slowly **(Burrell 1974)**.

Winks and Bailey (1965) described the experimental and commercial use of grain aeration in Australia and the beneficial effects of the cooling achieved on reducing the breakdown of malathion applied to the grain.

Elder (1972) in describing this technology and its attendant advantages pointed out that in Australia, of a total capacity of about 20 million tonnes, about 20% or 4 million tonnes, was aerated at that time. Aeration is a useful complement to grain protectant insecticides, prolonging the biological life of the insecticide and thus assisting in insect control.

Navarro (1974), in a comprehensive evaluation of aeration, drew on experience in Australia and in Israel but pointed out that at the time most of the experiments had been carried out in temperate climates and there was need for further work in hotter climates.

DeLima (1978a), in a review of the use of physical storage procedures in East Africa, pointed out that in the high altitude areas of Kenya the temperature conditions are sufficiently low to permit cooling of produce for better storage. He does not indicate whether aeration had been adopted there.

Quinlan (1979b) reported experiments in which corn stored in circular metal bins equipped with vertically placed aeration systems was treated with malathion thermal aerosol applied in the grain overspace. Though there was a considerable reduction in the number of live insects the distribution of malathion throughout the grain mass left much to be desired.

Hunter and Taylor (1980) showed how aeration ducting could be modified to enable the heat in the grain to be removed by refrigerated aeration using mobile or fixed refrigeration units of moderate capacity. The design and application of a system to a 1700 tonne cell is described. Wheat was cooled from 34°C to less than 10°C in 7 weeks and held there for a further 7 months. To be economic, the storage had to be thermally insulated and the cooled air recirculated. The average energy consumption was 2.7 kWh per tonne per month of storage.

Elder (1984a) has provided a brief description of the aeration process and its equipment including the automatic controls. He pointed out that aeration is complementary to the use of grain protectant insecticides particularly in prolonging the life of the insecticide deposit. In addition to assisting in the control of insects, aeration helps to control moulds and moisture migration and to preserve the desirable germination, milling and nutritive properties of grain. Some observers have commented favourably on the removal of odours by aeration of bulk stored grain.

Sutherland (1984) has summarised the physical and engineering principles which provide an understanding of temperature and moisture fronts in grain aeration. He pointed out that, in aeration, the majority of the grain bed is not cooled to the inlet air dry-bulb temperature, but comes to a temperature which is also dependent on the air and grain moisture contents. One of the major problems in aeration is the formation of a high-moisture zone near the point of air entry. To obviate this a humidistat is often employed in aeration control systems to limit the relative humidity of the entering air to an acceptable value. **Elder (1984b)** pointed out that in order to achieve maximum benefit from aeration it was necessary to instal automatic controllers in storages which cannot be supervised and monitored frequently. Such controllers are worth while even when regular supervision is available.

Longstaff (1984), in discussing the effect of aeration upon insect population growth, pointed out that aeration, by itself, is an inadequate insect control measure. Certainly it can reduce the rate of population growth but it cannot bring about the eradication of a pest population and, therefore, must be used as part of an integrated control strategy using, for example, low levels of grain protectant or fumigant. On the other hand, aeration effectively removes the heat produced by the metabolic activities of insects, particularly those species whose immature phase is spent within the kernel. The dispersal of hot spots and thermal gradients effectively swamps the impact of the insects.

Thorpe and Elder (1982) developed a mathematical model of the effects of aeration on the persistence of insecticides applied to stored bulk grain and they supported the validity of this model by experimental evidence. Aeration was shown to reduce the rate of degradation of pesticides applied to stored grain and to render the rate of decay relatively insensitive to initial grain conditions. In the temperate and sub-tropical wheat growing regions of Australia, aeration can reduce usage of the insecticide methacrifos by factors of 7 and 4 respectively.

Ghaly (1984) carried out an aeration trial to study the moisture content and quality changes in farm-stored wheat as a result of moisture uptake from the entering air. After 4 months of aeration the average grain temperature at the centre of the bulk for each month ranged from 11–21°C depending on the seasons. The average moisture content of one bulk investigated in detail increased from 9.5 to 11.4% after 18 months of storage. Germination and some physical dough properties revealed some deterioration of the wheat near the inlet duct. The rest of the bulk remained practically unaffected during storage and mainly free from storage fungi. The average insect fragment count at outloading was 6/100 g of flour which is considered to be a very low level.

Desmarchelier (1985) reviewed the work of himself and others on the high insecticidal potency of the vapours of slightly volatile insecticides and developed the concept of using aeration to distribute this vapour throughout a grain mass.

4.2.6 Heat

Recent work on the use of elevated temperatures to disinfest stored grain and cereal products has been largely concerned with radiant heating processes such as infra-red, microwave and dielectric heating (**Boulanger *et al.* 1971; Nelson 1972, 1973; Kirkpatrick and Tilton 1972, 1973; Kirkpatrick *et al.* 1972; Kirkpatrick 1975a, 1975b; Watters 1976a**). Essentially, these studies have evaluated the effectiveness of rapidly heating infested grain or flour to temperatures within the range 48°C to 85°C, by means of exposures ranging from several seconds to about 2 minutes, and then allowing it to cool passively to ambient temperatures over several minutes or, more typically, several hours.

Disinfestation processes which depend on the retention of heat for long periods, and which may therefore require insulated storage have little potential where a continuous-flow, in-line system capable of matching shipping rates approaching 2000 tonnes per hour is considered essential. **Dermott and Evans (1978)** and **Evans and Dermott (1979)** discussed fluidised-bed heating as a means of disinfesting wheat in an in-line system. Although still at the pilot stage, the process appears practicable and the cost comparable with the cost of re-circulatory fumigation. The process apparently does not affect the quality of the grain or its suitability for milling and bread-making.

Tilton and Shroeder (1963) demonstrated that infestations of insects within kernels of rice can be effectively controlled with infra-red radiation. The possibility of using infra-red radiation for grain drying has been recognised for many years. **Faulkner and Wratten (1969)** proposed a large scale infra-red source for drying rice; such a unit could also be used for insect control. Microwave radiation was proposed as an alternative method for obtaining insect control with high temperatures by **Hamid and Boulanger (1969)**.

Twelve species of adult stored-product beetles were treated in wheat with infra-red radiation at temperatures of 49°C, 57°C, and 65°C (**Kirkpatrick and Tilton 1972**). Considerable variation in mortality was evident among the species at the lowest temperatures. Mortality of the insects treated at 57°C ranged from 93 to 100% and mortality of insects treated at 65°C was 99.6% or more for each species tested.

Kirkpatrick *et al.* (1972) conducted studies to compare the efficiency of microwave and infra-red energies for insect control. They compared the effectiveness of the same temperature produced by microwave radiation and by infra-red radiation on all stages of *Sitophilus oryzae* infesting soft winter wheat. A temperature of 54°C was used. Infra-red heating gave greater insect control than did microwave by the following average differences in percentage: 13 for immature stages, 76 for adults and 56 for F_1 progeny of treated adults.

Kirkpatrick and Tilton (1973) took samples of wheat containing *Sitophilus oryzae* and *Rhyzopertha dominica* of known age and exposed them to temperatures of 39°C or 43.3°C with relative humidities ranging from 75 to 50% during a 4-day adult ovipositional period, and a continuous exposure to 39°C and 60% relative humidity during immature stages of development. Greater than 99% control was obtained at 39°C with a relative humidity of 60% for *Sitophilus oryzae* and at 43.3°C with a relative humidity of 50% for *Rhyzopertha dominica*.

Kirkpatrick (1975a) reported trials in which bulk quantities of wheat infested with *Sitophilus oryzae* and *Rhyzopertha dominica* were exposed to infra-red radiation that raised the temperature of the wheat to a maximum of 48.6°C. After 24 hours, 93 and 99% of these species, respectively, had been killed.

Kirkpatrick (1975b) reported that thermal energy obtained from either infra-red or microwave radiation or from heat by convection can be used to control insects affecting stored products. The time required to obtain temperatures that are lethal to insects depends not only on the insect species but also on the moisture content and the heat-retention characteristics of the grain or other commodities. The mature larvae and pupae of *Sitotroga cerealella, Sitophilus oryzae* or *Rhyzopertha dominica* were more resistant to microwave energy than either the young larvae or eggs. A temperature of 43°C and 50% relative humidity were required for control of *Rhyzopertha dominica* when using convected heat. Wheat quality appeared unaffected by storage temperatures of 49°C for either 1 or 2 months. However, the flour from this wheat did change colour especially when the wheat samples were stored for 2 months.

Watters (1976a) showed that the susceptibility of *Tribolium confusum* to microwave energy was a function of exposure time and wheat moisture content. Adult mortality was higher in insulated samples of wheat than in non-insulated samples. The order of susceptibility of stages was eggs, pupae, adults, larvae with eggs being the most susceptible.

Hurlock *et al.* (1979), reporting a short series of tests in the UK with a microwave machine generating 896 MHz, showed that a dose of 800 kw should kill *Oryzaephilus surinamensis, Tribolium castaneum* and *Sitophilus granarius* in stored grain, but killed only 30% of them in cacao beans. The mortality of *Ephestia cautella* on cacao beans was 80%. Supplementing this dose with hot air considerably reduced the amount of irradiation required. Slowing the rate of cooling almost eliminated a difference in susceptibility between adults of *Tribolium castaneum* and larvae of *Ephestia cautella* noted in treatments that did not use hot air. Based upon the results reported it was calculated that a microwave plant could be built to operate at a running cost considerably below that of fumigation.

Vardell and Tilton (1981), in a laboratory study, showed that grain temperatures of 62°C and 58°C completely controlled all developmental stages of *Rhyzopertha dominica* and *Sitophilus oryzae* respectively, in rough rice treated in a heated fluidised bed. It was not necessary to maintain these grain temperatures once they had been reached.

Tilton and Vardell (1982a) reported a laboratory study carried out in the USA that showed that populations of insects infesting stored grain could be reduced or completely controlled by a combination of microwave heating and partial vacuum. Infestations with *Sitotroga cerealella* were completely controlled in rye and maize and partially controlled (96.8%) in wheat after treatment at 0.25 par density units of microwave energy and a partial vacuum of 35 torr for 10 minutes. *Rhyzopertha dominica* was eliminated by this treatment in maize, whilst 99.4% control was achieved in rye and 95.6% in wheat. *Sitophilus oryzae* was completely controlled in rye, and 99.2% control was achieved in wheat. *Sitophilus zeamais* in maize was completely controlled by the treatment.

Tilton and Vardell (1982b) reported the performance of a pilot-scale microwave/vacuum grain dryer tested in the laboratory in the USA for use in the control of pests of stored grain. *Rhyzopertha dominica* and *Sitophilus oryzae* infesting wheat were completely controlled by treatment with the highest density of microwaves (1.0 par density units) while under 35 torr pressure. Lower microwave intensity at this pressure did not result in complete control.

Evans (1981) and **Evans and Dermott (1981)** studied the heat tolerance of various species of insects at different life stages. The immature stage of *Rhyzopertha dominica* appeared most tolerant. The time required to obtain 100% mortality of immature *R. dominica* depends on the temperature of the inlet air, the rate of flow of the air and the depth of the grain bed. Grain needs to be held at a temperature of 60°C or higher for about 3 minutes to achieve the required high level of insect kill. The process appears to be more effective with very dry (less than 11% moisture) grain than with high moisture (above 14% moisture) grain. These authors have discussed some of the biological and physical factors that influence the heat tolerance of these major pest species.

Evans *et al.* (1983) discussed some of the design characteristics which influence the performance of a continuous-flow, fluidised bed heating system and its effect on immature stages of *Rhyzopertha dominica* in stored wheat in Australia. The mortality obtained depends largely on maximum grain temperature, complete mortality being obtained consistently with maximum grain temperatures of 64.9°C or higher. In contrast, variation in mean residence time (2–4 minutes) had no significant influence on either maximum grain temperature or insect mortality.

Evans *et al.* (1984) reported the results of experiments with a grain disinfestation pilot-plant recently installed in Australia. This fluidised bed heating unit handled wheat at the rate of 150 tonnes per hour with a residence time of 2.2 minutes. When the maximal grain temperature was either 65°C or 70°C complete mortality of *Rhyzopertha dominica* was obtained in all trials but when the temperature was only 60°C the mortality ranged from 60 to 75%. The energy required to treat each tonne of grain was 1.13 kWh of electricity and 26.6 kWh of gas. In Australia, this is equivalent to a total energy cost of the order of $US0.63 per tonne using liquid petroleum gas and $US0.45 per tonne if natural gas is available. The performance of the unit was considered highly encouraging.

Zakladnoi (1984) stated that the measures used against stored product pests in the USSR included heat treatment of stored grain.

4.2.7 Manipulation of Both Temperature and Moisture

By altering the environment, one can alter the rate of growth of populations of insects, mites and mould. The economically optimum combination of relative humidity and temperature obviously varies from place to place, but a limit of 60–65% relative humidity, to control moulds and mites, is one commonly accepted goal. The temperature that is required to prevent the development of insects depends on the relative humidity and the species, and varies from about 10°C to about 21°C over the range of moisture contents (9–14%) on wheat **(Desmarchelier *et al.* 1979b)**. In some climates, insulation and refrigerated air are required to reach the desired temperature. In other climates, these are not required, but often grain has to be dried. Useful benefits are obtained even in environments that do not completely inhibit the development of progeny. Judicious use of insecticides can be made to supplement the control achieved by environmental means.

Most of the publications on the activity and fate of insecticides draw attention to the effect of temperature and/or moisture on the fate of the deposit applied to the commodity. Quite a few papers point to the advantage of manipulating either or both of these physical parameters. However, as these deal principally with the effect of temperature and/or moisture on the persistence of the insecticide they will be considered in the chapters dealing with the individual compounds and in chapters dealing with the effect of temperature and humidity (Chapters 28 and 29).

Howe (1965b) provided an extensive summary of estimates of optimal and minimal conditions of temperature and humidity at which 43 species of beetle, 9 species of moth and 1 mite can multiply sufficiently to become pests, and the range of temperature most favourable to each. An estimate of the maximum rate of increase for each species is also given. It is fairly obvious that the amount of stored foodstuff eaten and, therefore, the weight loss caused by a population of a species of insect is roughly proportional to the size of the population. Similarly, the amount of damage likely to be caused in the future will depend on how rapidly the population increases. This, in turn, depends mainly on the temperature and humidity of the environment and the nutritional value and moisture content of the food. A successful species must be able to multiply rapidly when conditions are favourable, and also to be able to withstand unfavourable conditions. The author collected data on both of these aspects for a number of species. A considerable number of references is supplied but no recommendations were made as to how to profit from this knowledge. Many authors have, however, used this information in planning their own control strategies.

Davey and Elcoate (1966), **Gough and Bateman (1977)** and **Gough and Lippiatt (1978)** reviewed the world literature for data on the relationship between equilibrium relative humidity and moisture content of stored products from the tropics. Results for cereals were selected according to specified criteria and are considered reliable but cannot be used for extrapolating from one variety of stored product to another. Additional valuable information concerning the equilibrium relative humidity of various cereal grains and similar commodities at different storage temperatures can be found in many other publications **(Coleman and Fellow 1925; Gane 1941; Gay 1946; Babbitt 1949; Karon and Adams 1949; Houston 1952; Breese 1955; Hubbard *et al.* 1957; Hall and Rodriquez-Arias 1958; Becker 1960; Juliano 1964; Ayerst 1965; Davey and Elcoate 1966; Chung and Pfrost 1967a, 1967b; Pixton 1967; Pixton and Griffiths 1971; Pixton and Warburton 1971a, 1971b; Berry and Dickerson 1973; Rangaswamy 1973; Neher *et al.* 1973; Pixton and Warburton 1975; Gough 1975; Gough and Lippiatt 1978; Gough and King 1980)**. **Figure 4.1** illustrates how the equilibrium relative humidity of wheat changes with the temperature and moisture content of the grain. **Figure 4.2** shows how the equilibrium relative humidity of wheat, rough rise and yellow flint maize differ from each other at different temperatures and grain moisture contents.

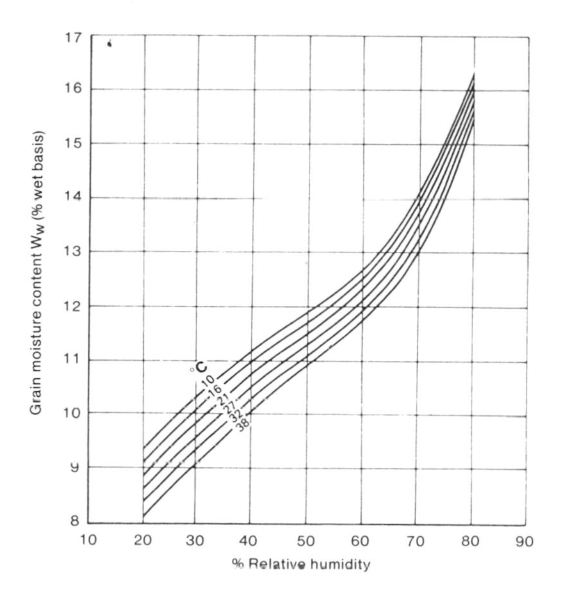

Figure 4.1. Hygroscopic equilibria of Australian wheat. Redrawn from **Gay (1946).**

Since the water activity influences not only biological activity but also chemical reaction, the understanding of these criteria is germane to a proper appreciation of the likelihood and consequences of infestation of stored produce. **Desmarchelier *et al.* (1979b)** discussed the desirability of combining temperature and moisture manipulation with the use of grain protectants. They were able show that this provides flexible procedures of pest control that are generally superior to, and cheaper than, the control given by cooling alone, drying alone or chemicals alone. This is especially the case when the strong point of one component covers the weak point of the other component, and when one component aids the other component. The effects of drying and cooling on both pest biology and the chemistry of protectants are outlined. The weaknesses and strengths of control of pests solely by manipulation of conditions are also discussed. It is argued that the control of mites and microflora by grain drying is by far the most useful and, perhaps, the only completely reliable manipulation suitable for an exporting country. In a series of tables the authors set out the conditions of relative humidity and temperature that, in theory, control the various species. There always seems to be one species or life-stage of a pest that is especially tolerant to any manipulated condition, whether cold, heat or aridity. To obtain complete control of all pest species by the manipulation of conditions, however, it is necessary to manipulate both relative humidity and temperature; such combined manipulations present practical difficulties. These investigators show, for example, how manipulation of relative humidity in Australia to 55% or below has proved effective in controlling mites and microflora. This success has probably been due to the acceptable grain moisture limit having been set at a level well below that required in theory.

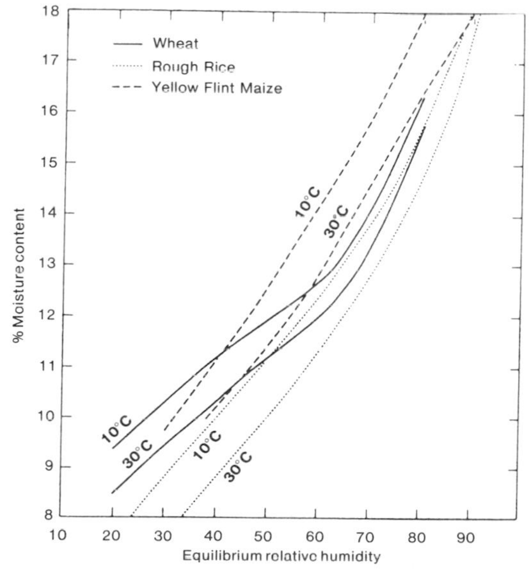

Figure 4.2. Hygroscopic equilibria of wheat, rough rice and yellow flint maize **(FAO 1982a).**

A combination of r.h. less than 40 to 50% and temperature less than 18°C is generally effective in controlling most pest species and such a combination frequently prevails in aerated Australian storages. Localised, and sometimes heavy infestations may nevertheless occur in the absence of grain protectants that provide control when, for example, pests breed in pockets of grain that are warmer or moister than the rest of the grain bulk.

Evans (1982) studied the influence of temperature and grain moisture content on the intrinsic rate of increase of *Sitophilus oryzae* and was able to show a sixfold difference in the weekly rate of population increase occurring at 21°C over that developing at 15°C irrespective of grain moisture content over the range 11.2%–14%. Though the effect of moisture content could be clearly demonstrated it was not nearly so marked as the effect of temperature over this range.

Evans (1983) studied the influence of relative humidity and thermal acclimation on the survival of adult grain beetles in cooled grain. He reported that survival at low temperatures differed between species and there was considerable interaction between the effects of temperature and humidity.

4.2.8 Hermetic (Airtight) Storage

In the hermetic storage process, grain is placed in a sealed enclosure and the natural respiratory activities of the grain and associated pests and fungi reduce the oxygen level in the system to a level which controls or eliminates the pests and fungi. The process is simple, self-regulating, and requires no added pesticide, inert gas or energy inputs. In its simplest form it relies entirely on locally produced materials.

The hermetic storage process has been used for conservation of grain stocks by man for at least 9000 years. This ancient and simple system is still in widespread use in subsistence agriculture. However, attempts to adapt the process to meet the constraints of modern large bulk handling systems have met with limited success. This has largely been because of the need for specialised structures and the lack of attention to important design problems, notably grain loading and unloading systems. Hermetic storage techniques have a great potential for developing residue-free storage methods using low energy inputs without reliance on imported pesticides and components and must be regarded as one of the grain storage methods of the future as well as the past.

Although the concept of airtight storage is quite ancient and the technique has been well studied and understood, it does not yet seem to have reached its full potential. It is well established in a number of temperate countries as a means of storage for high-moisture grain for feeding to livestock, the only possible use when the grain is damp. Under such conditions the oxygen level becomes too low for insects to survive but there is considerable danger resulting from the growth of micro-organisms, many of them harmful. There is a certain amount of airtight storage of relatively dry grain in large installations in France, but this development has not become widespread. However, the use of simple forms of airtight storage could be much extended where dry grain is available. In airtight storage, dry grain continues to respire until there is an equilibrium when the atmosphere contains 5–10% oxygen and 10–5% carbon dioxide. Experience in Australia shows that in bunker storage, where the oxygen level fell to 5%, *Rhyzopertha dominica* could survive and reproduce quite well. **Spratt (1979)**, who has been studying the effects of such atmospheres on insect development, proposes to publish further findings. The initial work indicated that the oxygen level must be well below 10% since at that level reproduction is only depressed by about 50%.

Hyde (1969, 1974) and **Hyde *et al.* (1973)** provided a comprehensive review of the scientific literature on airtight storage. The costs and logistics associated with airtight storage for extremely large quantities of grain appear to preclude its general adoption.

Ancient and currently used systems of hermetic storage are described in the papers of **Gast and Sigaut (1979). Hyde *et al.* (1973)** give descriptions of two large hermetic storage systems in current use and of the various traditional and experimental structures suitable for hermetic storage of grain at the subsistence level of agriculture. **Hyde (1974)** provides a detailed review of the practice of hermetic storage and its scientific basis. These publications should be consulted for details of the various systems and their operations. The section dealing with natural airtight storage in **Shejbal (1980)** is also a useful information source. In hermetic storage, the controlled atmosphere required for grain preservation is produced through biological respiration. At equilibrium relative humidities of greater than 70% in a system containing stored grain there may be a number of different types of living organisms, respiring aerobically. This action takes in oxygen from the atmosphere and releases CO_2 as the organisms oxidise carbohydrates, fats and other components for the production of usable energy for their life processes. Grain itself, at greater than 13% m.c., respires significantly and is always, in practice, associated with micro-organisms, such as moulds, which can respire and grow under such moist conditions. Insects and mites may also be present and if so, may contribute substantially to the overall rate of oxygen loss by respiration. Under dry storage conditions (less than 13% m.c. grain), oxygen loss occurs through insect respiration. Under very wet storage conditions (greater than 80% e.r.h.), yeasts and bacteria become active. These organisms produce CO_2 metabolically without uptake of oxygen (anaerobic respiration). They also produce acids and alcohols in the process and give rise to the characteristic fermented odour of wet grain stored under hermetic conditions **(Banks 1984d)**.

Where the leakiness of an enclosure is high, it may not be possible to generate proper hermetic storage conditions and insect infestation, moulding and grain deterioration will continue unchecked. However, with well-sealed enclosures the process is largely self-regulating.

The main problems associated with hermetic storage are:

- accessibility of stocks, loading and unloading;
- maintenance of the seal in storage structures;
- safety to workforce;
- moisture migration phenomena;
- protection of grain on removal from storage.

Hermetic storage is the only residue-free storage method which will ensure long-term preservation of grain without attention or energy input. It thus has great potential as a method of storing long-term food reserves against famine or in order to take advantage of high market prices. The more widespread adoption of the system is hindered by its inconvenience. Day-to-day stock movement is not possible and it is difficult to integrate into a modern system of handling because of the structural requirements of the most suitable storages. With engineering development, it may be possible to overcome these difficulties and produce a commercially viable system which can be applied in regions of the world where long-term food stocks are required and storage at present is difficult.

Hyde (1969) wrote of the hazards of storing high-moisture grain in airtight silos in tropical countries. She pointed out that the prevailing temperatures in tropical and sub-tropical countries were ideal for the development of moulds, not only those which cause unacceptable fermentation and taint but those which produced toxins or allergens. She listed 7 major factors which go against storing high-moisture grain in sealed silos in the tropics and concluded that this could be a very hazardous procedure. However, she agreed that it was not fair to condemn it out of hand as impossible but to wait the results of on-going experiments in Nigeria and Kenya which might provide the basis for the necessary developments.

Mitsuda *et al.* (1971) reported preliminary results of trials on the hermetic storage of cereals and legumes under water and under ground. In their work the commodities were stored in strong plastic or rubber bags. The idea was to combine a complete physical barrier to storage pests with temperature and environmental gas control. **Tani *et al.* (1972)** reported favourably on similar studies with underwater storage of brown rice.

Bailey and Banks (1974), in discussing the use of controlled atmospheres for storage of grain, introduced their subject with a brief review of the history of airtight storage. They concluded that though this concept had many fine features it did not lend itself to modern large-scale installation because of lack of convenience in grain handling.

Hyde (1974) has presented an extensive evaluation of all aspects of airtight storage including its use to control insects in dry grain, and to prevent mould growth in high-moisture grain. In discussing the modification of hermetic storage by addition of inert gas to produce 'controlled atmosphere' storage the author concludes that airtight storage is a simple method, and it may be best to let it remain so. This paper should be consulted for comprehensive information.

Banks (1984d) has made an extensive appraisal of most aspects of hermetic storage including basic principles, designs for storage systems, pest control in hermetic storages, problems and the future of hermetic storage.

Champ and McCabe (1984) reported that the storage of grain in plastic lined, earth-covered bunkers has been demonstrated to provide a low-cost method for maintaining quality in grain held for long periods. Various designs of bunker have been evaluated in Australia and the most satisfactory has been based on an above-ground structure with permanent earthen side walls and ends built on a slight slope to facilitate drainage. The walls and base are covered with a thin layer of sand and lined with polyethylene sheeting. After filling and trimming, the grain is covered with a woven polyethylene fabric, a layer of sand is spread over the cover sheets to protect them, and then soil is placed over the structure to a total depth of 1 metre. The method has been used commercially in Australia with bulks of wheat up to 25 000 tonnes in each bunker at total storage costs averaging $US4.83 per tonne. Although the conditions in such bunkers depress the reproduction and feeding of insects, for complete freedom from insects the grain should be treated with an appropriate insecticide whilst it is being loaded into the bunker, or fumigated when the bunker is full and sealed. Trials extending over 8 or more years have shown that bunker storage is a viable and cost-effective option for long-term storage of grain at safe moisture content. Some supplementary treatment, however, is necessary if the grain is to be kept free of infestation.

4.2.9 Storage in Controlled Atmospheres (CA)

Hermetic storage involving the use of an externally generated gas to displace the intergranular atmosphere shows promise as a means of insect control. Investigations involving the use of nitrogen and carbon dioxide are under way in many places. If the grain is sufficiently dry, controlled atmosphere storage has the advantage that it preserves the grain in a high quality, insect-free condition almost indefinitely if left undisturbed. It requires little or no attention for long periods, involves minimum technology and energy consumption but does require a hermetic structure which may be hard to obtain from existing installations. It has the disadvantage that grain handling is frequently a problem, disinfestation is slow, conditions of low oxygen must be re-established each time the structure is opened and the fabric must be maintained in perfect condition. Moisture migration and condensation may sometimes be a problem **(Hyde 1969)**. This form of storage appears limited to centralised storage systems with a high level of capital investment and where a high level of technology and management skill is constantly available. **Bailey and Banks (1974)** and **Banks and Annis (1977)** reviewed recent developments.

Bailey and Banks (1974), who had worked for a long time on the use of controlled atmospheres for the storage of grain, pointed out that much of the recent work had been carried out at the time when newer organophosphorus grain protectants were achieving an excellent result in many countries. During this time the grain industries showed little interest in controlled atmosphere storage as they believed their insect control problems had been largely solved for the foreseeable future.

The controlled atmosphere method of grain storage involves placing grain in a structure and altering the atmospheric composition within that structure to give a combination of the atmospheric gases, oxygen, nitrogen and carbon dioxide lethal to insect pests. It is a version of the process of hermetic storage but does not rely on the respiratory activities of grain, insects and moulds to alter the atmospheric composition in a gas-tight storage structure. In the present mode of use of CA storage the appropriate atmosphere is supplied externally and introduced into the structure. Thereafter additional gas is added to compensate for losses during the required exposure period.

Atmospheres of nitrogen containing less than 2% oxygen, a mixture of 60% carbon dioxide in air (i.e. about 8% oxygen) and low oxygen exhausts produced from the burning of hydrocarbons have all been shown to be lethal to all stored-product pests tested. They are thus each suitable for CA grain storage. Decision as to which is most suitable will eventually be made on economic grounds. The method is capable of refinement to give very low cost storages of large bulks, e.g. by use of atmospheres generated by burning hydrocarbons. Insect movement, insect-caused damage and insect reproduction ceases as soon as the commodity is placed in CA storage. CA storage is best suited to parts of the storage system where movement of grain is slow. Its slow action, particularly at low temperatures, makes it inconvenient in its present form as a disinfestation measure at terminals if a high through-put is required.

Bailey and Banks (1974) carried out field trials parallel with laboratory studies. Nitrogen was used as the gas because it was conveniently available and was best covered by supporting literature. The goal of the field trials was to maintain an atmosphere of less than 1.5% oxygen indefinitely at reasonable economic cost. This long-term storage approach contrasts with the work of other investigators who have all used an approach similar to fumigation. The toxic atmospheres in these studies was maintained only for a brief period with relatively high make-up rates of gas. Bailey and Banks used effective sealing to eliminate the possibility of the existence of local areas of high oxygen and to substantially reduce the gas required to be introduced on a continuing basis for maintenance of the atmosphere within acceptable oxygen limits. One of the limitations of this approach is the lack of storage facilities that can be made sufficiently gas-tight for the purpose.

Among the limitations that are inherent in the use of controlled atmospheres is the relatively long time for an existing insect population to be killed, especially at low grain temperatures.

It appears very unlikely that a strain of stored product insect will evolve which is capable of completing its life cycle in very low oxygen or oxygen-free atmospheres though resistance to high concentrations of carbon dioxide have been encountered **(Banks and Annis 1977)**.

It is interesting to consider whether any form of resistance by the insects to controlled atmospheres could develop. **Bailey and Banks (1974)** feel that an increase in the time required to kill could be selected over many generations or that some species might enter into a form of diapause. In either case effective control would still be achieved as reproduction would be extremely slow or would cease. The possibility of an increase in the time factor is a good reason to aim the program at continuous storage under controlled atmospheres, rather than looking upon its use as a form of fumigation.

Banks (1978), in reviewing recent advances in the use of modified atmospheres for stored-product pest control, pointed to the importance of the relevance of sealing of the enclosure to the general technique of modified atmosphere storage. Though sealing may sometimes be expensive, the cost can be offset against the saving of gas. The sealing of a storage also provides an insect-proof barrier. **Banks (1978)** reported experience from 8 large-scale field trials carried out in Australia with nitrogen generated from liquid nitrogen and from a series of other trials carried out in Australia, Israel, USA and elsewhere with carbon dioxide from cylinders and atmospheres generated by burning hydrocarbons. His conclusion was that it is possible to generate and maintain an insecticidal concentration of oxygen, nitrogen and carbon dioxide under a variety of field conditions. While efficiencies have been found to be high, there is scope for some improvement, particularly with nitrogen by altering the purging rate. Several trials carried out in Papua New Guinea under conditions similar to that to be expected in commercial use were highly successful.

Banks (1978) pointed out that much more data are needed on the response of stored-product insects to differing atmospheric compositions and particularly to the effect of a varying concentration. The effectiveness of the modified atmospheres at temperatures below 20°C is inadequately researched. Despite these limitations, there are many advantages of modified atmosphere storage.

Bailey and Banks (1980), in a review of recent studies of the effects of controlled atmospheres on stored-product pests, point to the conflicting literature on the relative susceptibilities of various species, the relative speed of action of nitrogen/oxygen mixtures and those containing carbon dioxide and the temperature dependence of the action of air/carbon dioxide mixtures. They reviewed the effects of controlled atmospheres on stored product mites and predators as also the possible development of tolerance to such atmospheres. They concluded that an accentuation of the delay in development observed when stored-product insects are treated with low oxygen atmospheres together with an increased overall tolerance would result in a form of resistance. There appears no reason why such resistance could not be selected for and why it could not eventually become significant in commercial practice, particularly if short exposures of 1 month or less are used. A similar form of tolerance could arise to high carbon dioxide atmospheres. They point to numerous studies where such a tolerance has been examined. The possibility of the eventual rise of resistance cannot be taken to invalidate the controlled atmosphere technique but must be borne in mind, as in fumigation, so that the technique is not used in situations where it is likely to be incompletely effective, thus providing a population of insects selected for tolerance to the control measure.

Bond and Buckland (1979) have shown that a high level of resistance to carbon dioxide can easily be developed in *Sitophilus granarius*.

Sharp (1979) and **Sharp and Banks (1979)** reported a series of trials, carried out in Australia, which resulted in the development of disinfestation procedures for wheat in freight containers. The containers must be selected to be of high gas-tightness. After loading the grain, about 60 kg of dry ice is added, half loose and half in an insulated box. The loose dry ice quickly sublimes to carbon dioxide gas, while that in the insulated box gives a slow release of gas over about 8 days, which compensates for gas loss by leakage. This technique was shown to be equally effective for bulk-loaded and bagged wheat and could no doubt be applicable for use with other grains.

Modified atmosphere techniques for grain storage have been under study in Australia for over 60 years but until recently have been widely used only during one brief period (1917–19). Full-scale testing of nitrogen-based atmospheres started in 1972 and has led to development of a process, based on tanker-delivered liquid nitrogen. Since, under dry Australian conditions, the quality of grains such as wheat is maintained adequately in storage, emphasis in the use of nitrogen-based systems has been on insect rather than quality control.

More recently the use of carbon dioxide has been investigated. Carbon dioxide, at present, is the preferred atmosphere for insect control as it is easy to apply and, unlike nitrogen atmospheres, does not require a continuous input of gas after the initial purge if the storage is well sealed. Carbon dioxide-based atmospheres are now in use in Australia for insect control in stored grain. **Banks *et al.* (1980)** published a summary of the experimental and commercial trials carried out with modified atmospheres in Australia. If insect control only is required, it should only be necessary to maintain the modified atmosphere until elimination of insects is achieved. The sealed fabric of the storage should then provide a barrier against reinfestation.

Kashi (1981) evaluated the prospects of controlling pests in stored grain with carbon dioxide. When the air in a storage structure is replaced with either the inert gas nitrogen, with a biologically-active gas like carbon dioxide, or with a low-oxygen atmosphere (for example one that can be produced by burning hydrocarbons so as to contain about 0.1% oxygen, 8.5–11.5% carbon dioxide, with the balance principally nitrogen) an environment which is unfavourable for the normal life of insects and other pests, such as rodents, is created.

Inert, low-oxygen atmospheres need to be maintained for extended periods and hence are only effective if the structures are extremely gas-tight and when provision is made for the frequent addition of inert gas. Carbon dioxide is only mildly toxic to insects and is effective only when concentrations in excess of 40% are maintained for a sufficiently long period. In the past, carbon dioxide has not been used successfully because techniques were not available for maintaining sufficiently high concentrations of the gas for extended periods in normal storages. The situation has now changed, however, and the use of carbon dioxide is currently receiving favourable attention **(Banks and Annis 1977; Banks 1978; Kashi 1981)** with the expectation that it may be adequately effective at concentrations below 40% when the atmosphere can be retained for sufficiently long periods.

Laboratory tests with the major stored-product insects have shown that most adults can be killed with pure carbon dioxide within 10 to 48 hours but it is not possible to use pure CO_2 for fumigation. Exposure times of more than 14 days are required to kill insects when the atmosphere contains 35–80% carbon dioxide. Adult insects recover completely from limited exposure to carbon dioxide. Mites are more difficult to kill, although their activity ceases in atmospheres having a high concentration of carbon dioxide.

Not surprisingly, the gas-tightness of structures has been identified as the most important criterion in ensuring effective and economical treatment with carbon dioxide. Good sealing of the structure prevents the loss of gas due to wind and temperature cycling in the headspace. The initial cost of sealing existing structures or fabricating new, gas-tight structures may appear prohibitive but the cost may well be compensated by the savings in gas. Reports indicate failure to kill insects with carbon dioxide in high dockage (i.e. dirty) grain.

The possibility of insects surviving under high levels of carbon dioxide, and thus developing resistance to it, cannot be overlooked, for the extensive use of carbon dioxide might be expected to promote the selection of resistant strains. **Bond and Buckland (1979)** showed that the treatment of successive generations of granary weevil (*Sitophilus granarius*) adults with carbon dioxide produced insects with a threefold increase in tolerance to the gas in seven generations. The most obvious disadvantages of using carbon dioxide for grain protection are the need for long exposure periods, special evaporators and large quantities of gas. Slow action may make its use inconvenient at terminals requiring a high throughput. Practical experience has shown that concrete structures rapidly absorb large quantities of carbon dioxide and it is not yet clear whether this may have deleterious effects on the strength and stability of concrete structures.

Banks (1981a), in reviewing the use of alternatives to contact insecticides pointed out that there are several important constraints to the adoption of an unfamiliar physical control system involving controlled atmospheres and insect exclusion:

1. Adoption of the process may involve training and substantial alteration of current practices.
2. The capital costs of the provision of suitable enclosures may be considerable.
3. The method is not well adapted to use with produce which is moved frequently.
4. Carbon dioxide may be unavailable. The technology of external inert atmosphere generation has not been developed for small-scale use and requires research.

None of these appears insuperable.

A comprehensive review of the effects of controlled atmosphere storage on grain quality was published by **Banks (1981b)**. He concluded that both low-oxygen and carbon dioxide-rich atmospheres do not have detrimental effects on the overall storability of grain.

In a later review **Banks (1984a)** pointed out that CA storage in its various forms is a relatively slow process for insect control, compared with conventional fumigation. With the artificially induced modified atmospheres, the process may take days, weeks or even months to produce complete mortality of stored-product pests. If the atmosphere is maintained either at a low oxygen (less than 2%, preferably 0.5% oxygen) or high-CO_2 (greater than 40%, preferably 60% CO_2), it provides a good storage atmosphere in which grain quality is preserved at least as well as, and for most grain, better than in air.

One of the essentials for successful use of CA storage is a gas-tight enclosure. Procedures for testing storage structures for gas-tightness were reviewed by **Banks (1981b)**. These tests are relatively easy to carry out, even with complex structures such as buildings. The generation of gases suitable for use as modified atmospheres was discussed by **Banks (1984c)**.

Connell and Johnston (1981) reported an investigation of the economics of introducing controlled atmosphere storage under Australian conditions. The study reported the costs of introducing these measures to existing grain storages and compared these costs with existing chemical costs.

Annis *et al.* (1984) told of successful experiments using carbon dioxide to control insects in stacks of bagged rice under a PVC-membrane enclosure. Bagged stacks of milled rice were fully enclosed in an envelope of PVC, consisting of a fitted membrane covering the top and sides of the stack, bonded with adhesive to the floor sheet. Addition of gas at a rate of 2.4 kg/tonne gave an average concentration of 70% within the stack. In all cases the rate of loss was such that an initial level of 70% of carbon dioxide would have taken longer than 10 days to decay to 35%, the regime required for complete insect kill. When, under humid tropical conditions in Indonesia, the covers were removed from the 4 treated stacks 28, 56, 86 and 133 days after the addition of carbon dioxide, the rice was in good condition and the initial infestation by a number of pest species had been eliminated.

Jay (1984) reported studies to control *Rhyzopertha dominica* with modified atmospheres at low temperatures. During laboratory studies carried out in the USA, eggs, larvae and pupae of *Rhyzopertha dominica* were exposed to 60 or 98% carbon dioxide, 99% nitrogen or normal air at temperatures of 1.7, 4.4, 10.0 or 15.4°C and 58% relative humidity for 1, 2, 3 or 4 weeks. The 4 atmospheres tested eliminated adult emergence almost completely at 1.7 and 4.4°C after 3 and 4 weeks, respectively. At 10°C, adult emergence was prevented by treatment with 60 or 98% carbon dioxide after 3 weeks, and at 15.4°C with 60% carbon dioxide after 2 weeks. At most of the temperatures tested, the atmosphere containing 60% carbon dioxide was the most effective.

Jay (1984a) reviewed recent advances in the use of modified atmospheres for the control of stored-product insects and concluded that this was a viable alternative to the use of fumigants. However, he believes that many more field studies are still needed to develop data applicable to commodity storage. This research should include further studies on application and distribution techniques, sealing, and the development of sound economic data so that industry can fully evaluate the method and determine if it fits into their particular marketing programmes.

4.2.10 Ionising Radiation

The fundamental principles and practical problems of using ionising radiation for the control of stored-product insects are well set forth in the publication by **Cornwell (1966)** which deals mainly with research carried out in the United Kingdom. **Cogburn *et al.* (1972)** and **Tilton *et al.* (1974)** provided extensive information on practical aspects of gamma-irradiation for control of insects in bulk wheat and wheat flour.

A distinction must be made between the objective of producing sterile insects for subsequent use in genetic control programs (producing fully vigorous but sterile individuals) and carrying out a total control program (producing sterile and incapacitated or preferably dead insects). The latter object is considered here.

Gamma-rays, accelerated electrons, natural *beta* particles **(Robertson 1974)** and neutrons **(Hooper 1971; Smittle *et al.* 1971)** have been used to kill insects or to render them sterile or debilitated.

Estimates of doses required for sterility and death vary widely between research workers and species. In general a dose of 50 krad (gamma rays) is required for sterilising Lepidoptera though Coleoptera may only require 25 krad. Small numbers of *Gnathocerus* have been recorded to emerge from 100 krad **(Brower 1974a)** and the acute LD/99 of *Corcyra cephalonica* is said to be 170 krad **(Loaharanu *et al.* 1972)**. Adult life span is much reduced at these levels. A dose of 150 krad was required to prevent reproduction in *Piophilo casei* **(Zuska 1973)** and larvae of *Ephestia cautella* survive up to 4 weeks after 100 krad dose **(Cogburn *et al.* 1973)**.

Watters and MacQueen (1967) demonstrated that there were large differences among insect pest species in radiation sensitivity and time of death after treatment, especially at lower radiation doses. In tests on 5 principal stored-product species, all species except *Rhyzopertha dominica* died in less than 3 weeks after irradiation at 50 krad. Irradiated *Tribolium castaneum* lived longer than non-irradiated beetles and this was the most resistant species at 6.25 krad. Though some individuals were temporarily sterilised for the first few weeks, they later regained fertility. Adults of other species that survived 6.25 krad were permanently sterilised. Survivors of all species continued to feed on wheat kernels.

Tilton *et al.* (1974) showed that when naturally or artificially infested wheat flour was packed in metal cans or multi-wall paper bags and treated with doses of gamma-irradiation ranging from 31.8 to 48 krad, populations of mixed stages and species of insects were not eliminated within 1 month. However, doses averaging 15 krad did cause 100% mortality within 3 months of irradiation among eggs and young larvae of those species most likely to be present in flour. The result showed that if mortality of all eggs and young larvae is to be complete by the end of 1 month, the dose selected must be higher than 50 krad.

The mortality produced by given doses is influenced by the dose rate and pre-irradiation holding temperature **(Singh 1973)**. High irradiation rates are more biologically effective against *Tribolium castaneum* eggs **(Brown and Davis 1973)** but lower rates produce higher mortality of adult *Rhyzopertha dominica* **(Singh and Liles 1972)**. Nitrogen atmospheres exert a slight protectant effect against the influence of gamma-rays **(Langley and Maly 1971; Wakid (1973)**. Fractionated doses give higher mortality than an equivalent single dose against adult *Rhyzopertha dominica* **(Liggins and Singh 1971)**. Adult longevity usually is reduced by irradiation but under special circumstances may be prolonged **(c.g. White and Hutt 1972)**. On occasions fertility has partly recovered from an initial sterilising dose **(Hutt and White 1973; Robinson 1973)**. It has been claimed that some benefit ensues from the combination of gamma-irradiation with microwave or infra-red treatments for disinfestation of grain **(Cogburn *et al.* 1971; Tilton *et al.* 1972; Kirkpatrick *et al.* 1973)** though no cost study was done.

On a pilot scale, gamma-irradiation has been tested for the disinfestation of wheat **(Cogburn *et al.* 1972)** and flour **(Tilton *et al.* 1972)**. The dosage requirements of 45 krad **(Calderon and Gonen 1971)** to ensure sterilisation of all stored-product pests severely limits the economic scale of any commercial bulk irradiation plant because it reduces throughput or requires a very large facility and thus high capital costs. Death of the insects may be much delayed (e.g. >3 months) even at this level **(Tilton *et al.* 1974)**. Even in 1975 the number of commercial bulk commodity irradiators in use was very few. A full-scale facility was constructed at Iskenderun, Turkey, but never commissioned. In 1980 an electron accelerator was established at Odessa Port's grain elevator (USSR) where it irradiates up to 400 tonnes an hour with a claimed efficiency of 100% **(Gilzin 1984)**.

The use of a low-intensity source placed permanently in a grain storage has been discussed but rejected because of the possibility of continuous reinfestation. It has been shown that chronic irradiation at 170 R/day causes slow population decline in *Tribolium* spp. **(Erdman 1973)**.

Atomic radiation at present is the only physical control measure used which requires tolerance levels on the applied dose. These have been established by several countries for bulk grain and packaged foods **(Watters 1972; Ahmad 1973; Tilton *et al.* 1974)** and are at a level which only permits a single 40 krad disinfestation. Since an irradiated commodity is as infestable after as before irradiation, it requires further subsequent protection. This places severe limitations on the applicability of the method to unpackaged foods.

Baker *et al.* (1953) wrote of the lethal effects of electrons generated by the Van de Graaff accelerator on insects infesting wheat and flour. They concluded that the biological and chemical effects of accelerated electrons are the same as those of X-rays and gamma-rays. The penetration of accelerated electrons into matter is less than that of X-rays of corresponding voltages, but the penetration is of sufficient magnitude to be considered. In a second paper **(1954) Baker *et al.*** provided further information on the effects of accelerated electrons on insects which infest wheat flour and beans together with information used to calculate dosage, penetration of electrons into wheat and flour, rate of treatment of product, cost of electrical energy for a given dose and a review of literature.

Cornwell (1964) used wheat labelled with P^{32} and found that *Sitophilus granarius*, irradiated at 16 krad, consumed half as much food as did untreated adults. Fourteen weeks after irradiation at 6.25 krad, *Tribolium confusum, T. castaneum, Cryptolestes ferrugineus* and *Rhyzopertha dominica* were placed on wheat of 13.5% moisture content to determine the extent of feeding. All species damaged wheat, especially in the germ region of the kernels; in some, the germ was completely removed while in other kernels the germs were partially eaten.

Lai *et al.* (1959) reported no damage to the milling and baking qualities of wheat irradiated at 100 000 rep (roentgens-equivalent-physical) but there was progressive damage to dough volume, crumb grain, texture and colour as the dose was increased to 1 million rep.

Cornwell *et al.* (1957) studied the lethal and sterilising effects of gamma-radiation on insects infesting cereal commodities. They irradiated immature stages of 12 species at 20 000 rep. There was a reduction in emergence of adults. Insects which survived the treatment were sterile. The vast majority of these emerged during the first 17 days after irradiation which suggests that growth of young larvae had been inhibited and that only pupae and late larvae completed their development. Irradiated larvae of *Ephestia* spp. remained alive for many weeks, but failed to pupate. These studies showed that irradiation prolongs the duration of larval stages; doses greater than 20 000 rep are required to prevent the emergence of adults from irradiated pupae. The irradiation of adults at the highest experimental level (50 000 rep) produces no immediate 'knock down' effect. Increasing the dose produces only a slight reduction in the time required to obtain 100% kill. While the insecticidal property of radiation exposes marked differences in the tolerance of the various species, irradiation with low doses has shown that 13 species are sterilised at or near 6000 rep. This would suggest that lethal doses evoke physiological processes which are more complex than those involved in destroying germ cells.

Banham and Crook (1966) showed that pupae of *Tribolium castaneum* and to a lesser extent, adults, when irradiated at low doses, survived better than the controls.

Tilton *et al.* (1966) exposed 4 major stored-product pests to various levels of gamma-radiation from a $cobalt^{60}$

source. All four metamorphic stages of each insect were used. In all species the adults were the most tolerant, followed by the pupae, then the larvae and then the eggs. No dosage used was high enough to produce immediate complete mortality of all species. All 4 species were sterilised by gamma-radiation in a single continuous dose of 25 krad or more.

Tilton (1975) reviewed the achievements and limitations of ionising radiation for stored-product insect control and showed that extensive tests conducted by many workers throughout the world had proved that doses of gamma-irradiation below 50 krad controlled insects in stored products. Practical methods of using gamma-radiation with most products had been developed. Combinations of infra-red heating with gamma-radiation gave excellent results controlling stored-product insects in wheat. Results with electron irradiation agree closely with those for gamma-irradiation. Disinfestation with ionising radiation is technically feasible, but the method is not used commercially for economic reasons. Some studies indicate that irradiation disinfestation has a favourable cost relationship to fumigation. These studies were based on the assumption that all the product to be handled by a facility will be treated and that the product will be available on a continuous uniform flow basis for treatment. Neither of these assumptions is correct: fumigation of commodities is usually highly selective and is used only when needed, and most products move through collection or shipping points with great fluctuations in volume. Another disadvantage that must be considered is the initial high cost of an irradiator, with most of the cost for several years of disinfestation being paid at the beginning of the operation.

Tilton (1979), in a later review of the status of irradiation for use in stored-product insect control, drew attention to the work of the Joint FAO/IAEA International Program in the Field of Food Irradiation (IFIP). Clearance for the use of irradiation of grain had been given in the USSR in 1959, in the USA in 1963, in Canada 1969 and in Bulgaria 1972. The difficulty in obtaining clearance from health agencies and the non-uniformity of regulations between countries is therefore not the major reason for current lack of practical use of irradiation. Electron accelerators for grain irradiation appeared to offer considerable economic and practical advantages.

Watters *et al.* (1978) discussed the potential of accelerated electrons for insect control in stored grain and pointed to two factors which have limited the use of ionising radiation for insect control. First is the apparent high cost of installation and operation; secondly, are restrictions that may be imposed due to possible quality changes in the irradiated product. Both of these factors have adversely influenced the widespread acceptance of irradiation for the direct control of stored grain infestation. Furthermore, the irradiated grain must be stored in uninfested bins, ship's holds or containers to protect it from further infestation until it reaches its final destination. This implies that insecticides and good sanitary practices must be employed to supplement radiation disinfestation of the grain. The authors were confident that in identifying and recognising the limitations of ionising radiation we may be in a better position to determine its proper role for the control of insects in stored grains.

Sethi *et al.* (1979) published the results of a series of studies on the control of stored-grain pests by gamma-radiations under different storage conditions. Studies carried out under laboratory conditions in India had shown that it was possible to halt reproduction in *Tribolium castaneum* infesting stored grain by release of sterile adults. This technique was also tested in galvanised iron bins and mud storage structures. When a further test was carried out in a warehouse, sterile adults of *Tribolium castaneum* were released in the ratio of 10 sterile insects to each normal individual. It was found that, though the population of *T. castaneum* was checked, populations of other stored products insect pests had increased. It was concluded that this method has comparatively limited chances of success under conditions where the grain is infested by more than one pest species.

Brower (1980a) studied the inheritance of partial sterility in progeny of irradiated males of *Ephestia cautella* and its effect on theoretical population supression. Adult males of *Ephestia cautella* were irradiated with 5–20 krad gamma-radiation and paired with untreated virgin females in the laboratory in Georgia, USA. Progeny were collected for 3 generations, and lines were established with all possible degrees of inbreeding and outbreeding. Although a high degree of population reduction occurred only in the F_1 generation that had doses of 15–20 krad, population models showed inbreeding of those insects that received the lower doses caused sufficient depression to reduce the abundance of a natural population greatly for several generations after a single inundative release of sub-sterilised males.

Zhbara and Shikrenov (1983) reported that laboratory studies carried out in Bulgaria showed that eggs and larvae of *Ephestia kuehniella* infesting stored grain could be controlled by gamma-irradiation. A dosage of 6 krad was necessary to kill all eggs. A dosage of at least 14 krad was necessary to protect grain against larvae, causing total mortality after 80 days. The shortest period in which complete mortality could be obtained was 60 days at 20 and 22 krad against first and second instar larvae and 50 days at 22 krad against third and fourth instar larvae. Puperal mortality ranged from 28.4% at 8 krad to 72% at 22 krad.

Gilzin (1984) described a 'radiation disinsectisation unit' which has been operating since 1980 in Odessa Port's grain elevator in the USSR where it irradiates up to 400 tonnes an hour. It is stated to be effective against all stages of insect development, 'its pests extermination efficiency being 100%'. By raising the temperature of the grain being irradiated, pests can be exterminated much more quickly. It is claimed, therefore, to be especially suited for use in the tropics.

After long and exhaustive studies involving the feeding of colonies of animals on irradiated wheat and wheat flour, it was proven that these products were wholesome and nutritious and were therefore given legal clearance for human consumption by the Governments of Canada, the United States of America and the Union of Soviet Socialist Republics. This means that in those countries irradiated wheat is given the same status as other foods that appear on the market **(Goresline 1973)**.

The US Food and Drug Administration has approved the irradiation of wheat and flour with absorbed doses of 20 000 to 50 000 rad from gamma sources providing energies of not greater than 2.2 MeV. Electron irradiation of energies not greater than 5 MeV is currently being considered for FDA approval.

Irradiation from cobalt[60] or caesium[137] sources or from electron-accelerating machines is certainly effective but the cost at present is very high and the process gives no subsequent protection to the grain. Furthermore, the treatment can only be given at a few major grain handling centres where very large quantities of grain are collected together. The cost of the process prohibits its use in country areas when the grain is first received from the grower. Thus chemical control would still be required if the buildup of large populations of insects is to be prevented **(Goresline 1973; Bailey 1979)**.

Bailey (1979), reviewing the potential for the use of irradiation of grain in an exporting country, concluded that a country, such as Australia, which lives by its exports and its international reputation, and which has a heavy investment in an existing storage system, would obtain no benefit, but would incur a great deal of extra cost, if it were to adopt the use of irradiation. He concluded that the real role of grain irradiation is for use as a quarantine measure by importing countries. A number of the problems faced by Australia would not confront importers. It would be feasible for grain to be passed through the irradiator as soon as it is received since it would be channelled through just a few import terminals. Because the effectiveness of the treatment would be under the control of the importing country, the delay before all pests died would be acceptable.

Watters (1984) reviewed the potential of ionising radiation for insect control in the cereal food industry and pointed out that high capital costs meant that an installation would have to operate for 4000 hours per year to approach the cost of fumigation. If it could only operate for 500 hours per year fumigation would be more economic. The decision on whether to install an ionising radiation device will depend on deficiencies of current control methods, the amounts of insecticides presently used, their frequency of application, comparative costs, the feasibility of integrating an irradiator into a cereal flow system, and customer or consumer acceptance of irradiated foods. Some of the answers cannot be determined in advance of industrial utilisation.

4.2.11 Inert Dusts

Inert dusts made from a variety of local materials such as ground rock or wood ash have been used to control insects in subsistence level grain stores **(Azab *et al.* 1971)**. These may be used at rates above 30% of the weight of grain. It is recognised that such materials protect grain by preventing movement of insects within the grain mass. Modern equivalents, such as silica aerogels, have been tested on a large scale but have not been adopted despite their efficacy. The large quantities of material required, frequently up to 1% by weight, render it inconvenient and lead to downgrading of the commodity through dust and lowered bulk density.

Dusts made from silica aerogels, various clays, diatomaceous earth, activated carbon, pyrophylite and a number of other silicates function as insect killers by absorbing or abrading the waxy layer from the insect cuticle, thus causing the insect to lose control of its rate of water loss. It dies of desiccation. The true mode of action of these dusts has long been in dispute. **Beaument (1961)** rejected absorption, favouring disruption of the spatial configuration of the insect epicuticle as the mechanism whereas **Ebeling (1969 and 1971)** provided evidence in favour of the importance of absorption. At this stage of our knowledge, it may be better to disregard theoretical considerations and to take at its face value the good correlation between the ability of dusts to absorb wax and their ability to increase the rate of water loss from living insects. The effectiveness of such sorptive dusts has been reported by **Cotton and Frankenfeld (1949), King *et al.* (1962), LaHue and Fifield (1967), LaHue (1970a), Ebeling (1971), and White *et al.* (1975)**. Performance is considerably higher under conditions of low humidity and in some instances failure has been reported when the sorptive dusts were incorporated in high moisture grain **(Ebeling 1971; LaHue and Fifield 1967)**. A number of such dusts are extremely effective in quickly killing a wide range of stored-grain insects when added to low-moisture grain at rates ranging up from 10 kg/tonne. At least one appears effective at 1 kg/tonne **(Ebeling 1971)**.

Many workers over several decades have endeavoured to exploit the capacity of diatomaceous earth to abrade and absorb insect cuticle and to cause dehydration of stored-product pests **(Strong and Sbur 1963; LaHue 1965, 1966, 1967, 1970a; LaHue and Fifield 1967)**. The application rates required are often unacceptably high.

Inert dusts do not entirely overcome the residue problem although their residues are not toxic. The dusts in the grain are unpleasant for workers handling treated grain. Some of the dusts are abrasive on machinery. Industry now spends considerable sums of money cleaning grain before sale or export and hence it does not welcome the addition of dust for any purpose. Attempts to introduce such dusts have resulted in downgrading of the grain with consequent loss of value and sales appeal.

Silica aerogels, because of their low cost and ease of application may lend themselves for on-farm use in developing countries where absolute freedom from toxic hazard is an important consideration. Diatomaceous earth dusts impregnated with low concentrations of grain-protectant insecticides appear to have particular promise for use in developing countries **(LaHue 1978).**

Golob (1981c) pointed out that frequently grain is stored on the farm for seed, in small quantities and protected by admixture with locally available 'inert' materials, such as ashes from the kitchen fire, sand, sawdust, vegetable oils, tobacco dust, etc. In this way the farmer can adequately protect seed at no cost to himself, even though the quantities required for admixture are very much larger than would be needed if synthetic insecticides were used.

Golob *et al.* (1982) set up trials under ambient conditions in southern Malawi to determine the effectiveness of admixing locally available powders with maize to protect the grain during storage. All of the powders, dolomite, wood ash, tobacco dust, sawdust and sand, restricted infestation. The effectiveness was directly related to dosage. The highest dose, 30% by weight, completely covered the grain. The protection afforded by woodash admixed at 30% by weight was of the same order as that provided by admixing pirimiphos-methyl at 8.8 mg/kg.

Golob *et al.* (1983) assessed the effectiveness of maize core ash, an additive traditionally used as a storage protectant in Tanzania in preliminary field trials to control *Prostephanus truncatus* which had the previous year been first recorded as an established pest of maize in Africa. They were able to show that 10% or 30% ash would reduce the percentage damage to grain by more than half.

Rheenen *et al.* (1983) showed that ashes gave some protection to bean seeds against bean bruchid beetles but the beetle control was not always sufficient.

Diatomaceous earth has been subjected to extensive evaluation as a grain protectant. Considerable work was done in the USA in the early days including many official trials **(Strong and Sbur 1963; LaHue 1965, 1966, 1967, 1970a, 1977c, 1978; LaHue and Dicke 1977; LaHue and Fifield 1967)**. Diatomaceous earth has been used as the carrier for various insecticides, particularly malathion, where it appeared to enhance the potency of the insecticide. However, such insecticidal dust formulations will not be considered here but will be dealt with in the chapters devoted to the individual insecticidal compounds.

LaHue (1967) evaluated diatomaceous earth along with malathion and pyrethrum as a protectant against insects in sorghum grain. The diatomaceous earth was applied at the rate of 1860 grams per tonne. It was thoroughly mixed with the sorghum grain by rotating in a steel barrel roller for 5 minutes. The treated grain was then poured into each of 5 bins of one half cubic metre capacity in a heated room which was heavily infested artificially with 5 species of stored-product pests. The diatomaceous earth treatment killed *Sitophilus oryzae* for 1 month only and had no adverse affect on *Tribolium confusum* in any of the tests. Repellency lessened with the increase in damage to the grain.

White *et al.* (1975) evaluated diatomaceous earth against silica aerogel dust and malathion for the protection of stored wheat from insects. The wheat was stored in 3250 bushell circular metal bins. Two diatomaceous earth preparations, 'Perma-guard' and 'Kenite 2–1' were compared with the standard treatment with malathion. Perma-guard proved most successful, comparable to malathion, in controlling insect infestations. Airborne dust was a nuisance when loading and unloading the bins, and increased problems with the machinery. **LaHue and Fifield (1967)** carried out a small-bin storage study in which a diatomaceous earth (Kenite 2–1) was applied to the wheat at a rate of 3.5 kg/tonne. This treatment proved equally effective to the standard malathion spray during a 15 month period when samples from the bins were tested with *Sitophilus oryzae* and *Rhyzopertha dominica*.

LaHue (1970a) evaluated diatomaceous earth against malathion, diazinon and silica aerogel in half cubic metre bins over a 12 month period. The diatomaceous earth was applied at a rate of 3.5 kg/tonne; it proved comparable to malathion in preventing damage from *Rhyzopertha dominica, Sitophilus oryzae* and *Cryptolestes pusillus*.

LaHue (1977c) reported that diatomaceous earth dust was very effective but its abrasive quality, largely due to silica content, damages all types of machinery used in commercial seed handling and farm planters.

When diatomaceous earth was mixed with wheat at 1785 g/tonne, grain-infesting beetles ranked in order of decreasing susceptibility as follows: *Cryptolestes pusillus, Sitophilus oryzae, Sitophilus granarius, Rhyzopertha dominica, Oryzaephilus surinamensis, Trogoderma parabile, Tribolium castaneum*, and *Tribolium confusum* **(Carlson and Ball 1962)**. In another investigation, when used against 6 species of grain pests diatomaceous earth at 1750 g/tonne of wheat prevented infestation for 6 months, 2680 g/tonne for 9 months and 3570 g/tonne for 12 months **(Strong and Sbur 1963). LaHue (1965)** found diatomaceous earth at 1785 g/tonne to be less effective after 1 year than malathion at 15 or even 10 g/tonne of wheat. In a later investigation 2 brands of diatomaceous earth at dosages of 1785, 3125 and 4460 g/tonne protected the quality of wheat better than malathion emulsion applied at the rate of 10 g/tonne and much better than 2 brands of unfluorinated silica gel at 670 g/tonne **(LaHue and Fifield 1967)**. Wheat treated with diatomaceous earth was downgraded because of decreased test weight, and when treated with silica aerogels it was classified as very low quality because of the presence of foreign substances.

Cotton and Frankenfeld (1949) reported effective control of stored-grain pests with silica aerogel but commented that treated grain would be subject to discount because of the residue. **King *et al.* (1962)** considered that the low mammalian toxicity and protracted residual effectiveness of silica aerogels would make them useful for protection of grain intended for farm consumption and for packaged food products. The good control of *Sitophilus oryzae* and *Tribolium castaneum* resulting from the use of this material led to further studies. Sorghum grain containing 12% moisture was treated with silica aerogel (SG-68) ranging up to 1% by weight. Excellent control of *Sitophilus oryzae* and *Oryzaephilus surinamensis* resulted from applications of silica gel at all concentrations used. Consequently a further series of trials with lower concentrations ranging up to 0.2% of the grain by weight was conducted. Effective control of *Sitophilus oryzae* was obtained with concentrations as low as 0.05%. Further work showed that satisfactory control apparently did not result when concentrations were below 0.05%.

In a small-bin storage study in the USA, **LaHue and Fifield (1967)** found a silica aerogel (Cab-O-sil) did not give satisfactory protection to wheat when applied at the rate of 670 g/tonne although the end results indicated that a higher dosage might be satisfactory.

Ebeling (1969) considered that dusts with a high surface area per gram of substance and with particles exceeding 20Å with low mineral sorptivity for water appeared to be the most effective.

LaHue (1970a) found that the silica aerogel Cab-O-sil at 890 g/tonne afforded nearly complete protection to wheat for 12 months.

Ebeling (1971) published a 55 page review of the subject of sorptive dusts for pest control and this paper is recommended for further study since it discusses in great detail the physical and chemical properties of silica aerogels and the epicuticle of insects and thereby draws conclusions about the mode of action of silica aerogels. It draws attention to the importance of incorporating ammonium fluorosilicate during the manufacturing process to increase the electrostatic charge and greatly improve silica aerogels as dust desiccants by causing particles of the aerogel to be attracted to insects which possess the opposite (negative) charge. When insects crawl over such charged particles they adhere to their cuticle in much greater quantity than uncharged particles and this results in a more rapid mortality. Such positive electrostatic charges are retained indefinitely when the powder is kept in a closed container but when the powder is exposed in the thin layers in the atmosphere it loses its charge within 2 to 3 months, depending upon humidity and the rate of air movement.

White *et al.* (1975) report evaluations of silica aerogel dusts, diatomaceous earth and malathion to protect stored wheat from insects. Wheat was stored in circular metal bins holding approximately 90 tonnes. Two silica aerogel materials, Cab-O-sil and SG-68, were compared with the standard treatment with malathion. They were applied as the bins were loaded. The wheat was sampled periodically to measure insecticidal effectiveness, physical changes that would affect grade and chemical residues. Both materials were effective but not superior to malathion. Airborne dust was a nuisance both when loading and unloading the bins.

Rodriguez *et al.* (1984) described experiments carried out to examine selected newer formulations of silica gels and compare these with other inert materials for use as coatings on cardboard and evaluate their efficacy as desiccants on a number of common stored-product insects. The authors were very enthusiastic about the results, particularly the relative efficacy of one product which they described as 'micronized silica'. However, the results are very poor indeed compared with those reported by **Ebeling (1971)** or **White *et al.* (1975)**. They bear no comparison with the performance of Dryacide reported by **Arthur (1981)**.

Silica aerogels, while highly effective in terms of insecticidal activity, have proved unmanageable because of their extremely low bulk density. The discovery of a process to coat diatomaceous earth with a film of silica gel has presented a practical method for utilising the insecticidal properties of silica gel for the treatment of grain. **Arthur (1981)** told how the new product, known as Dryacide, has been subjected to laboratory and field assessment. A grain treatment rate of 0.1% (1 kg/tonne) has been found to provide effective protection against the major insect pests where grain moisture has not exceeded 12%. *Sitophilus granarius* has been the most difficult of the test insects to control with Dryacide; other species could be controlled at treatment rates of less than 1 kg/tonne. In view of its mode of action, it will not control insects at the time they are developing within the grain. However, because of Dryacide's inert and stable nature it will, over a time, control active infestations and protect grain against further attack. Dryacide is currently (1986) being used on more than 500 000 tonnes of farm-stored grain and seed in Australia each year and the use is expanding rapidly.

4.2.12 Physical Shock and Disturbance

Both experimental studies and commercial experience have shown that insects and mites can be killed by physical stress and damage experienced during the handling and processing of grain and cereal products. The effectiveness of physical stress depends on particle velocity, distance travelled and force of impact. Death appears to be due to gross rather than minor physical damage such as abrasion of the cuticle resulting in excessive water loss or physiological disturbance.

Turning grain from one silo to another or from the bottom to the top within the one silo may have some value in reducing insect numbers but it is not possible to extrapolate from experience with one species to another let alone all pest species.

In the laboratory, percussion forces experienced by *Sitophilus granarius* impinging on solid surfaces at 45.7 metres per second killed all adults treated, but destroyed only 98–99% of the immature stages within the grain **(Bailey 1962)**. Some 20% of the grain was shattered by the impact. Repeated impacts of infested grain, moving at 12.8 metres per second, yielded lower kills of *Sitophilus* *granarius*, but did not damage the grain **(Bailey 1969)**. Adult *Cryptolestes ferrugineus* are more susceptible than either *Sitophilus oryzae* or *Sitophilus granarius* when dropped 14.1 metres at 16.6 metres per second **(Loschiavo 1978b)**.

Several workers **(e.g. Cogburn *et al.* 1972)** have reported considerable kills of stored-grain beetles in pneumatically conveyed grain (velocity greater than 7 metres per second). Conveying through augers or elevators has a less dramatic effect than pneumatic conveying but, nevertheless, may lead to appreciable mortalities, particularly of the immature stages. Sampling after turning may reveal a population that appears to consist only of adults (e.g. *Oryzaephilus surinamensis*).

Millers exploit the sensitivity of insects and mites to impact by using 'entoleters' in which infested grain is swung from the centre of a high speed rotor (about 1800–3600 revolutions per minute) on to fixed or mobile studs at the periphery of the machine. Insect fragments and broken grains are removed by aspiration and sieving **(Watters 1972).**

Watters and Bickis (1978) subjected wheat infested with *Cryptolestes ferrugineus* to mechanical movement through a farm auger to compare the extent of insect control with that obtained by treatment of augered wheat with malathion at 8 mg/kg. Assessment of insect populations for 34 weeks after treatment showed that the auger treatment, by itself, did not control adults or larvae.

4.2.13 Light

Insects respond to visible, ultraviolet and infra-red radiation **(Eldumiati and Levengood 1972; Marzke *et al.* 1973)** but the potential of this response for control has yet to be realised. It may be possible by artificial light regimes to alter the photoperiodic responses of insects, particularly moths, **(Schecter *et al.* 1971)**. This can lower frequency of mating or reduce fertility **(e.g. Lum and Flaherty 1969; Lum 1975)**, alter diapause responses **(Hayes *et al.* 1970)** or cause flights suitable for efficient control by chemical means. Females of *Plodia interpunctella* mated with males reared in continuous light have reduced numbers of spermatozoa in the spermatheca and produce fewer eggs **(Lum and Flaherty 1970)**. The effect is increased in the presence of carbon dioxide **(Lum and Phillips 1972)**.

More directly it may be possible to attract insects into an area in which another control measure is operating. Traps where an insect is attracted by long-wave ultraviolet light into a high tension electric grid e.g. 'Insect-O-cutor' **(Anon 1969)** are in use in many food retailing premises.

There have been studies to improve light trap design **(e.g. Lam and Stewart 1969); Stanley and Dominick 1970; Barrett *et al.* 1971)** and to predict the number of traps required to achieve pest population control **(Heartstack *et al.* 1971; Onsagher and Day 1973)**.

The influence of visible light on the behaviour and biology of stored-grain pests has been reviewed by **Lum (1975)**. Adult moths, in particular, are generally influenced by the manipulation of light/dark cycles (photoperiodism). The patterns of adult emergence, mating behaviour, oviposition and diapause can be modified by changes in the lighting and relative proportions of the light and dark periods in which the rhythmic behaviours of the moths are entrained. The underlying mechanisms are not yet well understood but it is possible that the manipulation of photoperiod could render stored-product moths, in particular, more amenable to other control methods.

Light traps emitting specific wave lengths have been used to trap insects for control as well as for survey purposes **(Watters 1972; Nelson 1975)**. *Plodia interpunctella* and *Ephestia kuehniella* are more attracted to green light than they are to ultraviolet. By contrast, the converse was true for several species of stored-grain beetles **(Soderstrom 1970)**. Once attracted, the insects can be trapped on sticky plates or killed in high voltage electric grids (Electr-O-Cuters) **(Gilbert 1984)**.

Insects that infest stored products respond, like many other species, to visible light wave lengths ranging from 350 to 770 nm. Responses other than phototactic responses are reflected in many types of rhythmic behaviour. In *Anagasta kuehniella, Cadra cautella* and *Plodia interpunctella*, visible light influences the development and emergence of adults, their reproductive behaviour and the development of reproductive organs. Adult emergence can be entrained to specific times of the day, and the light/dark cycle can be used to regulate oviposition, mating, and the 'calling' and release of pheromone by the females. Also, light, mediated probably through the neuroendocrine system, influences spermatogenesis in male *Plodia interpunctella* and *Anagasta kuehniella*. The many responses of insects to light provide opportunities for investigation of the physiological mechanisms that can be exploited to alter or manipulate behaviour of stored-product insects. Such an investigation would supplement existing physiological, chemical or biological data and could pinpoint ways to develop or improve control methods **(Lum 1975)**.

Interest in the use of short wave length (253 nm) ultraviolet (UV) radiation as a method of insect control has revived in recent years, partly because it is non-chemical and partly because it could be extremely effective in certain situations (e.g. insect rearing facilities). However, despite its documented lethal effects **(Wharton 1971; Beard 1972; Gingrich *et al.* 1977)**, UV radiation has played no part in control programs to date. There are several reasons. Undoubtedly, the primary difficulty is that UV radiation, at least at the shorter wave lengths, can be injurious to human cells although the underlying mechanisms are only partly understood **(Giese 1964)**. Any attempt, therefore, to employ high intensity sources of UV radiation in areas of human activity would probably be restricted. Lack of penetration of these light rays through some substances would certainly be a second reason. However, there are alternative ways of using UV: a pest species could be attracted to a shielded trap, a commodity could be irradiated for a very short time, or the phenomenon of photoreactivation (light repair of UV-induced damage) could be used. UV radiation should be viewed not only as a control method with lethal capabilities but also one which can provide new approaches to insect control: e.g. pheromone breakdown **(Goto *et al.* 1974; Bruce and Lum 1976)** and the converse, pheromone protection using UV-absorbing compounds, offer exciting possibilities **(Bruce and Lum 1978)**.

4.3 Fumigation

The decision to exclude fumigation from this review should not be interpreted as an indication that the author fails to recognise the importance of fumigants in controlling stored-product pests. Quite the contrary. However, fumigation is a major topic in its own right and it deserves to be considered in the light of the extensive literature that is available. Fumigants will be briefly mentioned wherever they have been used in conjunction with the use of grain protectant insecticides. It is, however, important to recognise that fumigants and insecticides are different, that each has its place, its advantages and its limitations. They need to be considered and used so that they complement each other.

One reason for the popularity of fumigation as a means of controlling stored-grain insects must surely be its flexibility. It is nearly always possible to modify application methods so as to meet the requirements of different climate and storage conditions. The most comprehensive information available on grain fumigants is provided in the *FAO Manual* by **Monro (1969)** and the subsequent revision by **Bond (1984)**. A good review is provided by **Harein and De Las Casas (1974)**.

Industrial fumigants can be applied singly or in combination to improve their effectiveness or decrease their potential hazard. It is generally recognised that the ideal fumigant has not been developed, especially when considering the wide variety of fumigation circumstances one may encounter. It is considered unlikely that any useful fumigants yet remain to be discovered. All fumigants are potentially lethal to humans before, during or after application. Considerable care must be exercised to reduce the hazard to acceptable levels.

Whilst in the past it has generally been assumed that fumigants do not give rise to residues in fumigated commodities, modern analytical techniques have revealed that a large number of fumigants give rise to substantial residues which have now become a matter of concern to health authorities.

To gain full value from the use of fumigants, insect-proof storage facilities are required because no fumigant confers any residual protection upon the fumigated commodity which can readily become re-infested if any insects remain in adjacent premises or in the vicinity.

Fumigants remain the materials of choice for disinsecting grain which has become infested prior to shipment as well as for ensuring that transport vehicles and vessels awaiting the loading of clean grain are themselves free of insect pest infestation.

5. Insecticides in Stored-Product Pest Control

5.1 Their Properties, Development and Regulation

Having considered some of the non-chemical means of controlling pests in stored grain it is now proposed to discuss the need and justification for the use of insecticides and to review the evidence which provides the requisite assurance of consumer safety.

The use of insecticides to control insects in stored products is usually not a matter of choice between methods of control, but a choice between use of insecticides and food losses. It devolves into a question of whether infestation and losses in commodities are to be tolerated and, if not, whether alternative control measures are practicable. Fortunately, because of the efficacy of the range of insecticides that may be used and their low toxicity when correctly used, chemical control is by far the lesser evil when compared with losses that may occur without their use.

Until comparatively recently there was a tendency to regard the presence of insects in close association with food as inevitable. However, in most industrialised countries the presence, in food, of any recognisable part of an insect is no longer regarded as acceptable. The highest degree of purity in our foods, including freedom from pests or their remains, is now expected. In many countries the required purity is legally enforced. The improvement in food hygiene which has been effected during the past 20 years can, to a very large extent, be attributed to the development of the hybrid science of pest control and, more particularly, to the development and usage of synthetic insecticides which pest control research has stimulated.

The use of insecticides on stored products has progressed in recent years from application of a few inorganic materials to the use of a large number of highly effective organic compounds and subsequently to the selection of a preferred few of these. Some very effective compounds have been discarded or displaced because of potential hazards to human health or the environment. Effective insecticides such as DDT and dieldrin have fallen into disuse because of residual stability and adverse effects on wildlife. Others have or may become obsolete because of toxic or mutagenic effects on animals. Still others may become obsolete for other reasons such as insect resistance. Some compounds have been eliminated or restricted because of the action of pressure groups and polititians.

Notwithstanding the shortcomings of synthetic insecticides, it has been proved that chemical pest control methods, if carried out intelligently and knowledgeably, can be both effective and safe. There is no doubt that insecticides contribute largely to the conservation of our food stocks and to the maintenance of their quality and purity. Therefore, unless there is to be an unexpected acceptance of foods contaminated by insects or until some alternative, practicable, economic and effective form of pest control, free from all hazards to operators and consumers, can be put into effect, we must rely on these insecticides.

Although the total quantity of insecticides used for grain protection is not known, it is certainly considerable and is increasing rapidly. **Herve (1985)**, quoting from statistics provided by the Wood-Mackenzie organisation in 1983, states that of US$675 million worth of pyrethroids sold for non-agricultural use in 1980, US $100 million worth (ex-manufacturer) were utilised for treatment of stored grain. This was equal to the value of pyrethroid insecticides used in public health and half the value used for veterinary purposes.

The criteria that determine whether or not a pesticide can survive for commercial use are many and they are changing. The requirements for registration are stringent and there is some indication that they may become even more stringent. Regulations are made to ensure that new compounds coming on the market will have minimal hazard to health and the environment. Since the possibility of new insecticides being discovered is small and the development of new materials is costly and slow it is essential that the insecticides we now have be used with the greatest care and effectiveness possible. Many questions can be asked about present policies regarding the use of insecticides on stored products: should our methods for screening, evaluating and registering insecticides change? Should we put increasing emphasis on new formulations, new methods of application and use? How important is the resistance problem and can we do anything about it? These are all questions that come to mind when we think of the great demand for food preservation and pest control in the future **(Bond 1975)**.

Chemical control must be placed in its correct perspective. It should be emphasised that synthetic insecticides must be regarded as an adjunct to good management, to reinforce hygiene and sanitation, to enhance the effectiveness of available storage facilities, to complement physical control methods and not to replace good warehouse-keeping or regular inspection for infestation or deterioration.

Grain-protectant insecticides should be regarded as

tools used in, rather than as substitutes for, well-planned storehouse management, for without such management all available pest control techniques will fail. They must be regarded as one of several options that might be considered in each situation where grain is to be stored, whether it be in the hands of the tribal system, the peasant farmer, village storekeeper, grain merchant or central grain authority.

Each of the many methods of insect control has its place and its advantages. There are many situations where 2 or more methods can be used in an integrated program. Not all methods, however, are, suitable for use in every situation. Grain-protectant insecticides offer many advantages, though admittedly, these advantages are more important in some grain-storage situations than in others. As the name implies, grain-protectant insecticides should be used to protect grain from attack, damage and infestation. They should not be looked upon as a primary means of destroying an already established infestation. They perform best when used in conjunction with a high standard of hygiene and sanitation and should never be seen as a substitute for these important measures.

Bond (1974), in a paper on future needs and developments to an international conference on stored-product entomology, appealed for a great deal more information on properties and uses of existing pesticides in order that they can be exploited most effectively. Whilst there is a pressing need for new pesticides to deal with some of our current control problems, the possibilities of new compounds being discovered and developed is not great. We will most likely have to rely on compounds already in use for most of our treatments. In addition, more basic information is required on the toxicology of pesticides, the response of the pest to treatment, persistence and degradation of pesticides and formation of residues is needed. We must know more about the materials we are using and the way pests respond to them so that the pesticide can be directed to the most vulnerable point of the organism's lifecycle and biochemistry. Information on mode of action is required to understand and deal with resistance with pesticides.

Winks and Bailey (1974) stressed the regulatory aspects governing the use of pesticides on stored products in Australia.

Minett (1975), in discussing some factors influencing the effectiveness of grain protectants, described experiments with non-uniform insecticide distribution which were found to significantly increase the effectiveness and persistence of malathion and dichlorvos.

Watters (1975) pointed out how the need for safer and better pesticides, greater understanding of the inter-relation between pest biology and pesticide action, and the application of techniques and practices for improving the effectiveness of pest control agents will provide wide scope for future research on stored-product pesticides. The combined use of physical and chemical controls should reduce the amount of chemicals needed for insect control and consequently reduce selection pressure which contributes to the emergence of resistant strains. Environmental factors such as temperature, grain moisture content and availability of food, influence insect locomotor activity and, consequently, the amount of toxicant picked up by insects. Thus, toxicants which stimulate insect activity will counteract environmental effects which tend to reduce insect activity. Public concern about contamination of foods suggests that more attention should be given to extension and public relations in the safe and efficient use of stored-product insecticides to counteract the unfavourable publicity given to pesticides.

Watters (1975) further emphasised that most chemical compounds supplied for testing have already been critically evaluated by the chemical industry or by various research institutes. Evidence concerning efficacy against stored-product insects is usually derived from a limited number of laboratory tests which form part of an overall screening program. Additional research therefore is often needed to evaluate those aspects of an insecticide's performance that are of concern for a particular climatic region, or for a special need. This usually involves tests of efficacy against one or more economically important species, and studies to determine persistence and rate of breakdown under certain environmental conditions. Comparisons with other insecticides should reveal where the candidate insecticides overcome some of the deficiencies inherent in currently used compounds. The acid test, of course, is whether the insecticide actually meets the expectation of a wide variety of users under a wide variety of conditions. Considering the many combinations of factors involved in the practical use of an insecticide it is little wonder that there are instances where a recommended insecticide use does not provide the control expected. Variations in dosages, application equipment, breeding surfaces, substrates, environmental conditions, the susceptibility of insect species, stages and strains and in the people who actually apply the insecticide, all contribute to wide divergencies in results. When all of these factors are considered it is indeed remarkable that there is any unanimity of opinion at all concerning the performance of an insecticide. It must not be forgotten also that of the people who are applying insecticide to control an infestation, some are sceptical of its performance and do a slipshod job, some do a token job to conform with instructions issued by management or a regulatory agency, while others are conscientious in trying to obtain the desired results.

The use of chemicals on or near foodstuffs for insect control or for prevention of infestations has to be considered in relation to the ultimate use of the food. Imported foods would be subjected to scrutiny both for chemical residues and for insects to comply with national standards of health and food sanitation. If the commodity is to be used immediately for processing, the requirements may be less stringent for the presence of insects but more exacting for chemical residues. By contrast, if the food is to be stored in reserve for several months or years, the presence of insects will not be tolerated but there will be a longer period for the breakdown of chemical residues. High food standards in developed countries place a heavy burden on developing countries in tropical and sub-tropical zones that are expected to export foods which are both uninfested and free of insecticide residues and other contaminants.

Golob (1977) edited an extension handbook, designed to assist the small farmer in storing grain for consumption, not for seed. It discusses the problems which farmers face after harvest, and suggests methods of using insecticides which apply to all types of village-level stores. It points out the simple basic needs in regard to insecticidal powders and powder sprinklers, gives notes on preparation of store and produce before applying the insecticide, describes the treatment of maize on the cob and of threshed grain and finally points out simple safety precautions.

Desmarchelier and Bengston (1979a) summarised the methods used to evaluate protectants suitable for any given set of conditions as follows:

1. laboratory tests on effectiveness, resistance patterns, etc;
2. laboratory tests on period of protection, under a range of conditions, against artificial re-infestations;
3. development of methods of analysis of residues;
4. trials in 2 commercial storages, 1 in a summer rainfall climate, 1 in a winter rainfall climate; chemical assays by up to 10 residue chemists, and biological assays with up to 20 strains in 8 species, are performed on samples of stored grain at intervals of 6 weeks for a period of 9 months;
5. extensive pilot usage trials in 20 commercial storages; in such trials, residues are monitored at regular intervals and the storages are regularly examined for possible insect infestation;
6. development of a predictive model of loss of residues, and comparison of laboratory predictions with commercial results;
7. organoleptic and residue tests on grain products such as cooked rice, malt and bread.

Desmarchelier and Bengston (1979a) described predictive models for loss of residues of 12 protectants on grains. From these models one can calculate the persistence of a chemical on any grain under any constant or varying conditions of temperature and moisture content.

Taylor and Webley (1979) discussed the constraints on the use of pesticides to protect stored grain under rural conditions in developing countries where damage to grain is often great. It is suspected that, though effective insecticides exist to overcome this problem, they may be being used only in a very small measure in some developing countries. There are many reasons this is so, for example the absence of a supplier or convenient package size, lack of government approval or an extension campaign and the high relative cost of chemicals to the rural user.

Snelson (1979b) discussed the topic of pesticide residues in stored grain, their significance and safety and described the measures taken to provide the necessary assurance to governments, traders and consumers.

Greening (1979b) pointed out that chemical treatments to control grain insects on farms in Australia are likely to be needed even if it were possible to convert the bulk grain handling system to using physical methods of insect control. Grain kept for stock feed may become a reservoir of insects that can spread to harvesting machinery and contaminate new grain at harvest. Methods of protecting and fumigating farm-stored grain and disinfesting harvesting machinery have been developed.

Pinniger and Halls (1981), summarising the recommendations of an international seminar on the use of pesticides for the control of stored-product pests in developing countries, stated that it was agreed that the subsistence farmer will conserve food by whatever means are at his or her disposal. He/she needs to reduce damage but not necessarily achieve complete control. When capacity to grow more is increased he/she needs to store more. The farmer may have to store higher yielding varieties which have greater susceptibility to insect attack. Then assistance will be needed to kill insects and prevent losses by use of pesticides. It was agreed that there was a great shortage of data on efficacy at the farm level, although many workers had experience of particular problems and insecticide usage. The collection and collation of data from different storage situations was recommended.

Taylor (1981) reported that the results of a survey initiated by **Taylor and Webley (1979)** indicated that the major constraints on the use of insecticides for farm-stored grain were non-availability, high cost and lack of appropriate small packs. Ignorance of modern insecticides among farmers was also an important consideration.

Webley (1981a) was critical of the poor quality of dilute insecticidal dusts available to small farmers. The results of a survey had indicated that many such dusts collected from the field had degraded completely, largely due to the use of unsuitable materials as carriers. He stressed that higher standards of stability are needed in developing countries and that a 2-year shelf life must be the minimum requirement.

Webley (1981b) pointed to the importance of improved insecticides since in every farm or village storage system in which there are susceptible crops being stored in conditions supporting insect pests some insecticidal measures may be required. The alternative traditional methods may not be safer and he questioned whether ash, oil or powdered leaves and roots applied at up to 30% of the weight of the grain can be safer than a few parts per million of well-tested insecticide. He pointed to the value of the work of the Joint FAO/WHO Meeting on Pesticide Residues and urged manufacturers not to recommend to developing countries methods which would not be used in the West and not to encourage countries to use chemicals in ways which do not strictly conform with the recommendations of the JMPR. He considers that we are currently fortunate in having some very good insecticides, which the farmer should be encouraged to make good use of, to everyone's advantage.

McCallum-Deighton (1981) examined the possible reasons for failure of farmers to use effective insecticides and suggested remedial measures. Although the potential market for stored-product pesticides in the developing countries is quite large there is such tremendous fragmentation into many countries and among many farmers, each producing relatively small amounts of cereals, that the interest to manufacturers is reduced accordingly. All the expenses involved in getting the small amount of insecticide dust into the hands of the farmer increase the cost to the point where the farmer is barely able to pay. The lack of education, incentives and perceived benefits among subsistence farmers in developing countries inhibits the adoption of effective insect control measures. Inappropriate formulations and techniques not suitable to the peasant farmer's way of life constitute another disincentive. The problems and delays involved in registering new pesticides for use with stored produce is yet a further complication. The level of literacy in many developing countries is low so that there is a language barrier in trying to educate the farmer.

Snelson (1984a, b, c,), in lecture notes for an international training course on preservation of stored cereals, outlined the general principles of the use of insecticides, the properties of insecticides and their preparation and application.

The report of discussions and recommendations of the seminar by the Group for Assistance on Systems relating to Grain After-harvest **(Anon 1981)** stated that whilst some progress has been made in identifying situations where simple treatments with pesticides will be effective and acceptable to the farmer, there is an urgent need for cost/benefit studies to be carried out to determine at what point the promotion of the use of pesticides and protectants at the farm level can be recommended. Such studies should also consider the important problem of ensuring the availability of suitable pesticides, both in terms of efficiency of the active ingredient and formulation. There is an urgent need for further studies on the comparative cost-effectiveness of different warehouse protective treatments in order to evolve treatment regimes which are more effective than the commonly practiced stack spraying.

Calverley (1981), in a keynote address to an international seminar on the appropriate use of pesticides for the control of stored-product pests in developing countries, pointed out that there are few situations in the storage of grain for which some technical solution cannot be found. On the other hand, we are very conscious that it is the inability to transfer existing knowledge which is the main hindrance to progress. There exists in too many developing countries a limited understanding of technology and restricted facilities or opportunities for training to improve this situation. He suggested that the seminar distinguish between treatments or pest control practices which have failed because the treatment was unsuitable and those for which the application was inadequate. He pointed to the very substantial improvements in pest control in Australia, enforced by recent grain surpluses and expanded export programs. He suggested that these have come from improved application and better control of long established treatments as much as from the application of new technology of insect control.

Based on the experience of the Joint FAO/WHO Meeting on Pesticide Residues and the Codex Committee on Pesticide Residues, **Snelson (1983)** discussed and presented recommendations on the quantity and quality of residue data required for the establishment and enforcement of maximum residue limits.

Harein (1982) pointed out that grain protectants have several advantages over fumigants; they persist for extended periods at concentrations lethal to the target insects; they are generally safer to apply; and they require little, if any, special application equipment. Protectants are applied to kill most indigenous insects but, of more importance, to prevent insects from establishing an infestation. One protectant application may be sufficient during a single storage season if the grain moisture content is low and especially if applied to grain as it warms above 15°C in early summer. Protectants can be used effectively in loosely constructed storage facilities that could not be fumigated successfully without extensive and expensive sealing.

Desmarchelier and Banks (1984) described some of the processes undertaken in the Australian grain storage industry which lead to optimum and appropriate use of protectants. These procedures are designed to meet a 'nil tolerance' specification for insects in grain. It should be possible to meet such a specification in any central storage system, given adequate management and laboratory back-up. Many of the monitoring procedures require a substantial level of training on the part of the investigators. Managers often lose sight of the fact that accurate application of pesticides at a parts-per-million level is a fairly difficult exercise in applied chemistry or chemical engineering and it is not surprising that proper control of such a process must include personnel with a reasonable understanding of these disciplines.

A summary report of discussions and recommendations at the GASGA (Group for Assistance on Systems relating to Grain After-harvest) Seminar held, 17–20 February 1981, at the Tropical Products Institute, Slough, England **(Anon 1984)** recommended that national governments, in collaboration with GASGA members, should undertake, as a matter of priority, a review of current post-harvest pest control problems in relation to:

1. materials and techniques currently used in the country;
2. the level of efficiency and success being achieved;

3. constraints affecting the use of pesticides, such as availability, suitability of formulations currently available and problems of registration;
4. the potential for improvements of existing pest control procedures;
5. the ability of locally available manpower to use the pesticides properly;

and formulate an action program to remedy any problems which emerge from the review.

Samson (1985) reviewed the effect of factors such as moisture and temperature on the biological efficacy of residual pesticides. Effects on efficacy reflecting availability of residues for pick-up by insects, differences in pick-up for reasons other than availability, and insect responsiveness after pick-up were outlined. Insect responsiveness is little affected by humidity. Its relationship with temperature, described by temperature coefficients, varies depending on the insecticide. With admixture treatments, efficacy is little affected within limits by distribution of the insecticide in the grain mass. Efficacy is reduced on small grains because of their greater surface area to volume ratio and perhaps insect movement is restricted. Availability of actual residues declines during storage, and is reduced irreversibly at high moisture contents. Residues on fabrics also lose their efficacy over time, particularly on absorbent substrates. The effect is lessened if insecticides are applied in wettable powder form. The availability of sorbed residues on hydrophilic substrates may be reversibly increased at higher humidities. Pick-up of insecticide is enhanced at higher temperatures because of increased availability and insect activity, but whether this is reflected in insect mortality depends upon the temperature coefficient of the insecticide.

Samson (1985) concluded that the basic understanding of the biological action of admixed insecticides may considerably improve their efficiency in practice. Insect control by means of mixing insecticide with grain represents a very 'finely-tuned' system. A balance must be struck between minimum effective concentrations for insect control on the one hand and maximum residue limits on the other. This balance is achievable because many of the factors influencing efficacy of the insecticides can be controlled: the substrate to be treated, viz. the grain, is very uniform; a sufficiently uniform distribution of insecticide in the grain mass is readily achieved; desired concentrations of insecticide can be applied accurately; and both temperature and moisture content are controllable to some extent. Under these circumstances, it is quite feasible that insecticides or dose rates can be varied, on the basis of results of basic research, to be appropriate for different storage conditions.

Snelson (1985b) described the national and international systems and practices for the regulation and registration of pesticides for use in grain storage systems. He described the many safeguards introduced for the protection of operators, consumers, livestock and trade and outlined the type of developmental work required to produce adequate evidence of efficacy, fate and residues of grain protectant insecticides.

Evans (1985) conducted laboratory tests using populations of *Tribolium castaneum, Tribolium confusum, Sitophilus zeamais, Callosobruchus maculatus* and *Zobrotes subfasciatus* from Uganda, firstly to determine whether they were resistant to malathion and/or lindane and secondly to measure the effectiveness and stability of pirimiphos-methyl, fenitrothion, etrimfos, permethrin and deltamethrin dilute dusts in protecting cereals and pulses from these insect pests. All the insect populations tested were resistant to malathion or lindane and some were resistant to both. Of the insecticides tested, deltamethrin at 1 mg/kg was generally the most effective. The organophosphorus compounds were only effective against the *Tribolium* species and *Sitophilus zeamais*. Permethrin was the least effective, only controlling *Callosobruchus maculatus*.

The following is a list of the main advantages which grain protectant insecticides offer:

(1) *Versatility*. From the extensive range of grain protectant insecticides available, it is possible to select one or more products which are ideally suitable for protecting any stored product from any, or all, stored-product pests.
(2) *Universal*. Grain protectants are applicable to all storage systems. Remarkable success has been achieved with simple insecticide powders applied to primitive maize cribs under tribal conditions in the least developed countries. Countries like Australia, which have a highly sophisticated central storage system for handling all types of grain, have used grain protectant insecticides as part of an integrated system for eliminating (virtually) all insects from many millions of tonnes of grain.
(3) *Practicable*. Grain protectant insecticides fit in with, but do not hamper, storage and handling practices. The insecticide, its formulation and method of application, are easily adapted to be compatible with, or complement, all such procedures.
(4) *Equally suitable for all grains* provided appropriate rates are used.
(5) *Predictable*. The fate of each grain protectant insecticide can be be predicted from a knowledge of the storage temperature and relative humidity. Treatment can be planned accordingly so as to ensure that an adequate level of protection is maintained for the full period of storage without leaving unnecessarily high levels of residue.
(6) *Simplicity*. The formulation and method of use can be designed to suit the level of sophistication of the particular grain storage system. Dilute dusts, pre-packed into small sachets, sufficient material to treat one bag, basket or other normal unit of grain can be provided for safe use by the

most unsophisticated communities. Emulsifiable concentrates or ultra-low-volume formulations are available for use in mechanised or automated central storage systems.

(7) *User safety*. Provided the product is chosen with due regard to the conditions under which it is to be used, there should be no risk of misadventure or harm to users.

(8) *Economical*. An extremely high level of protection can be obtained with absolute minimum of capital cost and an insignificant direct cost that is offset many times over by direct and indirect savings and increased return from treated produce.

(9) *Speed*. Treatment with grain-protectant insecticides can be carried out immediately since no preparation is required. Extremely effective applicators can be devised at little or no cost.

(10) *Do not interfere with fertility of seed*. This should be contrasted with fumigants and irradiation.

(11) *Preventative*. The use of grain protectants, particularly at the stage the grain is being taken into store, reduces the risk of infestation of premises, other stocks, transport and prepared food; also avoids the risk of damage to grain and contamination of grain with insects and their secretions.

(12) *Consumer safety*. All or most of the residues applied to the raw grain taken into storage, disappear or are removed before the cereal food reaches the consumer. This increases the high level of safety already provided by the extensive data available on the toxicology. The insecticides are approved as grain protectants by health authorities who are satisfied that there are no risks to consumers.

In several countries, where grain protectant insecticides are used exclusively or extensively for the protection of grain, total diet studies have revealed that the intake of residues of these insecticides by consumers is well within the amount that is deemed to be without hazard if consumed daily for an entire lifetime.

Tens of millions of tonnes of edible food grains have been treated successfully with grain-protectant insecticides under a wide range of conditions. These grains have entered, and have been accepted in, international trade, in spite of the enormous quantities of grain involved. Meticulous monitoring has revealed that it is possible to apply grain-protectant insecticides to the extent necessary to virtually eliminate insects from commercial grain shipments whilst complying completely with the rigid legal limits for residues.

In spite of this outstanding achievement, the use of chemical treatments for insect control in stored commodities has not been accepted universally. The following are believed to be some of the reasons:

1. general lack of understanding about the problems of tropical countries and countries which have not yet attained a high level of sophistication in the handling and storage of grain;
2. failure of many people and national authorities to recognise the need to use chemicals to protect valuable food, ensure the availability of staple commodities as a buffer against famine, maintain economy and meet the demand for food to feed the increasing population;
3. lack of knowledge about the limitations of available non-chemical measures to control stored-product pests;
4. fear of the unknown;
5. belief that man-made chemicals are somehow different to chemicals that occur in nature; failure to recognise that grain and similar stored products are, by nature, chemicals and that many of their natural chemical components are at least as toxic to laboratory animals as those approved for addition to grain for pest control;
6. traditions, sometimes backed by national legislation, that nothing which might be regarded as 'deleterious' should be added to food commodities;
7. the development of the 'natural food' cult and the attendant rackets in 'health foods';
8. political pressure by merchants, domestic producers and other self-interest groups to create misgivings about the safety of treated commodities, in order to produce non-tariff barriers to trade.

Grain protectants are insecticides which, when applied to grain, prevent infestations becoming established. They are not intended to control heavy infestations present in the commodity at time of treatment — such infestations should be controlled by fumigation as a separate operation. In practice, grain-protectant insecticides will control light infestations present at time of treatment, but, because this will accelerate the selection of resistant strains, it should be avoided.

Only those insecticides which have been specifically approved for use on and around grain should be used. The choice of insecticides which may be used is limited by the very strict requirements which must be enforced to ensure absolute safety of important basic food commodities. To qualify for selection as a possible candidate grain protectant for use on grain the insecticide must fulfil the following requirements:

1. it must be effective at economic rates of use;
2. it must be effective against a wide variety of insect pests;
3. it must be capable of being used without hazard to operators;
4. its use must be acceptable to health authorities;
5. it must not give rise to unacceptable residues;
6. legal maximum residue limits must be established;
7. it must present no hazard to consumers of grain and grain products;

8. it must not affect the quality, flavour, smell or handling of grain;
9. it must be acceptable in international grain trade;
10. it must not be flammable, explosive or corrosive;
11. its method of use must be compatible with established grain handling procedures.

Although the scientific literature contains many references to the effectiveness against stored-product pests of a large number of insecticides, the number which have been cleared for application to stored grain and for which maximum residue limits are established is strictly limited. Of these, only a few have yet been adopted commercially though the rate of adoption appears to be increasing. The following compounds have been cleared and are currently being used for treating stored grain and probably other stored commodities.

- bioresmethrin
- bromophos
- carbaryl
- chlorpyrifos-methyl
- dichlorvos
- fenitrothion
- lindane
- malathion
- pirimiphos-methyl
- piperonyl butoxide
- pyrethrins

The following compounds have been subjected to extensive study and several appear capable of fulfilling all criteria for acceptance as approved grain protectant insecticides:

- deltamethrin;
- diazinon;
- etrimphos;
- fenvalerate;
- iodophenphos;
- methacrifos;
- permethrin;
- phenothrin;
- phoxim;
- tetrachlorvinphos.

The acceptability of each of the above insecticides will be discussed in the light of the extensive knowledge of the following characteristics:

1. usefulness;
2. degradation;
3. fate in milling, processing and cooking;
4. metabolism;
5. fate in animals;
6. safety to livestock;
7. toxicological evaluation;
8. maximum residue limits.

5.2 Insecticide Residues — Their Measurement and Fate

Although extensive reference is made to analytical data on pesticide residues, a decision was made not to review analytical methods used to study the level and fate of insecticide residues on grain and cereal commodities. This was not because of any lack of appreciation of the importance of analytical methodology but rather that it was felt that justice could not be done to the subject without extending the text to an unacceptable length. It is hoped that a separate monograph dealing with many important aspects of analytical methodology can be produced in the near future.

A proper understanding of the performance and fate of grain protectant insecticides can be obtained only from chemical analysis data generated systematically during the lifetime of the deposit on the grain and the resultant residues on the products of subsequent processing. Although many studies have been conducted without the use of chemical control or residue analysis, such studies provide less than ideal understanding of the reasons for the biological performance which has been reported. It should be clearly understood, however, that residue analysis on its own does not provide a clear understanding of the bio-availability of the aged insecticide deposit.

Loss of insecticidal activity with time has been attributed, in part, to the decrease in availability of the insecticide to the insect (**Champ *et al.* 1969**). The biological evidence for decreased availability is not inconsistent with the chemical data reviewed by **Rowlands (1971)** on movement of insecticides from the outer grain layers. However, chemical and biological data have seldom been correlated. **Desmarchelier (1978c)** presented a mathematical evaluation of the availability to insects of aged insecticide deposits on wheat and he used his mathematics in interpreting data previously published by other authors.

People who are not trained as analytical chemists are inclined to regard analytical chemistry as an exact (precise) science which produces nothing but real numbers. However, the following important papers draw attention to the many factors which affect accuracy, precision and reliability: **Federal Working Group on Pest Management (1974); Desmarchelier *et al.* (1977); Thompson (1975); Cochrane and Whitney (1979); Horwitz *et al.* (1980); Cochrane *et al.* (1979); Gunther (1970); Carl (1979); Gunther (1980); Atallah (1981); Puschel (1981); Horwitz (1982);** and **Snelson (1982)**.

The results obtained with any analytical method tend to be sensitive to the skills and to the environment of the analysts (**American Chemical Society 1978**). A great deal has been written about the pitfalls of analytical practice (**Telling 1979**) and the many steps that need to be taken to ensure that the results finally reported reflect the true position.

Analytical reference standards are without doubt a key to good or bad analytical results (**Thompson 1975; Cochrane and Whitney 1979; Horwitz *et al.* 1980**). The precaution of checking the sensitivity and response of detectors is often neglected (**Cochrane and Whitney 1979; Cochrane 1979**), and frequently the need to check the percentage recovery through the whole procedure of preparation, extraction, clean-up and quantitation is forgotten (**Gunther 1970; Carl 1979; Gunther 1980; Frehse and Timme 1980**).

Many published data on insecticide stability on grains are of doubtful significance because of lack of replication.

The importance of careful, unbiased, and representative sampling by trained personnel cannot be over-emphasised. The validity or usefulness of analytical samples or residue analysis hinges upon an intelligent, realistic approach to the problem of obtaining a suitable sample. The analyst's residue data may be precisely determined but woefully inaccurate because of inadequate sampling. An erroneous result may be worse than none at all. Many workers know this fact yet not all are apparently aware that many of the discrepancies frequently encountered in residue studies derive from a failure to consider all the controlling parameters in planning to take such a sample **(Bates 1974). Snelson (1971)** presented extensive data to demonstrate the problems encountered in sampling bulk grain for the determination of pesticide residues and showed how the natural segregation of light particles to the side and top of grain bulks away from the heavier kernels which gravitated to the centre and bottom, produced a distinct separation of its insecticide residues. The segregation was such as to produce more then tenfold differences between the residue levels on different grab samples. **Minett *et al.* (1984)** were able to demonstrate that the insecticide residue levels throughout the bulk of wheat treated with conventional spray equipment and with gravity feed applicators varied through a tenfold range in spite of care and technical supervision. Small (40 g) samples drawn at various levels and from various segments of a vertical cylindrical silo were found to contain fenitrothion residues ranging from 2.8 to 28.8 mg/kg as indicated in **Table 5.1**. The residue distribution on individual grains varied through a much greater range as indicated in **Figure 5.1.**

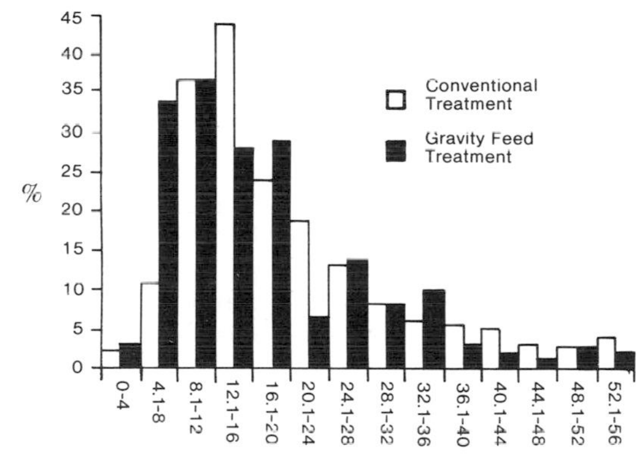

Figure 5.1. Distribution of fenitrothion on industrial wheat grains by conventional spraying and by gravity feed **(Minett *et al.* 1984).**

Desmarchelier and Elek (1978) demonstrated simple procedures which were suitable for use by untrained personnel for sampling bulk grain for pesticide residues and which gave results that were reproducible and compared favourably with results obtained by more elegant sampling methods. Several guidelines on sampling have been published **(Federal Working Group on Pest Management 1974)** and, following many years of

Table 5.1. Fenitrothion residues on bulk (40 g) samples taken after two weeks storage of conventionally and experimentally treated wheat **(Minett *et al.* 1984)**

Bin sampling point		Fenitrothion concentration (mg/kg) Gravity feed treatment		Conventional treatment
Centre 0 m		11.0		3.6
Centre 2		9.5		5.3
Centre 10		11.6		8.9
North 0		9.1		8.0
North 2		11.6		20.1
North 10		6.8		8.1
East 0		9.5		2.3
East 2		5.9		17.4
East 10		6.7		5.6
South 0		11.8		4.5
South 2		2.8		4.5
South 10		2.9		7.3
West 0		28.8		4.3
West 2		9.9		3.0
West 10		15.6		5.7
	Mean	10.2	Mean	7.2
	Range	2.8–28.8	Range	2.3–20.1
Coefficient of variation		60.5%		70.5%

international discussion, the **Codex Committee on Pesticide Residues (CCPR)** in 1979 published a set of recommended methods of sampling for the determination of pesticide residues. These methods have now been adopted for international use. **Wilson and Desmarchelier (1981)** showed how railway wagons could be used for obtaining representative samples of grain for residue analysis.

Not every investigator has the opportunity to conduct or confirm studies under conditions which truly represent the practical use situation. It is recognised that small-scale studies are both necessary and valuable but in many of the studies which have been reviewed it would appear that the investigators were either not sufficently familiar with the practical conditions encountered in field storage or that they were prepared to disregard the highly significant effects of temperature, humidity, surface effects and surface/bulk ratio.

The controversy which has existed for a long time concerning the relative relationship of results obtained in small laboratory studies to those obtained under the variety of practical conditions encountered in field practice will no doubt continue. Extensive Australian experience which embraces small-scale laboratory studies (100–1000 g), silo-scale studies (50–500 tonnes), and pilot-scale studies (multiples of 1000 tonnes) has shown that it is not justifiable to draw conclusions regarding the fate of insecticides applied to grain in large bulks from studies performed in small containers unless appropriate precautions have been taken when carrying out the laboratory studies, and even then the results are only approximately comparable. **Rowlands (1985)** described some of the pitfalls encountered in very small scale laboratory experiments where the ratio of grain surface to air, and grain surface to container surface can have disastrous effects upon the results. **Desmarchelier**

(1977a) drew attention to the effect of container size, shape and pre-treatment on the recovery of malathion and dichlorvos after application to small quantities of wheat. Losses exceeding 75% occurred within 48 hours at 25°C in small containers, but not in bulk grain.

One cannot fail to be concerned about the fact that the majority of papers dealing with the fate of insecticides applied to stored grain record deposits immediately after treatment to be well below the nominal or calculated (expected) concentration. Some investigators have wondered about this, others have expressed concern, but relatively few have apparently made any effort to find an explanation for the missing insecticide. Some have assumed it resulted from failure to land the spray on the target and undoubtedly this is a factor, especially with small-scale experiments. However, under practical conditions it should not represent a significant source of loss.

In the case of organophosphorus and carbamate insecticides, there seems no doubt that enzymes present in the seed-coat of freshly harvested grain cause rapid degradation of insecticide deposits **(Rowlands 1964, 1965a, 1965b, 1966c; Rowlands and Clements 1965a, 1965b, 1966; Andregg and Madisen 1983c)** showed that malathion was degraded in stored corn and wheat by fungi of the type regularly encountered on stored grains. However, these effects should be considered in the light of the publications by **Desmarchelier (1978a, 1980), Banks and Desmarchelier (1978)**, and **Desmarchelier and Bengston (1979a, 1979b)** on the influence of water vapour and temperature on changes in insecticide residue levels on the grain, particularly recognising that many investigators have failed to measure or record either temperature, moisture content or equilibrium relative humidity.

The losses which can occur between sampling and analysis can be highly significant, particularly in the case of freshly treated grain. Even where precautions are taken to avoid such losses, they can still be highly significant due to the large surface from which the insecticide can be co-distilled in the vapour from the film of water. The relatively large surface of the sample container makes for absorption or degradation of the insecticide. Glass rapidly degrades some organophosphorus compounds **(Rowlands 1985)** and plastics readily absorb a wide variety of insecticides from the solid as well as the vapour state. Tinplate can catalyse the degradation of some insecticides, especially when they are in the vapour state.

Australian experience has been that losses from freshly treated grain are highly significant and that they can be reduced, if not eliminated, by placing the sample immediately into solvent (methanol) at the sampling site, before attempting to transport the sample to the laboratory even by the quickest available route. As an example of the importance of this step, **Webley (1985)** reported that the apparent loss overnight due to delayed extraction of methacrifos in 55 out of 60 comparisons in one trial was 12–16%. The difference between the addition of the grain to methanol on site and addition in the laboratory the following day was significant at the 0.001 level (paired t test). In a similar situation where the insecticide was chlorpyrifos-methyl the addition of methanol on site gave a 23% increase in the analytical results initally but this decreased to zero when analysing samples taken from the bulk 24 weeks after treatment with the insecticide.

Desmarchelier (1977a) showed that the nature and pre-treatment of sample containers had a pronounced bearing on the loss of both malathion and dichlorvos from freshly-treated wheat samples stored in the containers for only one day. Loss was less when wheat was stored in acid-washed glassware that had been rinsed 10 times, than when stored in glassware that had been washed in dilute alkali prior to repeated rinsing. In the case of the alkali-treated containers, the loss of malathion from a wheat sample was 61% and that of dichlorvos was 72% in 24 hours.

Without doubt, one of the greatest errors introduced into the measurement of insecticide residues on grain comes from failure to extract and recover aged insecticide residues. The problem largely arises through the use of non-polar solvents which are either unable to penetrate the grain tissues or to dislodge the bonding between the residue and the substrate. The pattern appears to have been established by the Malathion Panel which published methods for the determination of malathion in cereals and oil seeds **(Anon. 1960)**. The Panel adopted carbon tetrachloride as the extracting solvent using the soxhlet method. The same solvent and extraction procedure was later adopted for the determination of dichlorvos residues. Although many residue reports published during the late 1960s indicated that a variety of non-polar and polar solvents, such as methylene dichloride, were used to extract the dichlorvos residue from the food commodity, these methods and the accompanying reports, were later found to be unreliable **(FAO/WHO 1971)**. In 1968, **Shell Chemicals** (personal communication) recommended the use of ethyl acetate or acetone for the extraction of dichlorvos residues from grain. **Minett and Belcher (1969)** and **Elms *et al.* (1970)** showed that dichlorvos cannot be recovered quantitatively from plant materials, especially grain, by the use of solvents even as polar as methylene dichloride. Water or water-miscible solvents, such as methanol or ethanol, are absolutely essential for the recovery of dichlorvos residues from plant materials. A study group in the U.K. **(Panel 1973)** later reversed the previous recommendation by proposing the use of water-miscible solvents.

Horler (1972) published a brief note of a study of the recovery of ^{14}C-labelled malathion from aged samples of wheat with different moisture contents ranging from 10% through 15% to 20%. He reported large differences between the amount extracted with n-hexane compared with that recovered by methanol. The difference was more than 50% with wheat of 20% moisture content. In all cases methanol extracted more than n-hexane. The low recovery with hexane from the high-moisture wheat emphasises the misgivings felt about some published data on residues of insecticides in processed grains and cereal-base foods in particular.

Horler and Clark (1974) reported on studies to determine the extractability of ^{14}C- malathion residues from wheat samples of different moisture content, freshly treated and aged for 1 year under cold and soxhlet extraction with hexane and methanol. Their results indicated little difference between cold and soxhlet extraction, except that at higher grain moisture contents malathion appeared to be broken down in the soxhlet procedure. Only with freshly treated samples was recovery complete in terms of unchanged ^{14}C- malathion.

Desmarchelier *et al.* (1977) conducted a collaborative study of five organophosphorus insecticide residues on wheat, involving the development of suitable methods and their evaluation. They found that methanol was by far the best solvent and that substantially complete extraction could be obtained by soaking the whole grain in methanol with only occasional agitation. They showed that the addition of methanolic wheat extract or acetic acid to the extracting methanol ensured that there was no break-down of the organophosphorus residue in the extract. Approximately 10% additional residue could be recovered by a second extraction. These workers warned against the snare and delusion of checking recoveries on fortified samples. There is no doubt that insecticide deposits applied within the last 1 to 2 days can readily be extracted even with hydrophobic solvents. This does not hold true for aged residues. The use of water-miscible solvents is even more vital for the recovery of residues from cooked cereal foods such as bread, pasta, noodles, rice etc.

For such reasons many descriptions of rates of loss of insecticides may be inaccurate, because they are based on analytical procedures not shown to be appropriate for aged residues.

Since the results obtained with any analytical method tend to be sensitive to the skills and to the environment of the analysts, the Codex Committee on Pesticide Residues, as part of the FAO/WHO Food Standards Program, has developed Guidelines for Good Analytical Practice **(Codex 1983)**. These define good practice in the analysis of pesticide residues and point out that there are 3 major components, the analyst, basic resources and the execution of the analytical procedure.

The typical multi-residue analysis of biological material as is currently conducted, is a highly complex procedure fraught with pitfalls throughout the operation. Generally, the procedure consists of several distinct steps; sample preparation; extraction of pesticides from the substrate; clean-up; gas-liquid chromatography; and interpretation of results. For the assurance of accurate final data, it is necessary to build in a set of controls designed to prevent or detect erroneous deviations. Without such a system, the procedure is reduced to a hit or miss situation wherein the final results may, or may not, estimate the actual pesticide content of the sample. Many papers have been written about analytical quality control in pesticide residue laboratories and about check-sample programs designed to provide quality assurance. Among these, the following references are most pertinent: **Cochrane and Whitney (1979); Carl (1979); Kateman and Pijpers (1981); Thompson and Mann (1975); Burke and Corneliussen (1975); Joint FAO/WHO Food and Animal Feed Contamination Monitoring Programme (1981); and Snelson (1983)**.

Reliability in analysis of insecticide residues on stored products is required for a number of reasons, including quality control, reliable comparison of data obtained by different workers and good agricultural practice. In order to draw attention to the deficiencies in analytical practice and, hopefully, to eliminate some of the defects, a number of collaborative studies were carried out and reported **(Snelson 1974; Snelson and Desmarchelier 1975; Desmarchelier *et al.* 1977b; Desmarchelier 1980b)**. These studies brought to light the fact that official government analysts with considerable experience, working in well-equipped laboratories and exercising above-average care, submitted results on identical samples which often varied by more than 100% between the lowest and highest values. The magnitude of such variations was reduced considerably when the extraction procedure was standardised, the quality of the reference standard was perfected and variations in equipment and reactions were eliminated as far as possible.

It is hoped that these comments which really do not do justice to the many complex issues surrounding residue analysis of grains and milled cereal products will at least alert readers to the need for care in conducting residue analysis and in evaluating analytical data.

6. Bioresmethrin

Bioresmethrin (5-benzyl-3-furfurylmethyl (+)−*trans*-chrysanthemate) is the (+)−*trans* isomer of the racemic (+)−*cis, trans*-mixture known as resmethrin (NRDC 104). It was one of the first successful synthetic pyrethroids developed by **Elliott and co-workers (1967a, b, 1969a, b)** at Rothampsted in the United Kingdom. It is one of the most potent broad spectrum insecticides currently available.

Bioresmethrin has a good knock-down performance against insects, particularly when compared with the organochlorine or organophosphorus insecticides.

Bioresmethrin, at low concentrations, is an effective killing agent against most insect pests affecting households, industrial premises and food storage. The principal uses to which bioresmethrin is currently being put are:

(a) in household aerosols and sprays formulated in combination with pyrethrum, bioallethrin, tetramethrin and piperonyl butoxide;

(b) as an insecticide for the control of pests in food premises;

(c) in grain disinfestation and protection.

Bioresmethrin has been tested in conjunction with a number of well-established synergists. Unlike natural pyrethrum, there is only a minimal increase in performance against flies, mosquitoes and cockroaches when piperonyl butoxide is added to bioresmethrin. However, in the case of a number of the more important insect pests of stored products, bioresmethrin is synergised to a significant extent with piperonyl butoxide, the factor of synergism ranging from 2 to 9 fold.

Bioresmethrin has an exceptionally high potency against some insect species, particularly *Rhyzopertha dominica* and it has proved useful when applied at the rate of 1 mg/kg in conjunction with selected organophosphorus insecticides, enabling the amount of organophosphorus insecticide to be reduced considerably without loss of effectiveness. Bioresmethrin is almost completely removed from grain in milling and processing and any residues which carry over into flour or milled cereals are destroyed in cooking or malting.

The effective performance of bioresmethrin and its novelty, being the first synthetic pyrethroid with properties comparable to natural pyrethrins, has stimulated independent scientists and government research workers to study the chemistry, metabolism, biological effects and fate of bioresmethrin. Much of this information is available in published literature. Because the (+)−*cis* isomer of bioresmethrin is acutely somewhat more toxic than bioresmethrin, there was interest in whether this isomer could be a significant component of technical bioresmethrin. In the manufacturing process the optical isomers of chrysanthemic acid are separated by procedures which include crystallisation and it is understood that substantially all of the (+)−*cis* isomer is removed before the esterification reaction **(FAO/WHO 1975, 1976)**.

6.1 Usefulness

Numerous authors have reported studies of the effect and usefulness of bioresmethrin for the control of stored-product pests and as a grain protectant **(Ardley 1973, 1976; Ardley and Desmarchelier 1975, 1978; Bengston 1979; Bengston *et al.* 1975, 1978; Bengston and Desmarchelier 1979; Bitran *et al.* 1980b; Carter *et al.* 1975; Chawla and Bindra 1976; Desmarchelier 1977b, 1978b; Desmarchelier *et al.* 1981a, 1981b; L'Hoste *et al.* 1969; Lloyd 1973; Wilkin 1975a)**.

The development of organophosphorus- and fumigant-resistant strains of many stored-product pests **(Champ and Dyte 1976; Champ 1976, 1977)** in most countries and in all continents points to the urgency of having available acceptable insecticides which have different properties and modes of action. Supplies of natural pyrethrum are not adequate to meet the increasing demands.

In Australia, the occurrence of malathion-resistant strains of *Rhyzopertha dominica*, after 12 years of extensive and intensive use of malathion, presented a severe threat to the grain export industry. Malathion was never entirely satisfactory for the control of this species, which shows a degree of tolerance to all organophosphorus insecticides that could possibly be used for the protection of stored grains. The outstanding efficiency of bioresmethrin in controlling this species is therefore important. **Bengston *et al.* (1975)** showed clearly that the high potency of bioresmethrin against *Rhyzopertha dominica* makes it an ideal insecticide to complement such organophosphorus grain protectants as malathion, fenitrothion, pirimiphos-methyl and chlorpyrifos-methyl.

Lloyd and Field (1969) reported results of tests with bioresmethrin and resmethrin on a number of stored-product pests when applied topically dissolved in cyclohexanone containing 0.5% butylated hydroxy-toluene. Against *Sitophilus granarius*, bioresmethrin was 3.6 times as toxic as pyrethrins. When synergised with piperonyl butoxide using an insecticide:synergist ratio of 1:10, a factor of synergism of ×5.8 was obtained which was increased to ×13.4 when the ratio was increased to

1:100. Against *Tribolium castaneum*, which proved to be less susceptible, bioresmethrin was found to be 1.9 times as toxic as pyrethrins. With this species a synergism factor of $\times$ 1.5 was obtained with piperonyl butoxide used in the ratio of 1:10.

Lloyd (1973) published an extensive account of the work in which the toxicity of pyrethrins was compared with that of 5 synthetic pyrethroids, including bioresmethrin, to *Tribolium castaneum* and susceptible and pyrethrum-resistant *Sitophilus granarius*. Bioresmethrin was by far the most toxic compound and depending upon the strain tested, bioresmethrin was up to 16 times more toxic than pyrethrins. The potency of bioresmethrin was synergised by the addition of piperonyl butoxide.

Wilkin and Hope (1973), who recorded the results of the evaluation of more than 20 pesticides against stored-product mites, found bioresmethrin moderately effective against *Acarus siro* and *Glycyphagus destructor* but when the concentration was doubled the otherwise mediocre effect against *Tyrophagus putrescentiae* became entirely satisfactory.

Ardley (1976) compared bioresmethrin with the standard malathion (12 mg/kg) treatment over 2 seasons in silo trials and found that the best treatment on an efficiency basis was 4 mg/kg bioresmethrin plus 20 mg/kg piperonyl butoxide. After 12 months, the grain treated with bioresmethrin and piperonyl butoxide controlled all species and strains exposed in bioassay. No comment was offered about the comparative cost. It was later found that the bioresmethrin treatment was more than 20 times more costly.

The Queensland Department of Primary Industries (1975) reported that insecticides being tested against malathion-resistant strains had proved effective against all species except *Rhyzopertha dominica*. Against this species bioresmethrin appeared necessary. It was effective at rates as low as 1 mg/kg when added to grain.

Ardley and Desmarchelier (1975) reported that for protection of grain in long-term storage the most cost-effective treatment was 4 mg/kg bioresmethrin plus 20 mg/kg piperonyl butoxide. After 12 months storage the treated grain controlled *Rhyzopertha dominica, Sitophilus granarius, Sitophilus oryzae, Oryzaephilus surinamensis, Plodia interpunctella* and *Ephestia cautella*. A much lower efficiency was observed in the control of *Tribolium* spp. and in this regard bioresmethrin parallels the documented insecticidal efficiency of pyrethrins.

Wilkin (1975a) reported that bioresmethrin plus piperonyl butoxide will control mites, including lindane-resistant strains, in the fabric of buildings and in stored grain.

Carter *et al.* (1975) published comparative observations on the activity of 6 pyrethroids against stored-product beetles. They concluded that synergised bioresmethrin was the most suitable pyrethroid for controlling stored-product beetles, except *Lasioderma serricorne* and *Stegobium paniceum*, and was of value against organophosphorus-resistant strains.

Bengston *et al.* (1975a), reporting extensive field trials to compare a range of grain protectant insecticides, showed that none of the 5 organophosphorus insecticides evaluated was adequate to control *Rhyzopertha dominica*, which lived and reproduced in all samples including those drawn immediately after insecticide application.

Bengston *et al.* (1978a, 1980a, 1983a, b) demonstrated, in commercial trials, the outstanding performance of several organophosphorus grain protectants when combined with bioresmethrin. Complete protection was obtained against all strains of all species of stored-product pests and reproduction was completely inhibited for many months.

Desmarchelier (1977b) showed that combinations of pyrethroids, which are particularly effective against *Rhyzopertha dominica*, and organophosphorus insecticides which are effective against *Tribolium castaneum*, controlled both species as long as the pyrethroid alone controlled *Rhyzopertha dominica* and the organophosphorus compound alone controlled *Tribolium castaneum*. Bioresmethrin was one of the pyrethroids used in constructing these experiments.

Ardley and Desmarchelier (1978) were able to show, by field trials, that a combination of synergised bioresmethrin and fenitrothion was completely effective against all the major pest species particularly those against which either insecticide alone was inadequately effective. From their results these authors suggested that there appeared a degree of potentiation and/or synergism in such combinations.

Amos *et al.* (1979) and **Minett *et al.* (1981)**, who had developed a gravity-feed technique for applying insecticide non-uniformly to grain, carried out an extensive trial to determine whether the technique would be commercially acceptable. An insecticide mixture of fenitrothion, bioresmethrin and piperonyl butoxide was applied to wheat so as to deposit 1 mg/kg of bioresmethrin and 10 mg/kg of piperonyl butoxide. The rate of application was adjusted automatically to allow for the variations in the rate of flow of wheat on the conveyor belt. Wheat samples for biological and chemical assay were taken from the bulk silos into which the grain was conveyed. Samples were taken immediately after treatment and approximately 6 months later. From the result of the assays the authors were able to predict that treatment would be fully effective against *Rhyzopertha dominica* for at least 6 months. The method of application offered many practical advantages.

Bengston (1979) reported that testing of synthetic pyrethroids available in the early 1970s established that bioresmethrin was clearly more potent than bioallethrin and tetramethrin. Bioresmethrin is decomposed by light but this has not been a problem in practice. In treated grain, it has a synergism factor of $\times 2$ with piperonyl butoxide at 10 mg/kg against *Rhyzopertha dominica*. Synergised bioresmethrin at 1 mg/kg provides effective protection against *Rhyzopertha dominica* for 9 months.

Bengston and Desmarchelier (1979) summarised the extensive laboratory and field trials that led to the introduction of the combination, fenitrothion plus bioresmethrin, as a grain protectant to control malathion-resistant insects in Australia.

Bengston *et al.* (1978b) showed that a combination of bioresmethrin and fenitrothion was outstandingly effective for the contol of malathion-resistant insects in stored sorghum.

Bengston *et al.* (1983a, b) reported the results of extensive field-scale trials in many parts of Australia using 3 of the most promising organophosphorus grain protectant insecticides with and without bioresmethrin synergised with piperonyl butoxide. The value of the bioresmethrin was clearly demonstrated by outstanding control of all stored-product pests including *Rhyzopertha dominica*. The value of the piperonyl butoxide synergist was confirmed by evidence that it potentiated the effectiveness of bioresmethrin by a factor of more than 2.

Desmarchelier *et al.* (1981b and in press) reported extensive pilot trials of grain protectants comprising combinations of bioresmethrin with a range of organophosphorus compounds. These combinations, under a wide variety of conditions, proved to be entirely satisfactory for commercial requirements.

Bitran *et al.* (1980b) evaluated the residual persistence of deltamethrin and bioresmethrin for the protection of maize and coffee during storage. Deltamethrin alone or synergised with piperonyl butoxide (1:4) and bioresmethrin synergised with piperonyl butoxide (1:4) were tested in Brazil for their effectiveness in protecting stored maize from attack by *Sitophilus zeamais* and stored coffee from attack by *Araecerus fasciculatus*. With maize, deltamethrin, alone or synergised, and synergised bioresmethrin were tested at 0.5 and 1 mg/kg; with coffee, deltamethrin alone at 1 and 2 mg/kg and synergised bioresmethrin at 4 mg/kg were used. Malathion at 8 mg/kg was used as a standard. The best results were obtained with deltamethrin, which at 1 mg/kg afforded excellent control of *Sitophilus zeamais* for at least 270 days and at 2 mg/kg gave excellent control of *Araecerus fasciculatus* for the same period. The addition of the synergist did not improve its effectiveness. Malathion gave good control of *Sitophilus* for 180 days and was superior to bioresmethrin. Neither malathion nor synergised bioresmethrin afforded satisfactory control of *Araecerus*.

Bengston *et al.* (1983a) reported duplicate experiments carried out on bulk sorghum stored in Queensland. Bioassays of treated grain, conducted during 6 months storage, established that fenitrothion at 12 mg/kg plus bioresmethrin at 1 mg/kg controlled typical malathion-resistant strains of *Sitophilus oryzae, Rhyzopertha dominica, Tribolium castaneum* and *Ephestia cautella*.

Desmarchelier (1983) demonstrated that the insecticidal efficiency of bioresmethrin was distinctly greater at 25°C than at 35°C.

Desmarchelier *et al.* (1986) reported the results of extensive field trials with methacrifos and the protectant combinations chlorpyrifos-methyl plus bioresmethrin, fenitrothion plus phenothrin and pirimiphos-methyl plus carbaryl. Each of these insecticides was applied to grain that was stored at 15 sites throughout Australia. Samples from 12 storages were taken for laboratory assays against *Rhyzopertha dominica* and *Tribolium castaneum*. In laboratory bioassay, the order of effectiveness against *Tribolium castaneum* was methacrifos superior to chlorpyrifos-methyl plus bioresmethrin, superior to fenitrothion plus phenothrin equal to pirimiphos-methyl plus carbaryl. The order of effectiveness against *Rhyzopertha dominica* was pirimiphos-methyl plus carbaryl equal to fenitrothion plus phenothrin greater than chlorpyrifos-methyl plus bioresmethrin equal to methacrifos.

6.2 Degradation

Rowlands (1975) reported that bioresmethrin appeared to be relatively stable under normal storage conditions, certainly more so than most other pyrethroids then available. Preliminary investigations indicated that bioresmethrin is, in fact, reasonably stable on stored wheat with 15% moisture but somewhat more susceptible to breakdown on freshly harvested wheat in which the ripening systems and microflora are perhaps still active. Initial breakdown was at the rate of about 30% per week for the first 1 to 10 days, thereafter levelling out. Interestingly, it seems that piperonyl butoxide may slow down this initial decline, if jointly applied.

Extensive commercial scale trials were carried out in Australia during 1976 at 20 sites involving 42 silos each containing from 2000–8000 tonnes of wheat. Bioresmethrin was combined with either fenitrothion or pirimiphos-methyl and the fate of the insecticide deposit was studied by bioassay and chemical analysis **(Desmarchelier *et al.* 1986; FAO/WHO 1976c)**. The data confirmed conclusions reached in previous laboratory scale studies, that the half-life of bioresmethrin in low-moisture stored grain is directly dependent on temperature. At 30–35°C the half-life in wheat of 11% moisture content is about 20 weeks while at 20°C it exceeds 26 weeks. This degree of stability is greater than that shown by malathion under similar conditions **(Desmarchelier 1977b; Desmarchelier and Bengston 1979b)**.

Desmarchelier (1978a) developed a mathematical model to serve as a means of predicting the fate of grain-protectant insecticide residues on grain of known temperature and moisture content. **Desmarchelier and Bengston (1979a, 1979b)** showed how these mathematical models had been used to reduce the application rates of bioresmethrin and certain other grain-protectant insecticides in trials and commercial usage in Australia. Their calculations showed that at 30°C and 50% relative humidity, the half life of bioresmethrin was 24 weeks. The addition of piperonyl butoxide at the rate of 20 mg/kg to the bioresmethrin (1 mg/kg) increased the half-life at 30°C to 38 weeks. These predictions were tested by further studies **(Desmarchelier 1980)**.

During the first season of commercial use some disappointing results were obtained. Failure to afford full protection of stored grain under commercial conditions was suspected to be due to decomposition brought about by physical, chemical and biological factors. **Desmarchelier (1978b)**, reporting the results of an extensive field survey into the cause of failure to protect grain, showed that the failure was entirely due to human error and mechanical failure leading to inadequate and inefficient distribution of the bioresmethrin throughout the bulk of grain. In no instance was it due to decomposition of bioresmethrin in the spray vat or of the residue on the grain. Correction of the causes of the mechanical failure have resulted in excellent performance from 1978 to the present.

Desmarchelier and Bengston (1979b) described how it is possible to predict the fate of residues of grain protectants from simple models. Loss of protectants in bulk grain depends entirely on grain equilibrium relative humidity, grain temperature and time of storage. They summarised these models for 12 protectants, including bioresmethrin, and showed how they had verified these predictions in commercial-scale applications. **Desmarchelier *et al.* (1980a)** determined the levels of bioresmethrin residues together with those of 5 other insecticides on unhusked rice, husked rice, polished rice and barley over a storage period of 6 months. The observed levels were close to levels predicted from the use of the above model which relates rate of loss of residue levels to a rate constant and only 2 variables, temperature and equilibrium relative humidity.

Desmarchelier *et al.* (1980a) also compared predicted and observed residues of bioresmethrin on rice and barley during and after storage. Unhusked rice, husked rice, polished rice and barley were studied under identical laboratory conditions. Loss of bioresmethrin during storage was compared with losses predicted from quantitative models **(Desmarchelier 1980)**. The observed levels were close to levels predicted from the model. The difference between the predicted and observed value on the 3 grades of rice was less than 5% but on barley the observed value was about 16% lower than predicted. This may be due to errors in sampling and analysis and the relatively low concentration of the deposit.

Bengston *et al.* (1980a) showed that field trials carried out on bulk wheat in commercial silos in Queensland, South Australia and Western Australia had demonstrated that bioresmethrin, applied at the rate of 1 mg/kg had declined to 0.4 mg/kg after 30 weeks by which time the initial temperature of 30–35°C had declined to 25–30°C.

Minett *et al.* (1981) reported a trial in which gravity-feed application of insecticide concentrate to wheat in a commercial silo was compared with the conventional application of a diluted emulsion applied at the rate of 1 litre per tonne. They applied bioresmethrin by both methods at the rate of 1 mg/kg. Chemical analysis revealed that both techniques applied substantially the same deposit initially and that after storage at 30–35°C for 6 months both lots of wheat contained levels of residue. The loss over this period was approximately 50%.

Desmarchelier *et al.* (1981b) provided the results of extensive pilot usage of grain protectant combinations containing bioresmethrin at 42 commercial storages throughout Australia in 1976. The amount of wheat in each storage ranged from 250 to 2500 tonnes and the moisture content from 9.5 to 11.5%. The temperature of the grain was recorded each month throughout the trial. Four of the 42 storages were aerated and in these the temperature eventually declined to within the range of 8–15°C. The temperature of the unaerated grain at the beginning of the trial was within the range of 30–35°C and after 8–9 months ranged from 20–30°C. The concentration of the bioresmethrin residues declined over 9 months from 1 mg/kg (0.8–1.1) to 0.35 mg/kg (0.25–0.45). The predictive model for bioresmethrin **(Desmarchelier and Bengston 1979b)** and the measured temperatures and moisture contents of grain were used to calculate, from the amount of residue measured in the first sample, the expected residue level at each site. The amount of bioresmethrin residue found by analysis agreed within 10% of the predicted residue level.

Desmarchelier *et al.* (1986) reported the results of pilot usage of grain protectant combinations in 63 storages throughout the 5 mainland States of Australia. Bioresmethrin was combined with chlorpyrifos-methyl in 15 of these storages. It was applied at the rate of 1 mg/kg and at the end of 6–9 months the average concentration of bioresmethrin in the 15 storages was 0.35 mg/kg. It was predicted from the work of **Desmarchelier and Bengston (1979b)** that the concentration of bioresmethrin after 6 months should have been 0.4 mg/kg and at the end of 9 months 0.3 mg/kg. The level of aged residue required to control *Rhyzopertha dominica* was found by bioassay to be 0.5 mg/kg. The concentration of bioresmethrin residues originally applied at the rate of 1 mg/kg fell steadily over the course of the trial so that at the end of 24 weeks the average was 0.55 mg/kg. The bioresmethrin used did not contain piperonyl butoxide.

6.3 Fate in Milling, Processing and Cooking

Ardley (1975) reported trials in which wheat treated with bioresmethrin at varying rates, with and without piperonyl butoxide, was subjected to standard milling and baking tests. The milling fractions and the bread were analysed. Residues of bioresmethrin and piperonyl butoxide in milling fractions and bread are summarised in **Table 6.1.**

Desmarchelier (1975a) took wheat which had been treated 6 weeks and 7 months previously and subjected it to standard milling procedures for the preparation of wholemeal and white flour. The following results **(Table 6.2)** indicate the distribution and fate of the bioresmethrin.

Studies to determine the fate of bioresmethrin following the processing of barley, oats, rice and wheat indicate

Table 6.1. Residues of bioresmethrin and piperonyl butoxide in milling fractions and bread (**Ardley 1975**)

Protectant[1] treatment	Recovered residue (i) and bioresmethrin (ii) piperonyl butoxide (ppm)							
	(i)				(ii)			
	bran	pollard	flour	bread	bran	pollard	flour	bread
a	0.3	4.0	4.0	nil	21.0	10.0	2.0	2.0
b	0.3	1.0	1.4	nil	0.5	0.3	nil	nil
c	0.5	1.0	0.5	nil	11.0	8.0	10.0	nil
d	1.1	1.2	0.5	nil	2.0	2.0	2.0	nil
e	nil	nil	nil	nil	nil	nil	nil	nil

1 a = 4 mg/kg bioresmethrin + 20 mg/kg piperonyl butoxide
b = 2 mg/kg bioresmethrin alone
c = 4 mg/kg bioresmethrin + 10 mg/kg piperonyl butoxide + 4 mg/kg anti-oxidant
d = 2 mg/kg bioresmethrin + 2 mg/kg piperonyl butoxide + 10 mg/kg anti-oxidant + 4 mg/kg fenitrothion.

Table 6.2. Distribution of bioresmethrin in milling fractions and bread (**Desmarchelier 1975a**)

Sample	A	B
Interval since treated	6 weeks	7 months
Residue in wheat	2.9 mg/kg	1.9 mg/kg
bran	5.2 mg/kg	1.7 mg/kg
pollard (shorts)	0.7 mg/kg	–
white flour	*	*
wholemeal bread	1.0 mg/kg	0.6 mg/kg
white bread	ND	ND
limit of determination = 0.05 mg/kg		

*Interference led to poor recoveries during assay of flour.

(**FAO/WHO 1976c**) that though the bulk of the bioresmethrin is removed by the processing and cooking the deposit is sufficiently stable to withstand this treatment to some degree. Thus a significant proportion of the original quantity applied is found in the processed and cooked cereals as is indicated in **Table 6.3.**

It should be recognised that the original rate of application, 7 mg/kg, is greatly in excess of that normally used in practice. If bioresmethrin were used as the only insecticide the maximum rate of application would be of the order of 4 mg/kg. When used in combination with an approved organophosphorus insecticide, bioresmethrin would be used in conjunction with piperonyl butoxide at the rate of 1 mg/kg.

Murray and Snelson (1978) carried out a study to determine the fate of fenitrothion and bioresmethrin residues in milled products from wheat processed through

Table 6.3. Fate of bioresmethrin in various grains processed after storage at 25°C (**FAO/WHO 1976c**)

Grain	Storage period (months)	Residue after storage (mg/kg)*	Processing	Residue after processing (mg/kg)*
Barley	3	3.5	primitive malting	1.5
Barley	6	1.75	commercial malting	0.35
Oats	3	3.5	boil 15 minutes	1.5
Rice in husk	6	1.5	husked	0.25
			milled/polished	0.1
Husked rice	3	3.5	boiled 15 minutes	2.5
			boiled 25 minutes	0.5
Polished rice	3	3.5	boiled 15 minutes	2.0
Wheat	3	3.5	bran	5.5
			shorts	3.5
			flour	2.5
			white bread	<0.1
			wholemeal bread	1.0

*Original rate of application = 7.0 mg/kg.

a commercial flour mill. They found that substantially all of the residue was contained in the bran and pollard fractions but that commercial flour will contain approximately 0.1 mg/kg. There was a significant accumulation of the residue in bran and germ. Their detailed results are given in **Table 6.4.**

Table 6.4. Residues of bioresmethrin (mg/kg) in wheat milling fractions* **(Murray and Snelson 1978)**

Sample number	Description	Bioresmethrin (mg/kg)
1	Wheat when received into mill	0.7
2	Wheat at first break roll before it enters the rolls	0.6
3	Crushed wheat at first break roll after it drops from the rolls	0.6
4	Flour sieved from the sifter fed from first break roll (clean stream)	<0.05
5	Flour sieved from the sifter fed from last break roll (dirty stream)	0.1
6	Flour sieved from the sifter fed by the first reduction roll (clean stream)	<0.05
7	Flour sieved from the sifter fed from the last reduction roll (dirty stream)	<0.05
8	Flour from all streams being fed into bulk flour bins or into packers	0.1
9	Pollard from all streams before it is mixed with bran	0.7
10	Bran from all streams before it is mixed with the pollard	1.0
11	Wheat germ	2.0

* Wheat treated at harvest with bioresmethrin (1 mg/kg) synergised with piperonyl butoxide (10 mg/kg) and subjected to milling through a commercial flour mill 3 months later.

Tempone (1979) studied the effect of various insecticides, including bioresmethrin, on barley malting. In the process, he studied the effect of malting on the level of the insecticide deposit on the malting barley. In one trial, bioresmethrin was applied at 4 and 8 mg/kg and in another at the approved commercial rate of 1 mg/kg, in conjunction with fenitrothion, pirimiphos-methyl and chlorpyrifos-methyl. No adverse effects upon the malting process, yield or malt quality were found. No residues of bioresmethrin were detected in any samples of malt analysed by a method with a limit of determination of 0.1 mg/kg.

Desmarchelier *et al.* (1980a) studied the fate of bioresmethrin on barley during and after storage and determined the effect of the malting process on the residue. To simplify the analytical work, the bioresmethrin was applied at the rate of 7 mg/kg. This deposit had declined to the level of 2.5 mg/kg after 6 months storage at 25°C and a relative humidity of 65%. After subsequent malting, the bioresmethrin residue was found to be 0.25 mg/kg representing a loss of 90% in the malting process.

In the same studies, **Desmarchelier *et al.* (1980a)** determined the effect of cooking husked rice and polished rice containing aged bioresmethrin residues. They showed that 50% of the residues on husked rice were destroyed by cooking. In the case of polished rice, the loss in cooking was 78%. Rice in husk was milled to produce husked rice and polished rice. Removal of the husk also removed 93% of the insecticide. A further 4% was removed by the polishing process. When the milled rice was subsequently cooked, about 80% of the remaining residue was destroyed. Thus, rice in husk treated with bioresmethrin at the rate of 7 mg/kg produced cooked, polished rice with less than 0.05 mg/kg of bioresmethrin.

Since all of these studies were carried out at highly exaggerated levels, it is clear that no significant residues of bioresmethrin will occur in foods at the point of consumption when bioresmethrin is used in combination with other insecticides for protecting grain against stored-product pests.

6.4 Metabolism

Extensive studies of the metabolism of bioresmethrin in animals have been carried out by many authors including the following: **Abernathy and Casida (1973)**; **Abernathy *et al.* (1973)**; **Farebrother (1973)**; **Foote *et al.* (1967); Jao and Casida (1974); Miyamoto (1975); Miyamoto *et al.* (1971, 1974); Suzuki and Miyamoto (1974); Ueda *et al.* (1975a, b); Weeks *et al.* (1972)**. The fate of bioresmethrin on plants and under the influence of sunlight has been studied by **Rosen (1972)**.

All this information has been reviewed by the Joint FAO/WHO Meeting of Experts on Pesticide Residues **(FAO/WHO 1976c, 1977b)**. The metabolic pathways proposed by **Ueda *et al.* (1975a, b)** are illustrated in **Figure 6.1.**

Most metabolism studies have been carried out on laboratory animals, mainly rats. Radioactivity measurements and radioautographs indicated that the compound was rapidly absorbed from the intestinal tract and distributed into various tissues where only a negligible amount of intact bioresmethrin was found. However, the radioactivity was excreted rather slowly and it took 3 weeks to recover all the radioactivity in the excreta (36% in urine and 64% in feces).

Neither urine nor faeces contained intact bioresmethrin or the ester metabolites; the predominant urinary metabolite being 5-benzyl-3-furoic acid amounting to approximately one-third of the radiocarbon recovered.

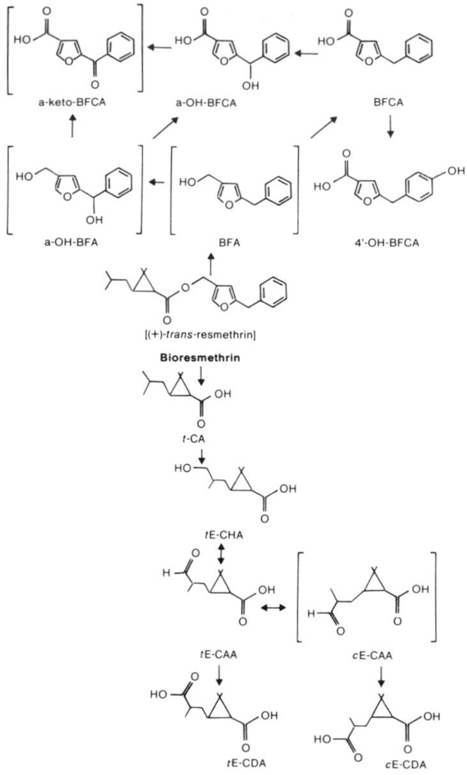

Figure 6.1. Metabolic pathway of bioresmethrin in rats **(Ueda *et al.* 1975a,b).**

Farebrother (1973), in a whole-body autoradiographical study in rats, showed that bioresmethrin (or at least its radiolabels) was absorbed from the gut and widely distributed in the body 2 hours after dosing. At 6 hours the distribution was similar but the concentrations were increased, particularly in fatty tissues. At 24 hours most tissues showed greatly reduced activity but the concentration in fatty tissues remained high. No studies appear to have been carried out to determine the level or nature of the metabolites in animal fat or other tissues.

Bioresmethrin is degraded by ester cleavage and the alcohol moiety is oxidised to 5-benzyl-3-furylmethanol (BFA), 5-benzyl-3-furoic acid (BFCA), 4'-hydroxy BFCA and *a*'-hydroxy BFCA (*a*-OH-BFCA). The chrysanthemate (acid) moiety undergoes oxidation from *trans*-chrysanthemic acid (*t*-CA) to 2,2-dimethyl 3-(2' hydroxymethyl-1'-propenyl) cyclopropanecarboxylic acid (*t*E-CHA). This is further oxidised through the formyl derivative (CAA) to the dicarboxylic acid isomers (*t*E-CDA and *c*E-CDA). It is at the CAA oxidation stage where isomerisation may occur through the proposed aldehyde (*c*E-CAA) intermediate to (*c*E-CDA) the *cis*-dicarboxylic acid **(Ueda *et al.* 1975b)**. This metabolic sequence may also account for the consideration of **Verschoyle and Barnes (1972)** that, as a delay in signs of poisoning was evident following *intravenous* administration, bioresmethrin may be converted *in vivo* to a toxic metabolite. The presence of (+)-*trans*-CA, BFA and BFCA as metabolites, which are more toxic than bioresmethrin, may account for their observation and inclusions. The metabolic sequence is very similar qualitatively to that observed with resmethrin although much less complicated because of the lack of isomeric products **(Miyamoto *et al.* 1971)** and because of the specificity of certain isomers to enzymatic degradation by selected routes as mentioned above **(Ueda *et al.* 1975a, 1975b; Abernathy and Casida 1973)**.

Soderlund and Casida (1977b) studied the effects of pyrethroid structure on rates of hydrolysis and oxidation by mouse liver microsomal enzymes. Having examined 24 pyrethroids they concluded that primary alcohol esters of *trans*-substituted cyclopropanecarboxylic acids are more rapidly metabolised, with hydrolysis generally serving as the major component of the total metabolism rate.

Rowlands (1975) reported that preliminary investigations in his laboratory indicated that bioresmethrin is reasonably stable on stored wheat at 15% moisture but somewhat more susceptible to breakdown on freshly harvested wheat in which the ripening systems and microflora are perhaps still active. The only degradation products found in detectable quantities were tentatively identified as the free benzylfurancarboxylic acid and a starchy conjugation product. Initial breakdown was at the rate of about 30% per week for the first 1 to 10 days, thereafter levelling out. **Miyamoto** *et al.* (1971) identified some 8 products of degradation, among which is the aforementioned benzylfurancarboxylic acid.

Leahey (1985) reviewed the metabolism and environmental degradation of resmethrin which holds true also for bioresmethrin.

6.5 Fate in Animals

No specific information on the fate of bioresmethrin in livestock has come to light. This is not at all surprising in view of the limited use of bioresmethrin in situations where animals could possibly be exposed to significant residues.

It is clear that much of the information on other synthetic pyrethroids would be applicable to bioresmethrin. The extensive information on the metabolism of bioresmethrin (Section 6.4) provides reassurance that though bioresmethrin is rapidly absorbed from the intestinal tract and distributed into various tissues only a negligible amount of intact bioresmethrin is found in such tissues.

Bioresmethrin is degraded by ester cleavage and alcohol oxidation. This degradation occurs rapidly though the products of degradation are apparently retained within the animal body for some time before being excreted.

Since the concentration of bioresmethrin applied to grain or available on milling offals seldom exceeds 5 mg/

kg, the potential for livestock receiving grain or feeds based on milling offals to accumulate significant amounts of bioresmethrin in their tissues is too remote to warrant concern.

6.6 Safety to Livestock

The available acute toxicity studies in rats, mice and chickens indicate that bioresmethrin has a very low acute toxicity as indicated in **Table 6.5**

Table 6.5. Acute oral toxicity of bioresmethrin

Animal	Sex	mg/kg body weight	Reference
Rat	M	8 800	**Glomot and Chevalier 1969**
		>8 000	**Elllot *et al.* 1973**
	F	8 000	**Verschoyle and Barnes 1972**
	F	7 071	**Wallwork *et al.* 1970**
Mouse	F	>10 000	**Wallwork *et al.* 1970**
	M	3 100	**Ueda *et al.* 1975b.**
Chicken	M & F	>10 000	**Wallwork *et al.* 1970.**

Signs of poisoning: After 2 or more hours following oral administration, tremors occurred; animals were sensitive to each other and aggressive. The final stages of poisoning consisted of convulsive twitching, prostration, coma and death normally between 3 and 24 hours.

Several short-term toxicology studies in rat and dog are available. Bioresmethrin is not a teratogen although at high levels it has been shown to induce some fetal abnormalities and fetal mortality. In short-term studies at high doses, thymic atrophy was noted over a 3-week test period.

No studies specifically designed to establish the safety of bioresmethrin to livestock have been located but in view of the low concentration used in grain protection, the feeding of treated grain to livestock or the use of milling offals in livestock feeds does not represent a significant hazard.

6.7 Toxicological Evaluation

Several special studies and short-term studies in a variety of animals including rats and dogs are available. Toxic manifestations were produced in those animals receiving extremely high levels daily for 90 days. The compound did not show teratogenic effects, although at high doses of intake it has been shown to induce some fetal abnormalities and mortalities. It is non-irritating and does not induce sensitisation reactions. In short-term studies at high doses, thymic atrophy was noted over a 3-week test period. This was accompanied by increased liver size. Liver dysfunction and fatty infiltration of the liver were observed in a 90-day study at doses of 1200 ppm and above. In dogs, several blood parameters were affected by a 250 to 500 mg/kg body weight dose administered daily over a 90-day period. No effects were noted at 80 mg/kg body weight. No observations in man were available for consideration.

Although short-term studies were available along with several special studies which did not specifically raise any unusual toxicological factors, the absence of data from long-term studies precluded the Joint FAO/WHO Meeting on Pesticide Residues from estimating an ADI for man. In concurrence with previous conclusions, the meeting expressed its need for evaluation of long-term studies in its consideration of an ADI for man. This was especially important in the case of bioresmethrin, owing to the fact that this synthetic pyrethroid was the first of a chemical class of pesticides projected for extensive use in future. Although bioresmethrin effects have been observed only at relatively high dose levels in short-term studies, the potential for adverse toxicological effects in long-term studies needs to be evaluated.

Since these decisions were taken in 1976 the necessary lifetime studies in rats have been commissioned and will be available for evaluation at some future date.

6.8 Maximum Residue Limits

In the absence of an acceptable ADI the Joint FAO/WHO Meeting of Experts on Pesticide Residues has not recommended maximum residue limits for bioresmethrin in raw grain and milled products from grain. However, it has recommended the following guideline levels as limits which need not be exceeded when bioresmethrin is used according to good agricultural practice **(FAO/WHO 1977c)**.

Commodity	Guideline level mg/kg
Cereal grains	5
Cereal products (milled)	5
Cereal products (cooked)	0.05*

* At or about the limit of determination

The residues are determined and expressed as bioresmethrin.

The methods recommended for the analysis of bioresmethrin residues are those of **Zweig (1978)** and **Baker and Bottomley (1982)**.

7. Bromophos

Bromophos is a non-systemic halogen-containing organophosphorus insecticide used on a variety of crops and animals to control biting and sucking insects. It is also used to protect stored products, as a seed protection agent for grain crops, and as a vector control agent in public health. The lipophylic nature of the compound causes it to penetrate the cuticular wax of certain crops which delays release and degradation.

First data encouraging the development of bromophos as an insecticide for the protection of stored grain were presented by **Immel and Giesthardt (1964)**. They emphasised the insecticidal potential, the low mammalian toxicity and the persistence on alkaline surfaces. During the following years quite a body of data on the efficacy against the main species of storage pests was accumulated from laboratory work. In addition some of the more specific problems of a storage pesticide were studied, e.g. persistence on various surfaces, influence of temperature and moisture content of the grain on performance and resistance behaviour.

As soon as the results of some of these studies revealed the potential of bromophos as a grain protectant, field evaluations were initiated in order to obtain practical recommendations for the treatment of empty storages and for the protection of stored grain in bag stacks as well as in bulk. Meanwhile the data needed on toxicology, metabolism, residue behaviour and taint problems were generated **(Kinkel *et al.* 1966; Rowlands 1966a; Eichler 1972, 1974; Eichler and Knoll 1974)**.

Some of the early trials revealed that bromophos produced an objectionable taint in treated grain. This was found to be due to impurities in the technical active ingredient which could be removed by processing. Since that time only the purified grade of bromophos has been available for grain treatment.

Bromophos is, in many respects, similar but superior to malathion. It is more effective on some species but, like malathion, weak against *Rhyzopertha dominica* and moths. It is stable on concrete and therefore useful for treating storage structures. Bromophos has proved effective in controlling some malathion-resistant species and the potency increases at higher temperatures. Bromophos penetrates grain rapidly; it must be used at rates in the region of 10–20 mg/kg. It has a moderate rate of degradation on stored grain and though a relatively high proportion of the deposit transfers to the flour, 85–90% is destroyed in baking. Trials have demonstrated that bromophos is destroyed in the simple processes for preparing and cooking cereal-based foods under tribal conditions in developing countries.

7.1 Usefulness

The original manufacturers of bromophos, C.H. Boehringer Sohn of West Germany, published results of laboratory studies, which showed that the application of bromophos to grain at the rate of 10 mg/kg was effective in controlling granary weevil, *Sitophilus granarius*. However, field studies showed that a somewhat higher rate in the range of 10–20 mg/kg, was necessary to cope with insect control in practice **(Boehringer 1965).**

Lemon (1966) made tests by applying bromophos topically to *Tribolium* spp. He reported it to be slightly less effective than malathion against both *Tribolium castaneum* and *T. confusum*. **Lemon (1967a)** showed bromophos superior to malathion but inferior to fenitrothion when applied to wheat of 12% moisture content. Bromophos was extremely persistent on concrete, whereas malathion broke down rapidly.

Tyler *et al.* (1967), after laboratory tests in which bromophos proved to be more persistent than malathion on moist grain, conducted a practical trial to assess the value of this insecticide as a grain protectant. Two lots, each 14 tonnes, of wheat of 15% moisture content and heavily infested by *Oryzaephilus surinamensis*, were sprayed with bromophos at the rate of 8 and 16 mg/kg respectively during turning into empty bins. All larvae and pupae were killed within 24 hours but the bromophos was too slow in action to prevent large numbers of adults moving to the surface of the grain where they congregated in the corners of the bins and survived for 1–2 weeks. Others left the bins and infested cracks and crevices where living insects persisted for the entire observation period of 5 months. After 19 weeks storage on the farm the bins ceased to be available for experimental work but bulk samples of each batch of grain were retained until 1 year after treatment. It was thought that a low temperature at the time of treatment was probably responsible for the slow rate of action of the bromophos against adult *Oryzaephilus surinamensis*. The air and grain surface temperature was about 12°C and laboratory experiments showed that 50% mortality of adult *Oryzaephilus surinamensis* exposed on grain freshly treated with bromophos at 8 mg/kg occured after 6.5 hours at 25°C and after 32 hours at 10°C.

Kane *et al.* (1967) carried out laboratory experiments to test bromophos as a long-term protectant. Clean, bagged wheat and barley grain, 4–5 months old, was fumigated to ensure freedom from living insects and then treated with aqueous emulsions of bromophos at concentrations graded to give doses of 8, 16 and 24 mg/kg active ingredient. Six 50 kg lots of wheat and barley were treated at each dosage rate; a similar number of untreated bags served as controls. They were stored on dunnage at 25°C and about 60% r.h. under conditions of heavy cross-infestation by *Oryzaephilus surinamensis* and *Sitophilus granarius*. At 2, 6 and 9 months after treatment, 2 bags of wheat and 2 of barley from each dosage rate were withdrawn and sieved. The numbers of live insects in the treated grain were compared with those in the control bags. At 2 months both species were breeding in the control grain but the only live insects found in the treated material were a small number of adult *Sitophilus granarius*. By 6 months, however, there was some breeding by both species of insects in all grain. A large measure of control of *Oryzaephilus surinamensis* was achieved at all dosage rates but the bromophos was less effective against *Sitophilus granarius*.

Joubert and de Beer (1968b) tested bromophos by direct application to maize at 2 dosage levels (5 and 10 mg/kg) in comparison with pyrethrum and malathion. The bromophos treatments survived a 50 week bulk storage period without an appreciable infestation or increase in damage to the internal structure of the corn grains.

Iordanou and Watters (1969) tested the effect of temperature on the toxicity of bromophos against 5 species of stored-product insects by impregnating filter papers to which the beetles were exposed for 24 hours, the mortality being assessed 72 hours after the beetles were returned to flour at the same temperature. Bromophos was 30 times more potent at 26.7°C than it was at 10°C against *Tribolium castaneum, Tribolium confusum* and *Oryzaephilus surinamensis*. *Tribolium confusum* was the most tolerant species and *Cryptolestes ferrugineus* was the most susceptible. Bromophos was considerably more susceptible to variations in temperature than either malathion or lindane. This temperature effect may account for variations in performance reported by other workers.

Dyte and Rowlands (1970) studied the effects of some insecticide synergists on the potency and metabolism of bromophos in *Tribolium castaneum*. They found that sesamex and SKF525A antagonised bromophos in its action against this species.

Green *et al.* (1970) applied bromophos at the rate of 8, 16 or 24 mg/kg to clean, bagged wheat and barley, of 14 and 13% moisture content respectively, which was exposed to heavy infestations of *Oryzaephilus surinamensis* and *Sitophilus granarius* and at 9 or 20 mg/kg to infested wheat of 15% moisture content stored in bulk on a farm. The treatments conferred some protection on the bagged grain and killed the insects in the farm grain. Bromophos acted slowly, however, so that both species oviposited on the bagged grain and larvae were present after 24 weeks even at 24 mg/kg. On the farm wheat, the treatments failed to immobilise the insects quickly and some escaped and survived; those remaining in the wheat were all killed at both dosage rates. Biological and chemical assays showed that bromophos residues broke down more quickly on infested, bagged wheat and barley stored at 25°C than on cooled wheat stored during winter on the farm. However, complete mortality was obtained of *O. surinamensis* and *S. granarius* exposed for 3 days on samples from bagged wheat and barley treated at 24 mg/kg and stored at 25°C for 24 weeks. Farm grain treated at 9 mg/kg was similarly effective against *O. surinamensis* for about a year.

The same authors examined the effectiveness of bromophos and malathion applied at 10 and 20 mg/kg to warm, moist grain in the laboratory against *O. surinamensis*. Applied to wheat of 18.5% moisture content, stored at 30°C, bromophos lost effectiveness after about 6 weeks and malathion after about 2 weeks. The higher dosage rate did not give longer protection with either compound.

Chawla and Bindra (1971) compared bromophos with malathion, iodofenphos and pirimiphos-methyl by application to low moisture wheat at the rate of 30 mg/kg and 50 mg/kg. When the wheat was stored in bags under normal storage conditions, pirimiphos-methyl was the most effective compound. The effect of bromophos appeared to dissipate at a rate comparable to that of malathion.

Coulon *et al.* (1972), comparing the activity of bromophos against 3 other insecticides for the control of *Sitophilus granarius*, found it to be distinctly inferior to pirimiphos-methyl.

Pieterse and Schulten (1972) compared 17 strains of *Tribolium castaneum* collected from warehouses and stores throughout Malawi for possible resistance to malathion. Eleven of the 17 strains were found to be resistant to malathion and also showed cross-resistance to bromophos. However, **Verma and Ram (1974)**, working in India with malathion-resistant *Tribolium castaneum*, showed that the development of malathion resistance did not result in any serious cross-resistance to bromophos in this species.

Wilkin and Hope (1973) studied the effectiveness of bromophos against 3 stored-product mites. Bromophos was not particularly effective against mites, producing only 50% mortality in *Acarus siro* in 14 days.

McCallum-Deighton (1974) reported that pirimphos-methyl was effective against susceptible and malathion-resistant strains of insects at dosage levels below those required for bromophos and a number of other insecticides.

Chawla and Bindra (1973) and **Bindra (1974)** tested 9 insecticides, including bromophos, for effectiveness against adults of *Rhyzopertha dominica*, *Sitophilus oryzae* and larvae of *Trogoderma granarium*. Bromophos proved promising and was further tested in a small-scale trial lasting one year. At comparable doses, bromophos provided protection for a shorter period than malathion.

Bansch *et al.* (1974) reviewed the then available information on the biological evaluation of bromophos for the control of storage pests. This included a considerable number of studies reported by personnel of Celamerck, Germany, as well as published studies. This experience resulted in recommendations for application rates for bulk treatment of small grains of 12 mg/kg and 8 mg/kg on maize and beans for protection periods of at least a year. For shorter storage periods, 6 mg/kg was recommended to be used on all stored commodities.

In a test reported from South Africa, maize and sorghum were sprayed with bromophos at different concentrations on a conveyor belt or on the flowing grain stream. It is noteworthy that the treatment prevented significant insect development and at the same time preserved the germination of the maize above 97%, whereas the untreated maize showed only 41% germination — no doubt due to insect damage **(South Africa Department of Agriculture 1975)**.

Watters (1976b) demonstrated the effectiveness of malathion and bromophos for the treatment of structural surfaces of granaries. Although there was no significant difference in the persistence of the deposits of the 2 insecticides on metal and timber surfaces, bromophos was considerably more persistent on concrete.

Chawla and Bindra (1976) tested the relative toxicity of 10 insecticides against 3 stored-product pests under controlled laboratory conditions. Malathion was employed as a standard. Bromophos proved comparable with malathion.

Tyler and Binns (1977) evaluated 7 organophosphorus insecticides and lindane against 18 species of stored-product beetles. They rated bromophos inferior to chlorpyrifos-methyl, fenitrothion, pirimiphos-methyl, phoxim and iodofenphos but superior to lindane and malathion.

McCallum-Deighton (1978), in reviewing insecticides which have recently been developed as grain protectants as replacements for malathion, ranked bromophos slightly inferior to malathion and greatly less effective than fenitrothion and pirimiphos-methyl. The minimum effective dose against 5 species is in the range 2–5 mg/kg but against *Rhyzopertha dominica* bromophos is relatively ineffective requiring greater than 20 mg/kg. It is outstandingly effective against *Cryptolestes ferrugineus*.

Fall *et al.* (1979) described trials in traditional storage of millet in country areas of Senegal. They described an insecticide-treated silo that was developed in Senegal for storing millet under farm conditions. It has 8 compartments of individual capacity 1.4–5 tonnes and total capacity of 19 tonnes. The grain is treated with liquid formulations of bromophos initially, and again every 3 weeks.

Mensah and Watters (1979b) compared 4 organophosphorus insecticides on stored wheat for control of susceptible and malathion-resistant strains of *Tribolium castaneum*. They considered bromophos a promising alternative to malathion, though distinctly less persistent than pirimiphos-methyl.

Bitran *et al.* (1979a) evaluated the residual persistency of 4 insecticides, including bromophos and malathion for the protection of sorghum and coffee. When used at equal rates, bromophos was slightly less persistent than malathion and considerably less persistent than pirimiphos-methyl and methacriphos.

Yadav *et al.* (1980) carried out a laboratory study in India to determine the toxicity of 7 organophosphorus insecticides to 7 major stored-product pests. The lethal concentration for the 7 species was determined and bromophos was considered inferior to phoxim and etrimfos.

Bansode *et al.* (1981) studied the toxicity of 4 organophosphorus insecticides to a malathion-resistant strain of *Plodia interpunctella*. The 227-fold resistance to malathion was suppressed by the synergistic action of triphenyl phosphate and the authors concluded that carboxyesterase appears to play a major role in the detoxification mechanism for malathion in this strain. However, the tolerance to bromophos was only ×1.3.

Tyler and Binns (1982) examined the influence of temperature on the susceptibility to 8 organophosphorus insecticides of susceptible and resistant strains of *Tribolium castaneum, Oryzaephilus surinamensis* and *Sitophilus granarius*. Based upon knock-down and kill, the effectiveness of all insecticides was greater at 25°C than at 17.5°C and was markedly lower still at 10°C. At 10°C bromophos was virtually ineffective against *Sitophilus granarius* at levels 50 times greater than those completely adequate at 25°C.

Zhang *et al.* (1982) tested the effect of bromophos on the quality of stored seeds and found that an application rate of 40 mg/kg had no significant effect on the germination of rice, wheat, maize, sorghum and barley. No adverse effects on germination of the seeds of wheat, maize and sorghum were found when treated with 7500 mg/kg bromophos. There was adverse affect on the germination of maize treated with 10 000 mg/kg of bromophos.

Peng (1983), working in Taiwan, determined the relative toxicity of 10 insecticides against 6 coleopterous insect pests of stored rice. Adults of each species were exposed to impregnated filter papers for 4 hours and mortality was recorded after 72 hours. Bromophos was rated as moderately effective.

Golob and Hodges (1982), Golob *et al.* (1983) report efforts to control the outbreak of *Prostephanus truncatus* in Tanzania. Bromophos dust was one of the materials tested. The results were disappointing and considerably inferior to pirimiphos-methyl. The authors questioned whether the dust formulations contained the full concentration of active ingredient.

Hasan *et al.* 1983 tested bromophos, 5 other organophosphorus and 2 organochlorine insecticides as dry films in a laboratory in India where their toxicity to *Sitophilus oryzae, Tribolium castaneum, Rhyzopertha dominica* and *Callosobruchus chinensis* was compared. In tests of the synergistic effects on bromophos of sublethal amounts of the other insecticides, only 3 compounds with a structure similar to that of bromophos (malathion, parathion-methyl and fenitrothion) had marked synergistic effects.

Srivastava and Gopal (1984) studied the toxicity of 7 insecticides to 5 stored-grain pests in a laboratory in India. Bromophos was not ranked among the more effective materials.

7.2 Degradation

Soon after bromophos became available for application to stored grain, **Rowlands (1966a, 1966b)** carried out a detailed study of the fate of bromophos applied to stored-wheat grains. It was shown that bromophos, applied to Manitoba wheat, breaks down very rapidly over 7 days. Degradation then stops for approximately 21 days. Thereafter, bromophos break-down continues for 7 days, followed by a further halt for approximately 21 days and so on. When the quantitative production of metabolites is related to this stepped phenomenon, it can be seen that cessation of bromophos metabolism coincides with a high level of desmethyl bromophos, which then degrades while the remaining bromophos is unattacked. The level of free dichlorobromophenol does not increase significantly until the degradation of desmethyl bromophos is well under way. Moreover, only trace amounts of dimethyl phosphorothionate were found throughout, though large amounts of monomethyl phosphorothionate were detected. It appears, therefore, that the production of phenol is probably from the degradation of desmethyl bromophos or bromoxon, rather than from bromophos. Only when the desmethyl bromophos has almost disappeared does degradation of bromophos continue with subsequent re-accumulation of desmethyl bromophos. When applied to autoclaved grains, bromophos does not show the stepped effect on storage and the only metabolites produced are dimethyl phosphorothionate and dichlorobromophenol. The stepping phenomenon is not restricted to any variety of wheat, neither is it particularly dependent on the age or moisture-content of the grain; an English wheat treated with bromophos at various times up to 6 months after harvest and at moisture contents from 10 to 22% showed the stepping effect under all conditions.

Rowlands (1966a) found that very little oxidation of bromophos to the more toxic bromoxon occurred, probably due to the rapid penetration of bromophos through the pericarp, which is the main location of oxidase activity. No bromoxon was detected after 1 month of storage and the highest level recorded (approximately 10% of the total residue) occurred 2 days after treatment, thereafter steadily declining. The main metabolite accumulating over 10–12 weeks was found to be dichlorobromophenol, which seemed to be stored unchanged in the endosperm.

Rowlands (1966c) showed that bromophos penetrated to the endosperm and germ with great rapidity, whereas malathion penetrated but slowly. Oxidase activity, capable of converting phosphorothionate insecticides to their P — O analogues, was found both in the seed coat and, to a certain extent, in the germ and did not seem to be affected by the total moisture content of the grain. Much of the oxidation during the normal life cycle of the wheat grain is associated with the enzymic oxidation of phenols occurring at the time of the ripening, tinting and hardening.

After laboratory tests in which bromophos proved to be more persistent than malathion on moist grain, **Tyler *et al.* (1967)** conducted a practical trial to assess the value of bromophos as a grain protectant. Two lots of wheat of 15% moisture were sprayed with 8 and 16 mg/kg bromophos, respectively, during turning into empty bins. Chemical analysis of samples showed that the initial dosage rates were 9.3 and 19.9 mg/kg respectively and that 2.9 and 7.3 mg/kg remained after 19 weeks storage and at the end of a year after treatment 2.5 and 3.9 mg/kg respectively. The grain temperature was about 12°C.

Kane *et al.* (1967) carried out an experiment in which wheat and barley that had been fumigated to ensure freedom from living insects was treated with bromophos at concentrations of 8, 16 and 24 mg/kg. The treated grain was stored at 25°C and 60% r.h. under conditions of heavy infestation of insects. Although there was no evidence of insects breeding in the treated grain at the end of 2 months, by 6 months, however, there was some breeding by both species of insects in all grain indicating there had been substantial degradation at each of the rates applied.

Rowlands and Clarke (1968) reported that visual examination of wheat treated 18 months previously with bromophos revealed that the grain was considerably less mouldy than that treated with malathion. Assay for bromophos showed no detectable residue of the intact compound but considerable levels of dichlorobromophenol. Since some trihalogenated phenols are fungicidal, it is intriguing to speculate that the phenolic residues resulting from bromophos application may act fungicidally.

Joubert and de Beer (1968b) studied the fate of bromophos applied to maize held in large bins containing up to 800 tonnes. The moisture content of the grain ranged from 9.7 to 12% and the temperature 28°C. The residue level on the maize treated with bromophos at the rate of 10 mg/kg fell to half (5 mg/kg) in about 27 weeks and to 1 mg/kg in 40 weeks. The maize treated at the rate of 5 mg/kg lost half of its deposit in about 17 weeks, after which there was a gradual reduction until, at the end of 40 weeks, the residue level was below 1 mg/kg.

Chawla and Bindra (1971, 1973) showed that bromophos applied to wheat grain with extremely low moisture content (8.6%) degraded at a rate comparable with that of malathion and considerably faster than iodofenphos and pirimphos-methyl when stored for a period of 6 months.

Eichler (1974) and **Eichler and Knoll (1974)** described a series of large-scale laboratory trials to determine the degradation of bromophos under different experimental conditions. Dust and emulsion formulations were used.

Two samples of grain, one with 15% moisture and the other with 13.5% moisture were used at an average temperature of 15°C. The wheat which had been treated at the rate of 8 mg/kg lost approximately 40% of its bromophos after 1 year, there being little difference between the dust and emulsion treatments. The wheat treated at the rate of 12 mg/kg showed 41 and 50% degradation at the end of 1 year for the dust and emulsion treatment respectively. The results showed that the main breakdown, amounting to about 35%, occurred during the first 6 months. Subsequent experiments by these workers showed that the rate of degradation increased with increasing temperature. At 26°C about 60% of the deposit was degraded in 12 months, whereas at 15°C only about 40% was lost. The results tended to suggest that, within the range of moisture levels used, the moisture content does not influence the rapidity of bromophos degradation.

Rowlands (1974) discussed the work of himself and co-workers on the rapid degradation of bromophos during the first few days following application to grain. He gave an example of the intended application of 10 mg/kg bromophos to 28 tonnes of wheat stored at 26–30°C and a moisture content of 18%. Analysis of samples drawn immediately after application revealed 9.3 mg/kg but within a day the residue of intact bromophos was down to 8.4 mg/kg and within 3 days to 7.6 mg/kg. He discussed whether these 'losses' could be due to experimental error but in view of the figure at the end of 7 days (6.8 mg/kg) he considered that there was unequivocal evidence of loss which was accounted for by the presence of 2.1 mg/kg of desmethyl bromophos. Laboratory studies at lower temperatures (20°C) showed a similar pattern.

Watters (1976b) studied the persistence of bromophos applied to various structural surfaces and the uptake of bromophos by wheat stored in contact with such treated surfaces. Bromophos remained effective on metal surfaces for 40 weeks, on plywood for 18 weeks, and on hardwood and concrete for 4 weeks. It was shown that the uptake of bromophos into wheat continued for up to 40 weeks though the residue levels were only of the order of 0.5 mg/kg. **Mensah *et al.* (1979a)** reported virtually identical studies and results. Their work revealed a much more pronounced tendency for bromophos deposits on wood to transfer to wheat, barley and corn stored in contact with the treated surfaces for a period of 1 week.

Mensah and Watters (1979a) applied an aqueous suspension of bromophos at 1.0 g/m^2 to the concrete floor and inside wall of an empty cylindrical steel granary to determine the uptake of bromophos into 30 tonnes of bulk wheat during 7 months of storage. The wheat was sampled at intervals as the granary was emptied. Insecticide uptake was determined on 50 g samples by chemical assay and bioassay with adults of *Cryptolestes ferrugineus*. No bromophos was detected in the upper 50% of wheat stored in the treated granary. The highest residue (0.68 mg/kg) was recovered in the final 0.5 tonne of wheat taken from the treated floor. Complete mortality of *Cryptolestes ferrugineus* occurred on wheat samples with 0.5 mg/kg or higher.

Mensah *et al.* (1979b) applied bromophos at 2 dosage rates to wheat of 12 and 16% moisture content. They showed that the deposit of 8 mg/kg declined after 6 months storage to 2.6 mg/kg in the case of the 12% moisture wheat and 1.4 mg/kg in the case of the 16% moisture wheat. Similar wheat treated at a dosage of 12 mg/kg was found to contain 4.5 and 1.5 mg/kg in the presence of 12% and 16% moisture respectively.

Watters and Nowicki (1982) showed by field studies that 11.3 tonnes of rape seed stored in a plywood-lined granary treated with bromophos at 0.5 g/m^2 had residues of 0.1 to 2.6 mg/kg after 16 weeks and 0.4 to 3.5 mg/kg after a further 36 weeks of storage. Laboratory studies of insecticide uptake by rape seed or wheat from wood or concrete surfaces treated with bromophos showed that uptake was higher in rape seed than in wheat and that uptake by rape seed or wheat was greater from treated wood than concrete.

7.3 Fate on Milling, Processing and Cooking

Eichler (1972) reported that wheat treated with bromophos dust at rates equivalent to 6 and 12 mg/kg was converted into wholemeal after being in store for 270 days. The residues in the meal were determined by a colorimetric method and were found to be 1.2 and 2.73 mg/kg respectively. This wholemeal was converted into wholemeal bread, which was analysed after baking. The residues found in the baked bread were less than half those found in the wholemeal flour when the results were compared on the basis of dry substance. He concluded that, after preparation for human consumption, only about 10% of the bromophos originally applied to the raw wheat was still detectable. The author drew attention to the fact that, in the milling of wheat for flour, the bran and pollard are removed and therefore the level of bromophos in white bread will be substantially lower still.

Eichler and Knoll (1974) and **Eichler (1974)** reported experiments designed to study the degradation of bromophos in stored wheat, milled cereals and cooked cereal foods. Samples of wheat typically used for baking were treated with bromophos in the form of both dust and emulsion. Both formulations were added to the grain in a mixer at concentrations of 8 and 12 mg/kg bromophos. Each series of experiments comprised 100 kg of wheat with a moisture content of 13.5%. The treated grain was stored at an average storage temperature of 15°C. Directly after treatment and 3, 6 and 12 months later, samples were prepared by mixing the grain withdrawn from different depths of the containers. These samples were set aside for milling and analytical assay. The wheat intended for baking was cleaned, conditioned and milled in a laboratory mill, giving a flour typical of commercial quality. Samples of the various milling fractions were retained. Crushed grain and flour prepared at four separate sampling times were converted into breads, which were submitted to organoleptic testing and residue analysis. **Table 7.1** shows the distribution of bromophos residues from the whole grain between coarse bran, semolina bran (pollard) and flour. For convenience only the results on the wheat treated at the rate of 12 mg/kg are presented.

Table 7.1. Distribution of bromophos (ppm) in milling products (application rate 12 ppm a.i.).

Months after applic.	whole[a] grain	coarse[b] bran	semolina[c] bran	flour
		application as dust:		
3	7.2	22.0	21.2	2.34
12	6.1	14.3	9.4	1.17
		application as emulsion:		
3	9.2	23.5	22.7	2.66
12	4.7	14.1	10.1	1.10

a crushed just before analysis in a grinder
b consists mainly of inner pericarp and seed coat
c consists besides inner pericarp, seed coat and flour, of main parts of aleurone layer and germ.

These experiments showed that after 3 months considerable amounts of bromophos can be found in coarse bran and pollard. These values gradually fell when the treated wheat was held in storage for a period of 12 months. The level of bromophos residues in wholemeal bread and white bread, prepared by milling the grains in this experiment, are recorded in **Tables 7.2** and **7.3**.

After the baking of crushed wheat, only about 25–35% of the residue originally present in the grain at the time of crushing was recovered. In bread made with white flour, 85–90% had been destroyed. Apparently, in the larger particles of crushed wheat, the residues of bromophos are protected from hydrolysis by the free moisture used in the preparation of the bread, whereas in the white bread the residues are more intimately in contact with moisture.

These studies by **Eichler and Knoll (1974)** revealed that the residues present in bread at the time of consumption, following the application of insecticidally-effective rates of bromophos 3–12 months previously, were, in the case of wholemeal bread, 1.4–2.5 mg/kg and in the case of white bread, 0.1–0.3 mg/kg. This represented only 2–3% of the bromophos originally applied to the wheat.

Chawla and Bindra (1973) reported that during the processing of treated grain into chapaties, the loss of bromophos was as fast as that of malathion, and considerably faster than that for iodofenphos and pirimiphos-methyl. Chapaties are prepared from crushed grain mixed with water to form a soft dough, which is cooked by spreading on a hot surface of a simple stove. In another study by the same authors, it was shown that the processing of grain treated with iodofenphos into chapaties and popcorn caused a loss of from 94–99% of the insecticide. Washing of the grain, followed by its drying, removed most of the residues of malathion, iodofenphos and bromophos. Addition of an emulsifier at the rate of 0.1% enhanced the residue-removal efficiency of water significantly, and sun-drying proved more effective than oven drying **(Chawla and Bindra 1971)**.

Mensah *et al.* (1979b) determined the level of insecticide residues in milled fractions of dry or tough (high moisture) wheat treated with bromophos and 3 other insecticides. Water-based emulsions of bromophos were applied at 2 dosage rates to wheat of 12 and 16% moisture content to compare residue distribution in fractions milled from wheat samples stored over a period of 6 months. The residue degradation of bromophos on dry and tough wheat and milled fractions were similar to that observed for malathion. Bromophos appeared to be more stable on dry wheat than on tough wheat as the age of the deposit increased. In fractions milled from dry wheat, high residues were found in the middlings during 3 months of storage. As high as 37.7 mg/kg bromophos was found in the middlings in contrast to 27.3 and 3.4 mg/kg in the bran and flour, respectively, 1 month after treatment. However, after 6 months, more bromophos was recovered from bran than from middlings or flour notwithstanding the fact that the concentration on the bran had declined by more than 30% in the intervening time.

Table 7.2. Comparison of bromophos residues (ppm) in flour and bread

Months after application	Application of dust[a]		Application of emulsion[a]	
	Flour	Bread	Flour	Bread
3	2.34	0.18	2.66	0.30
6	1.02	0.12	0.99	0.10
12	1.17	0.08	1.10	0.07

a Application of 12 ppm a.i.

Table 7.3. Comparison of bromophos residues (ppm) in crushed grain and bread

Months after application	Application of dust[a]		Application of emulsion[a]	
	Crushed grain	Bread	Crushed grain	Bread
3	7.2	1.59	9.2	2.48
6	6.7	1.73	6.2	2.05
12	6.1	2.23	4.7	1.40

[a]Application of 12 ppm a.i.

7.4 Metabolism

Rowlands (1966a) published an extensive account of studies on the metabolism of bromophos on stored wheat grains. This has been reviewed in the section on degradation and is best explained diagrammatically. **Figure 7.1** is taken from the Rowlands' study.

Rowlands (1966b), by identification using thin-layer and gas-liquid chromotography of the metabolites produced *in vivo*, showed that desmethyl bromophos was the major product during the first week of storage. Crude acid-phosphatase preparations from wheat grains, applied to bromophos as the substrate, showed that, of the metabolites produced *in vivo*, only desmethyl bromophos interferes with the phosphatase activity *in vitro*. Furthermore, while metabolism of bromophos is dormant the level of desmethyl bromophos achieved is steadily decreasing. When the desmethly bromophos has virtually disappeared, degradation of bromophos recommences with subsequent re-accumulation of the desmethyl derivatives.

Oxidative activity converting bromophos to its phosphate analogue was found in the seed coats and germs of wheat grains by *in vivo* and *in vitro* studies. Hydrolytic activity found in the germ and endosperm, detoxifying bromophos, was demonstrated *in vitro* and non-specific hydrolases were located by histochemical tests **(Rowlands 1966c).**

Rowlands and Clarke (1968) reported that after 18 months storage, wheat originally treated with bromophos was considerably less mouldy than that treated with malathion. Assay of the bromophos-treated samples showed considerable levels of dichlorobromophenol and since some halogenated phenols are fungicidal the authors speculated that the phenolic residues resulting from bromophos application may act fungicidally.

Stiasni *et al.* (1969), who studied the translocation, penetration and metabolism of bromophos in tomato plants reported that, in addition to unchanged bromophos, dichlorobromophenol was found as a main metabolite, and small amounts of bromoxon, monodesmethylbromophos, dimethyl thionophosphate, and inorganic phosphate were detected.

Dyte and Rowlands (1970), studying the effects of some insecticide synergists on the potency and metabolism of bromophos in *Tribolium castaneum*, reported that sesamex and SKF 525A antagonised the effect of bromophos on *Tribolium castaneum*, SKF 525A being the more effective antagonist. The main hydrolytic metabolite recovered from homogenates prepared from adult beetles

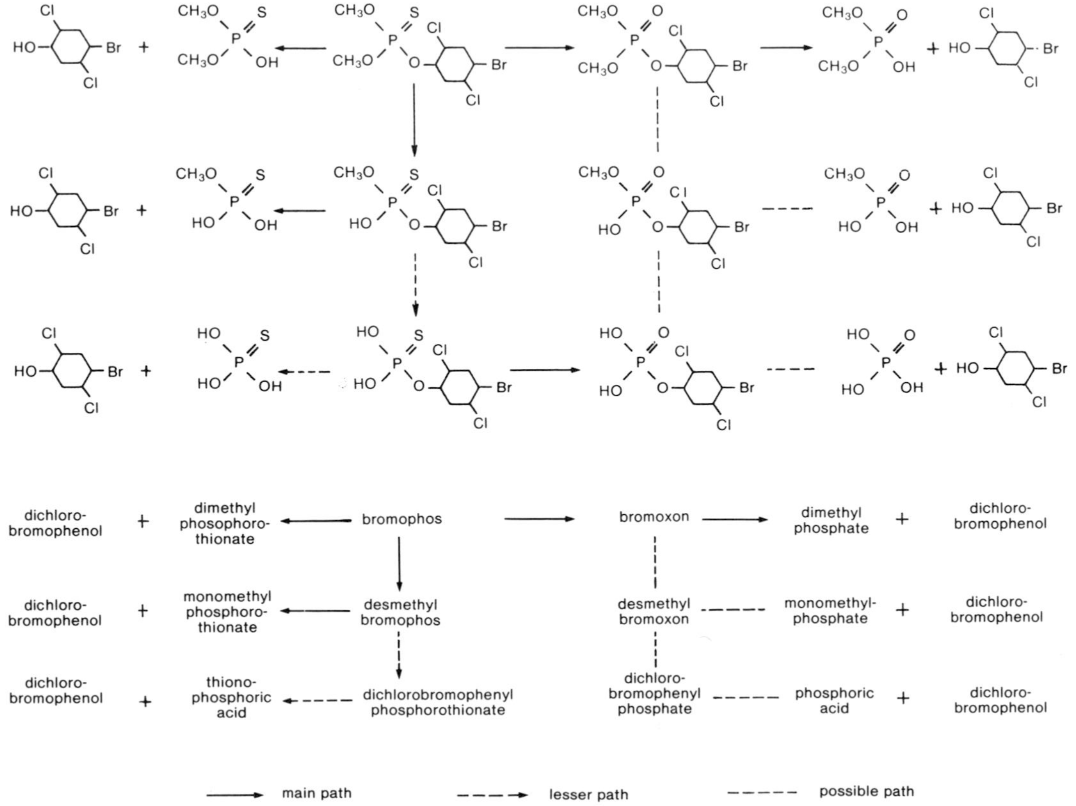

dichloro-bromophenol + dimethyl phosophoro-thionate ← bromophos → bromoxon → dimethyl phosphate + dichloro-bromophenol

dichloro-bromophenol + monomethyl phosphoro-thionate ← desmethyl bromophos; desmethyl bromoxon ---- monomethyl-phosphate + dichloro-bromophenol

dichloro-bromophenol + thiono-phosphoric acid ← ---- dichlorobromophenyl phosphorothionate; dichloro-bromophenyl phosphate ---- phosphoric acid + dichloro-bromophenol

→ main path ---→ lesser path ---- possible path

Figure 7.1. Suggested metabolism of bromophos by stored grain **(Rowlands 1966a).**

after topical treatment with bromophos was O-desmethyl bromophos, but O-O-dimethyl thiophosphoric acid, O-O-dimethyl phosphoric acid and a little O-methyl phosphoric acid were also found together with 2,5-dichloro, 4-bromophenol. Very little of either the oxygen analogue or dimethyl phosphoric acid were found. Sesamex or SKF 525A strongly inhibited the production of the O-desmethyl bromophos.

Eichler (1972) reviewed the then available studies. The first metabolic studies on animals were carried out by **Stiasni *et al.* (1967)** using rats. Five urine metabolites were found after administration of ^{32}P-bromophos, whereas three were found after administration of ^{3}H-bromophos. Analysis showed that the metabolites were phosphate, dimethylthionophosphate, monodesmethyl bromophos and dichlorobromophenol. Two metabolites were unidentified. Bromophos itself, its O-analogue, (bromoxon) and monodesmethyl-bromoxon were not found in urine.

The same workers found that, following oral, intravenous and intraperitoneal administration, the principal route of excretion was via the kidneys; more than 90% being eliminated within 24 hours. These studies led the authors to postulate that animal metabolism of bromophos proceeds according to **Figure 7.2.**

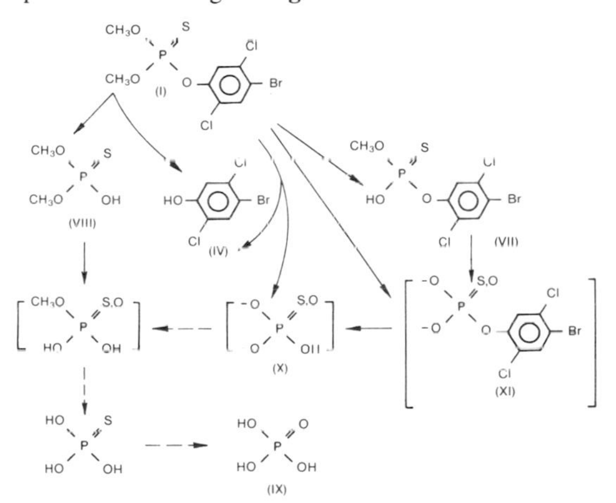

Figure 7.2. Suggested routes of metabolism of bromophos in the rat **(Stiasni *et al.* 1967).** (I) = bromophos.

Although bromoxon was not found in rat urine in the studies by **Stiasni *et al.* (1967) Muacevic *et al.* (1970)** detected extremely low concentrations of bromoxon in blood, and considered that this correlates with the inhibition of cholinesterase activity observed following administration of bromophos.

From these results we can conclude that bromophos is at least partially broken down in animals via bromoxon, but that the bromoxon is rapidly decomposed by hydrolysis so that it occurs only in extremely small amounts. It is also conceivable that this is only a secondary reaction and that the primary metabolic pathway is via direct hydrolysis of bromophos. **Sternersen (1969)** studied the metabolism of bromophos in plants and showed that it is much the same as that in animals.

Rowlands in his *Review of World Literature on the Metabolism of Contact Insecticides Used in Stored Grains* **(Rowlands 1975)** does not mention any new studies published during the period 1970–74.

Eichler and Knoll (1974) reported a series of studies on the degradation of bromophos in stored wheat, milling products and bread. When grain samples were examined 6 and 12 months after application of bromophos dust or emulsion no trace of bromoxon could be detected. This is not surprising as **Rowlands (1966a)** had found traces only within the first 3–4 weeks of storage. This coincides with metabolism studies in plants reported by **Eichler (1972).**

Calderon and Desmarchelier (1979) showed that the susceptibility of *Tribolium castaneum* adults to bromophos was increased measurably by pre-exposure to carbon monoxide (30% in air for 20 hours). However, an increase in susceptibility to bromoxon could not be shown. Carbon monoxide is well known as an inhibitor of the microsomal mixed-function oxidase system noted for its role in the metabolism of insecticides. Most oxidative reactions catalysed by this system are mediated by a carbon monoxide binding pigment designated as Cytochrome P-450. In view of the ever increasing importance of the insecticide resistance problem in insects, including stored-product insects, the reported study was undertaken in an attempt to provide further information on the *in vivo* inhibiting effect of carbon monoxide on the detoxification of several insecticides by *Tribolium castaneum*.

7.5 Fate in Animals

Stiasni *et al.* (1967), who carried out studies on rats with ^{32}P- and ^{3}H-labelled bromophos, showed it is well absorbed from the gastrointestinal tract. In the first 12 hours following oral administration, large quantities of radioactivity were found in the stomach, liver and kidneys of rats, but after 24 hours the majority was found in urine and feces. Bromophos showed no tendency to accumulate in any organ. The maximum blood level was found 7 hours after oral administration and the biological half life in rats was 14 hours. Approximately 96% of the activity of the 10 mg/kg dose of ^{3}H-bromophos was excreted in urine and 1% in feces during 24 hours following oral administration. Excretion was complete in 96 hours, a total of 2% of activity appearing in the feces. In the same period, administration of ^{32}P-bromophos resulted in the excretion of 63% of activity in urine and 16% in feces. During 8 hours following intraduodenal administration of 5 mg/kg of ^{3}H-bromophos, 25% of the activity was excreted by rats in bile. Since only 1% of ^{32}P-bromophos was excreted in the same period the biliary excretion was probably mainly dichlorobromophenol or its metabolites. Biliary excretion of ^{3}H is probably resorbed since only 2% of an oral dose was found in feces.

The relatively high ^{32}P content in feces may result from metabolism of bromophos in the intestine.

After spraying 20 mg of ^{32}P-labelled bromophos on the skin of lactating cows, labelled phosphate was detectable in blood and milk. The blood concentration calculated as bromophos was about 0.01 mg/kg, and was not in the form of the analogue. The principal metabolite, desmethylbromophos, was identified in blood and milk at concentrations of 0.4–0.7 mg/kg but these residues were not toxicologically significant **(Dedek and Schwarz 1969).**

7.6 Safety to Livestock

The available toxicology data from studies on laboratory animals suggests that bromophos is a typical organophosphorus insecticide which inhibits cholinesterase, plasma cholinesterase being considerably more sensitive than erythrocyte cholinesterase and brain cholinesterase not being inhibited at intake levels which cause a 20–50% reduction in acetyl (plasma) cholinesterase. Bromophos appears to inhibit cholinesterase activity indirectly through its metabolite, bromoxon.

Groups of 10 adult hens were administered 0, 12.5 or 125 mg/kg of bromophos per day in their food for 4 weeks. Clinically apparent neurological signs were reversible when treatment ceased and a neuropathological examination of 2 birds of each group showed no central nervous system changes **(Muacevic and Glees 1968)**.

In reproduction studies in rats, animals consuming 20 mg/kg bromophos/day were no different from control animals, no malformations were observed at any dosage or in any generation and the behaviour, appearance, food intake, results of haematological investigations and adult survival were also unaffected **(Leuschner *et al.*, 1967)**.

Studies on the acute oral toxicity of bromophos in several animal species are summarised in **Table 7.4.**

Table 7.4. Acute oral toxicity of bromophos in animals

Animal	Sex	LD/50 (mg/kg body weight)	Reference
Rat	M & F	3750–5180	**Jones *et al.* 1968**
	M & F	6000	**Novozhilov 1975**
	M	1600 (1322–1936)	**Gaines 1969**
	F	1730 (1373–2180)	**Gaines 1969**
	M	4000	**Kirkel *et al.* 1966**
	F	6100	**Kirkel *et al.* 1966**
Mouse	M	3700–5850	**Kirkel *et al.* 1966**
	F	2929	**Kirkel *et al.* 1966**
Guinea pig	M & F	1500	**Kirkel *et al.* 1966**
		>6000	**Muacevic 1967**
Rabbit	M	720	**Kinkel *et al.* 1966**
	M & F	720	**Muacevic 1964**
Fowl		9700	**Kinkel 1964**
Dog		>625	**Worth *et al.* 1967**

Observations on the acute toxicity of bromophos have been summarised by the manufacturers (Celamerck, Ingelheim, Federal Republic of Germany). Study of the symptoms of poisoning with bromophos after oral administration to 12 different animal species revealed that they agree in many respects with the reactions known to result with conventional organophosphate insecticides: over-activity of secretory glands in the form of increased salivation or lacrimation. There are typical signs of parasympathetic stimulation. The effect on the gastrointestinal tract was to cause diarrhoea, which may also be regarded as a result of acetylcholine imbalance. Dyspnoea, as a symptom of asthma-like conditions, is a symptom of poisoning with both bromophos and conventional organophosphate insecticides. The restlessness, tremors, myasthenia and infrequent convulsive paroxysms are similar to nicotine poisoning symptoms. The highest doses ultimately lead to loss of consciousness and collapse. Death occurs in coma, probably as a result of central and peripheral (curareminetic) paralysis.

On the whole, the toxic symptoms of bromophos do not distinguish this compound among organophosphates, since many well-known insecticides also produce all of these symptoms. However, bromophos poisoning differs from that of the conventional phosphoric acid ester insecticides in the following 4 ways:

(a) Dosages which lead to functional disturbances are unusually high. As the desired biological effects against insects may be attained at very low doses, this results in quite a broad therapeutic range, particularly in the control of animal parasites (ectoparasites and sometimes endoparasites).

(b) The course of the poisoning is unusually protracted. Whereas death generally occurs within minutes or a few hours after administration of high doses of the conventional organophosphate insecticides, death seldom occurs so soon after administration of lethal doses of bromophos. The majority of animals in the animal species tested did not die until 1 or even several days after administration.

(c) It is significant that a large gap exists between the doses which effect a marked inhibition of cholinesterase activity and those which lead to functional disorders. The conventional phosphoric acid esters, however, cause toxic symptoms and cholinesterase inhibition at the same or nearly the same doses. It is possible that this property of bromophos masks a toxicological mode of action which differs somewhat from that of the classic phosphoric acid ester insecticides, and which may also be the reason for its good tolerance.

(d) In contrast to poisoning from numerous organophosphorus compounds, bromophos poisoning responds relatively poorly to pyridine aldoxime methiodide (PAM). This too places bromophos in a special category.

After parenteral administration (local, intraperitoneal, subcutaneous, intracutaneous and inhalation) bromophos behaves like its more toxic congeners, but again with the peculiarity that toxic symptoms only develop at high doses and after a long period.

Bromophos is easily absorbed after all modes of administration and leads to toxic symptoms which correspond — as far as has been determined until now — to those observed after oral administration. It should be mentioned in particular that absorption through the epidermis is only slightly slower than through the mucosa of the gastro-intestional tract. Local reactions are slight or absent, and are usually due to the formulation used.

7.6.1 Domestic Fowl

Tests for determination of the acute toxicity of bromophos in chickens were carried out with a 1:1 mixture of bromophos and liquid paraffin. The symptoms observed were general debility, ataxia and paralysis of the extremities, some cases of diarrhoea and melaena, and pale combs. In cases of lethal poisoning, death occurred 4 to 19 days after administration.

Apart from cachexia, post-mortem examination of the animals which died did not reveal any marked macroscopic changes in the organs. In all the surviving animals the clinical symptoms were reversible within the observation period.

The LD_{50} was approximately 9700 mg/kg, the lowest toxic dose 500 mg/kg, and the first deaths were observed at 8000 mg/kg. Tests were carried out principally to establish the general poisoning symptoms and to reveal any possible neurotoxic symptoms. Two tests were carried out to establish the neurotoxic activity of bromophos; in 1 case a single administration of 300 mg/kg to 2 test animals resulted in mild toxic symptoms (temporary agitation and diarrhoea). After a dose of 200 mg/kg all the animals in the group developed diarrhoea, 4 animals had slight to severe ataxia, the combs became pale, they stopped laying and lost weight.

7.6.1.1 Four-week trial

(a) Fifty-four pure-bred hens (14 weeks old) were divided into 6 groups containing 9 animals each, and treated with 20, 50, 100, 200, 500 and 1000 mg bromophos/kg/day respectively for 4 weeks; administration took place 6 times a week. Ten other hens were used as controls. One-third of the test animals received bromophos alone, another third received atropine in addition, and the remaining third received PAM, atropine and bromophos.

(b) Even the lowest dose (20 mg/kg) led to debility after administration for 24 days. In contrast, the weight gain remained normal up to 200 mg/kg. From 500 mg/kg severe toxic symptoms developed after 4–21 days.

The walk became first atactic, then paretic. Eventually the animals were unable to stand. Flaccid paralysis developed, being particularly noticeable in the claws. Other symptoms were reduced circulation, hypothermia and melaena. All the animals receiving 500 and 1000 mg/kg died between the 11th and 21st day of treatment.

Histological examination of the liver revealed isolated patches of pigmentation (from 20 mg/kg). Otherwise nothing abnormal was detected.

Atropine and atropine in combination with PAM had no effect on the paralysis and mortality of the animals. However, after this combined treatment cases of debility were consistently absent or occurred considerably later at the higher doses. It must be mentioned that it was not made clear whether the optimal doses of atropine and PAM were used in these trials.

7.6.1.2 Unlimited trial with daily administration of 1 g bromophos/kg

(a) Ten hens approximately 12 months old and with an average weight of 1.3 kg received 1 g bromophos/kg/day orally in gelatin capsules. Administration was continued until signs of paralysis developed. The animals were then sacrificed, dissected, and the brain and the spinal cord were examined histologically.

(b) The animals began to refuse food after only a few days and their weight consequently fell sharply. Three days after the beginning of treatment weakness of the claws and legs was observed with subsequent paralysis. The feces contained blood. Eventually the animals became unable to raise themselves from lateral decubitus. This stage occurred after 12–56 days of treatment.
Dissection did not reveal macroscopically-detectable findings. Histological examination of the brain and spinal cord revealed changes which were very probably artifacts.

7.6.1.3 Four-week trial without administration of antidote

Ten 1-year old Leghorn-HNL-Hybrid hens were administered 12.5 mg/kg and 125 mg/kg in food daily for a total of 4 weeks. No clinical symptoms with the exception of a decrease in laying ability, were observed. The pathological findings indicated a degeneration of the ovaries of hens from both dosage groups. No significant difference was observed between the treated and untreated animals from a neurohistological examination.

7.6.2 Dog

In a range-finding test 2 males and 2 females (mongrels) received a single dose of 625 mg/kg. Two males and 1 female transiently showed signs of slight systemic poisoning in the form of salivation. Two animals vomited. Twenty-four hours later all the dogs had recovered and remained normal throughout the following 14-day observation period. One dog did not use one of his hind legs for some time without discernible reason.

In a range-finding test for chronic toxicity, mongrel dogs (3 males and 3 females) received 1–3 doses of 350 mg/kg at 24-hour intervals. After the third dose at the latest, all the animals developed vomiting, increased salivation, diarrhoea and anorexia. The dogs recovered rapidly after treatment was discontinued.

7.6.3 Systematic Observations on the Toxicity of Bromophos in Acute Tolerance Tests and in General Use in Veterinary Practice (Clinical Investigations)

7.6.3.1 Cat

The same investigator obtained contradictory results in a series of 4 experiments. Three males tolerated 1250 mg/kg without developing toxic symptoms. Two out of 3 females that had received 500 mg/kg were very unwell and refused to eat for 1 day. The third animal tolerated the dose without developing any symptoms. One out of 3 females that had received 750 mg/kg died.

7.6.3.2 Swine

Two investigators who examined 597 pigs of different ages (most of which had mange) found no clinically detectable side-effects after single and repeated doses of 50, 75 and 100 mg/kg. After administration of 150 mg/kg no symptoms were noted except for a reduced food intake (probably due to the altered taste). A dose of 200 mg/kg was tolerated without development of symptoms. The increased pruritus which occurred on the day following treatment was very probably due to the movements of the poisoned mites or to sensitisation produced by the substances contained in the mites. In a further test series bromophos was administered in the food at doses of 200 to 2000 mg/kg to 19 pigs, including some with severe clinical symptoms. No reactions were observed which were definitely or even probably attributable to the quantity of bromophos administered.

7.6.3.3 Goat

Eight female Angora goats infested with helminths tolerated 100 mg/kg body weight without any reaction.

7.6.3.4 Sheep

Four investigators administered bromophos orally to various breeds of sheep. One investigator observed slight toxic symptoms (polypnea) at as little as 15 mg/kg. In all other cases, no reaction was observed from doses up to 400 mg/kg. At 750 to 1000 mg/kg the animals were listless and appeared ill; a dose of 1500 mg/kg was lethal in 1 case (only 1 animal tested). The effects observed at sub-lethal doses (750 and 100 mg/kg) were ataxia, increased salivation with foaming at the mouth, and varying degrees of reduced food intake. Two investigators discovered that buccal application of 100–300 mg/kg bromophos to sheep with a nasal bot-fly infestation produced no toxic side-effects.

7.6.3.5 Cattle

The oral tolerance of bromophos in various preparations was tested by 5 investigators. Approximately 180 cattle received doses of 25 to 400 mg/kg as single, or in individual cases, repeated doses. In most cases no side effects were observed at doses up to 300 mg/kg/; only 1 investigator reported a reduced food intake in 5 out of 49 animals tested, after administration of 120 mg/kg. From 400 mg/kg side-effects were seen in the treated animals. After 4 to 6 hours, 2 animals developed salivation, drowsiness and fatigue. On the following days 1 animal showed loss of appetite, and another 1 slight tympanites and diarrhoea. The third animal (African zebu), which had received 2 doses of 400 mg/kg with an interval of 4 days, died on the ninth day of the trial in spite of an unspecified dose of atropine after its collapse on the sixth day. This therapy complicates the evaluation of the case, since the possibility of an overdosage of atropine cannot be excluded. The post-mortem findings on this animal were within normal limits.

7.6.3.6 Horse

One investigator administered 200, 400 and 600 mg/kg respectively to 3 horses by gastric tube. At the 2 lower doses no clinically-detectable behavioural disturbances were seen; slight diarrhoea occurred at 600 mg/kg. Inhibition of plasma cholinesterase activity occurred in the animals receiving the higher doses. The post-mortem on the 3 animals, which were infested with helminths, showed no evidence of organic changes due to the treatment.

7.7 Toxicological Evaluation

Two reviews have been made of the toxicology of bromophos **(FAO/WHO 1973, 1978)**. A reproduction study in rats showed no effect at 20 mg/kg body weight per day. Short-term studies in the rat showed no effects on plasma cholinesterase at 0.63 mg/kg body weight per day. A short-term study in the dog indicated no effect on plasma cholinesterase at 1.5 mg/kg body weight per day. A long-term study in the rat did not reveal any tumours, but survival was poor at termination of the study. At the same time the Committee evaluated a long-term feeding study in rats, special studies of neurotoxicity, effects on reproduction, teratology and reports of observations of effects on man.

An adequate study in the dog dispelled the previous concern about the effect on this animal species. Plasma cholinesterase was more susceptible to inhibition than erythrocyte cholinesterase; however, the previous study revealed the opposite. A no-effect level was demonstrated in the 20-ppm group. In a recent long-term specific tumourigenic study in mice, no compound-related effect was observed. A 28-day study in human volunteers showed no significant effect of bromophos at a level of 0.4 mg/kg body weight per day based on plasma cholinesterase inhibition. An ADI for man of 0.04 mg/kg body weight was established **(FAO/WHO 1978)**.

7.8 Maximum Residue Limits

Based on the above acceptable daily intake, the Joint FAO/WHO Meeting of Experts on Pesticide Residues recommended maximum residue limits for bromophos in raw cereal grains and milled products from grain **(FAO/WHO, 1976)**. These are as follows:

Commodity	Maximum Residue Limits mg/kg
Bran (wheat)	20
Raw grain (wheat, maize, sorghum)	10
White flour, wholemeal bread	2
White bread	0.5

The residues are determined and expressed as bromophos.

The methods recommended for the analysis of bromophos residues are those of **Pesticide Analytical Manual *(1979a; 1970b; 1979c); Methodensammlung (1982; XII-*3; S5; S8–10; S13; S17; S19); Abbott *et al.* (1970); Ambrus *et al.* (1981); Panel 1980**. The following *methods are also recognised as being suitable:* ***Methoden*-sammlung (1982; 210A); Zweig (1978); Eichner (1978); Krause and Kirchhoff (1970); Mestres *et al.* (1979b); Sussons and Telling (1970); Sprecht and Tilkes (1980).**

8. Carbaryl

Carbaryl is a methyl carbamate insecticide extensively used around the world on a variety of agricultural crops, ornamentals, turf, forests, livestock, and poultry as well as on certain non-agricultural pests. It has been in use since 1958 and the strongest influence on the use pattern of carbaryl in recent years has been a marked reduction in the general use of certain low-cost organochlorine insecticides for which it is often selected as a replacement.

Though not particularly effective against the broad spectrum of stored-product pests, carbaryl is quite toxic to *Rhyzopertha dominica* and therefore has been chosen for use in combination with approved organophosphorus insecticides. It is cheap, freely available, has a low acute and chronic toxicity and is backed by extensive world-wide use in general horticulture and agriculture. It is stable under grain storage conditions for a considerable period but needs to be used at rates in the region of 5 mg/kg to be fully effective in protecting stored products against *Rhyzopertha dominica*. There is almost no penetration of carbaryl into the grain. All the deposit remains on the hulls and bran so that milled rice and flour contain only about 1% of the level on the raw grain. This residue is mainly destroyed in cooking and malting processes.

8.1 Usefulness

One of the first references to the evaluation of carbaryl against stored-product pests is the paper of **Strong and Sbur (1961)** in which they present the results from a series of tests with varying dosages of each of 36 insecticides in acetone solution sprayed on wheat for protection against adults of *Sitophilus oryzae, Sitophilus granarius, Tribolium confusum* and larvae of *Trogoderma granarium*. Mortalities of adult insects were recorded after an exposure of 14 days and of *Trogoderma* larvae after an exposure of 28 days to treated grain. The rates used ranged through 8 steps from 1.25 to 200 mg/kg. Carbaryl was one of the least effective against all 4 species giving incomplete control even at 200 mg/kg. By comparison malathion was effective against 2 species at 2.5 mg/kg and the other two at 10 mg/kg. These experiments did not specifically indicate the effectiveness of carbaryl on *Rhyzopertha dominica*.

Parkin and Forster (1964) carried out tests with carbaryl dust to test its efficacy as a grain protectant. They found that 1 mg/kg was sufficient to supress breeding of *Cryptolestes surinamensis*, 32 mg/kg was needed for *Tribolium castaneum* but 128 mg/kg was insufficient to control breeding of *Sitophilus granarius* or *Sitophilus zeamais*. They indicated that *Rhyzopertha dominica* may be sufficiently susceptible to be controllable in practice with reasonably small doses.

Extensive trials carried out in the USA, South Africa, England, the Philippines, Argentina, Brazil and Uruguay in 1963–64 showed carbaryl to be effective against a wide spectrum of stored-product pests but a number were sufficiently tolerant to require unacceptably high concentrations to bring about adequate control **(Union Carbide 1976)**. It was shown by **Roan (1964)** and **Strong and Sbur (1961)** that the nature of the formulation applied was highly critical in obtaining the desired effect against stored-product pests.

Work carried out in Australia in 1963, 1965 and 1967 **(Greening 1976)** showed that carbaryl effectively controlled *Rhyzopertha dominica* at a concentration as low as 1 mg/kg in laboratory trials and at 10 mg/kg in field trials. Since none of the organophosphorus insecticides was completely effective in controlling *Rhyzopertha dominica* under trial conditions **(Bengston *et al.* 1975; Bengston and Desmarchelier 1979)** there is a need for an insecticide, which can be combined with approved organophosphorus insecticides to complete the spectrum of control of all stored-product pests, including *Rhyzopertha dominica*. Although bioresmethrin had been used successfully for this purpose for several years, these workers believed that additional insecticides, which are from different chemical groups, should be developed.

Trials in Australia clearly demonstrated the advantage of carbaryl for the control of *Rhyzopertha dominica* **(Davies 1976; Desmarchelier 1976b; Davies and Desmarchelier 1981; Bengston *et al.* 1980b)**.

Desmarchelier (1976b) showed that carbaryl, applied to wheat at rates ranging from 3 to 6 mg/kg and held at 25°C, gave 100% control of adult *Rhyzopertha dominica* and prevented reproduction of this species for more than 6 months. Although ineffective against other species at this concentration, the addition of normal rates of other organophosphorus insecticides produced complete control of all species for more than 6 months. **Davies (1976)** and **Davies and Desmarchelier (1981)** reported equally impressive results for periods of at least 31 weeks after treatment.

Carbaryl has a number of distinct advantages for use in combination with organophosphorus grain protectant insecticides: it is cheap; it is readily available; it has been widely used and evaluated as a general insecticide for more than 20 years; it is in a different class of compound from organophosphorus or pyrethroid grain protectants, which reduces the possibility of resistance developing.

Commercially acceptable control of all major stored-product pests and complete protection of stored grains can be obtained by the use of combinations of approved organophosphorus insecticides to which carbaryl is added at rates equivalent to 5 mg/kg of treated grain.

Bengston *et al.* (1978b) reported the outcome of extensive trials with a range of grain protectants for the control of malathion-resistant insects in stored sorghum. Carbaryl applied at a nominal 8 mg/kg controlled *Rhyzopertha dominica* for more than 30 weeks in a situation where bioresmethrin (1 mg/kg), phenothrin (1 mg/kg) and pyrethrum (2 mg/kg) were only partly effective.

Desmarchelier and Bengston (1979a) described the approach taken in Australia to achieve 9 months protection against stored-product pests by using combinations of insecticides tested for their effectiveness against particular species. They recommended carbaryl applied at the rate of 8 mg/kg or 5 mg/kg in aerated storage.

Al-Saffar and Kansouh (1979) carried out a survey of stored-grain insects at Mosul, Iraq and reported that *Trogoderma granarium* was present in more than 50% of the samples. In the light of toxicity tests on the various species they reported carbaryl was less effective than dichlorvos or chlorpyrifos.

Bengston *et al.* (1980b) evaluated the combination pirimiphos-methyl plus carbaryl as a grain protectant combination for wheat. Triplicate field trials carried out on bulk wheat in commercial silos were sampled at regular intervals over 9 months and the grain samples were submitted to bioassay using malathion-resistant strains of insects. Whilst these trials indicated that the combination was not as effective as a mixture of fenitrothion and phenothrin against 2 strains, the order of effectiveness was reversed against one other strain. Against 4 additional strains, the 2 treatments were equally effective, preventing the production of progeny.

Davies and Desmarchelier (1981) in a complex series of laboratory experiments treated wheat with pirimiphos-methyl or carbaryl or combinations of these 2 insecticides and then stored it at 25°C for bioassay at various intervals over a period of 39 weeks. The insects used were *Sitophilus granarius, Tribolium confusum* and *Rhyzopertha dominica*. Pirimiphos-methyl at 5.1 mg/kg effectively controlled the first two but was ineffective against *Rhyzopertha* of a strain showing malathion resistance. Conversely, carbaryl at 6.5 mg/kg was effective against *Rhyzopertha*, but ineffective against the other 2 species. Various combinations of the 2 insecticides were tested and the combination of pirimiphos-methyl at 4–5 mg/kg and carbaryl at 5–6 mg/kg was suggested where long-term storage is required. Bioassays conducted at 20, 25, 30 and 35°C showed carbaryl to be much less effective at the higher temperatures.

Hsich *et al.* (1983) evaluated the toxicity of 26 insecticides to *Sitophilus zeamais* and *Rhyzopertha dominica* in a laboratory in Taiwan by the admixture method. Carbaryl was one of 7 insecticides which were more toxic to *Rhyzopertha dominica* than to *Sitophilus zeamais*. The LC/50 of carbaryl to *Sitophilus zeamais* was found to be 1088 mg/kg.

8.2 Degradation

Srivastava *et al.* (1970) carried out laboratory studies in India on the effect of temperature and humidity on the persistence of carbaryl, malathion and lindane. Carbaryl, at concentrations of 2000 and 6000 mg/kg, which correspond to the LD/50 and LD/90 against *Tribolium castaneum*, was used to treat filter papers to which adults of this species were periodically exposed at different combinations of temperature (35°C and 45°C) and relative humidity (about 60% and 90%). The toxicity of the deposits was lost most rapidly at 45°C and 60% r.h. Carbaryl was the first to lose its effectiveness.

From laboratory trials in Australia, **Davies (1976)** measured the decline of active carbaryl residues on wheat by bioassay using *Rhyzopertha dominica*. From his data, illustrated in **Figure 8.1**, Davies calculated the approximate biological half-life of carbaryl in wheat treated with 5 mg/kg to be 40 weeks at 35°, 60 weeks at 30°C, 80 weeks at 25°C and much longer than 80 weeks at 20°C.

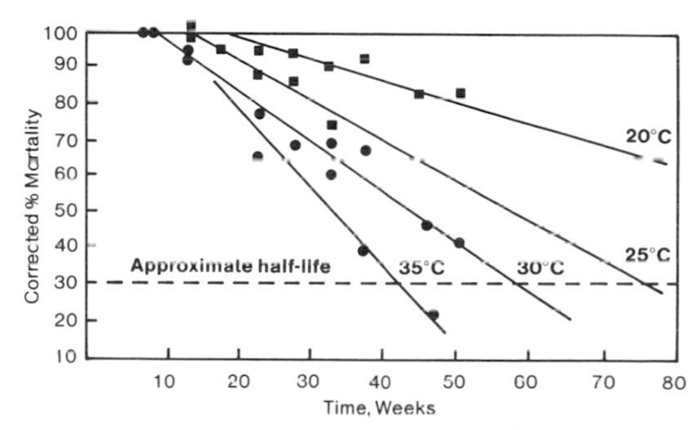

Figure 8.1. Decline of carbaryl residues in wheat after storage at various temperatures **(Davies 1976).**

Desmarchelier (1977c, 1979d, 1980) measured the rate of loss of carbaryl residues under 12 fixed sets of conditions of temperature and relative humidity in the laboratory and derived a general model relating the time required to degrade to half the former concentration to temperature and relative humidity. The model is indicated in a graph which is reproduced as **Figure 8.2**. From this model it is possible to predict the residue level after any interval of storage from a knowledge of temperature and the relative humidity of the inter-grain space and thus to determine the rate of application required to provide the requisite level of protection for the anticipated period of storage. Application of the model to extensive field trial data on wheat, barley and sorghum has confirmed the practical validity of the prediction. The use of this model

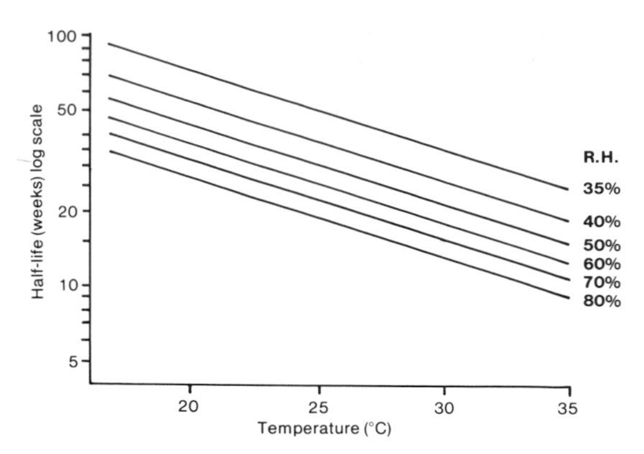

Figure 8.2. The relationship of half-life (log scale) of carbaryl on grains to equilibrium relative humidity and temperature **(Desmarchelier and Bengston 1979a).**

should enable residue levels to be maintained as low as possible consistent with obtaining adequate insect control.

Bengston *et al.* (1978b), reporting the results of commercial-scale trials, found the half life to range from 22–50 weeks in grain held at temperatures between 30°C and 22°C and an inter-grain relative humidity of from 35–49%.

Desmarchelier (1979d) compared his laboratory model with extensive data collected over periods of 6 to 10 months from 17 commercial storages containing wheat in 5 Australian States, in 2 commercial storages in Queensland containing sorghum and in 1 commercial storage containing malting barley. Half-lives were calculated for the mean of wheat residues in each State, for the mean of sorghum residues and for barley residues. These half-lives were compared with half-lives predicted from the model, and from the mean of measured field conditions of temperature and moisture content. Extremely good correlation was obtained, except in 2 cases where the number of measured residues was only 3.

Desmarchelier and Bengston (1979b) stated that chemical residues of grain protectants could be accurately predicted from simple models. Loss of protectants in bulk grain depends entirely on grain equilibrium relative humidity, grain temperature and time of storage. The basis for these models was originally published in 1978 **(Desmarchelier 1978a)**. The authors developed models for 12 protectants, including carbaryl, and these models have proved accurate in predicting residues on wheat, barley, sorghum, oats and rice in bulk. They showed how these models can be used to select the appropriate application rate for any given situation. The models demonstrated the saving of protectant which results from cooling or drying the grain. At 30°C and 50% relative humidity the half-life of carbaryl is stated to be 21 weeks. A chart relating the half-life of carbaryl residues on grain to equilibrium relative humidity and temperature was provided.

In 1980, Desmarchelier published further results of studies on the loss of bioresmethrin, carbaryl and phenothrin from wheat during storage. These studies were carried out to determine whether the stabilities of several protectants on grain, under different conditions of storage depend on only 2 variables, namely grain temperature and grain equilibrium relative humidity. Piperonyl butoxide increased the stability of bioresmethrin but had little if any effect on the persistence of carbaryl. Good agreement was demonstrated between the theoretical model and the model derived in an empirical manner from experimental data. The correlation co-efficient was generally greater than 0.9% **(Desmarchelier 1980)**.

Desmarchelier *et al.* (1980a) determined the level of residues of carbaryl and 5 other grain protectant insecticides on unhusked rice, husked rice, polished rice and barley over a storage period of 6 months. The observed levels were close to the levels predicted from the use of the model relating rate of loss of residues to a rate constant and 2 variables, temperature and equilibrium relative humidity. The barley, husked rice, unhusked rice and polished rice were treated with carbaryl at the rate of 10 mg/kg. The grain equilibrium relative humidity ranged from 55 to 65% and the storage temperature was 25°C. The carbaryl residue level at the end of 3 and 6 months was 6.5 and 3.5 mg/kg respectively on all 4 lots of grain. The deviation between the predicted and observed value of residues was less than 10%.

Bengston *et al.* (1980b) reported the results of 2 studies carried out on bulk wheat in commercial silos from which samples of treated grain were withdrawn at intervals for bioassay and chemical analysis. At these sites carbaryl was applied in combination with pirimiphos-methyl. The analytically determined residues compared very closely with those predicted from the work of **Desmarchelier (1980)** and were confirmed by the results of bioassay using 7 separate species and/or strains of stored-product insects. The initial deposit of 10 mg/kg carbaryl declined to 6 mg/kg over 30 weeks at a temperature of 27°C and an initial moisture content of 11.1%. The deposit of 5 mg/kg at the second site declined to 2 mg/kg at the end of 37 weeks on wheat having a temperature of 29.4°C.

8.3 Fate in Milling, Processing and Cooking

A number of studies are available to show the distribution and fate of carbaryl residues in various grains, milling fractions of grains and prepared cereals subjected to cooking **(FAO/WHO 1977d)**.

Desmarchelier (1976b) showed that wheat treated with carbaryl 3 to 5 months previously but still containing residues of carbaryl in the range 3–6 mg/kg, when converted into wholemeal, yielded a flour containing 1–3 mg/kg of carbaryl. When the same grain was milled for the preparation of white flour (approximately 70% extraction) the residue level in flour was reduced to about 0.1 mg/kg. The baking of wholemeal and white bread resulted in a further loss of about half the carbaryl content, bringing the final residue in wholemeal bread into the range of 1 mg/kg and in white bread to below 0.1 mg/kg. Thus the reduction in the residue between raw grain and white bread was 99%, and between grain and wholemeal bread about 75% (**Table 8.1**).

Table 8.1. Effect of milling and baking on carbaryl residues in wheat (**Desmarchelier 1976b**)

	Location N	Location B
Carbaryl applied, mg/kg	5	10
Carbaryl recovered, day 1, mg–kg	3.4	6.5
Wheat withdrawn after	19 weeks	13 weeks
Residues in wheat, mg/kg	3.1	6.0
Residues in mill fractions, mg/kg		
wholemeal flour	1.2	2.6
wholemeal bread	0.7	1.5
bran	7.0	14
shorts	1.5	3
white flour	0.07	0.15
white bread	<0.05	0.08
Reduction in residues, %		
wheat/wholemeal bread	77	75
wholemeal flour/wholemeal bread	42	42
wheat/white flour	97	97
white flour/white bread	50	50
wheat/white bread	98	99

In a series of experiments representing the most primitive type of processing to which husked rice, polished rice, oats and barley would be subjected, **Desmarchelier (1976b)** showed that simple boiling in minimal amounts of water for 15 minutes reduced carbaryl residues in husked rice, polished rice and oats by 84%. A primitive malting procedure reduced residues in barley by 77% (**Table 8.2**).

Tempone (1979) included carbaryl in extensive studies of the effects of insecticides on barley malting and vice versa. Carbaryl was applied at the rate of 10 mg/kg in combination with pirimiphos-methyl. No detrimental effects were observed in malts with this treatment. Residues were minimal in the malts initially and not detectable after 3 months storage.

Bengston *et al.* (1980b) reporting the results of extensive studies, which included a combination of pimimiphos-methyl plus carbaryl for the treatment of wheat, showed that the bulk of the carbaryl residues was removed in the bran and pollard — very little finding its way into white flour. Whilst there appeared to be some loss of carbaryl during the milling of wheat into wholemeal, the baking of wholemeal bread caused substantial further reduction in the carbaryl residue level. The details are given in **Tables 8.3 and 8.4.**

Desmarchelier *et al.* (1980a) studied the fate of residues of carbaryl and 5 other grain protectant insecticides applied to barley and to rice in husk, husked rice, and polished rice and the effect of malting of barley, milling of rice and cooking of the various grades of rice on the level of carbaryl residues. These studies showed that 90 to 95% of the residue present on milled and polished rice is destroyed by cooking, that about 95% of the residue present on rice in husk is removed by milling and most of the remaining residues are removed during the

Table 8.2. Effect of storage at 25°C, processing and cooking on carbaryl residues in treated grains (**Desmarchelier 1976b**)

Grain	Moisture (%)	Application rate, (mg/kg)	Residue (mg/kg) after storage for 3 months	6 months	Processed after (months)	Processed into	Residue (mg/kg) after cooking for –	15 min	25 min
Barley	13	10.6	6.5	3.5	3	Primitive malt	1.5		
					6	Commercial malt	0.2		
Oats	12	10.0	7.5	3.5	3	Rolled oats	–	1.2	
Rice in husk	13	10.0	7.5	3.5	6	Husked	0.4	0.2	
					6	Milled	0.07		<0.05
Husked rice	12.5	10.0	7.5	3.4	3	Cooked		1.2	
					6	Cooked			0.7
Polished rice	12.7	10.0	7.5	3.5	3	Cooked		1.2	
Wheat	12	10.0	7.2	6.3 (4.2 after storage for 9 months)	5	Wholemeal	2.5		1.5

Table 8.3. Mean carbaryl residue levels following milling and baking of wheat stored at Site B for 13 weeks or at Site D for 19 weeks after being treated with carbaryl suspension **(Bengston *et al.* 1980b)**

Site	Storage (weeks)	Carbaryl residue (mg–kg)						
		Wheat	Bran	Pollard	Wholemeal flour	White flour	Wholemeal bread	White bread
Site B	13	5-7	12-16	2-4	2.6	0.1-0.2	1.0-2.0	0.05-0.1
Site D	19	2.5-3.0	6-8	1-2	1.2	0.05-0.1	0.5-1.0	0.05

Table 8.4. Reduction in carbaryl residue content following milling and baking of wheat referred to in Table 8.3

Site	% reduction in residue content					
	Wheat to wholemeal	Wheat to white flour	Wholemeal to wholemeal bread	White flour to white bread	Wheat to wholemeal bread	Wheat to white bread
Site B	57	98	50	40	75	99
Site D	53	96	–	38	67	98

polishing process. Virtually all of the residue present on the rice in husk is lost when the milled and polished rice is cooked. More than 90% of the carbaryl present on barley is destroyed in the malting process. These results are set out in **Table 8.5.**

Bengston *et al.* (1983a), who carried out duplicate experiments on bulk sorghum stored in Queensland, found that carbaryl, applied in conjunction with pirimiphos-methyl at the rate of 8 mg/kg had declined to 5 mg/kg by the end of the fifth week after application. Thereafter the rate of decline in the concentration of carbaryl residues was quite smooth and regular so that at

Table 8.5. Residues of carbaryl on barley and rice before and after storage, processing and cooking **(Desmarchelier *et al.* 1980a)**

Grain	Application level (mg/kg)	Residues (mg/kg)			
		After 3 months	After 6 months	After cooking at 3 months	After cooking at 6 months
Husked rice	10	6.5	3.5	1.2	0.7
Polished rice	10	7.0	4.5	1.2	0.4

Grain	Application level (mg/kg)	After 3 months	After 6 months	Residues after 6 months storage and			
				Milling		Milling & cooking	
				Milled	Polished	Milled	Polished
Rice in husk	10	6.5	3.3	0.45	0.07	0.1	0.05

Grain	Application level (mg/kg)	After 3 months	After 6 months	Residues after malting	
				Home malting after 3 months	Commercial after 6 months
Barley	10	6.5	3.5	1.2	0.25

the end of 24 weeks the residue concentration was 0.6 mg/kg. The rate of degradation of carbaryl was considerably greater than 7 other compounds included in these studies. The grain temperature was 27°C and the average moisture content 12.3%.

8.4 Metabolism

There is an extensive literature on the metabolism of carbaryl in plants, animals and soil. Much of this has been evaluated by the JMPR on a number of occasions since 1969 and a comprehensive review is provided by **Kuhr and Dorough (1976)**.

The metabolic pathway **(Fig. 8.3)** for carbaryl in plants is identical whether the compound is introduced by stem injection or applied to the leaf surface. Laboratory or field conditions do not affect or alter the metabolic course. Only after it gains entrance into the plant tissue does carbaryl undergo biotransformation to its primary metabolites, which are similar to the ones formed by animals. The major metabolites in decreasing order of quantitative importance are: 1-naphthyl N-hydroxymethylcarbamate (or the methylol of carbaryl), 7-hydroxycarbaryl and 4-hydroxycarbaryl; the minor ones: 5-hydroxycarbaryl, 1-naphthol and 5,6-dihydro-5,6-dihydroxycarbaryl. These hydroxylated metabolites, which are less toxic than carbaryl itself, are conjugated by plants to form water-soluble glycosides. The formation of the primary metabolites appears to be the rate-limiting step and subsequent conjugation a rapid conversion, since the primary metabolites are not found free in plant tissue **(Dorough and Casida 1964; Kuhr and Dorough 1976).**

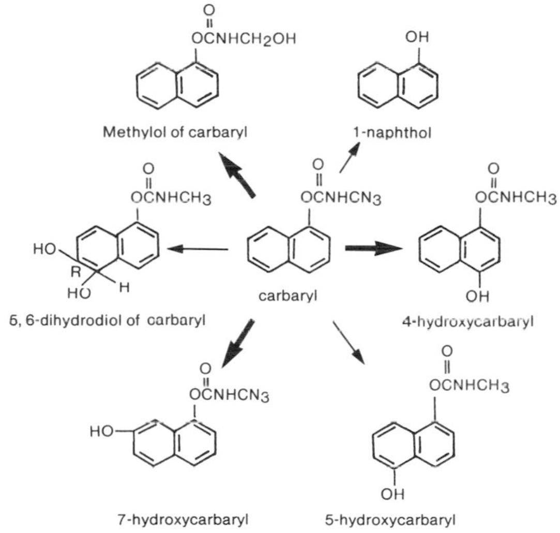

Figure 8.3. Primary metabolism of carbaryl in plants. The primary metabolites, which are hydroxylated compounds, are found not in the free state but conjugated as glycosides. Heavy arrows represent the major conversions; light arrows the minor conversions **(FAO/WHO 1970).**

8.5 Fate in Animals

Extensive information on the fate of carbaryl residues in domestic animals has been provided in the monographs in the 1966, 1968 and 1973 meetings of the Joint FAO/WHO Meeting on Pesticide Residues. From these monographs it is obvious that many separate investigations have verified that when carbaryl is included in the ration of cows only about 0.2% of the amount of carbaryl ingested is excreted in milk as carbaryl and metabolites. At least 7 separate metabolites have been identified in milk, most of them water-soluble. Carbaryl itself represents only about 5% of the total residue following continuous feeding at levels of 100 ppm in the ration (equivalent to about 1.5 mg/kg body weight). Since it is very unlikely that any dairy animal would ever consume as much as 100 mg of unchanged carbaryl per kg of the entire ration every day, actual milk residues would be negligibly small. Radiotracer studies have shown that when cows are fed the equivalent of 100 ppm of carbaryl in their ration, residues of all fragments do not exceed 1 mg/kg in kidney, 0.4 mg/kg in liver and 0.1 mg/kg in the muscle. Only about 3–17% of these residues is parent carbaryl.

Gyrisco *et al.* (1960) fed up to 450 ppm of carbaryl to cows for 2 weeks and found no residue in milk by a colorimetric method.

Dorough (1967) fed ring-labelled carbaryl to a lactating cow. A total of approximately 1 mg/kg, based upon a radioactivity measurement calculated as carbaryl, was found in the skim milk 6 to 12 hours after administration of 3.05 mg/kg body weight. Approximately one half of this residue was chloroform extractable, and one half was water extractable with a small unextractable fraction. At the end of 60 hours all the radioactivity (calculated to be 0.01 mg/kg as equivalents of carbaryl) was unextractable from the milk. In the 6-hour samples about 30% of the radioactivity in the milk was characterised as 5,6-dihydro-5,6-dihydroxy 1-naphthol N-methylcarbamate, and approximately 20% was an unknown metabolite. Neither of these materials, corresponding to 50% of the total milk residue (0.5 mg/kg), responded to the standard colorimetric method used for carbaryl.

No residues were detected in tissues of cattle fed carbaryl for 20 days **(Clayborn 1963). Whitehurst *et al.* (1963)**, using the colorimetric method of analysis sensitive to carbaryl, 1-naphthol and 1-naphthol conjugates concluded that, when cows were fed on a diet containing carbaryl, no residues, to the limit of the method, were found in the milk.

Dorough (1967) and **Baron *et al.* (1969)** fed radiolabelled carbaryl to lactating cows and found very small amounts of radioactivity in milk and this was proved to be not from carbaryl. **Baron *et al.* (1969)** found that the radioactivity was incorporated into the lactose of milk.

After oral administration of single doses of 1-naphthyl-^{14}C-carbaryl at levels of 0.25 and 3.05 mg/kg, approximately 0.35% of each dose was detected in the milk **(Dorough 1970)**. Maximum concentrations which, following the 2 treatments, were 0.063 and 0.95 mg/kg, respectively, were found in samples taken 6 hours after dosing. In another study, 1-naphthyl-^{14}C-carbaryl was fed to lactating cows at levels of 0.15, 0.43 and 1.35 mg/kg body weight (equivalent to 10, 30 and 100 ppm in the feed) for 14 days **(Dorough 1971)**. Equilibrium between intake and elimination was reached within 2 days of initiation of the treatment. At each feeding level, approximately 0.2% of the dose was secreted in the milk. The concentration of the total ^{14}C-carbaryl equivalents in the milk was 1/400 of that in the diet. Most of the ^{14}C-residues were in the aqueous phase.

Continuous feeding of 1-naphthyl-^{14}C-carbaryl to cows and a single oral dose of the same material demonstrated that carbaryl residues do not accumulate in the body tissues **(Dorough 1971)**. Ingested carbaryl is rapidly metabolised in cows and other animals, 70–80% being excreted in urine within 24 hours. In a continuous feeding study in cows, equilibration of total radioactive residues in milk, urine and feces occurred by the second day. Within 18 hours after the last of 14 days continuous feeding, the highest total ^{14}C-residues were found in the kidneys. The lowest residues were found in fat, indicating that metabolites are not stored in body tissues. The major components of the residue in tissues were carbaryl, naphthol, naphthyl sulphate, 5,6-dihydro-dihydroxycarbaryl and 5,6-dihydro-dihydroxynaphthol. Of the radioactivity appearing in the milk, carbaryl *per se* comprises less than 10%.

Saivaraj *et al.* (1977) studied the level and fate of carbaryl residues in maize fodder and after feeding the fodder to milch cows. Maize plants sprayed with a 0.1% solution of carbaryl showed an initial concentration of 53 mg/kg. This declined rapidly within 7 days with a half-life of 1.44 days. Carbaryl residues of up to 0.11 mg/kg were observed in milk from cows fed the carbaryl-treated maize at the rate of 60 kg/day. These residues declined rapidly and disappeared within 7 days. This is indicative of what might be expected when carbaryl-treated grain was fed to cattle.

Carbaryl in doses of 180 mg and 540 mg/kg of body weight per day was administered to laying White Leghorn hens for 60 consecutive days. No detectable residues of the insecticide or its naphthol metabolites were found in the meat when the low dose was administered. Residues in the meat and histopathological changes in various organs were found when the high dose was given. Residues were present in the fat tissue with both the high and the low dose **(Nir *et al.* 1966)**.

Following administration of 1-naphthyl-^{14}C-carbaryl to hens, total ^{14}C-residues reached a maximum and dissipated at a much faster rate in egg white than in egg yolk. **Paulson and Feil (1969)** showed that, following a single dose of 10 mg/kg, the maximum concentration of ^{14}C-residues in egg white was 0.12 mg/kg at day one and this dropped to trace amounts on the second day after treatment. The yolk residues reached a maximum at the fifth day (0.36 mg/kg) and had dissipated by the ninth day (0.03 mg/kg). Under continuous feeding conditions, the total residue in the yolk or white at each sampling time was dosage related **(Andrawes *et al.* 1972)**. Concentrations of ^{14}C-carbaryl equivalents reached a maximum (0.1 mg/kg from 70 ppm in feed; 0.025 mg/kg from 21 ppm in feed) in the white after 2–6 days and in the yolk (1.0 mg/kg from 70 ppm in feed; 0.3 mg/kg from 21 ppm in feed) after 6–9 days of dosing and remained level until the end of the treatment period. At plateau levels, the level of ^{14}C-carbaryl equivalents in the white was 1/10 that in the yolk; however, the total equivalents were in the ratio of 5:1 between yolk and white. The ratio of the concentration of carbaryl in whole eggs (white and yolk) to that in the diet was 0.006 at equilibration. After discontinuation of dosing, residues in the whites had a half-life of less than 1 day; for yolk residues the half life was approximately 2–3 days.

The distribution of carbaryl residues was determined in

Table 8.6. Acute oral toxicity of carbaryl to animals

Animal	Sex	Vehicle	LD/50 (mg/kg)	Reference
Rats	M	10% Tween 80	190	**Mellon Institute 1958**
	M	" + 0.75% NaCl	310	"
	M + F	0.25% agar	480–610	"
	F	Corn oil	560	"
	M	Corn oil	308	"
	M	Peanut oil	850	**Gaines 1960**
	F	Peanut oil	500	"
	M + F	Sunflower oil	515	**Rybakova 1966**
Guinea pigs	M	0.25% agar	280	**Mellon Institute 1958**
Dogs	M + F	Powder	>795	
Rabbit	M	0.25% agar	707	
Mouse	M + F	Sunflower oil	438	**Rybakova 1966**

hen tissues after continuous treatment with either 7, 21 or 70 ppm of 1-naphthyl-^{14}C in the diet (**Andrawes *et al.* 1972**). Tissue residues were directly proportional to the concentration of carbaryl in the diet. The highest residues were found in the blood and tissues of high blood content (liver, kidney, lung and spleen); body fat, brain and muscles contained the lowest residues.

8.6 Safety to Livestock

In view of the transitory nature of any effects on cholinesterase inhibition and the rapidity with which carbaryl is metabolised in animals, including livestock and poultry, the likelihood of adverse effects from the consumption of residues of carbaryl on grain or milling offals is so remote as to warrant no concern. The acute oral toxicity of carbaryl to laboratory animals is given in **Table 8.6.**

Carbaryl in doses of 180 mg and 540 mg/kg of body weight per day were administered to laying White Leghorn hens for 60 consecutive days. No noticeable effects were found (**Nir *et al.* 1966**).

Focal loss of striation and fatty infiltration of muscle of hens were observed at 3 g/kg administered subcutaneously. Transient leg weakness for 1–2 days occurred after 2 g/kg and a nephrotoxic action was observed after 2 g/kg or more. No demyelination was seen (**Carpenter *et al.* 1961**).

Ghadiri *et al.* (1967) and **Ghadiri (1968)** indicated that carbaryl fed to chickens at 75 to 600 ppm for 3 weeks caused embryonic deformities.

Khamelevskii (1968) fed carbaryl at levels up to 14 400 ppm to chickens for 6 months with no signs of toxicity. However, it tended to decrease cholinesterase activity at this level.

Gaines (1969) reported that the lethal dose of carbaryl to chickens was 800 mg/kg body weight.

Carbaryl gave no adverse effect when fed to chickens at levels as high as 500 ppm for 36 weeks (**Lillie 1972**).

De Witt and Menzie (1972) found that when carbaryl was fed to quail at 25 ppm or to pheasants at 1000 ppm it caused a reduction in hatchability.

Lillie (1973) fed caged White Leghorn pullets a breeder diet supplemented with 250 or 500 ppm malathion and/or carbaryl. Body weight changes, egg production, egg weights, specific gravity of eggs, feed consumption, mortality, fertility, hatchability, embryonic abnormalities and progeny performance were studied. In the progeny performance studies, progeny from hens fed 500 ppm malathion and/or carbaryl were fed a boiler diet supplemented with 500 ppm malathion or carbaryl for a 4-week period in batteries. The only significant differences resulting from malathion and or carbaryl supplementation in the feed were pullet weights and 4-week progeny weights. The pullet weight gains were significantly reduced by carbaryl with or without malathion. A significant growth depression was observed with the progeny fed carbaryl, irrespective of maternal diet. In a separate 4-week study, the incorporation of 500 ppm malathion and/or carbaryl in the caged Leghorn male diet exerted no significant changes in the fertility pattern or in the incidence of sperm and embryonic abnormalities.

Carbaryl was fed to hens at 0, 250 and 500 ppm in the diet for 36 weeks and to their progeny for 4 weeks at 0 and 500 ppm, either alone or in combination with malathion. Growth was affected in both parents and chicks, but reproduction and egg characteristics were not affected. A study at 500 ppm in males showed no effect on fertility over a 4-week period. In hens fed carbaryl at 500 ppm for 36 weeks, no effects on hens or progeny were observed except a slight weight loss and growth depression. Administration of carbaryl to male Leghorns resulted in no effects on reproduction (**Lillie 1973**).

Three groups of sheep commenced feeding on diets containing 0 (23 sheep), 100 (26 sheep), or 250 (23 sheep) ppm carbaryl, 4 days prior to pairing. Thirty sheep were lame at pairing, due to a viral polyarthritis, and thus all animals were treated with 450 000 units of benzathine penicillin just after pairing commenced. Incidence of pregnancy was comparable between groups and resulted in 25, 22 and 24 offspring (of which 15, 14 and 16 were delivered by Caesarean section prior to parturition) from the 0, 100 and 250 ppm levels respectively. Abnormalities were restricted to the 250 ppm group, where one Caesarean, and one normal parturition lamb were found to show intraventricular septal defects in the heart (**Panciera 1967**).

Carbaryl is a reversible cholinesterase inhibiter. In fact, the reversal is so rapid that unless special precautions are taken, measurements of blood cholinesterase of animals or persons exposed to carbaryl are likely to be inaccurate and always tending to appear normal. Pyridine 2 aldox ime methiodide (PAM), which is a good antidote for some organophosphorus compounds, is not effective in reversing cholinesterase inhibition by carbaryl. Atropine sulphate is effective in controlling symptoms.

The signs and symptoms of carbaryl poisoning in mammals can be appreciated by examining the results reported when swine were fed large doses of carbaryl (**Smalley *et al.* 1969, Smalley 1970**). Single oral treatments of 1000 or 2000 mg/kg bw produced salivation, vomiting, difficult breathing, central nervous-system depression, muscle tremors and cyanosis. When 13-week-old pigs were fed a diet containing carbaryl, 150 to 300 mg/kg bw per day, for up to 12 weeks, intoxication was evidenced by progressive myasthenia, inco-ordination, ataxia, and clonic muscular contractions. Continued exposure of the animals to carbaryl resulted in paraplegia and prostration.

8.7 Toxicological Evaluation

The toxicology of carbaryl has been reviewed 7 times by the Joint FAO/WHO Meeting on Pesticide Residues. Carbaryl is metabolised by a similar route in the rat, guinea pig, sheep, pig, monkey and man. However, it is metabolised in a different manner in the dog.

Several studies on the effect of carbaryl on reproduction were reviewed. No effect on reproduction was observed in Rhesus monkeys. Studies in several species of animals showed that administration by gavage is more likely to affect reproduction than administration in the diet. Numerous long-term feeding studies using mice, rats and dogs were considered to provide toxicological data sufficient for recommending an ADI for man of 0.01 mg/kg body weight. This was based on no effect levels determined in rat (10 mg/kg body weight per day) and man (0.06 mg/kg body weight per day).

8.8 Maximum Residue Limits

Based on the acceptable daily intake established some years previously and on the basis of information indicating that carbaryl was needed for the protection of grain, the 1976 Joint FAO/WHO Meeting of Experts on Pesticide Residues evaluated extensive information on the level and fate of carbaryl residues, resulting from both pre-harvest and post-harvest application of carbaryl to cereal grains and recommended the following maximum residue limits. These were designed to cover residues resulting from either pre-harvest or post-harvest use of carbaryl. They refer to carbaryl only, excluding metabolites:

Commodity	Maximum Residue Limits (mg/kg)
Bran	20
Barley, oats, rice (in husk and hulled), rye, wheat	5
Sorghum	10
Wholemeal flour	2
Wheat flour (white)	0.2

The residues are determined and expressed as carbaryl.

The level and fate of carbaryl residues in processed cereal products were summarised in the accompanying monograph but, in keeping with accepted practice, separate limits were not proposed for residues in such processed foods.

The methods recommended for the analysis of carbaryl residues are those of: **AOAC (1980b; 1980c); Pesticide Analytical Manual (1979c; 1979e); Manual of Analytical Methods (1984); Ambrus *et al.* (1981)**. The following methods are also recognised as being suitable: **Pesticide Analytical Manual (1979d); Methodensammlung (1982, 100); Cohen *et al.* (1970); Funch (1981); Lawrence (1977); Mestres *et al.* (1979a).**

9. Chlorpyrifos-methyl

Chlorpyrifos-methyl is a broad-spectrum organophosphorus insecticide of relatively low toxicity and moderate persistence. It shows reasonably good stability in stored products such as grains and dried fruits. In these products it controls a wide spectrum of beetles, weevils, moths and mites, including several such species which may have developed resistance towards other insecticides, e.g. malathion.

Chlorpyrifos-methyl is potent against all storage pests except resistant *Rhyzopertha dominica*. It is effective against moths which are not readily controlled by malathion. Deposits on grain are stable under most storage conditions. The deposit does not readily penetrate into the grain so that more than 90% is removed on the bran. Of the 10% carried forward into flour, much more than half is destroyed in cooking. This results in a 98% reduction between wheat and white bread and an 83% reduction between wheat and wholemeal bread. Only a small residue remains in malt after treated barley has been malted.

9.1 Usefulness

Chlorpyrifos-methyl was first introduced as a potentially useful grain protectant insecticide under the designation DOWCO 214. It has since been marketed under the trademark RELDAN.

Quite a lot of the early work was carried out in France and this led to the registration there of chlorpyrifos-methyl as an approved grain protectant.

Coulon (1972) published the results of studies involving 10 new substances belonging either to the organophosphorus or pyrethroid group. Chlorpyrifos-methyl appeared capable of destroying a high proportion of the insect population and of remaining effective over a considerable period.

Schulten (1973) considered chlorpyrifos-methyl effective for controlling *Ephestia cautella* in stacks of bagged maize in Malawi.

Morallo-Rejesus and Carino (1974), in a study to determine the residual toxicity of 5 organophosphorus insecticides against 2 pests of stored corn, reported chlorpyrifos-methyl more potent than tetrachlorvinphos, pirimiphos-methyl, malathion and MIPC against *Sitophilus* spp. However, against *Rhyzopertha dominica*, chlorpyrifos-methyl was not as effective as tetrachlorvinphos. In general, the residual toxicity and the percentage mortality increased with increase in concentration. Chlorpyrifos-methyl had the longest residual activity.

Coulon and Barres (1975) carried out laboratory studies on the biological persistence of chlorpyrifos-methyl against *Sitophilus granarius* at rates equivalent to 5 mg/kg in the presence of grain with normal moisture levels.

LaHue (1975a) evaluated chlorpyrifos-methyl, pirimiphos-methyl, malathion, fenitrothion as sprays and malathion as a dust for the control of *Sitotroga cerealella* in stored maize. Chlorpyrifos-methyl, pirimiphos-methyl and fenitrothion were all superior to malathion and gave good control and protection for up to 8 months.

Bengston *et al.* (1975) carried out a very extensive and detailed comparison of chlorpyrifos-methyl, bioresmethrin and pirimiphos-methyl as grain protectants against malathion-resistant insects in wheat. Chlorpyrifos-methyl proved more potent than pirimiphos-methyl and malathion against all species, although all 3 were relatively weak against *Rhyzopertha dominica*. *R. dominica* could only be controlled satisfactorily with bioresmethrin. The authors suggested the use of combinations of chlorpyrifos-methyl with bioresmethrin for use as a grain protectant, effective against all species.

The Queensland Department of Primary Industries (1975) announced that chlorpyrifos-methyl was being evaluated, along with a number of other insecticides, for the control of malathion-resistant insect pests, which pose a major threat to the grain industries. (These materials, including chlorpyrifos-methyl, are not considered adequately effective against *Rhyzopertha dominica* and the addition of bioresmethrin or other synthetic pyrethroid insecticide appears necessary.)

Bitran *et al.* (1975) studied the residual effect of 2 pyrethroids and chlorpyrifos-methyl in coffee.

LaHue (1976) compared the performance of chlorpyrifos-methyl, pirimiphos-methyl and fenitrothion with malathion as broad-spectrum grain protectants of seed corn over a 21 month period. For the first 4 months after treatment and for the eighth and sixteenth month, the seed corn was kept in cold storage to represent conditions prevailing throughout the corn growing area of the USA. Chlorpyrifos-methyl, applied at the rate of 6.7 mg/kg, was outstandingly effective against *Sitophilus oryzae, Sitophilus zeamais* and *Sitophilus granarius* over the first 16 months but thereafter the mortality of adult insects declined. However, no progeny were produced and no damage occurred in the corn seed.

LaHue and Dicke (1976a, b) evaluated chlorpyrifos-methyl, pirimiphos-methyl and fenitrothion against malathion for preventing insect attack on high-moisture sorghum grain. Six insect pests were involved. Chlorpyrifos-methyl gave excellent protection for 12 months, and it was the most effective treatment against *Rhyzopertha dominica* in bioassays conducted at the termination of the study. Pirimiphos-methyl was rated slightly superior against other species.

Morallo-Rejesus and Carino (1976a, b) evaluated the residual toxicity of 5 insecticides on 3 varieties of corn and sorghum under conditions prevailing in the Philippines. Application was at the rate of 10, 20, 30 and 50 mg/kg. Pronounced differences in the susceptibility of the test insects to the insecticides were noted. Chlorpyrifos-methyl was effective against *Sitophilus zeamais* and *Rhyzopertha dominica*, showing a longer residual activity on both maize and sorghum than the other insecticides.

Bengston *et al.* (1978b, 1979a, 1979b) reported the results of extensive field trials to compare chlorpyrifos-methyl against 4 other organophosphorus insecticides for the control of malathion-resistant insects infesting wheat in Australia. It proved particularly effective, comparing well with both fenitrothion and pirimiphos-methyl. However, none of the materials was considered adequately effective against *Rhyzopertha dominica* without the addition of an insecticide specifically effective against this insect.

Field tests carried out with a range of grain protectants for seed corn against eight common stored-product pests showed chlorpyrifos-methyl to be second only to pirimiphos-methyl in effectiveness when applied as an emulsion spray **(LaHue 1977c).**

LaHue and Dicke (1977) reported similar studies against stored-grain insects attacking wheat. The results were generally similar to those on maize.

Chlorpyrifos-methyl was among 6 insecticides evaluated in Malaysia against stored-product insects **(Loong Fatt Lim and Sudderuddin 1977)**. It proved particularly effective being more potent at 28°C than at 18°C.

LaHue (1977d) applied chlorpyrifos-methyl to clean wheat as a water emulsion at doses ranging from 1–10 mg/kg and checked the effectiveness of residues in samples withdrawn at intervals during a 12 month storage. These showed that initial applications of 3 mg/kg or more controlled *Sitophilus oryzae, Sitophilus granarius* and *Sitophilus zeamais*; however, 8 mg/kg was required to control 90% of *Rhyzopertha dominica*, 83% of *Tribolium confusum*, and 98% of *Tribolium castaneum*, 12 months after application.

Desmarchelier (1978c) used a series of mathematical expressions to describe and interpret biological and chemical residue data obtained from a series of experiments in which chlorpyrifos-methyl, pirimiphos-methyl and dichlorvos were applied, in a range of concentrations to wheat which was held at 20°C, 25°C and 30°C and tested over a period of 12 weeks. The experiments were designed to test 9 predictions derived from the mathematical expressions and related to the loss of insecticidal activity with time. The rapid initial loss of toxicity is attributed to rapid loss of chlorpyrifos-methyl from the surface of the grain kernels.

Quinlan (1978) compared chlorpyrifos-methyl and malathion as protectants for high-moisture stored wheat held in plywood bins. The wheat was sampled at various intervals over a 9-month period to examine for live insects present and for bioassay. Chlorpyrifos-methyl was found to be the more effective of the two and no live insects were found in grain treated with this compound during the 9-month observation period.

A laboratory evaluation of malathion, chlorpyrifos and chlorpyrifos-methyl, for use against beetles infesting stored wheat **(Williams *et al.* 1978)**, showed chlorpyrifos-methyl to be the most toxic compound to the 5 species tested.

In discussing the residual behaviour of chemicals on stored grain, **Desmarchelier and Bengston (1979a, b)** stated that chlorpyrifos-methyl at the rate of 10 mg/kg was satisfactory for all species except *Rhyzopertha dominica* and that the cost of such treatment is of the order of 8 cents per tonne.

Bengston and Desmarchelier (1979) stated that chlorpyrifos-methyl used at the rate of 10 mg/kg was effective against most species but not against resistant *Rhyzopertha dominica,* which required the addition of bioresmethrin (1 mg/kg) or similar pyrethroid.

Quinlan *et al.* (1979) evaluated the effectiveness of chlorpyrifos-methyl and malathion as protectants for high-moisture stored wheat in the USA and found that chlorpyrifos-methyl was more effective than malathion throughout a 9-month period in controlling insects on 14.6% moisture wheat. It also persisted longer on the wheat than did malathion.

Al-Saffar and Kansouh (1979) carried out a survey of stored-grain insects in wheat silos at Mosul, Iraq. The more abundant species in wheat was *Trogoderma granarium* which was present in more than 50% of the samples. In a series of toxicity tests with a variety of insecticides, chlorpyrifos-methyl proved to be the most toxic to larvae of *Trogoderma granarium.*

Coulon *et al.* (1979a) compared the activities of methacrifos and chlorpyrifos-methyl against *Sitophilus granarius*. Methacrifos was slightly more effective than chlorpyrifos-methyl in controlling *Sitophilus granarius* when applied to infested wheat at 2.5 mg/kg. When applied preventatively 2.5 mg/kg chlorpyrifos gave an 18-month protection whereas methacrifos protected the wheat only for 6 months.

Coulon *et al.* (1979b) studied the residual activity of methacrifos and chlorpyrifos-methyl applied to wheat and showed that at least 1.8 mg/kg of chlorpyrifos-methyl was required to be present on the grains to prevent infestation by *Sitophilus granarius*. To achieve this level at the end of 9 months requires the application of 4 mg/kg when the wheat is put into store.

Bengston *et al.* (1980a) reported extensive field trials with various insecticide combinations applied to bulk wheat in commercial silos throughout Australia. Samples of treated grain were subjected to laboratory bioassay at intervals over 8 months. Chlorpyrifos-methyl (10 mg/kg) plus bioresmethrin (1 mg/kg) proved effective against all species, completely preventing the production of progeny.

The effectiveness of chlorpyrifos-methyl, pirimiphos-methyl and fenitrothion was compared with that of malathion as protectants against insect pests of stored rice **(Cogburn 1981)**, and attention was drawn to the advantages of using an encapsulated formulation.

Viljoen *et al.* (1981a) reported the results of a series of trials to evaluate the protection against re-infestation of maize and groundnuts in bag stacks. Small bag stacks of maize and groundnuts in the open and in a shed in South Africa were sprayed with various contact insecticides directly after fumigation and thereafter at intervals of 4 weeks. A sample was taken to determine the degree of damage caused by insects to the kernels, and the rest of the sample was kept for 2 months in order to note the numbers of insects that developed in them. Chlorpyrifos-methyl was slightly less effective against *Ephestia cautella* and *Plodia interpunctella* than were tetrachlorvinphos and a mixture of diazinon with pyrethrins but more effective than these materials against *Tribolium castaneum*. These workers considered that chlorpyrifos-methyl was the ideal substitute for malathion for bag-stack spraying.

(Dow 1981) published a summary of some of the data available on the biological activity of chlorpyrifos-methyl against stored-product insects in Indonesia.

Barson (1983) reported the results of a series of experiments to determine the effects of temperature and humidity on the toxicity of chlorpyrifos-methyl to adult *Oryzaephilus surinamensis*. The toxicity, to a UK laboratory susceptible strain and a resistant strain originally from India, of water-dispersable powder formulations of chlorpyrifos-methyl under constant conditions of 25°C and 70% r.h. were compared to the toxicities when the pests were exposed to a diurnal cycle of 12.5°, 20°, 12.5°C and 70%, 50%, 70% r.h. to simulate storage conditions in the UK during spring and autumn. The insecticide was more effective at 25°C and 70% r.h. The LD/50 value for the susceptible strain was 4.4 and 1.4 mg/kg/m² at 12.5–20°C and 25°C respectively. The LD/50 values obtained from the two sets of environmental conditions for the resistant strain differed by a factor of 1.8. Toxicity tests were also made with chlorpyrifos-methyl under various constant conditions of temperature and humidity from 5–30°C at 5°C intervals and 30, 50, 70 and 90% r.h., and also at 0°C and 60% r.h. Chlorpyrifos-methyl was very effective and there was little or no cross-resistance to it in the resistant strain. From 15°C to 30°C, mortality was high and the differences in mortality at the LD/50 level for the various humidities were slight, but there was a decrease in the mortality with decreasing humidity at any one temperature, in particular at 5°C, 50% and 70% r.h. and at 10°C and 50% r.h. Chlorpyrifos-methyl was more toxic to both strains at the highest humidity (90%) throughout the whole temperature range. The LD/50 values for each strain decreased at each temperature as the water vapour concentration was increased.

Duplicate field experiments were carried out in commercial silos containing sorghum **(Bengston *et al.* 1983a)**. Five combinations of organophosphorus and pyrethroid insecticides were used, including chlorpyrifos-methyl (10 mg/kg) plus pyrethrum (1.5 mg/kg). Bioassays of treated grain conducted during 6 months storage established that chlorpyrifos-methyl (10 mg/kg) controlled prevalent strains of *Sitophilus oryzae, Tribolium castaneum* and *Ephestia cautella* but not *Rhyzopertha dominica*. Carbaryl proved an effective additive against *R. dominica* but at the rates used pyrethrum and the pyrethroids were only partially effective.

Golob *et al.* (1985) treated *Prostephanus truncatus* adults with chlorpyrifos-methyl by topical application, by exposure to treated filter papers, and by exposure to maize grain treated with dilute dust formulations. They compared its performance with that of 7 other grain protectant insecticides. Chlorpyrifos-methyl out-performed the other organophosphorus insecticides but was inferior to permethrin. Chlorpyrifos-methyl treated maize showed the lowest percentage weight loss at different sampling intervals up to 10 months after treatment.

9.2 Degradation

LaHue (1974) was one of the first to publish information on the degradation of chlorpyrifos-methyl deposits on grain. He showed that a deposit of 6 mg/kg of chlorpyrifos-methyl, applied to wheat with 12.5% moisture, stored at a temperature of 27°C, declined to 2.7 mg/kg in 6 months and 1.6 mg/kg in 12 months. This was substantially the same as the rate of decline of the deposits of malathion and fenitrothion in the same trial but considerably more rapid than pirimiphos-methyl. Parallel trials carried out on maize with 12.5% moisture, stored at 27°C under identical conditions showed a similar rate of decline for chlorpyrifos-methyl and malathion but, in the case of fenitrothion, the degradation was significantly more rapid. Deposits of the same insecticides, applied to sorghum with 17.6% moisture, decomposed at rates comparable to those found on wheat and maize, except in the case of malathion where the rate of degradation was considerably more rapid on the sorghum.

In his *Review of the World Literature on the Metabolism of Insecticides in Grain*, **Rowlands (1975)** noted that for the period 1970–74 there had been no published work on the metabolism of chlorpyrifos-methyl in grain. **Rowlands and Wilkin (1975)** found in laboratory tests that an initial treatment of 10 mg/kg applied to wheat of 14 and 18% moisture and stored at 20°C for 6 months had degraded by some 15 and 35% respectively. Breakdown products identified included dimethyl phosphorothioate,

3,5,6-trichloro-2-pyridinol, and the O-dimethyl derivative of the parent compound.

Bengston *et al.* (1975), carried out experiments in which treated wheat was stored under conditions typical of bulk storage in the field, following the application of chlorpyrifos-methyl at the rate of 2 mg/kg and 4 mg/kg. They estimated the half-life at 19.6 and 18.2 weeks respectively when applied to wheat of 12% moisture and an initial temperature of 30°C. Parallel experiments indicated the half-life of malathion to be half that of chlorpyrifos-methyl. Pirimiphos-methyl on the other hand showed a half-life of the order of 45 weeks.

Deahl and Tucker (1974,1975) reported results of a study in which 1000 tonnes of wheat were treated with chlorpyrifos-methyl at the rate of 6 mg/kg as it was put into storage at a silo in Queensland. Grain samples were drawn from 3 depths at 1, 6, 11, 16, 22, and 26 weeks after treatment. The residue, 1 week after treatment, was found to be 3.7 mg/kg declining gradually to 2.2 mg/kg at 16 weeks and 1.5 mg/kg at 26 weeks.

LaHue (1975a) reported the results of a study of the control of infestation in shelled corn by *Sitotroga cerealella* with chlorpyrifos-methyl in comparison with 4 other insecticides. Chlorpyrifos-methyl was applied as an aqueous emulsion at the nominal rate at 6.7 mg/kg. The grain was sampled after 24 hours and at the end of each month during 8 months storage. Analysis revealed that the deposit of 6.1 mg/kg at 24 hours had declined to 5 mg/kg after 1 month, 4 mg/kg after 3 months, 2.3 mg/kg after 6 months and 1.9 mg/kg after 8 months. This corresponded very closely to fenitrothion.

LaHue and Dicke (1976a) who evaluated selected insecticides applied to high-moisture sorghum grain, found that about 27% of the initial chlorpyrifos-methyl deposit remained on the sorghum at the end of 12 months.

Morallo-Rejesus and Carino (1976b) who tested the residual toxicity of 5 insecticides on 3 varieties of corn and sorghum under typical Philippine conditions reported that chlorpyrifos-methyl had the longest activity of any of the 5 insecticides on both maize and sorghum.

Quinlan (1978) studied the effectiveness of chlorpyrifos-methyl against insect infestation in 1.5 tonne lots of wheat (14.6% moisture) stored in plywood bins. The half-life of the chlorpyrifos-methyl residues was determined to be 4.4 months, whereas the half-life of malathion was 1.7 months.

Banks and Desmarchelier (1978) pointed out that the loss of chlorpyrifos-methyl and many other insecticides from grain follows pseudo first-order reaction kinetics. They stated that laboratory and field studies are in reasonable accord for all compounds.

Desmarchelier (1978c) made a mathematical examination of data from a number of extensive trials to measure the availability to insects of aged insecticide deposits on wheat. Chlorpyrifos-methyl was one of three insecticides which served as models in these studies. He came to the conclusion that the loss of insecticidal activity is, at least on occasions, more rapid than the loss of chemical residues. He proposed that the quantitative measurement of loss of activity, coupled with quantitative measurement of loss of residues, should be used as an extra tool in the evaluation of grain protectants. It was proposed that preference be given to protectants which showed a low loss of activity relative to loss of residues or alternatively, that several small applications of a protectant might replace a single large application so that a desired insecticidal activity could be achieved with a minimum chemical residue. The author came to the conclusion that chlorpyrifos-methyl at 2 mg/kg is superior to malathion at 4 mg/kg.

Desmarchelier and Bengston (1979a, 1979b) in discussing the residual behaviour of chemicals in stored grain, pointed to the desirability of increasing the residual effectiveness of grain protectants by partial drying and cooling. Chlorpyrifos-methyl was used as an example in these studies since it had a high insecticidal activity, moderately long residual life and a relatively high rate of change in degradation with respect to increasing temperature. Again the half-life at 30°C and 50% relative humidity was stated to be 19 weeks.

Studies of the effectiveness of chlorpyrifos-methyl and malathion as protectants for high-moisture stored wheat **(Quinlan *et al.* 1979)**, showed that chlorpyrifos-methyl residues declined from 7.3 mg/kg to 2.7 mg/kg over a period of 9 months when applied to wheat of 14.6% moisture content. From the data the authors calculated the half-life to be 18 weeks, the average temperature being 16.6°C.

Tempone (1979), studying the effects of grain-protectant insecticides on barley malting and vice versa, reported rather erratic residue data obtained by analysing barley treated with chlorpyrifos-methyl in a number of trials and held in storage for periods up to 9 months. The rate of loss appeared to be considerably greater than that reported for other grains by other authors and the studies were to be repeated.

Coulon *et al.* (1979a) compared the activities of methacrifos and chlorpyrifos-methyl used in the laboratory against *Sitophilus granarius*. They reported that chlorpyrifos-methyl, applied preventively at the rate of 2.5 mg/kg, gave a 18-month-lasting protection whereas methacrifos protected the wheat for only 6 months. **Coulon *et al.* (1979b)** reported that when chlorpyrifos-methyl was applied to wheat stored in silos, 35% of the original deposit had disappeared after 9 months of storage.

Desmarchelier *et al.* (1980b) studied the rates of decay of 5 insecticides, including chlorpyrifos-methyl applied to wheat, under field conditions and in commercial silos. The decay of each compound was studied over 6 months at 3 depths within the wheat bulk. The experimental results conformed closely to the results predicted from a mathematical model based on temperature of the grain and interstitial moisture with a constant applicable to each insecticide. The half-life of chlorpyrifos-methyl at 30°C was 18.6 weeks.

Bengston *et al.* (1980a) reporting the results of field trials with various insecticide combinations, carried out on bulk wheat in commercial silos in widely scattered regions of Australia, reported that the breakdown of residues agreed with and extended the results of the previous study **(Desmarchelier *et al.* 1980)**. The actual rates of breakdown of individual compounds, including chlorpyrifos-methyl, were consistent with earlier data when adjustments were made for prevailing grain temperatures and moistures according to the principles set out for fenitrothion by **Desmarchelier (1978a)**.

The Joint FAO/WHO Meeting of Experts on Pesticide Residues **(FAO/WHO 1976)** referred to a number of studies on the degradation of chlorpyrifos-methyl on stored grains but at the time the studies were reviewed in 1975 they were all in the form of unpublished reports. It seems as though all of the information has since appeared in open scientific literature.

Bengston *et al.* (1983a) who studied the effect and fate of a range of insecticide mixtures, applied to sorghum of 12.6% moisture content and at 28.4°C, held in bulk storage under semi-tropical conditions, found that residues of chlorpyrifos-methyl declined from 7 mg/kg to 2.8 mg/kg in 24 weeks. The authors reported that rates of breakdown on sorghum conform to the general pattern for other cereal grains.

Cogburn *et al.* (1983) showed that rice in husk treated with chlorpyrifos-methyl and stored in galvanised metal farm bins remained free of infestation from all species for 12 months whereas most other treatments allowed infestation to develop during early periods of storage.

Sun *et al.* (1984), studying the effect of degradation of insecticides on the survival and reproduction rate of *Sitophilus zeamais* reported that chlorpyrifos-methyl degraded slowly under conditions prevailing in Taiwan.

Rowlands (1985) reported that **Adams (1985)** had been able to show that individual grains treated topically at 4.5 mg/kg with ^{14}C-chlorpyrifos-methyl lost 80–90% of the dose from unsealed containers within 7 days; no radioactive materials were recovered from the walls of the glass vessels. He also showed that in a sealed container, chlorpyrifos-methyl was transferred from both grain and paper to the inner surface of the vial and that quite rapid degradation to 3,5,6-trichloro-2-pyridinol occurred there. He was further able to suggest that the glass was the major catalyst responsible for the breakdown.

Rowlands (1985), in discussing the techniques for studying the degradation of insecticides on grain, cited recent experience in the UK where a 25 tonne bulk of wheat was treated at 4.5 mg/kg with chlorpyrifos-methyl to compare directly the breakdown of the insecticide on this scale with that in 1 kg of the same wheat stored under laboratory conditions. The aim was to determine how far the laboratory study, which used radio-labelled insecticide, could reflect the practical scale of treatment. Samples were taken from the concurrent experiments at monthly intervals and were extracted and analysed using the same procedures. The apparent rate of degradation of chlorpyrifos-methyl proved similar in both barn and laboratory experiments with some 70% of the applied dose of insecticide still recoverable after 5 months storage. Of the apparent loss, very little could be detected as discrete metabolic products. Traces of the 3,5,6-trichloro-2-pyridinol were detected after 3 months and 0.12 mg/kg of this compound was detected in the field samples at the 5 month termination compared to 0.14 mg/kg in the laboratory. Unextractable or 'bound' radioactivity remaining in the grain tissues gradually increased throughout the storage period (15% at 5 months) and at the termination point (14 months) accounted for 29.4% of the total applied activity. The nature of this unextractable material is being studied but obviously it results from a significant process continuously taking place during storage. The levels of the free metabolites were insignificant and intact chlorpyrifos-methyl accounted for the rest of the applied dose which was recovered.

Wetters and McKeller (1985) presented the results of a study of the fate of chlorpyrifos-methyl residues in barley, oats, rice, sorghum and wheat which was stored at room temperature (temperature not stated) for 28 days after being sprayed with chlorpyrifos-methyl emulsion at the rate of 6 mg/kg. The amount recovered by analysis immediately following treatment ranged from 4.2 to 6.2 mg/kg, the mean being 5.2 mg/kg, 86% of the target rate. After 28 days storage the residues ranged from 3.6 to 5.2 mg/kg, the average representing a loss of 28% of the original deposit found by analysis. The records are summarised in **Table 9.1.**

Table 9.1. Residues of chlorpyrifos-methyl on grain after storage **(Wetters and McKellar 1985)**

Grain	Residues (mg/kg)	
	Day of treatment	28 days after treatment
Barley	4.2	4.1
Oats	5.2	3.6
Rice in husk	5.5	4.4
Sorghum	6.2	4.1
Wheat	4.8	5.2

9.3 Fate in Milling, Processing and Cooking

Morel (1975), working in France carried out a trial in which wheat was treated with chlorpyrifos-methyl at the rate of 1.25, 2.5 and 3.75 mg/kg. After 4 days, samples of the wheat were cleaned and milled with an extraction efficiency of 67.5%. The results of these studies are set out in **Table 9.2**. The white flour was converted into bread which, when analysed was found to contain chlorpyrifos-methyl residues only at the limit of determination.

Table 9.2 Residues* (mg/kg) of chlorpyrifos-methyl after milling and baking **(Morel 1975)**

Application rate	Deposit found in wheat	Flour (a)	Middlings	Bran	Bread (b)
1.25	0.52	0.16	1.40	0.96	0.01
2.5	0.92	0.22	2.17	1.65	0.01
3.75	1.05	0.30	3.30	2.21	0.02

*Limit of determination 0.01 mg/kg
(a) Efficiency of milling process = 67.5% extraction
(b) Bread contains 33% moisture after cooling

Bulla and LaHue (1975) in the USA determined the effect of milling and fractionation of wheat containing 6.3 mg/kg of chlorpyrifos-methyl. After 3 or 6 months storage most of the residue was found on the outside ('red dog' or bran) fractions. The residues in the flour contained less than 1 mg/kg (0.53–0.39 mg/kg) chlorpyrifos-methyl. Flour samples containing 0.53 and 0.39 mg/kg chlorpyrifos-methyl were baked into bread which contained 0.21 and 0.16 mg/kg chlorpyrifos-methyl, respectively, showing 60% reduction in residue due to baking.

In the data evaluated by JMPR **(FAO/WHO 1976)**, several studies revealed that over 90% of the residues on the whole wheat remained in the outer bran and coarse middlings and were removed during the preparation of flour. Bread baking reduced the remaining residues further by about 60%. Thus, grain with a chlorpyrifos-methyl residue in the range 6–7 mg/kg produced white bread containing 0.2–0.3 mg/kg of chlorpyrifos-methyl. These data are summarised in **Table 9.3.**

The data from **Table 9.3** had been used to determine the percentage reduction in chlorpyrifos-methyl residues in each step of the process of converting wheat to bread. These are set out in **Table 9.4.**

Table 9.4. Percentage reduction in chlorpyrifos-methyl residues on milling and baking

Process of converting		% reduction in residues
Wheat to bread	— Wholemeal	83
	— White	98
Wheat to flour	— Wholemeal	67
	— White	94
Flour to bread	— Wholemeal	47
	— White	37

Desmarchelier *et al.* (1980b) reported a collaborative study to determine the rates of decay and the fate of residues of 5 organophosphorus insecticides including

Table 9.3. Chlorpyrifos-methyl residues in milled wheat fractions and bread

Ageing period	Residues (mg/kg) in: Whole wheat	'Red dog'	Bran	Pollard (or middling)	Flour	Bread: Wholemeal	Bread: White
USA **(Bulla and LaHue 1975)**							
3 months	6.3 mg/kg*	6.5	5.0		0.53		0.21
6 months	6.3 mg/kg*	3.3	3.0		0.39		0.16
France **(Morel and Gallet 1975)**							
6 weeks			4.3	1.4	0.26		
9 weeks	1.6–1.8 mg/kg*		2.96	1.46	0.22		
14 weeks			3.92	1.60	0.29		
17 weeks			3.55	1.61	0.23		
Australia **(Bengston *et al.* 1975; Desmarchelier 1975a)**							
6 weeks	6.7–6.9 mg/kg	–	7.6–8.8	2.6–3.2	0.38–0.38	1.11–1.17	0.25–0.28
11 weeks	3.6 mg/kg	–	9.3	9.0	0.6	–	0.2
11 weeks	2.0 mg/kg	–	5.2	4.6	<0.10	–	<0.12
22 weeks	1.8 mg/kg	–	6.6	3.6	0.4	–	0.13

*Treatment levels

chlorpyrifos-methyl. Wheat held in bulk storage for 11 and 22 weeks, following treatment with chlorpyrifos-methyl at the rate of 5 mg/kg, was milled through a conventional roller mill and portion of the flour was converted into white bread. The residues determined in the various fractions are set out in **Table 9.5.**

Table 9.5. Residues of chlorpyrifos-methyl determined following milling and baking tests on wheat from two trials lasting 11 and 22 weeks **(Desmarchelier *et al.* 1980b)**

Milling fraction	Residues (mg/kg) in each fraction of wheat treated	
	11 weeks before milling	22 weeks before milling
Whole grain	2.0	1.8
Bran	5.2	6.6
Pollard	4.6	3.6
Flour	0.1	0.4
White bread	0.1	0.13

Tempone (1979) included chlorpyrifos-methyl with 6 other insecticides in studies designed to determine the effects of insecticides on barley malting and vice versa. However, the data on chlorpyrifos-methyl are difficult to interpret and the work obviously needs to be repeated.

Barley grain treated with chlorpyrifos-methyl at the rate of 6 mg/kg was stored at a temperature of 20–22°C for 28 days **(Wetters 1980)**. The grain was then malted and brewed into processed fractions, which were analysed for chlorpyrifos-methyl and 3,5,6-trichloro-2-pyridinol. The results are set out in **Table 9.6**. This indicates that virtually 100% of the chlorpyrifos-methyl deposit on barley is destroyed in the process of malting and brewing.

Table 9.6. Residues of chlorpyrifos-methyl and 3,5,6-trichloro-pyridinol in barley malting and brewing fractions following treatment with chlorpyrifos-methyl **(Wetters 1980)**

Commodity	Residues (mg/kg)	
	Chlorpyrifos-methyl	3,5,6-trichloro-2-pyridinol
Barley at application	6.0	–
Barley at malting	4.1	0.9
Malt	0.28	1.2
Spent grains	0.39	0.13
Filter aid	<0.01	0.22
Yeast	N.D.	1.1
Malt cleanings	0.23	1.8
Cleaner overs	2.4	<0.01
Cleaner thrus	19	22
Beer	<0.01	0.15

Wetters and McKeller (1985) presented results of laboratory trials with the processing of barley (to malt and beer), oats (to oat flakes), rice (to white rice) and wheat (to flour and cookies). When oats were converted into oat flakes (rolled oats), 76% of the residues were removed even prior to the cooking process to which oat flakes are always subjected **(Table 9.7)**. Likewise 86% of the residues on rough rice (rice in husk) disappeared in the preparation of white rice **(Table 9.8)**. This too would be subjected to wet cooking before consumption. Sorghum did not lose quite as much of the residue in the preparation of sorghum flour where 74% of the amount on the raw grain remained **(Table 9.9)**. In the case of wheat the residue was mainly removed in the bran, shorts and red dog fractions leaving only 8% in the white flour. This was further degraded by 50% in the baking of cookies **(Table 9.10)**. In a separate experiment it was shown that about 80% of the chlorpyrifos-methyl residues in white flour were destroyed in the baking of white bread. Some of the residue was converted to the pyridinol which remained in the bread **(Table 9.11)**. From these results it can be calculated that over 97% of the residue on wheat grain is removed or destroyed before the grain-based food reaches the consumer.

Table 9.7. Residues of chlorpyrifos-methyl and 3,5,6-trichloro-2-pyridinol on oats subjected to processing **(Wetters and McKeller 1985)**

Grain fraction	Yield %	Residues (mg/kg)	
		Chlorpyrifos-methyl	pyridinol
Oats	100	3.6	1.3
Hulls	23.4	10.0	N.D.
Groats	65.3	1.1	N.D.
Flakes	–	0.87	N.D.
Disc Rejects	5.2	1.2	<0.1
Light Oats	0.3	16.0	0.15
Dust	0.2	61	26

Table 9.8. Residues of chlorpyrifos-methyl and 3,5,6-trichloro-2-pyridinol on rice subjected to processing **(Wetters and McKeller 1985)**

Grain fraction	Yield %	Residues (mg/kg)	
		Chlorpyrifos-methyl	pyridinol
Rice in husk	100	4.4	0.16
Hulls		15	4.5
Brown rice		0.89	0.10
White rice		0.62	N.D.
Bran		2.0	2.0
Grits		2.2	N.D.

Table 9.9. Residues of chlorpyrifos-methyl and 3,5,6-trichloro-2-pyridinol on sorghum subjected to processing **(Wetters and McKeller 1985)**

Grain fraction	Yield %	Residues (mg/kg) Chlorpyrifos-methyl	pyridinol
Sorghum	100	4.1	0.25
Flour	68.4	1.1	N.D.
Bran	13.8	10	1.0
Screenings	0.1	31	24
Shorts	13.8	5.8	0.26
Germ	3.9	12	N.D.

Table 9.10. Residues of chlorpyrifos-methyl and 3,5,6-trichloro-2-pyridinol on wheat subjected to malting **(Wetters and McKeller 1985)**

Grain fraction	Yield %	Residues (mg/kg) Chlorpyrifos-methyl	pyridinol
Wheat	100	5.2	0.1
Flour	70.4	0.41	N.D.
Bran	25.4	11	2.4
Shorts	2.7	17	1.5
Red Dog	1.2	6.9	1.6
Germ	–	14	2.0
Cookies	–	0.22	N.D

Table 9.11. Residues of chlorpyrifos and 3,5,6-tri-chloro-2-pyridinol residues in wheat flour and bread **(Wetters and McKeller 1985)**

Product	Time from treatment to milling (weeks)	Residues (mg/kg) Chlorpyrifos-methyl	pyridinol
Flour	0	1.3	N.D.
		1.3	N.D.
	4	0.95	N.D.
		0.90	N.D.
Bread	0	0.22	0.26
		0.26	0.25
	4	0.17	0.18
		0.17	0.20

9.4 Metabolism

Three studies of the metabolism of chlorpyrifos-methyl in rats and sheep were evaluated by JMPR **(FAO/WHO 1976)**. It appears as though the metabolism follows identical pathways to that already demonstrated for chlorpyrifos.

The absorption, excretion and distribution of 2,6-^{14}C-ring-labelled chlorpyrifos-methyl in rats was studied by **Branson and Litchfield (1971)**. Single oral doses of 16 mg/kg (3.2 mg/200 g rat) were found to be rapidly adsorbed and eliminated. The highest blood levels (2–4% of dose) occurred 5 hours post-treatment. This blood level pattern suggested a rapid clearance from the body, 90–94% of the dosage was eliminated from the body in 72 hours. The principal routes of elimination were as follows: urine 83–85%, feces 7–9%, and respired air 0.23–0.43%. The maximum amount of the 16 mg/kg dose remaining in the tissues by 72 hours was 0.65–1.3% of administered ^{14}C. Levels of ^{14}C calculated as chlorpyrifos-methyl equivalent in 18 tissues, including fat, were low in all cases (less than 1 ppm). The experimental method was sensitive to 1 ppb ^{14}C-chlorpyrifos-methyl equivalent with radioactive recoveries around 93%.

About 0.23–0.43% of the administered ^{14}C-chlorpyrifos-methyl was eliminated as CO_2, which demonstrated *in vivo* ring breakage to small carbon fragments. The major urinary metabolite was 3,5,6-trichloro-2-pyridinol. The only other detectable radioactivity in autoradiograms of the rat urine was at the origin. The possibility of urinary conjugates was not studied **(Branson and Litchfield 1971)**.

In a study designed to determine whether chlorpyrifos-methyl accumulated in tissues, ring labelled ^{14}C-chlorpyrifos-methyl was fed to 10 rats at 2.0 mg/kg/day for 7 days. Treatment was stopped on day 7 as the animals had reached equilibrium on intake and output of ^{14}C activity. Analysis of tissues from 6 rats sacrificed on day 7 of treatment showed chlorpyrifos-methyl residues of 0.148 ± 0.024 ppm in hind leg subcutaneous fat, 0.135 ± 0.045 ppm in cervical brown fat and 0.052 ± 0.023 ppm in hind leg muscle. On the eighth day of the study, 24 hours post-exposure, analysis showed 0.012 ppm chlorpyrifos-methyl residue in the hind leg subcutaneous fat of one animal and no residues above the limit of detection (0.01 ppm) in the other tissues **(Branson and Litchfield 1971)**.

See Section 9.5 for a review of study in sheep **(Bakke and Price 1975)**.

Based on the results of these metabolic studies, there does not appear to be sequestering or accumulation of orally administered chlorpyrifos-methyl in the fat of experimental animals, nor is it stored as are certain chlorinated insecticides.

The principal metabolite of chlorpyrifos-methyl is the same as that for chlorpyrifos, namely, 3,5,6-trichloro-2-pyridinol. The metabolism of this pyridinol in rats and fish was summarised in **FAO/WHO (1973). (Smith *et al.* 1970).**

Nakajima *et al.* (1974) studied the behaviour of ^{14}C-labelled chlorpyrifos-methyl by means of whole-body autoradiographs of rats receiving single oral doses of about 24 mg/kg. Autoradiograms were taken periodically 0.5, 1, 3, 5, 7, 24, 72 and 120 hours after administration. The highest ^{14}C concentrations were usually in the blood.

Residues were noted in extremely low concentrations in the kidney, liver, intra-intestinal feces and fat at 72 hours and eliminated from the whole body 120 hours after administration.

Rowlands and Wilkin (1975), tentatively identifying the breakdown products of chlorpyrifos-methyl on wheat, included dimethyl phosphorothionate, 3,5,6-trichloro-2-pyridinol, and the O-desmethyl derivative of the parent compound.

Rowlands (1985) refers to work by his collegue **Adams (1985)** who has confirmed the free pyridinol and an unidentified more-polar compound as the only detectable metabolites from ^{14}C-chlorpyrifos-methyl during 14 months storage. A large amount (about 30% of the total radioactivity applied) of radio-labelled material that was not extractable by conventional solvent blending or acid digestion remains to be identified.

9.5 Fate in Animals

Johnson *et al.* (1974) reported an extensive trial in which the persistence of chlorpyrifos-methyl in corn silage and the effects of feeding dairy cows the treated silage were studied. Corn sprayed in the field was ensiled 1 day later and the residues of chlorpyrifos-methyl and its metabolites, determined in the silage through 83 days of ensiling, were 76% of that applied. Beginning 83 days post ensiling, silage was fed to cows for 42 days during which chlorpyrifos methyl averaged 1.85 ppm and was stable in the silage. Residues intake, amounting to 0.054 mg chlorpyrifos-methyl/kg body weight daily, failed to affect silage intake, milk production, blood cholinesterase activity, or body weight gains. Traces of chlorpyrifos-methyl (0.003 mg/kg or less) were found in milk from cows on the highest treatment. Milk from all cows fed treated silage contained traces of the pyridinol (0.011 mg/kg or less). No trace of the O-analogue of chlorpyrifos-methyl was found in any samples and all milk, urine and feces were free of residues within 1 week after the cows were withdrawn from treated silage.

The metabolism of chlorpyrifos-methyl in sheep was determined by **Bakke and Price (1975)** by use of radioactivity counts, infra-red spectra, mass spectra, and column, thin-layer and gas-liquid chromatography. Recoveries of ^{14}C from urine, feces, and tissues of sheep treated with 100 mg chlorpyrifos-methyl/kg totalled 69.2%, 12.5%, and 2.0%, in 2 days. No ^{14}C was found in respired air. Radioactivity in the urine consisted of 2 major components, one a glucoronide conjugate of 3,5,6-trichloro-2-pyridyl phosphorothionate (38%); and 3 minor components totalling 5.2% of which 0.5% was the hydrolyzed trichloro-2-pyridinol. The same 3 identified components were found in the plasma. The feces contained unchanged chlorpyrifos-methyl, its mono-demethylated derivative and its hydrolytic pyridinol derivative, which account for 89.3% of the ^{14}C in feces. Detectable amounts (0.32–11.8 mg/kg) of chlorpyrifos-methyl equivalent (including metabolites) were found in all tissues 96 hours after treatment. The highest levels by far were found in the visceral fat (as expected from the high fat solubility of chlorpyrifos-methyl and partitioning co-efficient between octanol and water).

Swart *et al.* (1976b) conducted a feeding study in Texas where 9 groups of 3 calves were fed levels of 0, 1, 3, 30, and 100 ppm chlorpyrifos-methyl in their rations for 28 days. All feeding was *ad libitum*. Samples for residue analysis were collected at 0 withdrawal for each level and at 7, 14 and 28 day withdrawal for the 100 ppm level. Tissues collected (blood, muscle, fat, liver, and kidney) were analysed and reported by **Kuper (1978a)**. Average residues of chlorpyrifos-methyl were 0.03 mg/kg (0.02–0.05 mg/kg) in kidney and 0.77 mg/kg (0.65–0.91 mg/kg) in fat of calves fed 100 ppm for 28 days with no withdrawal. Signficant residues were not detected in muscle or in liver under these conditions. At lower feeding levels residues were lower and, after 7 days withdrawal from treated feed, residues in all tissues were less than the validated level of quantitation.

Rowe (1978) set up a study to determine residues in milk and cream from cows fed chlorpyrifos-methyl. Chlorpyrifos-methyl was fed to 3 lactating dairy cows via supplementary rations in their feed at levels of 0, 1, 3, 10, 30 and 100 ppm based on total feed intake. The level of chlorpyrifos-methyl in the diet was increased at 14-day intervals, followed by a 14-day withdrawal period during which no chlorpyrifos-methyl was fed. Samples of milk and cream were collected at designated times during pre-treatment, treatment and withdrawal periods. Samples were analysed by **Kuper (1978c)** who found that residues averaged 0.03 mg/kg chlorpyrifos-methyl and 0.05 mg/kg 3,5,6-trichloro-2-pyridinol at the 100 ppm feeding level. Lower feeding levels resulted in lower levels of residue. Residues were below the validated level of 0.01 mg/kg at the 30 ppm feeding level for chlorpyrifos-methyl. No residue of either compound, greater than 0.01 mg/kg, was found after a 2-day withdrawal from the 100 ppm feeding level. Residues found in cream from cows fed 100 ppm chlorpyrifos-methyl averaged 0.43 mg/kg (0.31–0.50 mg/kg) chlorpyrifos-methyl. The concentration of chlorpyrifos-methyl in cream with respect to milk is approximately the same as the concentration of butterfat in cream versus milk.

Swart *et al.* (1976a) set up a chicken feeding study to determine residues of chlorpyrifos-methyl in eggs and tissues. Groups of White Leghorn laying hens were fed 0, 1, 3, 10, 30 and 100 ppm levels of chlorpyrifos-methyl in their rations for 28 days. Egg samples were collected every second day during the 7 to 11 day period prior to chemical feeding and through the treatment and withdrawal periods. Tissue samples of muscle and skin, fat, liver and kidney were collected from each group. The residues were determined by **Kuper (1978c)** who reported that the average residues of chlorpyrifos-methyl were 0.01 mg/kg in muscle, 0.08 mg/kg in fat and 0.02 mg/kg in eggs from birds fed 100 ppm with no withdrawal. No significant residues were detected in liver under these conditions. Residues of chlorpyrifos-methyl and pyridinol were below the validated level of quantitation for all tissues after a 7-day withdrawal from treated feed.

9.6 Safety to Livestock

In view of the extensive information on the fate of chlorpyrifos-methyl in cattle and chickens and the results of long-term feeding studies in mice, rats and dogs there are no grounds for believing that the continuous ingestion of chlorpyrifos-methyl residues, at levels likely to arise from the use of this insecticide for the protection of stored grain, could possibly have any adverse effect upon livestock. Furthermore the teratogenic studies, reproductive studies, neurotoxicity studies and various short and long-term studies in the rat, dog, mouse, monkey, Mallard duck, Bobwhite quail and Japanese quail provide considerable additional reassurance.

The numerous acute oral toxicity studies in various species indicate that the toxic dose of chlorpyrifos-methyl to various species of livestock is probably quite high. A summary of the acute oral toxicity to 5 species is given in **Table 9.12.**

Shellenberger (1970) conducted dietary tests with chlorpyrifos-methyl on 5- to 7-day old Mallards, Bobwhite and Japanese quail over a 5-day treatment and a 3-day post-treatment period (8-day protocol). Effects on mortality, feed consumption, weight gain, whole blood and brain cholinesterase were noted and no-effect levels are summarised in **Table 9.13.**

Table 9.12. Chlorpyrifos-methyl: acute oral toxicity

Animal	Sex	Vehicle	mg/kg body weight	Reference
Rat	F	Corn oil	>1000	**Olson & Taylor 1964**
	M	Corn oil	2140	**Olson 1964**
	F	Corn oil	1630	**Olson 1964**
	F	Glycerol/ethanol	>1600	**WHO 1966**
	M	Corn oil	2140	**Litchfield & Norris 1969**
	F	Corn oil	1090	**Litchfield & Norris 1969**
	F	Corn oil	1828	**Hasegawa, *et al.* 1973**
	M	Corn oil	2472	**Hasegawa, *et al.* 1973**
	F	CMC	3597	**Hasegawa, *et al.* 1973**
	M	CMC	3733	**Hasegawa, *et al.* 1973**
	M	Corn oil	1980	**US Army 1973**
	F	Corn oil	3600	**Esaki *et al.* 1973**
Guinea pig	M	Corn oil	2250	**Olson 1964**
Rabbit	M + F	Corn oil	2000	**Olson 1964**
Mouse	M	Peanut oil	1122	**WHO 1966**
	F	Corn oil	2032	**Hasegawa *et al.* 1973**
	M	Corn oil	2254	**Hasegawa *et al.* 1973**
	F	Corn oil	2440	**Esaki *et al.* 1973**
Chicken	M	Capsule	>2000	**Olson 1964**
Chicken	M	Capsule	>7950	**Olson 1964**
Chicken	M + F	Capsule	>8000	**Ross & Roberts 1974**
Chicken	M + F	Gavage	7532	**Ross & Roberts 1974**

Table 9.13. Dietary tests on birds with chlorpyrifos-methyl

Species	ppm in diet Lowest no-effect level — 5 days				
	LC50	Weight gain	Feed consumption	Cholinesterase	
				Blood	Brain
Bobwhite	1835	1250	312	–	–
Japanese quail	>5000	1250	1250	–	–
Mallard	2500–5000	>625	>625	39–78	>625

Ross and Sherman (1960) reported an extensive study of the effect of selected insecticides on growth and egg production when administered continuously to poultry in the feed. Chlorpyrifos-methyl was administered at 88 and 132 ppm in the feed of New Hampshire hens for 29 weeks. There was no mortality but the body weight gain was depressed, feed consumption indexes were increased by 2 to 10%, mean hen-day egg production was slightly higher. The weight gain ratios of chicks fed similar rates of chlorpyrifos-methyl were lower than controls but the final body weights after 29 weeks were unaffected.

9.7 Toxicological Evaluation

The toxicological data on chlorpyrifos-methyl were evaluated by the JMPR in 1975 **(FAO/WHO 1976).**

Teratogenic studies, reproductive studies, neurotoxicity tests and various short- and long-term studies in the rat, dog, mouse, monkey, Mallard duck, Bobwhite and Japanese quail, were considered to provide unequivocal assurance of the relatively low toxic hazard of this insecticide. In specific tumourigenic studies in the rat no compound-related effect or increase in tumour incidence was observed. No-effect levels were determined in the rat, dog, monkey and man on the basis of the most sensitive parameter, plasma cholinesterase inhibition. The level causing no toxicological effect was determined to be 0.1 mg/kg in all 4 species.

From these studies, the acceptable daily intake (ADI) was estimated to be 0.01 mg/kg/day.

9.8 Maximum Residue Limits

On the basis of the above ADI and extensive residue data, the JMPR **(FAO/WHO 1976)** recommended the following maximum residue limits in raw cereal grains and milled cereal products.

Commodity	Maximum Residue Limits (mg/kg)
Bran	20
Raw grain (wheat, maize, sorghum)	10
Flour, wholemeal bread	2
Bread (white)	0.5
Rice (pre-harvest treatment)	0.1

The residue is determined and expressed as chlorpyrifos-methyl.

Additional information is required to enable the JMPR to establish maximum residue limits in other raw grains, semi-processed milling fractions and prepared cereal food.

The methods recommended for the analysis of chlorpyrifos-methyl residues are those of **Desmarchelier *et al.* (1977b)** and **Mestres *et al.* (1979a; 1979b).**

10. Deltamethrin

Deltamethrin is one of the latest and most potent of the synthetic pyrethroids. The development of this family of products started from a recognised activity in natural pyrethrum extracts and classical chemical sleuthing by **Staudinger and Ruzika (1924)**. Based on this knowledge many scientists in many countries attempted to synthesise molecules which would reproduce the insecticidal properties of the *Chrysanthemum cinerariaefolium*. There were few successes other than the development of allethrin **(Schecter *et al.* 1949)** until bioresmethrin was developed by **Elliott *et al.* (1967a, b)**. Bioresmethrin represented a distinctive advance over the previous pyrethroids but stability in sunlight was not much improved. Decisive progress towards light stability was made by modifying the 2 photo-labile zones of the resmethrin structure and a combination of these two modifications led to permethrin **(Elliott *et al.* 1973)**. The introduction of a cyano group at the benzyl position increased insecticidal potency without reducing photostability. This led to the development of fenvalerate and cypermethrin.

A detailed study of the cypermethrin structure established that the great activity was obtained with esters of the *cis*-chain acids and that the dibromo was preferable to the dichloro grouping. Selection of the sterioisomer with maximum activity led to the development of deltamethrin **(Elliott *et al.* 1974)**. Deltamethrin is now considered the most potent insecticide against a wide variety of species. The crystalline nature of this insecticide undoubtedly greatly facilitated its isolation, as well as a very early study of its remarkable properties **(Barnes and Verschoyle 1974)**.

Deltamethrin is exceptionally potent against the whole spectrum of stored-product pests. It is effective at low dose levels in the range of 1–2 mg/kg. It is exceptionally stable on grain. It shows little or no tendency to penetrate the individual grains so it is expected to be removed on bran. The product is currently under development and there is little or no information available in open literature. Only limited data have been seen on the fate of deltamethrin residues on cooking but on a general knowledge of the compound it is believed that most of the residues finding their way into milled cereal products will persist, more or less undiminished, in foods presented for consumption. Field trials with conventional formulations have found them to cause irritation to the skin and mucous membranes of operators and new suspension concentrate formulations are currently being tested.

10.1 Usefulness

Coulon and Barres (1978), of the French National Agronomic Research Institute, carried out a very comprehensive study on the effects of deltamethrin on *Sitophilus granarius* taking bioresmethrin as the reference insecticide. The relative efficacy of bioresmethrin and deltamethrin was determined by bringing the insecticides into contact with adult insects placed on treated wheat and using the mortality values recorded 1 week after treatment as activity criteria. In this way, the concentrations required to kill 50% and 90% of the insects were found to be respectively 0.23 and 0.44 mg/kg for deltamethrin, against respectively 5.2 and 10.3 for bioresmethrin. The same tests were repeated with preparations in which the active ingredient was combined with piperonyl butoxide in a ratio of 1:4 and 1:10. The LC/50 and LC/90 values were found to have dropped considerably, especially for bioresmethrin, for which the synergistic effect is more pronounced. These authors also studied the effect of deltamethrin on the reproduction of *Sitophilus* and concluded that reproduction could be prevented by treating wheat with 0.5 to 1.0 mg/kg deltamethrin synergised with piperonyl butoxide in the ratio of 1:4 or by 0.25 to 0.5 mg/kg deltamethrin synergised in the ratio of 1:10.

Bitran and Campos (1978) of the Biological Institute of Sao Paulo, Brazil, provided preliminary results of trials with deltamethrin against stored-product pests of maize and coffee. **Carle (1979)** reported observations of the effectiveness of deltamethrin against *Sitophilus granarius* and *Tribolium castaneum*. Compared with bioresmethrin, deltamethrin, synergised with piperonyl butoxide (1:2.5) was reported to be from 5 to 30 times more potent.

Kalonji (undated) studied the effectiveness of a range of insecticides against *S. granarius* and reported that deltamethrin applied to wheat at 0.7, 0.85 and 1 mg/kg was considerably more persistent than deposits of iodofenphos, malathion and dichlorvos.

Bengston and Desmarchelier (1979), after reviewing the work of the Australian Wheat Board Working Party on Grain Protectants, stated that deltamethrin (2 mg/kg) plus piperonyl butoxide (10 mg/kg) had proved satisfactory for the control of all major species of stored-product pests, including malathion-resistant insects, in Australian grain during 9 months storage. However, emulsifiable formulations of deltamethrin had proved unacceptable because of irritation of the respiratory system and exposed skin of those working in grain storage systems during the application. A suspension concentrate formulation gave better results against insects without any observable effect on operators but when the suspension concentrate formulation was combined with piperonyl butoxide the irritancy recurred apparently due to the solubalisation of the deltamethrin by the piperonyl butoxide.

Bengston (1979) in reviewing the potential of pyrethroids as grain protectants pointed out that relatively high dosages are required to control *Sitophilus* spp. and *Tribolium* spp. and therefore these compounds are unlikely to be widely used against these species. In **Table 10.1** the relative tolerance of 3 species of grain beetles exposed for 3 days to 6 pyrethroid insecticides, including deltamethrin, is given.

Desmarchelier and Bengston (1979b) commented that deltamethrin (2 mg/kg) plus piperonyl butoxide (10 mg/kg) provided satisfactory protection for grain during 9 months storage in Australia.

Bitran *et al.* (1980b), in a series of tests to evaluate the residual persistence of deltamethrin and bioresmethrin on stored corn and coffee in order to control adults of the maize weevil, *S. zeamais*, applied deltamethrin and bioresmethrin alone and synergised with piperonyl butoxide (1:4) at the rate of 0.5 mg/kg and 1 mg/kg to the corn. They found that deltamethrin applied at the rate of 1 mg/kg was 100% effective for more than 270 days whilst the lower rate (0.5 mg/kg) was fully effective for 120 days. The treatments based on bioresmethrin were significantly less effective giving full control for only 120 and 30 days when applied at the rate of 1 and 0.5 mg/kg respectively. These findings appeared to contradict those of **Bengston (1979)** concerning the value of the synergist.

As part of studies on the protection against insects, especially *Sitophilus zeamais*, of maize cobs kept in farmers' stores, studies were carried out in Brazil **(Bitran *et al.* 1981)** to compare the effectiveness of treatment of the cobs with a deltamethrin dust at 1 mg/kg with the standard treatment with malathion at 8 mg/kg, with or without fumigation with phosphine. The cobs were examined after 4 and 8 months, and damage by *Sitophilus zeamais* and *Sitotroga cerealella* was assessed. Fumigation followed by treatment with deltamethrin afforded the best protection. Deltamethrin was superior to malathion when the cobs were not also fumigated. A significant correlation was found between the weight loss of maize and level of infestation by the 2 pests.

In laboratory studies in India, 4 synthetic pyrethroids, including deltamethrin, were tested alongside 4 organophosphorus insecticides against third-instar larvae of *Trogoderma granarium* **(Chahal and Ramzan, 1982)**. The pyrethroids were applied as 0.0125%, 0.025% and 0.05% solutions and the organophosphates at 0.025%, 0.05% and 0.1%. It was found that 1 day after spraying deltamethrin was the most effective, causing 100% mortality at 0.05%, while at 0.0125% it caused 26.5% mortality and at 0.025%, 66.2% mortality. The other compounds resulted in 0 to 16% mortality. Observations 2, 3, 5 and 7 days after treatment showed a progressive increase in mortality and after 7 days all compounds except malathion and cypermethrin caused 100% mortality of the larvae.

Residues of deltamethrin on wheat grain of 12% moisture content, stored in the laboratory at 25°C, remained constant within experimental error over a 15-month interval. However, the biological activity of the residues against *Sitophilus oryzae* declined to 78% of the initial value at 3 months and to 65% at 15 months after application. Inactivation of a proportion of the intact residues is indicated **(Hargreaves *et al.* 1982)**.

Bitran *et al.* (1983a) reported results of experiments with deltamethrin, compared to malathion, on maize and coffee, carried out in Brazil to control infestations of *Sitophilus zeamais, Acanthoscelides obtectus* and *Araecerus fasciculatus*. Treatments with deltamethrin synergised with piperonyl butoxide (1:10) applied directly to maize grains were superior to malathion, controlling *Sitophilus zeamais* for 9 months. Deltamethrin was also more effective than malathion in sack treatments, protecting maize and coffee for longer periods.

Bengston *et al.* (1983b) reported trials in which organophosphorothioates and synergised synthetic pyrethroids were used in duplicate field trials carried out on bulk wheat in commercial silos in Queensland and New South Wales. Laboratory bioassays using malathion-resistant

Table 10.1. LC99.9 of various insecticides obtained with adults deposited on freshly treated wheat **(Bengston 1979)**

Species	Strain	Bioresmethrin	D-phenothrin	Fenvalerate	Permethrin 75/25	Cypermethrin	Deltamethrin
S. oryzae	LS 2	15.8	10.7	9.8	3	1.5	2.9
	QSO 56	10.4	8.3	7.1	2.8	2.3	6.6
	QSO 231	352	52.5	108.1	11.5	19.7	35.7
T. castaneum	QTC 39	7.5	17.8	18.3	17.7	1.6	0.3
	QTC 34	26.1	62.8	43.8	13.0	1.6	0.5
R. dominica	QRD 14	0.6	0.9	0.7	0.7	0.1	0.05
	QRD 2	–	0.8	0.5	0.8	0.4	–
	QRD 63	0.6	2.5	0.7	1.7	0.2	0.05

strains of insects were carried out on samples of treated grain at intervals over 9 months. Residues of deltamethrin were shown to be highly persistent on stored wheat. There was no significant change in the concentration of deltamethrin over 9 months and at the end of 9 months the residues were still within the range of the experimental error.

Bitran *et al.* (1983b) presented results of biological tests carried to evaluate the residual action of some pyrethroids and organophosphorus insecticides on stored maize treated in Brazil to control *Sitophilus zeamais*. Insecticides were mixed with maize grains and evaluated for 9 months. Three of the pyrethroids showed high efficacy when synergised with piperonyl butoxide (1:5) but this effectiveness was reduced without piperonyl butoxide. Unsynergised deltamethrin was among the best treatments, being more persistent when applied as a dust.

Ishaaya *et al.* (1983) tested 6 synthetic pyrethroids as toxicants and inhibitors of weight gain in first and fourth instar larvae of *Tribolium castaneum*. Deltamethrin equalled *cis*-cypermethrin in being the most potent. Dosages that reduced larval weight also delayed pupation and emergence probably due to their anti-feeding activities. Three oxidase inhibitors including piperonyl butoxide added to the substrate at a concentration of 100 mg/kg synergised the toxicity of deltamethrin. On the other hand, an esterase inhibitor, profenofos, did not enhance the potency of any of the pyrethroids. The authors observed that oxidases appeared to be more important than esterases in pyrethroid detoxification by *Tribolium castaneum*.

Hsieh *et al.* (1983), working in Taiwan, tested the toxicity of 26 insecticides to *Sitophilus zeamais* and *Rhyzopertha dominica*. Only 7 compounds, including deltamethrin, were more toxic to *Rhyzopertha dominica* than to *Sitophilus zeamais*. Deltamethrin was the most toxic compound to *Rhyzopertha dominica* with an LC/50 of 0.07 mg/kg.

Longstaff and Desmarchelier (1983), in studying the effect of temperature on the toxicity of deltamethrin in comparison with pirimiphos-methyl, showed that deltamethrin was significantly more toxic to *S. oryzae* at 21°C than at 32°C. Pirimiphos-methyl, on the other hand, was considerably more active at the higher temperature. They showed there was significantly less reproduction at the higher temperature in wheat treated with deltamethrin at 0.05 and 0.1 mg/kg than at the somewhat higher temperature. On the strength of these findings these workers pointed to the advantage of artificial cooling of wheat treated with deltamethrin.

Bengston *et al.* (1984) reported a field experiment carried out on bulk sorghum stored for 26 weeks in concrete silos in south Queensland. No natural infestation occurred. Laboratory bioassays of treated grain, in which malathion-resistant strains of insects were added to grain samples, indicated that all the treatments were generally effective. Deltamethrin plus piperonyl butoxide (2 + 8 mg/kg), controlled typical malathion-resistant strains of *Sitophilus oryzae*, *Rhyzopertha dominica*, *Tribolium castaneum* and *Ephestia cautella*.

Arcozzi and Contessi (1984) tested 6 synthetic pyrethroids, including deltamethrin, in Italy in wheat held in small silos artificially infested with *Sitophilus granarius* at the rate of 4 to 5 adults per kilogram of wheat. The results indicated that deltamethrin at 0.5–1 mg/kg appreciably reduced the infestation from over 10 000 to below 1000 insects per kilogram of grain.

Bengston *et al.* (1984) applied synergised deltamethrin at the rate of 2 mg/kg to bulk sorghum held in concrete silos in south Queensland. Its performance was compared with that of combinations of fenitrothion with synergised fenvalerate, and synergised phenothrin and pirimiphos-methyl with permethrin. Deltamethrin controlled typical malathion-resistant strains of *Sitophilus oryzae*, *Rhyzopertha dominica*, *Tribolium castaneum* and *Ephestia cautella* for 26 weeks. Residue analyses indicated that its effectiveness should have continued well beyond the period of the trial.

Duguet and Wu (1984) evaluated deltamethrin against 2 bruchid species infesting grain legumes in the laboratory and in the field in Africa. They found that deltamethrin at 0.25 mg/kg was effective against *Callosobruchus chinensis* for more than 228 days in the laboratory and more than 6 months at 0.75 mg/kg under field conditions against *Callosobruchus maculatus*. A minimum of 5 days contact between the insects and the treated grain was required to produce maximum kill. Both species were equally sensitive to deltamethrin.

Duguet (1985a) reported experience and experiments with the protection of maize in village granaries and cribs in a hot climate with deltamethrin. He considered that deltamethrin powder applied at the rate of 1 mg/kg or deltamethrin spray (0.5 mg/kg and piperonyl butoxide 5 mg/kg) was effective for 8–10 months depending on the type of storage. He recommended that the sheath should be removed before treatment with insecticide and that the deltamethrin powder should be applied uniformly layer by layer as the cobs were placed in the crib.

Duguet (1985b) discussed the comparative effectiveness of deltamethrin against the 3 *Sitophilus* spp. and showed that in temperate climates 0.25 mg/kg deltamethrin and 2.5 mg/kg piperonyl butoxide provided good protection for wheat and rice against *Sitophilus oryzae* for 6 months or more. In tropical and subtropical climates where wheat and rice are attacked by limited numbers of *Sitophilus oryzae* and *Sitophilus zeamais* synergised deltamethrin at the rate of 0.5 mg/kg or deltamethrin alone at 1 mg/kg protects the commodity for 8–10 months. On maize infested with *Sitophilus zeamais* in the field it is necessary to increase the rate of unsynergised deltamethrin to 1.5 mg/kg to kill adults but this is not sufficient to prevent the development of a second generation. Under natural conditions of development *Sitophilus oryzae* is the most susceptible and *Sitophilus granarius* the most tolerant but *Sitophilus zeamais* develops in regions where the half-life of deltamethrin is shortest and where the harvested maize is already infested from the field.

Lessard (1985) reported on the acaricidal activity of deltamethrin formulations used for direct application to grains. Whereas bioresmesthrin reduced the population of *Tyrophagus putrescentiae* by 90% in 2 days, synergised deltamethrin at different dose rates reduced the mite numbers by only 33% to 61%. Resbuthrine reduced the numbers by 51% but only after 7 days contact. The results from pirimiphos-methyl and resbuthrine were no better than from bioresmethrin. The author concluded that synergised deltamethrin has a not negligible action against *Tyrophagus putrescentiae* but this activity is insufficient for practical use specifically for the control of mites in stored grain.

Nahal and Duguet (Undated) evaluated deltamethrin and malathion against 3 bruchid species. They found deltamethrin powder applied at the rate of 0.5 and 0.75 mg/kg was effective for more than 540 days for protecting broad-bean seeds and Egyptian clover seeds against *Callosobruchus maculatus, Callosobruchus chinensis* and *Bruchidus alfierii*. Malathion at 8 mg/kg was inferior in all tests.

Golob *et al.* (1985) tested deltamethrin by topical application to *Prostephanus truncatus* and found that it was from 5 to 50 times more effective than the most effective of 10 other insecticides. It was over 5000 times more potent than malathion.

10.2 Degradation

Results from a number of trials to determine the persistence of deltamethrin residues on stored wheat and maize conducted through the University of Montpellier, France, in Morocco, Belgium and Brazil were submitted for evaluation by the Joint FAO/WHO Meeting on Pesticide Residues in 1980 **(FAO/WHO 1981)**. The results suggest that difficulties were encountered in sampling the storages or in analysing samples because the analytical data were most inconsistent. However, it is clear that there was little or no degradation during 30 to 50 weeks of storage. There is a trend which suggests that the persistence of dust is somewhat greater than the corresponding emulsifiable formulation.

Halls and Periam (1980a) reported trials designed to determine the fate of deltamethrin on wheat during storage. Batches of English hard wheat were sprayed on a conveyor with diluted deltamethrin/piperonyl butoxide emulsion to give 250 kg batches of wheat containing 1 mg/kg deltamethrin plus 10 mg/kg piperonyl butoxide and 2 mg/kg deltamethrin plus 10 mg/kg piperonyl butoxide respectively. The grain was placed in 1 tonne capacity metal silos. Analysis of samples revealed that the actual treatment levels achieved were only about 40 to 70% of those nominated. Samples of the treated wheat were taken after spraying and thereafter at monthly intervals for 9 months and analysed for deltamethrin. Replicate samples were taken from each silo, analysed separately and the results expressed as the mean. The results are given in **Table 10.2**. Here again, the variability in analytical results suggest there were problems in achieving uniformity in distribution, sampling or analysis. It is reasonable to assume that the loss of deltamethrin from wheat under storage conditions in the UK is negligible.

Noble *et al.* (1982) carried out a carefully conducted laboratory experiment to determine the stability of 4 pyrethroid insecticides on wheat in storage at 25°C or 35°C and either 12% or 15% moisture content. Rates of loss were calculated from residue analysis of the wheat at 4 intervals during storage. Great care was taken to analyse the solution used to treat the wheat, to determine the recovery of pyrethroid from the treated wheat samples, to avoid instability problems and to measure the amount of pyrethroid on the surface of sample jars and laboratory equipment. As a consequence the analytical data are highly consistent **(Table 10.3)**. Pseudo first-order rate constants and half-lives were calculated for all of the storage conditions by assuming first-order loss of compound. Half-life values and 95% confidence limits for these values were calculated. These parameters are presented in **Table 10.4** for the 4 pyrethroids. These data show that the loss of these pyrethroids is slow, especially under milder conditions of temperature and moisture.

Table 10.2. Deltamethrin residues on wheat after indicated periods of storage in the UK[1] **(Halls and Periam 1980a)**

Target application rate (mg/kg)	Residue analysis (mg/kg deltamethrin)									
	Initial[2] (15.11.79)	1 month (13.12.79)	2 months (17.1.80)	3 months (11.2.80)	4 months (12.3.80)	5 months (21.4.80)	6 months (12.5.80)	7 months (10.6.80)	8 months (10.7.80)	9 months (6.8.80)
1.0	0.44	0.53	0.48	0.50	0.50	0.41	0.48	0.49	0.42	0.41
2.0	0.80	1.25	1.33	1.46	1.21	1.01	0.96	1.12	0.94	1.08

1 Analytical results are subject to overall confidence limits of 30%
2 Dates refer to day of sampling

Table 10.3. Concentration of four pyrethroids found on wheat stored for periods of up to 52 weeks at 25 or 35°C with 12 or 15% moisture content **(Noble *et al.* 1982)**

Initial concentration of pyrethroid (replicates) (mg/kg of moist grain)		Storage conditions: Temperature (°C)	Storage conditions: Moisture content (as % of net weight)	Storage conditions: Relative humidity (%)	Concentration of pyrethroid (mg/kg of moist grain) after storage for following periods: 13 weeks	26 weeks	39 weeks	52 weeks
Permethrin	0.96,0.92	25	12	54	1.00	0.89,0.89	0.81	0.87,0.91
	1.03,1.06	25	15	73	1.03	0.87,0.86	0.75	0.83,0.77
	0.95,0.98	35	12	57	0.91	0.77,0.77	0.60	0.74,0.68
	1.02,0.97	35	15	75	0.97	0.64,0.60	0.57	0.52,0.37
Phenothrin	2.11,1.73	25	12	54	1.79	1.62,1.64	1.56	1.07,1.15
	2.0,2.0	25	15	73	1.83	1.66,1.70	1.24	1.01,0.86
	1.92,1.90	35	12	57	1.40	1.21,1.30	1.08	0.72 –
	2.07,1.97	35	15	75	1.51	1.10,1.17	0.96	0.54,0.53
Fenvalerate	1.05,1.16	25	12	54	1.13	1.19,1.13	1.13	0.86,0.90
	1.14,1.08	25	15	73	1.16	1.19,1.22	1.14	0.88,0.81
	1.18 –	35	12	57	1,08	1.17,1.05	0.96	0.71,0.77
	1.14 –	35	15	75	1.06	0.87,1.06	0.87	0.64,0.68
Deltamethrin	2.07 –	25	12	54	1.91	1.89,1.77	1.66	1.51,1.42
	2.13 –	25	15	73	2.00	2.00,1.92	1.64	1.35,1.28
	2.01 –	35	12	57	1,83	1.67,1.66	1.38	1.17,1.26
	1.95 –	35	15	75	1.74	1.39,1.29	1.25	0.62,0.70

Table 10.4. Pseudo first-order rate constants and half-lives for four pyrethroids on wheat stored for 52 weeks at 25 or 35°C, with 12 or 15% moisture content **(Noble *et al.* 1982)**

Pyrethroid	Storage conditions: Temperature (°C)	Storage conditions: Moisture content (as % of wet weight)	Pseudo first-order rate constant $10^3k'$(week^{-1})	Half-life ($t½$) (weeks)	95% confidence limits on $t½$
Permethrin	25	12	2.7	252	161–582
	25	15	4.7	149	108–236
	35	12	7.8	89	69–127
	35	15	15.7	44	37–54
Phenothrin	25	12	9.6	72	58–93
	25	15	13.0	54	44–69
	35	12	17.8	39	34–45
	35	15	23.5	29	27–33
Fenvalerate	25	12	3.3	210	120–870
	25	15	3.8	182	93–1194
	35	12	6.7	104	73–183
	35	15	9.3	74	59–101
Deltamethrin	25	12	6.1	114	92–150
	25	15	7.7	90	65–148
	35	12	9.9	70	62–81
	35	15	20.1	35	28–44

Table 10.5. Comparison of three different criteria for assessment of biological activity (expressed as LC_{50} values) of deltamethrin on wheat **(Hargreaves *et al.* 1982)**

Treatment date	Time from treatment to bioassay (months)	LC_{50}[a] (mg/kg) assessed using following criteria: Mortality after 3 days	Mortality after 26 days	Reduction of F_1 progeny
19 July 1979	9	1.28	0.75	0.90
	12	0.93	0.56	0.62
18 October 1979	6	1.10	0.87	0.93
	9	1.10	0.63	0.79
10 January 1980	0	0.85	0.52	0.55
	3	1.24	0.84	0.73
	6	1.13	0.69	0.76
11 April 1980	0	0.92	0.64	0.66
	3	0.98	0.66	0.55

Regression equations: LC_{50} (26 days) = 0.54. LC_{50} (3 days) + 0.11 R^2 = 0.52. LC_{50} (F_1) = 0.86. LC_{50} (26 days) + 0.13 R^2 = 0.54. LC_{50} (F_1) = 0.65. LC_{50} (3 days) + 0.02 R^2 = 0.55.
[a]LC_{50} values determined from the mortality of *Sitophilus oryzae* (strain QSO56) on treated wheat at 25°C and 70% r.h.

The authors commented on the comparatively faster loss of deltamethrin than permethrin which is surprising because in various biological systems deltamethrin has been shown to be more resistant to degradation. Based on the work of **Soderlund and Casida (1977a)**, and the results of milling studies in their own laboratory, they concluded that when the pyrethroids are used post-harvest on grain, the bulk of the insecticide remains on the outside or at least in the outer layers of the grain, even after storage for 1 year. This suggests that the bulk of the degradation may be non-enzymatic and that chemical oxidative and hydrolytic processes would be more important in a system where light is essentially excluded.

Hargreaves *et al.* (1982) carried out a detailed study in which the biological activity of residues of deltamethrin on stored wheat was compared with the concentration of residues determined by chemical analysis. Two kg aliquots of conditioned wheat were placed in glass bottles of 4.5 litres capacity and an emulsifiable concentrate of deltamethrin, diluted with water, was applied accurately by a standardised method. The bottles were sealed immediately before being tumbled on a mechanical tumbler for 10 minutes. This procedure resulted in an insecticide-treated grain of 12% moisture content. The treated grain was held at 24°C before sampling for bioassay or for determination of residues. Ten treatment rates ranging from 0.125 to 16 mg/kg were used. The probable application rate for control of storage pests on grain in Australia is 1 mg/kg. Bioassays were carried out at 3-month intervals during a period of 15 months. At each sampling interval, new batches of grain were treated by the same technique to give fresh samples for comparative bioassays. Chemical assays were carried out on duplicate samples at 0, 8 and 15 months.

Variability in the distribution of applied deltamethrin was significantly low, the mean residues from 10 samples showing a relative standard deviation of 3.4%. By comparison, when using a slightly different technique, the relative standard deviation was 18.4%. The biological activity in the residues in treated grain was measured by the mortality of *Sitophilus oryzae* after 3 days. The biological activity was also assessed by determining the mortality after 26 days and the reduction of the progeny of the first generation. **Table 10.5** shows a comparison of the 3 different criteria for assessment of biological activity and **Table 10.6** shows the result of chemical analysis during 15 months.

Table 10.6. Chemical analysis by h.p.l.c. of duplicate samples of stored wheat for deltamethrin at various times after application of the insecticide **(Hargreaves *et al.* 1982)**

Deltamethrin applied (mg/kg)	Deltamethrin (mg/kg) found at following times after application: 0 month	8 months	15 months
0.125	0.14,0.12	0.10,0.09	NA
0.25	0.21,0.22	0.18,0.17	NA
0.3	0.27, –	0.27,0.26	0.29,0.29
0.5	0.46,0.43	0.41,0.45	0.46,0.47
0.75	0.75,0.75	0.66,0.74	0.80,0.84
1.0	0.84,0.84	0.95,0.89	0.93,0.95
2.0	1.7,1.5	1.5,1.6	1.9,1.9
4.0	3.1,3.4	2.9,3.2	NA
8.0	6.8,6.8	6.9,6.7	NA

NA = not analysed.

A comparison of the bioassay and the chemical assay data suggests that a significant process of inactivation occurred under these experimental conditions. This was clear from the mean loss of 22% of biological activity in the initial 3 months after treatment. Although significant, this loss was considerably less than has been reported for other protectants, such as pirimiphos-methyl. The deposit was found to retain 65% of the initial activity after 15 months. Whilst no attempt was made to investigate the mechanism of the process, the authors consider it is consistent with movement of the pesticide from exposed grain surfaces to within the bran layers of the grain. As discussed by **Rowlands (1971)**, studies in the authors' laboratories by **Simpson (unpublished data)** established that 67% of the residues of deltamethrin were present in the bran fraction 10 months after treatment while only 10% of the residue was in the flour. Adults of *S. oryzae* lay eggs through the bran layer of wheat into the endosperm and the insect completes its development within the kernel before emerging as an adult. The data are consistent with the hypothesis that most of the residues are in the outer bran layer and that once eggs are deposited in the endosperm they develop through the immature stages unaffected by the insecticide. When they become adults, they emerge from the endosperm region and are affected by the insecticide. The data for chemical residue levels and biological activity clearly indicate that deltamethrin should give prolonged residual action against susceptible species of grain pests. However, as little decay of the compound will occur during typical storage intervals, the initial application rates will need to be at or below the maximum residue limits.

10.3 Fate in Milling, Processing and Cooking

The results of a number of studies conducted through the University of Montpellier, evaluated by the JMPR in 1980, are not sufficiently consistent to provide a basis for valid conclusions. However, the report by **Halls and Periam (1980a)** of work done in the UK is more helpful. In this study 250 kg batches of English wheat treated at a nominal level of 1 and 2 mg/kg deltamethrin were sampled immediately after treatment and after 3, 6 and 9 months storage. The wheat was submitted to the Flour Milling and Baking Research Association, Chorleywood, Buckinghamshire, to be milled and baked.

The wheat was first cleaned of extraneous material then divided into 2 samples and each subjected to a different milling procedure. The first produced wholemeal flour, the second produced white flour, bran and fine offal. Two different samples of white flour were taken from this procedure, the first reduction flour (the cleanest flour mainly from the centre of the grain and used in the baking of cakes) and total flour (straight-run flour plus flour from the bran and fine offal fractions). These samples, together with the wholemeal flour, bran and fine offal were analysed for total deltamethrin content in both batches of treated wheat. In addition, bread was baked from both wholemeal and total white flours to produce wholemeal bread and white bread, respectively, and these loaves were also analysed for deltamethrin content.

Results of analysis for residues of deltamethrin on wheat, milling fractions and bread are given in **Table 10.7**. In both the deltamethrin treatments, levels were initially found by analysis to be only about 40% of the intended level of application. This could be due to, *inter alia*, incorrect spraying rates, inadequate extraction, loss of compound at some stage between nozzle and the grain or loss of residue from grain between sampling and analysis.

When the treated wheats were milled to produce white flour, relatively high deltamethrin residues were found in the bran fraction which is derived from the superficial

Table 10.7. Deltamethrin residues on milling fractions and bread made from freshly toasted wheat and from wheat after 3, 6 and 9 months storage[1] **(Halls and Periam 1980a)**

Target application rate (mg/kg)	Sampling period (months)	Wheat	Wholemeal flour	Wholemeal bread	Bran	Fine offal	First reduction flour	Total white flour	White bread
1.0	0	0.44	0.42	0.26	3.00	nd[2]	nd	0.09	0.10
	3	0.50	0.43	0.27	1.80	~0.5	nd	0.05	nd
	6	0.48	0.42	0.26	1.70	0.35	0.05	0.09	0.10
	9	0.41	0.36	0.36	1.90	0.46	0.03	0.03	0.05
2.0	0	0.80	0.73	0.63	5.40	nd	nd	0.20	0.29
	3	1.46	0.80	0.57	4.70	0.75	nd	0.20	0.15
	6	0.96	0.83	0.66	3.60	0.75	0.10	0.16	0.20
	9	1.08	0.95	0.70	4.40	1.36	0.05	0.17	0.11

[1]Analytical results are subject to confidence limits of ± 30%; [2]nd = not detectable (<0.03 mg/kg).

layers of the grain. Initially, deltamethrin residues on the bran were 6 to 7 times higher than those on wheat at both treatment rates. After 9 months storage the levels were still 4 to 5 times higher on the bran than on the wheat. This observation, together with the appearance of residues on the fine offal fractions after 3 months storage, and on the first-reduction flours after 6 months storage, suggests that although there is no change in the total deltamethrin level on wheat over a storage period of 9 months, there is a subtle change in its distribution. Thus, initially, as might be expected, the residues were found predominantly in the outer layers of the grain, mainly in the bran fraction, but also to a lesser extent in the total white flour fraction. The first-reduction flour, which tends not to come from the superficial layers of the grain, and the fine offal, consisting largely of wheatgerm, was uncontaminated initially but residues moved into these areas as the period of storage increased.

The deltamethrin levels on total white flour milled from treated wheat after different storage periods remained quite consistent at about 17% of the applied rate. Some fluctuation in this proportion was detectable at the lower application rate, but this was probably because, in these cases, the deltamethrin in the flour was at a concentration close to the limit of determination. When baked in white bread, there was no significant decline in the residue level showing that deltamethrin did not degrade during baking.

The following conclusions may be drawn from the results of this trial:

- deltamethrin is not degraded on wheat over a 9 month storage period,
- there is no degradation of deltamethrin when treated wheat is milled to produce either wholemeal or white flour;
- when wholemeal flour made from deltamethrin-treated wheat is baked, there is an apparent reduction in deltamethrin residues, but this is due to the greater moisture content of the bread;
- in the production of white flour and bread from deltamethrin-treated wheat, there is a reduction in the deltamethrin level to about 10 to 20% of the level applied to the wheat. This is because the deltamethrin is predominantly found on the bran fraction at a concentration of four- to sevenfold that in the whole wheat grain. There is no evidence of significant deltamethrin degradation when white flour is baked to produce white bread;
- over a 9 month storage period there appears to be a migration of deltamethrin from the bran fraction to the fine offal fraction, including the wheatgerm, and the first-reduction flour.

Bengston *et al.* (1983b) included synergised deltamethrin at the rate of 2 mg/kg in duplicate field trials carried out on bulk wheat in commercial silos in Queensland and NSW. At the end of 10 months wheat from one of the trials was milled to produce wholemeal flour and white flour, portion of which was made into bread. The milling fractions and bread were analysed and the results obtained are given in **Table 10.8.**

Table 10.8. Fate of deltamethrin residues following milling and baking **(Bengston *et al.* 1983b)**

Milling fraction	Residues (mg/kg) 'as is basis'	'moisture-free' basis	Distribution (%)
Wheat	1.82	–	66.5
Bran	8.4	–	23.2
Pollard	3.8	–	10.3
White flour	0.25	0.29	–
Wholemeal flour	1.8	2.1	–
Wholemeal bread	1.3	2.2	–
White bread	0.2	0.32	–

The work revealed that deltamethrin is highly persistent on stored wheat and that during milling, residues accumulate in the bran fractions so that relatively little appears in the white flour. Residues of deltamethrin are not significantly reduced during baking.

10.4 Metabolism

The following abbreviations have been used in this Section:

CA = chrysanthemic acid
PB = phenoxybenzyl
ITCA = 2-iminothiazolidin-4-carboxylic acid

On oral administration of ^{14}C- deltamethrin to rats, the radiocarbon was rapidly and almost completely eliminated from the body. After 8 days post-treatment, carcass and tissues contained only 1–2% of the dosed radiocarbon. With ^{14}C-labelled compound, approximately 20% of the radiocarbon remained in the animal body after 8 days, the highest concentration being in skin and stomach. No radiocarbon was determined in hair. Essentially all the radiocarbon in the stomach was thiocyanate. No noticeable $^{14}CO_2$ was obtained with either radio-active preparation. The major fecal residue was intact deltamethrin, accounting for 13–21% of the dosed radiocarbon, followed by 4'-HO- and 5 HO deltamethrin and a trace amount of 2'-HO-deltamethrin. Intact deltamethrin and the 4'-HO-derivative appeared not only as the administered *a*S-epimer, but also in parts as the *a*R-epimer, probably due to artificial racemisation on exchange of the *a*-proton in methanol. The metabolites from the acid moiety were mostly Br_2CA, and conjugated with glucuronic acid, with trace levels of the glycine conjugate of Br_2CA, and the glucuronide conjugate of the hydroxy acid probably on the methyl group *trans* to the carboxyl. The major metabolites of the aromatic portion of the alcohol moiety were PB acid derivatives and 4'-hydroxy derivatives. 4'-HO-PB acid sulphate accounted for about 50% of the $a^{14}C$- labelling, free and with small amounts of the glucuronide. PB acid was excreted either without conjugation or as the glucuronide and glycine conjugates. The CN group was converted mainly to thiocyanate and a small amount of 2-iminothiazolidin-4-carboxylic acid ITCA **(Ruzo *et al.* 1978)**.

Oral administration of the above alcohol-labelled or acid-labelled deltamethrin to male mice resulted in the complete recovery (ca. 99%) of the radiocarbon in the excreta after 8 days with 57–65% in the urine and 34–42% in the feces. The recovery of the cyano-labelled deltamethrin was lower, (93%, with 35.5% in the urine and 58% in the feces) with the remaining radioactivity in the skin and stomach. No expired radioactive carbon dioxide was detected. Smaller amounts of unmetabolised deltamethrin were excreted in feces in mice than in rats. The feces, but not the urine, contained 4-monohydroxy ester metabolites and 1-dihydroxy metabolite. Two unidentified ester metabolites were also detected in trace amounts. Major deltamethrin metabolites from the acid moiety were Br_2CA, *t*-HO-Br_2CA and their conjugates. As compared with rats much larger amounts of *t*-HO-Br_2CA and its conjugates were formed in mice. A major alcohol moiety metabolite was the taurine conjugate of PB acid in the urine which was undetectable in rats. Again, as compared with rats, more free PB acid as well as 4'-HO-PB acid and less 4'-HO-PB acid sulphate were excreted in mice. In mice, no ITCA was found **(Ruzo *et al.* 1979) (Figure 10.1).**

Intraperitoneal (ip) administration yields the same metabolites but in different ratios. Deltamethrin is hydrolised *in vitro* by esterases in blood, brain, kidney, liver and stomach preparations. Mice pretreated with piperonyl butoxide (PBO) or S,S,S-tributyl phosphorotrithioate (DEF) metabolise deltamethrin less rapidly than normal mice by oxidative or hydrolytic pathways, respectively. Equitoxic doses of deltamethrin administered orally or ip (with different vehicles for ip) to PBO- or DEF-treated animals yield similar levels (approximately 0.5 mg/kg) of deltamethrin in the brain. Severe poisoning symptoms result on introducing this level of deltramethrin into the brain by direct injection **(Ruzo *et al.* 1979; Soderlund and Casida 1977a)**.

The tissue distribution of toxic doses of ^{14}C-acid, ^{14}C-alcohol and ^{14}C-cyano-labelled deltamethrin after in-

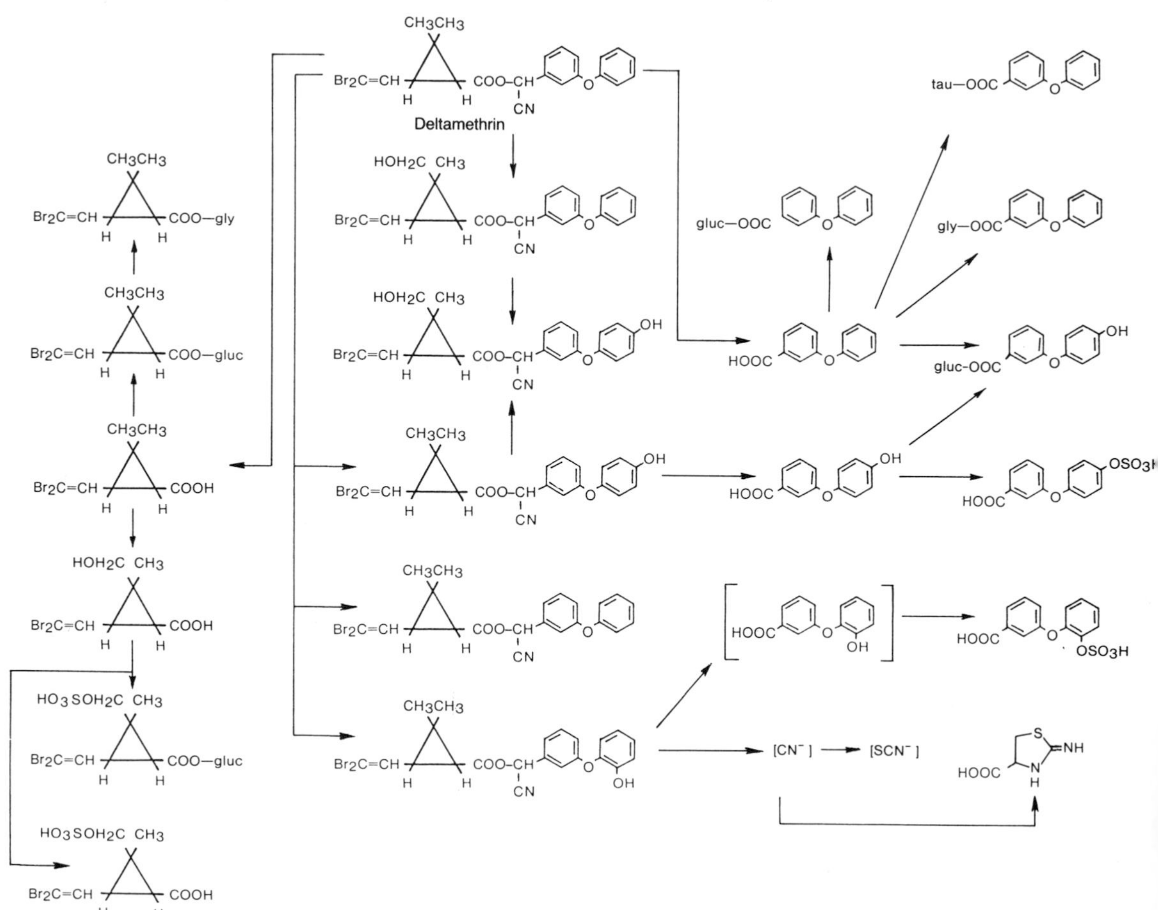

Figure 10.1. Metabolic pathways of deltamethrin in rats and mice **(Ruzo *et al.* 1979).**

travenous administration to rats has been studied **(Gray and Rickard 1981)**. All 3 radio-labelled preparations were found in every tissue examined 1 minute after injection. Peak central nervous system levels were achieved within 1 to 5 minutes but did not correspond to the onset of choreoathetosis.

Akhtar (1984) studied the metabolism of deltamethrin by cow and chicken liver enzyme preparations. He reported that deltamethrin is metabolised by an enzyme or enzymes in various fractions of liver homogenates. Studies that included acid- and benzyl-labelled insecticide showed the main metabolic pathway to be due to cleavage of the ester bond. The enzyme(s) responsible for ester bond cleavage was (were) located equally in both the soluble and microsomal fractions of a chicken liver homogenate. In cow liver homogenate, enzymatic activity was higher in the microsomal fraction. In a later paper **Akhtar *et al.* (1985)** described the metabolism of deltamethrin by Leghorn hens. Various metabolites were isolated and identified. The compounds identified indicated that the metabolic routes of deltamethrin in laying hens included hydrolysis of ester linkage, followed by hydroxylation of 1 or more *gem*-dimethyl groups, hydroxylation of the 2'-, 4'-, 5'-, or 6-position of the phenoxybenzyl moiety.

Leahey (1985) reviewed the metabolism and environmental degradation of deltamethrin and summarised the information by means of schematic diagrams.

10.5 Fate in Animals

The work of **Ruzo *et al.* (1978, 1979)** (see section 10.4) is relevant to an understanding of the fate of deltamethrin in domestic animals.

Studies were carried out on a single lactating Jersey cow, weighing 350 kg, giving milk of approximately 6.3% butterfat, using ^{14}C-deltamethrin labelled in the α-methine group of the alcohol moiety **(Wellcome Foundation Limited 1979)**. A dose of 0.27 g of labelled deltamethrin was injected intrarumenally as a solution in a sesame oil/alcohol mixture. Radioactivity was rapidly excreted, mainly in urine and feces (85.3%). Only 0.4% was found in whole milk, corresponding to peak residue levels after only 1 day of 0.045 and 0.92 mg/kg deltamethrin equivalents in whole milk and rendered butterfat respectively. The residue found in the peak butter sample was mainly (89%) unchanged deltamethrin. The half-life in milk and butter was 0.8 days. Omental fat and leg muscle biopsy samples, removed 2 days after treatment, contained 0.088 and 0.008 mg/kg deltamethrin equivalents respectively.

Akhtar *et al.* (1985) studied the metabolism, distribution, and excretion of deltamethrin by Leghorn hens. A dose of 7.5 mg/kg of ^{14}C-labelled (*gem*-dimethyl or benzyl) deltamethrin/hen/day was given on each of 3 consecutive days, and the elimination of radiocarbon in excreta and eggs was monitored for 5 days after the last dose. About 83% of the administered ^{14}C was eliminated during the first 24 hours after dosing. Tissue residues were generally very low with the exception of those for liver and kidney. Residues derived from the *gem*-dimethyl portion of the molecule tended to be higher than those from the benzyl moiety. Egg yolks contained considerably higher levels of residues than egg albumen. Residues in skeletal muscle were very low (traces to 0.21 mg/kg) during the entire period of the study. Residue levels as high as 3.95 and 6.85 mg/kg were found in the liver and kidneys, respectively, 18 hours after the last dose. The residue levels declined gradually during the withdrawal period. The residues in eggs reached a peak 4 days after treatment commenced with 0.5 mg/kg being found in yolks as compared to 0.2 mg/kg in albumen. The differences in the residue levels of albumen and yolks were probably related to the lipid content of yolks, as well as the manner in which the eggs are produced. The data suggested that once deltamethrin and its metabolites were incorporated in the yolks, there was very little exchange from the yolk to the other body compartments. Instead, the bird proceeded to encapsulate the preformed yolk containing ^{14}C residues with albumen and shell. Consequently, radiocarbon was detected in yolks for several days after the last dose of deltamethrin.

10.6 Safety to Livestock

No specific studies in which the oral toxicity of deltamethrin to livestock has been determined have been located. There are various publications concerning the acute oral toxicity of deltamethrin to laboratory animals, dogs, chickens, ducks and partridge. Although these studies indicate that the acute oral toxicity of the technical grade active ingredient dissolved in vegetable oils or suspended in polyethylene glycol results in a relatively high toxicity to rats and mice, the administration of deltamethrin formulated as wettable powder suspension or flowable (suspension) concentrate results in a relatively low toxic reaction in mice, rats, dogs and various poultry species **(Table 10.9)**.

Furthermore, numerous long-term feeding studies (refer Section 10.7) indicate that the rat, mouse and dog can tolerate relatively high concentrations in the diet daily for a long period without sustaining any observed toxic effect.

In view of the high biological activity of deltamethrin against stored-product pests, the concentration of deltamethrin residues likely to be found in grain or milling offals will be remarkably small, probably never exceeding 5 mg/kg. The likely intake by livestock is therefore not going to present either acute or chronic hazard.

10.7 Toxicological Evaluation

The toxicology of deltamethrin was evaluated by JMPR in 1980, 1981 and 1982 **(FAO/WHO 1981, 1982, 1983)**.

Extensive data from studies on biochemical aspects, mutagenicity, reproduction, teratogenicity, delayed neurotoxicity, potentiation, sensitisation, primary

Table 10.9. Acute oral toxicity of deltamethrin

Animal	Sex	Form[a]	mg/kg	Reference
Mouse	M	PEG 200	21	**Glomot & Chevalier 1976**
	M	sesame oil	33	**Glomot & Chevalier 1976**
	F	PEG 200	19	**Glomot & Chevalier 1976**
	F	sesame oil	34	**Glomot & Chevalier 1976**
	M + F	WP (2.5%)	>15 000	
Rat	M	PEG 200	67	**Glomot & Chevalier 1976**
	M	sesame oil	120	**Glomot & Chevalier 1976**
	F	PEG 200	86	**Glomot & Chevalier 1976**
	M	sesame oil	139	**Glomot & Chevalier 1976**
	M	peanut oil	52	**Kavlock *et al.* 1979**
	F	peanut oil	40	**Ray & Cremer, 1979**
	F	peanut oil	31	**Glomot *et al.* 1979**
	M	25g/L flowable	22 700	**Glomot *et al.* 1979**
	F	25g/L flowable	22 000	**Glomot *et al.* 1979**
	M + F	C.M.C	>5 000	**Glomot *et al.* 1981**
	M + F	WP (2.5%)	>15 000	**Glomot 1979**
Chicken		sesame oil	>1 000	**Roussel Uclaf 1976**
Hen	F	sesame oil	>2 500	**Ross *et al.* 1978**
	F	corn oil	>5 000	**Ross *et al.* 1978**
Mallard duck		corn oil	>4 640	**Beavers & Fink 1977**
Game duck	?	capsule	>4 000	**INRA 1976a**
Grey partridge	M + F	capsule	>1 800	**INRA 1976b**
Red partridge	M + F	capsule	>3 000	**INRA 1976b**
Beagle dog	M + F	PEG 200	>300	**Glomot *et al.* 1977**
Beagle dog	M + F	capsule	>300	**Glomot *et al.* 1977**
Dog	M + F	W.P 2.5%	>10 000	**Glomot *et al.* 1980**

a) The following formulations/media were employed in these studies:
PEG 200 — suspension in polyethylene glycol 200
WP (2.5%) — wettable powder formulation containing 25 g deltamethrin/kg
25 g/L flowable — suspension concentrate formulation containing 2.5 g deltamethrin/L
C.M.C — suspension in carboxy methylcellulose (0.25%) and polysorbate 80 (0.2%)
Sesame oil)
Corn oil) — solutions in the named vegetable oils
Peanut oil)

irritancy; acute toxicity to the mouse, rat, rabbit, chicken, hen, duck and dog; short-term toxicity studies in birds, rats and dogs; long-term toxicity studies in the mouse, rat, dog; carcinogenicity studies, together with information on effects from occupational exposure, were reviewed.

Acute toxicity studies on numerous species indicated that the formulation and medium of administration played a highly significant role in determining the toxic reaction in animals. Solutions in vegetable oils appeared to result in a much higher toxicity than suspensions from wettable powders or suspension concentrates. The latter formulations evidenced an acute toxic reaction only at massive doses. Short-term toxicity studies in rats and dogs served to define the manifestations of intoxication. Two reproduction studies in rats produced no evidence of adverse effect. Five teratology studies in mice, rats and rabbits showed no effects on reproduction or teratogenicity over a wide range of doses other than a significant dose-related incidence in the occurrence of supernumerary nibs in mice in one series of tests. Long-term studies in the mouse, rat and dog were considered satisfactory and the following no-observed-effect levels were determined.

Rat: 50 ppm in the diet, equivalent to 2.1 mg/kg bw
Mouse: 100 ppm in the diet, equivalent to 12 mg/kg bw
Dog: 40 ppm in the diet, equivalent to 1 mg/kg bw

On the basis of these studies an estimate of temporary acceptable daily intake for man of 0–0.01 mg/kg bw was determined.

10.8 Maximum Residue Limits

On the basis of the above toxicological evaluation, and the extensive residue data the JMPR **(FAO/WHO 1981,**

1982, 1983) recommended the following maximum residue limits in cereal grains and milled cereal products.

Commodity	Maximum Residue Limits (mg/kg)
Cereal grains	2
Wheat flour (wholemeal)	1
Wheat flour white	0.5
Wheat bran (unprocessed)	5

The residue is to be determined and expressed as deltamethrin.

The methods recommended for the analysis of deltamethrin residues are those of: **Ambrus *et al.* (1981); Baker and Bottomley (1982); Mestres *et al.* (1979a); Mourot *et al.* (1979).**

11. Diazinon

Diazinon is a broad spectrum organophosphorus insecticide which was first introduced in 1952 and which has been used against a wide variety of insect pests of households, public health, fruit, vegetables and grain crops, and external parasites of livestock. Though inherently unstable, its formulation may be stabilised to produce preparations of outstanding stability. Some of its effectiveness is derived from a moderate volatility and the high potency of its vapours. Diazinon residues have a high affinity for lipids and when dissolved in the fatty tissue of animals resist degradation and excretion for long periods.

Diazinon was widely tested 15 to 20 years ago. It proved more effective than malathion but the potency against some species left something to be desired. The use as a grain protectant was not fully developed and diazinon was never marketed as a grain protectant insecticide.

11.1 Usefulness

Although diazinon was introduced in 1952, one of the first indications of its usefulness for grain protection is the work of **Strong and Sbur (1961)** where it was shown to be 100% effective for killing adults of *Sitophilus oryzae, Sitophilus granarius* and *Tribolium confusum* and larvae of *Trogoderma granarium* after exposure for 28 days at rates of 10 mg/kg or less. *Sitophilus oryzae* proved to be particularly susceptible, 100% mortality being achieved at 1.25 mg/kg deposited on wheat. The performance of diazinon was somewhat better than malathion, dichlorvos and most other insecticides that could be considered for use on grain at that time.

A more extensive study **(Telford *et al.* 1964)** involving both dust and sprays applied to wheat of 2 moisture levels at rates equivalent to 1, 2.5 and 5 mg/kg and stored for 168 days in open jars at ambient temperature gave a critical measure of effectiveness against *Sitophilus granarius, Tribolium castaneum* and *Oryzaephilus surinamensis. T. castaneum* was most susceptible. The dusts were more effective than the sprays and the efficacy of both was reduced more in the high moisture (13.5%) wheat as compared with the wheat maintained at 10.6% moisture. The results in **Table 11.1** indicate there was a sharp fall in effectiveness after the treated grain had been in store for more than 7 days.

Strong and Sbur (1964a) studied the influence of grain moisture and storage temperature on the effectiveness of 5 insecticides as grain protectants. High-grade wheat with a graded moisture range of 10%, 13% and 16% was treated with diazinon (10 mg/kg) and naled (5 mg/kg), azinphos-methyl (5 mg/kg) and fenchlorphos (15 mg/kg). The influence of moisture on the insecticides was shown by reduced mortalities of *Sitophilus oryzae* and an increased number of progeny with increases in moisture content of grain. Likewise, the influence of storage temperature on insecticides applied to wheat of 10% moisture content and stored at graded ranges of 15°, 21°, 27°, 32°, and 38°C was shown by reduced mortalities of *S. oryzae* and increased numbers of progeny with increases in storage temperature. Under the conditions of the study, diazinon showed the best residual effectiveness, being greatly superior to the other 4 compounds. At the lower temperatures and moistures, it retained its full effectiveness for 12 months.

Strong and Sbur (1964b) evaluated diazinon along with 10 other insecticides as protective sprays against internal infestations of grain beetles in wheat. In laboratory studies in the USA, diazinon was applied to wheat containing internal infestations of *Sitophilus granarius, Sitophilus oryzae* and *Rhyzopertha dominica.* Wheat which contained all stages of development of the test insects normally found inside grain was sprayed with acetone solutions of insecticides to give deposits of 10 and 20 mg/kg. Of the compounds tested, dichlorvos was the most effective in reducing the total number of insects emerging from infested wheat, which is the criterion used to assess the effectiveness of treatments against internal infestations. Diazinon, like malathion, had little value against internal infestations but adult insects were killed when they came in contact with insecticide deposits after emerging from treated grain.

In a later study designed to show the inter-relation of moisture content, storage temperature and dosage on the effectiveness of diazinon as a grain protectant against *Sitophilus oryzae,* **Strong and Sbur (1965b)** found that though dosage levels of 2.5 mg/kg and higher killed 100% of adult *Sitophilus oryzae* in initial tests made 1 day after spraying, for full effectiveness over a long period of time, the dosage levels needed to range from 5 to 20 mg/kg. Less diazinon was required for protection of dry grain than grain stored at higher safe levels of moisture, and wheat stored at low temperatures required less diazinon for protection against the *Sitophilus oryzae* than wheat stored at higher temperatures. These results indicate the desirability of adjusting dosages of diazinon to compensate for adverse effects of moisture and temperature that may be found in actual storage of grain.

Lemon (1966) studied the relative susceptibilities of *Tribolium confusum* and *T. castaneum* to 16 organophosphorus insecticides. The insects were treated topically

Table 11.1. Percentage[a] mortality of stored grain insects exposed seven days to various dosages of diazinon-treated wheat after storage. May–November 1962. **(Telford *et al.* 1964)**

		ppm Diazinon													
		10.6% moisture							13.4% moisture						
		1.0		2.5		5.0			1.0		2.5		5.0		
Species	Days storage	Spray	Dust	Spray	Dust	Spray	Dust	Control	Spray	Dust	Spray	Dust	Spray	Dust	Control
STGB[b]	1	10.1	36.7	54.5	91.5	98.0	99.0	4.0	25.0	25.6	61.2	90.8	98.5	100	0.0
	7	5.0	10.4	41.1	59.3	84.9	95.3	4.0	3.5	11.8	25.0	57.0	72.2	94.0	.0
	31	0.5	1.5	5.5	27.6	37.6	57.1	0.5	0.0	0.5	4.4	13.9	35.7	54.4	.5
	84	.5	0.0	1.0	6.6	20.7	28.9	1.8	1.0	2.0	1.0	1.0	6.7	23.4	.0
	168	.5	1.0	2.0	4.5	6.5	5.5	0.0	0.5	1.0	0.5	0.0	2.5	5.5	.5
RFB	1	14.1	76.9	100	100	100	100	1.0	13.4	47.7	100	100	100	100	0.0
	7	2.0	–	82.8	–	100	–	0.0	3.3	6.0	67.6	–	100	100	5.3
	31	0.5	2.5	12.6	47.5	96.0	100	1.5	0.5	2.0	13.0	29.5	76.8	100	1.0
	84	.0	3.5	3.9	5.5	20.1	26.5	2.5	.0	2.5	2.0	10.6	20.1	22.3	2.5
	168	4.0	2.0	5.0	1.0	7.0	–	0.5	3.5	3.5	1.5	4.0	7.0	14.5	1.5
GW	1	0.5	12.4	62.7	96.0	100	100	0.0	1.5	5.5	38.1	95.4	99.5	100	0.0
	7	.0	1.5	7.9	48.2	87.5	97.5	.0	1.5	0.5	3.0	36.3	73.6	94.0	.5
	31	.0	1.0	2.0	3.5	10.6	46.7	.0	0.0	0.0	1.5	2.5	2.0	26.2	.0
	84	2.0	6.1	1.0	4.5	16.5	40.2	3.5	3.5	1.5	1.0	4.9	5.5	24.0	5.1
	168	1.5	2.5	3.5	3.0	8.6	33.2	1.5	0.5	6.0	2.5	4.0	2.0	28.0	1.5

[a]Average of 2 replicates of 100 insects each.
[b]STGB = saw-tooth grain beetle
RFB = red flour beetle
GW = granary weevil.

with 4 graded doses ranging from 0.016 to 0.43 micrograms insecticide per insect. Diazinon was more effective than malathion against *T. castaneum* but less effective against *T. confusum*. It was significantly less toxic to both species than fenitrothion but more effective than bromophos.

Strong *et al.* (1967) determined the relative toxicity and residual effectiveness against 17 species of stored-product insects of malathion and diazinon used for the protection of stored wheat. The insects were exposed to wheat which had been sprayed with solutions of each insecticide designed to achieve concentrations which would enable the lethal concentration to be determined. Diazinon was more effective than malathion against 12 of the 17 species and less effective against 5. Results from residual protectant tests designed to evaluate the relative effectiveness of various concentrations of insecticide residues on wheat in preventing infestations of 6 species of stored-product insects showed diazinon to be slightly more effective than malathion when considering counts of all insects, but the more important differences observed were in the response of species to the treatments involved. Diazinon was inferior to malathion against *Rhyzopertha dominica*.

Strong and Sbur (1968) evaluated the effectiveness of 48 insecticides applied directly to the adults as an acetone/olive oil solvent mixture. Diazinon, one of the standards used for comparing other insecticides, was superior to most, though not all, of the compounds tested.

Speirs and Zettler (1969) tested the toxicity of diazinon, other organophosphorus compounds and pyrethrins, to malathion-resistant *Tribolium castaneum*. The malathion-resistant beetles showed a 10-fold resistance to diazinon.

Using filter paper impregnated with malathion, diazinon, fenitrothion, dichlorvos and mixtures of dichlorvos with other insecticides, **Champ *et al.* (1969)** determined the responses of *Sitophilus oryzae* and *Rhyzopertha dominica* under laboratory conditions. The acute and residual toxicity of the 4 compounds and of the mixtures were checked over a range of moisture contents from 11% to 13%. Comparisons were then made under bulk storage conditions. Fenitrothion was the most promising insecticide. Treatment of immature stages with the 4 compounds showed that malathion was effective only against newly hatched larvae of *Sitophilus oryzae* and the free living stages of *Rhyzopertha dominica*. The activity of the other insecticides against later stage larvae and unemerged adults showed dichlorvos to be superior to fenitrothion which was somewhat more effective than diazinon. These compounds were not effective against mature larvae and pupae.

Strong (1970) compared the susceptibility of *Tribolium confusum* and *Tribolium castaneum* to 12 organophosphorus insecticides following direct application of spray solutions to adult insects. Diazinon was the least effective of the 12 insecticides against the two *Tribolium* species.

LaHue (1970a) evaluated diazinon against malathion, a silica aerogel and a diatomaceous earth, as protectants on wheat against attack by *Rhyzopertha dominica*. The treated grain was stored in bins each holding 110 kg. Diazinon was applied at the rate of 4 mg/kg and malathion at the rate of 10 mg/kg (8.3 mg/kg by analysis). Diazinon, at this concentration, did not give the desired protection. During the first 3 months, diazinon killed *R. dominica* adults in the grain but did not prevent development of their progeny. By the end of 12 months, considerable infestations of *R. dominica* and *Sitophilus oryzae* had become established in the wheat. Although the damage at 12 months was less in the diazinon-treated wheat than in the untreated wheat, it was twice that in the malathion-treated wheat and 6–15 times that in the dust-treated wheat.

Hussain and Qayyum (1972) compared the effectiveness of diazinon with malathion and dichlorvos for control of *Trogoderma granarium* in stored wheat. They found that malathion (0.1%) was slightly better than diazinon (0.1%) and both were better than dichlorvos (0.1%). All 3 insecticides were more effective at 75% than at 52% r.h. Although all three insecticides were initially more toxic at 35°C than at 25°C, the higher temperature decreased their residual effectiveness.

Weaving (1975) included diazinon along with 9 other insecticides in laboratory studies on maize designed to select grain protectants suitable for use under tribal storage conditions in Rhodesia. Diazinon applied to maize at the rate of 4 mg/kg conferred full protection against *Sitophilus zeamais* for 4 months but by the sixth month the mortality of the test insects after exposure for 6 days to treated maize had fallen to 96%. Thereafter, a considerable proportion survived so that at the end of 12 months the mortality had declined to 8%. Though this was superior to the performance given by malathion applied at the rate of 8 mg/kg, it was decidedly inferior to pirimiphos-methyl applied at the rate of 5 mg/kg or fenitrothion (8 mg/kg).

Weaving (1975) evaluated 10 insecticides for use as grain protectants under tribal storage conditions in Rhodesia. Although diazinon, when applied at a dosage of 4 mg/kg performed well for the first 4 weeks, it rapidly lost effectiveness after the sixth week and by week 12 was virtually ineffective. Whether a higher dosage would have been adequate is open to question. Fenitrothion applied at 8 mg/kg and pirimiphos-methyl at 5 mg/kg were still giving complete kill of *Sitophilus zeamais* at 12 weeks.

Attia (1976) reported studies on 3 strains of *Cadra cautella* collected from the field in Australia and which had displayed up to 260-fold resistance to malathion. These strains also showed low order resistance to other organophosphorus insecticides including diazinon. **Attia *et al.* (1980)**, using *Plodia interpunctella* from susceptible and multiple-OP resistant strains, showed that mixed-function oxidases had a significant role in resistance to a number of organophosphorus insecticides, including diazinon. S,S,S-Tributyl phosphorotrithioate (DEF) completely restored susceptibility of the OP-resistant strains to a range of OP insecticides.

11.2 Degradation

Gore (1958) found that diazinon applied as an emulsion concentrate to stored wheat was still present after 6 months at the levels 2.6, 4.6, 5.0, 6.2, 10.2, 17.3, 45.4 and 52.2 mg/kg of which he determined 0.9, 0.7, 1.8, 3.6, 1.8, 7.0, 16.0 and 12.9 mg/kg respectively, as remaining on the surface of the grains.

Telford *et al.* (1964) in making a laboratory evaluation of diazinon as a wheat protectant, analysed samples of wheat which had been treated with diazinon spray and dust after storage in open containers at ambient temperatures. Wheat with 2 different moisture contents was used and each was treated with 3 rates of diazinon, 1.0, 2.5 and 5 mg/kg. The analytical data presented in **Table 11.2** indicates only 25% recovery of the applied insecticide in samples taken immediately after treatment. There was not a gradual degradation as one would expect, the values found after 168 days and at several intermediate states being significantly higher than those found on the day of application.

Table 11.2. Diazinon residues in treated wheat after storage in open containers at ambient room temperatures, May-November 1986 (**Telford *et al.* 1964**).

% average moisture	Formulation	Days storage	Diazinon applied and recovered in ppm 1.0	2.5	5.0
10.6% ± .6%	Spray	0	0.23	0.81	1.44
		7	.26	.65	1.56
		31	.24	.94	1.51
		84	.57	1.07	2.30
			.31	*0.63*	*1.91*[a]
		168	.41	1.02	2.54
	Dust	0	0.51	0.73	2.29
		7	.39	.81	2.31
		31	.49	.97	2.37
		84	.94	2.20	4.29
			.62	*1.62*	*3.46*
		168	.49	1.52	2.56
13.4% ± .5%	Spray	0	0.25	0.55	1.32
		7	.25	.61	1.21
		31	.37	1.03	1.61
		84	.47	1.33	2.89
			.33	*0.68*	*2.11*
		168	.38	.95	2.47
	Dust	0	0.53	0.91	1.97
		7	.32	.88	1.89
		31	.53	1.17	2.31
		84	.73	2.40	4.00
			.45	*1.60*	*2.80*
		168	.29	1.23	2.83

a Italics indicate that the kernels were surface extracted. All other figures are from ground samples.

Strong and Sbur (1964a) studied the influence of moisture content and storage temperature on the residual persistence and effectiveness of selected doses of 5 insecticides in protecting stored grain against insects. The influence of moisture on insecticides, applied on wheat with a graded moisture range of 10%, 13% and 16%, and stored at 15°C, was shown by reduced mortalities of *Sitophilus oryzae* and increased numbers of progeny with increases in moisture content of grain. Likewise, the influence of storage temperature on insecticides applied to wheat of 10% moisture content and stored at graded ranges of 15°, 21°, 27°, 32° and 38°C was shown by reduced mortalities of *Sitophilus oryzae* and increased numbers of progeny with increases in storage temperature. Within the range of conditions likely to be found in grain storage, high moisture has less effect than high temperature. Under the conditions of the study, residual effectiveness of diazinon was considerably greater than azinphos-methyl, fenchlorphos, dichlorvos and dibrom.

Roan and Srivastava (1965) applied diazinon to wheat as an emulsion at the rate of 4.1, 8.2 and 32.8 mg/kg respectively. The treated grain was then stored in open containers for 130 days. Grain samples were removed at intervals to determine the external and internal diazinon residues on the grain, as well as the distribution in 3 milled fractions: bran, shorts and flour. Diazinon was found to penetrate to the interior of the grain with a stable concentration being reached at about 45 days. Residue dissipation beyond this period was extremely slow.

The work of **Rowlands (1966c)** concerned with the elucidation of the metabolic fate of the phosphorothionate insecticides, bromophos, malathion and dimethoate, when applied to stored grain showed that very little oxidation to the more toxic 'oxons' occurs. The insecticides were hydrolysed enzymatically to non-toxic phosphates and thiophosphates that could be incorporated into the normal metabolic processes of the grain. These studies have shown that the persistence of residues and their metabolic fate are dependent on the enzymes present and also on the rate and the route of penetration of the compounds into the grains. Metabolism may be affected by the moisture content, the age, the variety and the viability of the grains so treated and penetration may be determined by the polarity of the insecticide.

Weaving (1975) found that diazinon applied at the rate of 4 mg/kg to maize under tribal storage conditions in Rhodesia produced 100% mortality of *Sitophilus zeamais* for 4 months but by 6 months there were a small number (4%) of survivors. By 8 months more than half of the insects exposed survived. Although this performance is better than that given by malathion, it was inferior to fenitrothion (8 mg/kg) or pirimiphos-methyl (5 mg/kg). However, it must be conceded that the rate of application was hardly optimum for diazinon.

Kansouh (1975) studied diazinon metabolism in stored wheat grains during 3 months storage after treatment. He found the half-life of diazinon on wheat grains and glass beads to be of the order of 26–28 days but after 30 and 60 days only small amounts of the hydrolysis product 2-isopropyl-4-methylpyrimidin-6-ol were detected. The author postulated that the hydrolysis product probably degraded with the loss of ring ^{14}C, probably as $^{14}CO_2$ but little attention appears to have been given to the possibility that diazinon was lost by volatilisation.

Sun *et al.* (1984), in a laboratory study in Taiwan, reported that the toxicity of diazinon, phoxim and pirimiphos-methyl degraded relatively quickly, whilst malathion degraded slowly. Since this finding is contrary to the experience of so many other investigators it must be questioned.

11.3 Fate in Milling, Processing and Cooking

Telford *et al.* (1964) carried out a laboratory evaluation of diazinon as a wheat protectant. In preparation for a panel of organoleptic tests, the samples of the wheat which had been treated with diazinon at the rate of 1, 2, 5 and 8 mg/kg were milled and the diazinon residues in the bran and flour were determined. The results are presented in **Table 11.3.**

Table 11.3. Diazinon residues in flour and bran milled from emulsion-treated wheat employed in the second off-flavour tests, August 1962 **(Telford *et al.* 1964).**

Ppm diazinon applied	Ppm diazinon recovered	
	Bran	Flour
8.0	14.21	0.93
5.0	7.29	0.87
2.0	3.66	0.27
1.0	1.64	0.16
0	0.33	0.05

Note: Total period from application to extraction 9 days.

Roan and Srivastava (1965) applied diazinon to wheat as an emulsion giving a theoretical deposit of 4.1, 8.2 and 32.8 mg/kg. The treated grain was then stored in open containers for 130 days. Distribution of diazinon in the milled fractions showed more than half of the insecticide (57%) in the bran at 24 hours, while the shorts contained 26%, and the flour the least amount (17%). At the longer post-treatment intervals, diazinon residues decreased in the bran with a corresponding rise in the shorts and flour. Averages of all treatment rates indicated that about 70% of the theoretical application adhered to the grain and that cleaning, tempering and milling processes appeared to dissipate approximately half of the residue.

11.4 Metabolism

Perry (1960) reviewed the information then available on the metabolism of insecticides by various insect species. He recorded that diazinon-resistant houseflies exhibiting a 40-fold resistance, as compared with a normal strain, showed practically no difference in cuticle permeability following topical application. The level of diazoxon was 2.5-fold higher in normal than in resistant houseflies but these differences appear to be small in comparison with the magnitude of resistance demonstrated and therefore factors other than metabolism may play a vital part in the final toxicity.

One of the early workers **(Schrader 1963)** considered that diazinon can be broken down to diazoxon and monothiotetraethylpyrophosphate which are very potent cholinesterase inhibitors. Whilst this process can occur *in vitro* in technical grade active ingredient or formulations in the presence of small quantities of water, acids and sunlight, it does not occur in stabilized formulations even at temperatures of 100°C. There is little to suggest that this is a major pathway in animals **(Margot and Gysin 1957; Gysin and Margot 1958).**

Mucke *et al.* (1970), investigating the *in vivo* degradation of diazinon in the rat, characterised urinary oxidative metabolites of the pyrimidyl moiety following hydrolisis. Hydrolysis of the ester bond yielding 2-isopropyl-4-methyl-6-hydroxypyrimidine and oxidation at the primary and tertiary carbon atoms of the isopropyl side-chain were found as the main degenerative mechanisms.

Hastie (1963) reviewed available data on the metabolism and elimination of diazinon from animals and animal tissues. When orally administered, diazinon is degraded by the digestive enzymes before the lipid-soluble material can reach the fat depots. The more circuitous route involved following dermal application allows a certain amount of diazinon to by-pass the sites of degradation, thereby reaching the fat depots unchanged. Re-absorption into the digestive tract from these depots is a somewhat delayed process.

Machin *et al.* (1975) studied the hepatic metabolism of diazinon in the sheep, cow, pig, guinea pig, rat, turkey, chicken and duck. They incubated diazinon with liver microzomes from all of these animals and showed that the products of metabolism included hydroxydiazinon, iso-hydroxydiazinon, dehydrodiazinon, their oxons and diazoxon. An eighth metabolite was tentatively identified as the 6-aldehyde analogue of diazinon. Yields and rates of production of these metabolites varied greatly between species. Production of oxons was not generally correlated with susceptibility to diazinon poisoning, although it was lowest in the least susceptible animal, the sheep. The degradation of oxons by liver slices was too slow to explain the low toxicity of diazinon to the mammals. The relative importance of hepatic metabolism in determining toxicity to vertebrates was stressed.

11.5 Fate in Animals

Experiments carried out with ^{32}P labelled diazinon in a cow **(Robbins *et al.* 1957)** and a goat **(Vigne *et al.* 1957)** have shown that the ^{32}P is rapidly eliminated in the urine, since only a small proportion of the radioactivity can be detected in the blood, feces and milk after 24 hours. The urinary elimination products in the cow include diethyl thiophosphate and diethylphosphate. Similar results have been observed in a dog **(Miller 1963)**.

Investigations of the fate of diazinon in animals were initiated mainly by the widespread use of the product as an ectoparasiticide in ruminants. Residue studies in fat and milk of cows **(Bourne and Arthur 1967; Claborn *et al.* 1963; Derbyshire and Murphy 1962; Matthysse and Lisk 1968)** and of sheep **(Harrison and Hastie 1965)** were carried out after the insecticide had been applied regularly by spraying and dipping or by feeding the animals on pasture treated with the insecticide. Small amounts of diazinon residues have been found in fat and milk, whereas the other tissues were free of the insecticide.

Rai and Roan (1959) found no residues of diazinon in the milk of dairy animals given daily doses of diazinon at the rate of 1.06, 5.30 and 10.60 mg/kg of body weight over a 3 week feeding period. These administration rates were calculated to be 100, 500 and 1000 ppm on the basis of the grain fed, or 51, 290 and 500 ppm on the basis of hay consumed. Steers treated with 165 ppm and 825 ppm in daily oral doses calculated on the basis of grain fed showed traces of diazinon in blood, urine, muscle, liver and brain. Only in fat was a significant residue found, being 0.23 mg/kg at the maximum feeding level.

Because of the distinct difference in outcome of residues of diazinon applied topically to those administered in feed, **Hastie (1963)** reviewed the many papers reporting residues of diazinon following dipping or spraying. This review showed that there is no accumulation or storage of diazinon in any tissue of any species of livestock following oral intake. It is therefore clear that, should diazinon be applied to stored grain, the feeding of such grain or mill offals to livestock will not result in the deposition of detectable residues of diazinon or the excretion of diazinon in milk or eggs. However, because of the relatively high toxicity of diazinon to poultry and other bird species the question of danger to such livestock would have to be considered.

11.6 Safety to Livestock

The acute toxicity of diazinon to laboratory animals and domestic livestock is greatly influenced by the quality of the technical grade active constituent, the formulation and its correct storage. Early in the development of diazinon insecticides, **Gysin and Margot (1958)** and **Margot and Gysin (1957)** drew attention to the tendency to produce monothiotetraethylpyrophosphate (sulfotep) in concentrated solutions due to the action of acid, moisture or sunlight. Commercial diazinon insecticides were protected against this hazard by incorporating suitable specific stabilisers at the original manufacturing or formulation stage. Many of the earlier toxicological studies were carried out with unstabilised technical grade material and the results reflect the presence of small amounts of the highly potent cholinesterase inhibitor, sulfotep. Later studies reported diazinon to be only moderately toxic to mammals.

Birds of all types are highly sensitive to the effects of diazinon which should not be administered to, fed or applied to domestic poultry of any type. **Table 11.4** gives a selection of data on the LD/50 of diazinon to various species.

Radeleff (1958) reported that oral doses of 1 mg/kg of diazinon produced signs of toxicity in calves and that 10 mg/kg was a lethal dose. In steers and sheep doses up to 25 mg/kg and 30 mg/kg, respectively, produced toxic signs but not death. Doses of 10 mg/kg and 20 mg/kg, respectively, were non-toxic. It was later shown that these studies had been conducted with unstabilised technical diazinon which had been degraded by light and moisture and that it probably contained significant amounts of monothiotetraethylpyrophosphate.

Cows were given 1.06, 5.3 or 10.6 mg/kg-bw/day orally for 3 weeks and steers received 1.06, or 5.3 mg/kg-bw/day for 2 weeks. Blood cholinesterase was inhibited in all groups **(Rai and Roan 1960)**.

Fifteen calves were given 10, 25, 40 or 80 ppm of diazinon in the diet, starting at 1 week of age for 14 weeks. Blood cholinesterase inhibition was obvious in all the groups, with a dose-response relationship. However, it did not appear before weaning. In animals killed after the end of the treatment no pathological changes were found **(Geigy 1963)**.

Khera and Lyon (1968) conducted a series of experiments to ascertain whether chick and duck embryos could be used in the evaluation of pesticide toxicity. They found that eggs injected during the first 10 days of incubation were unsuitable but at mid-incubation there was a marked mortality response occurring primarily within 24 hours of injection and another period of mortality among the survivors was manifest near hatching. Diazinon showed a very high toxicity in chick embryos, comparable to that of parathion and mevinphos. In duck embryos it was apparently not quite so toxic. The authors calculated the LD/50 in avian embryos to be significantly lower than that quoted for rodents.

Groups of pigs were administered diazinon by capsule daily for periods up to 8 months at doses of 0, 1.25, 2.5, 5 and 10 mg/kg bw/day **(Earl *et al.* 1970)**. Mortality and cholinergic signs of poisoning were evident at 2.5 mg/kg/day and above. Although significantly increased myeloid/erythroid ratios were observed, no aplastic anaemia was evident.

Schlinke and Palmer (1971) studied the acute toxicity of diazinon to turkeys; groups of 5 birds per treatment were used. The maximum single oral doses that did not cause deaths among turkeys 6 and 16 months old were 1.0 mg/kg and 2.5 mg/kg body weight. Small oral doses repeated daily for 10 days had cumulative toxic effects. The maximum daily doses that did not cause deaths among birds 16 months old were 0.5 mg/kg body weight.

Table 11.4. Acute oral toxicity of diazinon

Animal	Sex	LD/50 (mg/kg)	Reference
Mouse	M	82	**Bruce *et al.* 1955**
	M + F	122	**Gasser 1953**
Rat	M	100–150	**Gasser 1953**
	M	108	**Gaines 1960**
	F	76	**Gaines 1960**
	M	250	**Gaines 1969**
	F	285	**Gaines 1969**
	M + F	466	**Boyd & Carsky 1969**
	M	203–408	**Edson & Noakes 1960**
Guinea Pig	–	320	**Gasser 1953**
Rabbit		143	**Gasser 1953**
Turkey		6.8	**Hazleton Laboratories 1954**
Chicken		40.8	**Hazleton Laboratories 1954**
Goose		14.7	**Hazleton Laboratories 1954**

All the dosages mentioned caused depression of plasma cholinesterase activity 24 hours after treatment, and some caused almost complete suppression.

11.7 Toxicological Evaluation

Four reviews have been prepared by the JMPR of the toxicology of diazinon, 1963, 1965, 1966, 1970 (**FAO/WHO 1964, 1966, 1967, 1971**). No signs of blood dyscrasias were noted in studies in dogs and pigs. No teratogenic or embryotoxic effects were observed in hamsters and rabbits. No-effect levels have been demonstrated in the rat (0.1 mg/kg body weight per day), dog (0.02 mg/kg body weight per day), monkey (0.05 mg/kg body weight per day), and man (0.02 mg/kg body weight per day) and an ADI for man of 0.002 mg/kg body weight was established.

11.8 Maximum Residue Limits

Available information on the level and fate of diazinon residues was evaluated by the JMPR on 7 occasions: 1965, 1966, 1967, 1968, 1970, 1975 and 1979 (**FAO/WHO**). None of the information dealt with post-harvest use on stored grain, apparently because no such uses had been registered in any country. For that reason no recommendations have been proposed for MRLs in cereal grain to deal with post-harvest applications. The MRLs that have been recommended are intended to cover residues resulting from pre-harvest application only.

The methods recommended for the analysis of diazinon residues are those of: **AOAC (1980a); Pesticide Analytical Manual (1979a, 1979b, 1979c); Manual of Analytical Methods (1984); Methodensammlung (1984 — XII-5,6; S5; S8; S10; S13; S17; S19); Abbott *et al.* (1970); Ambrus *et al.* (1981); Panel (1980)**. The following methods are recognised as being suitable: **Methodensammlung (1984 — 35B); Bowman and Beroza (1967); Carson (1981); Eichner (1978); Krause and Kirchhoff (1970); Mestres *et al.* (1977, 1979b); Sissons and Telling (1970); Sprecht and Tillkes (1980).**

12. Dichlorvos

Dichlorvos (dimethyl-2,2-dichlorovinyl phosphate, often referred to as DDVP) was introduced as an insecticide about 1955. It is widely used in households, food storage, for the treatment of companion animals, as an anthelmintic in pigs, poultry and horses and for the control of stored-product pests.

Dichlorvos is characterised among common insecticide compounds by its high vapour pressure and by high insecticidal activity in the vapour phase. The vapour pressure of dichlorvos at room temperature is approximately 1000 times greater than the figures quoted for parathion, malathion, diazinon and mevinphos at the same temperature.

Dichlorvos deposits are very active against the whole spectrum of stored-product pests; especially against the larval stages within the grain and also against moths. It is outstandingly effective in the vapour form. It performs best in storage systems which result in a physical or mechanical distribution of the vapour throughout the stored grain. Dichlorvos is useful for disinfesting grain brought into the storages with live insects. There is some evidence that the addition of dichlorvos increases the effectiveness and residual life of malathion. Dichlorvos has a short residual stability, especially at higher temperatures and higher moisture levels. It is readily metabolised and rapidly excreted. Because of the high penetrating effect, the residue on the whole grain readily transfers into flour, but these residues in flour degrade quickly and are completely destroyed in the cooking of bread or rice.

Shell Research Limited (1971, 1972, 1973, 1974) published a bibliography of 1510 references to the properties and uses of dichlorvos.

12.1 Usefulness

Strong and Sbur (1961) showed that dichlorvos was active against a range of insect pests of stored produce. Experimental work has since been carried out on its application for the protection of cereals and cereal products during storage and this has led to the widespread development of dichlorvos as a grain protectant. **Strong and Sbur (1964a)** studied the persistency of dichlorvos as a spray on wheat (16% moisture) at 10 mg/kg and found its toxicity to *Sitophilus oryzae* was lost after 2 weeks at high storage temperature; whereas when stored at 15°C, the treated wheat still killed these rice weevils for as long as 3 months.

Parkin *et al.* (1963) compared the susceptibility of several species of stored-product beetles to dichlorvos vapour derived from resin pellets containing 20% of this volatile insecticide. The exposures were made under rather variable conditions of room temperature and relative humidity. The vapour, at the high concentration used, caused all the beetles to be knocked down in from 5 to 60 minutes; *Anthrenus flavipes* and *Oryzaephilus surinamensis* were the most rapidly affected and *Sitophilus oryzae* and *Sitophilus granarius* the slowest. A mortality rate of 50% was obtained in 2 days with *Oryzaephilus surinamensis* and *Stegobium paniceum*; in 3 days with *Sitophilus oryzae, Sitophilus granarius* and *Lasioderma serricorne* and in 7 days with *Anthrenus flavipes*.

Strong and Sbur (1964a) studied the influence of grain moisture and storage temperature on the effectiveness of 5 insecticides as grain protectants. Under the conditions of the study, the residual effectiveness of dichlorvos ranked fourth among the 5 compounds tested.

Strong and Sbur (1964b) report laboratory tests made in a study to assess the effectiveness of 11 insecticides against internal infestations in wheat of 3 species of grain beetles: *Sitophilus granarius, Sitophilus oryzae* and *Rhyzopertha dominica*. Wheat samples sprayed with acetone solutions of insecticide to give deposits of 10 mg/kg and 20 mg/kg contained all stages of development of the test insects normally found inside grain. Dichlorvos was the most effective compound in reducing the total number of insects emerging from infested wheat, which is the criterion used to assess the effectiveness of treatments against internal infestations. The authors concluded that the combination of dichlorvos with a residual grain protectant would offer intriguing possibilities in a grain protectant program where infestations may originate in the field and storage facilities are unsuitable for fumigation.

Kane (1965) carried out experiments in the laboratory to evaluate dichlorvos for disinfestation of bagged grain heavily infested with *Oryzaephilus surinamensis* and *Sitophilus granarius*. The liquid dichlorvos was diluted with tetrachlorethylene and injected as 4 mL doses into the bags of grain. Some bags were given single, centre injections at the rate of 4 and 8 mg/kg and others were injected at each corner to give a total of 32 mg/kg dichlorvos. Counts of live and dead insects in spear samples taken from the bags indicated that there was little diffusion of insecticidal vapour through the intergranular spaces. Even at the highest dose, only 59% *Oryzaephilus surinamensis* and 14% *Sitophilus granarius* were dead in the sample taken from the mid-point of the 4 injections 24 hours after treatment and there was virtually no increase in kill after 4 days. The method is unlikely to be of practical value.

Green *et al.* (1965) treated a bulk of feed barley on the floor of a barn with dichlorvos emulsion at the rate of 4 mg/kg. The barley was heavily infested with *Oryzaephilus surinamensis* and was found to be heating. The treated barley was deposited in 2 small heaps, one being covered with gas-proof sheeting to create favourable conditions for fumigant action. The initial concentration of dichlorvos, as determined chemically, was 1.9 mg/kg in both heaps. No living insects were found in either heap 24 hours after treatment. Biological and chemical assays showed effective residues to be present for between 6 and 10 weeks.

Green and Tyler (1966) described a method for the protection and disinfestation of farm-stored grain in Britain by the admixture of water-based emulsions of insecticides, including dichlorvos. Heavy infestations of *Oryzaephilus surinamensis* in feed barley were quickly controlled by the application of dichlorvos at 4 mg/kg. Dichlorvos disappeared more quickly than did malathion or fenitrothion and conferred relatively short-lived protection.

Green *et al.* (1966a) conducted tests on dichlorvos-impregnated resin strips, which showed considerable promise for future extension of its use in the control of moths in grain stores.

Press and Childs (1966) applied dichlorvos in warehouses to control *Ephestia elutella*. Treatments applied weekly from mid-summer through autumn in 1959 reduced the number of moths caught in traps to nearly half that of 1958. Regular treatments starting in the spring as soon as moths began to appear resulted in a 99% reduction in 1960 and 90% in 1961. During the same time, the number of moths trapped in untreated warehouses increased.

Green *et al.* (1967) found that when dichlorvos was sprayed into an air-stream produced by a motorised knapsack sprayer this could be injected into bagged grain and a good distribution of the insecticide obtained. They carried out detailed trials to compare the effectiveness of injected dichlorvos against *Oryzaephilus surinamensis* and *Sitophilus granarius* infesting wheat and barley and, at the same time, compared the dosage rates on clean and holed grain for the control of *Oryzaephilus surinamensis*. Bags of wheat and barley (50 kg) were stored at 25°C and 45% r.h. Initially they were subjected to heavy cross-infestation by *Oryzaephilus surinamensis* which, because the grain was relatively undamaged, were confined to the intergranular spaces and the grain surface. After 15 weeks, when heavy breeding was evident, some bags were removed and dichlorvos was injected at rates of 2.5, 5, 10 and 20 mg/kg respectively. The remaining material was then also infested with *Sitophilus granarius* which, after a further 9 weeks were breeding inside the seeds and had made holes which provided shelter for many *Oryzaephilus surinamensis*. The treatments were then repeated on this material. Before the grain had been damaged by *Sitophilus granarius*, 100% kill of all stages of *Oryzaephilus surinamensis* was obtained with 5 mg/kg dichlorvos and about 99.8% with 2.5 mg/kg. After holes had been made in the grain by *Sitophilus*, however, very small numbers of *Oryzaephilus surinamensis* were able to survive at doses up to 10 mg/kg. With *Sitophilus granarius*, about 1% of adults, together with some other stages, were able to survive a dose of 20 mg/kg.

Vikhanski (1967) tested the efficacy of 8 organophosphorus compounds for controlling 5 stored-grain pests. Dichlorvos and fenitrothion possessed the highest contact toxicity against *Tribolium confusum, Rhyzopertha dominica, Sitophilus granarius, Oryzaephilus surinamensis* and tyroglyphid mites.

Harada (1967) tested a dichlorvos/PVC strip containing 12% dichlorvos against adult *Sitophilus oryzae*. The use of 10g of the preparation/m^3 was enough to kill all insects in 24 hours at 15°C. A similar efficiency was observed even after 40 days exposure of the preparation to air at 28°C.

Agarwal and Pillai (1967) determined the relative effectiveness of 21 organophosphorus insecticides against fully-fed larvae of *Trogoderma granarium* in the laboratory in India. The compounds were applied topically to the dorsal thoracic region, and mortality was recorded after 48 hours. Eight insecticides were more toxic, but dichlorvos was significantly less toxic, than malathion.

Scoble and Crawford (1967) described the properties and uses of dichlorvos fumigant strips for use in warehouses against stored-product moths. They recommended one commercial strip per one thousand cu. ft. (28 m^3.) of free air space for up to 3 months control of such pests as *Ephestia elutella, Anagasta kuehniella* but not *Sitophilus granarius*.

PVC strips impregnated with dichlorvos were tested in a small closed room and in a large grain storehouse in Nigeria (**Qureshi 1967**). In the closed room, laboratory-reared adults of 5 stored-product insects were exposed for 8 days to vapours from strips that had been hung there one day or 4–15 weeks previously. Mortality in all cases was greater than 99%, the highest occurring in the 4th week. The test insect most susceptible to the vapour was *Oryzaephilus mercator*, followed in order of decreasing susceptibility by *Callosobruchus maculatus, Tribolium castaneum, Dermestes maculatus* and *Trogoderma granarium*. In the storehouse, the grain was fumigated with methyl bromide to remove unwanted insects, dichlorvos strips were hung in position at the rate of 1 strip per 1500 cu. ft. (42.5 m^3) of free space, and adults of *Tribolium castaneum* in muslin bags containing groundnuts were introduced 4 and 6.5 weeks later and left suspended for 8 days. Mortality varied greatly in different parts of the storehouse, but total mortality occurred only in the hottest part under the corrugated iron roof; it is uncertain whether the heat caused death directly or whether it increased the effectiveness of the strips.

In the laboratory at Kanpur, India, the effect of constant temperature and humidity on the contact toxicity of dichlorvos and 3 other insecticides to adults of *Tribolium castaneum* was studied on the basis of their LC/50 at 2 levels of temperature (about 27 and 35°C), with about 75% humidity, and at 2 levels of humidity (70 and 90%), with a temperature of about 27°C. **(Teotia and Pandey, 1967).** The mortality rates obtained after exposure to residual films of the insecticides for 24 hours showed that dichlorvos gave better control at the higher temperature and that it was more effective at the higher, than at the lower, humidity.

Green *et al.* (1968) applied dichlorvos daily as an oil mist at the rate of 2.2 mg active ingredient/1000 L and demonstrated that it gave good control of *Ephestia elutella* in a London warehouse. The mist was dispensed automatically at night so as to minimise labour cost and interference with warehouse business.

Kirkpatrick *et al.* (1968) showed in laboratory tests that dichlorvos had limited effectiveness as a protectant when applied to wheat with 12% moisture content, but it was effective against *Sitophilus oryzae* when the insects were introduced immediately after treatment. In an unpublished study conducted by the USDA Plant Pest Control Laboratory **(Padget 1968)**, wheat was treated with dichlorvos prepared from an emulsifiable concentrate to give a deposit of 4, 6 and 8 mg/kg. The treatments were effective against *Graphognathus leucoloma* but the residues declined rapidly and by 14 days all dichlorvos residues were at, or below, 0.5 mg/kg.

The separate and combined action of 6 insecticides on adults of *Callosobruchus chinensis* exposed to residual films and kept for 24 or 48 hours at about 27°C and 75% relative humidity was studied in the laboratory at Kanpur, India. **(Teotia and Kewal Dhari, 1968)**. The results indicated that mixtures of 1 part carbaryl with 1 part dichlorvos had antagonistic action and were significantly less toxic to the beetles than the component insecticides, but that 1:1 mixtures of lindane and dichlorvos had a marked synergistic action.

Vikhanskii (1968) studied the potency and stability of some organophosphorus insecticides used for disinfestation of storehouses. DDT was used as the standard for comparison. Dichlorvos and fenithrothion were found to be 8 times more potent than DDT to *Tribolium confusum* larvae.

Coulon and Barres (1968) tested a number of methods of applying dichlorvos as aerosols, fogs and pressurised liquid sprays to the insides of barges used for transporting grain. The empty barge holds were artificially infested with *Sitophilus granarius, Tribolium castaneum* and *Acarus siro* and the insecticide was applied through open roof-panels of the holds. The treatments all reduced pest populations but did not prevent infestation of clean grain introduced without sufficient protection. The investigators concluded that for these methods to be effective in boats, holds must be made more airtight, the release of the insecticides improved so as to give more uniform coverage of the whole area, and the time of treatment prolonged to 24 hours in order to give the deposits time to vaporise and spread to sheltered refuges.

Champ *et al.* (1969) examined the effectiveness of dichlorvos, malathion, diazinon and fenitrothion for controlling *Sitophilus oryzae* and *Rhyzopertha dominica* which, in their larval stages, live within the grain and showed that dichlorvos was significantly more effective than the other materials against later stage larvae and unemerged adults.

Green and Wilkin (1969) used dichlorvos successfully to control insects in bagged grain. The insecticide was sprayed into the air stream from a motorised knapsack sprayer through a perforated lance which was inserted into the grain.

Vikhanskii and Sturua (1969) determined the toxicity of 5 organosphosphorus insecticides to the 4 life stages of *Tribolium confusum*. Their findings are summarised in **Table 12.1.**

Table 12.1. Toxicity of organophosphorus insecticides to *Tribolium confusum* **(Vikhanskii and Sturua 1969)**

Insecticide	LC/50 (mg/L)			
	adults	pupae	larvae	eggs
dichlorvos	167	443	194	150
fenitrothion	208	391	162	1 089
malathion	780	791	264	33 950

Mansur (1969) determined the concentrations of fumigants in the atmosphere that were toxic to *Sitophilus oryae* and *Tribolium confusum* held for 24 hours at 25°C and 65% relative humidity. The results are given in **Table 12.2.**

Dichlorvos was tested on hard winter wheat, shelled corn, and grain sorghum as a short-term protectant against *Sitophilus oryzae, Tribolium castaneum, T. confusum* and *Rhyzopertha dominica*. Evaluations of toxicity were made at intervals by the determination of mortalities after ageing of the deposits, by counting the number of progeny developing after toxicity test exposures, and by an assessment of progeny damage to the treated grain. Dichlorvos, at calculated deposits of 5, 10 and 20 mg/kg, was more effective against the adults of all four test-insect species on corn than on grain sorghum and wheat **(LaHue 1970b)**. Dichlorvos did not prevent the emergence of *Sitophilus oryzae* adults from immature forms present in infested material mixed with treated grain; however, the 20 mg/kg application prevented the establishment of an infestation in corn 14 days after treatment and in sorghum 7 days after treatment. Deposits of dichlorvos as high as 20 mg/kg did not prevent the emergence of a few *S. oryzae* adults from heavily infested corn, wheat and sorghum.

Coulon *et al.* (1970) showed that immature stages of

Table 12.2. Fumigating action of dichlorvos, chloropicrin and carbon tetrachloride **(Mansur 1969)**

Insecticide	LC/50 (mg/L)		LC/95 (mg/L)	
	S. oryzae	*T. confusum*	*S. oryzae*	*T. confusum*
dichlorvos	0.008	0.0054	0.024	0.012
chloropicrin	1.8	3.7	8.0	9.0
carbon tetrachloride	150.0	160	233	240

Sitophilus spp. are relatively insensitive to dichlorvos applied in the form of a diluted emulsion.

McGaughey (1970a) evaluated dichlorvos as a direct spray application at 5, 10, 15 and 20 mg/kg in comparison with malathion at 14 mg/kg for insect control in rough rice. Small-scale laboratory tests and small-bin tests were used. These revealed short-term protection by dichlorvos against *Sitophilus oryzae* and *Rhyzopertha dominica*. *Tribolium confusum* were controlled immediately after treatment but not at subsequent intervals. Dichlorvos was more effective than malathion in reducing emergence of *Sitophilus* spp. While high moisture content of the rice at the time of treatment adversely affected the duration of protection, it had no obvious immediate effect because nearly complete control of *Sitophilus* spp. was obtained at the time of treatment.

Strong (1970) compared the relative susceptibility of *Tribolium confusum* and *Tribolium castaneum* to 12 organophosphorus insecticides by direct application of spray solutions to adult insects. Dichlorvos was ranked seventh and eighth respectively.

Mukherjee and Saxena (1970) compared the relative toxicity of 14 organophosphorus insecticides as direct sprays or as films against the adults of *Tribolium castaneum*. Dichlorvos spray was found to be less toxic than fenitrothion but more toxic than all other materials. This was probably due to the volatility of dichlorvos causing its persistent toxicity to be low.

Girish *et al.* (1970a) measured the susceptibility of various developing stages of *Sitophilus oryzae* to dichlorvos. In decreasing order of susceptibility this was: adult, 0–48 hour-old eggs, 3–5 days-old larvae, 4–6 days-old eggs, 7–9 days-old larvae, 13–15 days-old larvae, 4–6 days-old pupae.

Merk (1970) described the properties of dichlorvos with regard to decontamination of stored grain. He advised that grain disinfestation is 100% effective with correctly formulated dichlorvos preparations uniformly dispersed at concentrations of 5–10 mg/kg.

Laboratory tests were made in Bulgaria with nine insecticides as possible alternatives to parathion for the control of *Sitophilus granarius* and *Sitophilus oryzae* in grain storages. When adults of these weevils were exposed to direct sprays of dichlorvos, total kill was obtained in 24 hours. When adults were released in glass dishes previously treated with dichlorvos, complete kill of *Sitophilus granarius* was obtained **(Shikrenov and Sengalevich 1970)**.

Two warehouses at Nairobi, Kenya, containing fumigated bagged wheat were treated with dichlorvos slow-release strips at the rate of 1 strip per 25 m^3 of free space and compared with neighbouring untreated warehouses during 18–20 weeks **(McFarlane 1970a)**. The treatment afforded a satisfactory and economic control of *Cadra cautella* throughout the observation period in an undisturbed warehouse with little ventilation and no time lapse between fumigation and application of the strips. In the other warehouse, with less favourable conditions, control was not so good, especially where there was a short delay of 2–3 days after fumigation before the dichlorvos strips were installed. The treatment was not effective against *Tribolium castaneum*.

Oshima *et al.* (1970) tested the insecticidal effect of dichlorvos/resin strips under semi-field conditions in Japan. In an airtight room the results against larvae of *Tribolium castaneum* and *Plodia interpunctella* were unsatisfactory, though their adults were sensitive even under ventilation.

Wyckoff and Anderson (1970) described the application of dichlorvos vapour as a preventative measure for controlling incipient warehouse infestations. During a 12-month test, experimental use of dichlorvos at a major military subsistence depot resulted in a 98% reduction of commodity tonnage infested.

LaHue (1971) exposed adult *Plodia interpunctella* to dichlorvos/PVC resin strips for 24 hours to obtain satisfactory control. The following year (1972), LaHue noted that female *Plodia interpunctella* deposited a greater number of eggs following exposure to dichlorvos. However, most adults emerging 3 to 4 weeks after exposure were killed before mating.

McGaughey and Cogburn (1971) attempted to utilise the effectiveness of the dichlorvos for rapidly disinfesting commodities to complement the residual effect of tetrachlorvinphos with encouraging results. They found that the combination could be used to treat infested grain to destroy the existing infestation and simultaneously provide long-term protection.

Qadri (1971) described a bioassay technique to estimate vapour toxicity of dichlorvos using the housefly (*Musca domestica*) which was 80.55 times as susceptible as *Tribolium castaneum* and 4.35 times as susceptible as *Sitophilus oryzae*. The method estimates about 20 μg of dichlorvos vapour per test chamber.

Reinhardt and Esther (1971) studied the sensitivity of some insects to organophosphorus insecticides throughout the day and measured the importance of light in determining sensitivity. The susceptibility of *Sitophilus granarius* to deposits of dichlorvos varied in the course of the day. In *Sitophilus granarius* the effect of dichlorvos varied according to the light conditions during exposure time. The susceptibility of the weevils was maximum when the exposure included the change from darkness to lightness and the reaction was minimal under the reverse conditions.

Coulon (1972) found that dichlorvos, when used at rates between 5 and 20 mg/kg, was effective in immediately destroying a high proportion of the population of *Sitophilus granarius* which was not immediately apparent because the larval, pupal and young adult stages were hidden within individual grains but it suffered from a rapid loss of residual activity. **Coulon *et al.* (1972)** rated dichlorvos at 10 and 20 mg/kg equivalent in residual effect to bromophos, iodofenphos and malathion used at 5 mg/kg.

Coulon *et al.* (1972a) studied the activities of different formulations of dichlorvos applied to grain infested internally by *Sitophilus granarius* and the influence of the amount and the homogeneity of the dispersion on emergence. The performance of the 4 formulations varied by a factor of 2 between the most effective and least effective. The reduction of emergences could be increased by up to 21% when the percentage of grains contacted by the insecticide was increased.

Bond *et al.* (1972a) studied the use of dichlorvos to control insects in empty cargo ships. Dichlorvos vapour applied to the empty holds was found to give effective control of *Sitophilus granarius* distributed throughout the holds. The insecticide was applied by vaporisation from impregnated resin pellets using a vapour dispenser and dispersion as an aerosol from a pressure cylinder. Both methods of application were found effective but the aerosol system was more versatile and easier to use.

Hussain and Qayyum (1972) tested the effectiveness of dichlorvos for controlling *Trogoderma granarium* in stored wheat. They found that 0.1% dichlorvos was inferior to similar concentrations of malathion and diazinon and that residues of the latter 2 insecticides were effective longer than dichlorvos residues at 25 35°C. Dichlorvos had a high initial toxicity and a short residual effect. All 3 insecticides were more effective at 75% than at 62% r.h.

McDonald and Press (1973) determined the toxicity of 8 insecticides to *Plodia interpunctella* adults. Though dichlorvos was less potent then pirimiphos-methyl and bioresmethrin it was more toxic than malathion.

McGaughey (1972a) evaluated dichlorvos, tetrachlorvinphos and a dichlorvos/tetrachlorvinphos mixture as protectants for stored rough rice. The study revealed that even when the dichlorvos was applied at twice the concentration of the tetrachlorvinphos, the residues of dichlorvos disappeared much more rapidly than those of tetrachlorvinphos, and, after 6 months storage, there were no detectable residues of dichlorvos.

Morallo-Rejesus and Santhoy (1972), who evaluated the toxicity of 6 organophosphorus insecticides against DDT-resistant strains of *Sitophilus oryzae*, found dichlorvos to be one of the less effective materials.

Harein and Rao (1972) tested dichlorvos and tetrachlorvinphos separately and in combination as protectants for stored wheat against *Sitophilus granarius*. Dichlorvos, applied at a concentration of 2 mg/kg, proved effective in controlling an existing infestation over an exposure period of 5 days. Dichlorvos at 4 mg/kg inhibited reproduction of the weevil. Generally, the mixtures became more effective as the percentage of dichlorvos in them increased.

Morallo-Rejesus (1973) compared dichlorvos with four other insecticides as protectants of shelled corn during storage in the Philippines. Though dichlorvos proved effective against a number of important pests, especially *Sitophilus oryzae*, its lack of residual protection was most apparent at the end of a 6-month storage.

Morallo-Rejesus and Nerona (1973) compared grain sacks impregnated with dichlorvos with sacks treated with other insecticides. Dichlorvos was shown to be unsuitable for this purpose.

McGaughey (1973) evaluated a dichlorvos/tetrachlorvinphos mixture for the control of *Sitotroga cerealella* moth in rough rice where it proved quite effective. **Harein (1974)** pointed to the uniform high potency of dichlorvos to the 5 most important stored-product moth pests.

Wilkin and Hope (1973) evaluated dichlorvos as an aqueous emulsion against 3 stored product mites. When applied at the rate of 2 mg/kg it was completely effective against *Glycyphagus destructor*, 50–75% effective against *Tyrophagus putrescentiae* but less than 25% effective against *Acarus siro*.

Lum *et al.* (1973) studied the fecundity and egg viability of stressed female *Plodia interpunctella*. They found that female moths that were stressed with carbon dioxide gas or dichlorvos vapour or by decapitation deposited viable eggs. Of the 3 stresses, carbon dioxide was the most effective in reducing egg production and egg hatchability. Dying females released viable eggs which could lead to reinfestation. The significant decrease in the hatchability of eggs laid by females treated with carbon dioxide were quite different from those produced by physical damage such as decapitation or by dichlorvos vapour.

McGaughey (1973a) studied the use of dichlorvos vapour for insect control in a rice mill. An average dose of 63.5g/100 m^3 dispensed over 6 hours twice a week for 2 weeks in each of 7 rooms of a rice mill caused higher mortalities among caged insects than a routine pyrethrum fog; 3–4.5 times higher among adult *Lasioderma serricorne;* 48–92 times higher among adult *Tribolium castaneum*; 3–5.5 times higher among adult *Rhyzopertha dominica;* and 5–7 times higher among larvae of *Cadra cautella*. These dichlorvos treatments caused a decrease in the insect population of the mill to one quarter the pre-treatment level.

Bengston (1976) developed a dichlorvos aerosol emission system that gave control of both naturally occurring and experimentally caged populations of *Ephestia cautella*, superior to that given by slow release dichlorvos strips in wheat storages in Queensland. Daily applications of 1 g dichlorvos aerosol per 100 m^3 were released at dusk using the system involving a time clock, timer and solenoids controlling simultaneous release from 15 aerosol dispensers in a storage 61 m x 15 m with a free airspace of 2100 m^3. Safety and reliability of the system were both satisfactory. The aerosol system is initially more expensive than the slow-release strips but costs are comparable over 3 summers and the aerosol system is cheaper thereafter.

Desmarchelier *et al.* (1977a), by means of bioassay tests, demonstrated the potency toxicity of dichlorvos vapour to *Sitophilus oryzae* and *Rhyzopertha dominica*. On freshly treated wheat at 20°C the lethal effect of dichlorvos was predominantly due to a mobile form of dichlorvos. This contrasts with the action of malathion. Dichlorvos vapour at toxic concentrations was moved through grain for up to four metres, using airflows comparable to those used in aeration of bulk-stored wheat. There was a marked increase in efficacy of dichlorvos under forced airflow conditions. These workers deduced that, were dichlorvos present entirely in the vapour form, an application of 0.0025 mg/kg would give a concentration in the inter-granular air space of more than 5.3 μg/*l*, a concentration sufficient to kill *Tribolium castaneum* after an exposure of 30 minutes. In these studies, *Rhyzopertha dominica* was controlled by an application equivalent to 0.7 mg/kg and the use of positive airflow, compared with failure to control at a conventional application rate of 18 mg/kg. This represented an increase in effectiveness of at least 25-fold, but the authors claimed it still fell far short of the potential of the system.

Desmarchelier (1977b) demonstrated that selective treatments using combinations of pyrethroid and organophosphorus insecticides including dichlorvos required less insecticide and gave superior results than any 'blanket' or all-purpose treatment. The net effect of lower temperature in reducing toxicity, but increasing persistence, was to increase the period of protection given by insecticides in cooled storages. It was suggested that the amounts of insecticides applied to grain could be considerably reduced if more consideration were given to grain conditions, if re-infestation pressures were more carefully examined and if less attention were given to surviving adults under conditions where progeny do not develop.

Loong Fat Lim and Sudderuddin (1977) demonstrated that, unlike a number of grain protectants, dichlorvos was more than twice as effective at 18°C against *Sitophilus oryzae* than at 28°C.

Zettler and Jones (1977) compared the toxicity of 7 insecticides, including dichlorvos, to malathion-resistant *Tribolium castaneum*. They found that all 7 insecticides were as toxic to the resistant strains as they were to the susceptible strain.

Desmarchelier (1978c), in experiments with caged and uncaged insects in wheat treated with dichlorvos, found that dichlorvos vapour was the dominant toxic agent as caged insects were killed to the same extent as uncaged insects. Therefore, dichlorvos vapour, not total dichlorvos, is the available form of the insecticide.

In a review of the formulation and use of 3 insecticides that have been developed for protecting stored products, **Saba and Matthaei (1979)** recommended dichlorvos for protecting grain, tobacco, skins etc. for short periods.

Singh (1979) evaluated malathion and dichlorvos applied pre-harvest for controlling insect infestation of stored rice. When rice in the field was sprayed with malathion 5 days before harvest the harvested rice exhibited almost complete resistance to the 4 main storage pests, and remained protected for 150 days. The residual effect of dichlorvos under similar conditions was inadequate.

Al-Saffar and Kansouh (1979), who surveyed the stored-grain insects in silos in Iraq and studied appropriate protective measures, considered that dichlorvos was the insecticide most toxic to the adults of *Tribolium confusum*. Dichlorvos could be used without significant effect upon the germination and vigour of seeds stored for more than 2 months. It was recommended for incorporation into grain held for planting.

Schmidt and Wohlgemuth (1979) investigated the efficiency of dichlorvos slow-release resin strips with active ingredient contents of 20, 28, 52, and 56 g per unit and different evaporation rates on *Ephestia elutella, Ephestia kuehniella* and *Plodia interpunctella* in 10 granaries in Berlin. In the first experiment to control *Ephestia elutella*, the moth count in the granary was reduced to one-fifteenth during 7 days. Most of the larvae of *Plodia interpunctella* introduced into the granary during the migratory phase shortly before pupation in the first 2 months after the strips had been placed there died either in the same stage or as pupae. Larvae that were still older when exposed to the insecticidal vapours were able to pupate, and 78% of them gave rise to sexually mature adults which oviposited, but none of the eggs hatched even in artificially ventilated experimental cages. In a trial for 3 years in 4 granaries, the moth count was steadily reduced until hardly any moths were observed at the end of the experimental period. The influence of the dichlorvos strips on the eggs of *Ephestia kuehniella* deposited on the grain and between sacks was tested. The effectiveness lasted 5 to 8 weeks. A sufficient mortality was guaranteed when the weight loss of the strips was about 20 mg/m^3/wk. Old larvae, ready to pupate, were killed during 4–8 weeks. Dichlorvos residues in the uppermost layers (0 to 10 cm) of the grain were about 0.2 mg/kg, one-tenth of the maximum residue limits of 2mg/kg.

Pagliarini and Hrlec (1982) studied the effectiveness of various acaricides against the stored-product mite *Tyrophagus putrescentiae* in the laboratory in Yugoslavia. Dichlorvos showed the highest initial effectiveness against the pest in treated wheat.

Gelosi (1982), in a review of the morphology, biology and control of *Sitophilus oryzae* in Italy, recommended that after cleaning of cereal storehouses and of containers and vehicles used for transportation, the structures and stored commodities should be treated with dichlorvos.

Bitran *et al.* (1982a) carried out trials in Brazil to evaluate malathion, dichlorvos and a malathion/dichlorvos mixture to protect corn cobs against *Sitophilus zeamais* and *Sitotroga cerealella*. Prior fumigation with phosphine increased the efficiency of the insecticides which had similar effects in reducing cereal losses. Later (1982b) the same authors tested the same insecticides singly or together for the protection of bagged maize against the same 2 pests. The best results were obtained with a 13:1 mixture of technical malathion and technical dichlorvos applied directly to the sacks at the rate of 1.78 ml/m^2. Monthly applications of the mixture afforded excellent protection even at high ambient temperatures. This treatment was as economical as the standard fumigation treatment.

Tsvetkov (1983) advised that the measures recommended for the control of insect pests of stored grain in Bulgaria include treatment with dichlorvos.

Peng (1983) carried out laboratory studies in Taiwan to determine the LD/50 and LD/90 of 10 insecticides to the adults of 6 species of insect pests of stored rice. Adults were exposed to impregnated filter papers for 4 hours and mortality was recorded after 82 hours. Dichlorvos was the most toxic to all 6 species; propoxur, malathion and permethrin were the least toxic.

Hasan *et al.* (1983) tested 6 organophosphorus and 2 organochlorine insecticides as dry films in the laboratory in India for their toxicity to *Sitophilus oryzae, Tribolium castaneum, Rhyzopertha dominica* and *Callosobruchus chinensis*. Dichlorvos was the most and malathion the least toxic to all 4 test insects, and *Sitophilus oryzae* was the most susceptible and *Rhyzopertha dominica* the most resistant species to all the test insecticides.

12.2 Degradation

Strong and Sbur (1964a) studied the persistency of dichlorvos as a spray on wheat (16% moisture) at 10 mg/kg and found its toxicity to *Sitophilus oryzae* was lost after 2 weeks at high storage temperatures; whereas when stored at 15°C, the treated wheat still killed the insects for as long as 3 months.

Green *et al.* (1965) carried out field trials on infested feed barley which was sprayed with malathion, fenitrothion or dichlorvos. Malathion (10 mg/kg) and fenitrothion (2 mg/kg) remained effective for at least 8 months but dichlorvos (4 mg/kg) disappeared so rapidly that within 2 weeks no insecticidal activity could be detected. In another trial, feed barley in a barn was sprayed with dichlorvos at the rate of 4 mg/kg. The initial concentration of dichlorvos determined by analysis was 1.9 mg/kg. Biological and chemical assays showed effective residues to be present for between 6 and 10 weeks, considerably longer than in the previous trial. This difference is probably attributable to the lower ambient temperatures prevailing.

Green and Tyler (1966) applied dichlorvos to farm-stored grain at the rate of 4 mg/kg and reported that the deposit disappeared quickly and conferred relatively little residual protection. Analysis of the samples taken during treatment revealed only 0.59 mg/kg dichlorvos. This level had fallen to 0.19 mg/kg at the end of 1 week and 0.07 mg/kg at the end of 2 weeks. These analytical results were confirmed by bioassay tests which showed complete failure to kill insects at the end of the first week.

Attfield and Webster (1966) concluded that dichlorvos, used in treating grain, had a residual effect of from 1 to 3 weeks.

Rowlands (1967b) carried out experiments designed to follow the metabolic degradation of dichlorvos, alone and in the presence of malathion, in wet and dry grain. His investigations were complicated by the formation of an unknown compound between dichlorvos, or one of its breakdown products, and wheat protein. The same complex could be formed by reacting dichlorvos with wheat gluten.

Harada (1967) reported that the efficiency of dichlorvos/resin strips against *Sitophilus oryzae* had not declined even after 40 days exposure of the preparation to air at 28°C.

Green *et al.* (1967), investigating a technique whereby dichlorvos was sprayed into an airstream produced by a motorised sprayer and injected into bagged grain found that immediately after treatment with a nominal dose of 20 mg/kg dichlorvos, the highest concentration recorded was 30 mg/kg on wheat and 24 mg/kg on barley. This fell in 6 days to 7 mg/kg on wheat and 1.7 mg/kg on barley. The grain had been stored at 25°C and 45% r.h.

In a laboratory study, **Kirkpatrick *et al.* (1968)** demonstrated that dichlorvos residues of 3.8–8 mg/kg decreased 89% and 76% respectively, in 28 days.

Green and Wilkin (1969) sprayed dichlorvos into the air stream from a motorised knapsack sprayer through to a perforated lance, which was inserted into bags of wheat and barley. Dichlorvos residues were higher on wheat than on barley and greatest in places where dust and frass had accumulated. When dichlorvos was injected at 50 mg/kg, the highest residue found 1 day after treatment was 30 mg/kg in a dusty pocket of wheat, and this fell to 5.3 mg/kg after 6 days. The range of residues reported for barley were 5.8–24 mg/kg after 1 day, falling to 0.9–1.4 mg/kg after 6 days.

Minett and Belcher (1970) determined the rate of loss of dichlorvos residues from wheat at 4 moisture levels and 2 storage temperatures. The critical moisture level

(**Minett *et al.* 1968**) observed in the case of malathion was not found for dichlorvos. It was, however, apparent that the disappearance of dichlorvos residues from wheat is dependent on both moisture content and storage temperature of the wheat. Wheat with moisture levels of 9.3, 11.1, 12.9 and 13.7%, treated with dichlorvos at 50 mg/kg and stored at –15°C, had levels of 49, 43 and 34 mg/kg respectively after 11 months storage. When the fate of dichlorvos, applied at an initial level of 50 mg/kg to wheat of a moisture level of 9.3%, was compared with malathion applied to similar wheat at a level of 10 mg/kg, it was found that after 14 days storage at 35°C the residue level of dichlorvos had fallen below that of malathion. When the storage temperature was 21°C, the concentration of dichlorvos fell to that of malathion only after 80 days. The effect of higher moisture levels and higher temperature were much more pronounced on dichlorvos residues than on malathion residues.

Merk (1970) advised that grain treated with dichlorvos at concentrations of 4–10 mg/kg complied with the International Maximum Residue Limit of 2 mg/kg within 3 weeks of treatment.

Sasinovich (1972) reported that the maximum amounts of residual dichlorvos found in grain and flour after storage was 0.25 mg/kg.

McGaughey (1970a) treated rough rice with dichlorvos at levels ranging from 5–20 mg/kg. After storage for 30 days, an initial concentration of 7.3 mg/kg (solvent extractable) at 1 day had degraded to 0.87 mg/kg. **McGaughey (1972)** applied dichlorvos in conjunction with malathion to rough rice of moisture content 13 to 14%, aiming to apply 20 + 10 mg/kg and 15 + 7.5 mg/kg respectively. After approximately 6 months storage, there were no detectable residues of dichlorvos.

Rowlands (1970a) studied the uptake of radio-labelled dichlorvos by the attachment and back regions of wheat grains. He found that uptake was very rapid from the attachment region at high moisture levels (18%). The situation found 1 hour after treatment was practically identical to that found after 7 days storage, except for the migration of metabolites from germ to endosperm. Penetration and internal distribution were slower from the back region at either 10.6% or 18% moisture. Dichlorvos, applied to the attachment region, took longer to penetrate grains of 10.6% moisture than those at 18%, although at both levels, equilibrium was reached after 24 hours.

Rowlands (1970b) showed that dichlorvos, applied to stored wheat, underwent rapid degradation to dimethyl phosphate and to fairly stable phosphorylated protein derivatives that were mainly water-soluble. These are probably mono- and dimethoxy phosphoryl derivatives of the hydroxy amino acids of the peptide chain. Dichlorvos prolonged the residual life of certain other organophosphorus insecticides if present in the grain and the effect may be reciprocal.

Rowlands (1971) reviewed the metabolism of contact insecticides on stored grains including information published during the period 1966–69.

Elms *et al.* (1972) studied the rate of breakdown of malathion and dichlorvos applied to wheat, utilising a range of mixtures containing from 0 to 20 mg/kg of each of the insecticides. Malathion and dichlorvos were applied together in one experiment and dichlorvos was applied 126 days after malathion in another. Dichlorvos was shown to retard malathion breakdown and malathion to influence, slightly, breakdown of dichlorvos. Results indicated that the presence of malathion residues in grain, treated with dichlorvos, would not appreciably influence the period required for the degradation of dichlorvos to acceptable residue levels under storage conditions approximating to 25°C and 12% moisture content. The results are described in **Figure 12.1.**

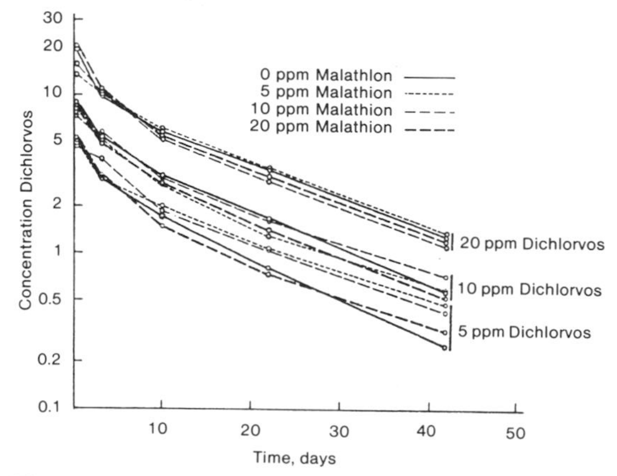

Figure 12.1. The breakdown of dichlorvos applied to wheat stored at 25°C and 12% moisture content in the presence of malathion residues. The malathion was applied at the same time as dichlorvos (**Elms *et al.* 1972**).

The authors pointed out that these results did not agree with those of **Rowlands (1970a)** but they attributed the difference to difficulties encountered in extracting dichlorvos from wheat for analysis. This point is also made by the JMPR (**FAO/WHO 1971, page 112**).

McGaughey (1973a) determined the residues of dichlorvos in various substrates following the use of dichlorvos vapour for insect control in a rice mill. After as many as 13 successive daily treatments, bulk products exposed in the mill had residues as follows: 4.48 mg/kg on rough rice, 1.29 mg/kg on brown rice, 1.18 mg/kg on milled rice and 8.26 mg/kg on bran. Milled rice and bran exposed in burlap bags had only half as much residue, milled rice exposed in cotton bags had about the same amount and milled rice exposed in paper bags had about one third as much residue. Spillage and sweepings exposed to single treatments had residues below 1 mg/kg.

Vardell *et al.* (1973) found that dichlorvos sprayed on wheat at 15 mg/kg resulted in deposits ranging from 2.4 to 6 mg/kg immediately after treatment. Residues decreased to 0.5 mg/kg or less during 6 weeks of storage in bins. Temperature of the grain was an important factor affecting residue-degradation rate. There was little or no loss of residues from wheat stored for three months at –18°C. The residue on wheat stored at 27°C for 1 week was about equal to that stored in bins at 18° to 12°C for 5–6 weeks.

Hall *et al.* (1973) applied tetrachlorvinphos plus dichlorvos at 26 + 13 mg/kg to wheat and at 22 + 11 mg/kg to maize. After 6 to 8 months storage on farm premises, they found no residues of dichlorvos in either cereal. As an instance of the rapid loss of dichlorvos, they stated that a theoretical application to wheat of 10 mg/kg resulted in the recovery of only 2.4 mg/kg some 8 hours afterwards. There are many pitfalls in such assumptions, as discussed earlier, because many investigators used solvent extraction methods that failed to recover dichlorvos from aged samples.

Desmarchelier (1977a), who developed a mathematical model of the breakdown of dichlorvos on grain, stated that the discrepancies in various reported models for decay of dichlorvos may be due to pronounced differences in surface effects observed in commercial and laboratory applications. He compared his predictive model of decay of dichlorvos residues and dichlorvos vapour concentrations with breakdown observed in grain stored in 6 railway trucks and 4 commercial silos. Decay rates were adequately modelled by simple kinetic treatments and no variations in rate of decay, caused by possible saturation of grain enzymes, were observed. One of the aims of this study was to assess the reliability of various laboratory models for predicting breakdown of dichlorvos in bulk grain.

The figure of a 32% loss of dichlorvos in the first day after application to wheat of 10.5% moisture content at 20°C **(Rowlands 1970b)** is reasonably close to the figures of 18% and 33% loss determined by **Desmarchelier (1977a)** who analysed samples of wheat and sorghum from the surface of treated bulks, but it vastly exaggerates the loss from the body of bulk grain which ranges between 3 and 4.4%. In general, laboratory data on residues obtained soon after application must be regarded as suspect because of possible errors arising from surface effects. It was recognised that about 10% of applied insecticide was lost by various processes including missing of target, volatility losses, specific enzyme reactions occurring only immediatly after application and that the proper understanding of the true position was often obscured by errors in chemical analysis due to incomplete extraction of aged residues **(Desmarchelier *et al.* 1977b)**.

One simple model that is consistent with data in open literature is that the time taken for residues to degrade below a certain level is the time predicted by use of first-order rate constants, assuming actual application of 90% of nominal concentration. By use of this very simple model, it was possible to predict, with reasonable accuracy, the time at which residues fell below 2 mg/kg after initial application of 50 mg/kg to wheat. The data presented in the paper by **Minett and Belcher (1970)** were compared with this mathematical model and the predicted and observed rate of decay are set out in **Table 12.1 (Desmarchelier 1977a)**.

It is possible that rapid initial decay of dichlorvos deduced from small-scale laboratory experiments could be due to vapour losses from freshly treated grain. Vapour losses from open containers were significant **(Champ *et al.* 1969)** and vapour losses from closed containers could be significant if dichlorvos was degraded or absorbed by container walls. Such vapour losses would be important only for small quantities of grain, or from surfaces of grain bulks, as there is a quasi or temporary equilibrium between dichlorvos in the vapour phase and dichlorvos bound to wheat, with less than 0.01% of the total dichlorvos being present in the vapour at any moment **(Desmarchelier *et al.* 1977a)**.

Formica (1977a) reported trials carried out in Switzerland on feed-grade wheat treated with various formulations of dichlorvos at a nominal dosage of 14 mg/kg. At the end of the first week after treatment the residues on the whole grain ranged from 2.5–4.5 mg/kg and after 14 weeks less than 0.12 mg/kg, the limit of determination.

Desmarchelier (1978c) published the results of a mathematical examination of availability to insects of aged insecticide deposits on wheat in which the fate of dichlorvos vapour was determined and discussed.

Mian and Kawa (1979) working in Pakistan measured the persistance of dichlorvos on wheat grains. Wheat grain treated with 10, 15 and 20 mg/kg of dichlorvos was analysed for insecticide residues after 8 months of storage. Dichlorvos was found to persist only in trace amounts.

Desmarchelier and Hogan (1978) reported a series of laboratory and pilot-scale experiments performed to reduce residues of dichlorvos in grain dust to levels below the maximum permitted for stock feed. Processing such as steaming or pelleting did not reduce residues sufficiently, but mixing grain dust with 3 alkaline materials, sodium hydroxide, sodium carbonate or ammonia, reduced residues considerably and this reduction was increased by subsequent steaming and pelleting. Reduction of residues was greater from use of liquid rather than solid alkalis, from use of the stronger (sodium hydroxide) rather than the weaker (sodium carbonate) base and was proportional to the amount of liquid sodium hydroxide that was used. Alkaline materials reduced pesticide residues in possible stock-feed combinations of grain dust plus straw and grain dust plus lucerne meal. As these results are consistent with general base-catalysed hydrolysis of organophosphorus esters, there is probably scope for considerable variations in procedure by use of different alkalis and different exposure times. In commercial applications amounts of base required (2% w/w) were less than those used to increase the food value of low-grade animal feed. In commercial practice it has been found that hydrated lime, incorporated as an aqueous suspension prior to pelleting, is likewise entirely satisfactory.

Desmarchelier and Bengston (1979a), in discussing the residual behaviour of chemicals on stored grain, pointed to the value of predictive models in enabling the grain industries to select rates appropriate for given circumstances. They pointed out that the effect of partial cooling on increasing the residual life of protectants was most pronounced for protectants with high temperature co-efficients and short half-lives. Thus, for 9 months protection with methacriphos, to leave a residue of 1 g/tonne to control *Tribolium* and *Sitophilus* spp., one would apply, at 50% equilibrium relative humidity, 22.5 g/tonne at temperatures of 30°C, 4.6 g/tonne at temperatures of 25°C and 2.2 g tonne at temperatures of 19–20°C. Though this example specifically refers to methacriphos, it could apply equally well to dichlorvos. There are enormous advantages in achieving even partial cooling; this cooling also reduces potential numbers of insects and enhances grain quality.

12.3 Fate in Milling, Processing and Cooking

Dichlorvos is readily soluble in water. It is therefore easy to understand why it is readily removed by washing and extensively degraded by cooking.

Processing and cooking removed large percentages of dichlorvos, which may be present on grain or other raw commodity. For example, wheat with 23.8 mg/kg of dichlorvos after treatment, produced flour with 4.6 mg/kg of dichlorvos. Three months later, the wheat had 2.8 mg/kg of dichlorvos and the white flour milled from it had 1.7 mg/kg. The dichlorvos residue in this flour after 14 days storage fell below the limits of detection **(Shell 1966)**.

In related studies, biscuits made from flour containing 0.35 mg/kg and 1.8 mg/kg of dichlorvos had 80% and 60% less dichlorvos, respectively. Flour with 9.5 mg/kg of dichlorvos, heated for 30 minutes at 100°C, 150°C and 200°C, had 0.2, 0.03 and 0.02 mg/kg of dichlorvos, respectively. Rice with 5.3 mg/kg of dichlorvos had only 0.06 mg/kg after cooking **(Shell 1966)**.

A study **(Shell 1970)**, using a number of cereal products with abnormally high dichlorvos residues, showed that residues can be almost completely removed by cooking. Rice, containing 4.5 mg/kg dichlorvos, lost 90% of the residue on boiling 20 to 30 minutes. When rice containing 19 mg/kg was similarly treated, 98% of the residue was destroyed. Samples of flour, containing dichlorvos residues in the range of 4.5 to 14 mg/kg, were found to lose about 90% of the residue on baking for 10 to 12 minutes at 230°C in the preparation of biscuits. Boiling the flour with water for two minutes, as in the preparation of gravy, was shown to decrease residue levels by 97%.

McGaughey (1970a), who evaluated dichlorvos for insect control in stored rough rice, found that residues on rough rice were as high as 7.3 mg/kg, 24 hours after treatment, but in 30 days they were 0.87 mg/kg on rough rice, 3.1 mg/kg in hulls, 0.41 mg/kg in bran, 0.03 mg/kg in brown rice and 0.02 mg/kg in milled rice.

Hall *et al.* (1973) studied the persistence and distribution of tetrachlorvinphos and dichlorvos in grain and grain products. A combination of the 2 insecticides was applied to corn and wheat held in farm storage for 8 months. Dichlorvos was rapidly lost from both grains, but it did exhibit some tendency to persist at trace residue levels. The milling products of corn, bran, fine bran and germ, contained dichlorvos residues that were only just detectable. The wheat bran contained 67% of the dichlorvos residue on the whole grain. In wheat treated with both compounds and immediately milled, the bran was found to contain 90% of the dichlorvos, the shorts about 3% and the flour 7%.

Vardell *et al.* (1973) investigated the carry-over of dichlorvos through the wheat milling process. Some 6 weeks after initial treatment at 15 mg/kg, the solvent-extractable residue in the total sample was 0.4 mg/kg. On milling samples from this bulk, the bran and shorts contained about 1 mg/kg and only trace amounts (less than 0.1 mg/kg) were found in the flour and in the bread baked from it.

A series of laboratory and pilot scale experiments was performed to reduce residues of dichlorvos and malathion in grain dust to levels below the maximum permitted for stock-feed **(Desmarchelier and Hogan 1978)**. Processing such as steaming or pelleting did not reduce pesticide residues sufficiently, but mixing grain dust with three alkaline materials, sodium hydroxide, sodium carbonate or ammonia, reduced residues considerably and this reduction was increased by subsequent processing. Reduction of residues was greater from use of liquid rather than solid alkalis, from use of the stronger (sodium hydroxide) rather than from the weaker (sodium carbonate) base and was proportional to the amount of liquid sodium hydroxide or ammonia that was used. Alkaline materials reduced pesticide residues in possible stock feed combinations of grain dust plus straw and grain dust plus lucerne meal. As these results are consistent with general base-catalysed hydrolysis of organophosphorus esters, there is probably scope for considerable variations in procedure by use of different alkalis and different exposure times to alkali. In commercial applications amounts of base required (2% w/w) were less than those used to increase the food value of low-grade animal feed. This experience has since been translated into commercial practice in Australia were quicklime (calcium oxide) and magnesite (magnesium carbonate) have proved to be effective and convenient to use.

12.4 Metabolism

In spite of its ready solubility in water dichlorvos is neither stored in the body, nor excreted in the milk to any appreciable extent in cows or rats, even when administered in doses that produce severe poisoning **(Tracey 1960; Tracey *et al.* 1960)**.

Perry (1960) reviewed the then available literature on metabolism of insecticides by various insect species recording that the rate of detoxication of dichlorvos by the housefly was slower than that of trichlorfon. Many separate pieces of information indicated that dichlorvos is mainly responsible for the toxic action of trichlorfon.

In whole homogenates of liver, kidney, spleen and adrenals from rat and rabbit, the principal labelled metabolite of dichlorvos-^{35}P was dimethyl phosphate (50–85%), with the remaining radioactivity appearing in demethyl-dichlorvos, monomethyl phosphate and inorganic phosphate **(Hodgson and Casida 1962)**.

In rats given 10 mg/kg dichlorvos-^{32}P orally, rapid absorption, distribution and hydrolysis took place. Female rats, treated every half hour with 4 mg/kg of dichlorvos-^{32}P for 2 hours and sacrificed one-half hour after the last dose, showed mainly hydrolysis products in the tissues **(Casida *et al.* 1962)**. Experiments in rats showed that the liver was highly efficient in detoxicating dichlorvos **(Gaines 1966)**.

Casida *et al.* (1962) studied the metabolism and residues of dichlorvos in rats, cows and a goat. They found that the insecticide was rapidly hydrolised in mammals to yield no persisting tissue residues and only trace levels of milk. The initial phosphorous-containing metabolites of dichlorvos, O,O-dimethyl phosphate and O-methyl 2,2-dichlorovinyl phosphate, are low in toxicity and rapidly excreted before being degraded. The 1-carbon of the 2,2-dichlorovinyl group in dichlorvos is excreted in urine predominantly as a conjugate of dichloroethanol, probably the glucuronide, in the feces as unknown derivatives, and in the expired air as carbon dioxide. Small amounts of dichloroacetic acid maybe formed, and some of the ^{14}C persists in liver, blood, and other tissues in an unidentified form. Limited metabolism studies of dichlorvos in plants and in bovine rumen fluid are also reported.

Rowlands (1967b) reported experiments designed to follow the metabolic degradation of dichlorvos, alone and in the presence of malathion, in wet and dry grain. This was found to lead to the formation of an unknown complex between dichlorvos, or one of its breakdown products, and wheat protein. This substance was extracted from crushed grain with the less polar metabolic products and it co-chromatographed with malathion under the GLC conditions used but was separable from malathion by TLC. The same (or a similar) complex could be formed by reacting dichlorvos with wheat gluten *in vitro* but was not produced if the gluten where first denatured by heat. Since the compound reacted with NBP reagent, it probably contains methoxy and phosphate groups. In the presence of dichlorvos, grain esterases were prevented from hydrolising malathion for some 3–4 weeks of storage and ultimate degradation of malathion proceeded via phosphatase products rather than carboxyesterase products. Under normal conditions, malathion degrades primarily to carboxyesterase products.

Rowlands (1970b), who studied the metabolic fate of dichlorvos on stored wheat grains, stated that it underwent rapid degradation to dimethyl phosphate and to fairly stable phosphorylated protein derivatives that were mainly water-soluble. These were probably mono- and di-methoxy phosphoryl derivatives of the hydroxy amino acids of the peptide chains. None of the phosphorylated protein fractions from dichlorvos-treated wheat was toxic to adult *Oryzaephilus surinamensis*.

Rowlands (1970b) postulated that dichlorvos significantly retarded the breakdown of malathion and vice versa but **Elms *et al.* (1972)**, who studied this phenomenon, found that the presence of malathion residues in grain, treated with dichlorvos, would not appreciably influence the period required for the degradation of dichlorvos to acceptable residue levels under storage conditions approximating to 25°C and 12% moisture content.

Page *et al.* (1971) studied the metabolic fate of ingested dichlorvos in pigs. After 4 hours, intact dichlorvos, desmethyl dichlorvos, dichloroacetaldehyde, dichloroethanol and dichloroacetic acid were found in the intestinal lumen, but only dichloroethanol was observed in blood. Even at 4 hours, degradation proceeded well beyond the chlorinated metabolites listed above. Following a single dose, considerable radioactive CO_2 was found in expired air and radioactive carbon in tissues even 14 days after dosing. It was concluded that the degradation of dichlorvos proceeds rapidly through the chlorinated metabolites described in the literature to dechlorinated compounds which enter many normal metabolic pools.

A comprehensive evaluation of the information on the metabolism of dichlorvos has been carried out by JMPR **(FAO/WHO 1967, 1971)**. Dichlorvos is rapidly broken down in mammals to products which are excreted or incorporated into natural biosynthetic pathways. Several studies have examined the potential for direct methylation of nucleic acids by dichlorvos. It has been shown to cause methylation of bacterial and mammalian DNA and RNA *in vitro*. Also, the urine from dichlorvos-treated mice and rats, was found to contain methylated purines. The alkylating properties of dichlorvos led to the suspicion that dichlorvos might be mutagenic and/or carcinogenic. However, no methylated 7-methyl guanine has been detected in the DNA or RNA from liver or kidney tissue or other soft tissue of rats and mice treated with ^{14}C-methyl-labelled dichlorvos. Although dichlorvos is a potential alkylating agent of DNA and RNA *in vitro*, this potential is apparently not realised *in vivo* owing to the rapid degradation of dichlorvos in mammals. These studies and the negative results in mammalian mutagenicity and carcinogenicity studies supported the view that dichlorvos has an extemely low potential for producing mutations or cancer in man.

Bull (1972) reviewed the metabolism of organophosphorus insecticides in animals and plants and described the reactions that activate or inactivate insecticides or their toxic metabolites. About 110 references are given.

Potter *et al.* (1973a), who determined the ^{14}C balance and residues of dichlorvos and its metabolites in pigs, concluded that the ^{14}C present in the tissues is the result of incorporation of C-1 and C-2 fragments from the vinyl moiety of dichlorvos into normal tissue constituents.

12.5 Fate in Animals

Dichlorvos is absorbed and degraded very rapidly following either oral or dermal administration. The rate of metabolism and detoxification of dichlorvos is so rapid in mammals that there is little or no storage in tissues. By the use of radio-tracer studies with ^{32}P and ^{14}C-labelled dichlorvos the distribution and fate of the resulting metabolites in rats has been well established. Cows fed diets containing up to 2000 ppm of dichlorvos in their rations for 9 days did not excrete sufficient dichlorvos or any toxic metabolite in the milk to be detected by fly bioassay. Calves suckling these cows were not affected. Dichlorvos is not excreted via the milk when the accrued dose exceeds by 2 or 3 times the LD/50 dose. Also, it has been shown that no storage occurs in animal tissues, either as dichlorvos or as toxic metabolites. **(Casida *et al.* 1962; Tracy *et al.* 1960)**.

Dichlorvos is very rapidly metabolised in mammals to products of reduced toxicity. Rapid detoxification is evidenced by the following:

- mammals survive repeated administration in nearly maximum sub-lethal levels;
- resulting tissues (except the stomach) are non-toxic to flies, irrespective of the accrued dichlorvos given to the test animals;
- offspring suckling animals fed rations containing dichlorvos sustain no illness or evidence of intoxication;
- studies with radioactive dichlorvos show rapid and complete *in vivo* detoxification in rats, cows and goats;
- *in vitro* observations **(Hodgson and Casida 1962)**.

Metabolic studies by various investigators have demonstrated the metabolism of dichlorvos and have indicated that the metabolic products are less toxic than the original compound. The fact that no pathology has been observed in animals administered dichlorvos shows that the metabolites formed do not cause any pathological effects or other unique manifestations. It is apparent that metabolites would be present in substantial amounts in animals fed high dosages of dichlorvos in view of its rapid rate of metabolism and detoxification in the animal's body.

The metabolite most likely, on *a priori* grounds, to be present is dichloracetaldehyde (DCA). Extensive analyses from animals given dichlorvos have failed to show the presence of DCA. Animals fed rations containing dichlorvos in a resin formulation at concentrations as high as 9540 ppm of dichlorvos have not shown DCA to be present in tissues. The fact that DCA is water-soluble, is highly reactive, and has a high vapour pressure lends support to the contention that it will not be likely to persist in the animal body any longer than the dichlorvos will. This is largely true for dichlorethanol and dichloroacetic acid which are 2 other possible metabolites of dichlorvos **(Arthur and Casida 1957; Casida *et al.* 1962; Hodgson and Casida 1962; Smith and Williams 1954)**.

The metabolites resulting from administration of dichlorvos are almost completely excreted within a few days of treatment and the phosphorus from the absorbed dichlorvos, as with other compounds in this class is probably incorporated into the normal phosphate metabolic pool. The metabolites in the urine and feces appear to be entirely hydrolysed material **(Casida *et al.* 1962)**.

Casida *et al.* (1962) studied the metabolism and residues of dichlorvos in rats, cows and a goat. They showed that it is rapidly hydrolysed in mammals to yield no persistent tissue residues and only trace levels in milk. The initial phosphorus-containing metabolites, O,O-dimethyl phosphate and O-methyl 2,2-dichlorovinyl phosphate, are low in toxicity and rapidly excreted or further degraded. The 1-carbon of the 2,2-dichlorovinyl group in dichlorvos is excreted in urine predominantly as a conjugate of dichloroethanol, probably the glucuronide, in the feces as unknown derivatives and in the expired air as carbon dioxide.

Pigs were fed daily at the high dosage of 9450 ppm dichlorvos for 90 days **(Singh 1964)**. Dichlorvos residues were not detectable in tissues and organs including spleen, kidney, small intestine, lung, heart, muscle and liver 0 and 1 day after last treatment.

Schwartz and Dedek (1965), who followed the breakdown and excretion of trichlorphon in the pig, were able to identify and measure dichlorvos, the breakdown product of trichlorphon, in the blood and intestinal contents but found that it had completely disappeared after 3.5 hours. The breakdown of dichlorvos occurred much more rapidly in the blood than in the intestinal contents because of the higher concentration of phosphate-splitting enzymes in the blood.

Lloyd and Matthysse (1966) tested polymer/insecticide feed additives for controlling flies breeding in manure. Rates of release of insecticide in the digestive tract of cattle were determined.

Lloyd and Matthysse (1970), who evaluated the effect of dichlorvos/PVC pellets fed to cattle, found that there were no detectable residues of dichlorvos in milk. The same authors (1971) fed dichlorvos in PVC pellets at the rate of 1.3, 1.8 and 2.6 mg of dichlorvos/kg body weight for 2 weeks. No insecticidal residues were found in milk samples collected at 1, 3, 7, 10 or 14 days.

Page (1970) has reviewed the extensive literature on the metabolism of dichlorvos and in the same paper has reported on extensive trials designed to show the metabolic fate and tissue residues of dichlorvos following oral administration. He believes that the data support the conclusion that when administered orally over a period of weeks, dichlorvos does not result in detectable toxic residues of the parent compound or product-related metabolites in tissues.

Page *et al.* (1971) studied the metabolic fate of ingested dichlorvos in pigs. They showed that even at 4 hours degradation proceeds well beyond the chlorinated metabolites. Complete degradation was demonstrated by the presence of radioactive CO_2 in expired air and the retention of radioactive carbon in tissues even 14 days after dosing. It was shown that the vinyl carbon entered glycine, serine and, at lower levels, glucose, cholesterol, fatty acids and RNA. The ^{36}Cl retained in tissues was identified as chloride ion. No chlorine-containing product-related metabolites could be found.

Three male pigs were fed ^{14}C-dichlorvos encapsulated in PVC at the rate of 40 mg/kg. After 14 days approximately 60% of the administered radioactivity was recovered in the escapsulated pellets in the feces. The expired carbon dioxide accounted for approximately 15% of the administered dose, 5% was excreted via the urine, 5% via the feces and the carcass contained approximately 10% of the administered dose. Of 20 tissues examined the highest level of radioactivity was found in liver tissues and the lowest in brain tissue **(Potter *et al.* 1973)**. Pregnant sows were given multiple doses of ^{14}C- or ^{36}Cl-labelled dichlorvos in PVC pellets at the rate of 4 mg/kg. In this experiment less of the administered ^{14}C activity was expired as carbon dioxide and more was released from the PVC pellets than in the feeding study with male pigs described above. Recoveries of ^{36}Cl from the sows comprised 60% recovered in pellets in the feces, 6% in feces, 25% in urine and 8% in the carcass. In newborn piglets, ^{14}C-tissue residues were increased in the femur, as were ^{36}Cl residues. This increase was apparent up to sacrifice at 21 days of age. The ^{36}Cl residues in the piglets were maximal for all tissues at 9 days of age **(Potter *et al.* 1973)**.

Potter *et al.* (1973a), who fed ^{14}C-dichlorvos formulated as slow-release PVC pellets to each of 9 young male pigs, reported that no dichlorvos, desmethyl dichlorvos, dichloroacetaldehyde or dichloroacetic acid were found in the tissues of the pigs.

Pym *et al.* (1984) reported an extensive study in which graded levels of malathion, dichlorvos and pirimiphos-methyl were given either separately or combined in the feed of laying hens over a 4 week test period. After 4 weeks on treatment, tissues and eggs were analysed. No dichlorvos residues were detected in any tissues.

12.6 Safety to Livestock

The effects of dichlorvos on many species of livestock have been studied by scientists over a long period. Many of these studies have involved topical application or inhalation but these studies are not necessarily relevant to the question of safety of treated animal feeds based on grain and milling offals.

Klotzsche (1955) reported studies where cows fed 20 ppm dichlorvos did not show symptoms of illness or depression of cholinesterase. At 100 and 200 ppm (dosages equivalent to 0.91 and 1.8 mg/kg-bw, respectively) slight depression of erythrocyte cholinesterase was observed after 8 days. At 500 and 2000 ppm (equivalent to 4.5 and 18 mg/kg-bw, respectively), severe depression of erythrocyte cholinesterase was observed after 8 to 12 days. A single dose of 3000 ppm (27 mg/kg-bw), caused severe symptoms and 100% depression of erythrocyte cholinesterase but not death. These data show that a high proportion of the acute lethal dose can be administered daily without toxic symptoms developing.

Tracy *et al.* (1960) showed that the cholinesterase activity of the blood from lactating cows fed dichlorvos at dosages of 4.5 mg/kg/day (equivalent to 500 ppm in the rations) for 12 days showed 80% to 90% depression. When the dosage was increased to 18 mg/kg/day (2000 ppm in the rations) for an additional 8 days, no further depletion of cholinesterase activity occurred. After 70 test days, and an accrued dosage of 164 500 mg, one of the test cows was given a single dose of 27 mg/kg (3000 ppm in the ration) which caused severe symptoms of intoxication. During this period, calves suckling these cows were not affected by any of the dichlorvos treatments. It is evident from these tests that dichlorvos ingested by the lactating cows was not transported *per se* or as toxic metabolites through their milk to the suckling young.

Horses tolerated a single oral dose of dichlorvos in feed at a rate of 50 mg/kg of body weight, but showed moderately acute poisoning when the compound was given by stomach tube at the rate of 25 mg/kg **(Jackson *et al.* 1960)**.

Casida *et al.* (1962) showed that single doses of 1 mg/kg of dichlorvos given to cows or goats (oral, subcutaneous or intraperitoneal to cows and intraperitoneal to goats) and a 20 mg/kg oral dose to cows did not yield a noticeable effect on lactation. At the 1 mg/kg dose, no depression of cholinesterase occurred. When a dose of 20 mg/kg was administered 7 days later, cholinesterase inhibition ranging between 20% and 30% was observed within 1 to 4 hours, followed by rapid recovery.

As is often the case with organophosphorus insecticides, birds appear to be more sensitive to dichlorvos than mammals. The acute oral LD/50 to adult Leghorn hens has been determined as 22.8 mg/kg **(Kettering Laboratory)**, while the acute oral LD/50 to male chicks has been determined as 14.8 mg/kg **(Sherman and Ross 1961)**. Chickens given multiple doses of dichlorvos in capsules at levels from 2.28 to 13.7 mg/kg were killed by successive doses above 2.5 mg/kg. No pathological changes were found in the viscera, muscles or peripheral nerves of the hens that survived. The hens that died exhibited only non-specific changes associated with acute poisoning; namely, haemorrhages in the lungs, and marked hyperaemia of the brain, liver and spleen. The peripheral nerves showed no pathological changes **(Kettering Laboratory 1964)**.

Batte *et al.* (1965) published a critical evaluation of dichlorvos as an anthelmintic for swine. They tested 9 formulations in addition to the technical active ingredient. Of the 262 pigs dosed with the different formulations, only 2 were killed by drug action. They were 2 of 5 pigs given 160 mg/kg of technical material; overt signs of organophosphate toxicity appeared in the other 3 pigs within 2 hours of drug administration, but were completely reversed within 24 hours without supportive treatment. None of the pigs receiving the commercial PVC/dichlorvos granules showed signs which could be interpreted as clear organophosphate toxicity. Histological examination of organs and tissues from selected animals revealed no evidence of pathological alteration. Consistent, statistically significant depression of blood cholinesterase was not obtained by treatment of pigs with different levels of dichlorvos.

Gaines *et al.* (1966) discussed the function of liver metabolism on anticholinesterase compounds in live rats in relation to toxicity. They pointed out that whereas isolan is considerably more toxic by dermal or intravenous administration than by oral intake, parathion is considerably more toxic by the oral route. It is clear that the efficient detoxification of isolan by the liver is one factor contributing to its relatively low toxicity by the oral, as compared to the dermal, route. The conversion of parathion to paraoxon *in vivo* results in the higher toxicity of parathion when administered orally. However, one or more other factors must be important because the detoxification of dichlorvos by the liver seems equally efficient, and yet dichlorvos is slightly more toxic by the oral than by the dermal route.

Dunachie and Fletcher (1966) studied the effect of some insecticides on the hatching rate of hen eggs following the injection of the insecticide in acetone directly into the egg yolk. Although the report is difficult to interpret it appears that dichlorvos, when injected at a rate of 5 mg/egg, equivalent to 100 mg/kg, reduced the hatch to 60% and when applied at the rate of 0.5 mg/egg or 10 mg/kg reduced the hatch to 80%. The combination of one part dichlorvos and 3 parts malathion injected at the rate of 5 mg/egg, equivalent to 100 mg/kg, did not depress the percentage hatch and produced only slight abnormalities in a few chicks.

Howes (1966) studied the effects of dichlorvos on the bovine but this reviewer has not sighted a copy of the M.S. thesis presented at the Washington State University.

Bris (1968) conducted experiments to determine some physiological responses of ruminants to dichlorvos. Three experiments were performed to evaluate the effect of dichlorvos with and without diethylstilbestrol (DES) on the performance and physiological responses of 187 cattle fed several different rations. A signifcant added effect was found between dichlorvos and DES when fed to cattle. It was concluded that a level of 1.6 mg dichlorvos per kg of body weight was the optimal dosage level with fattening cattle in conjunction with a 24 mg implant of DES. The average feed intake and feed efficiency were increased by dichlorvos. The performance of 20 sheep was not signifcantly changed by the addition of dichlorvos to the ration.

Bris *et al.* 1968) conducted 4 experiments involving 239 fattening steers used to evaluate the anthelmintic efficacy of dichlorvos. The optimum dosage was determined to be 1.6 mg/kg body weight equivalent to 55 ppm in the feed. Whilst there was evidence of an anticholinesterase reaction of the drug on the host animals at greater (unstated) dosage levels there was no evidence of such effect at 1.6 mg/kg body weight and, in the views of these investigators, the possibility of harmful effects of dichlorvos to cattle appear nil.

Singh *et al.* (1968) studied the effects of dichlorvos fed to gravid sows on the performance of their offspring to weaning. Sixty cross-bred gravid sows, ranging from 2 to 4 years in age, were used to study the effects of daily consumption of dichlorvos for 21 to 30 days before parturition. Groups of 12 sows received 200, 400 and 800 mg/head/day of dichlorvos in various forms. The average number of pigs farrowed alive, the average birthweight of pigs, and average number of pigs weaned at 35 days was not adversely affected by the administration of dichlorvos. Greater gains were observed for pigs from dichlorvos treated dams in comparison to gains from pigs from control or placebo treated dams.

Foster (1968) conducted 2 trials, involving 442 sows and gilts, to determine the influence of dichlorvos on sow reproductive performance and subsequent performance of the offspring. The dichlorvos treatment resulted in more live pigs farrowed; fewer stillborn pigs per litter; heavier litter birth weights; heavier litter weights at 4 weeks; and decreased mortality from birth to 4 weeks.

Khera and Lyon (1968) experimented with chick and duck embryos in the evaluation of pesticide toxicity. They found that dichlorvos was only slightly toxic to chick embryos when injected on the 10th day of incubation. It was considerably less toxic to duck embryos. It was concluded that the toxicity to avian embryos was similar to that in the rodent.

Bazer *et al.* (1969) studied the effect of dichlorvos and pregnant mares serum on reproduction in swine. Many parameters were recorded. There were no significant differences due to treatment.

Batte *et al.* (1969) studied the effects of dichlorvos on sow reproduction. 472 dams received dichlorvos in the ration at the rate of 400 or 800 mg per head per day for 3 weeks before breeding or from 3 to 6 weeks before parturition. Individual and litter weights at farrowing and at weaning were highly significantly greater in pigs from dichlorvos treated dams. The number of pigs weaned was also greater.

Frolov and Berlin (1969) reported on the toxicity of 8 insecticides to chicks. When given orally to chicks the LD/50 value of dichlorvos was 12 mg/kg.

Roger *et al.* showed that when dichlorvos was injected into the egg yolk teratogenic abnormalities were produced.

Mukhamedshin (1970) reported an investigation carried out in the Soviet Union in which dichlorvos emulsion concentrate was administered orally to hens by means of a probe at dosages of 0.0038–12.45 mg/kg-bw. Poisoning, generally resulting in death, occurred in birds given dosages of 0.042 mg/kg and higher, the speed with which the symptoms developed increasing with increases in dosage.

Lloyd and Matthysse (1970) who evaluated the use of dichlorvos/PVC pellets for feeding to cattle for control of flies breeding in manure reported that there was no detectable toxicity to the cattle receiving dichlorvos pellets in the ration.

England and Day (1971) reported a series of experiments involving 199 litters of pigs during a 2-year period in which dichlorvos was included in the ration at the rate of 800 mg/day from the 91st day of gestation through day of farrow. Litters had significantly heavier average birthweights than litters from control dams. Increase in average birth weight occurred and there were significantly greater weight gains for pigs in treated litters and a higher percentage of pigs gaining weight.

Loeffler *et al.* (1971) studied the metabolic fate of inhaled dichlorvos in pigs and reported that the metabolites found were desmethyl dichlorvos and methyl esters of phosphoric acid. Short-term inhalation trials did not show the presence of intact dichlorvos or desmethyl dichlorvos in blood or lung tissues. Even in the two 4-hour trials the degradation had proceeded to the stage where only methyl phosphate and phosphoric acid could be detected.

Collins et al. (1971) presented the results of a study of the effect of dietary dichlorvos on swine reproduction and viability of their offspring. Polyvinylchloride resin formulations of dichlorvos at concentrations up to 500 ppm of the active ingredient were fed to male and female breeding swine for up to 37 months and to their offspring for varying periods. The dichlorvos in the diet had no inhibitory effect on the numbers, viablity or growth rates of the offspring, nor were gross anatomic aberrances observed in any of the piglets. No pathologic changes suggestive of neoplasia were found in any of the young or adult animals examined at necropsy. A variable but dose-related depression of cholinesterase activity was observed in whole blood. These dietary levels had no significant effects on a wide range of blood parameters examined.

Macklin and Ribelin (1971) and **Ribelin and Macklin (1971)** conducted a survey to determine causes of abortion in dairy cattle in Wisconsin, USA. They reported that the feeding of 5 insecticides, including dichlorvos, to pregnant cows failed to cause abortion. In the opinion of these investigators pesticides do not contribute significantly to the incidence of abortion on Wisconsin dairy farms.

Lloyd (1973a) reported a case of accidental poisoning of poultry that gained access to manure from a horse that had received dichlorvos/PVC pellets for the control of helminths. It became apparent that the fowls were scratching over the dung and picking out the pellets.

Pym *et al.* (1976) having evaluated the effects of dichlorvos as a contaminant in the feed on the performance of laying hens concluded that 30 ppm in feed had a statistically significant effect on depressing egg production when fed over long periods. In a later, and more complex study **(Pym *et al.* 1983)** in which dichlorvos, malathion and pirimiphos-methyl were administered alone and in combinations to laying hens the authors concluded that 80 ppm dichlorvos alone in the feed had no significant effect upon feed consumption or egg production. Higher rates of dichlorvos and combinations with the other insecticides were clearly shown to reduce egg production. The lowered egg production figures were reproduced by hens receiving correspondingly lower levels of untreated feed, indicating that the effect upon egg production was mediated via a reduced feed intake.

Female pigs were fed dietary levels of 0, 200, 250, 288, 400, 500 or 750 ppm of dichlorvos for up to 37 months. The animals were first mated after they had received the test diet for 6 months, and the study was followed through 2 generations. Initially, a total of 22 females and 1 male were used, but after the first generation 4 more males were used in order to prevent inbreeding. The males received 0, 288 or 400 ppm of dichlorvos in their diets. Dichlorvos did not affect either the numbers of litters or their size and survival. The total of 490 piglets were examined and none showed anatomical abnormalities **(Singh and Rainier 1966)**.

Young *et al.* (1979) described a complex series of experiments in which dichlorvos was fed to parasitised and non-parasitised sows to study the joint effects of dichlorvos treatment and parasitism upon reproductive performance. Each treated sow was given dichlorvos (4 mg/kg/day) from gestation day 80 until parturition. The mean birth weight of piglets from dichlorvos-treated sows was 60 g heavier than that of the control, while the litter birth weight was 1.3 kg heavier. Treating the sows with dichlorvos increased the mean piglet weaning weight by 0.5 kg over that of the controls. Dichlorvos treated sows weaned litters 9.1 kg heavier than controls.

Wratthall *et al.* (1980), working in England, administered dichlorvos to pregnant sows at the rate of 8.5 mg/kg/day from day 41 to day 70 of pregnancy. Piglets appeared clinically and morphologically normal but the weights of some of their organs were heavier than those of piglets from untreated sows. These investigators observed that the effects of dichlorvos were quite different from those of the chemically related trichlorphon which, if fed to sows in mid-pregnancy, is teratogenic, causing severe cerebella hypoplasia with congenital ataxia in the progeny.

Ciba Geigy (1971) reported that the effects of dichlorvos on horses have been studied and that horses showed no clinical signs of poisoning after ingesting 10 mg dichlorvos/kg-bw, administered over a 48-hour period, but there was marked inhibition of whole-blood cholinesterase. Doses of 25 and 50 mg/kg were toxic to horses when administered as a single dose by stomach tube. Symptoms of organophosphate poisoning in horses differ from those in ruminants. The predominant signs in the horse are colic and digestive disturbances, together with muscular weakness, salivation and dyspnoea.

12.7 Toxicological Evaluation

The JMPR has evaluated the toxicology of dichlorvos on 5 occasions (1965, 1966, 1967, 1970, 1977). Acute toxicity studies were available on the mouse, rat, chick, horse, hen, pig and man. Short-term feeding studies on rats, dogs, monkeys, horses and man were evaluated. There was also at least one long-term feeding study on the rat and another on the dog.

The no-toxic effect level in the rat was 10 ppm in the diet, equivalent to 0.5 mg/kg body weight/day and in the dog 0.37 mg/kg body weight/day. From the short-term studies in man, the no-effect level was 0.033 mg/kg/day. An ADI for man of 0.004 mg/kg body weight has been established based on a level of 0.033 mg/kg/day, which caused toxicologically insignificant depression of plasma cholinesterase activity. The erythrocyte cholinesterase was not affected.

Special studies on reproduction in the chick, rat, rabbit and pig were also available, as were *in vitro* and *in vivo* studies of mutagenicity, carcinogenicity studies in the rat and inhalation studies in guinea pigs, mice and rats.

Dichlorvos has been shown to cause methylation of bacterial and mammalian DNA and RNA *in vitro*. Also, the urine from dichlorvos-treated mice and rats was shown to contain methylated purines. The alkylating properties of dichlorvos led to the suspicion that dichlorvos might be mutagenic and/or carcinogenic. In subsequent evaluation it has been observed that on the basis of recent and earlier studies, the metabolic patterns of dichlorvos in animals are clearly elucidated. It is rapidly broken down in mammals to products which are excreted or incorporated into natural biosynthetic pathways. Several new studies have examined the potential for direct methylation of nucleic acids by dichlorvos. However, no methylated 7-methylguanine has been detected in the DNA or RNA from liver or kidney tissue or other soft tissues of rats and mice treated with ^{14}C-methyl-labelled dichlorvos. Although dichlorvos is a potential alkylating agent of DNA and RNA *in vitro*, this potential is apparently not realised *in vivo* owing to the rapid degradation of dichlorvos in mammals. The presence of ^{14}C-labelled purines in the urine of animals treated with ^{14}C-methyl-labelled methylating agents is not conclusive proof of the methylation of DNA and RNA. At the polymeric levels, the ^{14}C may be incorporated into the free purine bases. Thus, animals treated with ^{14}C-labelled guanine excreted ^{14}C-labelled 7-methylguanine in the urine. These studies and the negative results of mammalian mutagenicity and carcinogenicity studies support the view that dichlorvos has an extremely low potential for producing mutations or cancer in man. None of the recent studies has changed the basis for establishing an ADI for man.

12.8 Maximum Residue Limits

The JMPR has reviewed the occurrence, nature, level and fate of residues of dichlorvos in food on 5 occasions (1965, 1966, 1967, 1970 and 1977). On the basis of these reviews, and the corresponding toxicological evaluation, the following maximum residue limits have been recommended for raw cereal grains and milling products.

Commodity	Maximum Residue Limits (mg/kg)
Cereal grain	2
Cereal products (milled)	0.5
Miscellaneous food items not otherwise specified	0.1

The residue is determined and expressed as dichlorvos.

For consistency, it would appear appropriate to also have a limit for dichlorvos in bran and wholemeal flour.

The methods recommended for the analysis of dichlorvos residues are: **Pesticide Analytical Manual (1979b, 1979c); Manual of Analytical Methods (1984); Methodensammlung (1982 — XII-3; S5; S13; S17); Abbott *et al.* (1970); Amburs *et al.* (1981); Panel (1973, 1977).** The following methods are recognised as being suitable: **Methodensammlung (1982 — 200); Dale *et al.* (1973); Draeger (1968); Eichner (1978); Elgar *et al.* (1970); Krause and Kirchhoff (1970); Mestres *et al.* (1979a, 1979b).**

13. Etrimfos

Etrimfos, a new organophosphorus insecticide developed in Switzerland, is O-(6-ethoxy-2-ethyl-4-pyrimidinyl) O,O-dimethyl phosphorothioate marketed for stored product protection under the Trade Name Satisfar. Six formulations designed for use on stored commodities are available. Etrimfos possesses contact and stomach activity on most stages of development of a broad spectrum of stored-product pests but has a low acute mammalian toxicity.

There is only limited information in open literature concerning the effect or fate of etrimfos as a grain-protectant insecticide. It appears to be fairly stable on grain, possibly because the deposit penetrates the grain, thus giving rise to significant residues in flour, where they appear to be carried through, to a considerable extent, into bread.

13.1 Usefulness

Stables *et al.* (1979) carried out a field trial on two 10-tonne lots of stored oil seed rape to control a widespread infestation of *Acarus siro* and *Glycyphagus destructor*. Dilute etrimfos emulsion was admixed with the rape seed. The mean initial level of etrimfos on the seed was shown by analysis to be 11.8 mg/kg. The etrimfos treatment controlled the mite infestation in 3 to 4 weeks and afforded protection against infestation for up to 24 weeks. Untreated oil seed rape stored nearby contained 50 000 to 100 000 mites per kg of grain at the end of the storage period.

Hall (1979) studied the effect of a range of temperatures on the knock-down speed of etrimfos against pests of stored grain. He included pirimiphos-methyl as a reference standard. The results obtained from his experiment showed that temperature does affect knock-down speed in that the speed was always slower at 2–5°C than at 20–25°C. This effect could be due to the reduced activity of the insects. The insects take up chemical either by feeding or through the cuticle. At low temperatures, insects enter a state of dormancy as observed in this experiment at 2–5°C. Whilst dormant, they do not feed or move around and their metabolic rate is very slow so that there is very little exchange of materials through the cuticle. Of the 3 species of insects tested *Cryptolestes* spp. proved least susceptible at 2–5°C.

Partington *et al.* (1979) presented the results of a series of trials in the laboratory and farm stores in the UK where etrimfos was evaluated for the control of arthropod pests of stored grain, including *Oryzaephilus surinamensis*, *Sitophilus granarius*, *Cryptolestes ferrugineus*, *Ahasverus advena*, *Tribolium confusum*, *Glycyphagus* spp., *Acarus siro*, *Trogoderma granarium* and *Plodia interpunctella*. As an admixture at concentrations of 5 and 10 mg/kg, etrimfos had a wide spectrum of effectiveness and outstanding persistence. Total control of *Ahasverus advena* and *Oryzaephilus surinamensis*, including a malathion-resistant strain, was still being obtained 6 months after treatment. Etrimfos applied at 0.5 g/m^2 also had a very good residual activity on all types of surface normally found in grain stores, including concrete, iron, hessian and brick.

Donat (1979a) determined the acute toxic dose of etrimfos and four other insecticides to *Sitophilus granarius* by exposure of the adult insects to treated wheat. He found etrimfos to be slightly less potent then pirimiphos methyl and chlorpyrifos-methyl but just slightly more toxic than methacriphos. The LD/95 values quoted in mg/kg active substance determined after 24 hours exposure were; pirimiphos-methyl, 0.86; chlorpyrifos-methyl, 0.95; etrimfos, 1.65; methacriphos, 3; dichlorvos, 54.

Donat (1979b) evaluated the effect of etrimfos upon the various development stages of *Sitophilus granarius* in wheat grain. When insecticides were applied at the rate of 5 and 10 mg/kg, etrimfos was less effective against eggs and young larvae than either methacriphos or dichlorvos but was somewhat more effective than pirimiphos-methyl. Against old larvae and pupae, etrimfos, pirimiphos-methyl and dichlorvos were all relatively ineffective while methacriphos was moderately active.

Yadav *et al.* (1980) determined the LD/50 values of 7 organophosphorus insecticides against 7 stored-product beetle pests. The insects were exposed to an insecticidal film in a glass dish at 27°C and 60–70% relative humidity. The mortality was counted after 24 hours. Etrimfos was the most potent of the 7 materials tested, *Sitophilus oryzae* being the most susceptible and *Callosobruchus maculatus* being the most tolerant. Phoxim had a comparable toxicity, bromophos, fenitrothion, iodofenphos, malathion and pirimiphos-methyl being significantly less potent.

Stables (1980) reported the results of screening 10 recently developed insecticides against 4 species of stored-product mites, *Tyrophagus longior*, *Tyrophagus putrescentiae*, *Acarus siro* and *Glycyphagus destructor*. The insecticides were applied as dusts or emulsion concentrates to wheat and the mortality of adult mites was evaluated 2 weeks later. Etrimfos at 20 mg/kg killed all mites of all species.

Prakash and Pasalu (1981) carried out a laboratory

evaluation of 5 insecticides, including etrimfos, as paddy rice seed protectants in order to minimise the emergence of adults of *Sitotroga cerealella* on paddy rice grains stored in glass jars. Etrimfos proved significantly more potent than fenitrothion, dichlorvos, permethrin and malathion. Even so these workers considered that a concentration of 30 mg/kg was necessary. Permethrin gave comparable control at this concentration.

Sandoz (1981) in a technical bulletin on Satisfar state that etrimfos has shown promising efficacy against stored-crop pests in Switzerland, England and France. The knock-down effect of etrimfos is stated to be outstanding, being more rapid at lower concentrations than comparison insecticides (unidentified). Admixture of etrimfos with wheat at the rate of 5 mg/kg is stated to result in 95–100% mortality of *Sitophilus granarius* adults after 6 hours. It is claimed that when etrimfos was admixed with wheat at 5 and 10 mg/kg a 96–100% effect was obtained against eggs of *Sitophilus granarius*. Against the larvae of the same insect the control ranges from 84–87%. Etrimfos was considered to be insufficiently effective against older larval stages and pupae of *Sitophilus granarius*. As persistence of etrimfos is good on the grain surface, all the insects are killed during hatching or immediately after hatching when they come in contact with the grain surface. Etrimfos is said to show a high level of efficacy against strains of stored-product pests known to be resistant to lindane, malathion and dichlorvos. The results of trials on which these claims are based have not been available to this reviewer.

In investigations in India, of the possibility of saving time, labour and money by applying insecticides and fungicides together for the protection of stored rice against *Sitotroga cerealella* and storage fungi such *Aspergillus* and *Penicillium* spp. and seed-borne fungi such as *Cochliobolus myabeanus, Alternaria padwicki* and *Fusarium* spp., only 4 of the 16 combinations tested were compatible, controlled both insect and fungi and improved seed viability. Among these was the combination of etrimfos with carbendazim **(Prakash and Kauraw 1983)**.

Prakash *et al.* (1983) evaluated etrimfos as a paddy seed protectant in Orissa, India. Etrimfos was applied in an aqueous solution at 9–21 mg toxicant/kg rice, either directly on to the grains by surface treatment or to gunny storage bags by dipping. Treated grains in untreated bags and untreated grains in treated bags were stored for 9 months and compared. Direct grain treatment proved more effective than bag treatment at all doses and throughout the storage period for protecting the rice against *Sitotroga cerealella, Rhyzopertha dominica* and *Sitophilus oryzae*. Even at 21 mg/kg etrimfos did not give complete protection for more than 3 months, but after 9 months, treated grain and grain in treated bags showed only 1.0% and 2.5% damage, as compared with 8% damage for no treatment. The treatments did not adversely affect germination.

Srivastava and Gopal (1984) studied the relative toxicity of organophosphorus insecticides against stored-grain pests in India. They tested 7 insecticides against 5 insects and on the basis of the LC/50, etrimfos was most toxic to *Sitophilus oryzae, Tribolium castaneum, Trogoderma granarium* and *Lasioderma serricorne*, while phoxim was most toxic to *Callosobruchus maculatus*. It was considered that phoxim applied at the LC/99 of 25.7 mg/kg and etrimfos applied at the LC/99 of 10.3 mg/kg against *Trogoderma granarium* and *Callosobruchus maculatus*, respectively, the least susceptible species, would be effective against all the other species. Both these compounds were considered suitable for field evaluation.

Bengston *et al.* (1984a) reported the outcome of silo-scale experiments in which etrimfos was applied at either 10 mg/kg, 7 mg/kg or 5 mg/kg along with synergised bioresmethrin (1 mg/kg) at 3 sites in Queensland and NSW. Etrimfos at 10 mg/kg controlled all typical strains of test species used in bioassays throughout the 9-month storage. Etrimfos at 5 mg/kg in combination with synergised bioresmethrin allowed some live progeny of one typical strain of *Sitophulus oryzae* to survive in samples 3 or 6 months after treatment. No treatment was completely successful against a highly resistant laboratory strain of *Sitophilus oryzae*. No natural infestations developed at any site.

Wohlgemuth (1984) carried out comparative laboratory trials with 14 insecticides under tropical conditions to determine the mortality of 5 stored-product beetles at various times during a 24 month storage of treated grain. At 36°C and 50% r.h. etrimfos, when applied at the rate of 15 mg/kg, was fully effective for more than 24 months against *Cryptolestes ferrugineus, Oryzaephilus surinamensis*, and *Sitophilus granarius* when applied as a dust but was effective for only 3 to 6 months when applied as an emulsion except against *Sitophilus granarius* where it appeared to be particularly potent. When applied at the rate of 5 mg/kg it was fully effective for a lesser period, the dust being more pesistent than the emulsion. However, against *Rhyzopertha dominica* both spray and dust were fully effective for only a short period even when applied at the rate of 15 mg/kg.

Chuwit Sukprakran (1984) presented a paper discussing the control of stored-grain insect pests in maize in Thailand, including tables showing the effect of insecticides against *Sitophilus* spp. However, the paper contained no information about experimental method. Maize treated with etrimfos at either 5 or 10 mg/kg showed significantly less damage over 4, 6 and 8 months than did similar grain treated with pirimiphos-methyl, chlorpyrifos-methyl, deltamethrin or cypermethrin (the latter 2 at 1 and 5 mg/kg).

Bandyopadhyay and Ghosh (1984) reported an experiment conducted in a storehouse in West Bengal, India, where stacks of wheat bags were sprayed with etrimfos (formulation not identified) at concentrations of

0.05% and 0.1%. The temperature ranged from 26° to 36°C, the relative humidity from 56% to 98% and the grain moisture content from 10.2% to 12.5%. Grain drawn from the bags at intervals over 3 months was examined and the percentage of damaged grains was determined. Bags treated with etrimfos at a concentration of 0.1% protected the contents from serious infestation for 3 months. Grain in bags treated with malathion suffered 4 times the rate of damage and that in untreated bags 7 times as much as that in etrimfos-treated bags.

Vyas (1984) presented a paper describing the evaluation of different indigenous materials and insecticides against *Rhyzopertha dominica* on wheat during storage. Etrimfos, applied either as dust or emulsion at 6 mg/kg, was more effective than all other materials tested in reducing the percentage of damaged grains 180 days after starting the experiment.

Ratnadass (1984), in an extensive report on entomological problems linked with tribal storage in the Ivory Coast, indicated that etrimfos, applied at the rate of 5 mg/kg to maize was highly effective in reducing the number of *Sitophilus* spp. infesting maize and statistically equal to pirimiphos-methyl (10 mg/kg) and deltamethrin (1 mg/kg). In another trial etrimfos (5 mg/kg) effectively reduced the weight-loss of maize and reduced the number of damaged grains to 7% of that occurring in untreated controls.

SEAMEO (1985) reported an efficacy trial with etrimfos emulsion against post-harvest pests in rough and milled rice in Indonesia. The insecticide was applied to the bag surface using a pressurised hand sprayer so as to apply a deposit of 0.25, 0.5 and 1 g a.i./m^2. The bags of rice were subject to natural infestation in a rice mill. The efficacy of etrimfos at the 1 g a.i./m^2 rate was equal to the standard treatment with pirimiphos-methyl at 1 g a.i./m^2. It was recommended that the application interval should not exceed 2 months.

Evans (1985) tested the effectiveness of various insecticides on some resistant beetle pests of stored products from Uganda. The discriminating-dose tests showed that all populations of insects received except *Tribolium confusum* were resistant to lindane. All populations of *Tribolium* spp. showed some resistance to malathion. None of the three populations of *Sitophilus zeamais* was resistant to malathion. Deltamethrin was the only insecticide to give complete control of *Zabrotes subfasciatus* and *Callosobruchus maculatus* after 24 weeks storage. Etrimfos gave complete control of these two species only when adults were exposed to treatments immediately after insecticide application. Large numbers of adults emerged from the etrimfos treatments after 24 weeks storage.

Rai *et al.* (1985) reported trials on the impregnation of jute bags with insecticides for protecting stored food grains. Etrimfos (0.5%) was the only one of 5 insecticides completely effective in killing *Rhyzopertha dominica* exposed on treated jute bags 90 days after treatment. The kill was only 94% at the end of 120 days. Complete kill of all other species was given by each insecticide 120 days after application. Etrimfos-treated bags were more effective than those treated with other insecticides for protecting wheat grain, reducing kernel damage and percent weight loss 4 months after exposure of the wheat in treated bags.

Bengston *et al.* (1986) showed, in a series of supervised trials with bulk wheat stored in 10 commercial silos at 8 sites in 4 Australian states, that etrimfos applied at the rate of 8 mg/kg was completely effective against 5 major species of stored-product pests. It provided complete protection against all strains, except multiresistant *Rhyzopertha dominica*, for 9 months storage. The temperature of the grain ranged between 30°C at time of treatment to 22°–20°C after 9 months storage. The moisture content varied between 10% and 12.5% depending on location.

13.2 Degradation

Very little information on the degradation of etrimfos has come into the hands of this reviewer.

The results of several trials conducted by the manufacturers in different countries on wheat, barley and corn were submitted to the 1980 Joint FAO/WHO Meeting on Pesticide Residues. The results published in the monographs **(FAO/WHO 1981)** are reproduced in **Table 13.1**. Practically all these results appear to come from small-scale trials and the quality of the data leaves much to be desired. Although it appears that the rate of degradation is slow the lack of information about storage conditions and obvious sampling problems make it difficult to pronounce with any degree of certainty on the rate of degradation or degradation constants for this insecticide.

Stables (1980) found little breakdown of etrimfos on oil seed rape stored at a mean temperature of 6°C (range 3.0–9.3°C) during 24 weeks.

Donat (1979c) measured the persistence of etrimfos on wheat under various climatic conditions by means of a bioassay with *Sitophilus granarius*. At 35°C and 70–80% r.h. it was inferior to malathion but superior to pirimiphos-methyl and methacrifos at the end of 27 days when applied at the rate of 10 mg/kg. When tested at 20–25°C and 50–60% r.h. etrimfos gave 99% mortality at the end of 119 days by which time the other insecticides had failed to kill insects. In these latter tests pirimiphos-methyl and methacrifos failed to give 100% mortality after 7 days exposure when applied at the rate of 5 mg/kg.

Donat (1979d) determined the persistence of various etrimfos dust formulations and compared these with etrimfos emulsion and malathion and pirimiphos-methyl dust. The etrimfos dust formulations were all superior to the etrimfos emulsion at similar rates and slightly superior to malathion dust. However, pirimiphos-methyl dust remained 100% effective at 10 and 12 months, being just slightly superior to the corresponding etrimfos treatments.

Sandoz Ltd (1984) reported the results of analyses of

Table 13.1. Residues of etrimfos (mg/kg) in stored grain treated after harvest **(FAO/WHO 1981)**

Commodity (country)	Application			Storage period (months)						Remarks
		Type	rate (mg/kg)	0 (24 h)	1–2	3–4	6	8–10	12–13	
Barley (England)	50	EC	5	2.3	3.2–3.6	2.1	0.8			Amount treated unknown; moisture content 15%; temp. 0–15°C
(Kenya)	0.5	Dust	2.25	2.1		0.6	0.3			Amount treated: 45 kg each, stored in bags conditions not reported
	1	Dust	4.5	3.0		1.3	0.8			
	50	EC	5	1.3		0.5	0.2			
	2	Dust	9	9.7		1.2–2.2	0.8–1.4			
	50	Dust	10	4.1		1.0–2.1	0.5–0.9			
Corn (maize)										
(Kenya)	0.5	Dust	2.75	0.6–0.8		0.5	0.3–0.4	0.3–0.4	–	Amount treated: 45 kg each, stored in bags; storage conditions not reported
	50	EC	5	1.5–1.6		1.6–1.7	0.8–1.2	1.0–1.1	0.8	
	1	Dust	5.5	0.9–2.6		0.7–1.5	1.0–1.6	0.8–0.9	0.9–1.3	
	50	EC	10	0.8–4.2		1.7–2.9	1.5–2.5	1.9–2.5	2.5–2.6	
	2	Dust	11	1.7–5.5		2.4–3.8	2.2–2.9	1.2–2.4	1.5–2.3	
Wheat										
(England)	50	EC	5	6.8	5.6–6.1	5.9	3.9			Amount treated: 10 t; 30 t on conveyer belt; moisture cont. 11%; 0–17°C
	50	EC	10	10.9	6.7–7.2	5.4	6.4			
(France)	50	EC	5		1.9–2.0	1.4–20	1.6			Storage conditions not reported: samples for analysis 10 g each
	50	EC	7.5		1.6–2.5	2.1–3.3	2.1–2.9			
	50	EC	10		3.3–5.4	4.2	2.6–3.8			
(Kenya)	0.5	Dust	2.5	1.2		0.7	01.0	0.8	0.4	Amount treated: 10 kg each, stored in bags; storage temp. about 15–26°C; further conditions not reported
	1	Dust	5	2.0		1.8	1.5	1.3	1.0	
	50	EC	5	1.6		1.0	1.0	0.8	0.5	
	2	Dust	10	3.7		2.8	2.7	2.1	1.8	
	50	EC	10	2.0		2.5	2.6	2.5	1.7	
(Switzerland)	50	EC	5	2.9	2.9	2.7	3.0	1.9	1.1	Amount treated; 25 kg each, stored in air-permeable fibre drums; moisture content 12–14% temp. about 16–20°C
	50	EC	10	5.4	5.8	5.5	5.6	3.4	1.7	
	50	EC	15	7.4	7.8	7.6	8.1	5.6	–	

samples from silos of wheat treated in Australia with etrimfos at the rate of 5 mg/kg and 10 mg/kg. These silos contained 520 and 550 tonnes, the temperature ranging from 29.6°C to 31.4°C. The results of analysis are presented in **Table 13.2**. No trace of the oxygen analogue of etrimfos could be detected in any sample.

Bengston *et al.* (1984a) measured the loss of etrimfos residues on wheat treated with 5, 7 and 10 mg/kg etrimfos in combination with synergised bioresmethrin (1 mg/kg) and held in bulk wheat silos at 3 sites in Queensland and NSW in 1982–83. The findings are given in **Table 13.3**. The grain temperature remained at or above 30°C for more

Table 13.2. Etrimfos residues in wheat stored in silos in Australia **(Sandoz Ltd 1984)**

Days after treatment	Residues (mg/kg) (recovery included)			
	5 mg/kg rate		10 mg/kg rate	
	Etrimfos	EEHP*	Etrimfos	EEHP*
0	4.1	n.d	5.9	n.d
49	3.2	trace	6.3	0.21
77	3.3	trace	5.2	0.14
126	3.1	trace	5.2	0.21
168	3.0	trace	4.8	0.14
252	2.4	0.14	4.0	0.31

*EEHP = 6-ethoxy-2-ethyl-4-hydroxy pyrimidine
n.d = not detectable

than 4.5 months and had only declined to 22°-24°C at the end of 7 months. The moisture content of the separate lots of grain ranged from 7.6% to 10.6% and remained constant throughout the trial. The concentration of the etrimfos deposit did not decline to 50% of the initial level even after 7 or 9 months storage so the half-life was not accurately estimated.

Bengston *et al.* (1986) extended their investigations to include commercial-scale trials in 10 silos at 8 sites throughout Australia. Some, but not all, of these sites were sampled regularly throughout 9 months storage. The analytical results, summarised in **Table 13.4,** are generally similar to those obtained in the previous silo-scale experiments **(Bengston *et al.* 1984a)**.

Table 13.3. Residues of etrimfos (mg/kg) on wheat held in bulk silos in Australia 1982–83 **(Bengston *et al.* 1984a)**

Location	Target Treatment	Calculated Rate	Approx. time after application (months) 0	1.5	3	4.5	6	7	9
M.Q.	5	5.2	3.55	3.50	3.16	2.85	2.77	–	2.78
Q.N.	5	5.6	4.35	4.36	4.15	3.58	3.02	3.20	–
M.Q.	10	10.0	6.14	5.85	5.06	4.87	4.79	–	3.61
P.N.	10	8.3	7.11	6.40	6.12	6.05	5.18	5.53	–
P.N.	7	6.3	4.98	4.50	4.50	4.20	4.03	4.00	–

Table 13.4. Residues of etrimfos (mg/kg) on wheat held in commercial silos in 4 Australian states 1984–85 **(Bengston *et al.* 1986)**

Location	Target Treatment	Calculated Rate	Approx. time after application (months)					
			0	1.5	3	4.5	6	9
Qld 1	8	9.3	7.3	–	–	–	–	5.4
Qld 2	8	7.8	7.2	–	–	–	–	5.8
Qld 3	8	8.3	6.4	–	–	–	–	5.1
Qld 4	8	8.1	7.7	7.4	8.1	5.6	5.8	5.2
Qld 5	8	8.1	7.4	–	–	–	–	5.6
NSW 1	8	7.8	11.0	10.4	7.5	6.8	5.6	–
NSW 2	8	6.8	8.5	–	–	–	6.3	–
SA 1	8	8.04	13.0	10.2	8.2	7.3	–	–
SA 2	8	8.0	14.0	7.3	6.3	–	–	–
WA 1	8	8.01	10.7	9.8	8.0	6.6	–	–

13.3 Fate in Milling, Processing and Cooking

Here again very little information is available.

Wheat grain was treated in 25 kg lots with etrimfos 50 EC at the rate of 5, 10 or 15 mg/kg **(Sandoz 1979)**. The grain was stored at room temperature (not defined) and at intervals of 2, 4, 6 and 12 months, samples of grain were milled and separated into different fractions. Bread was baked from the flour from these samples. Etrimfos residues were analysed in all samples. The results are given in **Table 13.5**. There appears to be a gradual penetration of etrimfos into the endosperm with a consequent increase in the residue concentration in the white flour. There is apparently a significant decrease in the residue concentration between white flour and white bread but part of this is due to dilution with water which occurs in bread making. From these results it would appear that the residues of etrimfos are more stable than all of the other organophosphorus insecticides under consideration as grain-protectant insecticides. However, before drawing firm conclusions it is imperative that adequate data from commercial-scale trials should be made available because experience has shown that commercial milling, with its higher extraction rates, transfers a greater proportion of the residue from the aleurone layer to the flour than occurs in laboratory milling.

Table 13.5. Etrimfos residues in milled fractions and flour of wheat and in bread **(Sandoz 1979)**

Treatment rate	Residues (mg/kg) 2–12 months after treatment							
	2		4		6		12	
	Etrimfos	EEHP	Etrimfos	EEHP	Etrimfos	EEHP	Etrimfos	EEHP
5 mg/kg								
Grain	2.92	n.d.	2.74	n.d.	3.02	0.20	1.89	0.36
Bran	10.47	1.10	9.94	0.39	9.78	0.38	3.50	2.10
Grits	8.67	0.62	8.64	0.62	6.83	0.27	3.48	1.02
White flour	0.82	n.d.	1.02	n.d.	0.73	n.d.	0.70	n.d.
White bread	0.23	0.05	0.24	0.10	0.21	0.11	0.18	n.d.
Whole wheat flour	n.a.	n.a.	2.66	0.16	2.33	0.11	1.97	0.37
Whole wheat bread	n.a.	n.a.	1.52	0.22	1.33	0.34	0.69	0.23
10 mg/kg								
Grain	5.84	0.54	5.50	0.08	5.59	0.34	3.38	1.20
Bran	15.22	1.92	16.27	0.66	15.83	1.14	6.20	3.34
Grits	41.29	1.27	15.08	0.73	13.14	1.06	7.74	1.97
White flour	1.89	0.14	1.75	n.d.	1.47	0.06	0.96	n.d.
White bread	0.50	0.13	0.61	0.24	0.51	0.20	0.41	n.d.
Whole wheat flour	n.a.	n.a.	5.59	0.39	5.07	0.24	3.70	0.69
Whole wheat bread	n.a.	n.a.	2.83	0.60	2.38	0.51	1.37	0.51
15 mg/kg								
Grain	7.80	0.20	7.61	0.20	8.11	0.46	5.57	1.52
Bran	16.60	1.50	21.65	1.56	19.70	1.51	9.66	4.34
Grits	19.41	1.53	20.84	0.93	19.33	1.06	12.14	1.97
White flour	2.04	0.25	2.62	0.09	2.48	0.10	1.50	n.d.
White bread	0.74	0.18	0.95	0.49	0.78	0.31	0.54	0.23
Whole wheat flour	n.a.	n.a.	7.99	0.41	7.84	0.40	5.90	0.71
Whole wheat bread	n.a.	n.a.	3.80	0.40	3.77	0.70	2.30	0.46

EEHP = 2-ethyl-4-ethoxy-5-hydroxy-primidine
n.d. = not detectable
n.a. = not analysed

Stables *et al.* (1979) stated that preliminary studies indicated that 95–100% of the etrimfos residues on oil seed rape was lost during a simulated commercial extraction and refinement of the oil. **Chamberlain (1981)** provided more information about these studies. Little loss of residue occurred during the neutralisation of the extracted oil; however, appreciable loss occurred during bleaching and 95–99.5% loss occurred after steam distillation which is a commercial process for the deodorisation of vegetable oils. The aqueous distillate contained 12% of the etrimfos present at the start of the distillation. The extracted meal contained no detectable etrimfos residues.

Wilkin and Fishwick (1981) studied the fate of 6 organophosphorus insecticides applied to wheat as grain protectants. The treated wheat was milled to produce wholemeal which was converted into bread. Wheat treated with 5.9 mg/kg etrimfos was stored for 36 weeks after which it was found to contain 4.4 mg/kg etrimfos and when converted into wholemeal flour and bread, 4.6 and 2.6 mg/kg, respectively. This means that 57% of the residue in the flour survived baking. This percentage is slightly higher than any of the other compounds including fenitrothion and pirimiphos-methyl. It is not clear whether any allowance was made for the dilution which automatically occurs through the addition of water in bread making.

Bengston *et al.* (1984a) reported a series of studies in which etrimfos-treated wheat from 5 separate bulk trial sites where etrimfos had been applied 9 months previously was milled through experimental roller mills at 2 different research institutes prior to processing into wholemeal bread and white bread. Samples of the milling fractions and the bread were analysed for etrimfos and the results tabulated. A selection of the data is provided in **Table 13.6**. These data have been presented on a moisture-free basis in **Table 13.7**. From this information the distribution of etrimfos residues in the wheat fractions was calculated to be:

white flour 22.7%
bran 30.5%
pollard 46.8%

This represents a relatively high degree of penetration of the insecticide into the endosperm fraction of the grain.

13.4 Metabolism

Female rats were given a single oral dose of 50 mg/kg bw etrimfos (^{14}C-labelled at ring-positions 4 and 6) dissolved in 80% aqueous polyethylene-glycol. Excretion of radioactivity occurred mainly via the urine, 84% within 96 hours, of which 68% was within 12 hours; 6.6% was found in the feces within 24 hours. One hour after administration, continuous excretion of radioactivity was observed in the bile and amounted to 6% within 24 hours. Only 0.012% was exhaled within 24 hours of which 0.011% was within the first 8 hours, with a peak-value around 1 hour after administration. For most tissues the peak concentrations were reached at 4 hours after administration with the exceptions of the liver (2 hours), blood (3 hours) and fat (8 hours). At peak time the fat and kidney had the highest concentrations (42 and 62 mg/kg respectively) whereas the value in the other tissues ranged from 6 to 22 mg/kg with 13 mg/kg in the blood. The concentrations declined rapidly to values of 0.01 mg/kg at 96 hours except in the fat and skin with 0.3 mg/kg. In the urine and feces collected for 96 hours, neither etrimfos (I) nor its oxygen analogue were detected. The main metabolite in both excreta was 6-ethoxy-2-ethyl-4-hydroxypyrimidine (III) in amounts of 48% of the dose in

Table 13.6. Etrimfos residues (mg/kg) in milled products and bread from wheat treated 9 months before processing (**Bengston *et al.* 1984a**)

Sample analysed	Mill	Treated at Malu, Qld 5 mg/kg	Treated at Malu, Qld 10 mg/kg	Treated at Quandary & Pucuwan, NSW 5 mg/kg	Treated at Quandary & Pucuwan, NSW 7 mg/kg	Treated at Quandary & Pucuwan, NSW 10 mg/kg
Wheat	Qld	2.5	4.1	3.2	4.0	5.5
	NSW	2.5	4.1	3.2	4.0	5.5
Bran	Qld	8.3	–	10.2	15.1	16.3
	NSW	11.8	8.5	7.6	11.9	14.4
Pollard	Qld	8.1	13.1	12.4	17.7	18.1
	NSW	11.6	8.1	9.7	5.9	14.9
Wholemeal	Qld	2.4	4.6	3.6	4.6	4.9
	NSW	4.3	3.4	3.4	2.6	4.7
White flour	Qld	0.89	1.09	0.85	1.18	1.11
	NSW	0.60	0.61	1.20	0.44	1.50
Wholemeal bread	Qld	1.13	1.78	0.89	1.76	1.91
	NSW	0.94	1.71	1.21	1.50	1.15
White bread	Qld	<0.2	<0.2	<0.2	<0.2	<0.2
	NSW	0.3	0.7	0.4	0.3	0.7

Table 13.7. Etrimfos residues (mg/kg) from Table 13.6 calculated on a moisture-free basis

Sample analysed	Mill	Treated at Malu, Qld 5 mg/kg	Treated at Malu, Qld 10 mg/kg	Treated at Quandary & Pucuwan, NSW 5 mg/kg	Treated at Quandary & Pucuwan, NSW 7 mg/kg	Treated at Quandary & Pucuwan, NSW 10 mg/kg
Wheat	Qld	2.7	4.5	3.5	4.4	6.0
	NSW	2.7	4.5	3.5	4.4	6.0
Bran	Qld	6.2	10.1	9.2	10.4	11.1
	NSW	13.4	9.7	8.6	13.6	16.3
Pollard	Qld	9.0	14.9	14.1	19.8	18.2
	NSW	12.9	9.1	10.8	6.5	16.6
Wholemeal bread	Qld	2.7	5.2	4.1	5.2	5.6
	NSW	4.8	3.8	3.8	3.0	5.2
White flour	Qld	0.80	1.23	0.97	1.34	1.27
	NSW	0.68	0.69	1.35	0.98	1.69
Wholemeal bread	Qld	1.58	2.17	1.28	2.48	2.72
	NSW	1.27	–	–	–	–
White bread	Qld	–	–	–	–	–
	NSW	–	–	–	–	–

the urine and 3% in the feces. However, part of this metabolite is possibly formed from (I) and desmethyletrimfos (II) during the procedure for analysis. Other metabolites found in the urine and feces **(see Figure 13.1)** were the hydrolysis product of III (IV, 9% and 0.4% respectively) and 3 hydroxylation products of III (V, 5% and 0.25 respectively and VI and VII, together 3% and 0.4% respectively). The metabolites III, IV, V occurred both free and conjugated, the metabolites VI and VII mainly as conjugates **(Karapally 1975 and 1977).**

In another study with rats, metabolites II and III were determined as the main metabolites in 12-hours urine and feces; 65% of the ^{14}C-radioactivity found was II and 30% III **(Ioannou and Dauterman 1978).**

Etrimfos is rapidly degraded to water-soluble metabolites, mainly desmethyletrimfos (II) and EEHP (III), when incubated with rat or mouse liver subcellular fractions. The oxygen analogue of etrimfos could not be found. Glutathione-transferase and to a lesser extent mixed-function oxyidases are the main groups of enzymes

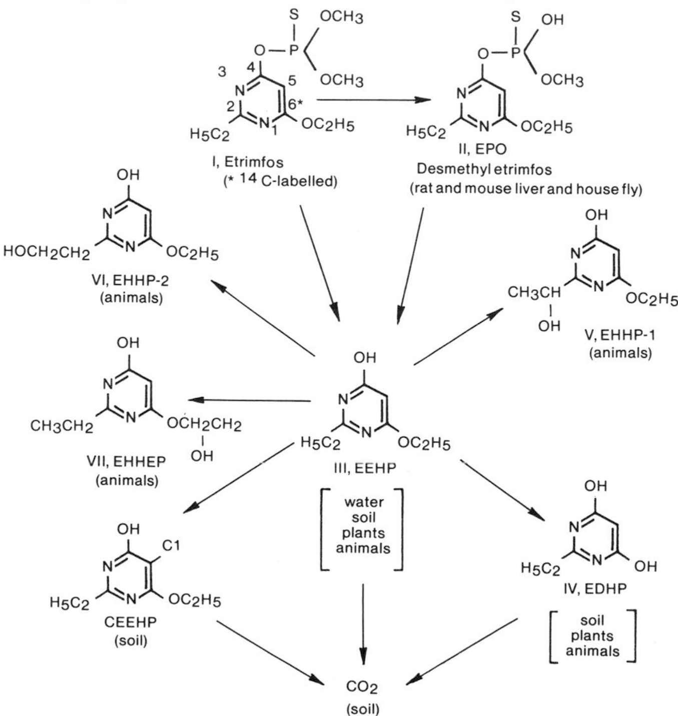

Figure 13.1. Metabolic pathways proposed for etrimphos in soil, plants and animals **(Karapally 1975, 1977)**

responsible for etrimfos metabolism **(Ioannou and Dauterman 1978)**.

The fate of etrimfos in bean and corn plants has been studied **(Akram *et al.* 1978)**. The primary leaves of bean and corn seedlings were treated with ^{14}C-etrimfos. The treated leaves and the untreated portions of the plant were sampled at 0, 3, 7, 14 and 21 days after treatment. The aqueous rinses and the extracts of the treated leaves were examined for the parent compound and its metabolites. Leaf remainder, and untreated portions, were analysed for total radio-carbon. Etrimfos rapidly volatilised from the leaf surface. Within the first 3 days after treatment it was lost with a half-life of approximately 3 days and during 3–15 days with a half-life of approximately 10 days from beans and approximately 5 days from corn. Three weeks after treatment the radioactivity, retained by the leaves after rinsing, ranged from 33% to 42% for beans, and from 25% to 42% for corn. No translocation to the root system was observed. A small percentage of the applied radio-carbon was present in the untreated foliage; 0.3–0.6% of the applied radioactivity became unextractable in bean leaves, and 0.1–0.3% in corn leaves. Leaf rinses of bean and corn contained mainly etrimfos and small quantities of EEHP. A small amount of the P = O analogue of etrimfos was also observed in the corn rinse. The leaf extracts of bean and corn contained etrimfos and minor quantities of EEHP and 5 unknown metabolites, of which 1 was tentatively identified as 2-ethyl-4,6-di-hydroxy-pyrimidine; 21-day bean leaf contained 33.5% of the applied radioactivity, of which 11.9% was etrimfos, 2.4% EEHP, 9.7% the unknown metabolites and 9.5% higher polar or conjugated metabolites. All single metabolites were less than 10% relative to applied etrimfos; 21-day corn leaf had about 24.3% of the applied radioactivity, of which 3.8% was etrimfos, 0.4% EEHP, 15.8% the unknown metabolites and 4.3% polar or conjugated metabolites. All single metabolites were less than 10% relative to applied etrimfos. The influence of metabolites on the toxicity of actual residue levels in practical field samples may be neglected.

13.5 Fate in Animals

A 3-level feeding study in dairy cattle with etrimfos incorporated in the diet at levels of 0.5, 1.5 or 5 mg/kg etrimfos was conducted **(Sandoz 1977)**. The milk production was not affected. The cows, sacrificed after 28 days of feeding, had only less than 0.01 mg/kg etrimfos in the muscle, liver, kidney and fat. Etrimfos residue in the milk throughout the study was less than 0.01 mg/kg. The residue in the milk solids separated from the twenty-first and twenty-eighth day samples also had only 0.01 mg/kg etrimfos.

Metabolism studies in the rat have shown that the P=O-pyrimidine ester bond in etrimfos is rapidly cleaved. This is the major detoxification process. Studies with rat and mouse liver homogenates **(Ioannou and Dauterman 1978)** have shown that the P-O-CH_3 and the P=O-pyrimidine ester bonds are rapidly cleaved. These studies clearly show that the triester molecule has only a transient existence in the animal system and that the phosphoric acid moiety exists either as the mono-methyl ester, or possibly as free phosphoric acid.

Re-esterification of the phosphoric acid moiety in the animal system, if it occurs, would not be something specific to etrimfos, but would be common to organo-phosphorus insecticides in general **(Karapally 1982)**.

13.6 Safety to Livestock

No studies dealing specifically with the effects of etrimfos on livestock have been located. The acute toxicity studies **(Table 13.8)** indicate that the compound has a relatively low acute toxicity.

Table 13.8. Etrimfos: acute oral toxicity to various species

Animal	Sex	Purity (%)	Vehicle	LD/50 (mg/kg)	Reference
mouse	M	97.2	water	470	**Hamburger & Klotzsche 1975**
	F	97.2	water	620	**Hamburger & Klotzsche 1975**
	M	90.5	corn oil	1120	**Anon. 1979b**
	F	90.5	corn oil	1100	**Anon. 1979b**
	M	–	corn oil	535	**Ioannou & Dauterman 1978**
rat	M	97.2	water	1800	**Hamburger & Klotzsche 1975**
	F	97.2	water	2354	**Hamburger & Klotzsche 1975**
	M	90.5	corn oil	1930	**Anon. 1979c**
	F	90.5	corn oil	1970	**Anon. 1979c**
	M	–	corn oil	2040	**Ioannou & Dauterman 1978**
Mallard	M & F	tech	–	1334	**Sandoz 1986**
ducks		tech	feed	405(1)	**Sandoz 1986**
chickens	M & F	–	50EC	240	**Sandoz 1986**
pigeons	M & F	–	50EC	500	**Sandoz 1986**

Note (1) 8 day LC/50 in Mallard ducks

Male and female rats were given feed containing 50, 250 or 1250 ppm etrimfos for 4 weeks. A decreased cholinesterase activity in erythrocytes was observed at 50 ppm and higher doses. Plasma cholinesterase was decreased in females from the 50 ppm level and in males from the 250 ppm and higher groups. Brain cholinesterase was decreased in females from the 250 ppm and males from the 1250 ppm groups **(Carpy and Klotzsche 1975a)**.

Groups of rats were fed etrimfos at concentrations of 3, 9 or 27 ppm for a period of 13 weeks. At the end of the study the glucose concentration in serum was increased in both males and females on the highest dose level. Microscopic examination showed no abnormalities attributable to the presence of etrimfos in the diet. After 8 and 13 weeks the plasma cholinesterase activity in the female rats in the 9 ppm group was slightly depressed and clearly depressed in the 27 ppm group. Cholinesterase activity in the erythrocytes and brain was within normal limits **(Carpy and Klotzsche 1975b)**.

Groups of 4 male and 4 female beagle dogs received 0, 4, 10 or 25 ppm etrimfos in their diet during 106 weeks. There were no signs of adverse behaviour and no mortality. Growth and food consumption were not clearly affected. Plasma cholinesterase activity was depressed in the 10 ppm and 25 ppm groups by 23% and 24% respectively. The cholinesterase activity of the erythrocytes was also decreased. Neither the weights of the organs nor the histopathological evaluation revealed indications for treatment-related changes in any organ or tissues. The no-observed-effect level in this study was 10 ppm **(Carpy and Klotzsche 1977)**.

13.7 Toxicological Evaluation

Etrimfos has been evaluated by the JMPR in 1980 and 1982 **(FAO/WHO 1981, 1983)**.

The Meeting evaluated *in vitro* and *in vivo* studies on mutagenicity and considered that etrimfos did not elicit a mutagenic response. A 3-generation, 2-litter per generation, reproduction study produced no adverse effects. Likewise, etrimfos was considered not to be teratogenic on the evidence of a study in white rabbits.

No delayed neurotoxic response was observed in chickens treated with etrimfos. Etrimfos was considered not to cause skin sensitisation.

The Meeting had access to short-term studies in rats and rabbits and long-term studies in rats and dogs. The no-effect levels in these studies were:

Rat: 6 mg/kg in the diet equivalent to 0.3 mg/kg bw/day

Dog: 10 mg/kg in the diet equivalent to 0.25 mg/kg bw/day

On the basis of this evidence the acceptable daily intake for man was estimated to be 0–0.003 mg/kg bw/day.

13.8 Maximum Residue Limits

On the basis of some rather inadequate studies of the nature and fate of etrimfos residues in grain and milled products the JMPR concluded that the residue levels listed below were suitable for establishing temporary maximum residue limits. The MRLs should remain temporary limits, irrespective of the status of the ADI, until adequate additional residue data from large-scale supervised trials on cereal grains and their products are made available.

Commodity	Maximum Residue Limits (mg/kg)
Bran (unprocessed)	20
Barley, maize, wheat	10
Wheat flour (wholemeal)	10
Wheat flour (white)	2
Rape seed	10
Rape seed oil (refined)	0.5
Rice	0.1

Residues are determined and expressed as etrimfos.

The methods recommended for the determination of etrimfos residues are: **Ambrus *et al.* (1981); Zweig (1980); Bowman *et al.* (1978).**

14. Fenitrothion

Fenitrothion is a broad-spectrum insecticide with a much lower acute mammalian toxicity than many similar insecticides. Its use is almost world wide for such crops as rice, fruits, vegetables, cotton, cereals, soy beans, coffee and tea. It is also used in public health, principally as a residual spray in houses for the control of mosquitoes. It also finds widespread application for the control of insects attacking forest trees and for the management of plague locust swarms. The composition of the technical material is apt to vary depending upon manufacturer, of which there are many, since the product was never covered by patents.

Fenitrothion has been widely used for grain protection in a number of countries for up to 10 years. It has a broad spectrum of effect against all species, though it is not fully effective against *Rhyzopertha dominica*. High potency and good stability mean that deposits in the region of 5–10 mg/kg are sufficient under most storage conditions to give complete protection for 9 to 12 months. When combined with pyrethrum or with synthetic pyrethroids, the effectiveness of fenitrothion is increased and the dosage level can be reduced. Fenitrothion products have proved suitable for use under tribal storage conditions where it is generally applied in the form of a very dilute dust. There is minimal penetration into the grain so that the deposit is mostly removed in bran of wheat and husks of rice. Less than 10% of the deposit is carried over into flour where it is degraded in storage, preparation and cooking until only 1–2% of the residue in the grain remains in white bread. The residues in bran are readily destroyed in the industrial processes used for preparing breakfast cereals. Residues on brown rice and milled rice are reduced 60–80% by cooking, leaving only a small residue in cooked rice.

14.1 Usefulness

Strong and Sbur (1965a) presented results from a series of tests with varying dosages of each of 46 insecticides, applied to wheat, for protection against 3 stored-product pests. Fenitrothion was more effective than malathion as indicated by dosages resulting in 100% mortality of all 3 species of test insects.

Lemon (1966) determined the relative susceptibility of *Tribolium confusum* and *T. castaneum* to 16 organophosphorus insecticides. Fenitrothion produced higher mortalities than those given by malathion.

Green and Tyler (1966) described a study to compare malathion, dichlorvos and fenitrothion under field conditions for the control of *Oryzaephilus* infesting stored barley. Fenitrothion, applied at the rate of 2 mg/kg, achieved complete control more quickly than when applied at 1 mg/kg and both rates remained effective for a long period after treatment, conferring good protection for as long as 8 months. This good performance was no doubt assisted by the fact that the grain cooled from 20°C to 10°C during the early stages of the trial. Fenitrothion proved superior to either malathion or dichlorvos.

Green *et al.* (1966b) carried out large-scale laboratory experiments to obtain information on fenitrothion as a grain protectant. The insecticide was applied as an emulsion to wheat and barley at the rate of 1, 2 and 4 mg/kg. The treated grain was stored in bags, together with untreated controls, at 25°C and about 60% r.h. under conditions of heavy cross-infestation by *Oryzaephilus surinamensis, Sitophilus granarius* and *Tribolium castaneum*. After 2 months, although some live adults could be found in the treated grain, there was no evidence of breeding. At 6 months limited breeding had occurred, although even at the lowest dosage rate a very high level of control was achieved of all except *Tribolium castaneum* on wheat. This species did not breed well on the grain, however, and relatively low numbers were recorded even in the control bags. In bioassay tests some breeding of *Sitophilus granarius* occurred on barley freshly treated at 1 and 2 mg/kg and on wheat at 1 mg/kg.

Green *et al.* (1966c) reported that fenitrothion applied at the rate of 4 mg/kg to wheat and barley, maintained under conditions of heavy cross-infestation by *Oryzaephilus surinamensis, Sitophilus granarius* and *Tribolium castaneum*, gave a very large measure of protection of wheat and barley for 10 months.

Lemon (1967a) compared the relative toxicities of malathion, bromophos and fenitrothion to 10 species of stored-product beetles by using a topical application technique. Fenitrothion was the most toxic compound to all species tested and when applied to wheat of 12% moisture content was the most effective of the 3 compounds as a protectant.

Cogburn (1967) found fenitrothion to be the most effective of 5 candidate insecticides, applied to rough rice, for the control of *Sitophilus oryzae, Tribolium confusum* and *Rhyzopertha dominica*.

Fenitrothion at 2 mg/kg and malathion at 10 mg/kg were compared as protectants on warm, moist grain in bins in a British farm granary **(Tyler and Green 1968)**. Both fenitrothion and malathion became ineffective after 2 weeks on moist, heating grain near the centre but the deposit remained effective for at least 12 weeks at the cooler, outer edges of the bins. At the doses used, the persistence of fenitrothion was similar to that of malathion on warm, damp grain and probably slightly superior on cooler, drier grain.

Strong and Sbur (1968) compared the effectiveness of 48 insecticides against 5 species of stored-product insects. Mortalities were recorded after 24 and 48 hours. Fenitrothion was among the most effective against all species.

Fenitrothion, applied as an emulsion to wheat and barley of 11% and 12% moisture content, respectively, or to the outsides of sacks, was highly effective for the protection of bagged grain stored at 25°C, under conditions of heavy cross-infestation pressure by 3 species of insects. Dosages of 2 mg/kg gave satisfactory protection for 6 months and 4 mg/kg for 10 months **(Kane and Green 1968).**

Strong (1969) determined the relative susceptibility of 5 species of stored-product moths to each of 12 organophosphorus insecticides by directly applying spray solutions to adult insects. Fenitrothion proved most effective against all 5 species, *Ephestia cautella* being the least susceptible.

A comparison of malathion, diazinon, fenitrothion and dichlorvos for the control of *Sitophilus oryzae* and *Rhyzopertha dominica* in wheat showed fenitrothion was the most promising material **(Champ *et al.* 1969)**. This result was confirmed by trials made under bulk storage conditions. A comparison study was made using malathion, diazinon and fenitrothion against 3 species of common storage pests in India **(Kashi 1972)**. Fenitrothion proved to be 3 times more toxic to *Tribolium castaneum*, 4 times more toxic to *Sitophilus oryzae* and 1.5 times more toxic to *Callosobruchus chinensis* than malathion. The author considered fenitrothion an ideal grain protectant insecticide for use in India.

The **Cyprus Agricultural Research Institute (1972)** evaluated fenitrothion among a number of low toxicity insecticides as grain treatments against selected species of stored-grain insects. By the third month, most of the chemicals had lost their effectiveness but fenitrothion was still effective 4 months after application.

Wilkin and Hope (1973), in an evaluation of 20 insecticides against 3 stored product mites, found that fenitrothion (2 mg/kg) was very effective against *Tyrophagus putrescentiae* and *Glycyphagus destructor* but not against *Acarus siro*. In a later report **Wilkin and Haward (1975)** advised that reducing the temperature decreased the effectiveness of the compounds tested against mites, including fenitrothion. There was only a very slow and incomplete effect at 5°C.

McCallum-Deighton (1974), in describing the use of pirimiphos-methyl, indicated that the lowest doses of pirimiphos-methyl, giving complete kill of susceptible and malathion-resistant strains of insects, was approximately the same as those for fenitrothion but lower than those for malathion, bromophos, iodofenphos, tetrachlorvinphos or dichlorvos.

Harein (1974) pointed out that fenitrothion is not as effective as some insecticides against stored-product moths, its effectiveness being only one-twelfth of that of dichlorvos.

Golob and Ashman (1974) studied the effect of a number of insecticides on 3 species of insects attacking rice bran. Fenitrothion at 3 and 6 mg/kg was compared with malathion at 16 or 20 mg/kg and pirimiphos-methyl at 4 or 8 mg/kg. At these rates, fenitrothion did not give adequate control of any species.

Weaving (1975) evaluated 5 insecticides for use under tribal storage conditions in Rhodesia. Fenitrothion was the most toxic to *Sitophilus zeamais*. At the rate of 8 mg/kg on maize and sorghum, fenitrothion showed good persistence for 12 months.

The **Queensland Department of Primary Industries (1975)** advised that fenitrothion had advantages over malathion but did not control malathion-resistant strains of *Rhyzopertha dominica*. Combinations with bioresmethrin appeared necessary.

LaHue (1975a) evaluated fenitrothion along with other insecticides for the control of *Sitotroga cerealella* in shelled corn. Fenitrothion gave good control and protection for up to 8 months, being superior to the standard recommended treatment of malathion.

LaHue and Dicke (1976a) evaluated selected insecticides, including fenitrothion, applied to high-moisture sorghum grain to prevent stored-grain insect attack. Fenitrothion, applied at the rate of 9 mg/kg, was effective in preventing damage to the grain during 12 months storage, though pirimiphos-methyl proved superior in reducing insect damage. Six of the more important insect pests were involved in these studies.

LaHue (1976) evaluated 5 broad-spectrum protectants of seed corn against stored-product insects over a 21 month period, involving almost 2 complete cycles of climatic changes. Fenitrothion, applied at the rate of 9 mg/kg, gave comparable performance to malathion applied at the rate of 12 mg/kg, but both treatments were inferior to pirimiphos-methyl and chlorpyrifos-methyl against 5 species.

LaHue and Dicke (1976b) reported fenitrothion to be inferior to pirimiphos-methyl and chlorpyrifos-methyl but equal to malathion when applied as a spray treatment for protecting shelled corn in small bins against stored-grain insects.

LaHue and Dicke (1977) evaluated the same insecticides as protectants for wheat against the same range of insect pests. Fenitrothion was generally more effective than malathion but inferior to pirimiphos-methyl and chlorpyrifos-methyl.

LaHue (1977c) reported the results of field tests under commercial conditions when the same insecticides were used as grain protectants for seed corn. When applied at similar rates to those used in the abovementioned studies, the general order of effectiveness of the treatments was the same.

Tyler and Binns (1977) determined the toxicity of 7 organophosphorus insecticides and lindane to 18 species of stored-product beetles. Fenitrothion was second only to chlorpyrifos-methyl in effectiveness.

Hyari *et al.* (1977) described the results of laboratory evaluations of emulsifiable and encapsulated formulations of malathion and fenitrothion on soft red winter wheat against attack by adults of 4 species of stored-product insects. Fenitrothion was superior to malathion against all 4 species at all rates examined. An initial concentration of 10 mg/kg appeared necessary to provide long-term protection, *Rhyzopertha dominica* being significantly more tolerant than other species. Encapsulated formulations were not significantly superior.

Desmarchelier (1977b) showed that bioresmethrin or pyrethrum supplemented the effect of fenitrothion and provided effective control of *R. dominica*. He showed that the ratio of the minimum effective dose to control *R. dominica* to the minimum effective dose to control *T. castaneum* was the same for fenitrothion and malathion.

Ardley and Sticka (1977) compared fenitrothion at 2.5, 5 and 6 mg/kg with malathion, applied at the rate of 12 and 18 mg/kg, for the protection of bulk wheat in vertical silos, and 10 mg/kg fenitrothion compared with 18 mg/kg malathion in horizontal bulk depots. In both types of storage the persistence of the insecticides could be correlated with the temperature and moisture content of the grain. Under the conditions of silo storage, both protectants remained effective against a range of pests for as long as 19 months, but under the more severe conditions of horizontal bulk storage the applications were effective for only 6–7 months. Fenitrothion at half the application rate of malathion appeared to give equivalent protection against a number of stored-product insect pests.

Bengston *et al.* (1978b) carried out studies in commercial silos holding sorghum, which they treated with combinations of fenitrothion (12 mg/kg) and bioresmethrin or phenothrin (1 mg/kg). Bioassay studies carried out on samples from each bulk storage indicated that complete protection was achieved against all species for at least 12 weeks and against some species for more than 22 weeks.

Ardley and Desmarchelier (1978) briefly reported the results of field trials with combinations of bioresmethrin and fenitrothion as potential grain protectants. The study demonstrated that low dosages of fenitrothion, in the presence of low doses of bioresmethrin, were effective in controlling many species and strains of insects, including a number known to have a high degree of resistance to malathion.

McCallum-Deighton (1978) compared the minimum effective dose (in mg/kg) for technical grade insecticides (8 compounds) against susceptible strains of 7 stored-product insects. Fenitrothion was shown to have a higher acute toxicity than all except pirimiphos-methyl which is marginally more potent. However, both compounds are relatively less effective against *Rhyzopertha dominica*.

LaHue and Kadoum (1979) compared the residual effectiveness of emulsion and encapsulated formulations of malathion and fenitrothion against 4 stored-grain beetles. The insects were exposed to treated wooden surfaces for 6, 12, 24, 48 and 72 hours at 1, 3, 6, 9 and 12 months post-treatment. Fenitrothion was more effective than malathion in comparative tests. The encapsulations were more persistent than the emulsion formulations.

Two articles in a UK trade journal **(Anon 1979a, d)** emphasised the importance of controlling insect pests of stored grain and recommended the admixture of fenitrothion with the grain. Recommendations and directions for the use of fenitrothion were given.

Bengston and Desmarchelier (1979) described extensive laboratory and field trials that led to the introduction of fenitrothion plus bioresmethrin as a grain protectant to control malathion-resistant insects in Australia. This combination has been effective in attaining and maintaining the grain export standard of a nil tolerance for live insects. Where complete protection is required for periods up to 9 months, fenitrothion is applied at the rate of 12 mg/kg and bioresmethrin, synergised with piperonyl butoxide, at the rate of 1 mg/kg. **Desmarchelier (1978c)** discussed the residual behaviour of chemicals in stored grain and provided a mathematical basis for the recommendations that have been officially adopted in Australia. These include the application of fenitrothion at the rate of 12 mg/kg to all classes of stored grain that is to be held in store for more than 6 months.

Amos *et al.* (1979) described a simple, controlled, non-uniform treatment of wheat with fenitrothion, with or without bioresmethrin, which has proved convenient and effective for protecting bulk wheat against a range of insect pests.

Small-scale laboratory trials carried out to determine the effectiveness of a range of insecticide dusts for use under tribal storage conditions in Zimbabwe showed that fenitrothion (8 mg/kg) possessed the greatest persistence and effectiveness against a range of insects attacking maize **(Weaving 1980).**

Margham and Thomas (1980) studied the weight/relative-susceptibility of *Tribolium castaneum* to fenitrothion. *Tribolium castaneum* adults, 27–29 days old were dosed topically with fenitrothion. The time to knock-down of each was determined, together with its dry weight. The times and the weights were significantly and positively correlated. The times for the females to be knocked down were on average about 13% higher than for the males, but the times adjusted for the beetle weight were the same for both sexes.

O'Donnell (1980) compared the toxicities of 4 insecticides to *Tribolium confusum* under 2 sets of conditions of temperature and humidity. He showed that all 4 insecticides were more effective at the higher temperature and that fenitrothion was more toxic than malathion and iodofenphos, but less toxic than pirimiphos-methyl.

Bengston *et al.* (1980b) carried out duplicate field trials on bulk wheat in commercial silos and tested the efficacy of the insecticidal deposit by bioassays on samples of treated grain at intervals over 9 months, using malathion-resistant strains of insects. It was shown that fenitrothion (12 mg/kg) plus phenothrin (2 mg/kg) was more effective than pirimiphos-methyl (6 mg/kg) plus carbaryl (10 mg/kg) against 5 important stored-product pests.

Hindmarsh and MacDonald (1980) reported field trials to control insect pests of farm-stored maize in Zambia. Severe infestations of *Sitophilus zeamais* and *Sitotroga cerealella* occur in maize cobs stored under farm conditions in Zambia causing up to 92% damage to grains 8 months after harvest. Shelling the maize reduced *Sitotroga cerealella* to low levels and also stabilised the moisture content of the grain providing a more suitable substrate for insecticide application and control of *Sitophilus zeamais*. Several insecticides applied at the rate of 12 mg/kg as dust admixture to shelled maize kept damaged grains below 10% up to 10 months after harvest. Fenitrothion applied at the rate of 2 mg/kg maintained these low damage levels for up to 8 months after harvest but there appears little doubt that a higher rate of treatment would have produced a result at least comparable to the other insecticides.

In a trial of grain protectants for use under tribal storage conditions in Zimbabwe, **Weaving (1981)** reported that fenitrothion applied as a dust at the rate of 8 mg/kg showed the greatest persistence and effectiveness in 2 small-scale storage trials for a 12 month period.

Viljoen *et al.* (1981a) included fenitrothion along with a number of other contact insecticides in field trials to protect maize and groundnuts against reinfestation. Fenitrothion was rated slower-acting and less effective than phoxim and iodofenphos against *Ephestia cautella, Plodia interpunctella* and *Tribolium castaneum*.

Desmarchelier *et al.* (1981b) reported the results of the use of grain protectant combinations, fenitrothion plus bioresmethrin and pirimiphos-methyl plus bioresmethrin applied to grain during 1976 in 21 commercial storages throughout Australia. Three storages became infested with *Rhyzopertha dominica* but all storages remained free of other insect pests. In 2 of the 3 infested storages, application of protectant was uneven, and the third became infested only after 8 months of storage. The bioresmethrin had been added specifically to control *Rhyzopertha dominica*.

Bansode *et al.* (1981) studied the toxicity of 4 organophosphorus insecticides to a malathion-resistant strain of *Plodia interpunctella* in North Carolina. This strain showed greater than 227-fold resistance to malathion but the tolerance to fenitrothion was only 1.6-fold.

Minett *et al.* (1981), who developed a gravity feed unit to apply insecticide concentrate to wheat being transferred on a conveyor belt, carried out extensive trials using fenitrothion and bioresmethrin and were able to demonstrate that the distribution and performance of the resulting deposit, as measured by chemical analysis and bioassay, was in no way different to the results achieved by conventional spraying equipment.

Desmarchelier and Wilson (1981) studied the distribution of fenitrothion applied to wheat by gravity feed and by spraying. They found that when fenitrothion was gravity fed as an undiluted concentrate, its distribution in 100 g samples of wheat taken from commercial storages was similar to the distribution after the use of conventional diluted sprays.

Cogburn (1981) compared the performance of fenitrothion and 3 other grain protectants against insect pests of stored rice and was able to show that encapsulated formulations of the insecticide offered a significant advantage.

Tyler and Binns (1982) determined the influence of temperature on the susceptibility of both susceptible and resistant strains of 3 stored product beetles to 8 organophosphorus insecticides. Based upon knock-down and kill, the effectiveness of all insecticides was greater at 25°C than at 17.5°C and was markedly lower at 10°C. Fenitrothion was ranked third amongst the most effective insecticides.

Kumar *et al.* (1982) carried out field trials with 5 organophosphorus insecticides, including fenitrothion, against insect pests of stored-food grains in godowns in various parts of India. Fenitrothion was superior to the standard treatment, malathion, but inferior to pirimiphos-methyl.

Jacobson and Pinniger (1982) reported an attempt to eradicate *Oryzaephilus surinamensis* from a farm grain store in the United Kingdom. The store, its equipment and grain were heavily infested with *Oryzaephilus surinamensis* but the objective was achieved and re-infestation was prevented during the following 2 storage seasons. The success of the treatment was attributed to a combination of the disinfestation, improved storage techniques and proper insecticide application to the store and the grain rather than any single factor. Fenitrothion was applied to structures and was admixed with the grain.

Chahal and Ramzan (1982) determined the relative efficiency of synthetic pyrethroids and some organophosphate insecticides including fenitrothion, against the larvae of *Trogoderma granarium*. Fenitrothion was applied as a spray at 0.025, 0.05 and 0.1% to third instar larvae of *Trogoderma granarium*. Although the kill after 1 day was low, a progressive increase in mortality was observed 2, 3, 5 and 7 days after treatment. At 7 days there was 100% mortality from all rates of application of fenitrothion.

Zhang *et al.* (1982) studied the effect of fenitrothion on the quality of stored seeds. They found, in laboratory tests in China, that fenitrothion at less than 30 mg/kg had no significant effect on the germination of rice, wheat, maize, sorghum and barley. Tolerance of the seeds of the 5 crops to the pesticide was in this order, maize more than sorghum, more than wheat, more than rice and barley. There was adverse effect on germination of maize treated with 10 000 mg/kg of fenitrothion.

Wilson (1983) reported the results of 2 surveys carried out in 1977 and 1980 in East Anglian (UK) farm grain stores. It was reported that though fenitrothion was used widely as a spray for structural treatments in 1976, by 1979 the farmers were showing a preference for pirimiphos-methyl. Only 6% of farms were using insecticidal admixture for grain protection. This did not include fenitrothion.

Kirkpatrick *et al.* (1983) studied the effectiveness of fenitrothion applied at the rate of 6, 10 or 15 mg/kg for the control of insects infesting stored maize grain in the USA. Malathion (11 mg/kg) was the standard reference treatment. Fenitrothion protected the grain from insect damage throughout the storage period and was superior to malathion in preventing such damage. A total of 9 species of insects was involved in the tests.

Bengston *et al.* (1983b) reported duplicate field trials carried out on bulk wheat in commercial silos in Queensland and New South Wales in which the grain was treated with combinations of organophosphorothioates and synergised synthetic pyrethroids including fenitrothion (12 mg/kg) + fenvalerate (1 mg/kg) and fenitrothion (12 mg/kg) + phenothrin (2 mg/kg); both treatments included piperonyl butoxide (8 mg/kg). Laboratory bioassays using malathion-resistant strains of insects were carried out on samples of treated grain at intervals over 9 months. These treatments controlled common field strains of *Sitophilus oryzae* and *Rhyzopertha dominica* and completely prevented progeny production in *Tribolium castaneum, Tribolium confusum* and in *Ephestia cautella*.

Bengston *et al.* (1983a) reported duplicate experiments carried out on bulk sorghum stored in south Queensland and in central Queensland. Bioassays of treated grain, conducted during 6 months storage, established that fenitrothion at 12 mg/kg + bioresmethrin at 1 mg/kg controlled typical malathion-resistant strains of *Sitophilus oryzae, Rhyzopertha dominica, Tribolium castaneum* and *Ephestia cautella*. Fenitrothion at 12 mg/kg + phenothrin at 1 mg/kg also controlled the test species except *Ephestia cautella*.

Barson (1983) examined the effects of temperature and humidity on the toxicity of 3 organophosphorus insecticides, including fenitrothion, to adult *Oryzaephilus surinamensis*. The toxicities, to a UK laboratory susceptible strain and to a resistant strain, originally from India, of *Oryzaephilus surinamensis* of water-dispersible powder formulations of fenitrothion were compared under constant conditions of 25°C and 70% r.h. with those obtained when the insects were exposed to a diurnal cycle simulating grain store conditions in the UK during spring and autumn. All insecticides, including fenitrothion were more effective at 25°C and 70% r.h. The LD/50 values obtained from the 2 sets of environmental conditions for the resistant strain differed by a factor of 7.3 for fenitrothion.

Wallbank (1983) studied the distribution of fenitrothion in a wheat bulk treated intermittently with a low volume formulation and compared this with the distribution when a standard water emulson of fenitrothion was sprayed continuously. Fenitrothion distribution was considerably more variable in the intermittently treated bulk, but the mean residues from grab samples of 200 g in both bulks was almost identical. All except 2 of the 100 samples taken during storage were within the range of half to double the anticipated level indicating good mixing of grains in the bulk. Up to 10 months storage, progeny of *Rhyzopertha dominica* developed in only 1 sample challenged in bioassays. This 'worst-case' sample had the lowest residue of 75 tested at that time.

Heather and Wilson (1983) noted that field resistance to fenitrothion in *Oryzaephilus surinamensis* was recorded for the first time from Queensland during 1979 in samples from 8 bulk storages where control failures had occurred. Resistance levels at the LC/99.9 ranged from x53 to x62 in tests with treated grain. For the strain most resistant overall, tests using the FAO impregnated-paper method and treated grain assays showed resistance at the LC/50 to fenitrothion (X159 and X53), malathion (X48 and X32), pirimiphos-methyl (X37 and X19), methacrifos (X1.8 and X5.7) and permethrin (X3.2 and X4.0).

Bengston *et al.* (1984) reported the results of a third experiment carried out on bulk sorghum stored for 26 weeks in concrete silos in south Queensland. Laboratory bioassays of treated grain in which malathion-resistant strains of insects were added to grain samples indicated that fenitrothion (12 mg/kg) + fenvalerate (1 mg/kg) + piperonyl butoxide (8 mg/kg) and fenitrothion (12 mg/kg) + phenothrin (2 mg/kg) + piperonyl butoxide (8 mg/kg) controlled typical malathion-resistant strains of *Sitophilus oryzae, Rhyzopertha dominica, Tribolium castaneum* and *Ephestia cautella*.

Wohlgemuth (1984) reported comparative laboratory trials with 14 insecticides under tropical conditions. Fenitrothion dust, at the rate of 10 mg/kg, proved outstandingly more effective than the fenitrothion emulsion applied at the rate of 12 mg/kg. This was noticable against all 5 species where the dust gave complete kill for 12 to 24 months whereas the emulsion treatment was fully effective for only 1 to 3 months.

Evans (1985) examined 10 specimens of 5 species of stored-product pests collected from the field in Uganda. All showed some degree of resistance to malathion, some being also resistant to lindane. Specimens of these insects were exposed to maize treated with fenitrothion at the rate of 10 mg/kg. All those exposed were killed initially and the treated maize remained 100% effective after 8, 16 and 24 weeks storage at 27°C and 70% r.h.

Desmarchelier *et al.* (1986) reported the outcome of the pilot usage of grain protectants including the combination fenitrothion + phenothrin on wheat stored at 63 storages at 15 sites throughout Australia. In laboratory bioassay, the order of effectiveness against *Tribolium castaneum* was methacrifos (22.5 mg/kg) superior to chlorpyrifos-methyl (10 mg/kg) + bioresmethrin (1 mg/

kg) being superior to fenitrothion (12 mg/kg) + phenothrin (2 mg/kg). Against *Rhyzopertha dominica* the fenitrothion combination was equal best.

14.2 Degradation

The persistence of insecticide residues on stored grain is known to depend partly on the moisture content and the temperature of the grain. It seems likely also that physiological and other changes taking place in grain in the early stages of storage after harvest may influence the rate of degradation of admixed insecticide. **Tyler *et al.* (1965b)** reported studies in which wheat and barley were obtained from the harvest field and stored so that the moisture contents (15.5% and 17.5%) and temperature (17°C) remained relatively constant. Batches of both cereals were treated with fenitrothion at 2 mg/kg after 2 days and similar batches at monthly intervals after harvest. Preliminary results indicated that fenitrothion breaks down more rapidly on newly harvested, than on stored grain. These preliminary experiments were later extended **(Tyler *et al.* 1966a)** and it was shown that the apparent loss of insecticide was always higher than expected particularly with newly harvested grain. This was shown to be due to an extremely rapid degradation of insecticide within the first few hours. As much as 40% of the applied dose of fenitrothion was broken down within 1 hour of application to newly harvested barley. After 1 month, however, residues did not differ significantly between stored and newly harvested grain. Bioassay using *Oryzaephilus surinamensis* confirmed the chemical analysis results.

As a result of field trials on barley which had been treated with fenitrothion at the rate of 2 mg/kg, **Tyler *et al.* (1966b)** showed that where the temperature and moisture content of the grain was highest, insecticide residues were soon undetectable by both chemical analysis and bioassay.

Fenitrothion applied to wheat and barley before bagging gave good protection during 6 months storage under conditions of heavy cross-infestation, high temperature and high moisture. Chemical analysis of the wheat showed that at 9 months only 0.2, 0.3 and 0.9 mg/kg fenitrothion remained of the original 1, 2 and 4 mg/kg respectively **(Green *et al.* 1966c)**.

Green and Tyler (1966) carried out trials in which fenitrothion was applied to barley (13–15% moisture) in farm bins at the rate of 1 or 2 mg/kg. The temperature of the grain ranged from 27°C to 32°C. Chemical analysis indicated that the residues did not decline to half the original deposit for at least 6 weeks, and bioassay indicated 100% kill of insects even at 19 weeks.

Green *et al.* (1966b) treated wheat and barley with a fenitrothion emulsion at the rate of 1, 2 and 4 mg/kg. The grain was 4 to 5 months old and had moisture contents of 10.9% and 13.7% for the wheat and barley respectively. After bagging, the grain was stored, together with untreated controls, at 25°C and 60% relative humidity. Samples of grain were withdrawn from the bags for chemical analysis at 0, 2 and 4 weeks after treatment and then at monthly intervals during the storage period. The persistence of fenitrothion residue was greater in wheat than in barley; however, the moisture content of the barley was greater than in the wheat.

Horler (1966) showed that fenitrothion applied to bagged wheat of 10.9% moisture content at 1, 2 and 4 mg/kg and stored at 25°C, had degraded to levels of 0.4, 0.8 and 3.4 mg/kg within 2 months and to 0.1, 0.3 and 2.5 mg/kg within 6 months. When applied at 1, 2 and 4 mg/kg levels to bagged barley of 13.7% moisture content and stored at 25°C, residues of 0.2, 0.5 and 0.6 mg/kg were detected at 2 months and no residues were found at 6 months.

Rowlands (1967a) observed that initially fenitrothion was degraded very rapidly to desmethyl fenitrothion in stored wheat and accumulation of free nitrocresol may be anticipated in the endosperm.

Tyler and Green (1968) applied fenitrothion at 2 mg/kg to moist, warm grain stored in bins in a British farm granary. They found there was a relationship between the persistence of the insecticide and the physical condition of the grain. Fenitrothion became ineffective after 2 weeks on moist, heating grain near the centre of the bins but the residue remained effective for at least 12 weeks at the cooler, outer edges of the bins and in those bins remaining cool after the grain was turned for treatment. In the case of the heating bins, where the temperature exceeded 50°C the residue level was found, at the end of 12 weeks, to have degraded to approximately 0.2 mg/kg, but in the bins where the temperature remained about 22°C the residue at the end of 12 weeks was 1 mg/kg (half the rate originally applied).

The **Cyprus Agricultural Research Institute (1972)** evaluated fenitrothion along with a number of organophosphorus insecticides as grain protectants. Whereas most of the insecticides had lost their effectiveness by the third month, fenitrothion was still effective 4 months after application.

LaHue (1974) applied fenitrothion to wheat, maize and sorghum, stored in small bins. The grain was analysed at regular intervals, post-treatment, and the results are set out in **Table 14.1**.

Weaving (1975) studied fenitrothion for its suitability for use as a grain protectant under tribal storage conditions in Zimbabwe. When applied to maize and sorghum at the rate of 8 mg/kg, fenitrothion showed good persistence for 12 months as indicated by bioassay with a number of insect species. With the exception of pirimiphos-methyl no other insecticide showed comparable stability.

LaHue (1975a) carried out trials to control *Sitotroga cerealella* in shelled corn. Fenitrothion was applied at the rate of 8.9 mg/kg and the residues were determined by analysis over 8 months storage. The results of analysis are given in **Table 14.2.**

LaHue and Dicke (1976a) applied fenitrothion at the

Table 14.1. Average residues of fenitrothion (mg/kg) in samples of grain stored in small bins at 27°C **(LaHue 1974)**

Grain	Moisture content %	Intended dosage	Post-treatment period (months) 24 hr	1	3	6	9	12
Wheat	12.5	8.3	6.2	4.9	3.4	2.6	1.6	1.3
Maize	12.5	8.9	6.2	4.8	3.4	1.7	1.1	0.8
Sorghum	17.6	8.9	6.1	4.0	3.0	2.0	1.3	1.1

Table 14.2. Residues (mg/kg) of fenitrothion on shelled corn during 8 months storage (mg/kg) **(LaHue 1974)**

Intended dose	Post-treatment period (months) 24 hr	1	2	3	4	5	6	8
8.9	6.2	4.8	4.2	3.4	2.2	2.1	1.7	1.5

rate of 8.9 mg/kg to high-moisture sorghum grain and examined its persistence by bioassay tests with 6 stored-product insect species. Fenitrothion was effective in preventing damage to the grain during 12 months storage, but damaging infestations were established in all replicates during the last month of the test. The grain was held at 27°C and 60% relative humidity throughout the tests.

Desmarchelier (1976b), in a study of the kinetic constants for breakdown of fenitrothion on grains in storage, found that the breakdown on wheat, rice paddy, oats and sorghum in storage at 4 temperatures, was first-order with respect to grain moisture. At a given moisture content, water activities were dependent upon, but rate constants were independent of, grain type. The effect of temperature on breakdown was in the form of the Arrhenius equation. A simple expression was derived from these basic concepts of kinetics, which accurately predicted residues under 24 different conditions set by altering grain type, moisture content and temperature **(FAO/WHO 1977)**.

Laboratory studies were carried out in Ghana **(Ofosu 1977)** on the persistence of fenitrothion applied in water-based emulsions to shelled maize against *Sitophilus zeamais* and *Tribolium castaneum*. At a rate of 4 mg/kg fenitrothion still gave 96% mortality of *Sitophilus zeamais* after a storage period of 24 weeks, but the compound lost its effectiveness against *Tribolium castaneum* after 12 weeks.

Desmarchelier (1977b) showed how selective treatments, including combinations of pyrethroid and organophosphorus insecticides, may be designed for the control of specific stored-product Coleoptera and how the nature and persistence of these treatments will vary according to temperature. **Figure 14.1** shows the decay of fenitrothion residues on wheat at 2 temperatures and 2 moisture levels.

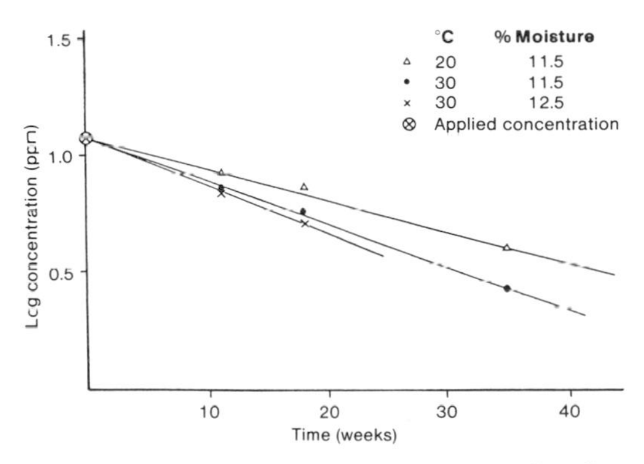

Figure 14.1. Decay of fenitrothion on Australian Standard white wheat (Desmarchelier 1977b)

Takimoto *et al.* (1978) treated rice grains containing 14% moisture with ^{14}C-fenitrothion labelled at the *m*-methyl group of the phenyl moiety, at a concentration of 6 and 15 mg/kg, and stored them at 15°C and 30°C for 12 months. Decomposition of fenitrothion proceeded more rapidly at 30°C than at 15°C, with the respective half-life of about 4 months and more than 12 months, independent of the applied concentrations. The nature, extent and distribution of radioactivity found in the rice grains at various periods over the 12 months storage are shown in **Tables 14.3 and 14.4**. Desmethylfenitrothion was the major product of decomposition in the early stages whereas 3-methyl-4-nitrophenol was formed with the lapse of time. Seven other metabolites were identified and determined but they represent only minor components of degradation.

Table 14.3. Decomposition of fenitrothion on/in rice grains stored at 15°C

Concen-tration	Month	Distribution of radioactivity extracted[a]											Bound[b]	KOH-Trap	Recovery
		Total	Fenitrothion	Fenitrooxon	Fenitrothion S-isomer	Desmethyl-fenitrothion	Desmethyl-fenitrooxon	Desmethyl-fenitrothion S-isomer	3-Methyl-4-nitrophenol	1-Methoxy-3-methyl-4-nitrobenzene	1,2-Dihydroxy-4-methyl-5-nitrobenzene	1,2-Dimethyoxy-4-methyl-5-nitrobenzene			
6 ppm	0	99.9	99.6	–[c]	–	0.2	–	–	0.1	–	–	–	0.8	–	100.7
	1	97.8	91.8	0.1	–	4.6	0.3	–	1.0	–	–	–	1.6	<0.1	99.4
	2	94.9	85.3	0.2	0.1	6.3	–0.7	–	2.3	–	–	–	2.0	0.1	97.0
	3	93.3	76.8	0.2	0.2	10.3	1.3	–	4.5	–	–	–	2.4	0.2	95.9
	4.5	98.5	76.3	0.2	0.1	13.2	1.7	–	7.9	–	–	–	3.4	0.3	102.2
	6	93.0	69.0	0.2	<0.1	12.5	1.9	<0.1	9.4	–	<0.1	–	3.0	0.3	96.3
	9	92.5	67.3	0.1	0.1	9.7	1.9	0.1	13.2		0.1		3.3	0.4	96.2
	12	96.0	64.7	0.1	<0.1	10.0	1.7	0.1	17.2	1.8	0.4	–	4.6	0.4	101.0
15 ppm	0	99.7	99.5	–	–	–0.2	–	–	–	–	–	–	0.2	–	99.9
	1	98.8	93.3	0.1	<0.1	4.2	0.3	–	0.9	–	–	–	1.8	<0.1	100.6
	2	99.8	89.0	0.2	0.2	7.3	0.7	–	2.4	–	–	–	2.0	0.1	101.9
	3	101.0	84.3	0.2	0.2	10.6	1.2	–	4.5	–	–	–	2.4	0.2	103.6
	4.5	96.7	77.5	0.1	0.1	10.9	1.3	0.1	6.7	–	–	–	3.3	0.2	100.2
	6	98.6	72.7	0.2	0.1	14.2	1.9	0.1	9.4	–	<0.1	–	3.0	0.2	101.8
	9	94.6	67.6	0.1	0.2	11.5	1.8	0.3	12.9	–	0.2	–	3.5	0.4	98.5
	12	95.7	65.0	0.2	<0.1	10.6	1.6	0.2	16.0	1.8	0.2	0.1	4.2	0.4	100.3

a Radioactivity applied to rice grains is referred to as 100%.
b Unextracted radioactivity in grains.
c <0.1%, if any.

Table 14.4. Decomposition of fenitrothion on/in rice grains stored at 30°C (***Takimoto et al.* 1978**)

Concentration	Month	Distribution of radioactivity extracted[a]: Total	Fenitrothion	Fenitrooxon	Fenitrothion S-isomer	Desmethyl-fenitrothion	Desmethyl-fenitrooxon	Desmethyl-fenitrothion S-isomer	3-Methyl-4-nitrophenol	3-Hydroxy-methyl-4-nitrophenol	1-Methoxy-3-methyl-4-nitrobenzene	1,2-Dihydroxy-4-methyl-5-nitrobenzene	1,2-Dimethyoxy-4-methyl-5-nitrobenzene	Unidentified compound	Bound[b]	KOH-Trap	Recovery
6 ppm	0	98.3	97.7	–[c]	–	–	–	–	0.6	–	–	–	–	–	0.7	–	99.0
	¼	100.1	91.1	<0.1	–	5.8	0.4	–	2.8	–	–	–	–	–	2.0	0.2	102.3
	1	94.7	67.4	0.1	–	16.8	1.4	0.1	8.9	–	–	–	–	–	3.5	0.4	98.6
	2	93.3	59.1	<0.1	<0.1	18.7	2.1	0.2	13.1	–	–	0.1	–	–	3.3	0.4	97.0
	3	95.5	58.4	<0.1	<0.1	16.8	2.4	0.2	17.5	–	–	0.2	–	–	4.4	0.4	100.3
	4.5	94.3	50.8	<0.1	–	15.6	2.6	0.1	24.7	–	–	0.5	–	–	4.7	0.5	99.5
	6	92.0	43.1	<0.1	–	16.6	2.2	0.3	28.8	–	–	1.0	–	–	4.1	0.6	96.7
	9	96.3	32.0	0.2	0.2	17.2	1.7	<0.1	36.0	0.2	4.0	1.9	1.0	1.9	4.5	0.9	101.7
	12	96.3	22.0	0.2	–	19.2	1.1	<0.1	38.0	0.2	9.1	2.9	2.6	0.8	5.0	1.2	102.5
15 ppm	0	102.1	101.2	–	–	0.3	–	–	0.6	–	–	–	–	–	0.6	–	102.7
	¼	100.6	90.4	<0.1	0.1	7.0	0.5	–	2.6	–	–	–	–	–	2.0	0.2	102.8
	1	94.0	71.6	0.1	0.2	14.1	1.0	0.1	6.9	–	–	–	–	–	3.3	0.4	97.7
	2	95.7	64.1	<0.1	0.1	17.9	1.8	0.2	11.5	–	–	0.1	–	–	4.0	0.4	100.1
	3	96.0	56.9	<0.1	<0.1	19.7	2.2	0.3	16.6	–	–	0.3	–	–	4.6	0.5	101.1
	4.5	95.3	45.1	<0.1	–	21.0	2.6	0.4	25.6	–	–	0.6	–	–	4.8	0.5	100.6
	6	95.7	40.8	<0.1	<0.1	18.1	1.1	0.2	30.1	<0.1	3.0	1.7	0.7	<0.1	4.2	0.7	100.6
	9	98.5	32.5	<0.1	<0.1	21.1	1.2	0.1	33.4	0.1	4.6	3.1	1.7	0.7	4.9	1.0	104.4
	12	96.3	26.3	<0.1	<0.1	18.1	1.1	0.3	38.0	0.2	6.7	3.2	1.9	0.5	5.6	1.0	102.9

a Radioactivity applied to rice grains is referred to as 100%.
b Unextracted radioactivity in grains.
c <0.1%, if any.

Loss of fenitrothion from post-harvest application to wheat, oats, paddy rice and sorghum followed a second-order rate process, with rate of loss being proportional, at a fixed temperature, to the amount of fenitrothion and the activity of water, which was obtained from the equilibrium partial pressure of water vapour **(Desmarchelier 1978a)**. The effect of temperature on loss was in the form of an Arrhenius equation. A chart relating half-life to temperature and relative humidity was presented in a form suitable for field use **(Figure 14.2)**, and a mechanism was proposed for loss of fenitrothion. The proposed mechanism is that an absorbed fenitrothion molecule is desorbed by replacement by a water molecule. The desorbed molecule is more likely to be degraded than an absorbed molecule, because it has a greater chance of collision with enzymes, metallic ions and other active molecules. The mechanism is consistent with the overall order of the rate process.

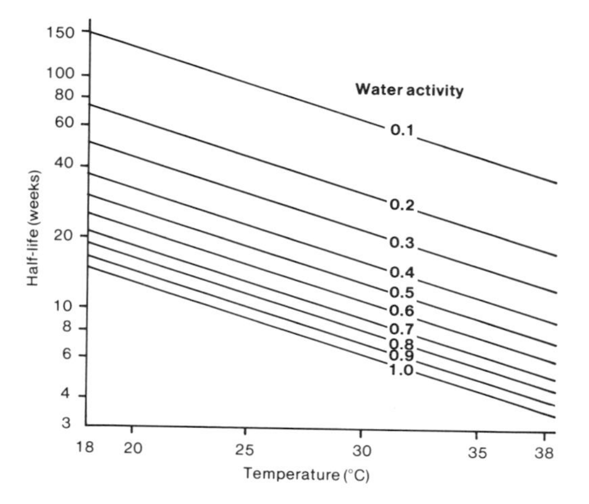

Figure 14.2. Half-life/temperature interaction with fenitrothion on grains as affected by humidity, calculated from a model based on simple chemical principles **(Desmarchelier 1978a).**

Desmarchelier *et al.* (1986) studied the effect and fate of fenitrothion applied at the rate of 12 mg/kg to wheat in 21 commercial grain silos in Australia in 1976. The fate of the fenitrothion was studied by chemical analysis and bioassay over a period of 9 months and the analytical results were compared with the values predicted from the mathematical model developed by **Desmarchelier (1978a)**. The high level of agreement between the predicted and observed values is given in **Figure 14.3.**

Banks and Desmarchelier (1978), in discussing the influence of water vapour and temperature on changes in pesticide residue levels with time, pointed out that a major factor obscuring the correct chemical interpretation of pesticide breakdown on grain is the widespread use of

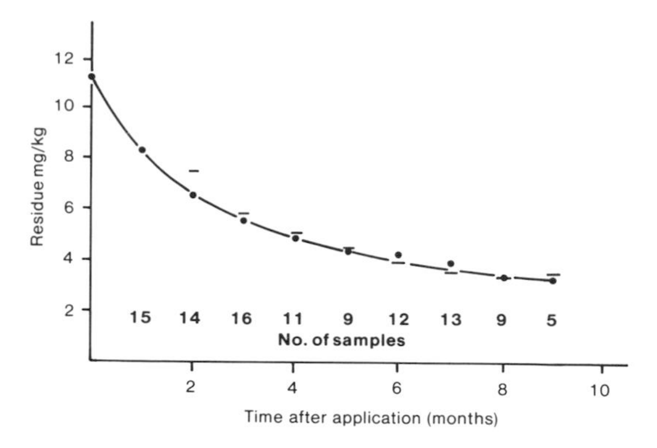

Figure 14.3. Mean observed (●) and predicted (—) residues of fenitrothion at intervals after application **(Desmarchelier 1978a).**

moisture content as a measure of the water present in grain. Moisture content, arbitrarily defined in terms of the weight loss on heating under specified conditions, is not linearly related to water activity for a particular grain and differs substantially between different commodities of the same water activity. Water activity, i.e. percentage equilibrium relative humidity is, however, easy to measure and compilations are available **(Hall 1963; Gough and Bateman 1977)** relating moisture content and water activity for the grain under test. **Banks and Desmarchelier (1978)** pointed out that the confusion arising from the widespread use of moisture content leads to statements such as the persistence of fenitrothion residue was greater in wheat than in barley; however the moisture content of barley was greater than that of wheat. According to these workers, however, if water activity is used as a measure of water present, the breakdown rate of fenitrothion and other organophosphorus insecticides is found to be independent of grain type **(see Figure 14.4)** and first order with respect to water activity.

The importance of this concept and its practical application was further described by **Desmarchelier and Bengston (1979a)** where they referred to fenitrothion as an example of how the rate and magnitude of the decomposition of a grain-protectant insecticide could be predicted in advance. This knowledge enables grain storage authorities to ensure adequate protection of the grain during its storage life but, by adjusting the initial rate of application to make certain that the residues will be minimal. The concept and its application were further expanded by **Desmarchelier *et al.* (1979b)** in a paper stressing the importance of manipulating the temperature and moisture of grain to achieve the maximum efficiency from the use of grain protectants such as fenitrothion.

Desmarchelier *et al.* (1980a) studied the levels of residues of fenitrothion and 5 other grain protectants on unhusked rice, husked rice, polished rice and barley over a storage period of 6 months. The residue levels determined were close to levels predicted from the use of

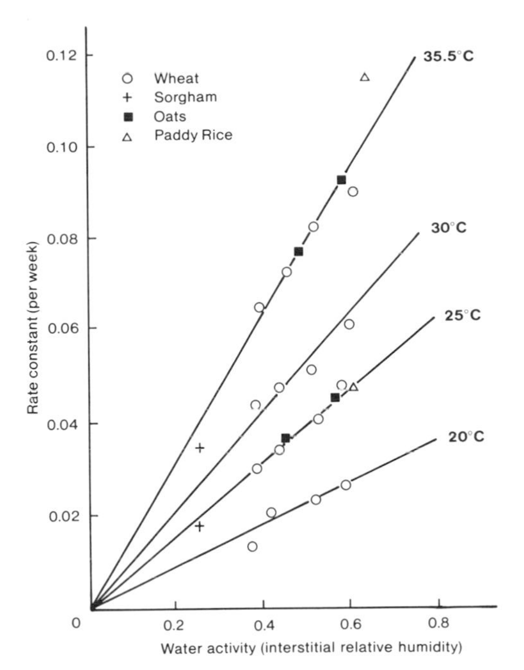

Figure 14.4. Variation on breakdown rate of fenitrothion on whole grains as a function of water activity and temperature **(Banks and Desmarchelier 1978).**

the model described by **Desmarchelier (1978a)**, there being only 8% deviation between the predicted and observed values for any commodity at the end of 6 months after initial application.

Desmarchelier *et al.* (1980b), reporting an extensive collaborative study of the rate of decay of residues of 5 grain-protectant insecticides on wheat carried out in 1976, quoted the half-life of fenitrothion at 30°C and 50% relative humidity as 16.2 weeks.

Weaving (1980), studying the performance and fate of various grain protectants for use under tribal storage conditions in Zimbabwe, found that fenitrothion applied at the rate of 8 mg/kg to maize persisted at effective levels for at least 12 months as indicated by bioassay with the 3 most important insect pests.

Bengston *et al.* (1980b) reported the outcome of duplicate field trials carried out on bulk wheat in commercial silos in Australia, noting that the residue levels as determined by chemical analysis throughout the experiment generally conformed with those previously reported (Desmarchelier and Bengston 1979a, 1979b).

Abdel-Kader and Webster (1980); Abdel-Kader (1981); Abdel-Kader *et al.* (1982) and **Webster and Abdel Kader (1983)** studied the degradation of fenitrothion in stored wheat at various temperatures following application to wheat at the rate of 8 mg/kg. This wheat was stored at –35° and –20° for 72 weeks after which it was found that less than 3% degradation had occurred. On treated wheat that had been stored at –5°C, 5°C, 10°C, 20°C, and 27°C, 18%, 35%, 56%, 90%, and 96% of the initial deposit had degraded by the end of 72 weeks **(Figure 14.5)**.

Hindmarsh and MacDonald (1980), who carried out field trials to control insect pests of farm-stored maize in Zambia, applied fenitrothion 1% dust at the rate of 2 mg/kg to the shelled maize. Samples were drawn for analysis once a month and the results indicate a half-life of more than 7 months.

Desmarchelier *et al.* (1981b) reported the outcome of extensive pilot use of grain protectant combinations including fenitrothion + bioresmethrin in 21 commercial storages throughout Australia in 1976. The fate of the residues on the grain were accurately described by predictive models. The results are given in **Table 14.5**.

Minett *et al.* (1981), who developed a gravity feed unit to apply insecticide concentrate to wheat being transferred into a silo on a conveyor belt, found that the rate of

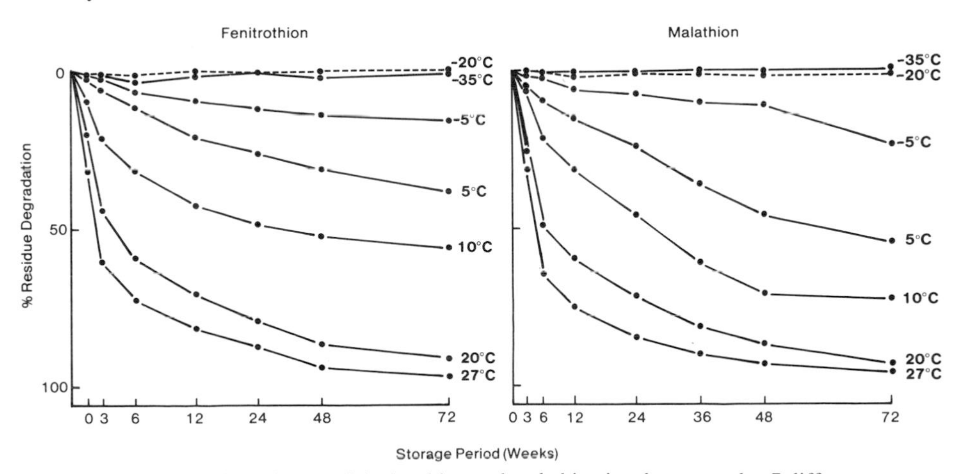

Figure 14.5. Percent residue degradation of fenitrothion and malathion in wheat stored at 7 different temperatures **(Abdel-Kader *et al.* 1982).**

Table 14.5. Fenitrothion residue in pilot studies with fenitrothion plus bioresmethrin, November 1975–September 1976 **(Desmarchelier *et al.* 1981b)**

Code number	Fenitrothion residues (mg/kg) measured during										
	Nov.	Dec.	Jan.	Feb.	Mar.	Apr.	May	June	July	Aug.	Sept.
22W		10.7[b]	6.4		4.7	4.7		3.2	2.7	2.5	
23W		11.8[b]	8.3		7.2			5.7	5.2	4.4	
24W		12.8[b]	6.7		5.3	5.2		3.9	3.2	2.8	
25W				12.0[b]	8.2	7.9		7.3	5.8	6.0	
26Q	9.8[b]	8.4	9.8	6.9	5.7	4.9	4.4	3.3	2.1	2.5	
27Q	11.9[b]	8.3	10.7	5.1	4.9	5.1	5.6	5.7	4.5	5.4	
28Q	10.8[b]	8.7	5.0	5.8	3.0	2.6	2.2	2.6	2.3	1.6	
29Q[a]	12.8[b]	8.5	10.1	8.4	6.1	5.6	6.4	5.5	5.2	4.6	
30Q	12.4[b]	6.0	7.0	5.2	6.0	6.6	5.3	5.7	5.9		
31Q		12.3[b]	9.8	6.2	5.6	6.0	5.4	5.2	5.0	4.1	
32N					11.7[b]	10.0			7.8		
33N					9.0[b]	7.8		5.7			
34N				8.5[b]	7.1	4.3	5.1				
35N				11.5[b]	12.2	6.8	10.2				
36N				10.2[b]		3.8	5.0				
37N				13.8[b]	9.4	8.4	6.6		4.6		4.0
38N				13.5[b]		7.0	7.0				2.2
39V			9.4[b]	Variable layers							
40V			12.8[b]	Variable layers							
41S[a]		10.4[b]		3.3	2.8	2.1	2.5	1.8	1.9	1.6	1.6
42S		10.4[b]		2.7	2.1	1.6	1.5	0.9	1.0	1.3	0.6

a Aerated.
b Calculated application rate; the recommended application rate for fenitrothion is 12 mg/ml.

degradation of the deposit was identical with that of the deposit from a dilute emulsion sprayed uniformly over most grains in the storage. In spite of the fact that the gravity feed unit applied the fenitrothion irregularly and, in a manner which meant that the deposit was not distributed as a fine film over a large surface area as is the case of when diluted emulsions are used, the concentration of the deposit on the grain from both forms of treatment fell from 11 mg/kg to 5.5 mg/kg over a period of 6 months, the grain temperature being 30–32°C and the moisture content 10–11%.

Kirkpatrick *et al.* (1983) studied the stability and effectiveness of fenitrothion on corn by treating stored maize grain with fenitrothion at 6, 10 or 15 mg/kg. During a 1-year storage at ambient temperatures at Savannah, USA, chemical analysis of the treated grain indicated that there was significant degradation of residues. Although residue levels decreased to 2, 4 or 6 mg/kg, depending upon the initial dosage, fenitrothion protected the grain from 9 species of insects involved in the tests.

Bengston *et al.* (1983a) reporting the outcome of duplicate experiments carried out on bulk sorghum stored in south Queensland and central Queensland during 6 months, showed that fenitrothion applied at the rate of 12 mg/kg degraded at rates consistent with the pattern for other cereal grains under corresponding moisture and temperature conditions. **Figure 14.6** shows how the fate of fenitrothion residues compares with that of other insecticides in the same trials.

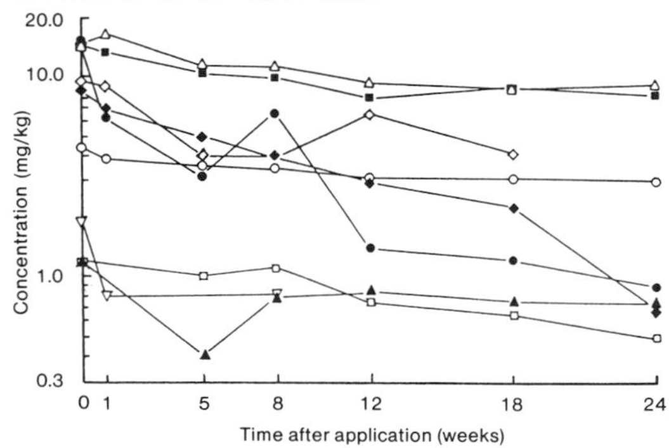

Figure 14.6. Calculated application rates at week 0, and analytically determined residues of insecticides at various times after application at the Central Queensland site. Insecticides: (●) methacrifos; (■) fenitrothion + bioresmethrin; (△) fenitrothion + (*d*)-phenothrin; (◇) chlorpyrifos-methyl; (◆) carbaryl; (○) pirimphos-methyl; (▽) pyrethrins; (□) bioresmethrin; (▲) (*d*)-phenothrin **(Bengston *et al.* 1983a).**

Bengston *et al.* (1983b) reported the results of duplicate field trials carried out on bulk wheat in commercial silos in Queensland and New South Wales in which insecticide combinations were applied to the wheat as it was placed in storage. The wheat was sampled at intervals over 9 months and **Table 14.6** gives the results of analysis for fenitrothion residues. The temperature of the grain at the time of treatment ranged between 30°C and 34°C and the initial moisture content of the grain between 10% and 11%.

operated grinding stone resulted in a much lower loss of fenitrothion. Less than 5% of the fenitrothion, applied to rice paddy, could be detected in polished rice milled 2 months later. This was obviously not due to degradation during storage because polished rice produced by milling the paddy rice immediately after application of the fenitrothion was found to contain less than 5% of the deposit on the paddy rice.

Chapatis (unleavened bread), prepared from the whole wheat flour, contained only 25–50% of the residue present in the wholemeal. A similar loss occurred when sorghum flour was converted into chapatis. Wet cooking methods used for the preparation of boiled rice, iddli (steamed, fermented rice), and boiled sorghum balls apparently destroyed all trace of fenitrothion residues because none could be detected by methods sensitive to 0.1 mg/kg **(Lockwood *et al.* 1974).**

Table 14.6. Calculated fenitrothion application rates and analytically determined residues on treated wheat stored in concrete silos at Natacha (Queensland) and Noonbinna/Eugowra (New South Wales) for varying times following spray application **(Bengston *et al.* 1983b).**

Site	Calculated rate (mg/kg)	Fenitrothion residue (mg/kg) at various times after treatment (months)						
		0.25	1.5	3	4.5	6	7	9
Natcha	10.9	9.3	6.2	5.8	4.9	4.4	–	4.8
Eugowra	13.6	20.8	–	10.2	11.6	9.3	8.0	–
Natcha	11.3	9.8	7.2	5.7	5.8	4.6	–	4.3
Noonbinna	11.1	–	–	7.8	8.2	9.0	7.4	0.7

Bengston *et al.* (1984) reported a third experiment carried out on bulk sorghum stored for 26 weeks in concrete silos in south Queensland. Combinations including fenitrothion were 2 of the 4 treatments involved. Chemical assays of samples drawn over 26 weeks revealed that the level of residue found at the end of the first week remained substantially constant for 26 weeks much the same as that reported by **Bengston *et al.* (1983a).**

14.3 Fate in Milling, Processing and Cooking

Preliminary work by **Parkin and Horler (1967)** indicated that fenitrothion residues on grain were reduced on processing. They subjected wheat, freshly treated with 5 mg/kg fenitrothion, to milling and baking tests. Before milling, 3.9 mg/kg of the applied dose was recovered and subsequently, a total of 3.9 mg/kg was determined for the bran, offal and flour fractions. Fenitrothion was not detected in the loaf after baking.

Lockwood (1973) and **Lockwood *et al.* (1974)** studied the degradation of organophosphorus pesticides on wheat, sorghum and rice paddy during milling and cooking in India. They found that when milling fenitrothion-treated wheat, using an electrically-powered disc mill, 62% of the deposit was destroyed and when milling sorghum through the same mill, the loss was only 28%. Minor adjustments to the clearance of the grinding discs made a noticeable difference in heating of the grain during grinding. Variation in residue breakdown may have partially resulted from such heating. Milling wheat with a hand-

Desmarchelier (1976b), studying the decay of 5 insecticides during the storage and malting of barley, carried out experiments in triplicate on samples of barley with a moisture content of 13% held at 25°C. After 3 months storage, some of the barley was malted by a primitive process and at the end of 6 months the remainder by a commercial process. In neither case was germination affected by any of the pesticides, including fenitrothion. An initial application of 15 mg/kg fenitrothion had declined to 6.8 and 3.1 mg/kg after 3 and 6 months respectively. When subjected to the primitive malting process, barley containing 6.8 mg/kg fenitrothion produced malt containing 2.9 mg/kg. The barley containing 3.1 mg/kg fenitrothion, when submitted to the commercial malting process, produced malt containing only 0.55 mg/kg of fenitrothion. More extensive studies were reported by **Tempone (1979)** using classical malting techniques. Fenitrothion was found to increase the malt extract and extract yield. Barley treated with fenitrothion at the rate of 12 mg/kg and malted immediately, yielded a malt with from 1 to 2 mg/kg of fenitrothion. When the barley, thus treated, was kept in storage for 3 months the residue in the prepared malt generally did not exceed

0.5 mg/kg. The wort (unfermented extract) from such malt did not yield any residues when analysed by methods with a limit of detection of 0.01 mg/kg.

Fenitrothion has been used for the treatment of all stocks of malting barley in Australia for more than 7 years and an extensive monitoring of malt, inspected prior to export, revealed no significant residue of fenitrothion in any of the many samples examined **(Snelson 1981a)**.

In a series of 3 trials, **Desmarchelier (1976c)** determined the fate of fenitrothion residues in wheat during milling and cooking. **Table 14.7** shows the results indicating that when wheat was milled for the production of white flour, only about 10% or less of the fenitrothion present in the raw grain was carried into the flour, the bulk being concentrated in the bran with a lesser amount in the pollard (shorts). When white bread is prepared from the flour, there is a further significant loss of fenitrothion so that the residue remaining in the bread was of the order of 1–2% of that present in the raw wheat. In the case of wholemeal bread, however, there was a significant carryover of the residue from the grain, equivalent to 20–25% of the original fenitrothion concentration on the wheat. In these trials, only a small proportion of the original residue remained in the form of the metabolite, 3-methyl-4-nitrophenol. Using the weight of the mill fractions obtained in these trials, the percentage reduction in the fenitrothion residue content was calculated for each step of the milling and baking process. The results shown in **Table 14.8**, with comparable results from the work of **Bengston *et al.* (1980b)**, indicate that the overall reduction of the residue during the various processes employed in converting the wheat to white bread, is of the order of 99%.

Bengston *et al.* (1980b) in a series of duplicate field trials carried out on bulk wheat in commercial silos in Australia, determined the residues of fenitrothion in the wheat and each milling fraction and in bread prepared from the flour. These results have been used to calculate the reduction in residue content at each stage of the process of milling and baking. These results are set out in **Table 14.8** for comparison with those obtained by **Desmarchelier (1976c)**.

Desmarchelier *et al.* (1980b) reported the results of a collaborative study of residues on wheat treated with fenitrothion and 4 other grain protectant insecticides and held in commercial silos at 2 sites in Australia. After the

Table 14.7. Fate of fenitrothion residues on wheat during milling and cooking **(Desmarchelier 1976c)**

	Trial 1	Trial 2	Trial 3	
			Fenitrothion	3-methyl-4-nitrophenol
Time after application (weeks)	19	13	10	–
Residue (mg/kg) in				
wheat	6.5	9.6	7.6	–
wholemeal flour	3.5	4.8	6.1	–
wholemeal bread	1.5	1.8	1.72	0.05
bran	19.5	22.6	20.8	0.25
shorts	5.8	11.0	3.3	0.12
white flour	0.53	0.74	0.25	0.06
white bread	0.09	0.21	0.08	0.06

Table 14.8. Reduction in fenitrothion residue content following milling and baking of wheat

Reference	Residue on wheat (mg/kg)	% reduction in residue content					
		Wheat to wholemeal flour	Wheat to white flour	Wholemeal flour to wholemeal bread	White flour to white bread	Wheat to wholemeal bread	Wheat to white bread
Desmarchelier (1976c)	7.6	67	94	47	37	83	98
Bengston *et al.* (1980b)	9.6	33	91	67	72	81	98
	6.5	46	92	57	83	77	99

grain was stored for 22 and 11 weeks respectively, a portion was subjected to milling and baking trials, which provided the results set out in **Table 14.9**.

Table 14.9. Fate of fenitrothion residues on wheat as a result of milling and baking (**Desmarchelier *et al.* 1980b**).

Fraction/Commodity	Mean fenitrothion residues (mg/kg)	
	Site M (22 weeks)	Site W (11 weeks)
Wheat	3.1	4.1
Bran	8.1	8.3
Pollard	6.0	6.1
White flour	0.7	0.3
White bread	0.2	0.2

Murray and Snelson (1978) studied the distribution of fenitrothion and bioresmethrin residues in milling fractions from wheat processed through a commercial flour mill. The results of their investigations are set out in **Table 14.10**. These show that the bran from all milling streams when blended together, contained fenitrothion residues 1.5 times as high as those present in raw wheat received into the mill. **Snelson (1979a)** also reported the results of monitoring bran from 13 flour mills processing wheat containing from 5 to 10 mg/kg of fenitrothion. The results given in **Table 14.11**, indicated that 100% of the bran samples contained less than 20 mg/kg but up to 90% contained up to 15 mg/kg. Only 23% of the bran samples contained less than 10 mg/kg. These findings contrasted with the residues found in processed bran which are very low (**Table 14.11**).

There is no wheat available in Australia without residues of grain-protectant insecticides. It was concluded that processed bran, available retail, would reflect the effect of processing on the residues present in raw bran. A survey was conducted in Australia (**Snelson 1979a**) in which 36 samples of processed bran, representing 12 different commercial products available in retail stores of 3 cities, located over a distance of 2000 kilometres, were analysed. In spite of the fact that all raw bran available for processing contained fenitrothion to a limit of 20 mg/kg, it was found that 80% of the processed bran available in the form of breakfast cereals contained less than 1 mg/kg of fenitrothion residue (**Table 14.11**). Of the samples, 8% contained more than 3 mg/kg but these represented products which were only partially processed. A total of 36% of the samples contained no detectable residues (less

Table 14.10. Residues of fenitrothion (mg/kg) in wheat milling fractions (and of bioresmethrin) (**Murray and Snelson 1978**)

Sample no. and description	Treated with 12 mg/kg fenitrothion (plus 1 mg/kg synergised bioresmethrin)		Treated at harvest with 12 mg/kg fenitrothion
	Laboratory A	Laboratory B	Laboratory A
1. Wheat before point of pesticide application	<0.5	–	N.A.
2. Wheat when received into the mill	8	7 (0.7)	7
3. Wheat at first break roll before it enters the rolls	4	4 (0.6)	3
4. Crushed wheat at first break roll after it drops from the rolls	5	4 (0.6)	3
5. Flour sieved from the sifter fed from first break roll (clean stream)	1	0.8 (<0.05)	1
6. Flour sieved from the sifter fed by the last break roll (dirty stream)	4	3 (0.1)	2
7. Flour sieved from the sifter fed by the first reduction roll (clean stream)	0.5	0.6 (<0.05)	0.5
8. Flour sieved from the sifter fed by the last reduction roll (dirty stream)	5	5 (0.3)	4
9. Flour from all streams being fed into bulk flour bins or into packers	2	2 (0.1)	1
10. Pollard from all streams before it is mixed with the bran	6	8 (0.7)	7
11. Bran from all streams before it is mixed with the pollard	12	11 (1)	11
12. Wheat germ	10	10 (2)	7

Values in brackets are for bioresmethrin.

Table 14.11. Fenitrothion residues on raw and processed bran (results from Australian survey) **(Snelson 1979a)**

Range of residues	Raw bran No.	%	Range of residues	Processed bran No.	%
<10	3	23	<0.1	13	36
10–12	5	38	0.1–0.5	7	19
12–14	5	8	0.5–1.0	7	25
14–16	3	23	1.0–2.0	3	8
16–18	0	9	2.0–3.0	1	3
18–20	1	8	>3.0*	3	8
	13	100		36**	100

* Maximum 4.0
**12 different brands.

than 0.1 mg/kg) of fenitrothion. In the light of the results of the survey, a supervised trial was carried out in a factory processing 3 bran products **(Snelson 1979c)**. The results of the trial are shown in **Table 14.12**. These indicate that substantially all of the fenitrothion residues are destroyed in the cooking process, which involves prolonged heating under pressure with live steam after the bran has been subjected to wet digestion with malt. The subsequent drying, extrusion and toasting are all vigorous processes employing high temperatures that would be expected to destroy any residues that were not destroyed in the cooking. The trial confirmed that normal commercial processing destroyed the bulk of the residues.

Snelson (1979a) observed that commercial flour mills could not comply with the national residue limit for fenitrothion in flour (1 mg/kg), particularly when wheat available had been treated within the previous 2 to 3 months with fenitrothion at the recommended level of 6 mg/kg. The matter was investigated by **Murray and Snelson (1978)** and the results, summarised in **Table 14.10**, indicate that although the flour sieved from the sifter fed by the first reduction rolls contained residues of the order of 0.5 mg/kg, this was later blended with flour from other streams containing slightly higher residues of fenitrothion, due to the higher degree of extraction, which occurs in these steps. As a consequence, the commercial flour contained up to 2 mg/kg of fenitrothion.

All other data available came from trials carried out with experimental milling equipment which, although designed to reproduce commercial practice, does not give the same high degree of extraction.

As part of a National Residue Survey in Australia, **Snelson (1979b)** reported that though 77% of raw bran samples available to breakfast cereal manufacturers contained fenitrothion at levels above 10 mg/kg, 92% of the samples of processed bran contained less than 3 mg/kg of fenitrothion. Those samples with fenitrothion residue levels above 1 mg/kg contained a proportion of raw bran along with processed bran.

Desmarchelier (1979c) and **Desmarchelier and Marsden (1979)** thoroughly investigated the transfer of fenitrothion residues from wheat via flour into wheat gluten and the fate of these residues during the cooking of many varieties of bread prepared from different recipes. They showed that although gluten represents approximately 10% of the content of raw grain, the residue in the extracted gluten is about 30% of that present in the raw grain, thus representing a threefold concentration above the expected level. This picture is not unique to fenitrothion, but appears to occur with many insecticides of quite different chemical groups. These investigations showed that, irrespective of the level of residues in the flour or in gluten, and independent of the type of bread, the recipe or the baking process, approximately 50% of the residue in gluten was destroyed in bread making.

Table 14.12. Effect of commercial processing on fenitrothion residues on bran **(Snelson 1979c)**

Processing Stage	Product A Rep. 1	Product A Rep. 2	Product B Rep. 1	Product B Rep. 2	Product C Rep. 1	Product C Rep. 2
1. Raw bran before cooking	5	4	3	4	23	21
2. Cooked bran taken directly from cookers	<0.05	<0.05	<0.05	<0.05	2	2
3. Bran taken after initial drying stage	<0.05	<0.05	<0.05	<0.05	1	2
4. Finished product taken after bran has been formed into final shape and dried/toasted through an oven	1	<0.05	<0.05	<0.05	1	1

Ito *et al.* (1976) carried out an extensive and detailed study on the fate of fenitrothion residues on rice. Following the application of fenitrothion to husked rice, the grain was stored for 12 months, then milled under standard conditions which yielded 87% polished rice and 13% bran. During the milling process, the major portion of the residue in the husked rice was removed in the bran and this feature was independent of the storage period or the amount of fenitrothion applied. Only a minor proportion of the fenitrothion remained in the polished rice and even this was reduced by washing the grains before cooking. The residue in the washed grain was reduced to about half by cooking. The higher the temperature and pressure applied during cooking, the greater was the reduction in the fenitrothion residue. **Table 14.13**, summarising this work, shows that the ratio of fenitrothion concentrations in polished rice and bran remain constant, irrespective of the amount of fenitrothion applied to the husked rice or found after prolonged storage. Only about 10% of the quantity occurring in the husked rice remains in the polished rice after milling. The residues remaining on polished rice, when washed and cooked are shown in **Table 14.14**. Three types of cooking were employed in these studies. Cooking condition (1) represents the typical method of cooking rice in the home and consists of boiling the rice in an approximately equal weight of water at atmospheric pressure for 15 minutes, followed by a further 15 minutes at a temperature of 80°C. Cooking condition (2) consists of boiling the rice in an equal quantity of water for 10 minutes at 110°C under a pressure of 1.5 atmospheres in an autoclave. Condition (3) consists of cooking for 10 minutes at a temperature of 120°C under a pressure of 2.1 atmospheres. The increased temperature and pressure had a slight but significant effect upon the destruction of the small quantity of fenitrothion remaining in the polished rice.

Table 14.13. Fate of fenitrothion applied to husked (brown) rice during milling and cooking (**Ito *et al.* 1976**)

Amount applied (mg/kg)	Storage period (months)	Residues (mg/kg) in Husked rice	Polished rice*	Rice bran*
15	0	9.38	1.02	65.0
2	12	0.61	0.09	4.03
6	12	1.66	0.25	11.7
15	12	4.39	0.55	31.6

*Ratio polished rice to bran = 87:13

Table 14.14. Effect of cooking on fenitrothion residues in rice (**Ito *et al.* 1976**)

Amount applied (mg/kg)	Storage period (months)	Residues (mg/kg) in Polished rice	Washed grains	Wash* water	Cooking** condition	Residue in cooked rice (mg/kg)
15	0	1.02	–	0.60	1	0.26
2	12	0.09	0.03	0.05	1	0.02
					2	0.01
					3	0.01
6	12	0.25	0.08	0.16	1	0.05
					2	0.03
					3	0.01
15	12	0.55	0.18	0.37	1	0.12
					2	0.07
					3	0.04

*Concentration expressed as mg/kg in grains before washing.
**For explanation see text.

Takimoto *et al.* (1978) studied the degradation and fate of fenitrothion applied to harvested rice grains with the aid of radio-labelled fenitrothion. The fenitrothion was applied to husked rice at the rate of 6.0 and 15 mg/kg and the treated grain was held in store at 30°C. When rice grains were cooked immediately after treatment with fenitrothion, the fenitrothion content in unpolished grains decreased to about 60%, accompanied by formation of desmethyl fenitrothion and 3-methyl-4-nitrophenol, as shown in **Table 14.15**.

In the case of rice grains that had been stored for some months after being treated, 30–40% of the fenitrothion originally present in grains was lost, with concomitant increase of desmethyl fenitrooxon (not desmethyl fenitrothion) and 3-methyl-4-nitrophenol. The level of other decomposition products decreased. When rice grains were polished and afterwards washed with water, the radioactivity remaining was only about 20% of that originally applied. When the washed rice was cooked there was a further reduction to about 5% to 6% of the applied

Table 14.15. Decomposition of fenitrothion and its decomposition products by cooking[a] (**Takimoto *et al.* 1978**)

Month	Distribution (%) of radioactivity extracted[a]																
	Total	Fenitrothion	Fenitrooxon	Fenitrothion S-isomer	Desmethyl-fenitrothion	Desmethyl-fenitrooxon	Desmethyl-fenitrothion S-isomer	3-Methyl-4-nitrophenol	3-Hydroxy-methyl-4-nitrophenol	1-Methoxy-3-methyl-4-nitrobenzene	1,2-Dihydroxy-4-methyl-5-nitrobenzene	1,2-Dimethyoxy-4-methyl-5-nitrobenzene	Unidentified compounds	Bound[c]	Ethanol trap	Other[d] fractions	Recovery
Unpolished																	
0	93.1	61.4	0.2	0.2	17.6	0.4	–[e]	12.1	–	–	–	–	1.2	2.1	4.4	–	99.6
6	102.2	25.4	0.2	0.2	19.1	8.4	0.4	45.8	0.3	0.2	0.6	0.1	1.5	1.6	1.7	–	105.5
12	97.5	18.4	0.3	0.4	11.8	10.5	1.0	51.5	0.4	0.2	0.6	0.1	2.3	2.0	1.0	–	100.5
Polished																	
6	18.2	6.1	<0.1	–	3.2	0.2	–	8.2	<0.1	–	0.2	–	0.3	2.5	0.5	82.7	103.9
12	17.9	5.1	<0.1	<0.1	3.8	0.2	–	8.4	–	–	0.2	–	0.2	3.3	0.3	77.4	98.9

a The rice grains treated with 15 ppm fenitrothion were stored at 30°C.
b Radioactivity applied to rice grains is referred to as 100%.
c Unextracted radioactivity in grains. See text for details.
d Including seed coat and germ fractions, and washings. See text.
e <0.1%, if any.

radioactivity, or a little less than the equivalent of 1 mg/kg fenitrothion in terms of uncooked rice. The residue content (fenitrothion, desmethyl fenitrothion and 3-methyl-4-nitrophenol) of boiled, polished rice decreased to half or less, compared with unwashed, polished rice.

Desmarchelier *et al.* (1980a), as part of larger study of the fate of fenitrothion and 4 other insecticides on rice and barley during storage, studied the losses of these insecticides during processing. After 3 and 6 months storage, husked and polished rice were cooked in a minimum amount of boiling water for 15 and 25 minutes respectively, and the residues were determined before and after processing. The results are given in **Table 14.16**. After 6 months storage, unhusked rice was milled and the husked and polished rice were cooked as described above, the results are included in **Table 14.16**. Generally, losses of residues from husked rice during cooking were less than those from polished rice. Residues in polished and husked rice after milling were only a fraction of the residues in unmilled rice.

Desmarchelier and Hogan (1978) performed a series of laboratory and pilot scale experiments on grain dust to reduce residues of dichlorvos and malathion to levels which would render grain dust acceptable as a livestock feed component. The residues of such insecticides often range up to 200 mg/kg, due to the preferential absorption of the insecticide spray on to the dust fraction of bulk grain. These studies showed that admixture of a small amount of alkali (in the form of caustic soda) rapidly resulted in the destruction of virtually all the insecticide residues. This information was later applied to grain dust containing fenitrothion residues and it was found that these could be destroyed with equal facility. Commercial processes were later developed, utilising the addition of a small quantity of caustic soda solution or finely powdered hydrated lime. The amount of base required was less than the amount used to increase the food value of silages or cereal straws and the process of pelleting reduced the alkalinity of the commodity to a neutral pH. It could be applied equally well to bran or other mill offals destined for livestock feeds. These workers also found that the level of residues on silo dust could be reduced considerably by not locating the spray nozzles adjacent to the dust extraction duct inlets.

Abdel-Kader and Webster (1980) who studied the degradation and translocation of malathion and fenitrothion in stored wheat at low temperatures found that, following the application of 8 mg/kg and storage for 6 months at 10°C and –5°C the residues on bran milled from the wheat were 20 and 26 mg/kg respectively. These residues were recognised as being higher than the legal limit. Residues recovered from fractions of wheat stored at 20°C were comparatively lower than residues found at 10°C or –5°C (**Abdel-Kader 1981**).

Wilkin and Fishwick (1981) studied the fate and distribution of 6 organophosphorus grain-protectant insecticides on wheat, wholemeal flour and bread produced from the treated wheat. Wheat treated with fenitrothion at the recommended dose of 6 mg/kg (actual dose 6.8 mg/kg) and stored for 4 weeks gave rise to residues of 3.3 mg/kg on the wheat, 3.0 on the flour and 1.6 on the wholemeal bread. This represents 53% of the insecticide in the wholemeal flour surviving baking.

14.4 Metabolism

Rowlands (1969) found fumigation with methyl bromide, that left residues of 20 mg/kg inorganic bromide after aeration, prevented desmethylation of fenitrothion applied 1 week later to the grain, presumably by methylating receptor sites of the desalkylation (phophatase) system. This level of methylation did not greatly affect the P-O-aryl phosphatase hydrolysis, however, nor

Table 14.16. Residues of fenitrothion on rice before and after storage, processing and cooking (**Desmarchelier *et al.* 1980a**).

Grain	Application level (mg/kg)	Residues (mg/kg) After 3 months	After 6 months	After cooking at 3 months*	After cooking at 6 months*
Husked rice	15	7.7	4.5	3.6	3.0
Polished rice	15	8.2	5.5	3.2	2.8

Grain	Application level (mg/kg)	After 3 months	After 6 months	Residues (mg/kg) after 6 months storage and Milling: Husked	Milling: Polished	Milling & cooking*: Husked	Milling & cooking*: Polished
Unhusked rice	15	7.7	3.5	0.62	0.27	0.39	0.13

* Residues expressed as if on rice, 65% r.h.

did more intensive fumigation that left residues of up to 120 mg/kg inorganic bromide.

Takimoto *et al.* (1978) studied the degradation of fenitrothion applied to harvested rice grains with the aid of radio-labelled fenitrothion. Rice grains with 14% moisture content were treated with ^{14}C-fenitrothion, at a concentration of 6 and 15 mg/kg, and stored at 15°C or 30°C for 12 months. Decomposition of fenitrothion proceeded more rapidly at 30°C than at 15°C, with half-lives of about 4 months and more than 12 months, respectively. The degradation was independent of the concentration applied. Desmethyl fenitrothion was produced at an early stage of storage, amounting to about 20% and 10% of the applied radioactivity at 30°C and 15°C respectively, whereas 3-methyl-4-nitrophenol was formed gradually and accounted for 38% and 17% at 30°C and 15°C after 12 months, respectively. A range of other minor decomposition products was also formed in small amounts. Fenitrothion and its decomposition products penetrated into grains about 100 μ from the surface, to the outer portions of the endosperm, after 12 months storage. Approximately 60% of the applied radioactivity was removed by milling, leaving 4 mg/kg of fenitrothion as the maximum in polished rice grains. Cooking unpolished rice grains treated with fenitrothion caused decomposition mainly to desmethylated derivatives and 3-methyl-4-nitrophenol. The degradation pathway for fenitrothion in rice grains is presented diagrammatically in **Figure 14.7.**

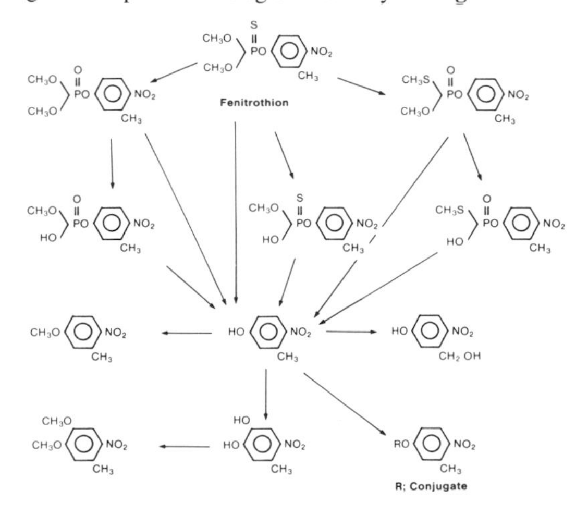

Figure 14.7. Degradation pathways for fenitrothion in rice grains **(Takimoto *et al.* 1978).**

Abdel-Kader (1981), Abdel-Kader and Webster (1981), Abdel-Kader and Webster (1982), and Webster and Abdel-Kader (1983) presented a series of papers on different aspects of the same studies involving the application of malathion and fenitrothion to wheat stored at various temperatures followed by a study of the distribution and quantification of metabolites. The major metabolites of fenitrothion degradation were demethyl-fenitrothion, 3-methyl-4-nitrophenol and dimethyl-phosphorothioic acid. Wheat treated with 12 mg/kg fenitrothion and stored at 20°C for 6 months contained 2 mg/kg demethyl-fenitrothion and 0.5 mg/kg dimethyl-hydrogen phosphorothioate. The level of 3-methyl-4-nitrophenol was still increasing slowly at 12 months, when its concentration was 0.8 mg/kg.

Various comprehensive studies in mouse, rat and guinea pig have dealt with the pharmacodynamic and biochemical aspects of fenitrothion and its metabolites. These were evaluated by the JMPR in 1969, 1974 and 1976 **(FAO/WHO 1970, 1975, 1977)**. A schematic presentation of the metabolism of fenitrothion in plants, mammals and bacteria was published by the **National Research Council of Canada (1975)** and is reproduced as **Figure 14.8.**

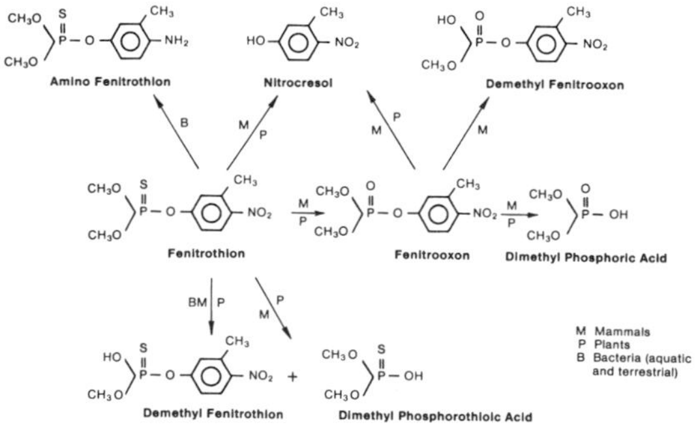

Figure 14.8. Metabolism of fenitrothion in plants, mammals, and bacteria **(National Research Council of Canada 1975).**

The metabolism of fenitrothion is rapid in several species of animals including rats, mice, rabbits, dogs, goats and birds, and the metabolites are substantially eliminated from the body like other organophosphorus insecticides. ^{32}P-labelled fenitrothion administered orally was rapidly metabolised in rats and guinea pigs and the phosphorus-containing metabolites were excreted into the urine within a few days post-treatment **(Miyamoto *et al.* 1963)**. The metabolites included desmethyl-fenitrothion, desmethyl-fenitrooxon, dimethylphosphorothioic acid, and dimethylphosphoric acid. Fenitrooxon was detected only after intravenous administration of a large amount of fenitrothion and not after oral administration **(Miyamoto *et al.* 1963; Miyamoto 1964)**.

Hollingworth *et al.* (1967a) studied the metabolism of fenitrothion in white mice. Of the radioactivity, 75% was recovered within 24 hours. The isolated products indicate that both P-O-alkyl and P-O-aryl bonds of fenitrothion and fenitrooxon are cleaved. No evidence was obtained indicating that the nitro group was reduced to form amino-fenitrothion or that the ring methyl group was oxidised. One of the main points of the their work was that some dearylation and dealkylation was dosage-dependent. At a small dose (17 mg/kg), desmethyl-fenitrothion, desmethyl-fenitrooxon, dimethylphosphorothioic acid and dimethylphosphoric acid were obtained as major metabolites. On the other hand, 200 mg/kg body weight doses decreased the amounts of dimethylphosphorothioic acid and dimethylphosphoric acid, and increased desmethyl-fenitrothion.

Further metabolic studies in mammals were carried out by using mainly 3-methyl-^{14}C-fenitrothion. Fenitrothion administered orally was readily absorbed from the gastrointestinal tract, and the radio-carbon was rapidly and completely excreted, mainly into the urine of rats (male and female), mice (male and female), rabbits (male and female) and dogs (male). Examination of 11 rat tissues, including fat and muscle, revealed that the concentration of fenitrothion was 0.004 to less than 0.001 mg/kg (except fat — 0.34 mg/kg) 24 hours after administration. Whole-body autoradiography also indicated rapid disappearance of the radio-carbon from tissues of mice. The cumulative excretion patterns were essentially the same among these animals over the range of dosage regimes tested (15 mg/kg-bw, 105 mg/kg-bw, or 15 mg/kg-bw radioactive fenitrothion given after pre-treatment of 15 mg/kg-bw of non-active fenitrothion every other day for five times). Thin-layer chromatographic analysis of the urinary metabolites showed the absence of intact fenitrothion and the presence of as many as 18 metabolites. Of these, from 92% up to 99% were identified. There were qualitative and quantitative differences in the composition of the metabolites among these animal species, and also between males and females of the same species (mostly quantitative in this case). The percentage of the metabolites retaining the P-O-aryl linkage varied with animal species, ranging from 15% (rabbits) to more than 55% (dogs). Most of these metabolites are demethylated products at the *o*-methyl position. Rats and mice tended to excrete greater amounts of demethyl fenitrooxon than did rabbits and dogs. 3-methyl-4-nitrophenol, free or bound with sulphate or glucuronic acid, constituted another group of major metabolites. With the exception of dogs, approximately 50–75% of the urinary radioactivity was accounted for by these 3 metabolites. Rabbits exceed other animal species in this respect. A trace amount of the oxidised phenols was present. Rabbits were exceptional in excreting fenitrooxon and amino-fenitrothion, though in small amounts. The urine of rabbits and, to a lesser extent, of rats contain several minor metabolites derived from amino fenitrothion or from 3-methyl-4-aminophenol. The minor products totalled approximately 20% of the excreted radio-carbon in rabbits and about 10% in rats. A few metabolites had unique structures resulting from reduction of the nitro group and oxidation of the 3-methyl group. An appreciable amount of the excreted metabolites from the 4 animal species was in the form of conjugates. It appears that rabbits are most active in this regard, and dogs least active. In the 4 species treated, higher amounts of sulfate conjugates than glucuronides were excreted. Thus, fenitrothion is metabolised through 2 major pathways, via O-demethylation and via cleavage of the P-O-aryl bond. Neither S-(3-methyl-4-nitro-phenyl) glutathione mediated by glutathione S-aryltransferase nor ring hydroxylation products have been positively demonstrated (**Miyamoto *et al.* 1976b**).

Although reduction of the nitro group in the fenitrothion molecule, presumably by intestinal micro-organisms, is a minor metabolic pathway in the above 4 animal species, it is by far the major metabolic pathway in ruminants. In fact, a study on female goats revealed that most of the urinary and fecal metabolites of fenitrothion are amino derivatives and are formed most probably in rumen fluid (**Mihara *et al.* 1978**). Jersey cows given 3 mg/kg-bw/day of fenitrothion for 7 consecutive days also secreted up to 0.003 mg/kg of amino-fenitrothion in fresh milk (**Miyamoto *et al.* 1967**). However, **Johnson and Bowman (1972)** reported that no fenitrothion or its metabolites were detected in the milk, urine or feces of lactating Jersey cows 7 days after feeding (for 28 days) on diets containing 1.84 mg/kg of the pesticide. By consecutive administration of the radioactive fenitrothion to a female goat for 7 days at the rate of 0.5 mg/kg/day, no fenitrothion nor fenitrooxon was secreted into milk, although 0.011 ppm, at the maximum, of radioactive metabolites was detected (**Mihara *et al.* 1978**).

The metabolism of fenitrothion in rats with experimentally induced hepatic lesions by dietary administration of diaminodiphenyl-methane (DDM), by feeding low protein-high fat diet (LPHF) or following intra-muscular treatment with carbon tetrachloride was also investigated. The *in vitro* rates of degradation of fenitrothion were significantly reduced in all preparations from livers with induced hepatic lesions, with the greatest reduction produced by LPHF followed by carbon tetrachloride and by DDM. The rate of degradation of fenitrothion and fenitrooxon was less severely affected than the rate of activation of fenitrothion because of the greater activity of glutathione S-alkyltransferase (**Miyamoto *et al.* 1977a**). However, the cumulative excretion patterns of radio-carbon were not altered by these 3 hepatic lesions after oral administration to the injured rats of either 15 or 50 mg/kg-bw ^{14}C-fenitrothion, or 15 mg/kg-bw of the compound after pretreatment with about 2.8 mg/kg-bw/day of nonactive fenitrothion for 4 weeks. Moreover, these injured rats metabolised fenitrothion equally well *in vivo*, as compared with the control animals, even though the pathway yielding demethyl-fenitrothion predominated. Higher or subacute doses of fenitrothion tended to reduce the metabolic differences observed between the injured and the control animals (**Miyamoto *et al.* 1977b**). These results may indicate that even very severe hepatic lesions such as those described here hardly influence *in vivo* detoxification of fenithrothion and excretion of its metabolites. The results also indicate that there is not much possibility of retention or storage of the parent compound and/or its toxic metabolites in the mammalian body even under such conditions.

14.5 Fate in Animals

Of the insecticide administered to test animals (rats, guinea pigs, mice, rabbits and dogs) 85–97% was excreted in the urine within 2–4 days (**Miyamoto *et al.* 1963; 1976a; Hollingworth *et al.* 1967**). Only up to 10% was eliminated in the feces. The radioactivity, excreted in the urine and feces combined, accounted for nearly 100% of the dose. No radioactive carbon dioxide was expired in rats. By consecutive administration of fenitrothion to mammals, no bioaccumulation of fenitrothion and its metabolites was observed (**Miyamoto 1977a**). The major metabolites eliminated in the urine included demethyl-fenitrothion, demethyl-fenitrooxon, 3-methyl-4-nitro-phenol (free and bound with sulphate and glucuronide), dimethyl-phosphorothioic and phosphorothioic acid and phosphoric acid. Among the animal species were observed some quantitative differences in the composition of these metabolites in the urine. In rat faeces, no metabolites other than those found in urine were present (**Miyamoto *et al.* 1976a**); fenitrothion, demethyl-fenitrothion, demethyl-fenitrooxon and 3-methyl-4-nitro-phenol were identified as major fecal metabolites.

Miyamoto *et al.* (1963) showed that orally administered ^{32}P-labelled fenitrothion was readily absorbed from the digestive tract of guinea pigs or rats and the major portion of the radioactivity was excreted in the urine. Neither fenitrothion nor fenitrooxon was detected and desmethyl fenitrothion, dimethyl phosphorothionate and dimethyl phosphate were eliminated in the urine. **Miyamoto (1964)** showed that following intravenous injection of radioactive ^{32}P-fenitrothion into guinea pigs and rats, fenitrothion rapidly disappeared from the blood. Fenitrothion and fenitrooxon were found in tissues and their amounts decreased rapidly. The desmethyl compound and the dimethyl esters mentioned above were found mostly in the liver and kidneys.

Excretion of metabolic products is rapid and chiefly in the form of 3-methyl-4-nitrophenol, the fenitrothion hydrolysis product (**Hladka and Nosal 1967**).

After oral administration of up to 40 g of fenitrothion per lactating cow, residues in the milk were as high as 0.4 mg/kg after 6 hours and below the limit of detection after 1 day (**Hais and Franz 1965**). Detoxication in bovine rumen was rapid owing to reduction of fenitrothion to the amino compound (**Miyamoto *et al.* 1967**).

Cows fed 3 mg/kg body weight of fenitrothion for 7 consecutive days produced milk having up to 0.002 mg/kg residue of fenitrothion on the second day, and no residue 1 day after administration was stopped. Less than 0.003 mg/kg amino fenitrothion and about 0.1 mg/kg 3-methyl-4-nitrophenol were detected during treatment, and no fenitrooxon was found. The residues diminished to undetectable levels 2 days after the last dosage (**Miyamoto *et al.* 1967**).

Thirty calves (1–1.5 years and average weight 243 kg) confined a pasture sprayed with 375 g/ha of fenitrothion (11.8 mg/kg initial residue on grass) were periodically sacrificed and breast muscle and omental fat analysed. On the first day residues in the meat and fat were about 0.01 mg/kg. No residue of fenitrothion was found in the meat from the third day on and only 0.004–0.007 mg/kg was found in the fat on the third day; these amounts decreased almost to control levels by the seventh day (**Miyamoto and Sato 1969**).

Silage prepared from corn treated with 1, 2 and 3 kg/ha of fenitrothion was fed to lactating Jersey cows for 8 weeks. Although traces (0.001–0.005 mg/kg) of amino-fenitrothion were found in the milk of cows fed the 3 kg/ha silage, no residues were found in the milk of cows consuming silage treated at lower levels (**Leuck *et al.* 1971**). The urine contained total metabolites averaging from 0.53–5.1 mg/kg but these consisted mostly of amino-fenitrothion and about 0.1 mg/kg or less of the parent insecticide and its cresol.

Jersey cows were fed on diets spiked with 0, 25, 50 and 100 mg/kg of fenitrothion for 28 days. Milk, urine and feces from cows fed as much as 100 mg/kg dietary fenitrothion contained neither fenitrothion, its oxygen analogue nor its cresol; but the amino analogue of fenitrothion in milk, urine and feces of cows fed the 25, 50 and 100 mg/kg diets averaged 0.002–0.17, 4.64–35.6, and 0.19–1.80 mg/kg respectively. Seven days after feeding the diets containing fenitrothion was terminated, residues could not be detected in milk, urine, or feces from any cow (**Johnson and Bowman 1972**).

Following administration of 10 or 3 mg/kg-bw/day fenitrothion to male rabbits for 6 months, blood, skeletal muscle and abdominal fat were analysed by gas chromatography for fenitrothion and fenitrooxon. In most cases, blood and muscle contained no detectable amounts of either compounds (detection limit for fenitrothion 0.005 or 0.002 ppm, and that of fenitrooxon, 0.01 ppm). An average of 0.131 mg/kg (0.243 ppm at the maximum) and of 0.045 ppm was measured in the fat of rabbits dosed at 10 and 3 mg/kg-bw/day respectively. No fenitrooxon was detected (**Miyamoto** *et al.* 1976a). Three male beagle dogs were treated orally with 5 mg/kg-bw/day of fenitrothion technical 6 days per week for 10 months. This dose significantly inhibited plasma and erythrocyte cholinesterases. After termination of the treatment, only the fat contained trace amounts of fenitrothion (to a maximum of 0.160 ppm) (**Tomita *et al.* 1974**).

To evaluate the possible retention of fenitrothion residues in animal tissues, ^{14}C-fenitrothion was administered orally to male Wistar rats at 15 mg/kg-bw/day for 7 days, and then at 30 mg/kg-bw/day for a further 3 days. A measurable amount of fenitrothion was found in abdominal fat and increased in concentration at the latter stages of administration. This amount, however, tended to disappear quite rapidly on cessation of the administration (**Miyamoto and Mihara 1977**).

After oral administration of 5 mg/kg-bw of fenitrothion to female Japanese quails, 99% of the radio-carbon was excreted in the first 24 hours (**Miyamoto 1977b; Mihara *et al.* 1977**). While white Leghorn hens dosed with 2 mg/kg-bw of fenitrothion for 7 consecutive days discharged in the excreta 95% of the radioactivity within the 6 hours following the last administration, analyses of the radio-carbon in the excreta revealed that among more than 12 metabolites, 3-methyl-4-nitrophenol (free and sulphate conjugate) accounted for 70% of the applied radioactivity in quails, and 50% in Leghorns. Demethyl-fenitrothion and demethyl-fenitrooxon were also identified together with a hydrolysis product, 3-hydroxymethyl-4-nitrophenol (free and bound with sulphate). The radioactivity in the hen egg white decreased sharply after the last dosage, with the highest concentration (0.02 mg/kg) recorded on the third administration day. The hen egg yolk showed a maximum radio-carbon of 0.10 mg/kg (fenitrothion, 0.006 mg/kg) after 1 day from the last dosage, followed by a prompt decline to 0.02 ppm after 1 week.

When lactating Japanese Sannen goats were treated orally with 0.5 mg/kg-bw of fenitrothion for 7 successive days, no residues of intact fenitrothion were found in the organs and tissues after 1 day, with trace amounts of amino-fenitrothion detected in digestive tracts (rumen, omasum and large intestine). The radio-carbon disappeared rapidly thereafter (**Mihara *et al.* 1978**). The administered radio-carbon was essentially quantitatively excreted during the week following treatment; 50% of the dose recovered in urine, 44% in feces and 0.1% in milk with a maximum of 0.011 mg/kg equivalent of the parent compound. The remaining radioactivity was further excreted, though gradually. The major metabolites included amino-fenitrothion, free in urine and sulphate conjugate in milk along with acetylamino-fenitrothion in feces. No intact fenitrothion nor fenitrooxon was found in milk, urine and feces.

Trottier and Jankowska (1980) carried out an *in vivo* study on the storage of fenitrothion in chicken tissues after long-term exposure to small doses. They used male chickens weighing about 500 g which were incubated with ^{3}H ring-methyl-labelled fenitrothion at regular intervals for 2 to 8 weeks. Some of the birds were slaughtered 8 or 16 days after they had received treatment for 0.5, 1 and 2 months. The amount of radioactivity was determined in a liquid scintillation counter. The findings suggest that chicken tissues retain no significant amount of fenitrothion or its metabolites even after a long-term exposure to small doses of this insecticide.

Goto *et al.* (1980) reported that, when fenitrothion was sprayed on drinking water and feeds or orally administered, trace quantities of fenitrothion (0.04–0.07 mg/kg) were detected in egg yolk and egg white.

The 1974 FAO/WHO Joint Meeting on Pesticide Residues stated that there is no tendency for fenitrothion or its metabolites to accumulate in animal tissues or food of animal origin following approved uses on pasture or forage crops or from feeding of food-processing wastes.

14.6 Safety to Livestock

Leuck *et al.* (1971) fed lactating Jersey cows for 8 weeks on silage prepared from corn treated with 1, 2 and 3 kg/ha of fenitrothion. Although feces of the cows contained low levels of residue (0.04–0.18 mg/kg, mainly as the amino compound) neither blood cholinesterase depression nor any abnormality of general health or milk production was noted.

The acute oral LD/50 values of fenitrothion to Mallard ducks, ringneck pheasants, Japanese quails (female) and hens were found to be 2550, 34.5, 140 and 523 mg/kg, respectively (**Kadota and Miyamoto 1975; Fletcher *et al.* 1971; Sumitomo 1971**). Despite marked differences among bird species, fenitrothion may be classified as an organophosphorus insecticide with a relatively low toxicity to birds.

Fletcher *et al.* (1971) reported that the results of the acute oral toxicity study conducted with fenitrothion in ringneck pheasants showed the acute oral medium lethal dose (LD/50) of the test material to be 34.5 mg/kg body weight with 95% confidence limits of 22.6 to 53.5 mg/kg.

Johnson and Bowman (1972) fed Jersey cows on diets spiked with 25, 50 and 100 mg/kg of fenitrothion for 28 days. Consumption of diets containing fenitrothion did not depress food intake, milk production or blood cholinesterase activity.

Fletcher (1974) conducted a toxicity and reproduction study with groups of Mallard ducks fed fenitrothion at dietary levels of 30 or 100 ppm. The body weight data collected during the investigation revealed no significant effects which could be attributed to the ingestion of fenitrothion. The test group consumed amounts of feed during the treatment period comparable to that consumed by the control group. There were no changes in the incidence of mortalities which could be attributed to the ingestion of fenitrothion. There were no abnormal behaviour reactions noted during the investigation. Autopsy of animals sacrificed either at the end of the treatment period or at the end of the recovery period revealed no gross pathologic effects. The weekly egg production, egg weight and egg quality were comparable to control groups. The percentage of infertile eggs and of eggs hatched was comparable to those in the control group. Likewise, the shell thickness was unaffected. The body weights and viability of ducklings hatched from eggs of treated birds were comparable to those of the control ducklings.

Fletcher *et al.* (1974) conducted a toxicity and reproduction study with groups of Bobwhite quail fed fenitrothion at dietary levels of 3 and 10 ppm. Body weights, food consumption, mortality and reaction, egg production, weight and quality, hatchability, body weights and viability of chicks and results of gross pathological examination were all comparable to those of unmedicated controls. There were no mortalities and no abnormal behavioural reactions during the investigation.

Kadota *et al.* (1974) found that the acute oral (single dose) LD/50 values of fenitrothion in male and female Japanese quails were 110 and 140 mg/kg, respectively. The sub-acute oral LD/50 of fenitrothion in male and female quails was found to be 21.5 and 13.5 mg/kg/day when fenitrothion emulsion was administered consecutively during 2 weeks. By dietary administration of as high as 50 ppm of fenitrothion for 4 weeks no adverse effects were observed with respect of behavioural reaction and body weight. The rate of egg laying was not affected by the administration of 15 ppm of fenitrothion in the diet.

Only slight depression on egg laying was observed at 50 ppm. At 150 ppm the rate of egg laying was inhibited to one-third of the control, but it recovered 2 weeks after completion of fenitrothion feeding. The cholinesterase in blood and brain were adversely affected at 15 ppm and above and blood cholinesterase was more susceptible than the brain enzyme. The recovery of the enzyme activity was rapid and complete after termination of the feeding. The maximum no-effect level of fenitrothion in this experiment estimated from the effect on cholinesterase was determined to be a little less than 5 ppm or 0.65 mg/kg/day in males and 0.69 mg/kg/day in females.

Joiner (1975) conducted a 5-day feeding study in Mallard ducklings and calculated the LC/50 to be 695 ppm. Apparent signs of toxic effects were noted at all levels of administration from 100 to 3000 ppm. Body weights of ducklings receiving fenitrothion at all levels were lower than control weights throughout the experimental period. Gains in average weight between weighing periods were lower for those receiving fenitrothion but they did recover to within the range of the controls during the 3 days they received the control diet. Food consumption was lower for all those dose levels than for the controls during the 5-day toxicant feeding period. However, all levels did recover to the consumption rate of the controls during the 3-day control feeding period.

Kadota *et al.* (1975) conducted an acute oral toxicity study with fenitrothion in adult hens and reported that treatment with fenitrothion and/or the formulation resulted in dose-dependent inhibition of red blood corpuscle and brain acetyl cholinesterase, plasma cholinesterase and hepatic and renal cholinesterase at doses above 1 mg/kg-bw/day. Recovery of plasma cholinesterase was very rapid but brain cholinesterase recovery took greater than 2 months. The calculated LD/50 for fenitrothion is given as 500 mg/kg (range 384–650). Toxic symptoms (motor ataxia, irregular respiration, decreased spontaneous motor activity) appeared early and then disappeared after 5–7 days and reappeared after the second dose. Slight body weight decrease was noted in treated hens. No paralysis of legs or abnormalities in sciatic nerves or spinal cords was observed.

Kadota and Kagoshima (1976) studied the delayed neurotoxicity of fenitrothion in hens. A dose of 500 mg/kg (the acute oral LD/50) was orally administered twice at the interval of 21 days to 16 adult white Leghorn hens, and the treated birds were observed for 3 weeks. Although 5 hens were dead of acute intoxication regardless of atropine and 2-PAM treatment, the surviving hens showed no delayed paralysis in their legs. In sub-acute administration where groups of 8 hens had been treated with 16.7 and 33.4 mg/kg of fenitrothion daily for 4 weeks with subsequent 3 weeks of observation periods, 15 surviving hens showed no clinical signs of delayed neurotoxicity. Histopathological examination revealed no structural changes in sciatic nerves and spinal cord. A second such report **(Sumitomo 1977)** evaluated by Canadian authorities presented almost identical findings.

Paul and Vadlamudi (1976) observed abnormalities, including dwarfism in Leghorn chicks hatched from eggs that had been injected with 0.1 mL of fenitrothion emulsions (0.1–1.0%). However, the experimental conditions are quite artificial and these dosages are much higher than the fenitrothion concentrations detected in the bird metabolism study **(Mihara *et al.* 1977)**.

Groups of young male and female Japanese quails were kept for 4 weeks on a diet containing 0, 1.5, 5, 15 and 50 ppm of fenitrothion **(Miyamoto 1977b)**. No abnormalities were found in behaviour, mortality, body weight gain and food consumption; egg production was slightly suppressed only at the 50 ppm level. The cholinesterase activity in whole blood and brain was significantly inhibited at the 15 and 50 ppm levels. The enzymes in females were more susceptible to fenitrothion. Reproduction studies in Bobwhite quails and Mallard ducks revealed that fenitrothion given sub-acutely at rates up to 10 ppm in feed (quails) or 100 ppm in feed (ducks) did not adversely affect the parental growth and reactions, egg production, egg weight and quality, hatchability of the eggs, and growth and viability of the young.

An indication of the acute toxicity of fenitrothion to various species of livestock can be obtained from the data on the acute toxicity to various laboratory animals provided in **Table 14.17**.

A number of studies in which the effect of repeated or continuous dosing of fenitrothion to avian species has led to the determination of the lethal concentration in feed have been published. These are summarised in **Table 14.18**.

Fenitrothion is, however, an inhibitor of cholinesterase. Therefore care should be taken when feeding livestock with milling offals containing relatively high levels of fenitrothion residues. These should be diluted with other feedstuffs or treated with alkali (caustic soda or quick lime) before using as a major source of livestock feed.

14.7 Toxicological Evaluations

Toxicological information on fenitrothion was evaluated by JMPR in 1969, 1974, 1977, 1982 and 1984. Extensive data from studies of mutagenicity; neurotoxicity; potentiation; reproduction; teratology; ocular toxicity; acute toxicity in many species; short-term feeding

Table 14.17. Fenitrothion: acute oral toxicity (mg/kg) to various species

Species — Strain	Sex	Vehicle	LD/50	Reference
Rat	M	–	740	**Gaines (1969)**
	F	–	570	**Gaines (1969)**
SD	F	–	386	**Sumitomo (1970)**
SD	F	–	456	**Sumitomo (1970)**
	F	–	433	**Sumitomo (1971)**
SD	M	–	330	**Sumitomo (1972)**
SD	F		800	**Sumitomo (1972)**
		Technical	490	**Rosival *et al.* 1976**
	M	Formulated	476	**Bioresearch (1982)**
	F	Formulated	406	**Bioresearch (1982)**
	M	Formulated	385	**Bioresearch (1982)**
	F	Formulated	441	**Bioresearch (1982)**
	M	Formulated	500	**Bioresearch (1982)**
	F	Formulated	476	**Bioresearch (1982)**
	M	Formulated	756	**Bioresearch (1982)**
	F	Formulated	868	**Bioresearch (1982)**
	M	–	940	**Benes & Cerna (1970)**
	F	–	600	**Benes & Cerna (1970)**
Mouse	M	–	1336	**Carshalton (1964)**
	F	–	1416	**Carshalton (1964)**
	M		880	**Sumitomo (1971)**
dd	M	–	1030	**Sumitomo (1972)**
dd	F	–	1040	**Sumitomo (1972)**
	–	Technical	870	**Rosival *et al.* (1976)**
Guinea pig	M	–	500	**Dubois & Puchala (1960)**
	M	–	1850	**Sumitomo (1971)**
	–	–	1850	**Miyamoto *et al.* (1963b)**
Pheasant	–	Technical	34.5	**Fletcher *et al.* (1971)**
Hen	F	Technical	500	**Kadota *et al.* (1975)**
Japanese quail	M	Technical	115	**Kadota & Miyamoto (1975**
	F	Technical	140	**Kadota & Miyamoto (1975**
	M	50% EC	85	**Hattori *et al.* (1974)**
	F	50% EC	74	**Hattori *et al.* (1974)**
Mallard duck	–		2550	**Fletcher 1971**
	M	Technical	1190	**Hudson *et al.* (1979)**
Redwinged blackbird	–	–	25	**Schafer (1972)**
Pigeon	–	50% EC	42	**Hattori 1974**
Cat	–	–	142	**Nishizawa *et al.* (1961)**
Dog		Technical	681	**IBT (1971)**
Sheep	M	Technical	770	**Sumitomo (1971)**
Milk cow	F	Technical	300	**Sumitomo (1971)**

Table 14.18. Lethal concentration (LC/50) of fenitrothion in the diet of avian species

Species	Age	Vehicle	Conditions	LC/50	Reference
Bobwhite quail	14d	technical	8d dietary	>5000 (no deaths)	**Hill *et al.* (1975)**
Japanese quail	14d	technical	8d dietary	>5000 (no deaths)	**Hill *et al.* (1975)**
Ring-neck pheasant	10d	technical	8d dietary	>5000 (no deaths)	**Hill *et al.* (1975)**
Mallard duck	10d	technical	8d dietary	>5000 (no deaths)	**Hill *et al.* (1975)**
Grackle		technical	5d dietary	78	**Grue (1982)**

studies in rats, dogs, cattle, sheep and pigs; long-term feeding studies in rats, together with information on effects in man from occupational exposure, were reviewed.

Results of studies on delayed neurotoxicity of fenitrothion in hens were negative. Results of tests for mutagenicity and 2 reproduction studies in rats indicated no adverse affects at doses below those toxic to parents. Short-term studies in rats and dogs showed that a depression of plasma cholinesterase was the most sensitive indicator of effects and was considerably more sensitive than brain cholinesterase inhibition. Studies of tumour incidence did not indicate carcinogenic action. No-effect levels have been determined in the rat (0.25 mg/kg body weight per day) and dog (0.3 mg/kg body weight per day) and an ADI for man of 0.005 mg/kg body weight was established. In 1982, when it was realised that some of the pivotal studies on fenitrothion had been conducted by Industrial Bio-test Laboratories, the ADI for man was reduced to 0.003 mg/kg body weight per day as an expression of the concern that these studies might not reflect the true toxicity of fenitrothion. In 1984 the JMPR received replacement studies for the IBT studies on which doubts had been expressed. The evaluation of the replacement studies removed all doubts concerning the toxicology of fenitrothion but the ADI was not restored to its former value.

Further information on rat teratology and observations in man have been requested to enable the ADI to be re-evaluated.

14.8 Maximum Residue Limits

On the basis of the above ADI, and extensive residue data, the JMPR **(FAO/WHO 1975, 1977, 1978, 1980)** recommended the following maximum residue limits in raw cereal grains and milled cereal products.

Commodity	Maximum Residue Limits (mg/kg)
Bran (processed)	2
Bran (raw)	20
Cereal grains	10
Flour (white)	3
Rice (in husk and hulled)	10
Rice (polished)	1
Wholemeal	5

The residue is determined and expressed as fenitrothion.

The methods recommended for the analysis of fenitrothion residues are: **Pesticide Analytical Manual (1979a, 1979b, 1979c); Manual of Analytical Methods (1984); Methodensammlung (1982) (XII-3, 6; S5, S8, S13, S17, S19); Abbott *et al.* (1970); Ambrus *et al.* (1981); Desmarchelier *et al.* (1977b); and Panel (1980)**. The following methods are recognised as being suitable: **Methodensammlung (1982) (58); Zweig (1974); Eichner (1978); Funch (1981); Krause and Kirchhoff (1970); Mestres *et al.* (1977, 1979b); Moelhoff (1968); Sissions and Telling (1970); Sprecht and Tillkes (1980); Takimoto and Miyamoto (1976).**

15. Fenvalerate

Fenvalerate, an ester closely related, and in many ways similar to the pyrethroids, is a highly active broad-spectrum insecticide. It is particularly effective as a contact and stomach poison with adequate stability on foliage, and relatively low toxicity to mammals. Fenvalerate is registered in many countries in Europe, Asia, the Americas and Australia. It is used against a wide range of pests and on a wide range of crops and in forestry. Fenvalerate has undergone laboratory and silo-scale field trials as a stored-grain protectant in Australia.

It has been shown to be effective at low doses against *Rhyzopertha dominica* and at higher doses against most species. It combines well with organophosphorus insecticides and its potency is synergised by the addition of piperonyl butoxide. Deposits on grain are stable, though the bulk of the deposit is removed with bran or hulls. Those residues which carry through to white flour or milled rice remain substantially undiminished following cooking.

15.1 Usefulness

Bengston (1979) pointed out that the main immediate potential of pyrethroid insecticides as grain protectants in Australia is as treatment to control *Rhyzopertha dominica* which is currently the most destructive species and is difficult to control with most organophosphorus insecticides. Bioresmethrin, synergised with piperonyl butoxide, has been used for over 10 years against *Rhyzopertha dominica* in conjunction with fenitrothion or other organophosphorus compounds. Fenvalerate is an effective alternative to bioresmethrin. Relatively high doses are required to control *Sitophilus* and *Tribolium* spp. and it is therefore unlikely to be used against these species.

Bengston and Desmarchelier (1979) indicated that fenvalerate, applied at the rate of 1 mg/kg, together with piperonyl butoxide at the rate of 10 mg/kg and an organophosphorus insecticide, will provide 9 months protection against all species and strains, including OP-resistant *Rhyzopertha dominica*.

Chahal and Ramzan (1982) studied the relative efficiency of synthetic pyrethroids and some organophosphate insecticides against the larvae of *Trogoderma granarium*. Each insecticide was tested at 0.0125, 0.025 and 0.05% sprayed on third instar larvae. Deltamethrin at 0.05% or higher was the only compound to give 100% mortality in 1 day but observations 2, 3, 5 and 7 days after treatment showed a progressive increase in mortality and by 7 days fenvalerate caused 100% mortality of the larvae.

Elliott *et al.* (1983) determined the selectivity of 20 pyrethroid insecticides between *Ephestia kuehniella* and its parasite *Venturia canescens*. Fenvalerate was ranked fourteenth out of 20 for potency against *Ephestia* but it required 8 times as much to kill the parasite. The ratio of the toxicity to host and parasite was much higher in 18 other compounds tested.

Joia (1983) in a PhD thesis wrote of the effectiveness of cypermethrin and fenvalerate against *Tribolium castaneum* and *Cryptolestes ferrugineus* in stored wheat. Wheat of 13.3% and 15% moisture content was treated with cypermethrin or fenvalerate at 8 or 12 mg/kg or malathion at 8 mg/kg. Treated wheat was stored at 25°C and –5°C for 60 weeks. Samples were removed for bioassay 6 times at intervals of 12 weeks. Bioassay studies revealed that cypermethrin was effective against both species. Although at 8 mg/kg, cypermethrin did not kill 100% of exposed adults of *Cryptolestes ferrugineus*, it prevented production of progeny. Fenvalerate failed to give effective control of *Cryptolestes ferrugineus* but was able to prevent progeny production of *Tribolium castaneum*. In contrast, malathion at 8 mg/kg was ineffective against both the species at 12 and 24 weeks after treatment.

Williams *et al.* (1983) compared the relative toxicity and persistence of 3 pyrethroid insecticides, including fenvalerate, on concrete, wood and iron surfaces for control of grain insects. Permethrin, deltamethrin and fenvalerate were applied at concentrations ranging from 0.01 to 0.5 g/m^2 to blocks of concrete, galvanised iron and wood which are materials commonly used in the construction of silos and other grain storage facilities. Their effectiveness was compared with the current recommended treatment of fenitrothion applied at 1 g/m^2. The treated surfaces were bioassayed with adults of organophosphorus-resistant strains of *Rhyzopertha dominica, Sitophilus oryzae* and *Tribolium castaneum* up to 32 weeks after the initial treatment. Fenvalerate was effective against *Rhyzopertha dominica* on iron and wood for 32 weeks and on concrete for 24 weeks. Against *Sitophilus oryzae* fenvalerate was effective on iron at its higher concentration for 24 weeks. On wood there was considerable variability, only the higher concentration being effective after 16 weeks. The effectiveness after 32 weeks ranged between 50% and 75%. Against *Tribolium castaneum* none of the pyrethroids was as effective as fenitrothion which was fully effective for 16 weeks. On wood all pyrethroids gave erratic results and generally produced lower insect mortalities than fenitrothion. On concrete, only the higher concentrations of deltamethrin and fenvalerate gave over 75% kill of *Rhyzopertha dominica* after 1 week; but against *Tribolium castaneum* and *Sitophilus oryzae* no pyrethroid or fenitrothion treatment was effective. The authors concluded that the higher concentrations of the 3 pyrethroids were significantly more effective than fenitrothion. The choice of an insecticide treatment for storage structures will inevitably depend on its cost-effectiveness.

Ishaaya *et al.* (1983) working in Israel tested the potency of 6 pyrethroids as dietary toxicants and inhibitors of weight gain in first- and fourth-instar larvae of *Tribolium castaneum*. Fenvalerate was third behind *cis*-cypermethrin and deltamethrin. Dosages that reduced larval weight also delayed pupation and emergence, probably due to their anti-feeding activity. Piperonyl butoxide did not synergise fenvalerate.

Hsieh *et al.* (1983) working in Taiwan determined the toxicity of 26 insecticides to *Sitophilus zeamais* and *Rhyzopertha dominica* following admixture with grain. Fenvalerate was one of the few compounds that was more toxic to *Rhyzopertha dominica* than to *Sitophilus zeamais*.

Bitran *et al.* (1983b) evaluated the residual action of some pyrethroid and organophosphorus insecticides in the control of *Sitophilus zeamais* in Brazil. Insecticides were mixed with maize grain and evaluated for 9 months. Fenvalerate plus piperonyl butoxide showed a high efficacy in control but this effectiveness was reduced when fenvalerate was used without piperonyl butoxide. Cypermethrin presented a better residual action than fenvalerate when applied without piperonyl butoxide.

Bengston *et al.* (1983b) evaluated a number of organophosphorothioates and synergised synthetic pyrethroids as grain protectants on bulk wheat in duplicate field trials carried out in commercial silos in Queensland and New South Wales. Fenvalerate at 1 mg/kg along with fenitrothion at 12 mg/kg and piperonyl butoxide at 8 mg/kg controlled common field strains of *Sitophilus oryzae* and *Rhyzopertha dominica* and completely prevented progeny production in *Tribolium castaneum, Tribolium confusum* and *Ephestia cautella*. In a field experiment carried out on bulk sorghum stored for 26 weeks in concrete silos in south Queensland, **Bengston *et al.* (1984)** found the same combination of fenvalerate controlled typical malathion-resistant strains of *Sitophilus oryzae, Rhyzopertha dominica, Tribolium castaneum* and *Ephestia cautella*.

Joia *et al.* (1985b) studied the initial effectiveness and residual activity of fenvalerate applied at 8 and 12 mg/kg to wheat of 13.3% and 15% moisture content using *Tribolium castaneum* and *Cryptolestes ferrugineus* as test species and malathion at 8 mg/kg as standard for comparison. Treated wheat was stored at 25°C and –5°C, and sampled at 6 intervals during a 60-week storage period. Bioassays using adult mortality and production of F_1 progeny as criteria revealed that fenvalerate caused only 33% mortality of *Cryptolestes ferrugineus*. At the 2 moisture contents, fenvalerate caused 63.3% and 66.7% initial mortality of *Tribolium castaneum* at 8 mg/kg and 90–100% at 12 mg/kg. Production of progeny was prevented under all conditions.

15.2 Degradation

Noble *et al.* (1982) studied the stability of 4 pyrethroids, permethrin, phenothrin, fenvalerate and deltamethrin on wheat in storage. The moisture content of 2 samples of wheat was carefully adjusted to ensure that they were exactly 12% and 15% respectively. Replicate samples were treated with diluted fenvalerate emulsion at a rate sufficient to deposit 1 mg/kg on the grain. The treated samples were maintained at either 25° or 35°C and after being sampled for analysis were maintained under these conditions for 52 weeks. Samples were withdrawn at 13, 26, 39 and 52 weeks and the results were as indicated in **Table 15.1**. From these data, the half-life was calculated to be as indicated in **Table 15.2**. It is clear that under high temperature and high-moisture conditions fenvalerate residues degrade more rapidly but nevertheless they are extremely stable. The stability of the fenvalerate deposit was considerably greater than that of permethrin.

Joia (1983) studied the fate of residues of fenvalerate on stored wheat of 13.3% and 15% moisture content treated at 8 or 12 mg/kg with fenvalerate and stored for 60 weeks at 25°C and –5°C. Residues degraded slowly indicating a half-life of 385 weeks for fenvalerate on wheat of 13.3% moisture **(Table 15.4)** stored at –5°C.

Table 15.1. Concentration of fenvalerate on wheat with 12% and 15% moisture content stored for up to 52 weeks at 25° or 35°C **(Noble *et al.* 1982)**.

Initial concentration (mg/kg)	Temp. (°C)	Moisture (%)	r.h. (%)	Concentration after storage for x weeks — (mg/kg) 13	26	39	52
1.05, 1.16	25	12	54	1.13	1.19, 1.13	1.13	0.86, 0.90
1.14, 1.08	25	15	73	1.16	1.19, 1.22	1.14	0.88, 0.81
1.18	35	12	57	1.08	1.17, 1.05	0.96	0.71, 0.77
1.14	35	15	75	1.06	0.87, 1.06	0.87	0.64, 0.68

Table 15.2. Pseudo first-order rate constants and half-lives for fenvalerate on wheat with 12% or 15% moisture stored at 25° or 35°C for 52 weeks **(Noble *et al.* 1982).**

Temp. (°C)	Moisture (%)	Pseudo fist-order rate constant (100 k/week)	Half-life (weeks)	95% confidence limits
25	12	3.3	210	120–870
25	15	3.8	182	93–1194
35	12	6.7	104	73–183
35	15	9.3	74	59–101

Bengston *et al.* (1983b) applied fenvalerate (1 mg/kg) along with pirimiphos-methyl (8 mg/kg) to bulk wheat in commercial silos in Queensland and New South Wales. Analysis of samples withdrawn over 9 months storage indicated a relatively slow degradation, as can be seen in **Table 15.3**.

Table 15.3. Fenvalerate residues on treated wheat stored in concrete silos at Natcha, Queensland and Eugowra, NSW **(Bengston *et al.* 1983b).**

Calculated application rate (mg/kg)	Fenvalerate residue (mg/kg) at various times after treatment (months)						
	0.25	1.5	3	4.5	6	7	9
Natcha — 0.9	0.86	0.70	0.79	0.73	0.92	–	0.74
Eugowra — 1.1	1.16	–	1.01	0.79	0.80	0.80	–

In a similar series of field trials with organophosphorus and synergised synthetic pyrethroid insecticides on stored sorghum, **Bengston *et al.* (1984)** showed that fenvalerate applied to sorghum of 11.8% moisture content held in 500 tonne silos at a mean grain temperature of 22.3°C suffered no significant degradation over 26 weeks other than that occurring in the first week. In addition to chemical analysis, bioassay confirmed that there was no significant degradation.

Sarkar *et al.* (1984) studied the residual toxicity and persistence of fenvalerate on jute bags and its persistence on wheat grains. Fenvalerate applied at 1 or 1.5 g/m^2 on jute bags was tested against *Tribolium castaneum* and its persistence on wheat grains inside the treated bags was investigated in New Delhi. Mortality varied from 100% after 1 month to 11.8% after 3.5 months from treatment, and the residues on the wheat grains decreased by up to 88% in the same period.

Joia *et al.* (1985b) reported fuller details of the study **(Joia 1983)** referred to above. Fenvalerate was applied at 8 and 12 mg/kg to wheat of 13.3% and 15% moisture content. Treated wheat was stored at 25°C and –5°C for 60 weeks and sampled each 12 weeks for residue analysis and milling. From an initial fenvalerate level of 8.2 mg/kg on wheat of 13.3% moisture content, the residue decreased to 6.1 mg/kg after 60 weeks at 25°C and to 7 mg/kg at –5°C. The level of 12.8 mg/kg fenvalerate decreased to 9.2 and 10.7 mg/kg after storage for 60 weeks at 25°C and –5°C, respectively. The rate of decline in fenvalerate residues on 15% moisture content wheat was faster than on 13.3% moisture content wheat. Thus, from an initial level of 8.1 mg/kg, fenvalerate residues declined to 4.3 mg/kg after 60 weeks storage at 25°C and to 4.5 mg/kg after storage for the same length of time at –5°C. Similarly, the residues from the level of 12.8 mg/kg decreased to 6.9 and 8.4 mg/kg, after 60 weeks storage at 25°C and –5°C respectively. The half-lives calculated from these data are given in **Table 15.4.**

15.3 Fate in Milling, Processing and Cooking

Bengston et al. (1983b) subjected wheat which had been treated with fenvalerate (1 mg/kg) and piperonyl butoxide (8 mg/kg) for 10 months in concrete silos in Queensland to standard milling and baking procedures. They found that 8.5% of the fenvalerate residues on the wheat carried through into the white flour and that none of this was apparently destroyed in bread making. The details results are given in **Table 15.5.**

Joia 1983 and **Joia *et al.* (1985a)** applied fenvalerate to 2 kg lots of wheat. Treated wheat was stored at 25°C for 4 weeks. Samples were milled to obtain white flour. Wholemeal flour was prepared by mixing the fractions obtained from milling. White bread and wholemeal bread was baked from these flours. As much as 88% (expressed on a moisture-free basis) of the fenvalerate in the white flour was found intact in the white bread and 87% in the wholemeal bread. No difference was observed in bread weight, volume, texture, and taste between treated and control breads. Residues were determined in wheat and milled fractions, *viz.*, bran, middlings and flour. It was observed that highest amounts of fenvalerate were present in bran and least in flour. Fenvalerate degraded on treated

Table 15.4. Half-lives of fenvalerate residues on wheat of 13.3 and 15% moisture content stored at 25°C or –5°C for 60 weeks **(Joia *et al.* 1985b).**

Treatment Level (mg/kg)	Temp. (°C)	Moisture (%)	Pseudo fist-order rate constant (1000/week)	Half-life (weeks)	95% confidence limits
8	–5	13.3	1.8	385	231–1155
12	–5	13.3	2.5	277	182–578
8	25	13.3	4.1	169	133–231
12	25	13.3	4.7	148	115–204
8	–5	15.0	8.9	78	59–114
12	–5	15.0	6.8	102	77–157
8	25	15.0	9.7	72	61–87
12	25	15.0	10.0	69	57–88

Table 15.5. Fenvalerate residues following the milling and baking of wheat stored in concrete silos for 10 months after application of insecticide **(Bengston et al. 1983b).**

Commodity	Mean residues (mg/kg) 'as-is basis'	Distribution (%)	Mean residues (mg/kg) 'moisture-free basis'
wheat	0.7	100	0.63
bran	3.3	68.5	3.0
pollard	1.5	23.0	1.35
white flour	0.08	8.5	0.09
white bread	0.06	–	0.10
wholemeal flour	0.70	–	0.80
wholemeal bread	0.49	–	0.73

wheat at slow rates. Half-lives of fenvalerate on grain ranged from 385 weeks on wheat of 13.3% moisture content stored at –5°C to 69.3 weeks on wheat of 15% moisture content stored at 25°C. The distribution of fenvalerate residues on wheat and milled fractions immediately after treatment at 8 and 12 mg/kg is shown in **Table 15.6.** The percentage residue degradation of fenvalerate in wheat and milled fractions during 60-week storage of wheat at 13.3% and 15% moisture content is illustrated in **Figure 15.1.**

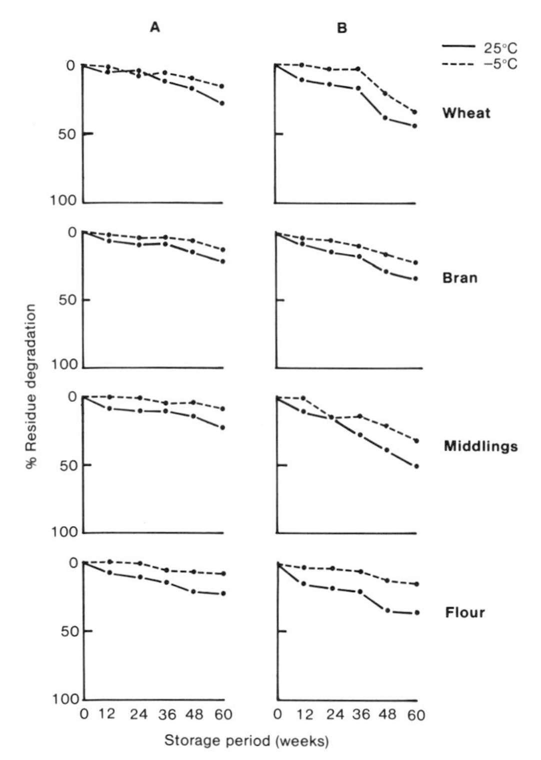

Figure 15.1. Percent residue degradation of fenvalerate (applied at approximately 12 mg/kg) in wheat and milled fractions during 60-week storage of wheat of 13.3% (A) and 15.0% (B) moisture content **(Joia *et al.* 1985a).**

15.4 Metabolism

The metabolic fate of fenvalerate has been examined in rodent species following acute and sub-acute oral and dietary administration. In all cases, the metabolism in both rats and mice and elimination of the metabolic components was rapid. Fenvalerate undergoes several major metabolic reactions; cleavage of the ester linkage, hydroxylation in the acid and alcohol moieties and conversion of the CN group to SCN and CO_2. The resulting metabolite acids and phenols were subsequently conjugated with glucuronic acid, sulphuric acid or amino acids. The reactions were similar in both rats and mice with differences being the nature of the conjugating material and the quantitative excretion of certain metabolites. Taurine was found to conjugate with 3-

Table 15.6. Distribution of fenvalerate residues on wheat and milled fractions immediately after treatment at 8 and 12 mg/kg **(Joia *et al.* 1985a).**

Moisture content of wheat (%)	Dosage (mg/kg)	Residues (mg/kg) Whole grain	Bran	Middlings	Flour
13.3	8	8.2	28.2	15.1	1.5
	12	12.8	39.7	23.2	2.1
15.0	8	8.1	27.1	11.1	1.1
	12	12.8	43.6	16.0	1.6

phenoxybenzoic acid, representing 10–13% of the dose in mouse urine. This conjugating mechanism was not observed with rats. In both species, major hydroxylation reactions were noted to occur in the 4' position of the phenoxybenzoic acid. Hydroxylation also has been noted in the 2' position. Several species differences in rats and mice were observed with respect to hydroxylation on the phenoxybenzoic acid. **Figure 15.2** gives a schematic of the metabolite's conjugating mechanisms observed in animals. Hydroxylated fenvalerate was detected in the feces of both rats and mice **(Boyer 1977b; Kaneko *et al.* 1979; Ohkawa *et al.* 1979; Kaneko *et al.* 1981a; Lee *et al.* 1985)**. The liver of rats fed fenvalerate for 28 days was analysed for residues and found to contain 3-phenoxybenzoic acid and the corresponding 4'-hydroxylated derivative **(Boyer 1977c)**. Sub-cellular fractions of rat liver have been shown to degrade fenvalerate, yielding a wide variety of products, many of which have been detected in urine and as conjugated products. The most widely noted were 3-phenoxybenzoic acid and its 4'-hyrdroxylated derivatives as well as the corresponding isovaleric acid **(Boyer 1976)**, which are less toxic than fenvalerate.

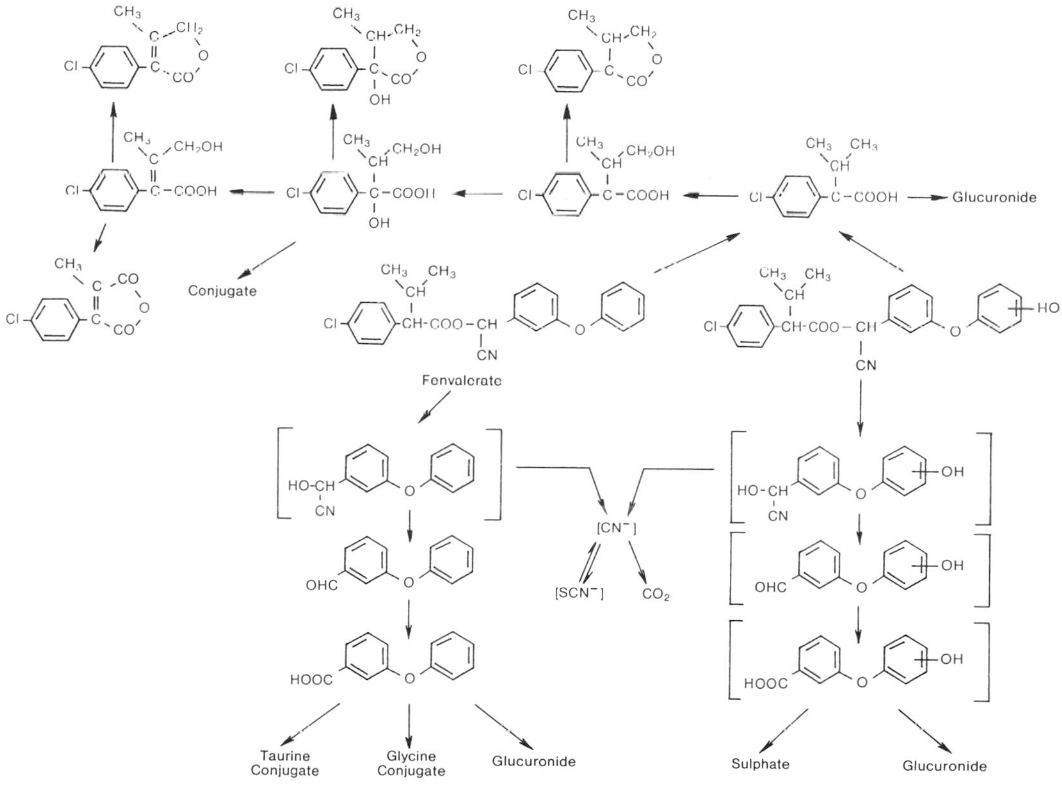

Figure 15.2. Fenvalerate metabolism with conjugating mechanisms in animals.

Casida *et al.* (1979) discussed the comparative metabolism of pyrethroids derived from 3-phenoxybenzyl alcohols covering most of the above points.

Miyamoto (1981) reviewed the chemistry, metabolism and residue analysis of synthetic pyrethroids, including fenvalerate, and dealt with its degradation in light, metabolism in plants, animals and soil and by soil micro-organisms. This extensive paper should be consulted for details.

Leahey (1985) reviewed the metabolism and environmental degradation of fenvalerate, presenting the information also in diagramatic form.

15.5 Fate in Animals

Fenvalerate, orally administered to rats and mice, was found to be rapidly absorbed, distributed to a variety of tissues and organs, metabolised and excreted from the body. The half-life for excretion in both rodent species was 0.5–0.6 days. Elimination of the CN-labelled fenvalerate was somewhat slower in both species, suggesting a different pattern of metabolism. Tissue residues following acute administration were extremely low with the highest concentration being absorbed in fat. High concentrations of the CN-labelled fenvalerate were noted in the hair, skin and stomach content which may account for the data showing that residues of this label were more slowly excreted from the body **(Kaneko *et al.* 1979, 1981a; Ohkawa *et al.* 1979)**.

Rats fed 20 ppm in the diet for 28 days were sacrificed and residues in adipose tissue were examined. Based upon chromatographic and mass spectral analysis, the residue in fat was characterised as unchanged fenvalerate containing both distereo isomers **(Boyer 1977a)**.

Male and female rats fed fenvalerate at the rate of 20 ppm for 28 days and placed on control diets for an additional 28 days were examined for tissue residues and/or depletion rates. Maximum residues were reached within 3 weeks of dietary administration. Of the tissues measured, adipose tissue contained the highest residue. Trace amounts were observed in other tissues after 28 days of treatment. Dissipation of residues from all tissues following the cessation of treatment was rapid, although from adipose tissue the dissipation was slower than with other tissues. At 28 days after the cessation of dietary fenvalerate administration, residues were still reported in adipose tissue, attesting to the slow clearance from this storage depot **(Potter and Arnold 1977; Potter 1976a).**

Kaneko *et al.* (1984) showed that following a single oral administration of ^{14}C-chlorophenyl-ring and ^{14}C-phenyl-ring-fenvalerate ((RS)-*a*-cyano-3-phenoxybenzyl-(RS)-2-(4-chlorophenyl)-3-methylbutylate) preparations to male dogs at 1.7 mg/kg, the radiocarbon from the acid and alcohol moieties was eliminated from the animal bodies within 3 days to the extent of 87.1% and 79.1% respectively. The biological half-life for fenvalerate in the blood was about 2.0 hr, and the level of fenvalerate decreased below the detection limit (0.01 ppm) in 48 hr after dosing. Fenvalerate was metabolised mainly by oxidation at the 4'-phenoxy position of the alcohol moiety and at the C-2 and C-3 positions of the acid moiety, cleavage of the ester linkage and conjugation of the resultant carboxylic acids, phenols and alcohols with glucuronic acid, sulphate and/or amino acid. Species differences were found between dogs and rodents such as rats and mice: 1) hydroxylation at the 2'-position of the alcohol moiety occurred in rats and mice, but not in dogs. 2) 3-phenoxybenzyl alcohol and 3-(4'-hydroxyphenoxy) benzyl alcohol were obtained from dogs to a considerable extent, but not detected in rats and mice. 3) 3-phenoxybenzoylglycine was the predominant conjugate of the alcohol moiety in dogs, but a minor one in rats and mice. 4) glucuronides of the acid moiety and its hydroxy derivatives were obtained from dogs to a larger extent than rats and mice.

Three lactating cows were fed daily doses of ^{14}C-fenvalerate at a level equivalent to 0.15 ppm in the total diet for 21 days and sacrificed 12 hours after receiving the last dose. Recovery of radio-carbon was about 65% of the administered dose of which 29% was recovered in the urine, 32% in the feces and less than 1% in the milk. The radio-carbon in the milk ranged from less than 0.0006 to 0.0019 mg equivalents/*l* and reached a plateau after 1 week on the treated feed. Nearly all the radio-carbon in the milk was present as unchanged fenvalerate. No detectable radio-carbon was found in the brain, fat, kidney, liver, lung and muscle. The limit of determination in this study was 0.007 mg/kg for fat and 0.004 mg/kg equivalents for other tissues **(Potter 1976a)**. In a separate study, 3 lactating cows were fed 10.9 ppm ^{14}C-fenvalerate in their total rations for 28 days. The ^{14}C-residues in the milk collected during the last 24 days of treatment ranged from 0.04 to 0.13 mg/kg equivalents. Fenvalerate content of the whole milk ranged from 0.037 to 0.082 mg/kg during the last 20 days of the test. The cows were sacrificed 12 hours after receiving the last dose. The highest tissue residues were found in the mesenteric fat (0.74–0.79 mg/kg equivalent). Muscle contained essentially no residues. One sample of 6 contained 0.06 mg/kg equivalents. A total of 92% of the ^{14}C-administered to the cows was recovered in the excreta and tissues **(Potter and Arnold 1977; DeVries 1976)**.

Three lactating Guernsey cows were fed daily doses of ^{14}C-fenvalerate at 0.15 ppm in their total ration for 21 days. The cows were sacrificed 12 hours after receiving the last dose. The ^{14}C in the milk ranged from 0.0006 to 0.0018 mg/kg equivalents. The ^{14}C residues in the milk reached a plateau after 1 week on the treated feed. Given the experimental error of the measurements all of the ^{14}C in the milk was present as fenvalerate. No ^{14}C was found in bone, brain, kidney, lung or muscle (limit of detection equals 0.004 to 0.007 mg/kg equivalents). Traces of ^{14}C were found in mesenteric fat and subcutaneous fat from 2 cows and in the liver from 1 cow **(Potter 1976b)**.

Lactating cows were fed an equivalent of 0.15 ppm ^{14}C-fenvalerate in their daily ration for 21 days. Residues of

0.01–0.02 mg/kg of fenvalerate were found in the cream by TLC and liquid scintillation counting. These residues were confirmed as fenvalerate by GLC-EC **(Potter 1976e)**.

Mature laying hens were fed 0.03 ppm of ^{14}C-fenvalerate in their total ration for periods up to 32 days. No detectable ^{14}C-residues were found in the fat, heart, gizzard, liver, meat, skin, egg whites and egg yolks **(Potter 1976d)**.

In a separate study, mature laying hens were fed 0.03 ppm of ^{14}C-fenvalerate in their total ration for periods up to 32 days. No ^{14}C-residues were detected in the light meat, dark meat, skin, gizzard, blood or plasma. The residues in egg yolks ranged from less than 0.002 to 0.003 mg/kg, and in fat samples less than 0.002 to 0.003 mg/kg **(Potter* and Sauls 1978)**.

McColl (1982) reported the results of a study in which dairy cows were fed with grain containing phenothrin and fenvalerate applied at the rate of 2 mg/kg. Analysis of the grain 48 hours after treatment indicated that only approximately 1 mg/kg of the 2 insecticides was present on the feed. The cows were supplied with this feed for 28 days by which time the residue was determined by analysis to have fallen to approximately 0.5 mg/kg. Milk samples taken from each cow at the end of the feeding period and 3 days after the cows returned to unmedicated food showed no trace of phenothrin or fenvalerate.

15.6 Safety to Livestock

No studies designed specifically to evaluate the safety of fenvalerate when fed to livestock have been located. There are many studies in which fenvalerate has been applied to livestock in the form of sprays and dips, and fenvalerate has been registered for the control of ectoparasites of cattle, sheep, pigs and hens in several countries.

The information reviewed in Section 15.4 and 15.5 indicates that when fenvalerate is orally administered to rats and mice it is rapidly absorbed, distributed to a variety of tissues and organs, metabolised and excreted from the body rapidly with a residence half-life in both species of the order of 0.5 days.

As indicated in **Table 15.7** the acute oral toxicity was influenced to a marked degree by the vehicle in which the insecticide was administered. When administered as a suspension in a mixture of polyethyleneglycol and water, the acute oral toxicity to rats, mice, dogs and hens was quite low.

15.7 Toxicological Evaluation

Toxicological information on fenvalerate was evaluated by JMPR in 1979, 1981 and 1984 **(FAO/WHO 1980, 1982, 1985)**.

Fenvalerate was studied for its effect on the reproduction of rats over 3 generations. With the exception of weight reduction in the third generation parents, there was no effect of fenvalerate on any parameter measured in the study **(Beliles *et al.* 1978; Stein 1977)**.

Teratogenicity studies were carried out in the mouse **(Khoda *et al.* 1976)** and in the rabbit **(van der Pauw *et al.* 1975)** but these showed no potential for inducing teratogenic events.

Studies on mutagenicity were negative. Studies on neurotoxicity have shown that, following high-level exposure, rats showed reversible clinical signs of ataxia. **(Hend and Butterworth 1976; Okuno and Kadota**

Table 15.7. Acute oral toxicity of fenvalerate to various animal species (mg/kg)

Species	Sex	Vehicle	LD/50	Reference
Mouse	M	DMS0*	200–300	**Walker *et al.* 1975**
	F	DMS0	100–200	**Walker *et al.* 1975**
		PEG†: water	1202	**Summit & Albert 1977b**
Rat		DMS0	451	**Walker *et al.* 1975**
		PEG: water	>3200	**Summit & Albert 1977a**
Chinese hamster	M	DMS0	98	**Walker *et al.* 1975**
	F	DMS0	82	**Walker *et al.* 1975**
Syrian hamster		PEG: water	760	**Hart 1976a (FAO/WHO 1980)**
Dog				
Dog		PEG: water or corn oil	Doses from 100 to 1000 mg/kg were emetic	**Hart 1976b**
Hen			>1500	**Milner and Butterworth 1977**

*DMS0 — dimethylsulphoxide
†PEG — polyethyleneglycol

1977). Microscopic examination of the sciatic nerve showed axonal swelling and myelin disruption, and biochemical studies revealed an increase in lysosomal enzyme activity. Fenvalerate, when administered to hens at high levels, did not induce signs of peripheral neuropathy **(Milner and Butterworth 1977)**.

Short-term and long-term studies have been performed in a variety of test animals. Fenvalerate is not a carcinogen and in short-term studies in dogs and long-term studies in rats and mice, dietary no-effect levels have been observed. The level causing no toxicological effect in the mouse was 30 ppm in the diet equivalent to 3.5 mg/kg body weight; 150 ppm in the diet of the rat equivalent to 7.5 mg/kg body weight; and 12.5 mg/kg body weight in the dog. Based on these studies the estimate of the temporary acceptable daily intake for man was 0–0.02 mg/kg body weight.

The JMPR desires the submission of the carcinogenicity studies with fenvalerate commissioned by IARC and NTP; the determination of the no-effect level in the dog with respect to granulomata formation; full details of the recently published 6-month feeding study of fenvalerate in dogs and further observations in occupationally exposed humans.

15.8 Maximum Residue Limits

Based on extensive data from published and unpublished studies on the effect and fate of fenvalerate residues in grain, the following maximum residue limits have been recommended for cereal grains and milled cereal products.

Commodity	Maximum Residue Limits (mg/kg)
Cereal grain	5
Bran	10
Flour (white)	0.5
Wholemeal	2

Residues are to be determined and expressed as fenvalerate.

The methods recommended for the analysis of fenvalerate residues are: **Pesticide Analytical Manual (1979a, 1979c); Ambrus *et al.* (1981)**. The following methods are recognised as being suitable: **Pesticide Analytical Manual (1979d); Baker and Bottomley (1982); Chapman and Harris (1978); Greenberg (1981); Lee *et al.* (1978); Reichel *et al.* (1981); Talekar (1977).**

16. Iodofenphos

Iodofenphos is a contact and stomach insecticide and acaricide for the control of flies and mosquitoes (adults and larvae) in public hygiene, indoor pests such as ants, bed bugs, cockroaches, fleas, spiders and warehouse pests. It is also recommended for farm use, especially in poultry houses against mealworm beetles, mites and flies.

This insecticide is widely mentioned in open literature as having been extensively evaluated as a grain protectant. It has an effect against a broad spectrum of stored-product pests and appears comparable with bromophos. Several years ago the manufacturers decided that the use of this insecticide as a grain protectant would not be further developed.

16.1 Usefulness

Strong and Sbur (1968) were among the first to publish concerning the effectiveness of iodofenphos against a range of stored-product insects. They tested the compound under the code number C9491 finding that sprays of the insecticide applied directly to adults killed 4 of the 5 species but were relatively ineffective against *Trogoderma variabile*. The performance was generally similar to malathion. On the other hand, iodofenphos proved to be more toxic than malathion to *Tribolium castaneum* adults **(Blackman 1969).**

Wilkin *et al.* (1970) applied iodofenphos emulsion to barley in 3 farm trials and a 5% dust in another trial in the UK and found it distinctly promising for protective treatment of grain. **Kane and Aggarwal (1970)** used iodofenphos emulsion on bagged wheat and barley at the rate of 5 and 10 mg/kg and found the treatment still effective after 9 weeks.

Girish *et al.* (1970) published studies on the comparative efficacy and residual toxicity of iodofenphos on concrete slabs and jute bags against *Sitophilus oryzae, Tribolium castaneum, Rhyzopertha dominica* and larvae of *Trogoderma granarium*. Four different doses of the insecticide were tested. The results indicated that iodofenphos has a fairly high contact toxicity for the 4 species and that it is more persistent on both the treated surfaces than malathion. When applied at the rate of 1.5 g/m^2 there is a good immediate kill of the 4 species and sufficient residual effect up to 1 month after treatment.

Wilkin *et al.* (1971) found iodofenphos less effective than malathion against a spectrum of grain pests but to have the advantage of greater stability on warm, damp grain.

Coulon *et al.* (1972) reported iodofenphos, along with bromophos and malathion, greatly inferior to pirimiphos-methyl in residual effect for the control of *Sitophilus granarius*.

In India, 9 insecticides including iodofenphos were screened for effectiveness against adults of *R. dominica* and *S. oryzae* and larvae of *Trogoderma granarium*. Iodofenphos, bromophos, phoxim and pirimiphos-methyl proved promising and were tested as protectants of wheat grain in a small-scale storage trial lasting 1 year **(Chawla and Bindra 1973)**. At comparable doses, the first 2 provided protection for a shorter period, whereas the last 2 provided protection for a longer period than malathion. The pattern of dissipation of iodofenphos was similar to that of malathion, but the rate of dissipation of pirimiphos-methyl was much slower.

Wilkin and Hope (1973) tested iodofenphos along with 20 other insecticides against 3 species of stored-product mites. Iodofenphos was effective against 2 of the 3 species only after 14 days exposure. Against the third species, *Acarus siro*, it was only 25% effective at the end of 14 days.

Rai and Croal (1973), working in Guyana, tested 9 insecticides for their duration of effectiveness on jute sacking. Iodofenphos was among the more effective though it was rather slow to act against the paddy moth, *Sitotroga cerealella*.

McCallum-Deighton (1974) quoting the minimum effective dose in mg/kg for technical-grade insecticides against susceptible and organophosphorus-resistant strains of *Tribolium castaneum*, indicated iodofenphos to be effective at 2–3 mg/kg against susceptible strains and 7.5–20 mg/kg against resistant strains. This is 4 to 10 times the minimum effective dose of pirimiphos-methyl.

Bindra (1974) reported trials in India where 9 insecticides, including iodofenphos, were screened for effectiveness against adults of *Rhyzopertha dominica* and *Sitophilus oryzae* and larvae of *Trogoderma granarium*. Iodofenphos proved promising and was tested in a small-scale storage trial lasting 1 year. At comparable doses it protected wheat for a shorter period than did malathion.

Chawla and Bindra (1976) compared the relative toxicity of 7 organophosphorus insecticides, including iodofenphos, in their commercial formulations, against adults of *R. dominica* and *S. oryzae* and larvae of *Trogoderma granarium* under controlled laboratory conditions. Iodofenphos was comparable to malathion. No significant change in relative toxicity of insecticides was noted when the exposure period was increased from 2 to 7 days.

Tyler and Binns (1977) reported studies on the toxicity of 7 organophosphorus insecticides and lindane to 18 species of stored-product beetles. Iodofenphos proved equal or superior to malathion in knock-down after 24 hours and in percentage kill it was superior to malathion against all species though it was not completely effective against *Rhyzopertha dominica* and *Ptinus tectus*. The authors ranked iodofenphos fifth in order of effectiveness.

Adesuyi (1978) determined the effectiveness of different doses of iodofenphos against *Sitophilus zeamais* on maize on the cob in cribs in Nigeria. It was found that iodofenphos, applied at 5, 10 and 15 mg/kg, was effective for 5, 8 and 10 months respectively.

Mensah and Watters (1979b) compared 4 organophosphorus insecticides on stored wheat for control of susceptible and malathion-resistant strains of *Tribolium castaneum*. Iodofenphos proved equal to malathion and bromophos in toxicity and residual effectiveness. It was more effective than malathion in preventing reproduction of the resistant strains.

Adesuyi (1979), in an assessment of new insecticides for use in maize storage at the farmer level in Nigeria, presented detailed bioassay results from field trials to show that iodofenphos applied to maize in farm cribs at rates equivalent to 5, 10 or 15 mg/kg was outstandingly effective in preventing insect damage for 8 months. The results of these trials showed that iodofenphos, applied at 10 mg/kg, can be used by Nigerian farmers to treat their maize stored in cribs as they do not generally store for longer than 8 months.

In evaluating grain protectants for use under tribal storage conditions in Zimbabwe, **Weaving (1980)** screened insecticide dusts as admixtures with maize in 2 small-scale storage trials over a 12 month period. Fenitrothion (8 mg/kg) and fenthion (8 mg/kg) showed the greatest persistence followed by malathion, iodofenphos and tetrachlorvinphos (all at 8 mg/kg) in decreasing order of effectiveness.

O'Donnell (1980) evaluated the toxicity of wettable powder formulations of malathion, pirimiphos-methyl, fenitrothion and iodofenphos to *Tribolium confusum* in constant conditions of 25°C, 70% r.h. This was compared to their toxicity when exposure was in a diurnal cycle of temperature and humidity varying between 5°C, 80% r.h. and 10°C, 79% r.h. All insecticides were more effective at the higher temperature. The effective dose (ED/50) values obtained in the 2 sets of environmental conditions differed by factors of less than 10 in the cases of malathion and pirimiphos-methyl, and more than 20 in the cases of fenitrothion and iodofenphos. The minimum doses required to kill all the insects tested in the colder conditions were as follows: pirimiphos-methyl 232 mg/m^2, fenitrothion 465 mg/m^2, malathion 1395 mg/m^2 and iodofenphos greater than 1395 mg/m^2. It is clear from the results that iodofenphos would be unlikely to be effective in practical conditions where the temperature is very low, for example, in a ship's hold in winter.

Hindmarsh and Macdonald (1980) conducted field trials to control insect pests of farm-stored maize in Zambia. Severe infestations of *Sitophilus zeamais* and *Sitotroga cerealla* occurred in maize cobs stored under farm conditions in Zambia causing up to 92% damaged grains 8 months after harvest. Shelling the maize reduced *S. cerealla* damage to low levels and also stabilised the moisture content of the grain, providing a more suitable substrate for insecticide application and control of *S. zeamais*. Iodofenphos applied at the rate of 12 mg/kg as a dust admixture to shelled maize in traditional mud-walled, timber and thatch storage cribs kept damaged grains below 10% up to 10 months after harvest.

Viljoen *et al.* (1981a) evaluated a range of contact insecticides for the control of pests of maize and groundnuts in bag stacks. Small bag stacks of maize and groundnuts in the open and in a shed in South Africa were sprayed with various contact insecticides directly after fumigation and further at intervals of 4 weeks. Iodofenphos, though somewhat slower acting, was very effective against *Ephestia cautella* and *Plodia interpunctella*. Though slightly less effective against *Tribolium castaneum*, iodofenphos was still vastly superior to malathion.

Weaving (1981) tested 7 insecticide mixtures under field conditions for protecting maize stored in traditional tribal grain storage bins in Zimbabwe. Iodofenphos proved to be one of the most effective and most persistent. The author makes the statement 'During the trials, iodofenphos was withdrawn for possible use as an admixture.' There is no explanation for the statement.

Tyler and Binns (1982) demonstrated the influence of temperature on the susceptibility of 3 stored-product beetles to 8 organophosphorus insecticides. The insecticides were applied as a range of deposits, from 10 to 5000 mg/m^2, to which the adult insects were exposed at 10°C, 17.5°C and 25°C. Based upon knock-down and kill, the effectiveness of all insecticides was greater at 25°C than 17.5°C and was markedly lower at 10°C. At 10°C iodofenphos was virtually ineffective against *S. granarius* even at 5000 mg/m^2. By contrast, at 25°C 100 mg/m^2 was adequate to give complete knock-down of all species.

Wohlgemuth (1984) tested iodofenphos dust on sorghum at the rate of 20 mg/kg under laboratory conditions designed to simulate tropical storage. The treated sorghum was divided into several samples, one for immediate investigation, the others for storage at 36°C and 50% r.h. for 1, 3, 6, 9, 12, 18 and 24 months. The efficiency was tested against 5 species of stored-product beetles. Iodofenphos was fully effective against some species for up to 24 months and was particularly effective in reducing reproduction. However, against *Rhyzopertha dominica* it was insufficently effective immediately and failed to prevent reproduction after 1 month.

16.2 Degradation

Chawla and Bindra (1971) reported a study in which iodofenphos was evaluated at 10, 20 and 40 mg/kg doses as a protectant of maize grain against 3 species of stored-product insects. Residues from 20 and 40 mg/kg applications were below 10 mg/kg after 5 and 6 months respectively, and from the 10 mg/kg dose these were below 5 mg/kg after 6 months.

Iodofenphos is reported to have greater stability than malathion on warm, damp grain **(Kane *et al.* 1971; Wilkin *et al.* 1971)**.

Mensah *et al.* (1979a) studied the translocation of malathion, bromophos and iodofenphos into stored grain from treated structural surfaces. Wheat, barley and corn were brought into contact with wood and concrete surfaces that had been previously treated with the insecticide at the rate of 1 g/m^2. Iodofenphos appeared to be somewhat less stable or less persistent that either malathion or bromophos when measured by the toxicity of the grain brought into contact with the treated surface. Alternatively, it is just possible that iodofenphos does not translocate from treated structural surfaces to the same extent as the other compounds.

Mensah *et al.* (1979b) studied the residues of iodofenphos, along with bromophos, malathion and pirimiphos-methyl in milled fractions of dry or tough wheat following storage for 1, 3 and 6 months. They found that iodofenphos degraded at about the same rate as malathion, more quickly than bromophos and considerably more rapidly than pirimiphos-methyl. Degradation was increased significantly by the higher moisture content of tough wheat (16% moisture).

Weaving (1980) measured many parameters in an extensive evaluation of 5 insecticides for use as grain protectants under tribal storage conditions in Zimbabwe. By all of these parameters, iodofenphos appeared to persist in a manner almost identical to that of malathion but significantly inferior to fenitrothion. The study extended over 12 months.

During extensive field trials to control insect pests of farm-stored maize in Zambia, **Hindmarsh and Macdonald (1980)** determined the insecticide residue level on shelled maize during 10 months storage in farm cribs. Over this period the iodofenphos residue declined steadily from 7.1 mg/kg to 4.0 mg/kg, the level after 6 months being 5.9 mg/kg. The mean moisture content of the maize varied from 8.8% to 14.3% according to prevailing climatic conditions. The iodofenphos was applied in the form of a 2% dust at a nominal concentration of 12 mg/kg.

16.3 Fate in Milling, Processing and Cooking

Chawla and Bindra (1971), who had treated maize grain at the rate of 10, 20 and 40 mg/kg with iodofenphos, found that when the treated grain was processed into chapaties and popcorn there was a loss of 94% to 99% of the insecticide. Washing of maize grain with water removed 70–88% of the residue. In another study **(Chawla and Bindra 1973)** it was found that washing of wheat grain followed by drying removed most of the iodofenphos residues. Addition of a surfactant at the rate of 0.1% enhanced the residue-removal efficiency of water significantly, and sun drying proved more effective than oven drying.

Mensah *et al.* (1979b) treated 2 batches of wheat (12% and 16% moisture content) with 2 rates of iodofenphos (10 mg/kg and 15 mg/kg) and after storage for 1, 3 and 6 months subjected portions to milling to obtain bran, middlings and flour. They analysed each fraction by GLC. In the case of dry (hard) wheat the residues in bran were about 50% higher than in middlings being from 3.3 to 6.3 times the concentration in the whole wheat. In moist (tough) wheat the residue concentration in bran was again about 50% higher than in middlings being from 4 to 8 times the concentration in the whole wheat. In all samples the concentration of iodofenphos in flour was less than 1.0 mg/kg being generally about 10% of the residue in the whole wheat. Though the residue on the whole wheat declined to about one-third between 1 and 6 months storage, the level in flour hardly changed at all suggesting that the degradation was occurring almost entirely in the surface layers of the kernel.

16.4 Metabolism

No information on the metabolism of iodofenphos was available.

16.5 Fate in Animals

No information on the fate of iodofenphos in animals was available.

16.6 Safety to Livestock

No information on the safety of iodofenphos to livestock was available.

16.7 Toxicological Evaluation

Iodofenphos has not been proposed for consideration by the Joint FAO/WHO Meeting on Pesticide Residues and so no independent toxicological evaluation is available.

16.8 Maximum Residue Limits

Iodofenphos has not been evaluated by the Joint FAO/WHO Meeting on Pesticide Residues and therefore no maximum residue limits have been recommended.

No recommendations have been made for methods for the analysis of iodofenphos residues.

17. Lindane

Lindane was the first synthetic insecticide widely used as a grain protectant. It was used extensively in many countries for many years and is still in use in some tropical countries. It is one of the most potent of all the insecticides considered as grain protectants. It is much more effective against mites and moths than the organophosphorus insecticides. Its efficacy is assisted by the biological activity of lindane vapour. Normal rates of application remain effective for long periods (2–5 years) but due to the exceptionally good penetrating effect, residues carry over into milled products. Much of the residue in flour or milled rice is lost in cooking. Use of lindane as a grain protectant is not favoured by some authorities because of questions about toxicity, but the high potency, reliability, stability and low cost have made lindane extremely valuable in many of the poorer countries where losses would otherwise be considerably higher. Doubts about toxicity have been largely removed by the latest evaluation of toxicological data.

An extensive monograph on the properties, uses, effect and fate of lindane has been published and a number of supplements to the original volume have been issued **(Ulmann 1972).**

17.1 Usefulness

Raucourt (1945) and **Slade (1945)**, more or less simultaneously, brought the insecticidal activity of benzene hexachloride (BHC) to the attention of the world. The original manufacture of the molecule by photochlorination of benzene is ascribed to Michael Faraday in 1825. The insecticidal effect appears to have been noticed in approximately 1939. Grant and Imperial Chemical Industries were granted a British patent for BHC as an insecticide in 1939, and Hardie a US patent in 1940. **Slade (1945)** made it clear that considerable information was known about the molecular structure of BHC, including recognition of the *gamma* isomer as the most highly insecticidal form. Later the *gamma* isomer in an isolated and purified state became referred to as lindane. BHC was assessed for use in the control of stored-product pests soon after its insecticidal nature was recognised. **Slade (1945)** referred to its use against *Sitophilus granarius* but reported *Ephestia kuehniella* to be tolerant.

Papers by **Cherian and Rao (1945), Burke (1946), Nasir (1946), Carpenter (1947), Ghosh (1947), Owen (1947), Srivastava and Wilson (1947), Block (1948a, b),** and **Smallman (1948)** all referred to the direct use of BHC either in food storage areas or in stored grain products for their protection.

The following references make it abundantly clear that in all climates, but particularly in the tropics, lindane has been widely tested, recommended, and used commercially for grain protection by application to packages, by use in thermal vaporisers and aerosols, and by direct addition to stored grains. A consensus appears to have emanated from the English laboratories in the 1950s that 10 mg/kg, in at least certain products, was a maximum permissible dosage. *Oryzaephilus surinamensis* is not in general reported as adequately controlled. *Trogoderma granarium* larvae are not readily controlled and resistance in other species has been shown to have developed. **O'Farrell *et al.* (1949)** found the persistence of BHC was shortened on alkaline surfaces but **Armstrong and Hill (1959)** reported the opposite effect. They stated that, applied to warehouse walls in a whitewash, it had prolonged effectiveness.

Srivastava and Wilson (1947) demonstrated that technical benzene hexachloride had a potent fumigant and contact effect against a variety of stored-product pests.

Hewlett and Clayton (1952) recommended 0.4% lindane as the upper limit for use in admixture with pyrethrins for *Ptinus tectus, Tribolium confusum* and *Tribolium castaneum* control but recognised that this level was not satisfactory for *Oryzaephilus surinamensis.*

Parkin and Bills (1953) reported on lindane's ineffectiveness against some species of grain insects.

Atkins and Greer (1953) reported successful protection of flour from treating bags with lindane. They found that flour held for 6 months at 20°C in jute bags impregnated with 0.1% by weight of lindane acquired a constant 4.2 mg/kg of lindane.

In Argentina in 1953 Cristobal described the use of synthetic insecticides, including lindane, for the preservation of stored grains.

Floyd and Smith (1953) and **Chao Yung-Chang and Delong (1953)** both demonstrated the usefulness of pyrethrum and lindane in the protection of corn and rough rice.

Fiedler (1954) used lindane dust for the protection of stored sorghum. He considered the effect to be largely a fumigant one enhanced by temperature.

Nasir (1954) reported a direct fumigant action from lindane and a tendency to transfer from treated surface to adjacent grain or flour in amounts which affected *Sitophilus, Sitotroga, Tribolium* and the eggs of *Sitotroga.*

Muthu and Pingale (1955) reported on lindane impregnated into jute bags for grain protection. They recommended lindane synergised with sesame oil applied at the rate of 15 mg/sq ft and stated that 25 to 50 mg in the bag did not contaminate grain but did contaminate wheat flour. Treatment gave protection but not control.

Starks and Lilly (1955) appeared to be among the first to demonstrate the usefulness and effect of lindane applied as a seed treatment on corn.

Floyd and Newsom (1956) used lindane-impregnated sawdust to protect stored corn from weevils in Louisiana. A dose of 1 mg/kg protected the corn against weevils for 12 months. A dose of 10 mg/kg gave control of *Sitotroga cerealella* and other pests as well.

Ida and Katsuya (1956) studied the effect of lindane fumigation on the optimum period to control insect injury to stored cereals.

Walkden and Nelson (1958) summarised tests by more than a dozen investigators of different disciplines on various aspects of the use of lindane as a grain protectant for stored wheat and shelled corn. Grain was treated in circular 90 tonne metal bins and sampled monthly. Wheat was treated with 2.5 to 5 mg/kg lindane. Control bins became weevily in an average of 11.2 months. Those treated with 2.5 mg/kg lindane were protected for an additional 9.3 months. A dose of 5 mg/kg was not quite as good as 2.5 mg/kg (sic!). The authors concluded that up to 10 mg/kg failed to provide adequate protection from all insect infestations for 1 year.

Gunther *et al.* (1958) treated wheat with lindane in ethylene dichloride to leave 4, 8 and 12 mg/kg of lindane on the grain. When the treated grain was stored in a sealed can in the dark there was no appreciable reduction of residue measured chemically, nor of effectiveness against *Sitophilus granarius* or *Rhyzopertha dominica*, over a 15 month period.

Kockum (1958), Passlow (1958), Walkden and Nelson (1958), and **Armstrong and Hill (1959)** reported on the use of lindane for the control of insects attacking stored grain in East Africa, Australia, USA, and the United Kingdom respectively.

A number of workers, including **Coaker (1959), Davies (1959, 1960), Prevett (1959)** and **Green and Kane (1960)** reported on the value of lindane for treating grain in storage in various parts of the African continent. **Coaker (1959)** proposed a novel method of injecting a small quantity of lindane emulsion into sacks of maize. Other workers investigated the usefulness of lindane in treating maize cribs under tribal conditions.

LePelly and Kockum (1954) reported 5 experiments in the use of insecticides against *Sitophilus oryzae* in stored maize in Kenya. Lindane at 1 mg/kg was completely effective for 16 months whilst 0.5 mg/kg had some protectant action but was not fully effective.

Parkin (1960) carried out a series of tests to determine the susceptibility of stored-product insects to contact insecticides including lindane. His results **(Table 17.1)** show that though some stored-product beetles are extremely susceptible to lindane, *Tribolium castaneum* and especially *Oryzaephilus surinamensis* are substantially tolerant.

Table 17.1. Susceptibility of stored-product insects to lindane **(Parkin 1960).**

Species	Approx. dose for 100% kill in 13 days (mg/kg)	Approx. dose to prevent breeding (mg/kg)
Oryzaephilus surinamensis	»20	20
Tribolium casteneum	>20	2.5–5.0
Rhyzopertha dominica	2.5	0.6
Sitophilus granarius	0.8	0.7–1.0
Sitophilus oryzae	0.6	0.6
Cryptolestes ferrugineus	0.6	0.4

Armstrong (1961) published a report of a trial in a Liverpool (UK) warehouse where lindane, in a commercial white-wash mixture, applied to walls and ceilings remained effective against *Ephestia elutella* during 7 months. There were indications that lindane/white-wash mixtures alone might control infestations of endemic *Ephestia elutella* in fully loaded warehouses.

Watters (1961) demonstrated the effectiveness of lindane for controlling *Ptinus villiger* in 8 commercial flour storages in western Canada. Lindane proved equally effective as methoxychlor, more effective than pyrethrins and considerably more effective than malathion.

Strong and Sbur (1961), evaluating insecticides as grain protectants against *Sitophilus oryzae, Tribolium confusum*, and *Tribolium granarium*, found lindane slightly less effective than malathion for the first 2 species.

Tsvetkov and Bogdanov (1961) recommended 12% *gamma* BHC at 4 kg/tonne for the protection of seed grain against *Sitophilus granarius*.

Lloyd and Parkin (1963) reported studies on a strain of *Sitophilus granarius* selected for resistance to pyrethrins. When the strain had a resistance of ×34, cross-tolerance tests were carried out with a range of insecticides. The strain was found to only have a resistance of ×5.5 to synergised pyrethrins and ×5.5 to lindane.

Champ and Cribb (1965) reported the results of a study of the first incidence of lindane resistance in *Sitophilus oryzae* in Australia. The level of resistance to lindane was ×80 but there was no significant resistance to other compounds.

Giles (1964, 1965) recommended that unthreshed sorghum grain, which was to be stored for more than 6 months in mud granaries, be treated with 0.8% lindane dust by the sandwich method at a rate of 10 mg/kg for threshed grain. Kochum (1965) demonstrated the usefulness of pyrethrum and lindane formulations for protecting maize in crib storage.

Perti *et al.* (1965) reported the LD/50 of topically applied lindane to be 1 to 31 times that of technical DDT to *Tribolium castaneum* and 23 times that of DDT to *Sitophilus oryzae*.

Champ (1965), Parkin (1965) and **Silva e Sousa (1965)** reported the onset of lindane resistance in several important species of stored-product pests in Australia, England and Spain respectively. In 1968 Champ proposed a test method for detecting insecticide resistance in *Sitophilus oryzae*.

The effectiveness of lindane dusts admixed with maize grains and tested as spray films on filter paper was reported by **Teotia and Singh Rajendra (1969)** as second only to malathion against adult *S. oryzae* and of similar effectiveness to malathion but slightly inferior to carbaryl for adult *Trogoderma granarium*. Even 100 mg/kg proved ineffective against *T. granarium* larvae.

Iordanu and Watters (1969) found lindane to be more toxic to *Tribolium castaneum* and *Oryzaephilus surinamensis* adults at 10°C than at 15°C but the opposite to be the case with *O. mercator* and *Cryptolestes ferruginus* adults.

Watters and Grussendorf (1969) applied wettable powders and oil solutions of lindane and methoxychlor to concrete, wood and metal surfaces and found them to be most persistent on wood surfaces. The uptake of these materials from treated structural surfaces by wheat and other grains was tested and proved rather high, especially the wettable powders.

Kockum (1965) showed that 1% lindane dust applied to maize cobs in cribs in Kenya was outstandingly effective against *Sitotroga cerealella* and *Sitophilus zeamais*.

Longoni and Michieli (1969), working in Italy, compared the effectiveness of phenthioate against that of malathion and lindane against insects affecting stored products. Lindane, used as a dust at 2 and 4 mg/kg, was fully effective for 1 year against *Rhyzopertha dominica, Sitophilus granarius* and *Sitophilus oryzae* and for a few months against *Tribolium confusum* larvae.

Seventeen samples of *Tribolium castaneum* collected from warehouses and stores throughout Malawi were tested for susceptibility to lindane and malathion. **(Pierterse and Schulten 1972)**. In all samples, beetles resistant to lindane were found and malathion resistance occurred in 11 of the samples.

Wilkin and Hope (1973) included lindane and a lindane/malathion mixture in trials against stored-product mites. Lindane was ineffective against *Tyrophagus putrescentiae* though it was effective against *Acarus siro* and *Glycyphagus destructor*. The mixture applied at the rate of 2.5 mg/kg lindane/7.5 mg/kg malathion was quickly effective against all 3 species.

Giles (1973) evaluated lindane in comparison with malathion, tetrachlorvinphos and pirimiphos-methyl. In maize crib trials in Nicaragua, the dust formulation was considered to be more convenient but the effectiveness of lindane was very disappointing since its residual protection was significantly less than the other materials.

Rai and Croal (1973) evaluated 9 insecticides for their duration of effectiveness against *Sitotroga cerealella* infesting rice paddy bags by applying the insecticide as a spray to jute sacking. Phoxim, pirimiphos-methyl, iodofenphos and lindane were more effective than the others. The times required to produce 50% knock-down were 27, 53, 102 and 107 minutes, respectively, when the insecticide was applied at a concentration of 0.1%.

Wilkin (1975a) evaluated a number of insecticides for effectiveness against a number of mites of stored products. The practical value of lindane or lindane/malathion mixtures was considered to be strictly limited due to widespread distribution of lindane-resistant strains of 2 of the mite species.

Wilkin (1975b) reported a number of trials in which farm-stored barley infested with *Acarus siro, Acarus farris, Glycyphagus destructor* and *Tyrophagus longior* was admixed with lindane plus malathion dust. The treatment controlled the infestation and protected the grain for 3 months in spite of the barley having a moisture content of more than 18%. Lindane dust or malathion dust, applied separately, were ineffective. Pirimiphos-methyl dust was more persistent.

Wilkin and Haward (1975) reported a series of laboratory tests on 3 stored-product mites to study the effects of temperature on the effectiveness of 5 insecticides. Lindane plus malathion was used as the standard treatment for reference. Reducing the temperature decreased the effectiveness of the compounds tested. With lindane plus malathion against *Acarus siro* the time required to obtain complete kill increased from 4 days at 25°C to 21 days at 10°C and complete kill was not obtained at 5°C within 21 days. These differences were even more marked with *Tyrophagus putrescentiae* and *Glycyphagus destructor*.

Tyler and Binns (1977) evaluated the toxicity of 7 organophosphorus insecticides and lindane to 18 species of stored-product beetles. Lindane was considered superior to malathion but distinctly inferior to fenitrothion, pirimiphos-methyl, chlorpyrifos-methyl and bromophos except against *Rhyzopertha dominica*. Lindane proved outstandingly superior to all other materials which are relatively ineffective against this species.

Adesuyi and Cornes (1978) in an advisory leaflet, issued by the Nigerian Stored Products Research Institute, recommended pirimiphos-methyl or lindane dusts for treating maize in cribs. When such treatments are applied to dried maize in cribs under conditions of good store hygiene, long-term protection is assured.

Yadav *et al.* (1979) tested the efficacy of DDT, lindane and malathion against larval stages of 3 species of moth pests in India. Eight to 10-day-old larvae of *Ephestia cautella, Corcyra cephalonica* and 1-day-old larvae of *Sitotroga cerealella* were used in the tests. Lindane was the most toxic to larvae of all 3 species, followed by malathion.

Yadav (1980) studied the toxicity of DDT and lindane against 13 species of stored-product pests and for lindane the decreasing order of toxicity was *Cryptolestes minutus, Stegobium paniceum, Sitotroga cerealella, Callosobruchus chinensis, Trogoderma granarium, Cadra cautella, Corcyra cephalonica, Callosobruchus maculatus, Rhyzopertha dominica, Lasioderma serricorne, Latheticus oryzae, Sitophilus oryzae* and *Tribolium castaneum*.

Bressani *et al.* (1982) reported studies on the control of insects to preserve stored maize in Guatemala. Treatment with lindane was effective in protecting the seeds of maize for 6 months.

Golob *et al.* (1985) found lindane to be moderately potent against young adult *Prostephanus truncatus* following topical application and following exposure to treated filter papers. However, in spite of its performance, lindane was not included in trials with treated maize grain.

Other papers dealing with the use and effectiveness of lindane for grain protection include **Coulon (1963), Tan *et al.* (1965)**, and **Pinniger (1975)**.

17.2 Degradation

Samples of whole Manitoba wheat were sprayed with an acetone solution of ^{14}C-labelled lindane to give an initial residue of 1 mg/kg **(Bridges and Trotman 1957)**. Samples stored in airtight containers showed no loss of the insecticide after 24 weeks. Samples under ideal airing conditions showed an initial fairly rapid loss, the residue falling to 0.6 mg/kg in the first 3 weeks, followed by a gradual loss to 0.4 mg/kg in the subsequent 33 weeks, the residue apparently being held in the fatty content of the wheat but no evidence was obtained of any breakdown of the lindane on storage or any reactions between it and the wheat constituents.

Walkden and Nelson (1958) treated wheat with a moisture content of less than 11% with lindane at the rate of 5 mg/kg in May 1953, and in August 1953 found 4.8 mg/kg lindane and in January 1956 they recovered 4.8 mg/kg, indicating no loss of lindane during storage. Likewise, maize treated with lindane at the rate of 5.4 mg/kg in August 1953 was found to contain 3.0 mg/kg in October 1953 but after a further 2 years in storage still contained 3.0 mg/kg lindane, indicating no loss during storage.

Gunther *et al.* (1958) determined the persistence of lindane used for the protection of stored wheat. They found that lindane was stable under the conditions of storage utilised in these studies (11.5% to 12% moisture stored at 20°C in darkness), very little loss occurring over a 15-month storage period.

Lindane labelled with ^{14}C was used to study the rate of loss of the insecticide from whole wheat and its distribution between the flour and bran fractions after milling. Loss from exposed wheat was rapid, but when it was stored in closed containers no loss was detected **(Bridges 1958a)**.

Lin (1960) showed in laboratory experiments that 50% to 70% of a 0.2% lindane dust applied to brown rice in Taiwan remained on the rice after 8 to 10 months storage.

Giles (1964) applied lindane to sorghum heads stored in jute bags inside a small warehouse. Three rates of application, 12.5, 25 and 50 mg/kg were used. Samples were taken at the beginning of storage and subsequently at intervals of 3 months. The lindane concentration on the sorghum heads substantially decreased with storage period, being less than one-tenth of the original application rate at the end of 12 months.

Lindgren *et al.* (1968), in an extensive review of the literature on the residues in raw and processed foods resulting from post-harvest insecticidal treatments, provide little or no information on the degradation of lindane on raw cereal grains or the effect of processing and cooking.

Srivastava *et al.* (1970) carried out laboratory studies on the effect of temperature and humidity on the persistence of lindane by applying the insecticide at the rate of 25 and 150 mg/kg (the LD/50 and LD/90 against *Tribolium castaneum*) to filter papers to which adults of this species were periodically exposed at different combinations of temperature (35° and 45°C) and relative humidity (60% and 90%). The toxicity of the deposits was lost most rapidly at 45°C and 60% r.h. and most slowly at 35°C and 90% r.h. Lindane was the last of three insecticides (carbaryl, malathion and lindane) to lose its effectiveness.

Rowlands (1971), in a second review of literature on the metabolism of contact insecticides in stored grain, noted that despite the continued use of lindane for application to stored grain, there seems to have been no further work on its metabolic fate in the grain. However, **Rowlands (1970b)** found that lindane was slightly dechlorinated in whole grains, and by the anaerobic lipoxidase system *in vitro*, to pentachlorocyclohexane and tetrachlorocyclohexane but no trichlorobenzenes were detected. **Rowlands (1971)** concluded that the lack of phytoxicity of lindane applied to stored grain may be taken as an indication that little or no aerobic degradation to the injurious trichlorobenzenes occurs.

Morrison (1972), in a most extensive review of the use and place of lindane in the protection of stored products from the ravages of insect pests, provides no indication of finding any studies dealing with the degradation of lindane on stored products.

Giles (1973) reported a series of trials in which the maize stored in cribs was treated with insecticidal dusts. Lindane was reported to give very disappointing results, providing protection for only a short period.

Rowlands (1975), reviewing world literature on the metabolism of contact insecticides in stored grain, concluded that where loss of residual toxicity occurred many authors attributed the loss to volatilisation of the intact insecticide. He concluded this to be a valid assumption generally. In a number of studies where lindane was applied at high concentrations to seeds to protect them from insect attack not only during storage but also after planting, the observed effect upon the depression of germination was apparently due to the production of trichlorobenzenes which had been shown capable of producing injury to the plant growth in wheat, oats and barley. It is not clear whether this conversion to trichlorobenzenes is due to the biological activity of the grain or to the effect of micro-organisms in the soil. **Rowlands (1975)** reported studies to determine the loss of ^{14}C-lindane applied to wheat of 15% moisture in the laboratory at 2.5 mg/kg and stored in either sealed (anaerobic or aerobic conditions) or unsealed jars at 20°C. From open jars, the residual half-life of lindane was 6 months. In the field samples stored for 6 months under anaerobic conditions, the loss (22%) was accounted for almost entirely by the production of pentachlorocyclohexane with trace amounts of trichlorobenzene and another substance, probably tetrachlorobenzene. In the sealed aerobic samples, loss was lower (12%) and was almost entirely due to production of trichlorobenzene. Similar amounts of trichlorobenzene, which is relatively stable under most physiological conditions, were obtained from samples in open jars, though overall loss of lindane was much greater in this case (50%), presumably due to volatilisation of intact lindane or breakdown products. When applied to very mouldy grain of 21% moisture content, lindane was almost entirely lost from open jars in 24 hours and only traces of breakdown products were found. Application to the mouldy wheat of radio-labelled pentachlorocyclohexane and trichlorobenzene suggested that they were both degraded via dichlorobenzene and subsequently lost from open jars, whereas they remained intact on normal non-mouldy wheat of 15% moisture.

Rowlands and Wilkin (1975) applied solutions of lindane and malathion to wheat grain of 12% and 18% moisture in the laboratory to give levels of 2 mg/kg of lindane and 8 mg/kg of malathion, and stood samples in the dark at 20°C. There was no indication of any metabolic or physical interaction between the 2 insecticides, both behaving as if applied separately. Typical half-lives obtained were lindane 6 months, malathion 4 months on the 12% moisture grain and lindane 4 months, and malathion 3 months on the 18% moisture wheat. In a field trial, the same workers found that metabolism accounted for only 25% of the lindane lost, most of it being accounted for by volatilisation.

Gupta *et al.* (1978) studied the fate of lindane residues in stored cereals. They treated stored wheat, barley and sorghum with 10, 20 and 30 mg/kg lindane and found by analysis that the deposits ranged between 5.2 and 9.5, 12.5 and 16.4, and 24.5 and 27.6 mg/kg respectively. There was slightly higher dissipation of lindane residues from low treatment doses than higher treatment dosages and the rate of dissipation from wheat was greater than from other cereals. The residues fell below 3 mg/kg in 30, 45 and 45 days from 10 mg/kg dosage; 45, 45 and 45 days from 20 mg/kg dosage and 75, 90 and 90 days from the 30 mg/kg dosage in the case of stored wheat, barley and sorghum respectively.

17.3 Fate of Residues in Milling, Processing and Cooking

Samples of whole Manitoba wheat were sprayed with an acetone solution of ^{14}C-labelled lindane to give an initial residue of 1 mg/kg **(Bridges and Trotman 1957; Bridges 1958a)**. During milling about 20% of the lindane was lost from the treated wheat. The residue that remained was almost equally divided between the bran and the white flour fractions.

Walkden and Nelson (1958) treated wheat with a moisture content below 11% with lindane at the rate of 5 mg/kg. It was found that lindane was carried over into the milling fractions, the largest amounts being found in the bran and shorts (11.0 to 23.0 mg/kg). Small amounts were found in the first and second clear flours (1.8 to 5.3 mg/kg), with less than 1 mg/kg in the patent flour. Baking reduced the residue only slightly in bread baked from flour containing known amounts of lindane (1 to 9 mg/kg); the amount of reduction ranged from 0 to 16%.

Schesser *et al.* (1958) treated wheat with lindane. The lindane was mixed with wheat flour and applied to wheat at a level of 5 mg/kg. Lindane was also applied as a spray at levels of 2.5 and 7.5 mg/kg. Wheat treated with lindane dust was stored for 18 to 24 months before tests were begun although wheat sprayed with lindane was evaluated about 10 days later. The results of analysis given in **Table 17.2** show the distribution in the various milling fractions. These data indicate most of the retained insecticide was in the pericarp and so was recovered in bran and shorts. Residues in flour were significantly lower than in the whole wheat. When lindane was applied as a spray about 10 days before wheat was milled, resulting flour con-

Table 17.2. Lindane residues in milling fractions of wheat after storage **(Schesser *et al.* 1958).**

Rate of application (mg/kg)	Residues (mg/kg) in: Whole grain, Before cleaning	Whole grain, After cleaning	Milling fractions: Shorts	Bran	Germ	Flour
5.0	8.6	7.8	13.0	49.5	11.8	2.3

tained 1.3 to 2.6 mg/kg. Highest concentrations were found in the outer layers of the kernel.

^{14}C-labelled lindane in acetone was sprayed **(Bridges 1985a)** on samples of hard wheat with a moisture content of 12.7% to give residues of 1.1, 4.1, 9.3 and 18.2 mg/kg. Samples of the wheat containing a residue of 1.1 mg/kg were placed in small tubes and continuously aerated. One such apparatus was kept at a constant temperature of 25°C and another at room temperature which varied between 15°C and 20°C. Samples of this wheat were also stored in stoppered glass containers. Wheat samples containing the 3 highest residues were exposed to room temperature spread out in a thin layer in a Petri dish. At intervals after the initial spraying, 0.5 g samples of the wheat were analysed for ^{14}C-radioactivity. Samples of wheat which had been treated with the labelled insecticide to give an initial residue of 1.1 mg/kg and which had either been aired continuously for 27 weeks or stored in closed bottles for 24 weeks were milled. The residue in each fraction was also determined. The rate of loss of lindane from treated wheat with initial residues of 18.2, 9.3, 4.1 and 1.1 mg/kg was fairly rapid under ideal airing conditions. Even with wheat containing the highest residue, 3 weeks was sufficient for the level of the residue to fall below 2.5 mg/kg. Complete loss by airing was not attained even after a prolonged period which suggests that some of the insecticide is strongly bound to the lipid portions of the grain. There was no significant difference in the rate of loss at 25°C compared with that at room temperature. When the wheat was stored in closed containers no loss of lindane occurred over 24 weeks. In practice, therefore, little loss of insecticide may be expected if it is intimately mixed with grain which is stored in bulk. If the insecticide is sprayed on to the surface of the bulk grain the loss from the surface may be quite rapid. The residue in the wheat which had been aired for 27 weeks was found to be 0.52 mg/kg whilst that held in the closed container for 24 weeks was 1.27 mg/kg. The distribution of the residue within the grain, as determined by the analysis of the milling fractions, is shown in **Table 17.3**. A comparison of the distribution of the residues in the various fractions obtained from Samples 1 and 2, shows a remarkable similarity, suggesting that absorption into the grain is governed by the concentration of the insecticide on the outer surface and that as this evaporates a redistribution occurs within the grain **(Bridges 1958a).**

Bridges (1958b) incorporated ^{14}C-lindane into wheat, starch, gluten and milled wheat. The activity retained by starch and gluten after heating for 1.5 hours to 180°C depended on the initial moisture content of the materials. Milled wheat, 10.4% to 17.3% moisture content, showed some loss. When made into a dough with water prior to heating a higher proportion of lindane was retained. The residue remaining after heating was 'locked-up' in desiccated starch granules and could not be extracted with acetone until the heated material was treated with water. The residue in heated starch was mostly unchanged lindane, but in flour it was a mixture of the breakdown products: the tri-,di- and mono-chlorobenzenes.

Feursenger (1960a) conducted large-scale experiments with lindane against weevils in grain and found that the flour made from the treated grain contained up to 1.8 mg/kg lindane and concluded that the use of lindane for the treatment of grain would have to be abandoned altogether.

Feursenger (1960a, 1962) is reported by JMPR **(FAO/WHO 1968)** as having indicated that lindane residues were reduced by cleaning and brushing for milling so that the residues in flour might be down to 50% of the original in the grain. Such cleaning would be more effective with dusts than with sprays and has been recognised as one of the advantages of specially formulated dusts **(Desmarchelier 1985)**.

Rohrlich *et al.* (1971) studied the distribution of lindane in wheat grains and in the processed products. They found the highest lindane concentrations in the germ, with considerably less in the husk (seedcoats) and virtually nothing in the endosperm. **Rowlands (unpublished data)** dissected and analysed individual wheat grains topically treated with ^{14}C-labelled lindane and found very rapid uptake from solutions in organic solvents or from dusts; this was surprising, since the majority of pesticides are only very slowly absorbed by grain from dust formulations. He confirmed Rohrlich's findings in that 85% of the radioactive lindane was present in the germ after 1 week, with some 12% in the seedcoat and 3% in the endosperm. Lindane, topically applied directly to the germ, was not so readily lost by volatilisation as it was from other sites of application, but a certain amount

Table 17.3. Distribution of lindane on wheat aired for 27 weeks (I) or stored for 24 weeks (II) before milling **(Schesser *et al.* 1958).**

Milling fraction	Sample I (0.52 mg/kg) residue (mg/kg)	Sample I (0.52 mg/kg) % lindane recovered	Sample II (1.27 mg/kg) residue (mg/kg)	Sample II (1.27 mg/kg) % lindane recovered
Coarse bran	1.20	21.8	5.06	25.1
Flour 100 mesh	0.67	11.9	2.34	18.6
Flour 100–200 mesh	0.40	10.4	1.07	15.5
Flour 200 mesh	0.21	32.0	0.59	30.7
Loss in milling	–	23.9	–	10.1

of initial redistribution to other tissues of the seed coat occurred during the 2 to 5 days subsequent to topical application. The same vapour-phase loss occurred, followed by an overall reconcentration in the germ tissues where a complex was formed with the ether-soluble lipid material **(Rowlands 1975)**.

Kanazawa (1973), in an article on pesticide residues in agricultural commodities in Japan, provided valuable information on the effect of cooking on BHC isomers, including lindane, present in rice grains. Approximately 50% of the BHC in the rice derived from pre-harvest treatments is lost by cooking either by simple boiling in an open pan or by cooking under pressure as is now popular.

Saha and Sumner (1974) studied the fate of radiolabelled lindane in wheat flour under normal conditions of bread-making. The wheat flour was treated with 0.26, 1.94 and 17.8 mg/kg lindane-^{14}C. Irrespective of the level of treatment, the baked bread retained 75–82% of the originally applied radioactivity. About 94% of these residues was present as lindane, the remaining radioactivity being represented by the following degradation products: gamma-pentachlorocyclohexane, 1,2,4-trichlorobenzene, and 3 isomers of tetrachlorobenzene.

17.4 Metabolism

Perry (1960) reviewed the then available information on the metabolism of insecticides by various insect species. He pointed out that there was extensive evidence that lindane is metabolised by the housefly into polar metabolites, which are rapidly excreted. In an early work, **Bradbury *et al.* (1953)** showed that pupae and adult houseflies originating from larvae reared in a lindane medium contained significant amounts of unchanged insecticide which indicated the absence of an efficient detoxication mechanism. On the other hand, the injection of lindane into resistant and susceptible houseflies **(Oppenoorth 1954)** resulted in rapid breakdown of the chemical in the resistant strain and slower metabolism in the susceptible strain. The enzyme, DDT-dehydrochlorinase, was not involved in lindane metabolism. Quantitative data on the metabolism of lindane in several insect species **(Bradbury 1957)** indicated that the housefly is unique in that it possesses an efficent detoxifying mechanism before any selection pressure is applied. It has been shown **(Bradbury and Standen 1959)** that alkaline hydrolysis of the metabolic product of lindane yields dichlorothiophenols. It was inferred that the first step in the metabolism of lindane involved the removal of one chlorine atom and the formation of a C–S bond, followed by further dehydrochlorination and the subsequent formation of dichlorothiophenol. When the procedure used for the measurement of DDT-dehydrochlinase in houseflies was followed in detail **(Sternburg *et al.* 1953)** it was shown that *in vitro* conversion of lindane into water-soluble metabolites requires reduced glutathione for activation of the enzyme. Thus glutathione might be the source of sulphur for the C-S bond.

The first stage of metabolism of lindane is probably dehydrochlorination to *gamma*-PCCH (pentachlorocyclohexane). This is followed by the cleaving off of further HCl groups leading, via the hypothetical intermediate tetrachlorocyclohexadiene, to 1,2,4-trichlorobenzene which is excreted in the urine in small quantities **(van Asperen and Oppenoorth, 1954; Bronisz *et al.* 1962; Coper *et al.* 1951; Grover and Sims 1965; Koransky *et al.* 1964; San Antonio 1959)**.

The main excretion products, however, are water-soluble conjugates of glucuronic acid and of sulphuric acid. After a single oral administration of 50 mg/kg or 100 mg lindane per kg body weight to rats, the excretion of glucuronic acid in the urine increased within 14 to 15 days by 1.5 mg/kg per day on average whereas the excretion of organic sulphur compounds increased over 13 to 14 days by 35% to 58% **(Rusiecki and Bronisz 1964)**. According to other experiments lindane and *gamma*-PCCH are metabolised by rats to 2,3,5- and 2,4,5-trichlorophenols and are excreted in the urine as free phenols, sulphates and glucuronic acid conjugates **(Grover and Sims 1965)**.

Such information as is available on the metabolism of lindane in grain is referred to in the previous section on degradation.

The metabolism of lindane in insects, especially the housefly, has been studied intensively. The metabolites identified before 1971 are summarised in **Table 17.4 (Sieper 1972)**.

Experiments on rats have shown that the metabolism of lindane in mammals is evidently very similar to that in insects. A scheme for the metabolic breakdown of lindane in mammals devised by **Grover and Sims (1965)** is given in **Figure 17.1**.

Macholz and Kujawa (1979, 1985) have reviewed the literature on lindane metabolism including more than 150 papers published from 1975 onwards. They refer to more than 80 lindane metabolites of different chemical structure and relevance including more than 70 metabolites found in warm-blooded animals. Until 1964 only 10 metabolites were reported, from 1965 to 1970 the number reported was 35, from 1971 to 1975, 118, and from 1976 to 1980, 238 (sometimes of the same metabolites).

17.5 Fate in Animals

Feeding studies designed to elucidate tissue distribution, physiological effects, excretion, etc of lindane **(Treon *et al.* 1951; Davidow and Frawley 1951; Claborn *et al.* 1953; Claborn 1956, Koransky and Portig 1963)** make it evident that lindane fed to animals is deposited in the tissues, especially the fat, and that after intake ceases the deposits are rapidly reduced by excretion.

Radeleff (1951) showed that lindane, as a contaminant of feed for beef cattle, was stored in the fat in proportion to the concentration fed. After 70 days, feeding 10 ppm in the feed produced 8 mg/kg in the fat, 100 ppm in the feed produced 98.5 mg/kg in the fat. Ten to 14 weeks after

Table 17.4. Lindane metabolites identified in insects till now

Test compound used	Identified metabolites
Lindane = γ-BHC	γ-PCCH
	iso-PCCH
	1,2,3-trichlorobenzene
	1,2,4-trichlorobenzene
	1,2,3,4-tetrachlorobenzene
	1,2,3,5-tetrachlorobenzene
	Pentachlorobenzene
	6 isomeric dichlorophenols (after hydrolysis)
	S-[2,4-dichlorophenyl]-glutathione
	Water-soluble metabolites
γ-PCCH	1,2,4-trichlorobenzene
	1,2,4,5-tetrachlorobenzene
iso-PCCH	1,2,4-trichlorobenzene
	1,2,3-trichlorobenzene
	1,2,4,5-tetrachlorobenzene
	1,2,3,4-tetrachlorobenzene
	Pentachlorobenzene

Abbreviations: γ-PCCH = γ-2,3,4,5,6-pentachlorocyclohexene-(1), iso-PCCH = iso pentachlorocyclohexen-

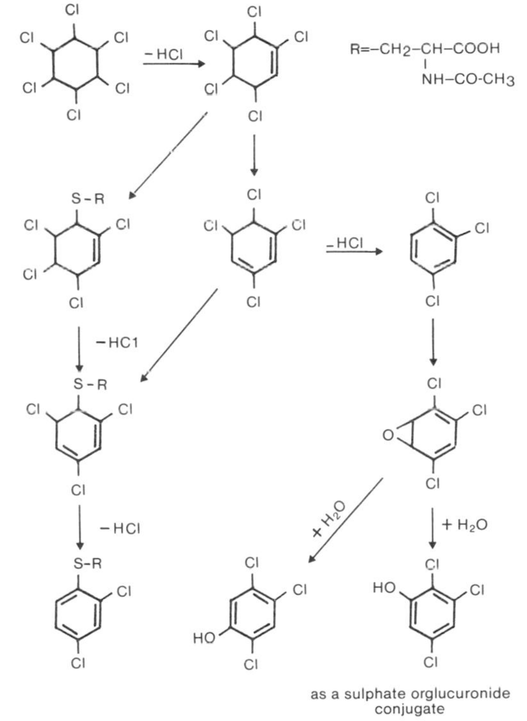

Figure 17.1. Scheme of the metabolic breakdown of lindane in mammals **(Grover and Simes 1965).**

feeding ceased the stored lindane had disappeared. No effect on health or weight gain was noted.

Knipling (1950) fed milk cows 7.5 ppm of lindane in their total diet and found little or no lindane in the milk. The limit of detection by his analytical method was 0.2 mg/kg.

Ely *et al.* (1952) fed lindane and soybean to cattle at daily intake levels of 0.07 to 6.22 mg/kg of body weight. Levels above 0.36 mg/kg resulted in increasing levels in the milk. When the intake ceased the levels in the milk dropped rapidly but lindane was still detectable in the milk after 30 days in animals which had originally received 2 or more grams daily.

Gyrisco *et al.* (1959) fed dairy cows on hay with lindane contents up to 10 ppm for up to 3 months. The maximum residues in the milk were 0.17 mg/kg from less than 1 ppm on the hay, 0.26 mg/kg from 2 ppm on the hay, 0.21 mg/kg from 4 ppm on the hay, and 0.17 mg/kg from 10 ppm on the hay.

Ware and Naeber (1961) fed caged laying hens 0.01, 0.10, 1.0 and 10 ppm of lindane in their daily rations for 60 days. Levels of 0.01 ppm resulted in detectable residues in the yolks of eggs but none in the albumen. The skin and visceral fats were the main storage areas and some residues occurred in the brain. Sixty days after feeding ceased lindane was still present in the eggs but not in the body tissues from the 10 ppm dosage.

Harrison *et al.* (1963) also observed the effects of feeding lindane to chickens. **Stadelman *et al.* (1964)** fed 0.015 ppm lindane to laying hens for 15 days and 15 ppm for 5 days. The lower level resulted in 0.3 mg/kg in the abdominal fat and more in the eggs. The higher level left 0.7 mg/kg in the fat and 0.4 mg/kg in the egg yolks one week after feeding had been discontinued, and no detectable residues after 11 weeks.

Ash and Taylor (1964, 1965) showed the presence of lindane in the flesh and the eggs of pheasants fed on lindane-treated grains. Residue disappearance was rapid when exposure ceased.

Cummings *et al.* (1966) fed laying hens mixtures of insecticides including 0.05, 0.015 or 0.45 ppm of lindane. Within 1 month after withdrawal lindane had disappeared from the eggs. **Cummings *et al.* (1967)** also reported on similar combinations of insecticides fed to hens. In this case tissue storage was monitored regularly. Breast and liver tissues never reached the 1.0 mg/kg level and again lindane residues declined rapidly after withdrawal.

Collett and Harrison (1968) grazed sheep on lindane-treated pastures and reported from 4 to 0.4 mg/kg in the mutton fat depending on how soon after treatment the sheep were placed on the pasture. Transferred to untreated pastures the fat content dropped to 0.5 mg/kg in 1 month and to 0.1 mg/kg in 2 months of subsequent grazing.

17.6 Safety to Livestock

Black *et al.* (1950) fed about 8 ppm of BHC for up to 6.5 months to chickens with no ill effects, and **Chen and**

Liang (1956) reported similar results from 10 ppm fed for 3 months. **Ware and Naber (1961, 1962)** reported no effects on egg production with up to 10 ppm of lindane in the feed. **Dahlen and Hougen (1954)** fed single acute dosages of lindane in gelatine capsules to 212 quail and 64 mourning doves. The medium lethal dose for Bob-white quail was 120 to 130 mg/kg lindane for males and 195 to 210 mg/kg for females. For doves the MLDs were 350 to 450 mg/kg.

Ash and Taylor (1964) attempted to feed pheasants on lindane-treated seed but it proved unpalatable. The same authors (1965) conditioned birds to low concentrations in the diet to overcome the unpalatability problem. Even then the birds consumed 14% less feed than the controls and when they started laying, reached a laying peak later than those on normal diets. Eggs contained 0.4 to 22 mg/kg of lindane. Hatching was not impaired. The residues in eggs remained constant on continued feeding and for 5 days thereafter, but then fell rapidly for 15 days. Within 9 days after ingestion ceased, 66% of the initial residue had been excreted.

Morrison (1972) reviewed the toxicity of lindane to vertebrates including laboratory and other small animals and this information shows that lindane is well tolerated at moderate doses by mammals and birds. Continuous feeding results in accumulation in fatty tissues but accumulated residues disappear quickly after exposure ceases.

17.7 Toxicological Evaluation

The toxicology of lindane has been evaluated by JMPR on 6 occasions (1963, 1965, 1967, 1971, 1973, 1977) **(FAO/WHO 1964, 1966, 1968, 1972, 1974, 1978)**. In all animal species tested lindane has proved to be a cumulative poison causing hepatic and renal lesions and disturbances of the central nervous system. The dog seems particularly susceptible to neurological effects.

No-effect levels were demonstrated in the rat (1.25 mg/kg body weight per day) and an ADI of 0.125 mg/kg body weight was established. Later, a no-effect level was demonstrated also in the dog (1.6 mg/kg body weight per day) and the temporary ADI for man was adjusted to 0.01 mg/kg body weight. In a 2-year study in dogs, hepatic lesions were evident at 100 ppm. Lindane at 100 ppm in the diet did not affect reproduction in the rat although at 50 ppm and above, hepatic lesions were observed in the F_2 generation. Lindane had no effect upon maintenance of pregnancy in the rabbit, rat or mouse and 2 tests currently used to evaluate mutagenic potential were negative. It was noted that with mice at high dietary intake, liver enlargement and nodular formation were evident but neither mouse nor rat studies, both undertaken with lindane of much greater purity than previously tested, provided any evidence of tumour or cancer induction following exposure to lindane. In the subsequent evaluation, the ADI for man previously established was confirmed on a fixed basis. It was considered that, in view of the relative rapidity with which lindane is degraded in mammalian organisms and in the environment to much less acutely toxic compounds, and the relatively high levels required to produce adverse effects in humans and laboratory animals, there appeared to be no toxicological objections to the continued maintenance of the previously established ADI for humans.

17.8 Maximum Residue Limits

Based on the extensive data available to it in 1967, the JMPR recommended the following maximum residue limits:-

Commodity	Maximum Residue Limits (mg/kg)
Cereal grains	0.5
Rough rice	0.5

Residues are determined and expressed as lindane.

The methods recommended for the analysis of lindane residues are: **AOAC (1980a); Pesticide Analytical Manual (1979a, 1979c); Manual of Analytical Methods (1984); Methodensammlung (1982 XII-5, 6; S1–5; S8–10, S12, S19); Ambrus *et al.* (1981); Greve and Grevenstuk (1975); Panel (1979); Telling *et al.* (1977)**. The following methods are recognised as being suitable: **Maybury (1980); De Vas *et al.* (1974); Greve and Heusinkveld (1981); Mestres *et al.* (1976, 1977, 1979a, 1979b); Porter and Burke (1973); Sissons and Telling (1970); Sprecht and Tillkes (1980).**

18. Malathion

Malathion has been widely used in many countries for over 20 years and there is now considerable experience and a wealth of scientific data on its properties, effects and fate. It is weaker against most stored-product pests than many of the other organophosphorus insecticides. It is virtually ineffective against stored-product moths and requires higher dosage to control non-resistant *Rhyzopertha dominica*. Because of the higher standards demanded today, malathion might not be acceptable if it were being proposed for the first time but it must be recognised that enormous quantities of grain and other stored products have been saved from destruction by stored-product pests because of the use of modest quantities of malathion sprays and dusts. In fact it is still performing an important service in many parts of the world, in spite of the development of strains of some pests that are resistant to malathion. Malathion is still used as standard for evaluating other insecticides. The amount of the malathion deposit which penetrates the individual grains is relatively small and therefore most of the deposit is removed in the milling of wheat and rice. More than 95% of the deposit on the raw cereal grain is removed or destroyed before the cereal food reaches the consumer.

In Australia the name 'Malathion' is a registered trade mark, the common name being 'maldison'. As this review is intended for use throughout the world the term 'malathion' is being used with the consent of the owners of the registered trade mark.

18.1 Usefulness

One of the first references to the performance of malathion against stored-product pests is that of **Lindgren et al. (1954)**, reporting laboratory trials which showed that malathion should be effective in protecting wheat from insect infestation. Wheat was treated with malathion either as a dust or spray. The effectiveness of the treatment was tested by confining 100 adults of each, *Sitophilus granarius, Sitophilus oryzae* and *Rhyzopertha dominica* in 250 g samples of treated grain intially on treatment and at monthly intervals thereafter. The insects were exposed for a period of 10 days and removed and checked for mortality. It was observed that the 3 species were killed at dosages as low as 2 mg/kg and after storage for 3 months, depending on the dosage used, the kill of these insects was relatively high especially when applied to the grain as a dust. Six to 7 months after application, 8 and 16 mg/kg were still effective.

Gunther *et al.* (1958) published results of studies of the biological effectiveness and persistence of malathion used for the protection of stored wheat and in 1958 Parkin prepared a provisional assessment of malathion for stored-product insect control in which he described not only its broad-spectrum insecticidal activity but the acceptability of such treatments because of the low mammalian toxicity of malathion. In a later paper **Parkin (1960)** discussed the variation in susceptibility of stored-product insects to various contact insecticides, including malathion. He provided the results of tests with 8 species of beetles exposed to malathion dusts on grain and this information is summarised in **Table 18.1**. His paper gives quite a deal of attention to the occurrence of resistant strains in wild populations and the selection of resistant strains.

Table 18.1. The susceptibility to malathion of several species of stored-product beetles 28°C and 70% r.h. **(Parkin 1960)**

Species	Approx. dose for 100% kill in 13 days (mg/kg)	Approx. dose to prevent breeding (mg/kg)
Oryzaephilus surinamensis	<0.5	0.5
Tribolium castaneum	4	0.5–1.0
Rhyzopertha dominica	»8 (10%)*	4
Sitophilus granarius	2	8
Sitophilus oryzae	2	4–8
Stegobium panicium	1	2
Lasioderma serricorne	4	4–8
Ptinus tectus	»8 (12%)*	8

* % kill at 13 days with 8 mg/kg malathion

Womack and LaHue (1959) determined the effectiveness of malathion emulsion sprays in protecting shelled corn from insect attack during an 8-month storage period. They used various application rates, ranging from 5 to 20 mg/kg.

Watters (1959a) showed that malathion applied to wheat at rates ranging from 2 to 8 mg/kg was effective in controlling *Tribolium castaneum* and *Sitophilus granarius* and preventing their reproduction but he drew attention to the importance of grain moisture content in promoting the degradation of the insecticide.

Strong and Sbur (1960) also stressed the importance of grain-moisture content and storage temperature on the effectiveness of malathion as a grain protectant. **Strong *et al.* (1961)** investigated the influence of formulation on the effectiveness of malathion sprays for stored wheat and **Watters (1961)** published an evaluation of the effectiveness of malathion and three other insecticides against the hairy spider beetle (*Ptinus villiger*).

Green and Kane (1960) compared monthly applications of DDT, malathion and lindane sprayed on to the surfaces of filled sacks for protection of bagged groundnuts against *Tribolium castaneum* under conditions of severe cross-infestations. All treatments were satisfactory but the results favoured malathion.

Watters (1961) reported experiments carried out in warehouses in western Canada to compare the residual effectiveness of lindane, malathion, methoxychlor and pyrethrins/piperonyl butoxide against *Ptinus villiger* which is the most prevalent pest of commercial flour storages in the region. Three of the warehouses treated with malathion were retreated after 6 weeks when high insect counts were noted at the first assessment. All other warehouses were retreated after 8 weeks. There were no significant differences among insecticides at the first assessment, 4 weeks after application, but between 4 and 6 weeks, malathion failed to give adequate control when compared with the other treatments. This result is not unexpected in view of the findings of **Parkin (1960)**.

Floyd (1961) published results of studies conducted to evaluate the protective value of malathion and pyrethrum grain protectants against damage to corn in Louisiana caused by *Sitophilus oryzae*. Both shelled and snapped corn were used in the studies. Malathion gave some protection to stored corn from progressive insect damage during storage; however, satisfactory protection was not obtained at any dosage level tested. Malathion was more effective than pyrethrum. Greater protection was obtained in shelled corn than was possible in snapped corn. Failure to protect corn in storage from insect damage with malathion or pyrethrum was very likely due to the high initial infestation incurred in the field and favourable conditions of temperature, relative humidity and grain moisture for insect development during much of the storage period.

Strong and Sbur (1961) presented results from a series of tests with varying dosages of each of 36 insecticides sprayed on wheat for protection against *Sitophilus oryzae, Sitophilus granarius, Tribolium confusum* and larvae of *Trogoderma granarium*. Insecticide deposits of 1.25, 2.5, 5, 10, 15, 25, 50, 100 and 200 mg/kg were used. Methyl parathion, phorate, parathion and mevinphos were more effective than malathion applied at 10 mg/kg as indicated by lower dosages required for 100% mortality of all species of test insects. If the effect against *Trogoderma granarium* was ignored many insecticides appeared to give a superior performance to that of malathion.

King *et al.* (1962), working in Texas, reported good control of coleopterous pests of stored grain was provided by malathion, fenchlorphos and silica aerogel. Further tests revealed that malathion did not remain effective on sorghum grain treated at moisture contents above 14% and that malathion does not kill any stages of *Sitophilus oryzae* by fumigation action.

Bang and Floyd (1962) reported that malathion dust or spray at 8 mg/kg on polished rice gave excellent protection from damage by *Sitophilus oryzae* during a 5-month storage period. A dose of 4 mg/kg gave complete protection from *Cryptolestes pusillus*. Less protection from damage by *Tribolium castaneum* was obtained. Effectiveness against *Sitophilus oryzae* of the residue from a 4 mg/kg treatment appeared to be decreasing at 3 months. Bioassay of malathion residue with *Sitophilus oryzae* on rice after 5 months storage showed that at 8 mg/kg a mortality of 100% was obtained after 10 days exposure.

Watt (1962) wrote on grain protection with malathion in Australia in 1962. By this stage, malathion was well established for the treatment of stored grain throughout the grain handling system in Australia.

Lloyd and Parkin (1963) reported that a strain of *Sitophilus granarius* selected for resistance to pyrethrins was found to have a resistance of x34 to pyrethrum and x5.7 to malathion.

Godavari Bai *et al.* (1964) studied the effect of concentration per unit area on the toxicity of malathion to *Tribolium castaneum* adults under laboratory conditions. Their studies confirmed that the insecticide distributed on the surface of the particles varies in its effectiveness according to the particle-size of the grain.

Strong and Sbur (1965a) presented results from a series of tests with varying dosages of each of 46 insecticides applied to wheat for the control of 3 insect pest species. Malathion was used as the standard insecticide in making comparisons.

Mookherjee *et al.* (1965), working in India, reported preliminary studies of the efficacy of malathion against *Sitophilus oryzae* and *Tribolium castaneum* and its persistence in stored wheat. They applied malathion dust at the rate of 8, 16 and 24 mg/kg to wheat with 13% moisture content and 28°C. The biological effectiveness was tested by 24 hours exposure of the 2 species immediately after the treatment of wheat and after 1, 2, 3 and 4 months storage. *Tribolium castaneum* proved much more tolerant with only 55% and 77% control being attained at the end of 2 months from wheat treated with 16 and 24 mg/kg malathion. Control of *Sitophilus oryzae* was incomplete at the end of 3 months. Longer exposure of the insects to the treated grain would have increased the mortality greatly but chemical analysis revealed that 96.6% of the malathion applied to the grain had degraded after 4 months.

Winks and Bailey (1965), in a review of treatment and storage of export wheat in Australia, pointed out that the significant change that had occurred in the cleanliness of Australian wheat had resulted from a demand by markets for certified freedom of the grain from insect pests. The outcome of trials conducted to 1961 to evaluate the efficiency of malathion as a protectant resulted in the adoption of malathion for this purpose. Since these trials, a steadily increasing quantity of grain had been treated with malathion and a significant change in attitude has resulted which now favours the use of protectants applied immediately after harvest. From 4 to 8 months protection can be expected from malathion depending on climatic conditions prevailing at the time of harvest and during the storage period. The normal rate of application of malathion is 10 to 12 mg/kg. This is considered adequate from the point of view of pest control and acceptable residue limits since it has been shown that up to 50% breakdown occurs in approximately 1 month after treatment at this rate.

Giles (1965), in a lengthy paper on the control of insects infesting stored sorghum in northern Nigeria, outlined some of the many problems which confront farmers and make the reduction of insect pest damage extremely difficult. A stable malathion dust based on flour, when applied at the rate of 15 mg/kg, was less effective than lindane dust at the rate of 10 mg/kg. As a result of this experience, it was recommended that farmers who wished to economically protect sorghum which is to be stored for more than 6 months in mud granaries should use 0.5% lindane dust applied to unthreshed sorghum as a 'sandwich' treatment, at a rate equivalent to 10 mg/kg threshed grain.

Tyler *et al.* (1966b) reported a field trial in which damp barley was treated with malathion (10 mg/kg) and stored in an infested farm granary. During storage after treatment, moisture contents ranged from 12% to 18% and exceeded 15% in 8 of the 11 bins. Two months after treatment, beetles were present in warm grain in all bins surrounding the control bin. Where the temperature and moisture content of the grain were highest, insecticide residues were soon undetectable by both chemical analysis and bioassay. The trial demonstrated that the long periods of protection conferred in other experiments are not obtained under the adverse conditions prevailing with heating grain. The trial demonstrates the importance of treating all grain in store rather than selected bins.

Green and Tyler (1966) described a method for the protection and disinfestation of farm-stored grain in Britain by the admixture of water-based emulsions of malathion, fenitrothion and dichlorvos. Malathion at the rate of 10 mg/kg quickly controlled heavy infestations of *Oryzaephilus surinamensis* and conferred good protection for as long as 8 months.

Lemon (1966) determined the relative susceptibilities of two *Tribolium* spp. to organophosphorus insecticides. Six compounds were found to produce higher mortalities than those given by malathion.

Lemon (1967a) determined the relative toxicities of malathion, bromophos and fenitrothion to 10 species of stored-product beetles, finding fenitrothion to be the most effective. This work was later extended to include 7 new organophosphorus insecticides, malathion being used as the standard **(Lemon 1967b)**.

LaHue (1967) reported the evaluation of malathion at 2 rates, synergised pyrethrum and a diatomaceous earth applied to sorghum grain to control an existing insect infestation and to prevent further insect damage. The treated grain was stored in 150-L Masonite bins for 12 months in a heated structure. Insect infestations were not adequately controlled by any of the treatments as damaging populations became firmly established in all bins during the 12-month period. The malathion treatment at 15 mg/kg gave the best protection. It killed nearly all *Sitophilus oryzae* for 6 months after the sorghum was treated and most of the *Tribolium castaneum* for the first 3 months. Treatment with malathion at the rate of 10 mg/kg afforded some protection, but did not kill a majority of the *Sitophilus oryzae* after 3 months, or of the *Tribolium confusum* after the first week. The *Sitophilus oryzae* neither preferred nor avoided the malathion-treated sorghum grain in food selection studies.

Strong *et al.* (1967) compared the toxicity and residual effectiveness of malathion and diazinon used for protection of stored wheat. The relative susceptibilities of 17 species of stored-product insects to malathion and diazinon deposits on wheat sprayed with acetone solutions of each insecticide were determined from LC values obtained in a series of toxicity tests. Mature larvae of 4 species of *Trogoderma* were used; all other test insects were adults. Diazinon was more effective than malathion against *Cryptolestes pusillus, Latheticus oryzae, Rhyzopertha dominica, Sitophilus zeamais, Sitotroga cerealella, Tribolium brevicornis, Tribolium castaneum, Tribolium confusum, Trogoderma inclusum, Trogoderma parabile* and *Trogoderma simplex*. Results from residual protectant tests designed to evaluate the relative effectiveness of various concentrations of insecticide residues on wheat in preventing infestations of 6 species of stored-product insects showed diazinon to be slightly more effective than malathion when considering counts of insects.

Joubert and De Beer (1968a) treated partially infested yellow maize with malathion at the rate of 8 mg/kg. A similar mass was treated with 1.75 mg/kg of synergised pyrethrum. In both treatments the initial infestation was eradicated after several weeks and thereafter the treated maize exhibited excellent resistance to reinfestation for 10 months. Internal damage in the treated maize remained below 2%, while 2 control masses were totally destroyed. During the period of the test 12 species of insects were collected in the silo.

Although the use of synergised pyrethrins had been approved in France in 1963 and the properties and effectiveness of malathion had been extensively studied **(Coulon 1968)**, it would appear that malathion was not approved for admixture with grain in France until after 1968.

The effectiveness of fenitrothion and malathion as grain protectants under severe practical conditions in a British farm granary was described by **Tyler and Green (1968)** and the usefulness of malathion for this purpose was discussed by **Green (1969). Strong and Sbur (1968)** evaluated the effectiveness of 48 insecticides against adults of 4 species and larvae of another stored-product pest, using malathion as the standard for reference.

Iordanou and Watters (1969) demonstrated the influence of temperature on the toxicity of 5 insecticides, including malathion, against 5 species of stored-product insects. They found that malathion had a positive temperature, co-efficient of toxicity (increasing toxicity with increasing temperature) against *Tribolium confusum* over the temperature range 10–27°C, though there was not much difference in the toxicity at 10°C and 15°C.

LaHue (1969) evaluated several formulations of malathion as a protectant of grain sorghum against insects in small bins. The standard malathion emulsifiable concentrate applied at twice the recommended rate as an undiluted low-volume spray was superior to the same rate applied as a water-diluted emulsion spray and significantly better than the standard rate applied as water emulsion. A diatomaceous earth admixed with the standard amount of malathion emulsion concentrate before application to the grain afforded excellent protection but reduced the test weight of the grain sorghum about 1 kg per bushel and consequently lowered the commercial grade.

Tyler *et al.* (1969) reported a complex series of experiments designed to determine whether the distribution of malathion among grain affected the biological activity and persistence of the insecticide. They were able to show that a high dose (200 mg/kg) on 5% of the grains in a bulk was likely to give a longer period of protection than an evenly distributed treatment and in the case of *Oryzaephilus surinamensis* a mean concentration of 2 mg/kg was sufficient to control the pest and give good protection against reproduction. The authors concluded that a chemical assessment of the insecticidal content of a whole sample of grain was not necessarily a reliable indication of the effectiveness of a treatment.

Champ *et al.* (1969) compared the acute and residual toxicity of 4 insecticides, including malathion, under laboratory conditions over a range of moisture contents from 11 to 13% and then under bulk storage conditions. Fenitrothion was the most promising material. The authors noted considerable variation in the effectiveness against various immature stages of *S. oryzae* and *R. dominica.*

The **Ministry of Agriculture (1969)** of the United Kingdom recommended malathion as well as fenitrothion and pirimiphos-methyl for the treatment of grain-storage structures and for admixture with grain.

LaHue (1970a) evaluated malathion and 3 other protectants against *Rhyzopertha dominica* attacking wheat. When applied at the rate of 8 mg/kg, malathion gave nearly complete protection against insect damage for 12 months.

Girish *et al.* (1970) studied the efficacy and residual toxicity of iodofenphos and malathion when applied to concrete slabs and jute bags. The results of the studies indicated that iodofenphos had a slightly higher contact toxicity and that it was comparatively more persistent on both the surfaces than was malathion.

Minett and Williams (1971) showed by laboratory experiments that treatment of a small proportion of a wheat bulk with high concentrations of malathion may prove a more effective method of application than attempting to treat all grains uniformly with the same overall level of insecticide. Treating 1 or 2% of grains gave as effective control as treating all grains, but control was inferior when only 0.1 or 0.2% of grains were treated.

Storey (1972) studied the effect of air movement on the biological effectiveness and persistence of malathion in stored wheat. He was able to show that vapour-phase activity of contact insecticides is an important aspect of storage-pest control.

Assessments of insecticides for control of *S. granarius* were reported from France **(Coulon *et al.* 1972)**. Malathion was judged to be similar in residual effect to bromophos and iodofenphos.

Experiments in Ecuador **(INIAP 1972)** demonstrated that malathion applied at the rate of 10 or 15 mg/kg gave control of *Sitophilus oryzae* for 143 days when applied to maize. Trials in the Philippines **(Morallo-Rejesus and Santhoy 1972)** showed that malathion was effective against the adults of DDT-resistant strains of *S. oryzae* and *Tribolium castaneum* collected in the Philippines, though it was inferior to several other insecticides. Work in Malawi **(Pieterse and Schultern 1972)** showed that malathion resistance occurred in 11 of 17 strains of *T. castaneum* collected throughout Malawi.

Srivastava and Dadhich (1973) carried out a laboratory evaluation of malathion as a protectant for the prevention of damage to stored gram (*Phaseolus mungo*) by pulse beetles. They reported that the addition of malathion at 10, 20 and 30 mg/kg to samples of gram provided absolute protection for 6, 7 and 8 months, respectively, against 2 common pulse beetle species, *Callosobruchus maculatus* and *Callosobruchus chinensis*, in that the insects died after 48 hours exposure to the grain.

In Nicaragua, **Giles (1973)** tested malathion and a number of other insecticides for application to maize cribs. The results observed over 4 months showed malathion was at least equal and often superior, to other materials tested.

In the United States, **Cogburn (1973)** reported preliminary studies with a number of protectant insecticides that might be used to counteract resistance to malathion among insects infesting stored rice. Malathion was used as a standard.

Trials reported from India **(Chawla and Bindra 1973)** to compare a number of grain-protectant insecticides against malathion for the control of a range of grain pests were designed to provide information against the day when widespread resistance to malathion would occur in India.

Bitran (1974) reported trials to evaluate the residual effect of malathion used to control insects of stored coffee in Brazil.

Wilkin and Hope (1973) published the results of a study of the effectiveness of 17 insecticides, including several mixtures, applied to grain against 3 stored-product mites. Malathion, applied at the rate of 10 mg/kg was ineffective against *Acarus siro* and slow to act against *Tyrophagus putrescentiae* and *Glycyphagus destructor*. Lindane at the rate of 2.5 mg/kg was completely inactive against the second species but the combination lindane 2.5/malathion 7.5 mg/kg was 100% effective and quick acting against all 3 species.

Studies in the Philippines **(Morallo-Rejesus and Carino 1974)** to determine the residual toxicity of malathion and 4 other insecticides on 3 varieties of corn against *Sitophilus* spp. and *Rhyzopertha dominica* found malathion to be the least effective of the five. The rate of application was 10, 20, 30 and 50 mg/kg, which contrasts greatly with the rates found adequate in temperate climates.

In Israel, **Carmi (1975)**, compared the effect of malathion and pirimiphos-methyl on *Ephestia cautella* at dosages of 2, 4 and 8 mg/kg mixed with wheat grains. Pirimiphos-methyl was superior. This confirms the observations of other workers and practical experience that malathion is not very effective against stored-product moths.

Trivelli (1974) discussed extensive experience with different types of stored wheat, their susceptibility to different species of insect pests and practical control measures applied in the Argentine. The value of malathion sprays and dusts was emphasised.

In India, the emergence of a malathion-resistant strain of *Tribolium castaneum* raised concern about the future of grain protection and **Verma and Ram (1974)** evaluated the performance of bromophos, tetrachlorvinphos, phoxim and pirimiphos-methyl and found there was no indication of cross-resistance, so these insecticides could be considered as replacements for malathion if malathion resistance spread.

Bindra and Udeaan (1975) reported studies to determine the dosage of malathion required for treatment of wheat in September (the cooler period) for rural storage in the Punjab, India. Dosages of from 10 to 30 mg/kg were admixed with wheat and in the 6-month period following treatment, losses due to insects in grain treated at 25 and 30 mg/kg were nil. It was concluded that a dosage of 25 mg/kg was adequate for treatment in September, as compared with 30 mg/kg recommended for treatment during May–June.

Bengston *et al.* (1975) examined the level of resistance to malathion in 3 species of insects collected in Australia. They found that *S. oryzae* had developed a ninefold resistance, *R. dominica* had developed a sixfold resistance and *T. castaneum* had a 39-fold resistance. Experiments were carried out by exposing wheat treated with bioresmethrin, chlorpyrifos-methyl and pirimiphos-methyl to typical conditions in the upper layers of a bulk grain store for intervals up to 25 weeks. These treated samples of wheat were then challenged with insects known to be resistant to malathion and it was found that chlorpyrifos-methyl and pirimiphos-methyl were adequately effective against the malathion-resistant strains, except *R. dominica* where the addition of bioresmethrin was indicated.

Insecticide dusts were screened as admixtures, primarily with maize in 2 small-scale storage trials designed to simulate tribal storage conditions in Zimbabwe. Fenitrothion (8 mg/kg) and fenthion (8 mg/kg) showed the greatest persistence, followed by malathion, iodofenphos, tetrachlorvinphos (all at 8 mg/kg) and pyrethrins (1.85 and 2.5 mg/kg) **(Weaving 1975)**.

LaHue (1975b) evaluated malathion low-volume spray and drip-on applications and a formulation on granular carbon as protective treatments against insect attack on wheat stored in small bins for 12 months. Damaging infestations of mixed populations of stored-grain insects developed in all untreated control bins during the first 4 months of storage from insects released in the storage room. The low volume malathion emulsion spray and the malathion granular carbon application gave excellent protection for 12 months.

Wilkin and Haward (1975) studied the effect of temperature on the action of 4 pesticides on 3 species of storage mites. They found that malathion plus lindane (7.5 + 2.5 mg/kg) was effective against all 3 species but that reducing the temperature decreased the effectiveness. The time required to obtain complete kill of *Acarus siro* increased from 4 days at 25°C to 21 days at 10°C and complete kill was not obtained at 5°C within 21 days. These differences were even more marked with *Tyrophagus putrescentiae* and *Glycyphagus destructor*.

DeLima (1976) published the results of an ecological study of traditional on-farm maize storage in Kenya and the effects of control measures. It was found that for a storage period of up to 3 months fumigation with phosphine was the best means of control. For longer periods of storage the application of 2% malathion dust at the rate of 10 mg/kg was superior.

LaHue and Dicke (1976a) evaluated selected insecticides applied to high-moisture sorghum grain to prevent stored-grain insect attack. They found that the addition of 0.2% propionic acid by weight to the high moisture sorghum to prevent spoilage did not influence insect infestations. Malathion was considered less effective than pirimiphos-methyl, chlorpyrifos-methyl and fenitrothion.

Watters (1976b) determined the persistence of malathion applied to granary surfaces to control *Tribolium castaneum* in Canada. Malathion emulsion was sprayed so as to give a deposit of 1 g/m^2 on concrete, metal, plywood and hardwood floor surfaces in a farm granary. Insecticide persistence was assessed by 24-hours exposure of adults of *Tribolium castaneum* from weeks 1 to 40 after treatment. Malathion provided 100% mortality on metal and plywood surfaces for 40 weeks and on hardwood for 4 weeks. The effectiveness of malathion on concrete was completely gone by week 10. It produced only 40% mortality by week 5.

Chawla and Bindra (1976) carried out a laboratory screening to determine the relative toxicity of 7 organophosphorus insecticides and 2 synthetic pyrethroids in their commercial formulations against adults of 3 stored-product pest species. Malathion was employed as the standard. Five of the 9 insecticides were comparable to malathion in effectiveness.

Cogburn (1976) found that a dosage of 14 mg/kg of malathion was ineffective in protecting unhusked rice against a malathion-resistant strain of *Sitotroga cerealella*.

LaHue (1976) tested pirimiphos-methyl, chlorpyrifos-methyl and fenitrothion as broad-spectrum protectants of seed corn against stored-product insects over a 21-month period. At the dosages used, the candidate materials were more effective than malathion emulsion. A similar study using high-moisture grain sorghum yielded similar results **(LaHue and Dicke 1976a).**

Morallo-Rejesus and Carino (1976b) tested the residual toxicity of 5 insecticides applied at 10, 20, 30 and 50 mg/kg to 2 species of adult insects on 3 maize and sorghum varieties. The residual toxicity increased with increase in concentration. Pronounced differences in the susceptibility of the test insects to the insecticides were noted. Malathion was more persistent on sorghum than on maize but the other materials were generally more effective.

Ardley and Sticka (1977) compared fenitrothion at 2.5, 5 and 6 mg/kg with malathion at 12 and 18 mg/kg for the protection of bulk wheat in vertical bin silos, and 10 mg/kg fenitrothion against 18 mg/kg malathion in horizontal bulk depots. In both types of storage, the persistence of the insecticides could be correlated with the temperature and moisture content of the grain. Under the conditions of silo storage, both protectants remained effective against *Rhyzopertha dominica, Sitophilus granarius* and secondary pests for as long as 19 months. Under the more severe conditions of horizontal bulk storage, the higher applications were effective for only 6–7 months. Fenitrothion at one half the application rate of malathion appeared to give equivalent protection against insect damage to stored wheat.

Desmarchelier *et al.* (1977a) demonstrated the potent vapour toxicity of dichlorvos to 2 species of insects and contrasted this with the action of malathion, which showed virtually no vapour effect.

Hyari *et al.* (1977) applied emulsifiable and encapsulated formulations of malathion and fenitrothion to wheat and after storing for 12 months determined the relative effectiveness of the residues against 4 insect pests. Fenitrothion was more effective than malathion against all species. No significant differences in relative effectiveness were found between the encapsulated and emulsifiable formulations.

LaHue (1977c), in field tests, showed that malathion was inferior to 3 other insecticides in controlling indigenous infestations of 7 species of grain insects infesting seed corn. **LaHue and Dicke (1977)** showed similar results when carrying out identical trials on wheat.

Loong Fat Lim and Sudderuddin (1977) found chlorpyrifos-methyl from 9 to 30 times more potent than malathion against 2 important stored-product pests in Malaysia.

Quinlan (1977) studied the effect of surface sprays of malathion in conjunction with the spraying of the outside of the silos with malathion for controlling insect populations in stored shelled corn. The results confirmed the value of the malathion surface spray treatment reported by **Quinlan (1972)** and demonstrated that it reduces insect populations in unaerated corn as well as in aerated corn. There was practically no value in spraying the outside of the silo. However, spray treatment of the inside walls may be of value.

Qayyum (1978) reviewed the importance of storing food grains in Pakistan and the measures which had been adopted to reduce losses. As a result of studies conducted on insect pests of stored products, malathion was found to be the best for protecting the grains from the ravages of insect pests. Structural treatment, surface treatment and admixture of malathion were recommended.

Williams *et al.* (1978) evaluated the relative toxicities of malathion and chlorpyrifos-methyl to stored-product beetles. Chlorpyrifos-methyl was the more toxic compound to the 5 species tested and was found to be equally effective against malathion-resistant and susceptible strains of *Tribolium*.

LaHue (1978) compared insecticidal dusts against sprays prepared from emulsifiable concentrates and showed that the performance of the dusts, prepared by adding the insecticide to diatomaceous earth was superior to the corresponding sprays. 2% malathion dust applied at the rate of 500 g/tonne gave complete protection for up to 8 months.

Watters and Bickis (1978), working in Canada, compared mechanical handling and mechanical handling supplemented with malathion admixture to control *Cryptolestes ferrugineus* in stored wheat. Wheat infested with *Cryptolestes ferrugineus* was subjected to mechanical movement through a farm auger to compare the extent of insect control with that obtained by treating augered wheat with malathion at 8 mg/kg. Augered grain treated with malathion markedly reduced the numbers of both adults and larvae whilst the auger treatment did not control adults or larvae.

Of 6 insecticides tested in the laboratory in India for protection of stored gram (*Phaseolus mungo*) seeds against *Callosobruchus chinensis*, 24 mg/kg malathion was the most persistent, based on mortality of the insects **(Dhari *et al.* 1978).**

Quinlan (1979a) studied the effect of malathion thermal aerosols applied to stored corn, soyabeans, wheat and sorghum using aeration. The 0.5 micron aerosol particles were introduced into the grain overspace at a rate of 2.5 mg/kg and then pulled down into the grain with the aeration fan. Samples of the grain were taken at 0.61 m depth intervals throughout the grain mass for residue analysis and bioassays. Screen test cages containing *Sitophilus oryzae* were placed at the same sample locations. Lethal amounts of malathion were found at all levels of the corn and soybeans. The malathion penetrated the wheat and sorghum only to the 3.05 m level. Complete insect mortality occurred only to the 1.83 m level both with the bioassay samples and the caged insects.

Calderon and Desmarchelier (1979) tested the susceptibility of *Tribolium castaneum* adults to some organophosphorus compounds after pre-exposure to carbon monoxide. They found that exposure to carbon monoxide in air for 20 hours reduced the susceptibility of *Tribolium castaneum* to malathion quite considerably. This is thought to be due to inhibition of the microsomal oxidation paths and inhibition of the formation of malaoxon.

LaHue and Kadoum (1979) compared the residual effectiveness of emulsion and encapsulated formulations of malathion against 4 species of stored-grain beetles. They found that when applied to plywood surfaces the encapsulations were much more persistent and more effective than the corresponding concentrations of emulsion.

Singh (1979) tested the value of a pre-harvest spray of malathion for controlling insect infestation of stored rice. Malathion emulsion was applied as a spray to the panicles of paddy rice growing in the field. The rate of application was adjusted to deposit 16, 24, 32 and 48 mg/kg on the rice in husk. The rice was harvested 5 days after spraying. It was then subjected to bioassay with *Sitophilus oryzae, Rhyzopertha dominica* and *Sitotroga cerealella*. The rice treated at the 2 higher rates (32 and 48 mg/kg) killed most of the insects after a 24 hour exposure period. Samples of rice from all treatments remained substantially free of infestation when exposed to conditions of high infestation pressure. Samples of rice from the 2 highest treatment rates remained free of living insects after 150 days exposure. The author concluded that this simple approach could replace fumigation treatment for pest control in stored grains.

Bengston and Desmarchelier (1979) reviewed the biological efficacy of a range of new grain protectants, by comparison with malathion, against a number of different strains of stored-product insects collected in Australia. These clearly showed why alternative grain protectants had given superior performance under Australian conditions. The emergence of malathion-resistant strains of several pests has further accentuated this.

Quinlan *et al.* (1979) showed in several laboratory and small-bin studies throughout a 9-month period that chlorpyrifos-methyl was much more effective than malathion in controlling a variety of insects in high moisture red winter wheat, stored in plywood bins.

Attia *et al.* (1980) studied the effect of a number of synergists with a variety of insecticides to ascertain whether it would be possible to potentiate their effect against organophosphorus-resistant strains of the Indian meal moth (*Plodia interpunctella*). None of the synergists increased the potency of the insecticides, including malathion, significantly.

Cherkovskaya *et al.* (1980), working in the USSR, discussed the value of malathion in protecting flour stored in sacks in warehouses. Following field tests in Uzbekistan in 1977 a rate of application of 0.3 g/m^2 was recommended for practical use. Sprayed stacks remained free from infestation for 9 months. The spraying of flour sacks with malathion was approved for general use in 1980.

Hindmarsh and Macdonald (1980) pointed out that severe infestations of *Sitophilus zeamais* and *Sitotroga cerealella* occur in maize cobs stored under farm conditions in Zambia causing up to 92% damaged grain 8 months after harvest. Shelling the maize reduced *Sitotroga cerealella* damage to low levels and also stabilised the control of *Sitophilus zeamais*. Malathion at 12 mg/kg as a dust admixture to shelled maize in traditional mud-walled, timber and thatch storage cribs kept damaged grains below 10% for up to 10 months after harvest.

O'Donnell (1980) determined the toxicities of 4 insecticides to *Tribolium confusum* under 2 sets of conditions of temperature and humidity and found that all 4 insecticides were much more effective at 25°C than at 5°C. The minimum dose of malathion required to kill all the insects tested at 25°C was 200 mg/m^2 but at 5°C this had to be increased to 1495 mg/m^2. Pirimiphos-methyl and fenitrothion were much less affected by the lowering of the temperature.

Weaving (1980) carried out further laboratory tests in Zimbabwe to screen insecticides for admixture with maize grain in 2 small-scale storage trials for a period of 12 months. Fenitrothion and fenthion at 8 mg/kg showed the greatest persistence, followed by malathion. *Sitophilus zeamais* was the principal pest causing most of the damage; there were 7 other species also present.

Borah and Mohan (1981) reported that the most effective treatment for reducing the natural infestations of *Sitotroga cerealella* and *Sitophilus oryzae* on stored paddy rice was obtained by fumigation with hydrogen phosphide followed by dusting with 5% malathion dust applied at the rate of 250 g/100 kg.

Schulten (1981), in a review of the use of pesticides at farmer and village level, pointed out that though malathion has many advantages, including safety in handling and residues, effectiveness against primary pests, rapid kill of adults etc., the disadvantage of malathion lies primarily in the increasing spread of resistance to this compound, its rapid breakdown on moist grains and the instability of the dust formulations.

Gelosi (1982), in a dissertation on the morphology, biology and control of *Sitophilus oryzae* in Italy, recommended thorough cleaning of cereal storehouses and of containers and vehicles used for transportation followed by treatment with malathion.

Salunkhe (1982) measured, by means of laboratory tests, the effectiveness of various concentrations of 6 insecticides as seed treatments for the control of *Tribolium castaneum* infesting stored sorghum in India. Test exposures of adult beetles to treated grain immediately, and 3 and 6 months after treatment, showed that malathion at a concentration of 25 mg/kg was the best treatment.

Storey *et al.* (1982a) in a critical appraisal of the incidence of insects and the use of malathion on wheat and maize exported from the United States pointed out that 28% of 2058 wheat samples and 8.4% of 2383 maize samples examined during a 2-year period, January 1977–December 1978, contained malathion and that the level of malathion residue was generally low (mean of 1.03 mg/kg on maize and 2.5 mg/kg on wheat). They did point out that few of the consignments found to have been treated with malathion actually contained live insects (0.5% for maize and 2.7% for wheat). Many of the treatments had been applied just prior to sampling but even so the incidence of malathion on grain arriving at the ports was only 11.6% for wheat and 7.7% for maize. The relatively low concentration of malathion in wheat (2.8 mg/kg) treated immediately before it passed through the sampler suggested that most port terminal elevator treatments do not result in recommended deposits of malathion on the grain. The authors concluded that the overall use of malathion in stored-grain insect control programs from farms to export was minimal.

Bressani *et al.* (1982) working in Guatemala, studied the best ways to protect harvested maize from insect damage while in storage. They found that burlap sacks previously treated with 2% malathion afforded the best protection and were practical and inexpensive.

Bitran *et al.* (1982a) carried out tests in Brazil under conditions simulating farm storage to compare various treatment programs to protect maize cobs from damage by *Sitophilus zeamais* and *Sitotroga cerealella*. Prior fumigation with phosphine increased the efficiency of treatment with malathion or a mixture of malathion and dichlorvos. The amount of malathion applied was equivalent to 20 mg/kg.

Bitran *et al.* (1982b) assessed the efficiency of a malathion/dichlorvos mixture for the protection of bagged maize in tests in Brazil. The best results were obtained with a 13:1 mixture of 95% malathion and 98% dichlorvos applied directly to the sacks at 1.78 ml/m^2 at monthly intervals in spite of high ambient temperatures. This treatment was as economical as the standard fumigation treatment.

Zhang *et al.* (1982) reported that 30 mg/kg malathion had no significant effect on the germination of rice, wheat, maize, sorghum and barley. There was adverse effect on the germination of maize treated with 10 000 mg/kg malathion.

Hasan *et al.* (1983), working in India, studied the synergistic effect on bromophos of sublethal amounts of other insecticides. Only 3 compounds with a structure like that of bromophos, including malathion, had marked synergistic effects.

Madrid *et al.* (1983) studied the effects of malathion dust on the development of infestations of *Plodia interpunctella* or *Ephestia cautella* on stored wheat. He found that young larvae failed to develop from eggs placed in the wheat treated with 2 and 8 mg/kg malathion. Fourth-instar larvae developed into adults in the wheat treated with 1 mg/kg malathion. Since insecticide degradation had occurred, both species of moths developed successfully when wheat, which initially contained malathion at 2 mg/kg, was reinfested after 2 weeks of storage. However, larval mortality in both species was high. For both species, none of the original larvae or those developing from eggs added after 2 weeks survived in wheat treated with malathion at 8 mg/kg.

Parker *et al.* (1983) described a controlled-release delivery system containing malathion for suppression of rice storage pests. The controlled release delivery system consists of plastic laminates in which the active ingredient is sealed in a layer between outer plastic layers. The active ingredient migrates continuously through one or more initially inert outer layers to the surface where it is rendered biologically available. The authors demonstrated by results of field experiments the protection of rice packed in sacks to which one or more of the controlled release delivery systems had been added. Chlorpyrifos appeared more efficacious than similar concentrations of malathion in reducing percentage increase of *Sitophilus oryzae* populations.

Mensah and White (1984) reported the laboratory evaluation of malathion-treated sawdust for control of stored-product insects in empty granaries and food warehouses in Canada. They found that the size of sawdust particles did not affect the residual activity of malathion on a given surface during 8 weeks at 25°C and 50% r.h. However, malathion on the sawdust decomposed more rapidly in contact with concrete surfaces than on wood surfaces. Sawdust, initially treated with 2% malathion, gave nearly 100% mortality of *Tribolium castaneum* for 4 weeks on concrete and 16 weeks on wood and steel.

Horton (1984) reported that discriminating-dose tests for malathion resistance indicated the widespread occurrence of malathion-specific resistance in strains of *Tribolium castaneum* and the lack of malathion resistance in strains of *Tribolium confusum, Sitophilus zeamais, Sitophilus oryzae, Sitophilus granarius* and *Rhyzopertha dominica* collected in the field in South Carolina.

Storey *et al.* (1984a) surveyed the use of pest management practices and their effectiveness in grain stored on the farm in the USA. From a survey of more than 8000 farms across 27 states they concluded that preventative and remedial actions reported by producers to maintain the quality of grain during storage on the farm was minimal. Application of malathion when the grain was put into storage was the most frequent action reported in wheat, aeration the most frequent in corn, and fumigation the principal action in oats. Malathion was found on only 14.6% of the wheat, 8.2% of the corn and 4.2% of the oats. When malathion was present, the incidence, number of different species, and/or density of most species was generally less than in untreated grain. The effectiveness of malathion decreased significantly at grain-moisture levels above 12%. Nearly two-thirds of the insects found in malathion-treated grain occurred in grain containing residues of 2 mg/kg or less.

Evans (1985) carried out laboratory tests using populations of 5 stored-product beetles collected in the field in Uganda to determine whether they were resistant to malathion. The three samples of *Sitophilus zeamais* were completely resistant to the discriminating-dose of malathion and the samples of *Tribolium castaneum* and *Tribolium confusum* exhibited varying degrees of resistance. There was also some cross-resistance to lindane.

Golob *et al.* (1985) tested a range of insecticides against young adults of *Prostephanus truncatus* by topical treatment and by exposure to treated filter papers. Malathion was relatively ineffective. When applied to maize grain at a rate of 24 mg/kg, malathion (as a dust) significantly reduced the damage to maize grain for between 4 and 6 months. This was confirmed in a field trial. However the weight loss of maize grain treated with malathion was considerably higher then maize treated with permethrin.

Wohlgemuth (1984) tested 18 commercial insecticides applied to sorghum under tropical conditions over a period of 24 months. Malathion powder was applied at the rate of 11.25 mg/kg and after storage at 36°C and 50% r.h., separate samples were infested with 30 adult insects of 5 species. The treatment was fully lethal against 3 species for 12 months and 90% effective against 2 species for 24 months. However, against adult *Rhyzopertha dominica* and *Tribolium castaneum* it was fully lethal for only 3 months and 1 month, respectively. Malathion proved effective in reducing reproduction but *Rhyzopertha dominica* and *Trogoderma granarium* were the most tolerant species.

Many further references which deal with the usefulness and effectiveness of malathion in grain storage have been located. Many of these deal with studies in which malathion was used as the reference material for the evaluation of other insecticides. Others deal with grain storage and management techniques which are compared against malathion treatments as a reference to the desirable degree of control. Seventy-eight such papers have been studied but it was decided not to include a detailed review merely in the interests of space. However, the references have been listed in **Table 18.2.**

18.2 Degradation

Gunther *et al.* (1958) determined the persistence of malathion on stored wheat of 11.5% to 12% moisture content, stored at 25°C in darkness. They found that malathion lost biological effectiveness with time and had a residual half-life of 5.6 months **(Table 18.3).**

Table 18.3. Residues of malathion on and in treated wheat (12% moisture content) **(Gunther *et al.* 1958)**

Months after treatment	Ppm found at various dosages			
	16 ppm	8 ppm	4 ppm	2 ppm
0	7.1	6.5	1.3	0.7
1	4.0	4.7	1.2	0.3
2	3.3	4.3	1.0	0.3
3	3.5	4.2	0.9	0.7
4	3.6	3.0	0.7	0.1
5	2.9	2.8	0.8	0.7
6	2.1	1.9	0.9	0.7

Malathion, applied at 2 mg/kg to wheat of 13.5% moisture content, caused 99% mortality of *Cryptolestes ferrugineus* 8 months after treatment, the wheat being stored in airtight containers at 9°C. A dosage of 16 mg/kg was required for comparable control in wheat of 15.5% moisture content but this dosage was ineffective in wheat of 18% moisture content at 5 months after treatment **(Watters 1959a).**

Womack and LaHue (1959) studied the fate of malathion-emulsion sprays applied to corn of 13.75% moisture content. At the end of 8 months, corn which had been treated at the rate of 5, 10, 15 and 20 mg/kg contained residues of 4.1, 5.8, 7.3 and 7.5 mg/kg of malathion respectively.

Following upon the realisation that available analytical techniques were unsatisfactory for determining malathion residues in cereals and oilseeds an expert Panel was set up in the UK to undertake collaborative studies of analytical methods for determining malathion residues in stored products: see report of the 'Malathion Panel' **(Anon 1960).** Experience showed that differences in techniques of different analysts affected results; the purity of reagents was important; the Soxhlet method gave the best recovery; and on wheat treated with malathion there was a measurable amount of breakdown of the chemical over a period of days. An average of 11.25% loss was found on sub-samples of wheat analysed by 8 Panel members on the fourth day after treatment with malathion at the rate of 4 mg/kg. The study showed that breakdown of malathion was influenced by temperature, the loss being 17.5% in 6

Table 18.2. Additional papers dealing with usefulness and effectiveness of malathion as a grain protectant

Al-Naji 1978	Lemon 1967b
Al-Naji *et al.* 1977	Longoni & Michieli 1969
Al-Saffar & Al-Iraqi 1981	Mensah & Watters 1979b
Armstrong 1961	Morallo-Rejesus 1973
Bansode *et al.* 1981	Morallo-Rejesus & Eroles 1974, 1976
Bengston *et al.* 1977	Morallo-Rejesus & Nerona 1973
Bindra 1974	McCallum-Deighton 1974, 1978
Bitran *et al.* 1979a; b; 1980b; c	McDonald & Gillenwater 1967
Bitran *et al.* 1983a; b	McGaughey 1972b
Blackman 1969	Pandey *et al.* 1979
Chahal & Ramzan 1982	Parkin 1966
Cogburn 1967; 1981	Peng 1983
Desmarchelier 1975b	Pinniger 1975
Desmarchelier & Bengston 1979a; b	Prakash & Pasalu 1981
Desmarchelier *et al.* 1981c; d	Quinlan 1978
Dhari *et al.* 1978	Quinlan *et al.* 1980
Ferguson & Waller 1982	Rai 1977
Girish *et al.* 1974	Raju 1984
Godavari Bai *et al.* 1960	Spiers & Zettler 1969
Golob & Ashman 1974	Srivastava & Gopal 1984
Graciet *et al.* 1974	Strong 1970
Green *et al.* 1970	Strong & Sbur 1964b
Greening 1980	Tyler & Binns 1977; 1982
Harein & De Las Casas 1974	Viljoen *et al.* 1981a; b
Joia 1983	Watters 1977
Kirkpatrick *et al.* 1983	Weaving 1981
Kumar *et al.* 1982	White 1984
LaHue 1965; 1966; 1970b	Wilkin 1975a; b
LaHue 1975a; 1975c; 1977a	Yadav *et al.* 1980
LaHue & Dicke 1976b; 1977	

weeks at 0°-5°C and 60% at 22–26°C. Likewise the degradation of malathion was promoted by grain moisture, a significant difference being found between wheat with 12.3% moisture and the same wheat when the moisture content had been increased to 15.3%. **Bates *et al.* (1962)** and **Bates and Rowlands (1964)** later contributed to improving the accuracy of the extraction and analysis of malathion residues and thus to the understanding of the degradation of malathion on stored commodities.

Godavari Bai *et al.* (1960) mixed malathion with flour to provide concentrations of 8, 16 and 32 mg/kg and the flour was stored in jute sacks at 24°C to 28°C and 68% to 80% relative humidity. The amounts of malathion remaining were 2, 5.6 and 12.8 mg/kg after 8 weeks and 0.9, 2.7 and 7.3 mg/kg after 12 weeks. Malathion caused off-odours in dough and unleavened bread prepared from flour containing 16 and 32 mg/kg of malathion. It is assumed that the malathion used must have contained significant quantities of mercaptans, which were subsequently eliminated in commercial production.

Barley with a moisture content of 15.5% was treated to give an average residue of 11.4 mg/kg and 80 days later was found to have an average residue of 8.9 mg/kg **(Papworth 1961).**

Strong *et al.* (1961) reported studies designed to determine the influence of formulation on the effectiveness of malathion sprays for stored wheat. Laboratory tests were made to determine the degree of toxicity and repellency to *Sitophilus oryzae* and *Tribolium confusum* of malathion (5 mg/kg) applied to wheat in sprays formulated as solutions, emulsions and wettable powder suspensions; the residues remaining on wheat at various intervals after application of sprays; the proportional amount of residue on the wheat related to each formulation that is removed by the usual handling and cleaning practices followed at milling and the residues finally resulting in the milled fractions. Differences between formulations in the effectiveness of malathion applied in sprays as a protectant against insect infestations in stored wheat were not clearly defined, the influence of moisture content and storage temperature on the persistence of biologically effective malathion deposits on wheat was recognised by reduced mortalities of test insects with increases in temperature and moisture content. Residues of malathion persisting in whole grain at various intervals after application of spray decreased with time. The influence of moisture content and storage temperature on the persistance of residues was indicated by the results from chemical analysis for residues on whole grain.

Parkin *et al.* (1962) treated samples of barley with a flour-based malathion dust at 4, 8, 16 and 32 mg/kg. The treated barley was stored at 20°C and 70% relative humidity. After 9 months storage in stoppered jars, the grain was found to contain 0, 0.3, 2.1 and 9.4 mg/kg malathion respectively. Larger quantities of barley were treated with 8 and 16 mg/kg malathion and stored outside in the United Kingdom in covered bins for the 9-month period from October to July. After 9 months, the barley contained 0.8 and 3.6 mg/kg respectively before malting and nil after malting.

Bang and Floyd (1962) treated polished rice of 12.7% moisture content with malathion as a dust or spray at 8 mg/kg. Analysis for chemical residue showed an actual 7.1 mg/kg in the spray treatment and 7.3 mg/kg in the dust treatment in samples taken immediately after treatments were applied. One month after treatment the malathion residues were 5.8 and 5.2 mg/kg, respectively, in the spray and in the dust treatments and after 3 months the residues were 1.0 and 0.7 mg/kg, respectively.

Maize and wheat treated with malathion was stored in sealed jars for 6 months and analysed monthly for malathion and its derivatives **(Rowlands 1964)**. Dimethyl phosphorothiolothionic acid, malathion mono-acid and malathion di-acid were identified, but malaoxon and its corresponding derivatives were not detected. The breakdown of malathion under these conditions appears to be hydrolytic rather than oxidative, or both.

Rowlands (1964) studied the metabolism of malathion in stored maize and wheat. The hydrolysis products, dimethyl dithiophosphoric acid; malathion monocarboxylic acid and malathion dicarboxylic acid were found in maize during 6 months storage and dimethyl dithiophosphoric acid and malathion dicarboxylic acid were found in wheat. No oxidation products were detected in either cereal but later work **(Rowlands 1965a)** showed that some slight oxidation to malaoxon occurred in wheat of unknown age during storage for 1 week, the highest level being reached at 1 to 2 days after treatment. Malaoxon was degraded to dimethyl phosphorothioate; dimethyl phosphate; and mono- and dicarboxylic acid derivatives. Later work **(Rowlands 1966c)** showed that when freshly harvested wheat, of 18% moisture content, was treated at the rate of 10 mg/kg with malathion and stored at 20°C, production of malaoxon was initially rapid and continued for 3 to 4 weeks, although the highest level of malaoxon (0.9 mg/kg) was observed after 2 to 3 days. A sample of the same wheat, treated at 1 week after harvest, produced malaoxon over 1 week only and treatment of similar samples at 2 to 8 weeks after harvest showed scarcely any oxidative activity at all, slight amounts of malaoxon being produced during 2 to 5 days following treatment. Triphenyl phosphate was degraded to di- and monophenyl phosphates by stored wheat **(Rowlands 1965b)**.

Alessandrini (1965) determined the persistence of malathion on wheat during storage. From the results, she concluded that malathion residues decreased fairly rapidly in treated wheat during storage and that the decrease is quicker on some varieties of wheat than on others. Examination of the data suggests that the difference in the rate of degradation was more directly related to free water content than to wheat variety, being much more rapid on wheat which had been washed prior to treatment than on the corresponding unwashed wheat.

Rowlands and Clements (1965a) studying the degradation of malathion in rice brans concluded that malathion, applied to rice bran, degraded during storage, both by chemical and enzymatic routes. The chemical route appeared to depend on the level of free fatty acids present from the hydrolysis of rice bran oil, and the enzymatic route involved a phosphatase-type mechanism.

Williams and Gradidge (1965), after studying the application of malathion spray to wheat on conveyor belts at a major storage installation, found that a significant amount of the spray deposited on the grain was quickly lost if the grain had to travel any distance between the point of application and the point of storage.

Mookherjee *et al.* (1965), who studied the efficacy of malathion dust against *Sitophilus oryzae* and *Tribolium castaneum* and its persistence on stored wheat, found that it was necessary to add malathion at a rate of 24 mg/kg to wheat of 13% moisture content held at 30°C in order to ensure that over 80% of both species were killed at the end of 3 months. Chemical analysis had indicated that by the end of 3 months after treatment the concentration of malathion had fallen from 24 mg/kg to 3.8 mg/kg. By the end of 4 months over 96% of the original deposit had degraded so that the remaining residues were below 1 mg/kg.

Rowlands (1965a), reporting *in vitro* and *in vivo* oxidation and hydrolysis of malathion by wheat grain enzymes, indicated that wheat grain esterases were shown to oxidise malathion to malaoxon which had no lasting inhibitory effect on those hydrolytic enzymes involved in the degradation of malathion. Breakdown of malaoxon proceeded by routes similar to those for malathion and it was concluded, from the small amount of malaoxon produced in a short storage trial, that oxidation played only a minor role in the degradation of malathion by wheat grain esterases.

Consideration of the enzymology of grain and the associated microflora suggests hydrolytic attack as the main metabolic pathway of phosphates in stored grain. Triphenyl phosphate (TPP) was degraded to di- and monophenyl phosphate by stored wheat **(Rowlands 1965b)**. It was shown that TPP prevented hydrolysis of malathion or malaoxon, by wheat germ and human serum preparations *in vitro*, at the carboxyester and the thiolo-carbon linkage but not at the thiolo-phosphorus bond. It was thought, therefore, that TPP, which has very low mammalian toxicity, might possibly prolong the effective life of malathion applied to damp grain, and **Rowlands and Clements (1966)** determined that TPP at 10:1 synergist:insecticide ratio did, in fact, decrease the breakdown of malathion by grain or frugal enzymes during storage. Unfortunately the inevitable hydrolysis of TPP to diphenyl phosphate results in increased inhibition of mammalian blood cholinesterase *in vitro* and TPP may not therefore be of any practical use.

Rowlands and Clements (1965b) treated maize and wheat with malathion which was then stored in sealed jars for 6 months and analysed monthly for malathion and its derivatives. The degradation of malathion was, as anticipated, much slower in autoclaved wheat where it was assumed no enzymic action could occur; of the 250 μg originally present in a jar, only 40 μg were lost in the first 3 months against 98 μg in the living sample. There was less breakdown in the maize than in the living wheat.

Tyler *et al.* (1965a) showed that malathion applied at 10 mg/kg breaks down more rapidly on newly harvested grain than on stored grain. Experiments showed this applied equally to wheat and barley.

Tyler *et al.* (1965b) found that grain which had become heated to 29°C before treatment with malathion (10 mg/kg) lost all its malathion and became infested with *Oryzaephilus surinamensis* within 5 weeks whereas the same grain which had cooled to 16.5°C still contained 2.2 mg/kg malathion and was effectively protected.

Godavari Bai (1965) who studied the persistence of malathion on treated food grains under different conditions of storage and processing showed that the moisture content of grain, especially at the extremes of the range, probably has a significant effect on the ability of the grain tissues to metabolise malathion to non-toxic products, and at these extremes the increased or decreased metabolism of malathion may affect significantly the location of the intact pesticide and its degradation products within the grain of the bulk stored wheat.

Green and Tyler (1966) in an extensive field comparison of malathion, dichlorvos and fenitrothion for control of *Oryzaephilus surinamensis* found that malathion applied to barley with a moisture content within the range 13–15% and an initial temperature of 25°C, brought the infestation under control and remained biologically fully effective for 22 weeks by which time the average residue level had fallen from the initial 8.53 mg/kg on the day of treatment to 1.9 mg/kg. At the end of 22 weeks the temperature of the grain had fallen to 10°C.

Rowlands and Clements (1966) extended the work begun by **Rowlands (1965b)** on the effect of triphenyl phosphate on malathion metabolism in wheat. They found that triphenyl phosphate added to wheat at 100 mg/kg either before or simultaneously with malathion treatment at 10 mg/kg slowed down the rate of malathion degradation considerably over a period of 6 months. The residual malathion was the same at 6 months when the grain was pre-treated or simultaneously treated with triphenyl phosphate; however pre-treatment kept the level of malathion at a higher value for a greater period of time. Examination of the malathion metabolites verified the expectation that triphenyl phosphate would interfere with the esterase attacking at the carboxyester and thiolophosphate linkages.

Rowlands (1966c) found oxidative activity converting malathion to its thiolate and phosphate analogues in the seed coats and germs of wheat grains by *in vivo* and *in vitro* studies. Hydrolytic activity found in the germ and endosperm and capable of detoxifying malathion was demonstrated *in vitro* and non-specific hydrolases were located by histochemical tests. **Tyler *et al.* (1966a)** confirmed preliminary experiments which showed that malathion breaks down more rapidly on newly harvested grain than on grain which had been stored up to 9 months before treatment. The apparent loss of insecticide was always greater than expected however, particularly with the newly harvested grain. This was shown to be due to an extremely rapid degradation of malathion within the first few hours. After 1 month, however, residues did not differ significantly between stored and newly harvested grain. Bioassay confirmed the chemical analysis.

Tyler *et al.* (1966b) reported further observations on clean grain, protectively treated with 10 mg/kg malathion, that high-moisture content and high temperature rapidly degrade malathion deposits. Wherever temperature and moisture content of the grain were highest, insecticide residues were soon undetectable by either chemical analysis or bioassay.

Rowlands (1967b) found that in the presence of dichlorvos, grain esterases were prevented from hydrolysing malathion for some 3–4 weeks of storage and ultimate degradation of malathion proceeded via phosphatase products rather than carboxyesterase products. Under normal conditions, malathion degrades primarily to carboxyesterase products.

Rowlands and Horler (1967) studied the penetration of malathion into wheat grains and found that in the first few days after application most of the intact malathion was in the endosperm. Moisture content did not seem to affect movement between pericarp and endosperm although the insecticide only penetrated into the germ at moisture contents of 14% or lower.

Horler and Rowlands (1967) showed that the penetration of malathion applied topically in hexane to the dorsal side of wheat grains is very little affected by the moisture content of the grain. At the attachment region there is a large increase in the amount of malathion entering the germ as moisture contents fall below 14%. Application to the attachment region leads to the very rapid appearance of malathion in the endosperm, irrespective of moisture content, whereas application to the back leads to slow penetration to the endosperm. It is in the endosperm that most of the breakdown products are eventually encountered.

LaHue (1967) treated sorghum with an initial moisture content of 13.5% with malathion at 2 rates, 11.4 and 17.1 mg/kg. The temperature in the infestation room ranged from 20–35°C during the summer and from 10–30°C during the winter. Moisture content of the grain ranged from 13.5% to 9.5% over 12 months. Over this period the residues recovered by chemical analysis fell from 4.4 mg/kg immediately after treatment to 0.56 mg/kg after 12 months in the case of the lower treatment and 11.0 to 1.14 mg/kg in the case of the higher rate.

Hill and Border (1967) set up a series of trials in order to dispel misgivings about the acceptability of barley, treated with malathion, for malting, brewing and distilling. They showed that there was a very substantial loss of malathion at each step in the malting process until the barley, after germination and drying (stage known as malt), contained less than 1% of the original level on the barley. In the trials where barley was treated at the rate of 10 mg/kg, the malt contained 0.08 mg/kg of malathion. The authors pointed out that in commercial practice the barley would contain less than 5 mg/kg at the stage it was utilised for malting. These studies confirmed the work of **Papworth (1961).**

Tyler and Green (1968) studied the effectiveness of malathion as a grain protectant under severe practical conditions on farms in Britain. Though good control of insects was achieved, the deposit degraded rapidly on moist barley in granaries; the concentration degrading from 10 mg/kg to approximately 1 mg/kg after 12 weeks. In grain which spontaneously heated to 60°C, no residues could be detected at the end of 12 weeks.

Lindgren *et al.* (1968) prepared a review of world literature on residues in raw and processed foods resulting from post-harvest insecticidal treatments in which they devote 11 pages to residues of malathion in various stored grain. These authors dealt with a number of papers which were not available for this review.

Tyler and Rowlands (1968) showed that the addition of sodium carboxymethyl cellulose to a malathion spray can delay the degradation of the residual insecticide. It was found that the addition of 0.5% sodium carboxymethyl cellulose to malathion emulsion applied as a grain protectant to wheat of 12% moisture content delayed the degradation of the malathion residue to such a degree that the residue on such wheat was higher at the end of 8 months than the residue from wheat treated with the malathion emulsion alone after only 5 months storage.

Kadoum and LaHue (1969) studied the effect of hybrid, moisture content, foreign material, and storage temperature on the degradation of malathion residue in grain sorghum. Malathion was applied to grain sorghum from 6 hybrids at the rate of 11 mg/kg. There was little effect of variations between hybrids on malathion degradation during 12 months storage at 27°C and 60% r.h. Grain moisture content and storage temperature significantly affected residue retention, but foreign material had no effect on the rate of malathion disappearance.

McGaughey (1969) reported preliminary results of intermediate scale tests of effectiveness and persistence of malathion applied for the protection of rough rice. When malathion was applied to rice in husk at 14, 21 and 28 mg/kg the residues only just exceeded 8 mg/kg on the whole rough rice 30 days after treatment.

Studies of malathion residues in imported cereals arriving in the UK were reported by **Thompson and Hill (1969)**. Forty-five samples from 19 shipments of Australian wheat were examined. Eleven samples contained more than 5 mg/kg, 6 of these from a shipment part of which cargo was known to have been treated twice with malathion, once at the up-country collecting centres and again at the port silos. The maximum found was 7.4 mg/kg, minimum 1.6 mg/kg and the mean 4.1 mg/kg. There was little difference between these and the figures for the previous 12 months, when 46 samples from 19 shipments were examined. In that period 5 exceeded 5 mg/kg, the maximum being 8.3 mg/kg, the minimum 0.4 mg/kg and the mean 3.2 mg/kg.

Koivistoinen and Aalto (1970) presented a review of information concerning malathion residues and their fate in cereals but they contributed no new information. However, they outlined a proposal to identify the chemistry of the degradation mechanisms in cereals using radio-labelled malathion. The outcome of these studies has not been sighted.

LaHue (1970a) evaluated malathion along with other protectants on wheat in small bins. He found that the malathion residue degraded from 7.5 mg/kg to 2.7 mg/kg during 12 months. The grain had a moisture content of 11.8% and the trials were carried out at an average temperature of 25°C.

McGaughey (1970b) found that malathion, applied to protect rough rice at a dosage of 20 mg/kg, left a residue at the end of 3 months of less than 8 mg/kg. The initial deposit on hulls was 80 mg/kg but this did not fall below 8 mg/kg for 11 months.

Kadoum and Sae (1970) showed *in vitro* that malathion and its breakdown products do not inhibit sorghum grain esterase, but observed that they could not exclude possible interactions at sites other than the active sites of enzymatic proteins.

In India, **Chawla and Bindra (1971)** showed that washing of maize grain with water removed 70–88% of the malathion residue. In another study by the same authors, washing of wheat grain followed by its drying, removed most of the residue. Addition of an emulsifier at the rate of 0.1% enhanced the residue-removal efficiency of water significantly, and sun-drying proved more effective than oven-drying **(Bindra 1974).**

Elms *et al.* (1972) studied the breakdown of malathion and dichlorvos mixtures, applied to wheat. Dichlorvos was shown to retard malathion degradation and malathion to decrease slightly the degradation of dichlorvos. Degradation curves for decay of malathion and dichlorvos were presented for different combinations of the insecticides.

Horler (1972) published a brief note of a study of the recovery of ^{14}C-labelled malathion from aged samples of wheat with different moisture contents ranging from 10% through 15% to 20%. He reported large differences between the amount extracted with n-hexane compared with that recovered by methanol. The difference was more than 50% with wheat of 20% moisture content. In all cases methanol extracted more than n-hexane. If one were to use the hexane extract as an indication it could be concluded that approximately 50% degradation occurred in 2 months on wheat of 15% moisture content (temperature not given). However, the same wheat extracted with methanol showed only 30% loss in the same time.

Bindra and Sidhu (1972) studied the dissipation of malathion residues on maize grain in relation to dosage, storage conditions and baking. They did not identify the products of breakdown but observed about 75% loss during the cooking of maize flour.

Kadoum and LaHue (1972) studied the degradation of malathion on viable and sterilised sorghum grain during 9 months storage. The migration of malathion residue to the endosperm and embryo of sterilised sorghum grains was greatest during the first month of storage; however, the amounts recovered during that time accounted for only 41% of the remaining total malathion residue. After 3 months storage, the total residue gradually decreased until 34% of the initial total deposit remained at the end of 9 months. At all time intervals, during the 9 months storage, more malathion residue was detected on, and in, the sterilised sorghum kernels than on the live viable kernels, but the pattern of residue decline was similar. Only 14% of the initial malathion deposit remained on the viable sorghum after 9 months storage. These findings indicated that during a normal storage period, enzymes contributed to the breakdown of malathion applied to sorghum grain as a grain protectant.

LaHue (1974) reported residue data obtained during a 12-month period from hard winter wheat, shelled maize and sorghum grain, treated with malathion emulsion and malathion dust and analysed at regular intervals throughout the storage period. **Table 18.4** shows the residue level found on wheat, maize and sorghum during the 12-month storage period. Malathion residues degraded more slowly on wheat treated with a dust formulation than on wheat treated with the emulsion. These results are similar to those obtained on maize. The residues degraded very rapidly on the 17.6% moisture sorghum. This corresponded with a loss in effectiveness against stored-product insects in the second month of storage.

The rate at which malathion penetrated and degraded on wheat, corn and sorghum was determined during a 6-month storage period. Analyses made 24 hours after treatment to determine initial residue deposit showed that 85% or more of the total residue remained on the exterior of the kernels of all 3 grains. During the first month of storage, residues increased internally but decreased pronouncedly on kernel exteriors. During the remainder of the storage period, malathion residue disappeared from the exterior of the kernels more rapidly than from the interior **(Kadoum and LaHue 1974).**

Bengston *et al.* (1975) made a detailed comparison of bioresmethrin, chlorpyrifos-methyl and pirimiphos-methyl against malathion-resistant insects in wheat. After determining the lethal concentration in sprayed grain assays, the workers subjected the treated wheat stored under bulk storage conditions to bioassay and to chemical analysis. Malathion, with an estimated half-life of 9.5 weeks, degraded so rapidly and to such low levels that it was quite obvious why malathion was no longer effective in controlling malathion-resistant strains of several species. The level of resistance was, in some instances, so high that the residues would have needed to be 100-fold greater to provide reasonable assurance of control.

Table 18.4. Average malathion residues in mg/kg on grain stored for 12 months in small bins after application as an emulsifiable concentrate (E) or as a dust (D) **(LaHue 1974)**

Grain	Moisture (%)	Formulation	Intended dosage (mg/kg)	Post-treatment periods					
				24 hours	Months 1	3	6	9	12
Wheat	12.5	E	10.4	8.6	5.2	3.8	3.0	2.0	1.4
		D	10.4	8.2	6.7	4.6	3.9	2.6	2.0
		D	10.4	8.8	7.3	5.3	4.0	2.8	2.3
		D	10.4	8.5	7.0	5.0	3.9	2.7	2.3
Maize	12.5	E	11.2	9.0	6.1	4.0	3.2	2.2	1.4
		E	11.2	7.3	4.5	4.1	2.0	2.3	1.6
		D	11.2	7.7	6.8	5.8	5.3	3.8	3.0
		D	11.2	8.4	7.7	5.0	4.6	4.7	4.4
		D	11.2	8.8	8.5	6.0	6.4	3.3	2.8
Sorghum	17.6	E	16.7	13.4	3.8	1.6	1.5	0.9	0.8
		D	11.2	9.1	0.5	0.3	0.2	0.2	0.1
		D	11.2	7.5	0.5	0.3	0.2	0.2	0.1

Rowlands (1975) published an extensive review of the metabolism of the contact insecticides in stored grains but this does not throw further light on the degradation of malathion on grain.

Williams *et al.* (1975) described how the application of malathion non-uniformly to wheat decreased its degradation. By treating 1% of the grain with 750–800 mg/kg malathion before returning it to the bulk of the grain going on to storage it was possible to achieve a blend of grains that contained an average of 10 mg/kg. This protected the grain from insects for a longer time than did the usual practice of distributing the insecticide as evenly as possible throughout the whole mass of grain. This technique protected a batch of wheat from *Sitophilus oryzae* infestation for 10 months whilst another batch treated with the same amount of malathion in the conventional way was infested in 6 months.

LaHue and Dicke (1976a) treated sorghum of 17.4% moisture content with malathion-emulsion spray and malathion dust. Residues resulting from the application of the malathion emulsion degraded rapidly from the initial deposit of 13.4 mg/kg to 3.8 mg/kg during the first month of storage, but thereafter a gradual reduction occurred. The residues resulting from the malathion dust application degraded from the initial deposit of 9.1 mg/kg to 0.5 mg/kg during the first month of storage. The moisture content of the sorghum decreased to 13% during the second and third month during of the 12 months storage.

Minett and Williams (1976) found that the breakdown of malathion in wheat treated conventionally with a diluted emulsion designed to give as uniform a deposit as possible proceeded more rapidly and to a greater degree over a period of 30 weeks than did a non-uniform deposit for undiluted concentrate applied to similar wheat under similar conditions.

Watters (1976b) showed that malathion, coming in contact with concrete surfaces of granaries, degraded much more rapidly than did similar deposits on wood.

LaHue (1977a) showed, in a study where hard winter wheat was treated in bulk bins with malathion and pirimiphos-methyl, the malathion degraded so that there was only 16% of the original deposit remaining at the end of 12 months storage, whereas 83% of the original pirimiphos-methyl deposit remained after the same period.

Kadoum and LaHue (1977) studied the degradation of malathion in wheat and milling fractions in Kansas, USA. They treated hard winter wheat with malathion emulsion at the rate of 10 mg/kg and recovered by analysis 9.4 mg/kg 24 hours after application when the moisture content averaged 12.1%. The malathion residues degraded gradually with 14.7% of the initial deposit remaining on the whole grain after 12 months storage. The grain had been stored at 26°C and 60% r.h.

Bitran and Olivera (1977), working in Brazil, took coffee beans with which malathion had been mixed at the rate of 8 mg/kg and artifically infested them with *Corcyra cephalonica* and *Araecerus fasciculatus* and stored them for 6 months in small jute sacks inside screened cages in a warehouse where the relative humidity remained throughout at about 90%. At the end of the first period, the moisture content of the coffee beans initially 12.9 — 13.2% had risen to more than 15%. Under these conditions malathion had not remained stable and the development of the two pests had been in no way inhibited being comparable with the development in untreated beans stored under the same conditions.

Quinlan (1978) carried out tests in Kansas, USA to determine the effectiveness of malathion applied as a protectant against insect infestations in 1.5-tonne lots of clean, 14.6% moisture, hard red winter wheat stored in plywood bins. The wheat was sampled before treatment and at various intervals over a 9-month period for inspection of live insects, bioassay, moisture content and determination of insecticide residues. The malathion-treated grain contained an average of 55 insects/kg of wheat, while untreated wheat contained an average of 156 insects/kg. The half-life of the malathion residues was 1.7 months.

Al-Nagi (1978) studied the effectiveness of malathion on stored-grain insects and the fate of the insecticide residues on the grain and milling products. Malathion residues were analysed in treated wheat of 12.5% moisture content, held for 365 days. It was found that 85% of the malathion degraded by the end of the study period.

Banks and Desmarchelier (1978), in discussing the influence of water vapour and temperature on changes in pesticide levels on grain with time, pointed out that, in the case of malathion, which had been used very widely for many years as a grain protectant in Australia, field experience had shown that a half-life of 3 months could be expected. Their laboratory-determined half-life correlated well with hard-won industry experience. **Figure 18.1** compares the average values for malathion residues in Queensland wheat in 1973 with the semi-logarithmic decay profile found in the laboratory. However, until about 1976, the grain industry had been using a 'linear decay with time' model. This is based on a half-life of 6 months; although in this instance, because the decay is not semi-logarithmic, the term half-life cannot be used correctly. This model was known to predict a slower loss rate than was observed. Furthermore, even the original estimate of the 'half-life of 6 months', although based on a correct semi-logarithmic model, was incorrect and was obtained from the data points shown in **Figure 18.1**. The scatter is obviously too great for an accurate estimate of the rate constant. The calculation is further complicated in this instance by poor analytical recovery of malathion, even from samples taken directly after application. **Figure 18.2** also shows the curve expected from a 'linear' model with a halving of the residue in the first 6 months and a semi-logarithmic model based on an incorrect 'half-life'

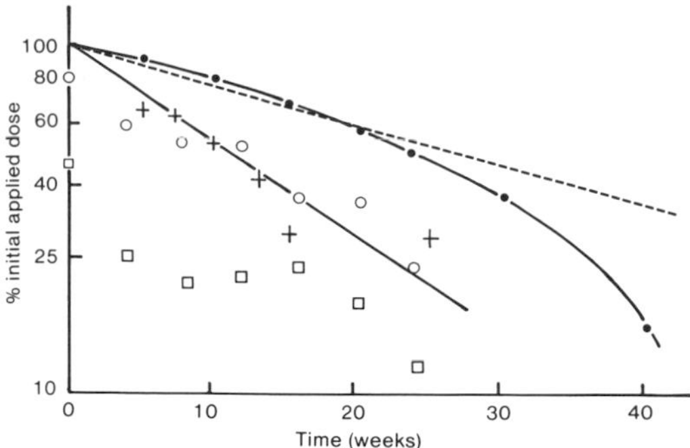

Figure 18.1. Decay of malathion on warm dry grain (≈ 30°C, 0.5-0.6 water activity). Field results (+) compared with pseudo first order decay model (−) as determined in the laboratory (half-life, 3 months). Decay profiles predicted from an incorrect half-life (6 months) (---) and a 'linear' decay model for 12 mg/kg (●) as used commercially until recently. Also shown, transformed data points for 8 mg/kg (○) and 16 mg/kg (□) applied dose on which an estimate of a 6 months half-life was based **(Banks and Desmarchelier 1978).**

of 6 months. In both cases, there is a substantial discrepancy between observed and predicted results.

A model of loss following pseudo first-order reaction kinetics applied to all pesticides on grain so far tested **(Desmarchelier 1977a)**.

Laboratory and field studies are in reasonable accord for all compounds **(see Figure 18.2)**. Further studies have shown that this model holds true, not only for each pesticide on wheat, but on all other grains provided the calculation is based on water activity (interstitial relative humidity) rather than absolute moisture content **(Desmarchelier and Bengston 1979a, 1979b; Desmarchelier 1978a, 1980).**

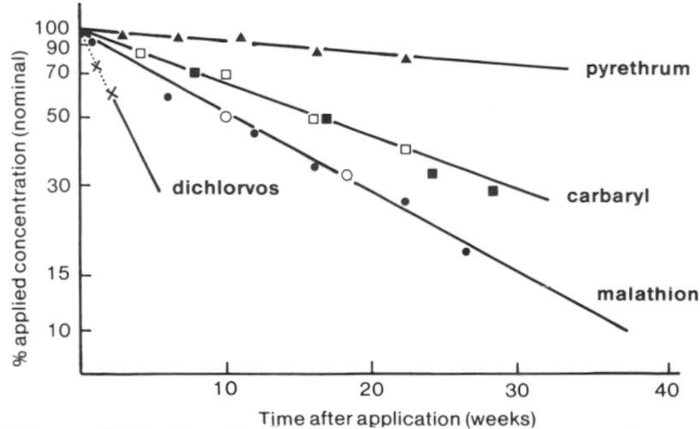

Figure 18.2. Semilogarithmic decay rates for pyrethrum (30°C), carbaryl (30°C), malathion (30°C), and dichlorvos (20°C) on wheat at 0.6 water activity with field results for malathion (●) and carbaryl (■) under similar conditions **(Banks and Desmarchelier 1978).**

Kadoum and LaHue (1979b) studied the effect of grain-moisture content on malathion-residue degradation. During studies carried out in Kansas, residue data were obtained over a 12 month period from stored maize and wheat with 4 moisture levels, following application of malathion-emulsion sprays. The sequence of malathion residue losses did not vary greatly between the different moisture levels when based on the initial deposits 24 hours after treatment, but the ratios of degradation were distinct for each moisture level. After 12 months storage, 42.7, 21.9, 9.1 and 4.7% of the initial residue deposit remained on the wheat of 10, 12, 14 and 16% moisture, respectively, and the corresponding residues for maize were 35.1, 26.1, 11.0 and 8.8%.

Watters and Mensah (1979) measured the stability of malathion applied to stored wheat for control of *Cryptolestes ferrugineus*. In studies in a farm granary in Manitoba, Canada, emulsion sprays of malathion were applied to wheat having 11.7 and 16.8% moisture contents to give concentrations of 8 and 12 mg/kg malathion in the grain in order to determine the stability of the insecticide at different temperatures during storage for 9 months. Bioassay with adults of *Cryptolestes ferrugineus* and residue analysis showed that there was a gradually increasing rate of breakdown of the insecticide on dry or moist grain stored in the granary and in the laboratory and that the breakdown rate was higher at 20° than at 10°C and highest at 30°C.

Quinlan *et al.* (1979) studied the effectiveness of malathion as a protectant for high-moisture stored wheat. Tests were conducted with 1.5-tonne lots of cleaned, untreated hard winter wheat held in plywood bins at 25°C. Samples were drawn at 8 intervals over 9 months for the determination of malathion residues. The calculated dosage applied (10.4 mg/kg) apparently declined to 8.6 mg/kg 1 day after application. By the fifth day this had declined to 4.1 mg/kg and by the twelfth day to 2.6 mg/kg. Thereafter there was little apparent degradation before the end of the second month. By the ninth month the amount of malathion found in the grain was only 0.5 mg/kg. The half-life of malathion under these conditions was calculated to be 1.7 months.

Kadoum and LaHue (1979a) studied the degradation of malathion on wheat and corn of various moisture contents. Residue data were obtained during a 12-month period from corn and wheat of various levels of moisture following 10 mg/kg application of malathion-emulsion spray. The sequence of malathion-residue losses did not vary greatly among the different moisture levels 24 hours after the initial treatment; however, the ratios of degradation were distinct for each moisture level. After 12 months storage, 41, 21, 7 and 3% of the initial deposit remained on wheat containing 10, 12, 14 and 16% moisture respectively. For corn of similar moisture contents the deposits remaining after 12 months were 34, 24, 8 and 5% respectively.

Desmarchelier *et al.* (1979b) described how temperature and moisture manipulation can be combined with the use of grain protectants to provide flexible procedures of pest control that are generally superior to, and cheaper than, the control given by cooling alone, or drying alone, or chemicals alone. Their argument applies particularly to the use of malathion since high temperature and high moisture plays such a powerful part in degrading the deposit.

Hindmarsh and Macdonald (1980) reported 2 trials to control insect pests of farm-stored maize in Zambia in which malathion applied as a dust admixture at the rate of 12 mg/kg to shelled maize in traditional mud-walled timber and thatch storage cribs kept the damage to grains below 10% for up to 10 months after harvest. The grain-moisture content fluctuated between 8.4% and 16%, the relative humidity between 40 and 87% and the temperature between 18 and 32°C. The malathion residue level fell from 5.8 mg/kg soon after application to 2.2 mg/kg 10 months later. In a second season the decline was much less (from 4.7mg/kg to 3.0mg/kg) over 8 months.

Abdel-Kader and Webster (1980), Abdel-Kader *et al.* (1980), and **Abdel-Kader (1981)** reported various aspects of a study to determine the low temperature degradation of malathion in stored wheat. Malathion was applied as a water-based emulsion to provide a deposit of 8 mg/kg on wheat. The treated grain was stored at –35°, –20°, –5°, 5°, 10°, 20° and 27°C in a study to determine the influence of Canadian prairie temperatures on the rate of malathion degradation. Less than 3% breakdown of the insecticide residue occurred on wheat that had been stored at –35°C and –20°C for 72 weeks. As the temperature increased, malathion residues decreased, the rate depending upon the age of the deposit. At the end of 72 weeks, 74, 59, 26, 5 and 4% of the initial deposit remained on wheat stored at –5°, 5°, 10°, 20° and 27°C respectively.

Desmarchelier *et al.* (1980b) provided the results of a collaborative study of the rates of decay of 5 grain protectant insecticides, including malathion, on wheat. These studies provided much of the basic information from which the model **(Desmarchelier 1978a, 1980; Desmarchelier and Bengston 1979a, 1979b)** was developed.

This study is closely related to the publications by **Desmarchelier *et al.* (1977b)** and **Desmarchelier (1980a)** and needs to be evaluated in the light of the Report of the Malathion Panel **(Anon 1960)** and the papers by **Bates *et al.* (1962)** and **Bates and Rowlands (1964)** (see also Chapter 5). The Australian studies revealed that incomplete extraction of aged residues with hexane resulted in models that overemphasise the loss of malathion. These workers showed quite conclusively that the loss of insecticide, including malathion, from samples collected immediately after treatment superimposed on the effect of failure to extract all available residues has created highly significant errors in deducing the rate of degradation of malathion (and numerous other insecticides) applied to stored grain. They were successful in deducing the actual application rate and confirming that this agreed with the calculated application rate in spite of indications to the contrary that were obtained from analytical data. The validity of this deduction was confirmed by placing the samples of wheat immediately in a glass sample jar containing methanol for transport to the laboratory. This had been shown to prevent the 20–30% loss that otherwise occurred particularly with freshly treated grain samples.

Webster and Abdel-Kader (1983) described analytical methods that were developed to analyse residues of malathion and its metabolites on stored wheat. When malathion was applied to stored wheat at 12 mg/kg and the grain was stored at 20°C for 12 months, the concentration of malathion mono-acid rose to a maximum of 2 mg/kg at 6 months, that of dimethyl hydrogen phosphorodithioate to 1 mg/kg at 3 months and malathion diacid to 0.7 mg/kg at 12 months.

Anderegg and Madlsen (1983a) studied the effect of insecticide distribution and storage time on the degradation of ^{14}C-malathion in stored wheat. The degradation of ^{14}C-malathion in stored wheat grain in which 5% or 100% of the kernels had been treated with the same overall amount of insecticide was determined over a 12-month period in studies in the USA. There was no difference in the degradation of the insecticide in either treatment. However, over the storage period, quantities of chloroform-soluble radio-carbon decreased, whereas those of water-soluble and unextractable residues increased. Quantities of volatile radio-carbon reached a peak after 6 months. *Tribolium castaneum* developed more slowly in the presence of unextractable residues than in their absence, although such 'bound' residues were not acutely toxic to adults and did not affect the number of progeny.

Anderegg and Madisen (1983b) studied the degradation of ^{14}C-malathion in whole wheat containing 0, 2.5, 5 or 10% dockage in the form of ground wheat. The effect of storage time on the degradation of ^{14}C-malathion on wheat containing 2.5% dockage was also studied. The total quantity of ^{14}C-malathion residues recovered in the dockage fraction increased significantly as the ratio of dockage to whole grain increased and as the storage time increased. The large proportion of the radio-carbon recovered from the dockage fraction was in the form of unextractable residues. As the proportion of dockage in the grain increased the recovery of volatile ^{14}C compounds decreased.

Anderegg and Madisen (1983c) also studied the degradation of ^{14}C-malathion in stored corn and wheat inoculated with *Aspergillus glaucus*. Grains free of internal storage fungi were surface-sterilised before treatment with ^{14}C-malathion and fungal inoculation. After 6 months, inoculated corn contained 74% of applied radio-carbon, compared with 92% in sterilised controls. Similarly, inoculated wheat contained 86% of applied radio-carbon, compared to 93% in sterilised control wheat. The low quantities of radio-carbon recovered from the inoculated grain may have resulted from losses by volatilisation or $^{14}CO_2$ evolution, both of which were determined in this study. Although malathion was degraded in both sterilised and inoculated grain, sterilised grain contained significantly higher amounts of ^{14}C-malathion than inoculated grain after 6 months.

White and Nowicki (1985) carried out a study on rapeseed containing up to 1 mg/kg malathion and about 10% moisture to determine the time required to reach residue levels of 0.1 mg/kg. The rapeseed was stored for up to 34 weeks at 10°, 20° and 30°C. Malathion levels below 0.1 mg/kg were found only in seed at 20°C, 7.7% moisture by 32 weeks and at 30°C, 7.1% moisture content by 16 weeks. Manipulation of temperature in the region of 60°C for periods of a few hours was ineffective in reducing malathion residues to 0.1 mg/kg.

Many more papers dealing with some aspect of the degradation of malathion residues on grain have been examined. Many of these reflect the degradation by the loss of biological activity, whilst others have gone quite far in efforts to unravel the process of metabolism of malathion on stored grain. Such references are recorded in **Table 18.5.**

18.3 Fate in Milling, Processing and Cooking

Schesser *et al.* (1958) were among the first to publish information on the level of malathion residues in milling fractions of wheat treated for the control of insect pests. Using wheat which had been treated as dosages of 2.5 to 7.5 mg/kg, these workers found that the highest residues were in the bran and shorts. Very little insecticide carried over into the flour, especially when the grain was milled shortly after treatment.

Milling and baking tests were carried out by **Watters (1959a)** on samples of untreated wheat and wheat treated with malathion at 4 and 8 mg/kg. The test samples were neither cleaned nor scoured and before milling were tested for odour. The samples were milled to approximately 72% extraction. Bakings were made to a typical bread formula. There was no adverse effect on the milling or baking quality of wheat treated with malathion at 4 or 8 mg/kg.

Godavari Bai *et al.* (1960) mixed malathion with flour to provide concentrations of 8, 16 and 32 mg/kg and the flour was stored in jute sacks and sampled at intervals. Malathion was reported to have caused off-odours in dough and unleavened bread prepared from flour containing 16 and 32 mg/kg respectively.

Witt *et al.* (1960) reported on the fate of malathion during storage of barley, on the retention of residual malathion during the malting process, on the effect of the residues on mashing, fermentation, and on the physiology of the yeast, and the effect on the initial taste and shelf-life stability of the final beer. Barley retained about 7.7 mg/kg of the protectant after 14 to 60 days storage. The malt, exposed for 14 days before processing, retained 5 mg/kg but the malt which had been in store for 60 days showed less than 0.2 mg/kg residues. Conventional laboratory analyses indicated that treated malts were somewhat better modified than the controls. This characteristic was translated into pilot-plant brewery performance and

Table 18.5. Further references which deal with the degradation of malathion residues on stored grain.

Al-Naji *et al.* 1977	Longoni & Michieli 1969
Al-Saffar & Al-Iraqi 1981	Masud *et al.* 1976
Banks 1972	Mensah *et al.* 1979b
Bindra 1974	Morallo-Rejesus & Carino 1976b
Bindra & Sidhu 1972	McFarlane & Harris 1964
Chawla & Bindra 1973	Oden & Sahin 1964
Cogburn 1981	Parker *et al.* 1983
Cogburn *et al.* 1983	Parkin 1966
Desmarchelier 1975b	Quinlan *et al.* 1980
Desmarchelier *et al.* 1977a, 1980b, 1981d	Redlinger & Womack 1966
Giles 1973	Rowlands 1966b, 1967a
Girish *et al.* 1970	Rowlands & Clarke 1968
Godavari Bai 1965, 1968	Rowlands *et al.* 1967
Golob & Ashman 1974	Schesser *et al.* 1958
Harein 1982	Schulten 1981
Joia 1983	Stankovic *et al.* 1965
Kadoum 1969	Sun *et al.* 1984
Kawamura *et al.* 1980	Udeaan & Bindra 1971b
Kirkpatrick *et al.* 1983	Viljoen *et al.* 1981b
LaHue 1965, 1966, 1975a	Weaving 1975, 1980
Lai *et al.* 1981	Wilkin 1975b
Lockwood *et al.* 1974	Winks & Bailey 1965
	Witt *et al.* 1960

treated malts yielded higher extracts. The beer, made from the malt held 14 days after malting, retained 2.4 mg/kg, whereas that made from malt held for 60 days before brewing, showed only a negligible residue. No difference was noted in taste, flavour or clarity between the beers.

Strong *et al.* (1961) studied the fate of residues remaining on wheat after application of malathion sprays, formulated as solutions, emulsions and wettable powder suspensions. In milled samples, the larger amounts of malathion residues were recovered from bran (including germ), shorts and wheat middlings. Very little malathion carried over into the flour. The solution formulation led to somewhat higher residues on the bran and shorts.

Watt (1962) reported tests carried out by the Bread Research Institute of Australia, as well as overseas work which showed that the milling quality of wheat, or the baking quality of the resultant flour, was unaffected even when excessive amounts of malathion were used.

Bates and Rowlands (1964) drew attention to the interference by extractives in the determination of malathion residues in rice bran and recommended chromatographic purification of the extracts on Fuller's earth.

McFarlane and Harris (1964) studied the uptake by 6 feed ingredients, citrus pulp, soybean meal, cotton seeed meal, coconut meal, rice bran, and groundnut meal, or malathion from malathion wettable powder applied to jute sacks for the control of stored product pests. They found that commodities with high oil-content are likely to be more susceptible to contamination, and that uptake increased with increasing temperatures. Such information is relevant to the question of the distribution of malathion into the lipid-rich protions of the grain kernel, particularly the bran and germ.

Alessandrini (1965) determined the persistence and fate of malathion in or on wheat during storage and subsequent milling and baking or cooking of the products made from treated wheat. When treated wheat was milled, most of the residue remained in the bran and shorts and relatively little appeared in the white flour, though the residue level in wholemeal flour was comparable to that occurring in the whole wheat. The bread, prepared from flours containing malathion, either added or coming from treated wheat, contained about 8% to 16% of the amount in the flour. With the low-malathion residue in flour, milled from wheat treated with dosages used in commercial practice, the residue in the bread was considered negligible. The residue in uncooked pasta was about 10% to 11% of the amount of malathion present in the flour used to prepare it. The residues in cooked pasta were not detectable.

Rowlands and Clements (1965a), who studied the degradation of malathion on rice bran, concluded that malathion, applied to bran, degraded during storage, both by chemical and enzymatic routes. The chemical route appeard to depend on the level of free fatty acids present from the hydrolysis of rice bran oil, and the enzymatic route involved a phosphatase-type mechanism.

In Yugoslavia, **Stankovic *et al.* (1965)** treated wheat with malathion grain-protectant dust at a rate resulting in a deposit of 10 mg/kg of malathion active ingredient. Malathion residues were determined 51 days and 203 days later on the wheat and on flour and bran produced from the treated wheat. On the basis of the amounts of residue found, the following conclusions were drawn:

(1) Wheat treated with malathion at the rate of 10 mg/kg may be used for human or domestic animal nutrition only if it is washed first.
(2) Flour produced from wheat treated with malathion at the rate of 10 mg/kg may be used for human or domestic animal nutrition only if the wheat is washed before milling.
(3) Bran produced from wheat treated with malathion at the rate of 10 mg/kg may be used for a limited period of time if the wheat is sieved before milling and for an unlimited period of time if the wheat is washed before milling.

Acton and Parouchais (1966) conducted chemical analyses to determine the amount of malathion which might eventually be found in flour and bread derived from wheat treated with malathion at the rate of 8 mg/kg. Samples of malathion-treated wheat were collected from various wheat receival points throughout South Australia. They drew three conclusions from the results obtained.

(1) Loss of malathion occurred during the milling process and the higher the initial dosage of malathion, the greater was the proportionate loss.
(2) Since the malathion was applied to the outside of the wheat grain, there was naturally more malathion in wholemeal flour than in white flour. However, even at low concentrations of malathion in the wheat, some malathion was found in white flour. It is not likely that much malathion would be transferred to the flour purely by mechanical contact with the bran during the milling process, so it would seem more probable that malathion found its way to the endosperm by penetration through the bran after the spray treatment, particularly over long storage periods.
(3) Within the limits of detection of the experiment (0.3 mg/kg) no malathion was found in white bread, even from wheat containing 26 mg/kg malathion. Malathion may be present in bread baked from malathion-treated wheat, but for white bread the amount of malathion is not likely to exceed 0.3 mg/kg. In the case of wholemeal bread, however, more than 1 mg/kg malathion may still be present, when made from wheat treated with no more than the generally accepted level of 8 mg/kg malathion.

Bressau (1966) (see also FAO/WHO 1978) conducted a trial with rye grain in two 3-tonne lots, each at 13.2% moisture content. One lot was treated with malathion dust and the other with emulsifiable-concentrate formulation at 10–11 mg/kg and stored in hoppers for 34 months. The temperature was initially 20°C and varied between 20°C and 4°C during storage. After 11 months storage, the grain was mixed and a portion was removed, cleaned by aspiration and milled. Malathion levels were determined in different milling fractions and in bread made from 2 types of flour. There was no significant difference in the malathion level of the 2 lots of rye treated with dust and emulsion, respectively. The aspirated grain was converted into 2 grades of flour, designated grey flour, type 1150 and white flour, type 815. Though the level of malathion in the grey type flour was approximately twice that in white flour, there was no significant difference between the level in the flours prepared from either lot of grain treated with dust or emulsion respectively. The shorts and pollards contained approximately 10 times the level in the white flour and the bran contained approximately 15 times that in the white flour. Flat (unleavened) bread, made from flour containing 1 and 2 mg/kg malathion contained residues between 0.2 and 0.6 mg/kg after baking for 15 minutes and drying. No off-flavours in grain, milling fractions or bread were observed.

Hill and Border (1967) published a report on the breakdown of malathion during the malting process in order to allay misgivings by maltsters, brewers and distillers, that barley, treated with malathion, might not be suitable for malting, brewing and fermenting. They treated 2 lots of barley with malathion to leave a deposit of approximately 10 mg/kg. They found that approximately 70% of the residue was removed or destroyed by the first steeping liquor. The second steeping liquor removed 50% of the remaining residue. During the 8 days germination, 90% of the residue present at the commencement of germination had been destroyed so that barley after germination and drying at 70°C for 7 hours contained no more than 0.08 mg/kg of malathion. Preparation of a wort from the final samples was not carried out because the malathion content had been reduced to such a low level in the dry malt that a determination of malathion in the brew would have been impossible by the analytical methods employed. In commercial practice, barley is unlikely to contain malathion at levels approaching 10 mg/kg and therefore the amount finding its way into wort and beer will be extremely low indeed.

Lindgren *et al.* (1968) reviewed the information available at that time on residues in raw and processed foods resulting from post-harvest insecticidal treatments and they have brought together much valuable information.

Alessandrini *et al.* (1968) reported that, during the preparation of flour from treated wheat, much of the residue is removed so that the residues remaining in the flour are in the order of one-tenth of those in the hull of wheat.

McGaughey (1969) studied the fate of malathion applied to rice in husk (rough rice) both during storage and in subsequent milling. He treated several lots of rice at rates equivalent to 14, 21 and 28 mg/kg respectively. Samples of rice were drawn from each lot for residue analysis. **Table 18.6** presents the average residue levels found on rough rice and each milling fraction after 30 days in storage under conditions typical of those in the Gulf Coast region of the USA.

Table 18.6. Malathion residues (mg/kg) determined in milling fractions of rice 30 days after application to rice in husk **(McGaughey 1969).**

Treatment level	Rice in husk	Hulls	Bran	Milled rice
0 mg/kg	0	0	0	0
14 mg/kg	4.1	13.7	13.9	0.14
21 mg/kg	6.5	23.6	20.9	0.24
28 mg/kg	8.7	33.5	28.2	0.32

It is noteworthy that the residue level on milled rice was only 1% of the treatment level applied 30 days previously. Subsequent studies **(McGaughey 1970b)** showed that malathion deposits of about 17 to 20 mg/kg protected rice for 6 to 12 months under Gulf Coast conditions. Initial deposits on rough rice of 20 mg/kg decreased to levels below 8 mg/kg within 3 months. Initial deposits on hulls were 80 mg/kg and these remained above 8 mg/kg for 7 to 11 months. Residues on bran and milled rice increased for 3 months and then declined. Residues on bran reached a maximum of 30 mg/kg but fell below 8 mg/kg in 6 to 10 months. The author noted significant differences in the concentration of residues in the bran between varieties of rice and deduced that this was dependent upon the relative surface area and the thickness of the hull covering kernels of each variety.

Udeaan and Bindra (1971a) obtained extensive data on the residues of malathion in different fractions of treated food grains and their finished derivatives. It was found that the smallest loss (62–77%) of malathion residue during processing of grain into chapatis was on pearl millet, a small grain. About 73% of the malathion in maize flour was dissipated during its processing into chapatis, while more than 90% of the malathion residue on rice in husk was removed during the dehusking and most of the remaining residue was lost during cooking.

McGaughey (1972b) investigated the amounts of residues present in the milled fractions of 3 varieties of malathion-treated rice. He found that treating at levels between 10 and 20 mg/kg caused excessively high residues (exceptionally 80 mg/kg) in the bran, though the low proportion of this tissue in the whole grain has to be taken into account and that these high residues persisted throughout 7 to 11 months of storage. In other experiments the concentration in the seed coat of the rice rose to a maximum of about 30 mg/kg during 3 months, thereafter declining to 8 mg/kg when the product was milled after a storage period of 6 to 10 months. This slow fluctuation within the tissues points to movements of insecticide within the grain. Overall residues in the rice milled after 6 to 10 months storage were about 0.3 mg/kg, relative to the 8 mg/kg present in the much lighter weight bran fraction comprising seed coat and husk.

Mukherjee *et al.* (1973) studied the fate of malathion residues in milled materials during processing and showed that 14–15% of the malathion in whole wheat flour survived the preparation into chapatis.

Bindra (1974) in reviewing grain protection and pesticide residues in India pointed out that local studies by **Chawla and Bindra (1971)** revealed that the washing of maize grain with water removed 70–88% of the residue. In another study, washing of wheat grain, followed by its drying, removed most of the residue of malathion, iodofenphos and bromophos. Addition of an emulsifier at the rate of 0.1% enhanced the residue-removal efficiency of water significantly, and sun-drying proved more effective than oven-drying **(Chawla and Bindra 1973).**

Lockwood *et al.* (1974) studied the degradation of organophosphorus pesticides in cereal grains during milling and cooking in India. Samples of paddy (rough rice), wheat and sorghum were treated with different concentrations of malathion for protection against insect infestation during storage. Subsequently, samples were milled, cooked and analysed by gas-liquid chromatography to determine residues remaining in the final products. Malathion treatments ranged from 10 to 80 mg/kg. During machine milling, residues on wheat and sorghum decreased by from 30% to 70%. Because residues were concentrated in the hull and bran of the paddy, relatively little remained on polished rice after milling. Even following treatments of paddy rice with 80 mg/kg, residue on polished rice was never more than 4 mg/kg. Milled products from each grain were cooked, using traditional Indian methods. Preparations involving boiling or steaming (wet cooking) appeared to result in complete degradation of residues as far as could be detected with methods sensitive to less than 0.1 mg/kg. Dry cooking methods, used in preparing chapatis from wheat and sorghum resulted in average losses of 43–75% of residues. The simple motorised mill used for grinding the wheat and sorghum was typical of that available in Indian villages. Minor adjustments in the clearance of the grinding discs made a noticeable difference in heating of the grain during grinding. Variation in residue loss may have partly resulted from that heating. Milling of wheat with a hand-operated grinding stone resulted in a much lower loss of malathion. Residue losses during power grinding ranged from 30% to 70%. That, combined with losses of 40–75% during the preparation of chapatis from wheat and sorghum flour, made the total residue loss during the processing of whole grain to chapatis 60–93%. With rice products, wet cooked, the residue totally degraded. This confirmed other studies **(Godavari Bai 1968; Udeaan and Bindra 1971a, 1971b; Godavari Bai *et al.* 1960).**

Golob and Ashman (1974) treated de-oiled rice bran with malathion and stored it at 27°C for 7 weeks. Unfortunately, they did not assay the dose actually applied, and their first analytical figures are for 8-day-old samples, wherein they found 18.1 and 13.8 mg/kg of malathion in identical bran samples. After 6 weeks storage at 27°C these had decreased to 13.7 and 11.4 mg/kg malathion.

Rowlands and Bramhall (1977) studied the uptake and translocation of malathion by stored wheat grain using gas-liquid chromatography and radio-labelled malathion. They reported that malathion residues are found chiefly in regions of high lipid content in the individual wheat kernel, mainly in the germ and scutellum, and seldom in the starchy endosperm. Malathion applied directly to the endosperm is rapidly transported to other regions.

Kadoum and LaHue (1977) studied the degradation of malathion in wheat and milling fractions by obtaining data on the malathion residues in whole wheat and wheat fractions during 12 months after treatment with the recommended 10 mg/kg application of malathion emulsion. Tempering the wheat to 15% moisture content required for milling, reduced the residue. Residues on the wheat fractions decreased with time. Residues in the fractions decreased in the following order: shorts, bran, red dog, flour. The detailed data are set out in **Table 18.7.**

Table 18.7. Average malathion residue in parts per million on indicated fractions of hard winter wheat after a 10 ppm application of a malathion emulsion spray [a] **(Kadoum and LaHue 1977)**

Ageing period	Whole wheat		Shorts	Bran	Red dog	Flour
	When sampled	At milling				
Days						
7	8.6	7.6	18.4	16.5	6.0	2.1
Months						
1	7.4	6.4	16.2	13.4	5.1	1.5
2	6.6	5.3	16.0	10.7	5.3	1.6
3	5.4	4.0	[b]	7.1	8.7	1.0
6	3.0	2.1	[b]	6.1	7.8	0.5
9	2.1	1.1	8.0	5.5	2.1	0.6
12	1.4	0.6	5.4	2.2	1.3	0.2
Control	<0.1	<0.1	<0.1	<0.1	<0.1	<0.1

a All untreated controls (at all ageing periods) contained <0.1 ppm malathion.
b Red dog and shorts combined in milling process.

Magallona and Celino (1977) studied the effect of cooking on malathion residues in rice. One kg of milled, polished rice was sprayed with malathion emulsions to obtain concentrations of 10, 20 and 30 mg/kg in the grains. The samples were air-dried and stored at room temperature for 1 week. Representative samples were taken for cooking. For each fortification level, 2 processing procedures were evaluated, (1) washing before cooking and (2) cooking without washing. Malathion residues were reduced by 85% with washing and cooking compared to 61% for cooking only. It must be appreciated that these experimental levels are considerably higher than occur in practice where malathion residues on polished rice are usually in the range of 0.5–1 mg/kg.

Al-Naji (1978) studied the fate of malathion residues on grain products. She found that greater residues of malathion were found in the bran and shorts and very small amounts in the flour. The losses of residues during milling were 5% to 7%. There were no changes in water absorption of flour, dough mixing tolerance, and the dough development time between samples treated with malathion and untreated samples. Chlorine bleaching at the rate of 125 g per 100 kg of soft wheat flour resulted in 32% to 46% loss in malathion residues.

Desmarchelier and Hogan (1978) who studied methods for destroying organophosphorus insecticide residues in grain dust (removed during the storage and handling of grain) found that aqueous alkali, such as caustic soda solution, when incorporated into the dust, would destroy virtually all of the residue within a few days. It was later found that hydrated lime, which was more economical and more readily blended with the dust, was equally effective. This finding was subsequently adopted for industrial use in processing grain dust and milling offal to render them suitable for use in livestock feeds.

Zemanek (1979), working in Czechoslovakia, reported that rye grain treated against storage pests with 10 mg/kg malathion contained 3.3 mg/kg after 11 months storage. Flour and bread made from this rye contained 1.2 and 0.2 mg/kg malathion respectively.

Mensah *et al.* (1979b) studied the level and fate of malathion residues in milled fractions of dry and tough wheat in Manitoba, Canada. Water-based emulsions of malathion were applied at 2 dosage rates to wheat of 12% and 16% moisture content to compare residue distribution in fractions milled from wheat samples over a period of 6 months. Comparatively smaller amounts of insecticide residue were found in the flour than in the bran or middlings. Residues recovered from fractions of wheat of 16% moisture content were generally lower than residues found in wheat fractions of 12% moisture content. The data are given in **Table 18.8.**

Kawamura *et al.* (1980) studied the distribution of malathion residues from milled wheat and the disappearance of malathion in flour during storage. **Table 18.9** shows the distribution of the various milling fractions, the concentration of malathion in each fraction and the proportion of the total malathion residues occurring in each fraction.

Wilkin and Fishwick (1981) reported a laboratory study in which malathion emulsion was applied to wheat to leave a deposit of 10 mg/kg. The treated wheat was stored for 4 weeks at ambient temperature before being milled to produce wholemeal flour. The actual deposit on the treated wheat before storage was found to be 8.2 mg/kg and after storage 5.4 mg/kg. The wholemeal flour prepared from this wheat contained 4.1 mg/kg and the wholemeal bread 1.6 mg/kg. This is largely similar to the findings with respect to 5 other grain-protectant insecticides.

18.4 Metabolism

The fate of radio-labelled malathion was extensively studied in the laying hen and, for comparative purposes, in the white mouse and American cockroach **(March *et al.* 1956a)**. Hens were treated by feeding 100 ppm ^{32}P-malathion in the mash, by spraying with 0.5% ^{32}P-malathion water emulsion, and by intraperitoneal injection. Residues were determined in the droppings, eggs and various tissues. In the experiments with fed hens, 60% of the consumed malathion was eliminated in the droppings in 2 to 4 days and 75% in 5 to 6 days. Of the radio-active compounds excreted, 97% to 98% was in the form of water-soluble metabolites and degradation products. The maximum total residues found in any of the

Table 18.8. Mean ± SE malathion residues (ppm) on dry (12% moisture content) and tough (16% moisture content) wheat, and milled fractions after 1, 3, and 6 months of storage of wheat treated at 8 or 12 ppm **(Mensah *et al.* 1979b)**

Dosage (ppm)	Storage period (months)	12% moisture content: Whole wheat[b]	Bran	Middlings[c]	Flour	16% moisture content: Whole wheat[b]	Bran	Middlings[c]	Flour
8	1	6.24±0.10	19.42±0.95	19.01±1.71	1.35±0.05	4.97±0.26	15.58±1.41	9.35±0.74	0.62±0.03
	3	3.19±0.14	15.57±1.29	14.33±0.18	1.39±0.12	1.97±0.06	14.78±0.67	6.68±0.05	0.41±0.03
	6	2.62±0.20	6.59±0.33	8.55±0.55	1.42±0.10	1.19±0.13	2.99±0.18	3.07±0.38	0.30±0.03
12	1	10.06±0.64	29.01±3.61	24.79±3.09	2.52±0.15	8.50±0.32	23.34±0.19	13.45±0.68	1.36±0.08
	3	6.59±0.16	25.42±2.00	22.06±0.20	1.73±0.15	4.57±0.19	19.95±0.91	11.31±0.04	0.51±0.03
	6	4.52±0.38	9.34±0.04	12.70±0.74	1.50±0.13	1.45±0.08	3.99±0.49	3.45±0.19	0.34±0.02

a Mean of 3 replicates
b Ground wheat
c Shorts, wheat germ, and coarse particles of flour

Table 18.9. Distribution of malathion in wheat fractions **(Kawamura *et al.* 1980)**

	Sample 1			Sample 2			Sample 3		
	a	b	c	a	b	c	a	b	c
Wheat		2.90			2.36			5.16	
B 1	5.8	0.26	0.5	6.9	0.24	0.6	5.7	0.38	0.5
B 2	7.2	0.36	0.9	7.9	0.36	1.1	7.5	0.53	0.7
B 3	2.2	0.94	0.7	1.9	0.69	0.5	2.3	1.60	0.8
M 1	43.1	0.33	5.0	41.5	0.36	5.7	47.1	0.75	7.6
M 2	9.7	0.95	3.2	12.1	0.92	4.1	9.3	2.22	4.4
M 3	2.4	1.60	1.3	2.5	1.45	1.4	2.0	3.66	1.6
Large bran (obusuma)	21.7	8.85	66.4	21.6	8.55	72.8	21.3	15.3	69.9
Small bran (kobusuma)	7.9	8.06	21.9	5.6	6.55	13.8	4.8	13.9	14.4

a ratio to fraction weight (%); b malathion concentration (mg/kg); c ratio to malathion content (%); B break system fraction; M middling system fraction

tissues or eggs examined were less than 3% of the concentration being fed. These residues decreased after the hens were removed from the radioactive feed. In the case of the sprayed hens, the largest concentration of radioactivity was excreted in the droppings in the first 24 hours after treatment, the amount decreasing throughout the remainder of the experiment. The maximum concentration in the droppings was only about one half that in the droppings of the fed hens. Less than 12% of the applied dose was absorbed and eliminated in 32 days; this suggests that malathion was poorly absorbed through the skin and unavailable for absorption from the feathers. About 90% of the compounds excreted was in the form of water-soluble metabolites and degradation products. The residues in tissues and eggs of sprayed hens were approximately one-tenth of those found in the hens receiving malathion in their feed.

When a massive dose was injected intraperitoneally into a hen, more than 50% was eliminated in 3 hours, and nearly the entire dose in 24 hours. During the period of greatest elimination, 1 to 3 hours after the injection, nearly twice as much radioactivity was eliminated as unchanged malathion and chloroform-soluble metabolites as was eliminated as water-soluble metabolites and degradation products. At other times, 90% or more was excreted as water-soluble metabolites and degradation products. The metabolism of malathion in the laying hen and white mouse is complex, probably involving progressive hydrolysis of the ethyl ester moieties to produce more ionic and water-soluble compounds, which apparently exhibit a low order of toxicity. It probably involves, as well, the oxidation of the thiono-sulphur moiety to produce a series of thiolophosphates. One of these is the dicarboxylic analogue of malathion and this may be the principal metabolite showing marked anti-cholinesterase activity and toxicity **(March *et al.* 1956a)**.

The metabolism of malathion in the American cockroach is apparently less extensive and complex than it is in warm blooded animals. The less effective metabolism in the insect may explain the great differential in toxicity for malathion between insects and warm blooded animals **(March *et al.* 1956a). March *et al.* (1956b)** studied the fate of radio-labelled malathion and metabolites in Jersey heifer calves after 2 spray applications (each 2 weeks apart) of 500 m*l* 0.5% ^{32}P-malathion water emulsion. The malathion was rapidly absorbed and metabolised and was eliminated in the urine principally as water-soluble metabolites and degradation products (96–99%). The amount eliminated was greatest during the first hours and then gradually decreased. The 2 calves were sacrificed 1 and 2 weeks after the second spray application, and residues in 10 cuts of meat and in tongue, brain, spinal cord, thymus, thyroid, pancreas, kidney, liver, heart, rumen, suet, bone, marrow and hide were determined radiometrically. Residues in all tissues except the hide were in the form of water-soluble metabolites and degradation products, no unchanged malathion or chloroform-soluble metabolites being found. In the hide, only 2.7% remained as unchanged malathion and chloroform-soluble metabolites, 2 weeks after the second spray application. Total residues in the meat cuts were low, ranging from 0.05 to 0.15 mg/kg. Higher residues (0.22 mg/kg) were found in thymus, thyroid, pancreas, liver and bone. The highest residues (3 to 18 mg/kg) were found in the hide. Chemical analysis for malathion in tissues of the fore leg, hind leg and rump of the two heifers showed no detectable malathion (less than 0.2 mg/kg).

Kruger and O'Brien (1959) attempted to account for the selective toxicity of malathion on the basis of differences in its metabolism by various species. Levels of metabolites were found in the German cockroach, American cockroach and house fly, and compared with the level in the mouse. Most metabolites were identified. Degradation of malathion is much more extensive in the mouse than in the insects and malaoxon production is correspondingly lower; these effects account satisfactorily for the low toxicity of malathion to the mouse. The low toxicity of topically applied malathion to the German cockroach is attributed to lack of penetration through the integument.

Perry (1960) reviewed the then available information on the metabolism of insecticides by various insect species. Malathion appears to be degraded and rapidly excreted by the housefly through intermediates involving hydrolysis of the diethyl succinate nucleus and hydrolysis of the P–S and S–C bonds. In the cockroach, activation (oxidation) is more rapid than hydrolysis and the oxidation product, malaoxon, presumably accumulates to a lethal level **(O'Brien 1957)**. With comparable doses of ^{32}P-malathion there seems to be little difference between susceptible and malathion-resistant houseflies in rate of excretion of non-toxic metabolites, but the tissues of susceptible houseflies contain a higher amount of a toxic compound that is not malathion. It is presumably malaoxon **(US Public Health Service 1958).**

Smith *et al.* (1960), using radio-labelled malathion, studied the fate of residues in meat and milk of a cow receiving 8 ppm malathion in its ration for 3 weeks. There was no excretion of the insecticide in milk, nor was it detectable in the blood, brain, liver, kidney, round or rib-eye muscle, when the animal was slaughtered at the conclusion of the experiment.

O'Brien *et al.* (1961) also studied the metabolism of orally administered malathion by a lactating cow. It was found that malathion was rapidly excreted, principally via the urine, which accounted for 90% of the excreted material. About 23% of the dose was not excreted over 3 weeks. As in non-ruminants, the major metabolite was produced by carboxyester hydrolysis; the principal faecal metabolite, however, was dimethyl phosphate. Milk contained no malathion or malaoxon, but had 0.11 mg/kg of radioactive materials, most of which could not be identified. Blood metabolites were also examined. When malathion was injected into mice, or into dogs, it was degraded rapidly, the principal reaction being hydrolysis of the carboxyester bonds **(Knaak and O'Brien 1960; Kruger and O'Brien 1959)**. In insects, the degradation is slower, and ethyl ester hydrolysis is somewhat less important, cleavage of a phosphate thioester bond being correspondingly more significant **(Kruger and O'Brien 1959).**

Pasarela *et al.* (1962) also studied the fate of malathion during feeding to cattle by determining residues in milk and tissues. No malathion was found in milk of cows fed up to 800 ppm of the insecticide based upon a 5 kg daily ration of dairy supplement. Hay was permitted *ad libitum*. Similarly, no malathion was detected in blood, liver, kidney, heart, muscle or fat of ruminating calves fed 200 ppm of the insecticide, based on total food intake for 41 to 44 days. Some malathion was found in the liver of 2 calves that were sacrificed after 14 days at the same level of intake. No ready explanation of why malathion should be present after the shorter interval of feeding is apparent.

Rowlands (1964) studied the metabolism of malathion in stored maize and wheat. The hydrolysis products: dimethyl dithiophosphoric acid, malathion monocarboxylic acid and malathion dicarboxylic acid, were found in maize during 6 months storage and dimethyl dithiophosphoric acid and malathion dicarboxylic acid were found in wheat. No oxidation products were detected in either cereal but later work **(Rowlands 1965a)** showed that some slight oxidation to malaoxon occurred in wheat of unknown age during storage for 1 week, the highest level being reached at 1 to 2 days after treatment. Malaoxon was degraded to dimethyl phosphorothiolate (chiefly), dimethyl phosphate, and mono- and dicarboxylic acid derivatives. Direct hydrolysis of malathion to dimethyl phosphorothionic acid, rather than dimethyl dithiophosphoric acid, was shown in *in vitro* studies with crude grain homogenates, whereas only dimethyl dithiophosphoric acid was derived *in vivo*. *In vitro* studies showed that malaoxon had no apparent inhibitory effect on carboxyesterase hydrolysis of diethyl maleate by a grain homogenate. Later work **(Rowlands 1966c)** showed that when freshly harvested wheat of 18% moisture content was treated at 10 mg/kg with malathion and stored at 20°C, production of malaoxon was initially rapid and continued for 3 to 4 weeks, although the highest level of malaoxon (0.9 mg/kg) was observed after 2 to 3 days. A sample from the same bulk wheat, treated at 1 week after harvest, produced malaoxon over 1 week only and treatment of similar samples at 2 to 8 weeks after harvest showed scarcely any oxidative activity at all, slight amounts of malaoxon being produced during 2 to 5 days. Similarly, much older wheat produced only a little malaoxon during 2 to 5 days.

Rowlands and Clements (1966) showed that when triphenyl phosphate was added to wheat at 100 mg/kg, either before or simultaneously with malathion treatment at 10 mg/kg, the rate of malathion degradation slowed down considerably over a period of 6 months. The residual malathion was the same at 6 months whether the grain was pre-treated or simultaneously treated with triphenyl phosphate; however pre-treatment kept the level of malathion at a higher value for a greater period of time. Examination of the malathion metabolites verified the expectation that triphenyl phosphate would interfere with the esterase attacking at the carboxyester and thiolophosphate linkages.

Rowlands (1966b) carried out studies on dissected wheat grains that showed that the seed coat is the main source of oxidases, the germ and endosperm contain the acid phosphatases, and the esterases are located in the seed-coat and germ. The rate of penetration and the routes followed by insecticides entering grain are important in relation to the persistence of residues and the toxic compounds that may be produced. For example, it has been found that malathion penetrates the seed-coat of wheat grains more slowly than does bromophos and, since the main oxidase activity is in the seed-coat there is more opportunity for malathion to be oxidised to a toxic thiolate than for bromophos to be converted to a toxic phosphate.

Rowlands *et al.* (1967a) showed that whereas crude preparations of oxidases from the seed-coats of wheat grains catalysed oxidation of malathion to malaoxon *in vitro*, a purified tyrosinase isolated from the seed-coat did not effect any such conversion, either in the presence or absence of $NADPH_2$. Similar experiments with a commercial tyrosinase from mushrooms were also unsuccessful in this respect. Both grain and mushroom enzymes were active towards *p*-cresol, the model substrate.

Rowlands (1967b) showed that in the presence of dichlorvos, grain esterases were prevented from hydrolysing malathion for some 3–4 weeks of storage and ultimate degradation of malathion proceeded by phosphatase products rather than carboxyesterase products. Under normal conditions, malathion degrades preliminarily to carboxyesterase products. One of the degradative products complexed with a wheat lipid fraction, giving a compound which also co-eluted with malathion on GLC; this compound contained a methoxythiophosphate grouping.

Calderon and Desmarchelier (1979) tested the susceptibility of *Tribolium castaneum* to malathion after pre-exposure to carbon monoxide. Carbon monoxide is an inhibitor of the microsomal mixed-function oxidase system but has little or no toxicity to *Tribolium castaneum*. The exposure to carbon monoxide at all concentrations tested increased susceptibility to malathion, the ratios ranging from 1.3 to 1.7 times.

Koivistoinen and Aalto (1970) reviewed the chemistry of malathion transformation, degradation and metabolism by animals, plants and micro-organisms. Most of the information in their review has already been covered above.

Rowlands (1971) reviewed the open literature on the metabolism of contact insecticides in stored grain for the period 1966–69 but found no new published information on malathion during this period.

Elms *et al.* (1972) studied the rate of breakdown of malathion and dichlorvos, applied together and separately to wheat. In one experiment, dichlorvos was applied 126 days after malathion. Dichlorvos was shown to retard malathion breakdown and malathion to influence slightly the breakdown of dichlorvos. Results indicated that the presence of malathion residues in grain, treated with dichlorvos, would not appreciably influence the period required for the degradation of dichlorvos to acceptable residue levels under storage conditions approximating to 25°C and 12% moisture content.

Rowlands (1975) found very little information had been published on the metabolism of malathion in grains during the period 1970–74.

Gupta and Paul (1977) studied the biological fate of radio-labelled malathion in poultry. Following a single oral dose of 394 mg/kg, the birds showed characteristic signs and symptoms of organophosphorus poisoning and the results suggested that the compound was rapidly absorbed from the gastro-intestinal tract, significant quantities being detected in plasma, one half hour after ingestion. Highest concentration of the radio-label was present within 6 to 8 hours of administration. At 6 hours, the radioactivity was highest in liver, followed by other organs. With the lapse of time, the concentration of radioactivity in various organs decreased and at 48 hours it was not detected, except in liver, kidney, lung and spleen, where only traces were observed. The cumulative urinary and faecal excretion study revealed that within 24 hours, 90% was rapidly excreted, mainly via the urine, with only small amounts in the faeces. Metabolism studies showed that the compound was quickly metabolised. Because of the rapid turnover of the compound, this study indicated that the accumulation of malathion in the body system is unlikely.

Hansen *et al.* (1981) reported the characterisation of an alteration product of malathion detected in stored rice. This was shown to be O-methyl-O-ethyl S-(1,2-bis-carboxy) ethyl phosphorodithioate, apparently resulting from the environmental alteration of malathion.

Abdel-Kader and Webster (1981) reported the development of analytical methods for the quantitative determination of malathion metabolites in stored wheat. **Abdel-Kader (1981)** and **Webster and Abdel-Kader (1983)** reported that when malathion was applied to stored wheat at 12 mg/kg and the grain was stored at 20°C for 12 months the concentration of malathion mono-acid rose to a maximum of 2 mg/kg at 6 months, and those of dimethyl hydrogen phosphorodithioate to 1 mg/kg at 3 months and malathion diacid to 0.7 mg/kg at 12 months.

Anderegg and Madisen (1983c), by using ^{14}C-malathion applied to wheat inoculated with *Aspergillus glaucus*, a common grain storage fungus, showed that there was significant degradation by the inoculated wheat which did not occur in sterilised controls.

18.5 Fate in Animals

Malathion does not leave detectable residues if fed to poultry at low levels or if used as a 5% dust **(Herrick *et al.* 1969; Smetana 1969)**. However, if fed at levels of 600–32000 ppm, residues appear in eggs and tissues **(Ghadiri *et al.* 1967; Khmelevskii 1968).**

Gupta and Paul (1977) studied the metabolism and biological fate of malathion in poultry birds using ^{32}P-malathion. Each bird was given a single oral dose equivalent to 394 mg/kg of radio-labelled malathion. Significant quantities of radioactivity were detected in plasma half an hour after ingestion. Higher concentrations of ^{32}P were present 6 to 8 hours after administration. At 6 hours ^{32}P was highest in liver followed by other organs. Metabolism studies showed that the compound is quickly metabolised. Because of the rapid turnover of the compound this study indicated that accumulation is unlikely in the body system.

Pym *et al.* (1984) fed hens for 4 weeks with rations which contained 100 ppm of malathion. There were no residues detected in eggs by a method sensitive to 0.01 mg/kg. However, residues within the range 0–0.07 mg/kg were recovered in the fat of birds receiving 100 ppm malathion in their rations.

March *et al.* (1956a) studied the fate of ^{32}P-malathion in the laying hen. Hens were treated by feeding 100 ppm ^{32}P-malathion in the mash. Residues were determined in the droppings, eggs and various tissues. Of the consumed malathion, 60% was eliminated in the droppings in 2 to 4 days and 75% in 5 to 6 days. Ninety-seven to 98% of the radioactive compounds excreted was in the form of water-soluble metabolites and degradation products. The maximum total residue found in any of the tissues or eggs examined was less than 3% of the concentration being fed. These residues decreased after the hens were removed from the malathion containing feed.

March *et al.* (1956b) studied the fate of ^{32}P-malathion sprayed on Jersey heifer calves. Two spray applications were made, each 2 weeks apart, of 1 pint 0.5% malathion emulsion. The malathion was rapidly absorbed and metabolised and was eliminated in the urine principally as water-soluble metabolites and degradation products (96 to 99%). The amount eliminated was greatest during the first hours and then gradually decreased. The calves were sacrificed 1 and 2 weeks after the second spray application and residues in 10 cuts of meat and in 12 tissues, bone marrow and hide were determined radiometrically. Residues in all tissues except the hide were in the form of water-soluble metabolites and degradation products, no unchanged malathion or chloroform-soluble metabolites being found.

Smith *et al.* (1960) studied the residues of malathion in milk and meat of cows fed for 3 weeks on rations containing 8 ppm malathion. No trace of malathion could be found in the milk, nor was it detectable in the blood, brain, liver, kidney or muscle when the animals were slaughtered at the conclusion of the experiment.

In order to acquire knowledge of the possible excretion of malathion in milk or of its accumulation in tissue at higher levels of cattle feeding than had previously been investigated, **Pasarela *et al.* (1962)** fed milk cows up to 800 ppm of malathion in a 5 kg daily ration of dairy feed supplement. Hay was permitted *ad libitum*. No malathion was found in the milk; similarly, no malathion was detected in the blood, liver, kidney, heart, muscle or fat of ruminating calves fed 200 ppm of malathion, based on total food intake, for 41 to 44 days. Some malathion was found in the liver of 2 calves that were sacrificed after 14 days on this level of intake. Statistical analysis permits the conclusion that, at a 95% level of confidence, 95% of such samples will contain less than 1 mg/kg of malathion. No ready explanation of why malathion should be present after the shorter interval of feeding is apparent.

18.6 Safety to Livestock

The toxicity of malathion has been the subject of a program of investigations that were commenced by the manufacturers in 1949. As manufacturing experience with malathion accumulated, the degree of purity of the product increased. Early samples of the technical material contained 65% malathion, while that currently offered for sale is greater than 99% w/w equivalent to 1030 g/L. It was observed that an increase in the grade of purity was associated with a decrease in acute oral toxicity as expressed by the LD/50. The acute oral toxicity of malathion is also influenced by administration of the dose in a solvent rather than as the undiluted product. The use of solvents such as propylene glycol and vegetable oils results in a numerically lower LD/50 (greater toxicity) presumably because the solvent facilitates gastro-intestinal absorption of the compound. The history of the development and the effects of the above factors can be seen in **Table 18.10 (Schaffer 1955)**.

In one study, when rats were administered malathion, males were more susceptible than females **(Hazleton and Holland 1953)**. However, this difference in susceptibility between the sexes was not shown in another study **(Gaines 1969)**. Gaines states that the majority of pesticides tested by the oral route were more toxic to female than male rats. The reason for these reported differences is not clear. Young animals appear to be more susceptible to malathion than older animals **(Brodeur and DuBois 1963)**. The dietary protein concentration also influences the acute oral toxicity of malathion. It was shown that as the amount of casein in the diet of rats is decreased, the acute toxicity is increased **(Boyd 1969)**. Thus, much of the variation in acute LD/50 values can be attributed to differences in experimental techniques.

The toxicity of malathion may be affected by other organophosphate compounds. **Frawley *et al.* (1957)** observed that the simultaneous administration of two organophosphate compounds produced a higher toxic effect in some instances than was to be expected, based on the known toxicity of each compound. This was a potentiation effect, and the toxicity of malathion has been shown to be influenced by other, but not all, organophosphate compounds **(Kimmerle and Lorke 1968)**.

Further developments have eliminated the impurities which at one stage significantly increased the acute oral toxicity of technical malathion. Information from the principal manufacturers indicates that the LD/50 to male albino rats of the RH Wistar (Royal Hart) strain of a typical batch of commercial malathion is 1650 (1438–1898) mg/kg. Modern practice is to dose animals which have fasted for 18 hours with the undiluted technical product, thus avoiding the interference due to solvents, surfactants and diet **(American Cyanamical Co. 1977)**.

Table 18.10. Single oral dose LD/50s of technical malathion **(Schaffer 1955)**

Grade	Species	Solvent	LD/50 (mg/kg) Male	Female
65% Technical (1949)	Albino mice	Propylene glycol	930	940
65% Technical (1950)	Albino rats	Propylene glycol	300	600
90% Technical (1951)	Albino mice	Vegetable oil	720	–
90% Technical (1951)	Albino rats	Propylene glycol	940	–
90% Technical (1951)	Albino rats	Vegetable oil	390	–
99%+ Technical (1951)	Albino mice	Undiluted	3300	–
99%+ Technical (1951)	Albino mice	Vegetable oil	2700	–
99%+ Technical (1951)	Albino rats	Undiluted	4700	–
99%+ Technical (1951)	Albino rats	Vegetable oil	1500	–
95%+ Technical (1954)	Albino rats	Undiluted	2100	–
95%+ Technical (1954)	Chickens[1]	Vegetable oil	>850	–
95%+ Technical (1954)	Dairy calves[2]	Undiluted	80	–
95%+ Technical (1954)	Dairy cows[3]	Undiluted	–	560

[1] New Hampshire
[2] Less than 3 weeks of age; animals of both sexes used
[3] Pregnant and non-pregnant animals included

A summary of the acute oral toxicity of malathion to mice is shown in **Table 18.11**. The vehicle and the composition of the formulations have a considerable influence on absorption following oral administration to mice. Mice appear to be more resistant to malathion than rats. Male and female mice appear to be about equally susceptible to malathion **(Hazleton and Holland 1953)**. The symptoms of toxicity of mice exposed to toxic doses of malathion are those due to cholinesterase inhibition. These symptoms include excessive salivation, depression and tremors. The less severe symptoms are usually of short duration and unless death occurs within several hours, recovery is rapid and apparently complete **(Golz and Shaffer 1956)**.

Table 18.11. Acute oral toxicity of malathion to mice.

Grade	LD/50 (mg/kg) male	female	Reference
Technical 99%	3321		**Hazleton and Holland (1953)**
Technical 90%	886		**Hazleton and Holland (1953)**
Technical 65%	1260	1158	**Hazleton and Holland (1953)**
Technical 99%	3330		**Golz and Shaffer (1956)**
Technical 90%	720		**Golz and Shaffer (1956)**
Technical 65%	930	940	**Golz and Shaffer (1956)**

The intraperitoneal toxicity of malathion varied with the age of the animals. The LD/50 for adult rats was 750 and the LD/50 for weanling rats was 340 mg/kg **(Brodeur and DuBois 1963)**.

Although exposure by the intraperitoneal route is of less importance than by some other route in characterising the potential health hazard of the compound, it is important in giving information as the inherent toxicity of the compound. The intravenous administration of malathion to rats represented the most toxic route and the sub-cutaneous toxicity was comparable to that of the oral. The acute intravenous and subcutaneous LD/50 values are 50 mg/kg and 1000 mg/kg, respectively. The dermal LD/50 value of malathion is 4444 mg/kg. Exposure of rats to saturated vapours of the compound caused no mortality and the only symptoms noted were laboured breathing and depression **(Spiller 1961)**.

Radeleff *et al.* (1955) reported that the minimum oral toxic dose for baby calves appeared to be between 10 and 20 mg/kg. They reported, however, that sheep tolerated 50 mg/kg doses but were poisoned by 100 mg/kg and all higher doses. These findings need to be judged in the light of improvements in the quality of technical malathion which have occurred over the intervening years.

Schaffer (1955) summarised a trial carried out by Hazleton Laboratories where a group of day-old New Hampshire chicks, both sexes included, was maintained on a feed containing 10 ppm malathion (95% technical).

After 2 weeks, this group was subdivided at random into 3 groups of 10 birds each, each group containing both males and females. These 3 groups were furnished feed containing 100 ppm, 1000 ppm, and 5000 ppm, respectively, of malathion 95% technical. Feeding of the compound at these levels was continued for 10 weeks. Plasma cholinesterase activity was determined after 2 weeks at the higher levels, and again at 6 and 8 weeks and at termination of the study (10 weeks). There were no deaths attributable to feeding of malathion in either the 100 or 1000 ppm groups. There was some evidence suggestive of an adverse effect on food consumption and growth, as well as an inhibition of plasma cholinesterase activity at 1000 ppm; however, these indications were not sufficiently clear-cut to be considered significant. Appearance and behaviour of the birds were normal. At 5000 ppm there were definite signs of toxicity such as retardation of growth, slow feather development, soft droppings, weakness of legs and paralysis. Two of the 10 birds in this group died during the second week, one during the fifth week and one during the sixth week. There was significant inhibition of plasma cholinesterase activity of survivors at 6 and 8 weeks. **Golz and Schaffer (1956)** provided this same information and indicated that the mean daily dosage of malathion 95% technical over the period of feeding was calculated as approximately 7 mg/kg, 90 mg/kg and 450 mg/kg body weight for the 100 ppm, 1000 ppm and 5000 ppm groups, respectively.

Gaafar and Turk (1957) disagreed with the findings reported by **Golz and Schaffer (1956)** after they carried out studies in which they determined the LD/50 for young chickens, 3 weeks of age, and reported it to be between 200 and 400 mg/kg and for older fowls (1 year old), between 150 and 200 mg/kg. They pointed out that almost all the poisoned birds recovered if they survived the first 16 hours. The signs of toxicity shown by the chickens in the poisoned groups included drowsiness, inco-ordination, reluctance to move, assuming a sitting position on the hocks, excessive salivation with mucus hanging from the beaks, slight blueness of the skin, diarrhoea, brownish droppings tinged with blood, coma, and death. At no time did the chickens show signs of nervous excitability or uneasiness.

Keller (1957) reported studies in which New Hampshire hens and roosters were fed diets containing technical malathion at levels of 250 and 2500 ppm for 104 weeks. A diet of 2500 ppm produced growth suppression in the hens; the roosters did not appear to show a marked retardation in weight gain. No signs of systemic toxicity were noted in any of the chickens. Survival among the experimental hens was comparable to the controls. The survival among the roosters fed 2500 ppm was below that of corresponding controls but because of the small number of chickens in the groups, this may not be meaningful. Egg production appeared to be slightly influenced by the ingestion of malathion at the 2500 ppm level since the hens came into production later than the control hens and laid slightly fewer eggs, at least for the period from the sixth through the eighteenth month. The hens fed 250 ppm malathion showed a normal production and were comparable to the controls throughout the study. Inclusion of malathion in the diet did not appear to decrease fertility of the hen or rooster and did not appear to affect hatchability of the developing chick. No deformities were noted among the chicks from eggs laid by hens receiving 2500 ppm malathion. The results of autopsies performed on chickens which died or were sacrificed indicate no gross pathology which may be attributed to the inclusion of malathion in the diet. Likewise, the inclusion of malathion in the diet did not produce observable microscopic pathology.

Ross and Sherman (1960) carried out an experiment in which they measured the food consumption, weight gain, egg production and survival of hens and chickens fed malathion in their rations for 29 weeks. Hens 4 years old consumed significantly less feed containing 45 ppm malathion than did similar hens on unmedicated feed during the first 4 weeks of the experiment but when the concentration of malathion was raised to 90 ppm for 4 weeks and 230 ppm for 20 weeks the food consumption was not significantly depressed. Egg production was depressed slightly but the quality of the eggs was comparable to controls. Chicks 3 weeks old receiving the same rations displayed a significantly lower weight gain initially but after the first 4 weeks their weight gain approached that of the controls. However, the final bodyweight at the end of 29 weeks was slightly less than the control group. There was one death among the hens and one among the chicks but the significance of the mortality is not clear.

McDonald *et al.* (1964), working in Australia, found that malathion at levels up to 100 ppm did not affect growth or feed conversion of young chickens over a period of 6.5 days. Malathion, at the rate of 15 ppm in the diet of growing chickens, fed from 3 to 10 weeks of age did not produce any effect on mortality, weight gain or feed conversion. In the diet of laying hens it did not affect average egg production, hen day egg production, mortality, egg size, egg specific gravity, albumen quality or incidence of blood and meat spots. Yolk colour of eggs from birds receiving malathion was significantly more intense than in eggs from control birds. Eggs from pullets receiving 15 ppm malathion hatched as well as eggs from control pullets. There were no differences in numbers of chickens culled at hatching, sex ratio or mortality during the first 14 days after hatching. The authors concluded that the fowl can tolerate 15 ppm malathion in the diet for a considerable period and growth is not affected by higher levels for short periods.

Ghadiri *et al.* (1967) and **Ghadiri (1968)** presented data which indicated that malathion fed to chickens at 75 to 600 ppm for 3 weeks produced embryonic deformities and reduced hatchability. These authors were able to demonstrate the presence of malathion residues in chicken tissues and eggs when feeding at 600 ppm. The presence of residues in tissues and eggs of hens receiving malathion at rates ranging from 600 to 32 000 ppm were confirmed by **Khmelevskii (1968).**

Khera and Lyon (1968) investigated the use of chick and duck embryos for the evaluation of pesticide toxicity. Malathion was injected into the embryos at mid-incubation of both species (10 days in chick and 13 days in duck embryos). Approximately 85% of the chick embryos and 95% of the duck embryos survived the administration of 1 mg of malathion. The authors considered the results with malathion to be variable.

Smetana (1969), working in Western Australia, demonstrated that malathion was not toxic to chickens when fed at up to 15 ppm. **Rehfeld *et al.* (1969)** and **Rehfeld (1971)** carried out feeding trials which included levels as high as 5000 ppm. Whereas very little toxicity was evident at 2500 ppm, the 5000 ppm level was extremely toxic to day-old chicks.

Sauter and Steele (1972) studied the effect of low-level pesticide feeding on the fertility and hatchability of chicken eggs. Hens were fed a commercial breeder ration to which malathion was added at the rate of 0.1, 1 or 10 ppm. Other groups received corresponding amounts of DDT, diazinon or lindane. All pesticides used except 0.1 ppm malathion significantly reduced hatchability. Embryonic mortality was increased throughout the incubation period. Egg production was reduced by all pesticides fed. These results are at variance with many other studies but the reason and significance are difficult to interpret.

Lillie (1972) reported studies of the reproductive performance and progeny performance of caged white Leghorns fed malathion. The hens were fed a breeder diet containing 250 or 500 ppm malathion. The traits studied in a 36-week period were body weight changes, egg production, egg weights, specific gravity of eggs, feed consumption, mortality, fertility, hatchability, embryonic abnormalities and progeny performance. The addition of malathion caused no significant changes in any of these parameters. In a separate 4-week study the incorporation of 500 ppm malathion in the caged Leghorn male diet exerted no significant changes in the fertility pattern or the incidence of sperm and embryonic abnormalities.

Page and Bush (1978) reported feeding trials conducted at the University of Georgia, USA, where birds were fed diets spiked with graded levels of malathion (2.5, 5, 10, 20 ppm). The results were reported to indicate that:

(1) feed levels up to 20 ppm malathion do not significantly affect egg production;
(2) feed levels as low as 5 ppm malathion tend to reduce hatchability and fertility of broiler hatching eggs;
(3) feed levels as low as 2.5 ppm malathion significantly reduce hatchability of Leghorn hatching eggs;
(4) malathion depressed hatchability of Leghorn eggs more severely than broiler hatching eggs.

It should be noted that details of these studies have not been sighted: the information has been obtained from reports in 2 poultry trade magazines.

Lillie (1973) fed caged white Leghorn pullets a breeder diet supplemented with 250 or 500 ppm malathion and/or carbaryl. Body weight changes, egg production, egg weights, specific gravity of eggs, feed consumption, mortality, fertility, hatchability, embryonic abnormalities and progeny performance were studied. In the progeny performance studies, progeny from hens fed 500 ppm malathion and/or carbaryl were fed a broiler diet supplemented with 500 ppm malathion or carbaryl for a 4-week period in batteries. The only significant differences resulting from malathion and or carbaryl supplementation in the feed were pullet weights and 4-week progeny weights. The pullet weight gains were significantly reduced by carbaryl with or without malathion. A significant growth depression was observed with the progeny fed carbaryl, irrespective of maternal diet. In the separate 4-week study, the incorporation of 500 ppm malathion and/or carbaryl in the caged male Leghorn diet exerted no significant changes in the fertility pattern or in the incidence of sperm and embryonic abnormalities.

Rajini and Krishnakumari (1981) reported an acute oral toxicity study with malathion in poultry. Technical malathion and its 50% EC formulation were evaluated for their acute oral toxicity in chickens. The computed LD/50 values were 948 and 1195 mg/kg for the technical grade and formulation respectively. Livers and kidneys of treated birds showed marked hypertrophy. Histological observations revealed cellular infiltration and necrotic changes in liver and moderate cellular infiltration in kidney.

Pym *et al.* (1984) carried out 3 experiments to study laying performance in hens given graded levels of malathion, dichlorvos and pirimiphos-methyl, either separately or combined in the feed over a 4-week test period. Results conclusively demonstrated interaction between dichlorvos and malathion as measured by depressed food consumption and egg production. Combining the 3 insecticides at levels which, when given separately, had no effect severely depressed food consumption and egg production. Plasma acetylcholinesterase levels were reduced by 30% with malathion at 100 mg/kg. There was no indication of potentiation between insecticides as measured by plasma acetylcholinesterase inhibition, and effects upon food consumption and egg production appeared unrelated to plasma acetylcholinesterase activity. The relationship between food consumption and egg production was similar in groups receiving dichlorvos/malathion mixtures and in those receiving graded levels of untreated food, indicating that the insecticide's effects upon egg production were mediated via a reduced food intake.

Vadlamudi and Paul (1979) studied the acute oral toxicity of malathion to the Indian buffalo. In 1–2 year old male buffalo calves administered malathion at the rate of

50, 75 and 100 mg/kg body weight in feed, marked excitement and salivation occurred within 10–25 minutes, suggesting rapid absorption from the gastro-intestinal tract. Muscular fasciculations, inco-ordination, rigidity of limbs, convulsions and dyspnoea followed 10–45 minutes later.

18.7 Toxicological Evaluation

The toxicology of malathion and its residues was evaluated by the JMPR in 1963, 1965 and 1967. The meeting had before it considerable information on biochemical aspects and a number of studies on the acute toxicity to the rat, mouse, chicken, calf and cow. These did not indicate any significant inter-species differences though it was quite clear that the premium grade technical malathion (99% purity) was significantly less toxic to all species examined than were the 60% or 90% technical material available and studied in earlier years.

It is important to realise that the grade of malathion used commercially for grain protection has been the 'premium' grade which not only has a lower acute toxicity but is substantially free of the mercaptan impurities which give rise to objectionable odour.

The JMPR also considered a number of short-term feeding studies on the mouse, rat, chicken and man, together with potentiation studies and long-term feeding studies in the rat. Some groups of animals in some of the long-term feeding studies received up to 5000 ppm malathion in the diet and these high levels produced more or less complete inhibition of erythrocyte cholinesterase activity.

There was noticeably less inhibition of plasma and brain cholinesterase activity produced by the premium grade technical malathion than was produced by the 90% technical malathion (**American Cyanamid Co. 1955; Hazleton and Holland 1953**). The no-effect level in the rat was determined to be 100 ppm in the diet, equivalent to 5 mg/kg body weight/day; and in man 16 mg/day equivalent to 0.2 mg/kg body weight/day, the only effects noticed being cholinesterase depression.

On the basis of these observations, the acceptable daily intake for man was estimated to be 0.02 mg/kg body weight.

18.8 Maximum Residue Limits

Based on reviews made by JMPR of extensive data in 1965, 1966, 1967, 1968, 1969, 1970, 1973, 1975 and 1977 and the ADI for man of 0.02 mg/kg body weight, the following maximum residue limits have been recommended for various stored products.

Commodity	Maximum Residue Limits (mg/kg)
Bran of rye and wheat	20
Cereal grains	8
Lentils, dried beans	8
Wholemeal and flour from rye and wheat	2

Residues are determined and expressed as malathion.

The methods recommended for the determination of malathion residues are : **AOAC (1980a); Pesticide Analytical Manual (1979a, 1979b, 1979c); Manual of Analytical Methods (1984); Methodensammlung 1982 — XII-3, 5, 6; S5, S8, S10, S13, S17, S19); Abbott *et al.* (1970); Ambrus *et al.* (1981); Desmarchelier *et al.* (1977b); Panel (1973, 1977, 1980)**. The following methods are recognised as being suitable: **Methodensammlung (1982 — 72); Bowman and Beroza (1967); Carson (1981); Eichner (1978); Krause and Kirchhoff (1970); Mestres *et al.* (1976, 1979a, 1979b); Sissons and Telling (1970); Sprecht and Tillkes (1980)**.

19. Methacrifos

Methacrifos is an organophosphorus insecticide discovered in Switzerland. The compound acts as a contact, vapour and stomach poison against all important arthropod pests of stored products. It is also highly effective against major malathion- and lindane-resistant insects. It does not suffer the disability of many other organophosphorus insectides of being particularly weak against *Rhyzopertha dominica*. Methacrifos will fill a widening gap in the possibility of grain protection in view of the world-wide occurrence of resistance in grain pests to malathion and lindane. It is particularly useful where grain temperature can be regulated and where aeration of the grain mass can take advantage of the high potency of methacrifos vapour. It was originally developed under the code name CGA 20168 and is now marketed under the trade name Damfine.

Methacrifos penetrates the individual grains fairly rapidly and is therefore effective against larval stages within the grain. It is extremely potent at lower temperatures and has a pronounced vapour action. Methacrifos degrades rapidly at high temperature and under high humidity conditions but persists exceptionally well at lower temperatures and lower humidities. There is a very steep increase in the rate of degradation with increase in temperature. Therefore, there are considerable economies if the grain can be cooled before, or soon after, treatment with methacrifos. Methacrifos is metabolised by demethylation to yield non-cholinergic compounds that are mineralised in grain and rapidly in animals. When fed to animals, methacrifos is rapidly degraded and excreted, though some fragments appear to be incorporated in life processes. The bulk of the deposit is removed on hulls and bran when milling rice and wheat, and any residues which are carried over into flour or milled rice are completely destroyed in baking or boiling of food prior to consumption. The malting of barley results in destruction of methacrifos with no carry-over of residues into beer.

19.1 Usefulness

Hart and Moore (1975) sprayed methacrifos on to brown and white rice at the rate of 10 mg/kg as the rice was bagged in 25 kg sacks. The sacks were sewn and subsequently held in stacks of 36 bags for 24 weeks. Biological assays using malathion-resistant *Sitophilus oryzae* and chemical analyses were undertaken at regular intervals on samples withdrawn from identified bags. Full insect control was obtained in the samples of treated rice for the full period of the trial. However, no satisfactory control was achieved against the moth, *Ephestia cautella*.

The report of the **Queensland Department of Primary Industries (1975)** indicated that methacrifos had been tested in their program to develop alternative grain protectants to take the place of malathion against which resistant strains were creating major problems. It was stated that methacrifos had proved effective against all species.

Hart and Moore (1976a) carried out a study in which 760 tonnes of grain sorghum was treated with either 15 mg/kg or 7.5 mg/kg of methacrifos. After treatment, the sorghum was stored in vertical concrete silos and sampled for chemical analysis and bioassay with insecticide-resistant *Sitophilus oryzae, Tribolium castaneum*, 2 strains of *Rhyzopertha dominica* and an insecticide-susceptible strain of *Plodia interpunctella*. A few progeny of 1 strain of *Rhyzopertha dominica* were produced in bioassays but the remaining insects were unable to breed. No insects were detected in the grain when it was outloaded from the silo 21 weeks after treatment.

Wyniger *et al.* (1977), in a presentation to the British Crop Protection Conference, pointed out that the lethal dose of methacrifos for several strains of a range of different species is at least 10 times less than either malathion or lindane. In addition to this advantage, methacrifos showed considerable activity against internal stages of grain weevil in wheat.

Bengston *et al.* (1977) carried out a series of field trials in which methacrifos was compared against chlorpyrifos-methyl, fenitrothion, pirimiphos-methyl and malathion for the control of malathion-resistant insects infesting wheat in Australia. Although methacrifos was less persistent than chlorpyrifos-methyl and pirimiphos-methyl it gave a superior performance against the whole spectrum of stored-product pests, several of the other materials being ineffective against *Rhyzopertha dominica*.

Renfer (1977) carried out a biological evaluation of methacrifos as a grain protectant under field conditions in Switzerland. Methacrifos was admixed with 15 tonnes of wheat, 15 tonnes of barley and 15 tonnes of oats at a target rate of 10 mg/kg and with 15 tonnes of barley at the rate of 15 mg/kg. Methacrifos gave effective control against adults and immature stages of 4 major insect pests, the protection period varying from 1 month to 2 years, according to insect species and development stages. The investigator considered that the cereal variety can influence the protection period afforded by methacrifos. The trial showed that a dosage increase in barley from 10 to 15 mg/kg doubled the protection period against *Rhyzopertha dominica*.

Renfer *et al.* (1978) described the spectrum of effectiveness of methacrifos against 36 species of beetles, moths, flies, roaches and mites present in stores. They also compared the toxicity of methacrifos with that of malathion, lindane, pirimiphos-methyl and chlorpyrifos-methyl against each of 8 major species, including a number of strains of each known to show a degree of resistance against one or more currently used insecticides. In every instance, except against 3 strains of *Rhyzopertha dominica*, where lindane was significantly more potent, methacrifos was noticeably more toxic than the other four compounds.

Bitran *et al.* (1979a), working in Brazil, evaluated methacrifos against pirimiphos-methyl, bromophos and malathion for protecting maize against *Sitophilus zeamais*. Over a storage period of 9 months, they rated methacrifos, applied at a rate of 10 mg/kg, equal to pirimiphos-methyl 5 mg/kg and considerably superior to both bromophos and malathion at 10 mg/kg. Only pirimiphos-methyl, applied at the rate of 10 mg/kg, was superior at the end of 9 months but methacrifos gave a comparable performance for 8 months.

Bengston and Desmarchelier (1979) discussed extensive laboratory and field trials with various grain protectants to control malathion-resistant insects in Australia and stated that methacrifos, used at the rate of 22.5 mg/kg in unaerated storage and 15 mg/kg in aerated grain, gave complete protection against all species for 9 months. They provided tabulated data indicating the LC/99.9 values for candidate grain protectants in newly-sprayed grain, showing that methacrifos was significantly more potent than all other materials examined, except in its toxicity to various strains of *Rhyzopertha dominica*, where bioresmethrin was outstandingly potent and chlorpyrifos-methyl slightly more potent than methracifos. Against malathion-susceptible strains, methacrifos was from 3 to 4 times more potent than malathion.

Desmarchelier and Bengston (1979b), discussing 12 grain protectants which have been evaluated in Australia in the light of recent knowledge about the predictability of the fate of their residues, showed that the rate of application of methacrifos could be substantially reduced if grain could be cooled at the time of application and that this reduction in the application rate would be greater for methacrifos than for any other protectant so far investigated.

Desmarchelier and Bengston (1979a) pointed out that methacrifos offers considerable advantages where it is desired to obtain long-term protection but where it would be an advantage to have minimum residues at the time the grain was taken out of storage. In support of their beliefs, they pointed to the advantage of controlling temperatures in storage and showed that in order to use methacrifos and achieve the required high level of protection for 9 months one would apply, at 50% equilibrium relative humidity, 22.5 mg/kg at temperatures of 30°C. This would ensure that at the end of 9 months, the grain contained a residue of 1 mg/kg, sufficient to control *Tribolium* and *Sitophilus* spp. If, however, the grain temperature could be reduced to 25°C, the initial application of methacrifos could be reduced to 4.6 mg/kg. Where grain temperatures were reduced to 19°C, the amount of methacrifos applied could be reduced to 2.2 mg/kg. There are clearly enormous advantages in achieving even partial cooling; this cooling also reduces potential numbers of insects and enhances grain quality.

Coulon *et al.* (1979a), working in France, found that methacrifos was equally effective as, or even more effective than chlorpyrifos-methyl in controlling *Sitophilus granarius* in stored wheat, when each was applied to infested wheat of 2.5 mg/kg. However, when applied preventatively, 2.5 mg/kg chlorpyrifos gave an 18-month lasting protection whereas methacrifos applied at the same rate protected the wheat for only 6 months. **Coulon *et al.* (1979b)** reported that at least 3.85 mg/kg methacrifos was required to prevent infestation by *Sitophilus granarius*. For grain being stored for 9 months, this corresponded to an application of 10 mg/kg.

Desmarchelier *et al.* (1979b) discussed the theory and practice of combining temperature and moisture manipulation with the use of grain protectant insecticides. They showed how the effects of drying and cooling on both pest biology and the chemistry of protectants can be turned to advantage. A quantitative assessment was made of the use of grain drying and cooling with a labile insecticide such as methacrifos compared with the use of drying and cooling with low application rates of persistent insecticides. The value of this concept in slowing down the selection for resistance is also discussed.

Greening (1980), in discussing the methods for controlling insects which infest harvest machinery and grain stored on farms, states that methacrifos is as effective as dichlorvos for this purpose, both insecticides gaining their effect because of the potency of their vapour.

Bitran *et al.* (1980a), reporting experiments carried out in Brazil on the protection of maize stored on the cob in farm storage systems, found that phosphine fumigation to destroy existing infestations, plus methacrifos to provide protection against *Sitophilus zeamais* and *Sitotroga cerealella*, was the most efficient treatment. Methacrifos, applied at the rate of 8 mg/kg, when used in conjunction with fumigation, provided a high level of protection for 10 months.

Bengston *et al.* (1980a), reporting extensive field trials in which methacrifos was compared with various pesticide combinations on bulk wheat in commercial silos in Queensland, South Australia and Western Australia, said that methacrifos, applied at the rate of 15 mg/kg, was comparable to the best combination of organophosphorus insecticide with bioresmethrin, but inferior only to a high rate of bioresmethrin against *Rhyzopertha dominica*. The biological efficacy of methacrifos was greater and the rate of degradation lower in aerated than in non-aerated storage.

Desmarchelier *et al.* (1981d) reported extensive field trials in which methacrifos, applied at the rate of 22.5 mg/

kg, was compared with 3 grain protectant combinations that were each applied to grain stored at 15 sites throughout Australia. Grain remained free of insect infestation at all sites and bioassays, carried out on samples taken at various intervals throughout the storage period from all sites, showed methacrifos to be the most effective against *Tribolium castaneum*, but least effective against *Rhyzopertha dominica*.

Bansode *et al.* (1981) found that a field-collected strain of *Plodia interpunctella* was greater than 227-fold resistant to malathion. Since the resistance was suppressed by the synergistic action of triphenyl phosphate, carboxyesterase appeared to play a major role in the detoxification of malathion. Tests revealed that the strain showed only a low level of tolerance to other organophosphorus compounds, including methacrifos, where the resistance factor was 1.5.

Pagliarini and Hrlec (1982) working in Yugoslavia, showed that methacrifos was slightly less effective initially than dichlorvos in controlling the mite *Tyrophagus putrescentiae* but it had a high residual activity irrespective of formulation. They determined that an application rate of 5 mg/kg was sufficient to control the mite.

Kumar *et al.* (1982) tested the effectiveness of methacrifos against insect pests of stored food-grains in godowns in various parts of India. Malathion was used as the standard for comparison. Methacrifos proved superior to malathion against *Sitophilus oryzae, Tribolium castaneum, Rhyzopertha dominica, Oryzaephilus surinamensis, Cryptolestes pusillus* and *Cadra cautella*.

Kirkpatrick *et al.* (1982) studied the effect and fate of methacrifos applied to corn for control of stored-product insects. Although large numbers of a wide variety of pests invaded the untreated controls and the malathion-treated corn samples, only a small number of a few species penetrated the corn treated with methacrifos. Weight loss from insect feeding damage was 56% in the untreated corn samples, 21% in the standard malathion-treated samples and a mean of less than 5% in corn treated with 3 rates of methacrifos.

Bengston *et al.* (1983a) carried out field trials on bulk sorghum stored in commercial silos in south Queensland and central Queensland. Grain samples taken at intervals during 6 months storage were tested by bioassay against *Sitophilus oryzae, Rhyzopertha dominica, Tribolium castaneum* and *Ephestia cautella*. Methacrifos applied at the rate of 15 mg/kg controlled all the tested species except *Ephestia cautella*. The authors recommend that if this species were a problem supplementary measures such as drying, aeration or cooling as advocated by **Desmarchelier and Bengston (1979a, 1979b)** would be required.

Wohlgemuth (1984) carried out comparative laboratory trials with commercial insecticide formulation under tropical conditions. Wheat treated with methacrifos dust and emulsion concentrate at a rate of 10 and 15 mg/kg respectively was stored at 36°C and 50% r.h. for 24 months, samples being artifically infested with 5 stored-product beetles after 1, 3, 6, 9, 12, 18 and 24 months. Although methacrifos, at these rates, produced 100% mortality of all species at the first month it appeared to lose its effectiveness quite rapidly under these conditions in comparison with some other materials. It was slightly more effective in reducing reproduction than in killing adult insects. *Rhyzopertha dominica* was the most tolerant of the 5 species.

Desmarchelier *et al.* (1986) reported the results of 2 field trials, carried out on bulk wheat in commercial silos, involving laboratory bioassays on samples of treated grain at intervals over 26 weeks using malathion-susceptible and malathion-resistant strains of 8 species of stored-product pests. Methacrifos was used at the rate of 18 mg/kg, in comparison with chlorpyrifos-methyl (5 mg/kg), pirimiphos-methyl (6 mg/kg), fenitrothion (7 mg/kg) and malathion (18 mg/kg). Methacrifos proved to be the best or equal to the best against all species except *Ephestia cautella*, against which it was inferior to all except malathion.

19.2 Degradation

The information available on the degradation of methacrifos on grain in storage is so extensive and so comprehensive as to overshadow the information on all other grain protectants. This reviewer has had the full text of no less than 63 detailed reports on studies, ranging from small but carefully controlled laboratory studies through comprehensive small-scale trials to field trials involving many thousands of tonnes of treated grain. In addition to these 63 studies, there is a considerable number of additional studies on other stored products, including oil seeds, coffee, cocoa, peanuts etc. The quality of the data produced from these studies is quite outstanding.

The first comprehensive study reported **(McDougall 1973, 1974)** was designed to determine the effect of moisture and temperature on the persistence of methacrifos in wheat. It is interesting to consider the author's summary 'for the moisture range 9.5–13% and the temperatures 22°C and 35°C, the half-life of methacrifos in wheat varies from 238 days to 19 days; the half-life decreasing as moisture and temperatures increase'. The detailed residue data provided in **Tables 19.1 and 19.2** adequately support this conclusion. All of the subsequent studies have only served to confirm these facts.

It is not possible for the reviewer to consider any of these studies in isolation from the work of **Desmarchelier (1977a, 1978a)** referring to dichlorvos and fenitrothion respectively, but these basic mathematical models apply equally well to methacrifos as was demonstrated by **Desmarchelier *et al.* (1979b, 1980a)**. If this line of reasoning had been available to other investigators at an earlier date, no doubt many of the empirical studies would have been considered unnecessary or they would have been pointed in somewhat different directions. The knowledge that the degradation of methacrifos follows a second-order reaction depending upon temperature and equilibrium relative humidity, rather than absolute moisture content, and that methacrifos had a coefficient of variation with respect to temperature that was considerably higher than all other grain-protectant insecticides yet evaluated, could not be ignored when evaluating the results of the otherwise excellent studies.

Table 19.1. Residues of methacrifos in wheat **(McDougall 1973, 1974)**

Moisture (%)	Temperature (°C)		Methacrifos residues Times (days) 0	6	14	21	28	42	70	103
		ppm CGA-20168	% Initial	% Initial	% Initial	% Initial	% Initial	% Initial	% Initial	% Initial
9.6	22	13.6	100	99	97	–	88	90	86	71
10.5	22	11.9	100	95	92	–	87	78	70	52
11.4	22	13.6	100	90	85	–	79	74	55	38
12.3	22	13.5	100	91	79	–	70	66	47	27
13.0	22	13.5	100	74	67	–	59	53	39	22
9.6	35	11.9	100	82	78	64	61	52	35	24
10.5	35	13.3	100	70	61	49	44	31	22	13
11.4	35	14.3	100	69	54	42	39	24	15	8
12.3	35	13.9	100	60	47	35	28	18	10	5
13.0	35	13.8	100	44	44	36	28	17	9	4

Table 19.2. Effect of temperature and moisture on the half-life of methacrifos in wheat **(McDougall 1973, 1974)**

Moisture (%)	Temperature (°C)	Half-life (days)	Time to 50% of day 0 (days)
9.6	22	238	238
10.5	22	124	124
11.4	22	75	75
12.3	22	58	58
13.0	22	45	45
9.6	35	52	44
10.5	35	39	22
11.4	35	30	17
12.3	35	20	11
13.0	35	19	11

Moore and McDougall (1974a), who studied factors affecting the degree of dissipation of methacrifos in wheat, showed that the half-life of the compound decreased as moisture and temperature increased. They found that increased application rates and repeated application did not influence the half-life of the chemical but they did find some evidence to suggest that enzymatic activity in the grain was at least partially responsible for the initial rapid breakdown of methacrifos in wheat. They developed a mathematical formula for expressing the half-life of the compound as influenced by moisture and temperature. Though their data on wheat fitted this formula fairly well, it could not be used to predict the fate on other grains where the equilibrium relative humidity was different to that of wheat at the same moisture content.

Moore and McDougall (1974b), who participated in the extensive collaborative studies to determine the rate of decay of methacrifos and other grain-protectant insecticide residues on wheat **(Desmarchelier *et al.* 1980b)**, estimated the half-life of methacrifos on the wheat stored in large silos under field conditions to be 9 weeks under the conditions that prevailed. After evaluating a second lot of data obtained in these collaborative studies but from another site, **Moore and McDougall (1974c)** calculated the half-life of methacrifos to be 12 weeks. Although the moisture content of the grain at the 2 sites was comparable, the temperature of the grain at the second site was somewhat lower.

Hart and Moore (1975) sprayed methacrifos on to brown and white rice at the rate of 10 mg/kg as it was bagged into 25 kg sacks. Residue levels, which were monitored frequently, declined at a similar rate and to the same degree on both grades of rice, the half-life being in the region of 18 weeks. Temperature of the grain throughout the study ranged from 21°C to 13°C. No doubt this somewhat lower temperature was the cause of the lower decay rate.

Formica (1975) reported the outcome of studies designed to determine the rate of degradation of methacrifos in wheat and maize at 3 temperatures, 15°C, 25°C and 35°C. At 15°C the residues declined to the same level in 9 months that were reached in less than 1 month at 35°C.

Moore and McDougall (1975), reporting results of studies carried out in collaboration with the Australian Wheat Board, stated that aeration of wheat decreased the rate of degradation of methacrifos and thus extended its effective life as a grain protectant. Though the moisture content of the grain in the aerated silo was, adventitiously, higher than that in the non- aerated silo, the methacrifos deposit degraded much more slowly in the aerated silo, no doubt due to the significant reduction in temperature brought about by the controlled aeration which was installed for that purpose. As a measure of the difference between the 2 systems, the authors calculated the time taken for the concentration to fall from 50% of the initial concentration to 25%. In the unaerated silo this was 8 weeks, but in the aerated silo 18 weeks.

McDougall (1976a) reported a series of studies in which wheat, maize, barley and sunflower seed, treated with methacrifos at the rate of 10 mg/kg, were kept at a constant temperature (26°C) in small silos which were subsequently sampled and analysed regularly over a storage period of 8 months. The residues dissipated twice as fast on barley as on sunflower seed, which in turn dissipated the residues significantly more rapidly than wheat or maize. The differences in moisture content of the grains would indicate a much more significant difference in equilibrium relative humidity, the importance of which was not appreciated at that stage. From these data the half-life of methacrifos on the grain was calculated. The results are given in **Table 19.3.**

Table 19.3. Degradation of methacrifos in grains held in mini-silos at 26°C **(McDougall 1976a).**

Grain	Moisture (%)	Half-life (days)
Wheat	10.2	105
Maize	12.8	100
Barley	12.9	44
Sunflowers	7.2	90

In a second experiment, **McDougall (1976a)** stored 5 varieties of grain in 120 L drums in a constant temperature room at 32°C. One lot of oats was stored in a sealed plastic bag. Each grain except the oats was treated at 2 different rates. The grain was sampled frequently over 210 days of storage and from the data the half-life was calculated. The results obtained are given in **Table 19.4.**

Table 19.4. Degradation of methacrifos in grains held in 120 L drums at 32° C **(McDougall 1976a).**

Grain	Moisture (%)	Treatment rate (mg/kg)	Half-life (days)
Wheat	10.3	5	70
	10.3	10	94
Maize	10.1	10	85
	10.1	10	87
Soybeans	7.2	5	95
	7.2	10	120
Oats	8.0	10	64
Oats (sealed)	8.2	10	56
Peanuts	7.4	10	55
	7.4	20	55

In the third study, **McDougall (1976b)** measured the dissipation of methacrifos in white, brown and paddy rice. Rice was stored in a mini-silo (paddy) or in 120 L cardboard drums held in a constant temperature room at 32°C. Each type of grain was treated at 2 rates (5 and 10 mg/kg) and samples were withdrawn at frequent intervals for analysis of methacrifos residues. From the data the rate of degradation was calculated and is expressed in terms of the half-life in **Table 19.5**. The moisture content of the grain declined steadily over the storage period.

Hart and Moore (1976a) reported trials in which methacrifos was applied to grain sorghum stored in large commercial silos in order to determine the residual fate and biological performance of the deposit over a period of 21 weeks. The temperature was 27°C and the moisture content was 13%. Two rates of application, 15.4 and 6.8 mg/kg, were used. The half-life under these conditions was approximately 98 days.

Hart and Moore (1976b) treated 2 lots of barley held in 300 tonne concrete silos with methacrifos at 12 or 20 mg/kg. The grain temperature varied between 20°C and 30.5°C and the moisture content of the barley from the 2 silos was 8.7% and 9.8%. Regular sampling for residue analysis produced data from which these workers calculated the half-life of methacrifos under the conditions of the trial to be approximately 98 days.

Table 19.5. Degradation of methacrifos in different grades of rice stored in a silo or in cardboard drums **(McDougall 1976b).**

Grain	Moisture (%)	Treatment rate (mg/kg)	Storage/ temp. (°C)	Half-life (days)
Paddy rice	11.8	10	Silo/26°C	110
	11.8–8.8	5	Drum/32°C	50
	11.8–8.8	10	Drum/32°C	52
Brown rice	12.3–9.5	5	Drum/32°C	72
	12.3–9.5	10	Drum/32°C	76
White rice	12.8–9.9	5	Drum/32°C	72
	12.8–9.9	10	Drum/32°C	70

Table 19.6. Degradation of methacrifos on 5 varieties of grain **(Desmarchelier 1976d).**

Grain	Moisture (%)	Temp. (°C)	Application rate (mg/kg)	Residue after storage for 3 months (mg/kg)	6 months (mg/kg)
Barley	13	25	18	4.8	1.4
Paddy rice	13	25	18	7.0	3.1
Brown rice	12.5	25	18	7.1	4.8
White rice	12.7	25	18	7.5	3.8
Oats	12	25	18	4.7	1.3

Desmarchelier (1976d) applied 5 grain protectant insecticides to 5 varieties of grain which he analysed after 3 and 6 months storage at 25°C. The results of analysis for methacrifos residues are given in **Table 19.6.**

Blattmann (1978) studied the degradation of methacrifos on wheat using ^{14}C-labelled insecticide. From 84% to 94% of the applied radioactivity was recovered at all time intervals, indicating the limited volatility of methacrifos and its degradation products during the storage period. After 1 year about 75% of the radioactivity was present in the form of degradation products. Methacrifos was degraded in wheat, primarily by hydrolytic processes to cholinesterase-inactive demethylated derivatives. These compounds eventually were mineralised. Isomerisation or oxidation reactions leading to strongly cholinesterase-inhibiting oxo-analogues were not observed.

Formica (1978a) determined the degradation of methacrifos on barley stored in Switzerland and treated at a dosage rate of about 20 mg/kg. The barley had been in store for 2 years and was infested with 3 species of stored-product pests. The formulation was applied as an aqueous emulsion at a rate calculated to apply 22.3 mg/kg methacrifos. The grain temperature ranged between 2°C and 25°C and the relative humidity between 40% and 80%. Samples were taken during the treatment and periodically during the storage. The results of residue analysis indicate that methacrifos dissipated from 13.8 to 0.86 mg/kg within 560 days of treatment. The analysis of the samples taken from separate sections of the silo demonstrate that the dissipation of methacrifos varied according to the temperature existing in the different sections of the silo. Higher levels were found in samples taken from the coolest, sun-sheltered northern section of the silo.

In an experiment carried out in Spain, **Formica (1978b)** determined the degradation of methacrifos on wheat and barley stored in commercial silos. The methacrifos was applied as a dilute aqueous emulsion at a rate corresponding to 10 mg/kg of grain. The moisture content of the wheat was 11% and of the barley 11.6%. The inside temperature was measured periodically with a thermacouple and ranged between 13°C and 26°C. The results of residue analysis demonstrate a dissipation of methacrifos in wheat from 9.4 to 1.0 mg/kg and in barley from 4.0 to 1.7 mg/kg within 430 days of treatment.

Formica (1978c) reported a study in which Swiss barley and Swiss oats held in wooden bins in a warehouse were treated with methacrifos at a rate equivalent to 10 mg/kg. The temperature within the grain ranged from 6°C in winter to 28°C in summer. Samples were taken during the treatment and periodically during the storage, the last being taken 74 weeks after treatment. The results of analysis demonstrated dissipation of methacrifos from about 12 to 2.5 mg/kg of barley and from about 10 to 3.7 mg/kg of oats within 17 months of treatment.

Formica (1978e) studied the fate of methacrifos applied to wheat stored for 21 months in Switzerland. Wheat was treated with sufficient methacrifos to deposit 11 mg/kg and samples were taken from the surface and from 2 metres depth on 24 occasions over a period of 736 days. Although the residue on the sample of grain collected on the day of treatment was, for unexplained reasons, less than half of the intended value, the rate of degradation was slow and the deposit remaining at the end of 736 days (approximately 1 mg/kg) would have been sufficient to control many species of insects. The temperature of the grain ranged from 5°C in winter to about 22°C in summer.

Formica (1978d) reported the results of a large-scale trial undertaken to establish the persistence of methacrifos sprayed on wheat at the rate of 10 mg/kg prior to storage and to determine the effect of air circulation on the breakdown of the active ingredient. The wheat was aerated by being turned through the conveyor system back into the same cell. This form of aeration had little or no influence on the rate of dissipation of the methacrifos residue.

Formica (1978f) studied the fate of methacrifos residues on barley stored in a covered, cylindrical, aluminium silo standing in the open. Analysis of the individual samples taken from different sections of the silo demonstrated a different dissipation rate, depending upon the temperature existing in the different sections of the silo. Highest residues were found in samples taken from the coolest, sheltered section of the silo.

Formica (1978g) reported an attempt to study the dissipation of methacrifos applied as a 2% dust to maize cobs held in a crib under sub-tropical storage conditions. Though extensive data were produced there were so many irregularities introduced by difficulties in applying the dust evenly and of drawing truly representative samples that it was not possible to draw reliable conclusions on the rate of dissipation. In spite of this, it was clear that no drastic degradation of the active ingredient occurred during the 4 months storage.

Desmarchelier *et al.* (1979b), in stressing the importance of manipulating temperature and moisture in order to enhance the effect of grain protectants, pointed out that for methacrifos, cooling by 5, 10 and 15°C, for a given initial application, increases the period of protection by factors of 2, 4 and 8 respectively. The value of aeration or of the use of mobile cooler units to achieve such reductions in temperature and thus minimise the amount of methacrifos applied or, alternatively, extend the period of protection obtained, are both practical and economical.

Coulon *et al.* (1979a) indicated that methacrifos, applied preventively at the rate of 2.5 mg/kg, will protect wheat for 6 months.

Thorpe (1979) discussed the effect of heat and moisture transfer on the degradation of methacrifos applied to stored grain. A mathematical model was postulated to describe this effect in aerated wheat. This model indicates that blowing cold night air through grain results in considerable savings of methacrifos. Aeration for 50% of the time during the first months of storage would lead to more uniform distribution of methacrifos.

Giannone and Formica (1979a) applied methacrifos to maize cobs stored in open wired cribs under artificially produced tropical conditions (28 ± 3°C and 45 ± 10% r.h.) during a period of 231 days after first treatment. The water content of the maize cobs decreased from 38% to 10% within 54 days and remained constant thereafter. The results of analysis of samples from the cribs indicated that methacrifos dissipates in maize cobs from 10–12 mg/kg 7 days after treatment to 3–4 mg/kg in about 50 days. The same behaviour was demonstrated following a second treatment although the water content of the maize cobs was distinctly lower than at the beginning of the experiment. During further storage, no significant reduction of methacrifos residues was demonstrable. Some of the maize cobs from this experiment were studied **(Giannone and Formica 1979b)**. They were analysed to demonstrate the degradation of methacrifos to demethylated metabolites. It was found that there was 2 to 3 times as much demethylated metabolites as parent compound present 230 days after the first treatment.

Schnabel and Formica (1979a) treated 28 tonnes of Swiss wheat with methacrifos at the rate of 10 mg/kg and after storage for 7 months in a concrete cell sampled the wheat to determine the proportions of unchanged methacrifos and its demethylated metabolites. They found that this grain contained 2.8 mg/kg methacrifos and 1.7 mg/kg demethylated metabolites.

Desmarchelier and Bengston (1979a), discussing the value and use of predictive models that enable the grain industries to select rates of application appropriate for given circumstances, used the example of methacrifos in showing the effect of partial cooling on increasing the residual life of protectants. Thus, for 9 months protection with methacrifos, to leave a residue of 1 mg/kg to control *Tribolium* and *Sitophilus* species, one would apply, at 50% equilibrium relative humidity, 22.5 mg/kg at temperatures of 30°C, 4.6 mg/kg at temperatures of 25°C, and 2.2 mg/kg at temperatures of 19–20°C. There are clearly enormous advantages in achieving even partial cooling when using a grain-protectant insecticide such as methacrifos.

This principle was further enunciated by **Desmarchelier *et al.* (1979b)** in a paper dealing with manipulation of temperature and moisture to improve the usefulness and efficacy of grain protectants. **Figure 19.1** illustrates the level of residues required to give 10 months protection for 3 constant temperatures (30°C, 25°C and 20°C) and in aerated grain cooled in 2 steps from 30 to 20°C.

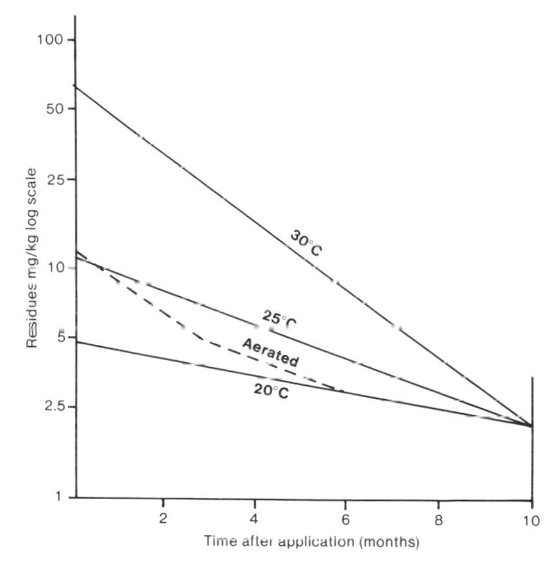

Figure 19.1. Semilogarithmic plot of methacrifos residues required to give 10 months protection at three constant temperatures (solid lines) and in aerated grain cooled in two steps from 30°C to 20°C (broken line) against time after application **(Desmarchelier *et al.* 1979b).**

Bengston *et al.* (1980a) presented the results of field trials with methacrifos and various pesticide combinations carried out on bulk wheat in commercial silos in Queensland, South Australia and Western Australia. They reported that the biological efficacy of methacrifos was greater and the rate of degradation lower in aerated than in non-aerated storage. Their findings are illustrated in **Figure 19.2.**

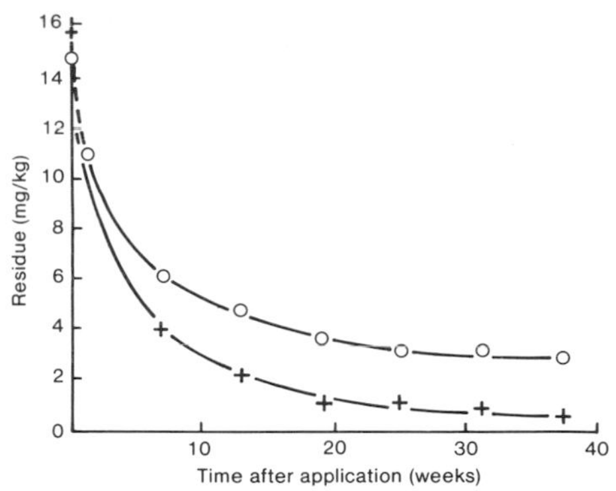

Figure 19.2. Calculated application rates (week 0) and mean residues of methacrifos throughout storage: aerated storage (○); unaerated storage (+) **(Bengston *et al.* 1980a).**

Thorpe and Elder (1980), in discussing the use of mechanical refrigeration to improve the storage of pesticide-treated grain, cited theoretical considerations to show that the loss of methacrifos on treated wheat, initially at 30°C and 11% moisture content, could be reduced during 6 months storage from 90% to about 30% by prompt cooling to 19–20°C. The energy required for such cooling would be only 1.5 kWh/tonne grain.

Giannone and Formica (1980a) reported a study on 60 tonnes of barley and 80 tonnes of maize treated to determine the fate of methacrifos and its dealkylated metabolites. The results of analysis of the samples of barley and maize stored during 370 days in closed concrete silo cells showed a gradual degradation of the residues of the parent methacrifos corresponding to a parallel increase in the residues of the dealkylated metabolites. However, the sum of the determined compounds (methacrifos and metabolites) did not seem to decrease significantly with time in the barley, whereas a slow degradation of the total residue was evident for maize. The results are given in **Table 19.7.**

Desmarchelier *et al.* (1980a) studied the fate of methacrifos and 5 other grain-protectant insecticides on rice in hull, husked rice, polished rice and barley and from the accumulated data, showed that the level of residues determined by analysis of these different grains compared closely with the values predicated by a model derived from other grains.

Giannone and Formica (1980b) studied the composition of the residues in the samples of wheat and barley from silo-scale trials conducted in Spain by **Formica (1978b)**. They found that wheat held in storage for 381 days still contained 20% of the applied nominal amount of methacrifos but it also contained 33% of this amount in the form of dealkylated metabolite. By the end of 430 days the methacrifos residues had declined to 10% of the applied nominal amount and the dealkylated metabolites to 24% of this value. In the case of the barley treated at the same time, the amount of methacrifos was just marginally higher and the proportion of dealkylated metabolites just slightly lower. However, these values must be accepted as being within the experimental error. Similar studies were carried out on wheat in Switzerland **(Giannone and Formica 1980c)**, sorghum in Mali **(Giannone and Formica 1980d)** and on wheat in Morocco **(Schnabel and Hormann 1980).**

Kirkpatrick *et al.* (1982) studied the fate of methacrifos applied to corn under ambient conditions in Savannah, Georgia, USA, during 12 months. No significant degradation of methacrifos was noted during the storage season. An average of only 20% of the residues from the 3 dosage levels applied was recovered after 12 months. The rate of degradation was remarkably similar to that of malathion at the same rate included in the trial. The residue data obtained are given in **Table 19.8.**

Bengston *et al.* (1983a) reported duplicate experiments carried out on bulk sorghum stored in southern Queensland and central Queensland. This was treated with methacrifos at the rate of 15 mg/kg. The grain had a mean moisture content of 12.5% and a temperature in the range of 25–20°C. At the end of 1 week after application the methacrifos residue on the sorghum was found to be approximately 10 mg/kg and thereafter it declined steadily

Table 19.7. Residues of methacrifos and its dealkylated metabolites in barley and maize grain after treatment with the formulation SO 050 **(Giannone and Formica 1980a)**

		barley			maize		
Sampling dates	days after treatment	t(°C)*	methacrifos (mg/kg)	metabolites (mg/kg)**	t(°C)*	methacrifos (mg/kg)	metabolites (mg/kg)**
Sep.21,1979	6	18	5.1	0.60	19	11.4	0.58
Nov.30,1978	76	8	4.2	2.0	9	6.4	1.6
Jul.12,1979	300	18	3.0	4.2	14	4.1	2.0
Sep.20,1979	370	24	1.5	3.8	26	0.7	5.0

*Temperature registered at 2 metres depth
**Sum of methacrifos and CGA 73712 formed after methylation

Table 19.8. Residues of methacrifos on corn (13% moisture) at selected intervals after treatment **(Kirkpatrick *et al.* 1982)**

Intended rate	Mean residues (mg/kg) after treatment (months) 0	1	3	6	9	12
6	7.2	5.1	4.9	3.0	1.2	1.2
10	10	9.1	7.9	4.7	2.7	2.4
15	14	12.5	12.3	5.3	3.8	2.8

The trial commenced in mid-winter — temperatures not given.

so that at the end of 24 weeks, 5 mg/kg remained at the southern Queensland site whereas at the central Queensland site it had declined to 1 mg/kg in the same time. This difference does not seem to be explained by the data on temperature or moisture.

19.3 Fate in Milling, Processing and Cooking

McDougall (1976b), who carried out a trial to determine the rate of dissipation of methacrifos from paddy rice, brown rice and white rice over a period of more than 6 months, submitted the paddy rice to milling and polishing, likewise the brown rice to polishing, in order to measure the distribution of methacrifos on the various portions of the rice. These experiments showed that 92.9% of the methacrifos was on the hulls of the rice and only 7.1% of the deposit remained on the brown (hulled) rice after removal of the hulls. Most of this deposit was removed with the bran in the production of the white rice. Detailed results are given in **Tables 19.9 and 19.10.**

Desmarchelier (1976d) treated 5 different cereal grains with methacrifos at the rate of 18 mg/kg and stored them for 3 and 6 months at 25°C. Brown rice, white rice, oats and barley were sampled for further processing and/or cooking after 3 months storage. Paddy rice and barley were taken for processing after 6 months storage. The methacrifos residues were determined before and after processing/cooking and the results are set out in **Table 19.11**. This shows that there was a substantial loss. The amount of methacrifos applied to the brown rice and white rice is considerably higher than would occur in practice.

Hart and Moore (1976b) treated barley in 300 tonne concrete silos with methacrifos at 12 or 20 mg/kg. Portion of the bulk was withdrawn for malting by a commercial maltster. This barley was found by analysis to contain 10–10.8 mg/kg methacrifos before malting. After malting the residues were undetectable (less than 0.25 mg/kg). The maltster reported that this grain germinated normally and there were no adverse effects on the quality of the malt produced.

Table 19.9. Distribution of methacrifos in paddy rice as determined by milling and polishing **(McDougall 1976b)**

Fraction	Fraction weight (g)	Residue (ppm methacrifos)	Total μg methacrifos	Methacrifos as % of total*
Paddy rice	1200	3.0	3600	–
Brown rice	934	0.36	336	7.1
Hulls	240	18.4	4416	92.9
White rice	818	<0.2	<164	1.7
Bran	95	2.7	257	5.4
Waste	40	–	–	–

*Total is brown rice + hulls

Table 19.10. The distribution of methacrifos in brown rice as determined by polishing **(McDougall 1976b)**

Fraction	Fraction weight (g)	Residue (ppm methacrifos)	Total μg methacrifos	Methacrifos as % of total**
Brown rice	935	4.3	4021	–
White rice	778	0.30	233	8.0
Bran	103	25.9	2668	92.0
Waste	50	–	–	–

**Total is white + bran

Formica (1978f) reported a trial conducted in Switzerland in which wheat was treated with methacrifos at 6 rates between 2.5 mg/kg and 20 mg/kg. The treated grain was stored in closed metal barrels at a constant temperature of 25°C and after 1 month a 10 kg sample was taken from the centre of each barrel and processed. The results of analysis of the methacrifos residues at the time of treatment, immediately before milling and in the 3 milling fractions, flour, low-grade flour and bran and in bread prepared from the flour are given in **Table 19.12**. This shows that there was virtually a complete loss of methacrifos as a result of milling and baking.

Blattmann (1978) studied the degradation of methacrifos in wheat during storage, processing and cooking. He used radio-labelled methacrifos to study the level and composition of the deposit at various times up to 385 days after treatment. Samples of the wheat were milled after 3 and 314 days in storage and the methacrifos equivalents determined in the whole grains, semolina, bran, flour, and bread and noodles made from the flour. After milling the grains, most of the radioactivity was found in the bran fractions. The flour fraction, amounting to about 70% of the total weight of the grains, contained less than 15% of the radioactivity. During further processing to bread or noodles, about 90% of the residual insecticide in the flour was degraded primarily to the dealkylated derivatives. This pattern seemed to follow whether or not the wheat was processed shortly after treatment or after a substantial period in storage.

Formica (1978d) reported the results of a large-scale trial undertaken with methacrifos under practical conditions in Switzerland with the aim of establishing the persistence of methacrifos following application at the rate of 10 mg/kg on wheat prior to storage. The level and fate of residues on processed products were determined after a storage period of 7 months. He reported that the

Table 19.11. Fate of methacrifos in various grains during storage and processing. Application rate 18 mg/kg **(Desmarchelier 1976d).**

Processing stage	Residues (mg/kg) in grain				
	Paddy rice	Brown rice	White rice	Oats	Barley
After storage — 3 months	7.0	7.1*	7.5*	4.7*	(1) 4.8*
— 6 months	3.1+	4.8	3.8	1.3	(2) 1.4+
After processing					
husked rice — raw	0.65+				
— cooked	0.23				
polished rice — raw	0.12				
— cooked	<0.02				
After cooking — 5 min		4.4*	4.1*	1.6*	
— 15 min		2.7*	2.6*	0.9*	
— 25 min		0.75	–	–	
After malting — primitive					(1) 0.1*
— commercial					(2) 0.09+

* and + indicate corresponding samples

Table 19.12. Residues of methacrifos in wheat and its products one month after treatment (mg a.i./kg)

Material	Rate of application (mg a.i./kg)						
	20	15	10	7.5	5	2.5	untreated
Wheat*	14.4/15.8	16.0/17.4	10.4/10.0	7.6/7.2	5.2/5.4	1.8/1.7	<0.02
Wheat**	8.0/7.4	7.8/7.8	4.2/3.8	2.6/2.9	1.7/1.8	0.75/0.74	<0.03
Flour	0.9/0.85	0.6/0.6	0.36/0.36	0.2/0.4	0.08/0.12	<0.03/0.06	<0.03
Low grade flour	3.1/3.2	2.5/2.7	1.6/1.7	2.4/2.4	0.6/0.5	0.3/0.4	<0.03
Bran	15.6/18.0	10.4/11.2	7.6/6.4	7.6/8.0	3.1/2.6	2.6/1.7	<0.03
Bread	0.03/0.04	<0.03/<0.03	<0.03/<0.03	<0.03/<0.03	<0.03/<0.03	<0.03/<0.03	<0.03

*Immediately after treatment
**Shortly before milling

deposit had declined from 10 to approximately 3 mg/kg over a period of 7 months, and the flour milled from the wheat contained 0.57 mg/kg methacrifos, the bran 9.5 mg/kg and 4 different types of bread baked from this flour contained no more than 0.01 mg/kg, representing virtually complete destruction of the residue prior to consumption.

Tournayre (1978) carried out a study in which 3 separate lots of wheat were treated with methacrifos at the rate of 5, 10 and 10 mg/kg and the fate of the residues was studied during 252 days of storage. Portion of the grain was removed at days 42, 84 and 252 and it was then milled and processed into bread. None of the bread prepared from grain that had been in store contained any detectable residues of methacrifos (limits of determination 0.04 mg/kg). Details of the data generated in these studies are contained in **Table 19.13.**

Table 19.13. Fate of methacrifos residues on wheat during storage and following milling and baking **(Tournayre 1978).**

Applied fraction (mg/kg)		Residues (mg/kg) after increasing no. of days						
		0	42	63	84	127	168	252
5	Whole wheat	1.2	0.6	0.5	0.72	0.52	0.66	0.26
	Bran	1.12		1.48			0.8	
	Bran/flour		1.48		1.0			0.57
	Flour	0.10		0.11			0.08	
	Bread		<0.04		<0.04			<0.04
10	Whole wheat	3.06	2.35	1.70	1.31	1.18	1.27	0.75
	Bran	2.4		2.94			2.58	
	Bran flour		3.51		2.52			1.91
	Flour	0.30		0.28			0.26	
	Bread	<0.04		<0.04			<0.04	
10	Whole wheat	3.15	3.47	2.3	2.25	2.75	2.04	2.0
	Bran		5.29		5.83		6.12	
	Bran/flour		4.08		3.98			4.25
	Flour	0.46		0.53			0.50	
	Bread	<0.04		<0.04			<0.04	

Tempone (1979), who carried out a study to determine the effect of a large variety of grain-protectant insecticides on barley malting, reported that methacrifos, applied at the rate of 15 mg/kg, improved the quality and yield of malt. Barley malted immediately after treatment produced malt with a trace of methacrifos residue (0.05 mg/kg) but the same barley kept in storage for 3 months before malting (by which time the residue level had declined from 13 to 2.4 mg/kg) yielded malt without a trace of methacrifos residue.

Schnabel and Formica (1979b) carried out a study to see whether methacrifos, added to malt, could find its way into beer produced from such malt. They added the methacrifos to malt at the rate of 5 and 10 mg/kg and the next day used the malt to brew beer. About 20% of the methacrifos, added to the malt, was recovered in the brewer's grains and less than 5% was recovered from the wort before boiling. However, no trace of methacrifos could be found in the yeast or the beer (limit of detection 0.01 mg/kg) prepared from the wort.

Giannone and Formica (1979c) reported a similar study in which the methacrifos was added to the malt at the rate of 5 and 10g/kg and the malt was held in storage for 44 days. At the end of this time, no trace of methacrifos could be found (limit of determination 0.007 mg/kg) but dealkylated metabolites were found at levels not exceeding 1 mg/kg. No trace of methacrifos could be found in beer brewed from the treated malt but following partitioning and methylation it was possible to determine 2 metabolites at a combined total concentration of less than 1 mg/kg.

Schnabel and Formica (1979a) carried out a study in which 28 tonnes of Swiss wheat were treated with methacrifos at a target rate of 10 mg/kg and, after remaining in storage for 7 months, the wheat was milled and portion of the flour was converted into 4 types of bread prepared by 2 different recipes. The results that are given in **Table 19.14** indicate not only the level of methacrifos but also the content of demethylated metabolites. This study provides the most comprehensive picture of the residues of methacrifos and its metabolites in grain, milling fractions and bread. The methacrifos residue in bread was little more than 1% of that occurring in the wheat grain. Even the inclusion of the biologically inactive metabolites brings the total residue to only about 15% of that occurring in the whole grain at the time of milling, or less than 5% of the rate applied as the grain was put into storage.

Desmarchelier *et al.* (1980a) published results of a number of studies of the fate of methacrifos and 5 other

Table 19.14. Gas chromatographic determination of total residues of methacrifos in wheat and its derivatives seven months after treatment with the formulation of SO 050 **(Schnabel and Formica 1979a)** (Results not corrected for recoveries)

	unchanged methacrifos (mg/kg)	methylated metabolites to methacrifos (mg/kg)	methylated metabolites to CGA 73712 (mg/kg)	methylated metabolites sum (mg/kg)	total residue calculated as methacrifos (mg/kg)
Wheat grain	2.91	0.57	1.06	1.63	4.5
	2.72	0.62	1.13	1.75	4.4
Bran	7.97	1.63	1.76	3.39	11.3
	8.40	1.55	1.57	3.12	11.5
Low grade flour	7.13	3.00	4.74	7.74	14.8
	7.06	2.89	4.03	6.92	13.9
Flour	0.50	0.14	0.31	0.45	0.95
	0.49	0.13	0.22	0.35	0.84
Bread large*	0.035	0.101	0.42	0.52	0.56
small*	0.034	0.060	0.23	0.29	0.32
Bread large**	0.040	0.105	0.55	0.66	0.70
small**	0.055	0.087	0.47	0.56	0.61

*long fermentation time
**short fermentation time

insecticides on rice and barley after storage, processing and cooking. Data presented in this publication is similar to that in **Desmarchelier (1978c).**

Desmarchelier *et al.* (1980b) reported that any methacrifos residue occurring in flour milled from treated wheat was destroyed in the baking process, being undetectable in bread by methods sensitive to 0.03 mg/kg.

Wilkin and Fishwick (1981) reported a study in which wheat treated with 6 different grain-protectant insecticides was processed into wholemeal flour after being in store for between 4 and 36 weeks. This study indicates that 39% of the methacrifos present on the wheat after storage could be recovered from the wholemeal bread.

Bull (1983a) conducted a special study in which prime hard wheat of the grade used specifically for the production of starches and gluten was treated with methacrifos at a concentration of 20 mg/kg and stored in a silo. The grain was milled 9 weeks after treatment at which stage it contained 15 mg/kg methacrifos which was reduced to 11 mg/kg by the conditioning process prior to milling, with about one-quarter of the total residue being present as the desmethyl metabolite (CGA-90953). The greatest proportion of the chemical residue was retained at the grain surface with the bran and the pollard fractions accounting for 61% of the total residue in the whole grain. The actual residue levels of methacrifos and CGA-90953 in the gluten prepared from the flour and in the milled fractions of the wheat are given in **Table 19.15.** Although the metabolite CGA-90953 is a phosphorous ester it is not biologically active. It has no insecticidal activity; neither has it any ability to inactivate cholinesterase *in vitro.*

Table 19.15. The distribution of methacrifos and its desmethyl metabolite (CGA-90953) in wheat, milling fractions and gluten **(Bull 1983a).**

Cereal fraction	Methacrifos (mg/kg)	CGA-90953 (mg/kg)	Total residue (mg/kg)
Whole wheat	16	–	20
Aged wheat	15	–	–
Conditioned wheat	10.8	3.1	13.9
Bran	34.9	6.7	41.6
Pollard	27.4	7.5	32.6
Flour	2.4	0.82	3.2
Gluten	3.9	2.0	5.9

19.4 Metabolism

The degradation of methacrifos in stored wheat was studied by **Blattmann (1978)** who applied radio-labelled methacrifos to grain at a concentration of 10 mg/kg. The grains were stored in glass cylinders fitted for the collection of $^{14}CO_2$. Grain samples were removed for analysis after 0, 23, 140, 294 and 385 days. From 84 to 94% of the applied radioactivity was recovered at all sampling times, indicating very limited volatility of methacrifos or its degradation products. After about 1 year, about 25% of the then recovered radioactivity was present as parent compound. The fact that little volatilisation of the radioactivity occurred during the storage indicated that methacrifos and its degradation products were strongly absorbed to the wheat grain. Less than 15%

of the recovered radioactivity could be washed off the surface of the grain with methylene chloride. The bulk of the radioactivity could be extracted only after grinding the grain. The non-extractable portion of radioactivity was low at the early sampling dates but rose to 9% and 14% after 140 and 385 days, respectively. Low but continuous evolution of $^{14}CO_2$ during the experiment demonstrated that the insecticide was metabolised by the whole grain. The organosoluble radioactivity on or in the grains consisted mainly of the parent compound. The oxon was not detected. Mono- and di-demethylated methacrifos were detected by ethylation followed by GLC analysis. Results give strong evidence for the degradation pathways shown in **Figure 19.3.**

about 50% of the nominal value between day 20 and day 126 after treatment. This suggested that a competitive reaction degraded the dealkylated metabolite to some extent as it was being formed.

Giannone and Formica (1980d) carried out a similar study on sorghum grain treated and stored in Mali; the results were generally similar.

Ifflaender and Mucke (1975) studied the distribution, degradation and excretion of methacrifos in the rat, using material labelled with ^{14}C in the carboxy group. After a single oral dose of approximately 5 mg/kg, the radioactivity was rapidly excreted; the excretion half-life times being less than 8 hours for both sexes. Within 5 days 54.7% and 52.7% of the applied radioactivity were

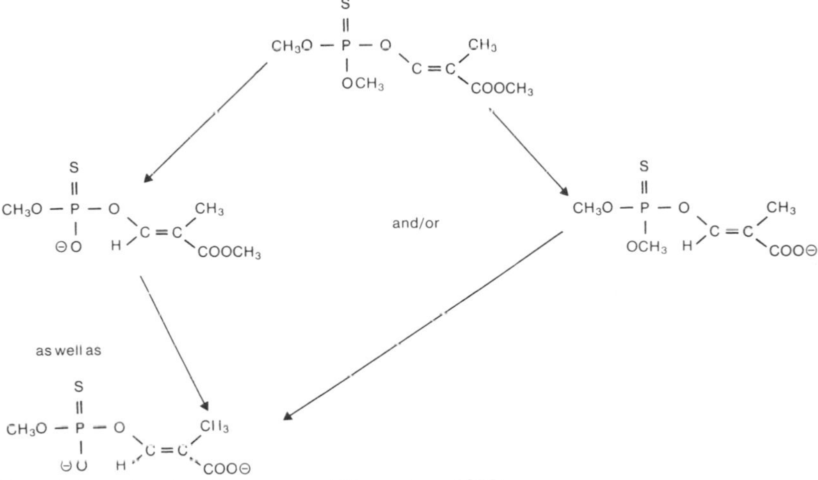

Figure 19.3. Degradation of methacrifos in grain **(Blattmann 1978).**

The decrease in the protective effect of methacrifos against insects is closely related to the decrease in the amount of the unchanged methacrifos still present. This suggests that the degradation products are significantly less toxic to insects than the parent compound. Most of the radioactivity was found in the bran fraction, the flour fraction amounting to about 70% of the total weight of the grains contained less than 15% of the radioactivity. During further processing of the flour into bread or noodles, about 90% of the residual insecticide in the flour was degraded primarily to the dealkylated derivatives **(Blattmann 1978).**

Schnabel and Hormann (1980) measured the level and fate of residues of methacrifos and its dealkylated metabolites in wheat grain over a period of 149 days by means of gas-chromatography using the flame photometric detector. Results indicated that the methacrifos residues declined from 80% of the nominal value to 30% within 126 days. The amount of the dealkylated metabolites, calculated as a percentage of the total residue, was found to gradually increase from 17 to 54% in the time between 20 and 149 days after treatment. The total residue was found to gradually decrease from 98% to

excreted with the expired air, 30.2% and 43.8% in the urine and 10.2% and 8.7% with the faeces by male and female rats, respectively. When the animals were killed 5 days after dosing, the tissues investigated showed the following amounts of ^{14}C were incorporated (in mg/kg methacrifos equivalents for males and females):

liver	0.31 and 0.29	mg/kg;
fat	0.11 and 0.04	mg/kg;
kidney	0.09 and 0.09	mg/kg;
muscle	0.04 and 0.02	mg/kg;
blood	0.04 and 0.04	mg/kg;
brain	0.03 and 0.02	mg/kg;
spleen	0.05 and 0.05	mg/kg;
testes	0.03	mg/kg;
ovary	0.04	mg/kg.

These residues were probably due to incorporation of ^{14}C- fragments into natural products. No unchanged methacrifos was present in the urine, indicating complete degradation of the compound.

A similar study by **Hambock (1978)** showed that 41% of an oral dose of 25 mg/kg ^{14}C-labelled methacrifos was excreted by the kidney within 48 hours, 38% appeared as $^{14}CO_2$ in the expired air and 10% in the faeces. The main urinary metabolite, representing 12% of the dose, was isolated and identified to be N-acetyl-S-(2-methoxy-carbonylprop-1-enyl) cysteine. In a separate experiment, it was demonstrated that this cysteine conjugate, when injected into rats, was not metabolised to CO_2 but was excreted mainly unchanged via the kidney. Based on the information obtained, a partial metabolic pathway of the insecticide in the rat was proposed. This is illustrated in **Figure 19.4.**

skin were taken and analysed by a method with a limit of determination of 0.01–0.03 mg/kg. Only in the birds slaughtered 6 hours after administration of the methacrifos were any residues determined. These ranged from less than 0.01 in muscle to 0.42 mg/kg in gizzard. Fat contained the next highest residue at 0.22 mg/kg.

Formica (1977b) studied 2 lactating cows given 30 mg methacrifos daily for 10 days. The dose of 30 mg methacrifos was calculated on the basis that cows consuming a daily ration of 3 kg of cereals treated at the rate of 10 mg/kg would consume 30 mg of methacrifos as a maximum. The cows were milked mechanically, twice daily and milk samples were taken daily for analysis. No residues of methacrifos could be found in the milk before, during or after the administration of methacrifos. The method used had a limit of determination of 0.001 mg/kg.

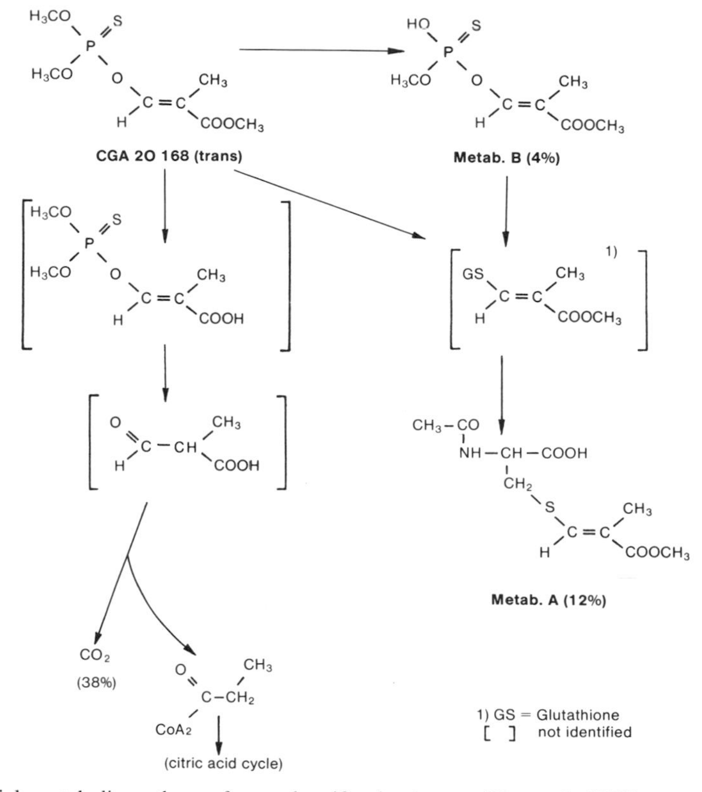

Figure 19.4. Partial metabolic pathway for methacrifos in the rat **(Hancock 1978).**

19.5 Fate in Animals

Formica (1974) reported a study in which 384 day-old chicks which were divided into 5 groups and were fed *ad libidum* during 63 days with feed fortified with methacrifos at graded levels. The concentration of methacrifos in the feed was checked by analysis. The fortification levels were 10, 100, 250, and 500 mg/kg. At the end of the feeding time, 20 chickens per group were slaughtered and the combined muscle, skin, kidney and liver used for residue analysis. None of the tissues from birds in any group contained methacrifos residues above the limit of determination (0.01 mg/kg).

Moore and McDougall (1974d) conducted a study in which 14 week-old cockerels were given a single oral dose of 20 mg/kg technical methacrifos and afterwards slaughtered 6, 12, 24 and 48 hours after treatment. Samples of fat, muscle, liver, kidney, heart, gizzard and

Schnabel and Formica (1979c) gave 45 laying hens held in battery cages feed containing 10 and 20 mg/kg methacrifos for 28 days. Eggs collected before, during and after the administration of methacrifos were analysed by a method capable of detecting 0.01 mg/kg methacrifos. None of the eggs contained any residues at this level.

Beef cattle were given feed prepared from wheat grain containing methacrifos at a level of 15 or 30 mg/kg **(Bull 1983b)**. The medicated feed, consisting of 5 parts wheat to 2 parts chaff, was made available to the cattle *ad libidum* for a maximum period of 14 days. Some animals were killed after 7 days exposure, some after 14 and the third group was given a recovery period of 7 days on non-medicated feed after the 14 day exposure period before they were killed. Neither methacrifos nor its dealkylated metabolite could be detected in the fat, liver, kidney or muscle of cattle exposed to medicated feed. The limit of detection was 0.01 mg/kg in all substrates.

19.6 Safety to Livestock

Gfeller (1974) fed 4 groups of 4 pigs (2 males and 2 females) methacrifos at concentrations of 0, 10, 100, and 1000 ppm for 4 weeks. With the exception of a severe inhibition of the cholinesterase activity of the blood at 1000 ppm, particularly the acetyl cholinesterase activity, and a slight inhibition of the brain cholinesterase at 1000 ppm as well, all the other parameters examined (clinical signs, mortality, body weight development, feed consumption and feed conversion, haematology and blood chemistry) were within the normal limits. At 100 and at 10 ppm no reaction to treatment was noted. At autopsy and upon histopathological examination no changes were seen which could be associated with the treatment.

Japanese quail (3 males and 7 females per group) were fed concentrations of 0, 1000, 6000 and 10 000 ppm of methacrifos for 5 days and observed for a further 3 days. The '8-day LC/50' was found to be approximately 1000 ppm. The birds of the 6000 and 10 000 ppm groups refused food during the entire feeding period. Therefore, a correct determination of the LC/50 was not possible. Food consumption and body weight were decreased in the 1000 ppm group during the treatment, but showed a recovery during the 3-day observation period. All female birds of the treated groups stopped their egg production from the third experimental day onwards **(Sachsse and Ullmann 1974b)**.

Methacrifos was given in the diet to 6 groups of chickens at concentrations of 0, 10, 100, 250, 500, and 1000 ppm for 63 days. The percentage of mortality recorded in the groups was as follows: control 1.6%, 10 ppm 0%, 100 ppm 1.6%, 250 ppm 3.1%, 500 ppm 37.6% and 1000 ppm 98.2%. There was no difference between males and females in mortality. Most animals died during the first 2 weeks. The dying animals of the 500 and 1000 ppm group showed signs of ataxia, inappetence, somnolence and ruffling up of plumage. Body weight development and food consumption were significantly below the controls in the 500 and 250 ppm groups; also the 100 ppm group showed a significant reduction in body weight development. The cholinesterase activity in the brain showed a slight (100 and 200 ppm groups) to moderate (500 ppm group) inhibition with a dosage related trend. In the 10 ppm group no reaction to treatment was observed. Gross pathological and histopathological examinations revealed no substance-related changes **(Strittnatter and Gfeller 1975)**.

Strong (1985) reported on the reaction of young cattle used in a residue study **(Bull 1983b)** with wheat medicated with methacrifos. Wheat containing 15 and 30 ppm methacrifos was crushed in a hammer mill and mixed with oaten chaff in the ratio of 5 parts wheat to 2 parts chaff by weight. Each level of medication was then used in feeding cattle for 7 or 14 days after which some of their number were slaughtered. One-third of the animals were retained on unmedicated feed for a further 7 days. Body weight was maintained. At no stage during the period of the trial did any of the calves show any signs of toxicity.

19.7 Toxicological Evaluation

The toxicology of methacrifos was evaluated by JMPR in 1980 and 1982 **(FAO/WHO 1981, 1983)**. Methacrifos shows moderate acute toxicity following oral administration to various animal species. Signs of poisoning are typical of cholinesterase inhibitors. The acute toxicity was found to be potentiated by several other organophosphorus esters.

Methacrifos did not induce a delayed neurotoxic response in hens. In short- and long-term studies methacrifos caused signs and symptoms of cholinesterase inhibition, which were associated with reduction of body weight gain and food consumption. Brain cholinesterase activity was inhibited at 100 ppm. In rats, at high dose levels, effects on several haematological parameters were observed and recovery of cholinesterase activity was relatively slow.

In long and short term studies in the rat, cholinesterase depression was observed at 10 ppm and above. A no-effect level of 1 ppm was observed.

Because of parental mortality in a 3-generation reproduction study in rats and fetal death in rat teratology bioassays, it was not possible to fully evaluate the potential effects in these areas.

Mutagenicity studies with bacteria, rats, mice and hampsters, a mouse carcinogenicity study and a rat and rabbit teratology studies were all negative. Acceptable data were available to allow the establishment of a no-effect level in 2 mammalian species and a temporary ADI was allocated. The level causing no toxicological effect is:

Rat:	1 ppm in the diet equivalent to 0.05 mg/kg body weight
Dog:	1 ppm in the diet equivalent to 0.025 mg/kg body weight.

From this information, the temporary acceptable daily intake for man was estimated to be 0.0003 mg/kg body weight.

Further studies, including a human intake study, which will be considered by JMPR in 1986, should enable the temporary ADI to be converted to a full ADI at a considerably higher level.

19.8 Maximum Residue Limits

Based on the extensive data available to it in 1980, the JMPR recommended the following temporary maximum residue limits.

Commodity	Maximum Residue Limits (mg/kg)
Bran	20
Cereal grains	10
Flour (white)	2
Wholemeal	10

Residues are determined and expressed as methacrifos.

The methods recommended for the determination of methacrifos residues are **Desmarchelier *et al.* (1977b)** and **Wallbank (1981)**.

20. Methoprene

Methoprene [1,isopropyl (2*E*,4*E*)-11-methoxy-3,7,11-trimethyldodeca-2,4-dienoate], available under the trademark Altosid, is the most commercially advanced example of the class of biorational insect growth regulators (IGRs) with juvenile hormone activity generally referred to 'juvenoids' (**Henrick *et al.* 1973; Henrick 1982**).

In order to appreciate the difference between methoprene and conventional insecticides used for stored-product pest control some understanding of juvenile hormones is desirable. I am indebted to Dr Clive A. Henrick for the following introduction.

The juvenile hormone (JH) of insects is synthesised and secreted by the *corpora allata* and has several functions in the life of the insect. To exert morphagenetic activity it acts in concert with the moulting hormone to suppress adult differentiation and to maintain the immature character of the developing insect. Thus, although the moulting hormone initiates all moulting processes and introduces metamorphosis, the JH modifies the outcome of the moult. In at least several orders of insects JH also has a gonadotrophic effect in that it is necessary for yolk formation in certain adult females and for development of the accessory glands in the males. The JH also exerts several other effects during the insect's life cycle (**Staal 1975; Slama *et al.* 1974; Sehnal 1976; Edwards and Menn 1981**). For example, insect eggs are sensitive to JH, especially in the release stages of embryogenesis, and the application of JH prevents the development of viable larvae. Also, in some species the treatment of adult females with JH results in the development of eggs that fail to hatch. The growth of immature larvae requires the presence of JH, but in order for mature larvae to undergo metamorphosis into adults the production of the hormone must decline. Contact with JH at this stage can cause the death of the insect due to the derangement of development.

The possibility that analogues of the natural JH would have potential as insect control agents was first recognised by **Carroll Williams (1956, 1967)**. The search for safer and more selective insecticides has prompted extensive research during the past 18 years into the area of compounds showing insect JH activity (juvenoids). The promise of practical utilisation of juvenoids has been realised with the discovery and commercialisation of methoprene, hydroprene and kinoprene. The advantages and disadvantages of the use of juvenoids for the management of insect pest populations and the successful commercial applications of these have been reviewed (**Staal 1977; Edwards and Menn 1981; Menn *et al.* 1981**).

The modes of action of JH are not fully understood, and the diverse effects of juvenoids suggests that their modes of action may not be identical in all cases. The general modes of action and the various effects of juvenoids on different groups of insects have been reviewed (**Slama *et al.* 1974; Staal 1975; Sehnal 1976; Vogel *et al.* 1979**). Juvenoids may be used to interfere with essential life processes, such as metamorphosis and adult emergence, in the later part of immature development. The effects of the application of juvenoids at these sensitive periods in the insect's life cycle may include the occurrence of larval-pupal or pupal-adult intermediates, defective reproductive organs, or abnormal embryogenesis. The abnormal morphological and developmental consequences of such treatment are largely irreversible and often lethal to the target insects. The biological activity of a juvenoid is usually measured only at a specific time during the sensitive stage of the insect's life cycle, although more detailed evaluations over more than 1 generation may reveal additional subtle and delayed effects that contribute to insect control.

Juvenoids include all chemicals showing the same qualitative physiological activity against an insect species as the natural JH. A wide variety of chemical structures can evoke a JH response in some insect species, including compounds with little or no structural resemblance to the natural JH's. The JH activity of some compounds is specific to a particular order or even family of insect species, whereas other juvenoids show activity against a broad spectrum of species. Some insect species respond to a diverse range of structures, while others are sensitive to very specific structures. Structure/activity relationships observed for 1 taxon of insects are not necessarily valid for another, and generalisations to many species, even within 1 order, are difficult to make. **Henrick (1982)** has prepared an 80 page review of juvenile hormone analogues and their structure/activity relationships. This contains an extensive bibliography on the subject.

This chapter is confined to methoprene because it is the first juvenoid which has reached the stage of commercial development. There is an extensive literature on related compounds, other juvenoids, unrelated insect growth regulators and chitin inhibitors. It has been decided that this is not the place to review these other compounds.

20.1 Usefulness

Schaefer and Wilder (1972) obtained favourable

results in their work on mosquito control with methoprene, a novel juvenile hormone analogue with a trans, trans-dienoate structure without the epoxide function essential for activity in compounds described earlier. Hydroprene, another dienoate without the epoxide function, was highly active against khapra beetle, *Trogoderma granarium* (**Metwally *et al.* 1972**). **Henrick *et al.* (1973, 1976)** surveyed the chemical and biological characteristics of this dienoate group of compounds with special emphasis on the effectiveness of hydroprene and methoprene against pests of economic importance.

These reports and references cited in them give an indication of a potentially important new concept in insect-pest control ranging from basic chemical and physiological studies to applied laboratory and field investigations. Obviously, synthetic organic chemicals acting as juvenile hormone analogues, rather than as toxicants in the manner of presently used insecticides, will present special problems in developing acceptable pest-control programs. However, the safety of these compounds, as known at present, makes them highly desirable candidates for use as post-harvest protectants of stored agricultural commodities.

Strong and Diekman (1973) reported results from a series of exploratory tests with 15 candidate juvenoid compounds admixed with appropriate foods to which each of the 12 species of stored-product insects were exposed. The products hydroprene and methoprene were the most active juvenoid compounds on all species tested. Neither hydroprene nor methoprene affects the fecundity of adults, when applied at that stage. However, reproduction was inhibited when the juvenoids were present at a critical time before metamorphosis to the adult stage. Direct ovicidal effects have also been observed and can contribute to the inhibition of reproduction if adults are present at the time of application. Hydroprene was effective at 5 mg/kg against *Trogoderma inclusum*, and *T. variable*; and at 10 mg/kg against *Rhyzopertha dominica*; and at 50 mg/kg against *Sitophilus granarius*. Hydroprene demonstrated a high level of activity at 10 mg/kg in tests with *Tribolium confusum* and at 50 mg/kg in tests with *Sitophilus oryzae*, but it did not completely prevent the emergence of adults of *Sitophilus oryzae* at these concentrations. Methoprene was effective at 5 mg/kg against *Plodia interpunctella, Lasioderma serricorne, Rhyzopertha dominica, Oryzaephilus surinamensis, Oryzaephilus mercator, Tribolium castaneum* and at 10 mg/kg against *Ephestia cautella* and *Tribolium confusum*. These authors concluded that while hydroprene and methoprene did not control parent insects, this should not hinder active investigation of these compounds as commodity protectants. Their obvious value in preventing development of large populations of insects rendered them potentially suitable for insect-pest management programs in stored commodities. Such programs are dependent upon sanitation procedures to eliminate sources of infestation and thus prevent immigration of untreated adults. The use of insecticides to eradicate existing infestations in and around storage facilities prior to storage of commodities is essential.

Williams and Amos (1974) showed that methoprene was active at concentrations of 5 mg/kg and above by either inhibiting or impairing development of *Tribolium castaneum*.

Amos and Williams (1975) determined the effect of methoprene on the development of *Tribolium castaneum* by placing eggs on wheat flour treated with methoprene to give concentrations of 0.001, 0.01, 0.1, 1, 5 and 20 mg/kg. The viability, reproductive performance and productivity of morphologically normal and deformed adults was determined. The productivity was only marginally reduced in deformed females and the re-productivity of males was greatly reduced irrespective of morphological deformity. The authors concluded that there was some degree of physiological, as well as morphological, impairment to successful breeding and that such sub-lethal effects may confer an advantage for use of methoprene as a protectant for stored products by retarding or even preventing the establishment of an insect infestation.

McGregor and Kramer (1975) found that when hydroprene and methoprene were applied to wheat and corn kernels at concentrations ranging from 2 to 10 mg/kg, they effectively prevented the metamorphosis of *Plodia interpunctella* and effectively reduced the F_1 populations of *Rhyzopertha dominica, Tribolium confusum* and *Oryzaephilus surinamensis*. However, they were not effective against *Sitophilus oryzae* since substantial numbers of progeny developed in wheat kernels treated at a rate of 10 mg/kg with either of the compounds.

Hoppe and Suchy (1975) reviewed the status of research on insect growth regulators for the protection of stored grain. They considered the pest insects of stored food to be an ideal target for juvenoids because, in addition to low mammalian toxicity, most exhibit very long persistence in stored commodities. The rapid development of most pest insects of stored food, resulting in a high number of generations in 1 year, provides an ideal opportunity for harnessing the particular properties of these compounds.

Loschiavo (1975) reported tests with 4 synthetic growth regulators with juvenile hormone activity including hydroprene and methoprene, against 7 species of stored-product insects.

Tan (1975) studied the effects of methoprene and hydroprene on *Ephestia kuehniella* and *Ephestia cautella*. When applied topically to mature larvae both juvenoids resulted in the production of super-larvae with invariably prolonged larval life and larval/pupal intermediates. When migrating last-instar larvae were treated with hydroprene and methoprene, larval/pupal intermediates and pupal mortality were induced. However, when applied topically, methoprene appeared more effective than hydroprene. Both analogues prevented adult emergence when topically applied to the migrating larvae at doses between 28–52 ng. One-day-old pupae were most susceptible while older individuals became less sensitive with age. When larvae pupated in corregated cardboard rolls treated with hydroprene those of both *Ephestia cautella* and *Ephestia kuehniella* failed to emerge. At an estimated dose of 179 mg/cm^2, methoprene prevented 77.6% *Ephestia cautella* and 100% *Ephestia kuehniella* larvae from emerging as adults.

Baker and Lum (1976) noted significant differences in the response of *Sitophilus granarius* and *S. oryzae* to methoprene. They proposed that the bacterial symbiont could be responsible for the difference. Actually, all *Sitophilus* spp. have been noted to be less susceptible to insect-growth regulators than many other stored-product insects.

Loschiavo (1976) studied the effects of methoprene and hydroprene on survival, development or reproduction of 6 species of stored-product insects. He used food treated with methoprene and hydroprene at 1, 5, 10 and 20 mg/kg. Both compounds at 20 mg/kg prevented emergence of pupae of *Tribolium castaneum* and substantially reduced those of *Tribolium confusum*. At 5 mg/kg or higher, they inhibited oviposition of both species. Larvae that failed to pupate in treated food continued to moult, and those that survived to 120 days or longer after emergence were larger than normal. Methoprene at 1 mg/kg or higher prevented emergence of adults of *Oryzaephilus mercator* and *O. surinamensis* in treated rolled oats or cornmeal. Hydroprene produced a similar effect in cornmeal treated at 5 mg/kg or higher. Both compounds caused morphogenetic malformities in flour beetles and grain beetles. None of the pupal/adult intermediates or malformed adults survived. Hydroprene at 10 and 20 mg/kg in wheat almost completely inhibited progeny production by *Sitophilus granarius*. Both compounds caused decreasing productivity in *S. oryzae* with increasing concentration but not enough to produce useful control.

Amos and Williams (1977) published additional research on methoprene and activity data on hydroprene. *Rhyzopertha dominica, Sitophilus oryzae* and *S. granarius* responded with reduced reproductivity. Parental mortality on wheat appeared greater with methoprene than with hydroprene. The authors concluded that ventilation was one of the factors influencing the sensitivity by *Sitophilus* spp. to juvenile hormones. It is clear that the lack of sensitivity of *Sitophilus* spp. has been paramount in hindering the otherwise potentially valuable development of these juvenoids as grain protectants (see however, **Edwards and Short 1984**).

Marzke *et al.* (1977), Long *et al.* (1978) claimed methoprene very effective in controlling *Lasioderma serricorne* on tobacco.

Amos *et al.* (1978) studied the sterilising ability of methoprene and hydroprene in *Tribolium castaneum*. The productivity of *Tribolium castaneum* adults, previously reared in flour incorporating either juvenoid, was found to be impaired depending upon the concentration of the juvenoid in the flour, whether or not the individual was morphologically deformed and its sex. The objective of these workers was to determine whether exposure to sub-lethal concentrations of the juvenoid could induce a degree of sterility that may be an advantage particularly when the juvenoid is used as a commodity protectant because such sub-lethal effects would extend the period of protection by retarding or even preventing a build-up of an infestation. They found that viability was significantly reduced in all male adults previously reared in flour containing 0.1 mg/kg methoprene, and in only the deformed males reared in flour with 0.01 and 0.1 mg/kg hydroprene. The viability of morphologically deformed female adults reared in either methoprene or hydroprene treated flour was markedly reduced at all concentrations whereas that of morphologically normal female adults was comparable to control viability. Productivity of female adults reared from methoprene treated flour was comparable to that of controls, and in the males a reduction occurred only in those reared in flour containing 0.1 mg/kg methoprene. The manner in which these juvenoids interfere with the reproductive processess is not understood. Nonetheless, the fact that both methoprene and hydroprene possess sterilising ability further enhances their potential as commodity protectants.

Gonen and Schwartz (1978) investigated the possibility of taking advantage of the fundamental biological difference between *Ephestia cautella* and most other stored-product insects, i.e. the fact that fully grown larvae, as well as the deposited eggs of *Ephestia cautella*, come in contact with the surface of the infested bulk. Eggs were therefore brought into contact with surfaces sprayed with graded doses of methoprene and larvae were caused to come in contact with small jute sacks which had been sprayed with methoprene at a range of concentrations. No meaningful reduction in the number of adults that developed from eggs that were in contact with a sprayed surface could be detected up to a dose of 14.7 mg/m^2. The higher doses tested reduced the number of developing adults by about 60%. A dose of 44–75 mg/m^2 was required to achieve 96–98% control when applied to the surface of jute bags. Complete mortality at larval or pupal stages was induced by direct spraying of last instar larvae with a dose of 6.6 mg/m^2. This treatment also caused a delay of up to 3 months in the pupation process. The results obtained in these experiments demonstrated that the concept of treating surfaces with which susceptible life stages of the insect come into contact, could be feasible.

Pardo and Nordlander (1979) tested the effect of methoprene-treated surfaces on *Tribolium confusum*. Methoprene was applied in graded doses to cardboard to which young and full-grown larvae, young and old pupae and adults of *Tribolium confusum* were exposed. Prolonged contact with methoprene caused high mortality of fully-grown larvae and pupae, while shorter periods of exposure were not sufficient for effective control of fully-grown larvae. No direct effects on adults or their fecundity were noted.

Tan and Tan (1980) studied how the degradation of methoprene and hydroprene by light and high temperature would affect the biological activity against *Ephestia* *kuehniella* under storage conditions in Malaysia. From the work they concluded that the half-life of methoprene would cover 5 generations of *Ephestia kuehniella* and that it would be suitable for practical control during short-term storage provided that the ambient temperature was not too high. They found that methoprene applied to pupation sites in semi-darkness was effective at 160 mg/m^2. In contrast, hydroprene was effective at 80 mg/m^2 and persisted for 8 generations of *Ephestia kuehniella*.

Long *et al.* (1980) showed methoprene to be effective for controlling *Lasioderma serricorne* under severe infestation conditions.

Hoppe (1981) tested insecticide-resistant strains of *Tribolium castaneum* for susceptibility to methoprene. Although the multi-resistant strain showed the highest cross-resistance to methoprene which could be demonstrated by any of the used criteria, no clear relation between the degree of insecticide resistance and the response to methoprene was observed. On the contrary 2 malathion-resistant strains were more susceptible to methoprene than the standard susceptible strains.

Edwards (1981), in a paper concerning alternative chemicals for use in grain protection, discussed the advantages which 'third-generation' pesticides such as methoprene offer for the protection of stored grain, but in doing so drew attention to the lack of incentive for industry to research, develop and promote such products or for users to develop the necessary sophistication and ability to pay for such materials. The article appealed to governments to conduct and sponsor basic research, to encourage industry to develop such methods and products perhaps by re-examining the requirements necessary for registration of such new compounds.

Mian and Mulla (1982a) studied 4 insect growth regulators, including methoprene, applied at 5 mg/kg to wheat flour or grain against *Oryzaephilus surinamensis, Tribolium castaneum, Rhyzopertha dominica* and *Sitophilus oryzae*. Methoprene, besides its ovicidal action against *Oryzaephilus surinamensis* and *Rhyzopertha dominica*, caused substantial mortality in the fully-grown larvae of the 3 species. It also affected the subsequent progeny production of parent adults pre-exposed to methoprene treated feed for 2 weeks. Methoprene had no adverse effect on the development of *Sitophilus oryzae* inside grain kernels infested with parent adults 1 day after grain was treated. However, in grain infested with *Sitophilus* after 12 months of post-treatement storage, the development of this species was arrested significantly. Methoprene showed activity from egg to pre-emergence adults of *Sitophilus oryzae*. The authors pointed out that when testing insect growth regulators against internal feeders, it is important to allow time for penetration of the active ingredients into the inner portions of grain kernels before assessing effects.

Methoprene was evaluated for residual activity against 2 internal feeders, *Rhyzopertha dominica* and *Sitophilus oryzae* in stored grains for 1 year **(Mian and Mulla 1982b)**. At 1–10 mg/kg it gave effective control of *R. dominica* in stored wheat, barley and corn for more than 12 months. The residual activity of methoprene at 0.1 and 0.5 mg/kg did not result in appreciable control of *R. dominica*. At 5 and 10 mg/kg, methoprene gave 100% control at 2 weeks to 12 months post-treatment. At 1 mg/kg its residual activity (93.7%) 2 weeks after application gradually increased to as high as 99.8% at 12 months post-treatment. Methoprene tested at 1, 5 and 10 mg/kg against *S. oryzae* did exhibit residual action which increased with time. Its activity, however, was low and even at 10 mg/kg it gave only 80% to 93% control at various intervals during the test period.

Manzelli (1982) pointed out that methoprene has demonstrated the capability of preventing the larval form of *Lasioderma serricorne* and *Ephestia elutella* from developing into normal pupae or adults. When applied directly to tobacco materials at a concentration of 10 mg/kg, just before warehouse storage, methoprene has prevented the emergence of adults of *Lasioderma serricorne* for a period of 4 years.

Ambika and Abraham (1982) studied the effect of applying methoprene at different concentrations to eggs and larvae of *Corcyra cephalonica* in laboratories in India. The results showed that spray application of a dose of 0.1 μg/egg gave the highest egg mortality. Topical application at a dose of 1 μg per fourth instar larvae gave 100% larval mortality by the forty-second day after treatment, while that of 0.5 μg/larva gave 100% larval mortality within 90 days. Such larval treatment resulted in the formation of supernumerary larvae, which died within 7–9 days of the supernumerary moult.

Abdel-Aal and Hussein (1983) studied the adult emergence and survival of the cow pea seed beetle, *Callosobruchus maculatus*, after treatment of 1- and 3-day-old eggs with methoprene. Adult emergence was reduced considerably by methoprene treatment, and the reduction in survival increased with the concentration applied. The compound was most effective when applied to newly-laid eggs, especially at the higher concentrations.

Mian and Mulla (1983b) studied the effects of methoprene on the germination of stored wheat at various intervals after the grain had been treated with aqueous solutions at rates to give 1, 5 or 10 mg/kg on the grain. Grain viability was significantly reduced by all the treatments when it was assessed 1 week and 1 month after treatment. Losses in germination were significantly higher than the allowable limit of 7%. Evaluations of organic solvents for their impact on grain viability showed that 70% ethanol was the most active compound inhibiting germination, followed by acetone, benzene and hexane.

Minor *et al.* (1983) reported a large-scale evaluation of methoprene for control of *Lasioderma serricorne*.

Edwards and Short (1984) tested 3 compounds with insect juvenile hormone activity as protectants for wheat grain against the 3 *Sitophilus* spp. Although methoprene and JH 1 did not reduce the number of progeny until the concentration reached 50 to 100 mg/kg, fenoxycarb (Ro

13–5223) showed a pronounced effect at concentrations as low as 1 and 5 mg/kg and well over 95% suppression of emergence at 10 mg/kg. These results confirm the work of **Kramer *et al.* (1981)** who found that Ro 13–5223 possessed high relative activity against the 3 *Sitophilus* spp.

Because of the spread of insecticide resistance in stored-product insects **(Champ and Dyte 1976)** and because cross-resistance to juvenile hormone analogues had been reported in some stored-product species, e.g. *Tribolium castaneum* **(Dyte 1972), Edwards and Short (1984)** included 2 insecticide-resistant strains of *S. granarius* in their tests. Their results have indicated that the 2 strains of *S. granarius* resistant to conventional insecticides may be more sensitive to juvenile hormone analogues than the corresponding susceptible strains.

It has been suggested that the reason for the low activity of juvenile hormone analogues against *Sitophilus* spp. is that the developing larvae are protected from the effects of these compounds which are unable to penetrate into the grain in which the larva is enclosed **(Bathnagar-Thomas 1973; Hoppe 1976; Kramer and McGregor 1978)**. However, **Rowlands (1976)** demonstrated that three juvenile hormone compounds rapidly penetrated wheat grains and accumulated in the region of the aleurone layer. Since final instar larvae are often found in close proximity to the aleurone layer, **Rowlands (1976)** suggested that the insensitivity of *Sitophilus* spp. might be due to factors other than the protective effect of the grain. Subsequently, several workers have examined the role of metabolism **(Edwards and Rowlands 1978)**, the influence of gut flora **(Baker and Lum 1976)** and the effect of ventilation **(Amos and Williams 1977)** as factors influencing the sensitivity of *Sitophilus* spp. to juvenile hormones. Although these studies have provided no simple explanation for the insensitivity of *Sitophilus* spp. to juvenile hormone analogues, it is clear that this problem has been paramount in hindering the otherwise potentially valuable contribution of juvenile hormone analogues as grain protectants.

It is likely that *Sitophilus* spp. may have an intrinsic insensitivity to at least some of the commonly used juvenoids and penetration deficiencies simply intensify this insensitivity.

Bengston *et al.* (1986a), having found that methoprene was specially effective in preventing progeny production in *Rhyzopertha dominica*, undertook silo-scale experiments to evaluate methoprene, as a replacement for bioresmethrin, in combination with fenitrothion. They found that methoprene emulsifiable concentrate applied at the rate of 1 mg/kg completely prevented the development of progeny of *Rhyzopertha dominica* over the 9 month storage period, as determined by bioassay using several strains of this pest species. No field infestation developed at either site in spite of known sources of these insects in the area. The temperature of the grain ranged from 31.4°C at the time of treatment to 21.8°C after 9 months. The moisture content of the two lots of wheat was 12% and 10.8%.

20.2 Degradation

Quistad *et al.* (1975b) studied the photodecomposition of methoprene as a thin film on glass or silica gel, as an aqueous emulsion, and as a methanolic solution with added photosensitiser. The most abundant degradation product after illumination of a thin film or aqueous emulsion was 7-methoxy citronellal (9–14%). Methoprene was quite stable in methanolic solution although a slow reaction with singlet oxygen occurred. Isomerisation of the 2–ene double bond to a mixture of *cis-trans* isomers was easily promoted.

Rowlands (1976) studied the uptake and metabolism by stored wheat grains of 3 synthetic juvenile hormone mimics, JH-1, Bowers' 2b and methoprene. He found that all 3 penetrated rapidly through the outer layers of wheat grains. After 2 days, the internal distribution was similar in all cases: highest residues occurring in the aleurone layers, much less in the germ, and hardly any in the endosperm or outer seed coats. Residual half-lives at 20°C in freshly-harvested wheat of 19% moisture were: JH 1, 1–2 weeks; Bowers' 2b, 5–6 weeks; methoprene 2–3 weeks. Metabolic changes observed included opening of epoxide groups and ester cleavage. The compounds used in this study all have a significant vapour pressure and their persistence may have been impaired by the fact that this work was carried out on a micro-scale where losses by evaporation are possible. These estimates of persistence may therefore not apply where grain is stored in bulk.

Tan and Tan (1980) found that when methoprene was applied to rolled corrugated paper its biological efficiency decreased by 23% after 50 days of storage at 30°C in semi-darkness. The loss of effectiveness was slightly greater when the rolls were stored under continuous fluorescent light. It was concluded that the half-life of methoprene would cover 5 generations on *Ephestia kuehniella* provided that the ambient temperature was not too high.

Miller (1981) reported only a slow decrease in the level of methoprene from the original application rate over a period from a few months up to 1 year on peanut kernels and peanut hulls. **Miller (1983a)** reported that methoprene applied to shelled corn showed no significant losses with time, even up to 200 days after treatment. Likewise no significant decline in methoprene residues was noted on cacao beans over a 50 day period at 25°C **(Miller 1983b)**.

Although the methoprene residues found by analysis were only about one-half the application rate in 2 studies, no significant decline in residues occurred for up to 150 days after application **(Miller 1983a, 1983c)**.

Mian and Mulla (1983a) reported the results of a laboratory study of the persistence of 3 juvenoids in stored wheat. The persistence of residues of methoprene in stored wheat was studied over a period of 12 to 23 months. The residues of methoprene in stored wheat were determined by gas-liquid chromatography. The loss of residues at 1, 5 and 10 mg/kg were 61, 66 and 62%, respectively, after 12 months. The details are shown in **Table 20.1**.

Table 20.1. Residues of methoprene in treated wheat grain at various post-treatment intervals* **(Mian and Mulla 1983a)**

Interval	Conc. (ppm)		
	1	5	10
0	0.99a	4.97a	9.92a
1 week	0.99a	4.60a	9.39a
	(0)	(7)	(5)
1 month	0.94a	3.93a	6.62a
	(5)	(21)	(33)
4 months	0.82a	3.49ab	5.53bc
	(17)	(30)	(44)
8 months	0.80a	1.78bc	4.07c
	(19)	(64)	(59)
12 months	0.38b	1.71c	3.73c
	(61)	(66)	(62)

* Mean of four analyses. Means followed by the same letter(s) in a column are not significantly different from one another (Duncan's multiple range test. $P = 0.05$). Values in parentheses below each mean represent percent loss in residues at indicated post-treatment interval.

These studies confirm that methoprene can persist in wheat grain at levels above 1 mg/kg, sufficiently to protect the grain against insect infestation (both external and internal feeders) for almost 2 years. These data confirm the biological results reported **(Mian and Mulla 1982a, 1982b)**.

The apparent loss of methoprene active ingredient from the whole grain with time is more than offset by the significant increase in efficacy reported by such workers as **Edwards and Short (1984)**, apparently due to the penetration of the active ingredient into the grain kernel.

Bengston *et al.* (1986a) applied methoprene to bulk wheat at the rate of 1 mg/kg in conjunction with fenitrothion (12mg/kg) at 2 sites in Australia and followed the degradation of the deposit by sampling and analysing the wheat on 6 occasions over 9 months storage. The temperature fell from 31.4°C to 21.8°C over the 9 months and the moisture content was 12% and 10.8% at the 2 sites. As indicated in **Table 20.2** there was little or no change in the concentration of the methoprene residue over this period.

20.3 Fate in Milling, Processing and Cooking

Only limited information has been located on the effect of milling, processing or cooking on the level and fate of methroprene on cereal grain. This is not unexpected because this material is only now undergoing field trials on cereal grains. Methoprene has, however, undergone bulk trials on peanuts, cocoa beans and several cereal commodities over recent years. Methoprene has been in commercial use for treating stored tobacco since about 1976.

A trial was carried out in Australia **(Zoecon 1984a)** in which methoprene was applied to wheat at the rate of 10 mg/kg. After storage for 2 weeks the wheat was milled through an experimental Buhler mill to provide the following fractions:

Fraction	Percentage by weight of each fraction
Bran	23%
Pollard (germ, bran, semolina)	3%
White flour	74%
	100%

A separate milling produced wholemeal.

The fractions were sampled and the samples were forwarded to USA for analysis. The results of analysis were as follows:

Fraction	Methoprene (mg/kg)
Wheat	8.9
Wholemeal	6.7
Flour	1.8
Bran	11.4
Pollard	17.7

The laboratory analysis was verified by use of an internal standard.

A separate trial carried out at Kansas State University **(Zoecon 1984b)** involved treatment of 2 lots of wheat with methoprene at the rate of 10 mg/kg followed, 9 days later, by milling through a Ross Walking Flour Mill to

Table 20.2. Residues of methoprene (mg/kg) in wheat stored in vertical concrete silos prior to assay **(Bengston *et al.* 1986a)**

Location	Target rate	Calculated rate	Approx. time after application (months)					
			0	1.5	3	4.5	6	9
QLD	1.0	1.1	0.61	0.57	0.57	0.54	0.52	0.61
NSW	1.0	1.0	1.0	1.0	1.0	1.0	0.78	–

produce the following fractions:

Fraction	Percentage by weight of each fraction	
White flour	70.2	70.1
Bran	14.1	14.4
Shorts)	12.7	12.4
Red dog) pollard	2.4	2.5
Germ	0.6	0.6
	100.0	100.0

A separate milling through a pin mill produced wholemeal.

The fractions were sampled and the samples were forwarded for analysis. The white flour and wholemeal were used to prepare bread. The results of analysis were as follows:

Fraction	Methoprene (mg/kg)	
Wheat (whole)	5.0	5.7 (1)
White flour	2.8	2.8
Wholemeal flour	4.1	5.5
Bran	8.7	10.2
Shorts)	9.6	9.6
Red dog) pollard	9.5	11.0
Germ	8.7	10.7
White bread (33% moisture)	0.41	
Wholemeal bread (33% moisture)	0.88	

Note (1): It appears as though a significant proportion of the original deposit (10 mg/kg) was lost, possibly by evaporation, between application and analysis. Spiked samples analysed at the same time yielded the results expected.

Bengston *et al.* (1986a) took wheat which had been treated with 1 mg/kg methoprene and which had been stored in concrete silos and subjected it to a standard milling through a pilot-scale roller mill. The methoprene content of the grain, milling fractions and bread was determined and the results are given in **Table 20.3.**

Table 20.3. Methoprene residues (mg/kg) in milled products and bread from wheat treated with methoprene (1 mg/kg) 9 months prior to processing **(Bengston *et al.* 1986a)**

Sample analysed	Methoprene residue (as-is basis)	Methoprene residue (moisture-free basis)
wheat	0.61	0.69
bran	1.35	1.53
pollard	2.33	2.59
wholemeal	0.53	0.60
white flour	0.17	0.19
wholemeal bread	0.06	0.09
white bread	<0.01	<0.01

20.4 Metabolism

An extensive review of the metabolism of methoprene and other insect growth regulators is provided by **Hammock and Quistad (1981)**.

Methoprene painted on the leaf surface or injected into the stem of the dwarf lima bean is essentially not translocated **(Bergot 1972)**. When injected into the stem of the plant, the parent compound is partially metabolised to the corresponding hydroxy and dihydroxy derivatives and a polar conjugate. Topical application results in more rapid metabolism after 24 hours' incubation.

Methoprene is very susceptible to photolytic decomposition under environmental conditions. It is degraded to a plethora of photo products, which are present in relatively low yield. As a thin film on glass, the half-life was 6 hours and in aqueous solution (0.01–0.5 mg/kg) was less than 1 day. In the dark, sterile aqueous solutions of methoprene at pH 5, 7 and 9 were found to be stable over the 4 week test period **(Quistad *et al.* 1975b; Zoecon 1984c)**.

The degradation of methoprene on soil was studied under aerobic and anaerobic conditions **(Schooley *et al.* 1975)**. On aerobic sandy loam soil, an initial half-life of 10 days was shown. Extensive breakdown is apparent from the significant evolution of CO_2. There is rapid incorporation into humic and fulvic acids as well as the humin fraction.

The metabolic fate of methoprene was studied on alfalfa and rice as a function of time **(Quistad *et al.* 1974a)**. Five primary non-polar products were isolated. The combined yield of these products was 2% after 7 days. The initial rapid loss of the methoprene radio-label from plants was attributed to evaporation of methoprene and 7-methoxy citronellal.

Quistad *et al.* (1974b) studied the metabolism of methoprene by a steer over a period of 14 days. It was found that the radio-labelled methoprene was converted into cholesterol, cholic acid and desoxycholic acid. Cholesterol and desoxycholic acid were degraded chemically by Kuhn-Roth and Barbier-Wieland oxidations to show catabolism of (5-^{14}C) methoprene to (2-^{14}C) acetate.

Quistad *et al.* (1975c) studied the metabolism of methoprene in a Hereford steer and a Jersey cow over a period of 14 days and 7 weeks respectively. A large percentage of the radio-label was incorporated in the tissues and respired by the animals (15% for cow). A total of 60.4% of the applied dose was recovered in the urine and feces of the steer. 7.6% was recovered in the milk of the cow and 19.8% in the urine and 30.2% in the feces of the cow. In the urine and feces, a small amount of radio-label was metabolised into free primary metabolites, somewhat more was incorporated into single glucuronides and a considerable quantity of radio-label was found in polar compounds, probably complex conjugates or polar biochemicals. No methoprene was found in the urine, but approximately 40% of the radio-label in feces was contributed by unmetabolised methoprene. The methoprene was extensively metabolised by the lactacting dairy cow to acetate which was isolated from blood as randomly labelled acetic acid. Radioactive acetate incorporated into milk fat which was degraded to radio-labelled saturated, monoenoic and dienoic fatty acids. Also isolated from milk were radioactive lactose, lactalbumen and casein. The presence of ^{14}C-cholesterol (free and esterified) was confirmed in blood, in agreement with a previous study in a steer.

A degradation study of methoprene in the bovine showed that it was converted to cholesterol and related natural products **(Quistad *et al.* 1975a). Chamberlain *et al.* (1975a, 1975b)** studied the absorption, excretion and metabolism of methoprene in a guinea pig, a steer and a cow. This showed that the material is rapidly absorbed, metabolised and incorporated into life systems. Similar results were obtained from studies on chickens **(Quistad *et al.* 1976a)** and in bluegill fish **(Quistad *et al.* 1976b).**

Rowlands (1976) studied the uptake and metabolism of 3 juvenoids by stored wheat grains, using radio-labelled compounds applied to wheat grains of 19% moisture content. Uptake and penetration data were obtained by dissecting the individual wheat kernels. The breakdown rate was determined on small lots stored in sealed jars in the dark at 20°C. All 3 compounds (JH-1 Bowers'2b and methoprene) penetrated rapidly, irrespective of method of application or the vehicle used to apply the compound. The quantities sorbed differed between the vapour and topical treatments. However, the results obtained for uptake after topical application or after exposure to the vapours were similar to those observed with many conventional insecticides **(Rowlands 1971)**. The only quantitative difference observed was that the proportion absorbed by the germ region from the vapour phase was initially greater than from solutions but this was only a transient effect. Two days after treatment, the distribution pattern was virtually the same in all cases; most radioactivity, i.e. intact compounds, being found in the aleurone layers, much less in the germ and virtually none in the endosperm proper or outer seedcoats. The residual half-lives in freshly harvested wheat of 19% moisture were found to be approximately: JH–1, 1–2 weeks; Bowers'2b, 5–6 weeks; and methoprene, 2–3 weeks. In older wheat of 12% and 18% moisture content the respective half-lives were JH–1 3–4 weeks and 1–2 weeks; Bowers'2b, 7–8 weeks and 5–6 weeks and methoprene 6–7 weeks and 3–4 weeks. Some 5 weeks after treating freshly-harvested wheat with 10 mg/kg methoprene there was still some 2 mg/kg intact, of which about 60% was present in the aleurone region. In dryer, older wheat, there appeared to be 3 times that quantity present.

Quistad *et al.* (1976a) studied the metabolism of methoprene in chickens. Treatment of Leghorn chickens with a single oral dose of 5-^{14}C-methoprene resulted in residual radioactivity in tissues and eggs. The chemical nature of the residual radio-label in tissues (muscle, fat, liver), eggs and excrement was thoroughly examined at several doses (0.6 to 77 mg/kg). Although a high initial dose (59 mg/kg) resulted in methoprene residues in muscle (0.01 mg/kg), fat (2.13 mg/kg) and egg yolk (8.03 mg/kg), these residues of methoprene represented only 39% and 2% of the total ^{14}C label in fat and egg yolk respectively. Radio-labelled natural products from extensive degradation of methoprene were by far the most important ^{14}C residues in tissues and eggs, particularly at the lower dose of 0.6 mg/kg where ^{14}C-cholesterol and normal ^{14}C-fatty acids, as triglycerides, contributed 8% and 71% of the total radio-label in egg yolk. Novel minor metabolites of methoprene were observed in lipid depots, resulting from saturation of the dienoate system. These minor metabolites were conjugated to glycerol and/or cholesterol.

Quistad *et al.* (1976b) also studied the metabolism of methoprene in bluegill fish. The fish, in a dynamic flow-through system, acquired moderate residues of largely unmetabolised methoprene when continuously exposed to about 30 times anticipated environmental levels of methoprene, but residues were rapidly eliminated (93–95%) within 2 weeks when fish were transferred to flowing uncontaminated water. When bluegill were treated with methoprene in a model aquatic ecosystem, fish showed a highly misleading accumulation of ^{14}C residues since residual radioactivity was found almost exclusively in radio-labelled natural products, including cholesterol, free fatty acids, glycerides and protein. Less than 0.1% of the total radioactivity in fish could be attributed to unmetabolised methoprene or its primary metabolites; thus, simple radio-assay procedures in ecosystem studies can be severely compromised unless coupled with more sophisticated analytical techniques.

Davison (1976) studied the distribution and elimination of carbon-14 given to chickens as ^{14}C-methoprene. When about 4 mg of methoprene was given in a single oral dose to colostomised chickens, elimination of carbon-14 was greatest in exhaled air; however, when 105 mg or 107 mg or methoprene was given, elimination of carbon-14 was greatest in urine. Up to 19% of the carbon-14 from a single dose of methoprene was eliminated over a 14 day period in the eggs of laying hens, and carbon-14 was detected in all tissues and organs examined. The nature of the metabolites or their biological significance was not determined.

Hammock and Quistad (1976) have reviewed the metabolism of methoprene in insects by comparing its degradation to that of other juvenoids.

A paper by **Edwards and Rowlands (1978)** on the metabolism of synthetic juvenile hormone JH-1 in 2 strains of the grain weevil was not available for review but was considered not relevant to the metabolism of methoprene.

20.5 Fate in Animals

Methoprene has been developed for administration to cattle for control of horn flies (*Haematobia irritans*) and face flies (*Musca autumnalis*). It is administered in mineral blocks, in drinking water, as a sustained-release bolus and as a feed additive (**Harris *et al.* 1974; Beadles *et al.* 1975; Miller *et al.* 1976; 1977a, 1977b, 1978).**

Quistad *et al.* (1974b, 1975a) administered [5-^{14}C] methoprene to a steer which was slaughtered 2 weeks later. Samples of fat, muscle, liver, lung, blood and bile were analysed for radioactive residues. No primary methoprene metabolites could be characterised but the majority of the tissue radioactivity was positively identified as cholesterol. A total of 72% of the bile radioactivity was contributed by cholesterol, cholic acid and deoxycholic acid. Radioactivity from catabolised methoprene was associated with protein and cholesteryl esters of fatty acids.

A cow balance/metabolic study with [5-^{14}C] methoprene was conducted on a 338 kg Jersey cow administered 207 mg/kg of radio-labelled methoprene in a gelatine capsule (**Chamberlain *et al.* 1975b**). This is 40 fold the proposed dose for fly control. The study proceeded for 7 days after the administration of the methoprene. It was found that 16.4% of the radioactivity was excreted as CO_2 between 0 and 170 hours. Peak excretion of CO_2 occurred just prior to 24 hours after dosing. A total of 30.3% of the applied dose was eliminated via the feces approximately 50 hours after dosing and 19.7% was eliminated via the urine in 30 hours. By the termination of the experiment 17.6% of the radio-label was excreted via the milk. The radioactivity was measured in 30 tissue samples after the cow was sacrificed 7 days after dosing. The total tissues represented 20% of the applied dose, intestine and contents contributing 5.96% and fat 4.59%. The organs of metabolism, lung, liver and kidney contributed a further 1.6–8%. The presence of radio-carbon in tissues was attributed to steroidal derivatives in which labelled acetate from the methoprene was anabolically re-incorporated into natural body constituents.

The results of the above studies and another on a guinea pig were compared in a publication by **Chamberlain *et al.* 1975a**. It was reported that a rather large percentage of the radio-label was incorporated in the tissues and respired by the animals. A small proportion of the radio-label was incorporated into free primary metabolites, a greater amount into glucuronides and a considerable proportion into polar compounds, possibly complex conjugates or polar biochemicals. No methroprene was found in the urine, but approximately 40% of the radio-label within feces was methoprene. This explains the activity of the administered methoprene against fly larvae developing in dung pats. The formation of conjugates and complex metabolites of methoprene was more extensive by the steer than by the guinea pig. Comprehensive information was provided on the radio-label profiles in tissues, organs and components of the animals.

Quistad *et al.* (1975c) showed in a further metabolism study in a lactating dairy cow that methoprene was extensively metabolised to acetate which was isolated from blood as randomly labelled acetic acid. Radioactive acetate incorporated into milk fat which was degraded to radio-labelled saturated, monoenoic and dienoic fatty acids. Also isolated from milk were radioactive lactose, lactalbumin and casein. The presence of ^{14}C-cholesterol (free and esterified) was confirmed in blood, in agreement with a previous study in a steer.

Davison (1976) investigated the distribution and elimination of ^{14}C given to chickens as ^{14}C-methoprene. When about 4 mg of methoprene was given in a single oral dose to colostomised chickens, elimination of ^{14}C was greatest in exhaled air; however, when 105 or 107 mg of methoprene was given, elimination of ^{14}C was greatest in urine. Up to 19% of the ^{14}C from a single dose of methoprene was eliminated over a 14 day period in the eggs of laying hens, and ^{14}C was determined in all tissues and organs examined.

Quistad *et al.* (1976a) showed that treatment of Leghorn chickens with a single oral dose of 5-^{14}C-methoprene resulted in residual radioactivity in tissues and eggs. The chemical nature of the residual radio-label in tissues (muscle, fat, liver), eggs and excrement was thoroughly examined at several doses (0.6 to 77 mg/kg). Although a high initial dose (59 mg/kg) resulted in methoprene residues in muscle (0.01 mg/kg), fat (2.13 mg/kg) and egg yolk (8.03 mg/kg), these residues of methoprene represented only 39% and 2% of the total ^{14}C label in fat and egg yolk, respectively. Radio-labelled natural products from extensive degradation of methoprene were by far the most important ^{14}C residues in tissues of eggs, particularly at the lower dose of 0.6 mg/kg where ^{14}C-cholesterol and normal ^{14}C-fatty acids (as triglyceride) contributed 8 and 71% of the total radio-label in egg yolk. Novel minor metabolites of methoprene were observed in lipid depots, resulting from saturating of the dienoate system. These minor metabolites were conjugated to glycerol and/or cholesterol.

Ivey *et al.* (1982) determined residues of methoprene in omental fat taken by biopsy from cattle at 30, 60, 90 and 180 days after either 1 or 2 boluses containing 1% methoprene had been placed into the reticulum. Residues of methoprene ranged from 0.02 to 0.0159 mg/kg. At 90 days post-treatment, only 2 animals contained residues in fat above the lower limit of detection (0.02 mg/kg), and no residues were detected at 180 days post-treatment.

When chickens were given feed treated with methoprene at 0, 25, 50 and 100 ppm for 14–63 days, residues in poultry meat and eggs were less than 0.1 mg/kg **(Zoecon 1984d).**

20.6 Safety to Livestock

Although the amount of direct evidence pointing to the safety of methoprene to livestock is limited there is more than sufficient indirect evidence to indicate that the feeding of grain and milling offals containing residues derived from the use of methoprene as a grain protectant could not possibly represent a hazard to livestock.

The information available on the acute oral toxicity of methoprene to various animal species indicates that it has an extremely low toxicity to all species tested. A summary of the available figures is given in **Table 20.4.**

Table 20.4. Methoprene: acute oral toxicity to various animal species

Species	Sex	LD/50(mg/kg)	Reference
rats	M & F	> 5 000	**Calandra (1972)**
dogs	M & F	> 5 000	**Hill (1972)**
rats	M & F	>34 600	**Jorgensen & Sasmore (1972)**
Mallard	M	> 2 000	**Hudson (1972)**
Mallard	M & F	>10 000	**Fink (1972a)**
Bobwhite quail	M & F	>10 000	**Fink (1972b)**
chickens	M & F	> 4 640	**Fink (1973)**

The extensive information on the metabolism of methoprene (Section 20.4) indicates that it is rapidly absorbed and excreted in bovines and poultry and that the portion which is absorbed is rapidly metabolised into simple substances which are incorporated into life procesess. Under the circumstances, there is no reason to believe that the continuous long-term feeding of grain and milling products that have been treated with methoprene for insect control could lead to the accumulation of significant levels of methoprene in livestock.

Fink (1973) has determined the 8-day dietary lethal concentration to young chickens to be greater than 4640 ppm in the feed. Likewise, **Fink (1972a)** determined the 8-day dietary lethal concentration to Mallard ducks to be greater than 10 000 ppm in the feed. Similarly the 8-day dietary lethal dose to Bobwhite quail was determined by **Fink (1972b)** to be greater than 10 000 ppm.

In a study to evaluate the potential for reproductive impairment in Mallard ducks when exposed to dietary levels of technical methoprene, **Fink and Rero (1973a)** fed groups of male and female Mallard ducks with feed containing 3 and 30 ppm methoprene during the 19-week duration of the study. The ducks were bred according to a standard protocol and the eggs were incubated. There were no signs of toxicity or behavioural abnormalities observed during the study. All birds appeared normal throughout the study and there was no mortality. Statistical analysis of body weight, food consumption, eggs layed, eggs cracked and egg shell thickness revealed no difference between the control and test birds. The study revealed only significant impairment in the 14-day-old survivors per hen at the 30 ppm test level. The investigators concluded that this statistical parameter was due to a reduced number of eggs layed in that test group, and therefore does not constitute meaningful reproductive impairment.

Likewise in a reproduction study with Bobwhite quail **(Fink and Rero 1973b)**, no differences occurred between the control and test birds with regard to body weight, food consumption, eggs layed, eggs cracked, eggs embryonated, 3-week embryo survival, normal hatchlings, 14-day-old survivors or egg shell thickness. There were no observable signs of toxicity or behavioural abnormalities during this study. The authors concluded that methoprene fed in the diet at levels of 3 and 30 ppm during a generation reproduction study had no adverse effect on the reproductive success of Bobwhite quail.

20.7 Toxicological Evaluation

Methoprene was evaluated by the Joint FAO/WHO Meeting on Pesticide Residues in 1984. Based on a 78-week mouse study, a 2-year rat oncogenicity study, a 3-generation reproduction study in rats and a 3-month dog study the levels causing no toxicological effect were determined to be:

Rat: 500 ppm in the diet, equivalent to 25 mg/kg body weight

Dog: 500 ppm in the diet, equivalent to 12.5 mg/kg body weight.

From these parameters a temporary acceptable daily intake for man was estimated to be 0–0.06 mg/kg body weight. For the estimation of a full ADI the Meeting required a six-month feeding study in dogs, a two-generation reproduction study in rats and adequate teratology studies.

20.8 Maximum Residue Limits

A petition has been submitted to the US Environment Protection Agency proposing that permanent tolerances be established for methoprene in the entire cereal grain crop grouping as defined by the EPA at the level of 10 ppm. This grouping includes barley, buckwheat, corn, millet, milo, oats, rice, rye and wheat.

The Joint FAO/WHO Meeting on Pesticide Residues in 1984 made recommendations for temporary maximum residue limits as follows:

Commodity	Maximum Residue Limits (mg/kg)
Bran	20
Cereal grains	10
Flour (white)	5
Wholemeal	10

The residue is determined and expressed as methoprene. The method recommended for the determination of methoprene residue is that of **Miller *et al.* (1975)**.

21. Permethrin

Permethrin was originally synthesised by **Elliott *et al.* (1973)** as one of the successful outcomes of the work at Rothamsted in the UK. It was originally known under the code number NRDC143 and it is generally described as 3-phenoxybenzyl(±)*cis, trans* 3-(2,2-dichlorovinyl)-2, 2-dimethylcyclopropane-1-carboxylate.

The technical grade permethrin contains 4 stereoisomers deriving from chirality of the cyclopropane ring at the C-1 and C-3 positions. The nomenclature standards do not prescribe the ratio of isomers in 'permethrin'. **Glenn and Sharpf (1977)** have shown that the ratio of *cis* to *trans* isomers varies with the method of synthesis. It is desirable to produce different *cis/trans* ratios for certain insecticidal applications (e.g. lower *cis/trans* ratios for animal health products). It is therefore important to note the isomer ratios in products used in the supervised trials and metabolism studies. *Cis* permethrin is more insecticidally potent and has a higher mammalian toxicity **(James, 1980)** than the *trans* isomer. The isomers differ significantly in rates of photolysis and hydrolysis, in biotransformation and bioaccumulation. Most of the information in this review refers to uses of technical grade permethrin containing *cis/trans* isomers in approximately a 40/60 ratio. Permethrin is moderately stable in the environment. **Elliott *et al.* (1973)** reported it to be 10–100 times more stable than earlier synthetic pyrethroids. The increased resistance to photolysis is attributable to substitution of the dichlorovinyl moiety for the isobutenyl group of chrysanthemic acid found in natural pyrethrins and other synthetic pyrethroids.

Permethrin has developed rapidly in world-wide agricultural usage even though it was commercially introduced only recently. It is a stomach and contact insecticide with adulticidal, ovicidal and larvicidal activity against a wide range of insects. The compound shows no systemic or fumigant activity, and has very little value in soil treatments because of rapid degradation in soil and lack of systemic action. The principal agricultural and horticultural uses are in repeated spray programs. Permethrin also has potential for animal health applications. It has been evaluated for post-harvest use for the control of stored-product insects.

Although effective against a wide range of stored-product pests, permethrin is most valuable when used in combination with organophosphorus insecticides since it is particularly effective against *Rhyzopertha dominica*. It is weak against *Tribolium* spp. Its efficacy is increased significantly by combining with piperonyl-butoxide. Permethrin deposits on grain are stable, 70% remaining after 15 months at 25°C. It is rather insensitive to moisture. The metabolism of permethrin is well understood. It degrades to biologically inactive compounds which are readily eliminated by animals. Milling trials have shown that 88% of the residue remains in the bran and pollard and only 12% carries over into the flour. Very little of the residue in the flour is further degraded by cooking so that bread will contain about 10% of the deposit originally added to raw grain. However, when combined with an approved organophosphorus insecticide, permethrin is usually applied at the rate of 1–2 mg/kg so that the level of residue in bread or cooked rice is generally of the order of 0.1 mg/kg or less.

21.1 Usefulness

Permethrin has an important application in the protection of stored grain, being particularly useful when applied in conjunction with organophosphorus insecticides to control insect species which are tolerant or resistant to organophosphorus compounds.

Gillenwater *et al.* (1978) evaluated the potential of permethrin as a protectant for wheat and corn against *Sitophilus oryzae* in the USA. Stored wheat and shelled corn treated with permethrin at 5, 10 and 20 mg/kg and with malathion at 10 mg/kg as the standard for reference were subjected to infestation by adults of *Sitophilus oryzae* at intervals during a 12 month period. Both malathion and permethrin were more effective protectants for corn than for wheat. However, the permethrin treatment of 10 mg/kg on wheat and 5 mg/kg on corn were much more effective protectants than malathion.

Bengston (1979), in discussing the potential of pyrethroids as grain protectants, pointed out that the main immediate potential is as treatment to control *Rhyzopertha dominica* which is currently the most destructive species in Australia and is difficult to control with most organophosphorus insecticides. Although relatively high doses are required to control *Sitophilus* spp. and *Tribolium* spp., permethrin and other pyrethroids are important in dealing with organophosphorus-resistant species. *Sitophilus oryzae* is the species most difficult to control with permethrin. Based on extensive Australian studies Bengston recommended the use of permethrin at the rate of 1 mg/kg, synergised with piperonyl butoxide at the rate of 10 mg/kg, to be used in conjunction with an organophosphorus insecticide.

Permethrin has been applied at 1 and 2 mg/kg in large-scale trials in Australia. At this level, a single application of permethrin, synergised with piperonyl-butoxide, adequately controls the common strains of the lesser grain borer (*Rhyzopertha dominica*), one of the most destructive of all stored-grain insects. When permethrin is used in conjunction with 4–6 mg/kg pirimiphos-methyl the common strains of storage insects are controlled for at least 9 months. Pirimiphos-methyl does not control some strains of *Rhyzopertha dominica* adequately. On the other hand, at 1 mg/kg, permethrin does not control *Tribolium* spp. Laboratory studies have shown that when used at 4–5 mg/kg, a single application of permethrin does give adequate control of all the commonly occurring strains of storage-insect pests when the grain is dry and cool. This indicates a future option of varying pyrethroid and organophosphorus protectant rates, in a complementary manner, to overcome future problems due to insect resistance, absence of a synergist and/or lack of a complementary insecticide (**Bengston *et al.* 1979a, b, c**).

Taylor and Evans (1980) evaluated dilute dusts of pirimiphos-methyl and permethrin (25:75) for control of bruchid beetles attacking stored pulses. Dilute dust formulations of pirimiphos-methyl, applied at the rate of 2.5, 5 and 10 mg/kg, together with permethrin at 2.5 and 5 mg/kg on pigeon peas and haricot beans, kept at 25°C and 70% relative humidity, gave rapid control of 2 important bruchid species (*Callosobruchus chinensis* and *Acanthoscelides obtectus*) infesting the samples. Damage to pulses was almost entirely prevented for up to 24 weeks.

Berck (1980) tested the insecticidal effectiveness of permethrin emulsion on various building materials and on wheat by both chemical analysis and bioassay with *Tribolium castaneum* and *Tribolium confusum*. He found *Tribolium confusum* the more tolerant of the 2 species. On absorptive surfaces, including wheat grain, insecticidal effectiveness at the same rates was markedly lower.

Davies (1981) reviewed the use of permethrin as an alternative insecticide for the control of *Rhyzopertha dominica* and demonstrated the outstanding effectiveness by results of bioassay conducted over a period of 9 months. When used at the rate of 1.5 mg/kg the control of *Oryzaephilus surinamensis* was most satisfactory though less spectacular than *Rhyzopertha dominica*. Unfortunately *Sitophilus granarius* was only controlled to a limited extent and *Tribolium confusum* barely affected at this level. However, in combination with pirimiphos-methyl, complete control of all species could be obtained for at least 9 months.

Prakash and Pasalu (1981) evaluated 5 insecticides against *Sitotroga cerealella* in stored paddy rice but the use rates ranged from 10 to 150 mg/kg. Although permethrin was the most effective in checking the development of *Sitotroga cerealella* in stored paddy, the rate (30 mg/kg) seems unusually high. The authors claim that much higher rates of all insecticides are required on paddy rice.

Halls (1981b) assessed the results of small-silo trials carried out in Australia. The target treatment levels of 0.5 mg/kg permethrin (25:75) and 1 mg/kg permethrin (both in combination with 10 mg/kg piperonyl butoxide and 12 mg/kg fenitrothion) and 2 and 5 mg/kg permethrin (both with and without addition of 10 mg/kg piperonyl-butoxide) were applied to wheat held in 5 tonne silos which were sampled soon after treatment and thereafter at monthly intervals for 9 months. The samples were subjected to bioassays, using several strains of each of 3 of the more important stored-product insects. Susceptible strains of *Sitophilus oryzae* and *Rhyzopertha dominica* and an OP resistant strain of *Rhyzopertha dominica* were controlled by 2 mg/kg permethrin. A level of 5 mg/kg was necessary to give complete control of *Tribolium castaneum* adults, although progeny production of this species was suppressed at 2 mg/kg. Addition of piperonyl-butoxide improved the activity of permethrin.

Adesuyi (1982), working in Nigeria, carried out trials on cob maize stored in cribs and found that 1% permethrin dust was effective in preventing insect damage and maintaining the viability of seed grain. A treatment level of 5 mg/kg permethrin (25:75) gave significant protection from insect damage over a storage period of 8 months for maize harvested in the rainy season, whilst the treatment level of 2.5 mg/kg gave similar protection for 5 months to maize harvested in the dry season.

Chahal and Ramzan (1982) tested the relative efficacy of synthetic pyrethroids and some organophosphorus insecticides against the larvae of *Trogoderma granarium*. They used 4 pyrethroids including permethrin and 4 organophosphorus insecticides. Whilst deltamethrin gave 100% mortality within 1 day when sprayed at a concentration of 0.05% the mortality resulting from the use of permethrin sprays of 0.0125, 0.025 and 0.05% increased progressively and was complete by 7 days.

Elliott *et al.* (1983) compared the selectivity of pyrethroid insecticides between *Ephestia kuehniella* and its parasite *Venturia canescens* with a view to identifying compounds with selectivity favouring the survival of beneficial parasites rather than pest hosts. They found that the ratio of the toxicity of permethrin to the pest and its host (14:1) justified further attention being focused on beneficial insects of more immediate and practical importance.

Williams *et al.* (1983) tested the relative toxicity and persistence of 3 pyrethroid insecticides on concrete, wood and iron surfaces for the control of stored-product insects. Permethrin was effective against *Rhyzopertha dominica* on iron, wood and concrete though it began to lose its effectiveness on concrete after 24 weeks. It was effective against *Sitophilus oryzae* when applied to iron but less so when applied to wood and it was relatively ineffective even shortly after application to concrete as were the other compounds tested. The effectiveness was considerably less against *Tribolium castaneum*.

Hsieh *et al.* (1983) tested the toxicity of 26 insecticides against *Sitophilus zeamais* and *Rhyzopertha dominica*. Permethrin was one of only 7 insecticides that was more effective against *Rhyzopertha dominica* than against *Sitophilus zeamais*.

Bengston *et al.* (1983b) applied 4 combinations of organophosphorothioates and synergised synthetic pyrethroids to bulk wheat in commercial silos in Queensland and New South Wales. Laboratory bioassays using malathion-resistant strains of insects were carried out on samples of treated wheat at intervals over 9 months. The combination of pirimiphos-methyl (4 mg/kg) plus permethrin (1 mg/kg) plus piperonyl butoxide (8 mg/kg) controlled common field strains of *Sitophilus oryzae, Rhyzopertha dominica* and prevented progeny production in *Tribolium castaneum, Tribolium confusum* and *Ephestia cautella*.

Desmarchelier (1983) demonstrated the action of higher temperature on the effectiveness of permethrin at 2.5 mg/kg in preventing reproduction of *Tribolium castaneum* in wheat where he showed that the insecticide was several times more effective at 25°C than at 35° due to the lower reproductive capacity of the insect and the greater potency of the insecticide at the lower temperature.

Peng (1983) determined the relative toxicity of 10 insecticides against 6 coleopterous stored-rice insect pests by exposing adults to impregnated filter papers for 4 hours and recording the mortality after 72 hours. Permethrin was 1 of the 3 least toxic compounds.

Bitran *et al.* (1983b), working in Brazil, evaluated the residual action of several pyrethroid and organophosphorus insecticides in the control of *Sitophilus zeamais* on stored maize. Permethrin showed a high efficacy but this effectiveness was reduced without the use of piperonyl butoxide.

Bengston *et al.* (1984) evaluated combinations of organophosphorus and synergised synthetic pyrethroid insecticides as grain protectants for stored sorghum in field experiements in concrete silos in south Queensland. Samples were drawn for laboratory bioassay at regular intervals over 26 weeks. Although most of the combinations were completely effective against typical malathion-resistant strains of *Sitophilus oryzae, Rhyzopertha dominica, Tribolium castaneum* and *Ephestia cautella*, pirimiphos-methyl (6 mg/kg) plus permethrin (1 mg/kg) plus piperonyl butoxide (8 mg/kg) allowed some survival of adults and progeny production by *Sitophilus oryzae* after 12 weeks, and by 1 strain of *Rhyzopertha dominica* throughout.

Bodnaryk *et al.* (1984), working in Canada, studied the potential synergistic activities of chlordimeform, DEF (S,S,S-tributyl phosphorothioate), and piperonyl butoxide by exposing adults of *Tribolium castaneum* to treated filter paper discs. Each of the chemicals was individually of low toxicity to *Tribolium castaneum*. However, when tested in various combinations, chlordimeform: DEF (1:1 and 1:4) exhibited a strong synergistic interaction, but other pairs (chlordimeform:piperonyl butoxide, 1:4, or DEF:piperonyl butoxide 1:4) exhibited no synergism. When tested at experimentally determined optimum ratios with permethrin, synergism was observed for permethrin: chlordimeform (1:2) and for permethrin:DEF (1:2) but not for permethrin:piperonyl butoxide (1:2). The triple mixture permethrin:chlordimeform:DEF (1:2:2) was twice as toxic as permethrin:DEF (1:2) but only slightly more toxic than permethrin:chlordimeform (1:2). Although chlordimeform and DEF interact synergistically, and each synergises permethrin by unrelated mechanisms, these interactions were not expressed independently in the mixture of all 3 substances tested against *Tribolium castaneum*.

Hodges and Meik (1984), studying the infestation of maize cobs by *Prostephanus truncatus* in East Africa, found that dipping the exposed ends of maize cob cores in 0.1% permethrin aqueous dispersion or 0.5% dust was very effective in preventing infestation by this pest. Such a technique is more acceptable than the treatment of shelled maize because of traditional storage practice. **Golob *et al.* (1983)** found that permethrin dust (25:75) applied at the rate of 3 mg/kg to shelled maize was effective in preventing attack by *Prostephanus truncatus*.

White (1984) studied the residual activity of permethrin and other insecticides applied to wheat stored under simulated western Canadian conditions. He found that permethrin was more toxic to *Tribolium castaneum* than to *Cryptolestes ferrugineus* and that knock down was consistently greater than mortality when the insects were allowed a 3-day recovery period.

Wohlgemuth (1984) carried out comparative laboratory trials with commercial formulations of 14 insecticides under tropical conditions. Wheat was treated with permethrin dust at the rate of 6 mg/kg and it was then stored at 36°C and 50% r.h. for 24 months. Samples were infested with 5 stored-products beetles after 1, 3, 6, 9, 12, 18 and 24 months. Permethrin was outstandingly effective against *Rhyzopertha dominica*, causing complete kill and preventing reproduction for the whole 24 months. Against the other species it was either ineffective or irregular in its performance.

Golob *et al.* (1985) determined the susceptibility of *Prostephanus truncatus* to 11 insecticides under laboratory conditions. Permethrin was one of the most toxic materials when applied topically and it was second only to lindane when applied to filter papers to which the adult beetles were exposed. When applied to maize grain at the rate of 2.5 and 5 mg/kg permethrin was fully effective for 24 months against adults. No live adult progeny remained 50 or 100 days after the initial exposure of parent adults. Maize cobs treated with permethrin dust at the rate of 5 mg/kg remained practically undamaged for 8 months and showed the lowest number of damaged grains after 10 months. Permethrin dust applied at the rate of 2.5 mg/kg suffered only 4% damaged grains after 10 months whereas in the case of 5 other treatments the damage was greater than 80%. The weight loss after 10 months (5.5%) compared well with some other treatments which allowed more then 30% weight loss.

Evans (1985) tested the effectiveness of 6 insecticides on 10 specimens of 5 species of insects received from

Uganda and which were subsequently shown to be substantially resistant to lindane or/and malathion. Permethrin, applied in the form of a dust at the rate of 2 mg/kg, was relatively ineffective against *Tribolium castaneum, Tribolium confusum, Sitophilus zeamais, Zabrotes subfasciatus* but it was particularly effective against *Callosobruchus maculatus*.

Semple (1985) quotes **Pranata *et al.* (1983)** as having demonstrated in Indonesia the efficacy of a simple application of permethrin dust (25:75) applied at 5 mg/kg, which effectively controlled *Rhyzopertha dominica, Sitophilus oryzae* and *Sitophilus zeamais* in paddy and milled rice for 6 months. *Tribolium castaneum* were only slowly killed but reproductivity was almost completely suppressed. However, *Liposcelis* spp. remained abundant, being seemingly unaffected by the insecticide and enjoying the lack of competition with the other pests.

21.2 Degradation

Gillenwater *et al.* (1978), working in the USA, applied permethrin to wheat at the rate of 10 mg/kg and to corn at 5 mg/kg. They found that whereas the residues of malathion degraded rapidly there was little or no degradation of residues of permethrin during the 12 month test period.

Simpson (1979) reported a study in which 2 samples of permethrin containing respectively 40% *cis* and 60% *trans* isomer and 25% *cis* and 75% *trans* isomer were admixed with wheat and held in storage for 39 weeks. Samples were drawn on 6 occasions throughout this period and were analysed to determine the *cis* and *trans* isomers separately. The small change found in the isomer ratio was considered to be within the experimental error. **(Table 21.1.).**

Permethrin has undergone extensive laboratory studies and silo-scale trials in the development of its post-harvest use as a grain-protectant insecticide in Australia. Since 1977, some 10 000 tonnes of grain have been treated in silo-scale pilot trials in 5 of the Australian States. All the residue studies showed that permethrin was persistent on grain under the conditions of temperature and moisture content prevailing in Australia storages **(Table 21.2) (Desmarchelier *et al.* 1979)**. Initial residues on grain were about 20% below the level expected from the amount applied. The residue levels thereafter hardly declined during 9 months storage; after 6–9 months about 80% of the initial (1 month) residue in grain remained. This level of persistence was found consistently in studies on wheat, barley and sorghum and probably can be generalised for all stored grain **(Bengston *et al.* 1979a, b, c; Desmarchelier *et al.* 1979).**

Halls and Periam (1980b) reported the results of a trial in which 2 permethrin liquid grain protectant formulations were applied to wheat, which was then stored for 9 months in metal silos in the UK. The wheat had a moisture content of 15.5–16%. The target application rates were 2 mg/kg and 5 mg/kg of permethrin. Both treatments included piperonyl-butoxide at the rate of 10 mg/kg. The treatments were carried out by spraying the grain on a conveyor with diluted permethrin/piperonyl-butoxide liquid. Chemical analysis revealed that the actual treatment levels achieved were only about 35–55% of the target application rates. The results given in **Table 21.3** show that for both permethrin application rates there was no detectable rate of breakdown of permethrin on wheat over a 9 month storage period. The fluctuation in levels between different samples was within the inherent errors of sampling and analysis. Moreover, the measurements of the contribution of the separate *cis* and *trans* isomers to the total permethrin content at 3-monthly intervals revealed that there was no detectable deviation from the expected 25:75 *cis:trans* ratio over the 9 months storage period. The stability of permethrin on wheat, found in these trials, agreed with the results obtained by **Gillenwater *et al.* (1978)**, who found little or no degradation of permethrin residues on laboratory-treated wheat or maize over a storage period of 12 months. In this respect, permethrin differs from many organophosphorus grain protectants, such as fenitrothion **(Desmarchelier 1978a).**

Table 21.1. Residues of *cis* and *trans* isomers of permethrin in stored grain; percentage proportions in stored wheat and sorghum **(Simpson 1979).**

Storage conditions			% Residues						
% moisture	Air temperatures* (°C)	Original proportion (nominal)	Period of storage (weeks) 1	6	13	17	20	26	39
10	32	40	37	36	35.5	–	38.3	38.8	36.3
		60	63	64	64.5	–	61.7	61.2	63.7
11.6	21.2	25	20.7	24.6	22.5	21.2	–	21	
		75	79.3	75.4	77.5	78.8	–	79	

Table 21.2. Permethrin residues in treated wheat and sorghum in concrete silos (Australian large-scale post harvest treatments*)

(a) Wheat data **(Bengston *et al.* 1979a,b; Desmarchelier *et al.* 1979)**

	Grain conditions		Amount applied (calc'd) (mg/kg)	Residues (mg/kg) Period of storage (months)									
	% moisture	Air temperatures** (°C)		1	2	3	4	5	6	7	8	9	15
range	8–12	14–35	0.94–1.15	0.70–0.90	0.65–0.83	0.6–0.8	0.70–0.78	0.75–0.80	0.50–0.80	0.75–0.80	0.63–0.65	0.57	0.60–0.63
mean			0.99	0.83	0.75	0.70	0.74	0.74	0.66	0.78	0.64	0.57	0.65
s.d				0.17	–	0.06	–	–	0.14	–	–	–	–
c.v (2)			21	–	9	–	–	21	–	–	–	–	

(Bengston *et al.* 1979a,b; Desmarchelier *et al.* 1979)

s.d = standard deviation; c.v = coefficient of variation

*Trials included treatment of 30 tonnes of barley apart from some 10,500 tonnes wheat.

**Initial temperatures were in the range 25–33°C, at 6 months they were 14–27°C, at 8 months they were 17–22°C.

Initial moisture contents were in the range 9–12%, at 6 months they were 8–12% and at 8 months they were 9–11%.

Data are from 1977/78 and 1978/79 large scale trials. Grain stored in 16 silos (30–1400 tonnes grain) in 5 Australian States.

(b) Sorghum data **(Bengston *et al.* 1979c)**

	Grain conditions		Amount applied (calc'd) (mg/kg)	Residues (mg/kg) Period of storage (weeks)				
	% moisture*	Air temperatures* (°C)		1	6	12	17	26
range	11.6	21.2	1.06	0.70–0.80	0.56–0.58	0.62–0.71	0.56–0.57	0.53–0.62
mean				0.75	0.57	0.67	0.57	0.58

Data from grain stored in a silo (430 tonnes) at one site. 1978 trial.

*Initial

Table 21.3. Permethrin residues on wheat after indicated periods of storage in the UK **(Halls and Periam 1980b)**

Target application rate (ppm permethrin)	Residue analysis (ppm permethrin)									
	Initial (15.11.79)	1 month (13.12.79)	2 months (17.1.80)	3 months (11.2.80)	4 months (12.3.80)	5 months (21.4.80)	6 months (12.5.80)	7 months (10.6.80)	8 months (10.7.80)	9 months (6.8.80)
2.0	0.82 (27:73)*	0.95	0.86	0.96 (26:74)*	0.72	0.72	0.95 (27:73)*	0.84	1.0	1.09 (25:75)*
5.0	1.74 (28:72)*	2.05	2.20	2.04 (25:75)*	2.04	2.35	2.32 (27:73)*	2.32	2.11	2.14 (24:76)*

Dates refer to day of sampling
**cis:trans* isomer ratio
The above analytical results are subject to overall confidence limits of ± 20%

Halls and Periam (1981) continued the studies begun in 1980 in order to determine the fate of permethrin residues on wheat after storage for 12 and 15 months. The same wheat, mentioned above, was analysed after 12 and 15 months storage and the authors reported that the residues did not show any signs of degradation over a 15 month storage period. This view is supported by the results in **Table 21.4.**

Table 21.4. Permethrin residues on wheat after indicated periods of storage in the UK **(Halls and Periam 1981)**

Target application rate (ppm permethrin)	Residue analysis (ppm permethrin)		
	Initial (15.11.79)	12 months (13.11.80)	15 months (26.2.81)
2.0	0.82 (27:73)*	0.81 (25:75)*	0.90
5.0	1.74 (28:72)*	2.22 (25:75)*	2.38

Dates refer to day of sampling
**cis:trans* isomer ratio

As part of the evaluation of permethrin (25% *cis*:75% *trans* isomers) as a grain protectant, trials were set up in Australia to examine the persistence of the compound on wheat. In these trials, permethrin was applied to wheat on its own or in combination with piperonyl-butoxide or in combination with piperonyl-butoxide and fenitrothion **(Halls 1981a)**. The rationale behind these treatments was that permethrin's insecticidal activity is generally synergised by piperonyl-butoxide and the addition of fenitrothion allows lower levels of permethrin to be used and yet still achieve broad spectrum control of stored-product insects. Each segment of the trial involved 5 tonnes of wheat stored separately in elevated steel silos. Target treatment levels were 0.5 and 1 mg/kg permethrin (both in combination with 10 mg/kg piperonyl-butoxide and 12 mg/kg fenitrothion) and 2 and 5 mg/kg permethrin (both with and without addition of 10 mg/kg piperonyl-butoxide). At the commencement of the trial, the grain temperature was in the range of 22–25°C. The temperature fell gradually so that after 5 months it reached a minimum in the region of 10–13°C and thereafter gradually rose again until at the end of 9 months it had returned to the original temperature. Residue levels for permethrin and, where appropriate, piperonyl-butoxide and fenitrothion are shown in **Table 21.5**. For all treatments, the level of permethrin found on the wheat stayed fairly constant over the 9 months storage period, demonstrating that the decay of residues of this compound was very slow.

Davies (1981) carried out laboratory experiments in which permethrin was admixed with wheat at the rate of 1.5 mg/kg initially at 12% moisture content and stored at 25°C and 65% R.H. Samples analysed at 1, 4 and 9 months after treatment showed that approximately 80% of the deposit initially applied remained after 9 months storage.

Noble *et al.* (1982) studied the stability of 4 pyrethroids, including permethrin, on wheat stored for 52 weeks at 25°C or 35°C, and either 12 or 15% moisture content. Rates of loss were calculated from residue analyses of the wheat at 5 intervals during storage. Calculated half-lives for permethrin at 25°C and 12% moisture were 252 weeks and at 35°C and 15% moisture, 44 weeks with intermediate values for the other 2 conditions. The importance of temperature and water activity on the stability of the deposit was quite clear. It was found that the *trans*-isomer degraded faster than the *cis*-isomer. The authors suggested that a hydrolytic mechanism may be one important pathway in the degradation under the conditions of their experiment.

Table 21.5. Residues of permethrin, piperonyl butoxide and fenitrothion detected on wheat after indicated periods of storage in Australia **(Halls 1981a)**

Compound	Target application rate (ppm)	Residue analysis (ppm)						
		Initial	1 month	2 months	3 months	6 months	8 months	9 months
PM	0.5	0.38	0.35	0.38	0.28	0.39	0.39	0.36
PB	10.0	3.9	4.5	4.4	5.0	5.7	3.8	–
FEN	12.0	7.2	7.2	7.7	6.9	6.9	4.3	4.3
PM	1.0	0.93	0.99	0.88	0.89	0.80	0.75	0.75
PB	10.0	7.9	4.6	4.1	5.5	6.7	4.0	–
FEN	12.0	8.9	7.6	6.6	8.0	7.0	5.8	5.2
PM	2.0	1.90	2.00	2.05	1.59	1.64	1.60	1.68
PM	2.0	1.82	1.76	1.95	2.02	1.88	1.75	2.15
PB	10.0	4.2	3.4	3.0	4.0	4.0	3.5	–
PM	5.0	4.5	4.4	4.9	4.5	5.0	3.6	2.6
PM	5.0	5.1	5.8	6.0	5.1	5.9	5.6	4.5
PB	10.0	5.6	7.3	6.9	6.6	6.4	5.9	–

PM = permethrin
PB = piperonyl butoxide
FEN = fenitrothion
– = not analysed

Bengston *et al.* (1983b) applied permethrin at the rate of 1 mg/kg along with pirimiphos methyl (4 mg/kg) and piperonyl butoxide (8 mg/kg) to bulk wheat in commercial silos in Queensland and New South Wales. Analysis of samples drawn at 6 times over the storage period of 9 months showed that the level of permethrin had declined 13–20% during this period. The initial moisture content of the grain was 10% and the initial temperature 32°C.

Bengston *et al.* (1984) reported a field experiment carried out on bulk sorghum stored for 26 weeks in concrete silos in south Queensland in which permethrin (1 mg/kg) was used in conjunction with pirimiphos-methyl (6 mg/kg) and piperonyl butoxide (8 mg/kg). Samples submitted to chemical analysis at the end of 1, 6, 12, 17 and 26 weeks indicated that the concentration of permethrin had declined by only about 15% in 26 weeks. The amount recovered by analysis at the end of 1 week's storage indicated a signficant loss during application or during first contact with the newly-harvested grain.

21.3 Fate in Milling, Processing and Cooking

During the processing of treated whole wheat, the permethrin residue is retained mainly in the bran component (62%) although a significant proportion (12%) remains in the white flour. Permethrin residues in flour from treated whole grain are carried over into bread baked from that flour. There is no reduction in residue level on a commodity weight basis. Whilst white bread prepared from treated grain would have a residue of about 0.15–0.2 mg/kg, the corresponding level in wholemeal bread would be about 0.7–1 mg/kg when permethrin is used, as it is in Australia, at the rate of 1 mg/kg on wheat going into storage **(Table 21.6) (Simpson 1979).**

Wheat from the trials by **Halls and Periam (1980b)**, 3, 6 and 9 months after treatment with permethrin (25:75, *cis:trans*), was submitted to the Flour Milling and Baking Research Association in the United Kingdom to be milled and baked. The wheat was first cleaned of extraneous material, then divided into 2 samples each of which was subjected to a different milling procedure. The first produced wholemeal flour and the second produced white flour, bran and fine offal. Two different samples of white flour were taken from this procedure, the first-reduction flour (the cleanest flour mainly from the centre of the grain and used in the baking of cakes) and total white flour (straight-run flour plus flour from the bran and fine offal fractions). These samples, together with the wholemeal flour, bran and fine offal, were analysed for total permethrin content for both batches of permethrin-treated wheat. In addition, bread was baked from both wholemeal and total white flours to produce wholemeal and white bread, respectively, and these loaves were also analysed

Table 21.6. Distribution of permethrin residues in whole wheat fractions and consequent residue levels in bread **(Simpson 1979)**

(a) Distribution of residue between whole grain fractions

Fraction	% whole grain	Permethrin residue	
		mg/kg	% distribution
Flour (white)	75.1	0.13	12.1
Pollard	11.3	1.9	26.3
Bran	13.6	3.7	61.7

Grain sample 1.5 kg. It had a residue level of 0.65 mg permethrin/kg after being stored for 9 months.

(b)Residue in bread baked from the ground flour from (a)

Commodity	Permethrin residue (mg/kg)
White flour	0.13–0.15
White bread	0.13–0.19
Wholemeal flour*	0.81–0.93
Wholemeal bread	0.67–1.00

*reconstituted from the fractions [in (a)] in the original whole grain proportions. The processing operations simulated those used in commercial practice; the unbaked bread recipe included potassium bromate and benzoyl peroxide. Wholemeal flour was formulated after reconstituting the wheat fractions in the original ratio.

for permethrin content. The milling data for wheat freshly treated with permethrin and from wheat after 3, 6 or 9 months storage are given in **Table 21.7**. The analytical data for the permethrin residues on milling fractions and bread, made from freshly treated wheat and from wheat after 3, 6 or 9 months storage, are given in **Table 21.8**. There was no detectable degradation of permethrin by the processes of milling or baking over the storage period, except possibly in the case of the baking of wholemeal bread from wheat stored for 6 months or more. The trials were continued and the results reported later **(Halls and Periam 1981b)** indicated that there was no significant degradation 15 months after treatment.

Bengston *et al.* (1983b) subjected grain from duplicate field trials on bulk wheat in commercial silos to a standard milling process to produce bran, pollard, white flour and wholemeal flour. The flour was later processed into bread. The permethrin residues were determined in each milling fraction and in the bread. The results are expressed both on a 'as received' basis and on a 'moisture free' basis in **Table 21.9.** This shows that 62% of the residue is in the bran and only 12% in the flour but that there is little or no loss when the flour is converted into bread.

21.4 Metabolism

A wealth of information is available on the absorption, distribution, metabolism and excretion of permethrin. Much of this information was evaluated by **JMPR** in **1979** and **1980.** Since then, considerably more information has become available. It is not possible to review all this information at the present time. Attention is drawn to the

Table 21.7. Milling data for wheat freshly treated with permethrin and from wheat after three, six or nine months storage **(Halls and Periam 1980b)**

Target application rate (ppm permethrin)	Sampling period (months)	Clean wheat		Bran		Fine offal		Total white flour	
		weight (kg)	prop. of total (%)	weight (kg)	prop. of total (%)	weight (kg)	prop. of total (%)	weight (kg)	prop. of total (%)
2.0	0	3.840	100	0.598	15.57	0.300	7.81	2.852	74.20
	3	4.857	100	0.735	15.13	0.433	8.91	3.603	74.18
	6	3.700	100	0.616	16.65	0.300	8.11	2.689	72.68
	9	3.700	100	0.625	16.97	0.295	7.97	2.691	72.73
5.0	0	3.764	100	0.591	15.70	0.284	7.55	2.789	74.10
	3	4.725	100	0.725	15.34	0.398	8.42	3.474	73.52
	6	3.700	100	0.630	17.03	0.305	8.24	2.694	72.81
	9	3.700	100	0.635	17.16	0.293	7.91	2.717	73.43
0 (Control)	0	2.010	100	0.312	15.52	0.175	8.71	1.525	75.87

There is a loss of up to 3% in the laboratory milling process

Table 21.8. Permethrin residues on milling fractions and bread made from freshly treated wheat and from wheat after three, six or nine months storage[1] **(Halls and Periam 1980b)**

Target application rate (mg/kg)	Sampling period (months)	Wheat	Wholemeal flour	Wholemeal bread	Bran	Fine offal	First-reduction flour	Total white flour	White bread
2.0	0	0.82	0.80	0.52	4.40	nd[2]	nd	0.27	0.19
	3	0.96	0.65	0.67	3.70	–1	nd	nd	0.15
	6	0.95	0.83	0.47 (25:75)[3]	2.90	0.74	0.12	0.17	0.06 (23:77)[3]
	9	1.09	0.78	0.24	4.00	1.04	0.12	0.25	0.12
5.0	0	1.74	1.60	0.95	10.20	nd	nd	0.55	0.51
	3	2.04	2.25	1.20	10.30	2.40	nd	0.42	0.30
	6	2.32	2.19	0.68 (25:75)[3]	5.90	1.79	0.23	0.34	0.18 (23:77)[3]
	9	2.14	2.21	0.64	8.00	2.60	0.23	0.60	0.20

1 Analytical results are subject to confidence limits of ± 20%;
2 nd = not detectable (<0.05 mg/kg);
3 *cis:trans* isomer ratio

Table 21.9. Fate of permethrin on wheat subjected to milling and baking **(Bengston *et al.* 1983b)**

Grain/milling fraction	Permethrin residue (mg/kg) as received	moisture free	% Distribution
wheat	0.82	0.93	100
bran	3.7	4.25	61.7
pollard	1.9	2.14	26.3
flour (white)	0.13	0.15	12.0
wholemeal bread	0.67	1.0	–
white bread	0.13	0.19	–

more important references and to the excellent reviews by **Hutson (1979)** and **Miyamoto (1981)**. Meanwhile, readers are referred to **1979, 1980** and **1981 JMPR** evaluations: **Elliott *et al.* (1973, 1976); Edwards and Iswaren (1977); Edwards and Swaine (1977); Gaughan *et al.* (1976, 1977, 1978a, 1978b); Hunt and Gilbert (1977); Leahey *et al.* (1977a, 1977b, 1977c); Miyamoto (1981); Ohkawa *et al.* (1977); Swaine *et al.* (1980a, 1980b); Shono *et al.* (1979); Ussary and Braithwaite (1980a, 1980b, 1980c, 1980d).**

A schematic metabolic profile for permethrin is given in **Figure 21.1.**

Oxidative and hydrolytic mechanisms play a major role in the metabolism of permethrin. Hydroxylation of a methyl group occurs more readily in the 1R- than in the 1S-permethrin isomer **(Soderlund and Casida 1977b)**. In general, it has been recognised that the lower toxicity of the *trans*-isomer relative to the *cis*-isomer of permethrin is associated and consistent with its greater ease of biodegradation both *in vivo* and *in vitro*.

Ohkawa *et al.* (1977) studied the metabolism of permethrin in bean plants. They found that both permethrin isomers and their metabolites hardly moved from the application site to other parts of the plants. When applied to the leaf surface of bean plants, (+)-*trans* and (+)-*cis* permethrin were readily metabolised. The half-life was about 7 days and 9 days, respectively, for *trans* and *cis* permethrin. Both isomers of permethrin underwent ester cleavage, oxidation at the phenoxy group of the alcohol and probably at the geminal dimethyl group in the acid, and conjugation of the resulting carboxylic acids and alcohols, the action proceeding somewhat more rapidly on the *cis*- than the *trans*-isomer.

Hutson (1979) undertook an extensive review of the metabolism of pyrethroid insecticides, including permethrin.

Casida *et al.* (1979) pointed out that various combinations of oxidation, hydrolysis and conjugation lead to more than 50 identified metabolites of the permethrin isomers in various insects and mammals. In addition, the

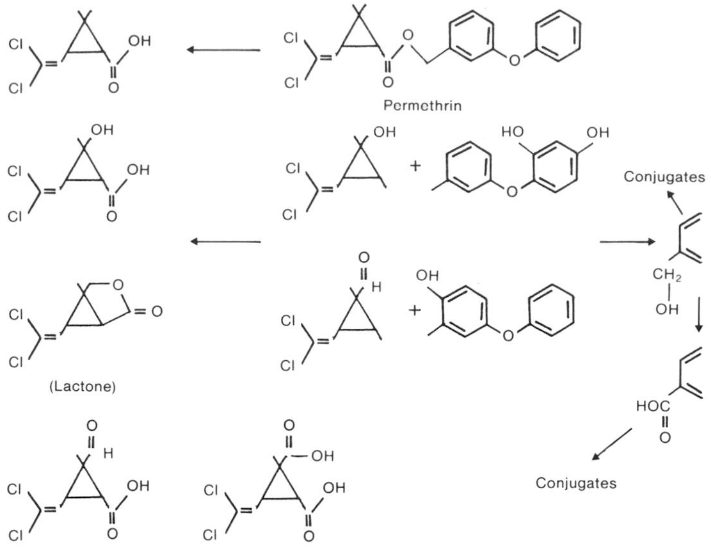

Figure 21.1. Schematic metabolic profile for permethrin.

degradation observed *in vitro* by the action of subcellular oxidative enzymes of rat, mouse and insects was described by **Shono *et al.* (1979)**.

Miyamoto (1981) prepared an extensive review of the chemistry, metabolism and residue analysis of synthetic pyrethroids including permethrin and this paper is recommended for those seeking further details.

Leahey (1985) reviewed the metabolism and environmental degradation of permethrin and other pyrethroids and summarised the available information in a series of diagrams which are most useful.

21.4.1 Rats and Mice

The sites of metabolic attack on permethrin include: ester cleavage (which appears to be more rapid or complete for the *trans-* than for the *cis*-isomer), hydroxylation of the *gem*-dimethyl group of the cyclopropane carboxylic acid, hydroxylation of the 4'-position of the 3'phenoxybenzoic acid and subsequent conjugation of both the phenolic and carboxylic acid substituents. Following oral administration to rats, the metabolic pathway for both *cis-* and *trans*-permethrin was reported by **Elliott *et al.* (1976)**, and **Gaughan *et al.* (1977)**.

Adult male rats were orally administered permethrin as a solution in corn oil at a dosage rate of 10 and 100 mg/kg. Within 24 hours, approximately half of the administered dose was excreted in urine and feces. Analysis of the urine and feces was performed in an effort to see if the rat produced the cyclopropane dicarboxylic acid metabolite observed as the plant metabolite. Low levels of this product were observed in both urine and feces. At least 2 of the 4 possible diastereo-isomers were also detected in this experiment **(Bewick and Leahey 1978)**.

Gaughan *et al.* (1976) discussed permethrin metabolism in rats, cows, bean plants and cotton plants. When administered orally to male rats at 1.6–4.8 mg/kg the [1R, *trans*], [1RS, *trans*], [1R, *cis*], and [1RS, *cis*] isomers of permethrin are rapidly metabolised, and the acid and alcohol fragments are almost completely eliminated from the body within a few days. *Cis*-permethrin is more stable than *trans*-permethrin and the *cis* compound yields 4 fecal ester metabolites which result from hydroxylation at the 2'-phenoxy, 4'-phenoxy, or 2-*trans*-methyl position or at both of the latter 2 sites. Other significant metabolites are 3' phenoxybenzoic acid (free and glucuronide and glycine conjugates), the sulphate conjugate of 4'-hydroxy-3-phenoxybenzoic acid, the sulphate conjugate of 2'-hydroxy-3-phenoxybenzoic acid (from *cis*-permethrin only), the *trans-* and *cis*-dichlorovinyldimethylcyclopropanecarboxylic acids (free and glucuronide conjugates), and the 2-*trans-* and 2-*cis*-hydroxymethyl derivatives of each of the aforementioned *trans* and *cis* acids (free and glucuronide conjugates).

21.4.2 Cows and Goats

In general, the permethrin isomers, although fat-soluble, are readily metabolised and excreted by cows and goats **(Gaughan *et al.* 1978a; Hunt and Gilbert 1977)**. **Leahey *et al.* (1977a)** found that when permethrin is administered to goats, a small amount of unchanged permethrin is excreted in the milk fat but the *cis:trans* ratio is changed from approximately 4:6 to 2:1.

Ivie and Hunt (1980) showed that *cis*-[1RS]- and *trans*-[1RS]-permethrins, radio-labelled with ^{14}C in either the acid or alcohol moiety, are rapidly metabolised and excreted after oral administration to lactating goats. Twenty-six metabolites of the permethrin isomers are fully characterised. The identified metabolites arise through hydrolysis of the ester linkage, hydroxylation at the *cis*- or *trans*-methyl of the geminal dimethyl group, and hydroxylation at the 4' position of the phenoxybenzyl moiety. Certain of these products are further oxidised and/or conjugated with glycine, glutamic acid, glucuronic acid, or other identified compounds before excretion. Unmetabolised permethrin and certain ester metabolites are found in feces, milk, and fat from the treated goats, but only metabolites arising from ester hydrolysis are seen in urine. GLC/mass spectral data are reported for the *cis*- and *trans*-permethrin isomers and 40 of their metabolites and analogues.

21.4.3 Hens

Gaughan *et al.* (1978b) investigated the metabolic fate of permethrin in hens following oral administration of a dose of 10 mg/kg/day for 3 consecutive days. The overall metabolic pathway was similar to that noted with mammalian species. Permethrin was extensively hydrolysed and oxidised with the *trans*-isomer more extensively degraded. In egg yolk, permethrin and *trans*-hydroxymethyl *cis*-permethrin were detected as residues. Extensive detoxication via hydrolitic, oxidative and conjugative reactions, is probably responsible for the relative insensitivity of avian species.

21.5 Fate in Animals

21.5.1 Cows

Bewick and Leahey (1976) conducted a study in which cows received a single oral dose of 40:60 *cis*:*trans* ^{14}C-permethrin at 2.5 mg/kg body weight, equivalent to approximately 80 ppm in the diet. Levels of radioactivity in milk reached a maximum of 0.13 mg permethrin equivalents/kg after 1–2 days. These declined to less than 0.02 mg/kg after 7 days. Levels of radioactivity in the fat were 0.12–0.18 mg permethrin equivalents/kg after 7 days and 0.05–0.08 mg/kg after 14 days, indicating that the small residues in fat are also not maintained on cessation of dosing.

Edwards and Iswaren (1977) conducted a residue transfer and toxicology study in which groups of 3 barren, Friesian cows, yielding 9–13 litres of milk per day were maintained on diets containing radio-labelled permethrin at approximately 0.2, 1, 10 and 50 ppm. The permethrin was absorbed on grassnuts. After 28–31 days, 2 cows in each group were sacrificed. The third was returned to control diet for 7 days before sacrifice. Samples of milk and meat tissues were analysed for permethrin residues. At the 0.2 and 1 ppm rate, permethrin residues in milk were less than 0.01 mg/kg. Residues in kidney, liver, muscle and subcutaneous fat were also less than 0.01 mg/kg and in peritoneal fat less than 0.05 mg/kg. The higher dietary levels of 10 and 50 ppm resulted in low residues in milk of 0.01–0.06 mg/kg (mean 0.02 mg/kg) and 0.03–0.2 mg/kg (mean 0.1 mg/kg) respectively. These levels were approximately 0.2% of the corresponding dietary levels. Residues did not accumulate over the period of the study and they declined rapidly on returning the animals to control diet, to below 0.01 mg/kg (the limit of detection) within 7 days. Permethrin residues in muscle, liver and kidney were below 0.1 mg/kg. Residues in peritoneal fat were again higher than in subcutaneous fat. Intact permethrin was also the major component of the small residue determined in adductor, pectoral and cardiac muscle. The residue levels present were approximately 0.1% to 0.2% of corresponding dietary inclusion levels. As found separately by **Leahey *et al.* (1977a)** in the goat, DCVA (dichlorovinyl cyclopropane carboxylic acid), 3-phenoxybenzyl alcohol and 3-phenoxybenzoic acid were major constituents of the residues in liver and kidney. As with milk and fat, residue levels in muscle, liver and kidney declined rapidly on cessation of exposure to permethrin.

Gaughan *et al.* (1978a) administered ^{14}C-acid- and ^{14}C-alcohol-labelled preparations of *trans*- and *cis*-permethrin to lactating Jersey cows for 3 consecutive days at about 1 mg/kg body weight and found that the radio-carbon was largely eliminated from the body within 12 or 13 days after the initial treatment. Milk and fat residues, although relatively low, are higher with *cis*- than with *trans*-permethrin and consist almost entirely of unmetabolised permethrin. *Cis*-permethrin, hydroxylated at the methyl group *trans* to the ester functionality, also appears in trace levels in milk. Major excreted metabolites (each 8–28% of the administered radio-carbon) from both isomers are: the esters hydroxylated at the *trans*-methyl group; the acid moieties hydroxylated at the *cis*-methyl group and the corresponding *gamma*-lactones; 3-phenoxybenzyl alcohol; the glutamic acid conjugate of 3 phenoxybenzoic acid. An additional 13 excreted metabolites of *trans*-permethrin and 10 of *cis*-permethrin where tentatively identified. As found separately by **Leahey *et al.* (1977a)** in studies with goats, **Swaine *et al.* (1980a)** found that metabolites rather than permethrin itself constituted the major part of the residues in liver and kidney. The levels declined rapidly on cessation of exposure.

21.5.2 Goats

Hunt and Gilbert (1977) dosed goats orally with either the *cis*- or *trans*-isomers of ^{14}C-labelled permethrin at a rate equivalent to approximately 6 ppm in the diet for 10 days. Total radioactive residues in the milk reached a plateau of 0.023–0.05 and less than 0.01–0.01 mg/kg permethrin equivalents/kg, respectively, for the *cis*- and *trans*-isomers. The goats were sacrificed 24 hours after receiving the final dose, when levels of radio-carbon in meat tissues were measured. Total radioactivity in the fat of animals receiving the *cis*-isomer was 10 times higher than those receiving the more readily hydrolysed *trans*-isomer.

Leahey *et al.* (1977a) dosed goats orally with 40:60 *cis*:*trans* ^{14}C-permethrin at a rate equivalent to approximately 10 ppm in the diet for 7 days. Total radioactive residues in the milk reached a plateau of 0.02–0.03 mg permethrin equivalents/kg after 5 days. Of this radioactivity 30–50% was associated with the butterfat fraction of the milk in which total radio-active residues were 0.13–0.27 mg permethrin equivalents/kg. Where alcohol-labelled permethrin was used, approximately 70% of the ^{14}C in kidney tissue was 3-phenoxybenzoic acid plus 3-(4'-hydroxyphenoxy)benzoic acid. Approximately 30% of the ^{14}C in the liver was due to 3-phenoxybenzyle alcohol plus 3-(4'-hydroxyphenoxy)benzyl alcohol. A further 15% was due to 3-phenoxybenzoic acid plus 3-(4'-hydroxyphenoxy)benzoic acid. Where acid-labelled permethrin was used, approximately 10–15% of the label in liver and kidney was due to the *cis*- and *trans*- 3-(2,2-dichlorovinyl) cyclopropanecarboxylic acids, principally the *trans*-isomer.

Ivie and Hunt (1980) showed that the metabolic fate of permethrin in the goat is similar to that in the cow. Radioactivity deriving from the oral administration of the *cis*-isomer of permethrin is excreted mainly in the feces, whereas that deriving from the more readily hydrolysed *trans*-isomer is excreted mainly in the urine. Radioactivity in feces, in the case of the *trans*-isomer, is due principally to permethrin and, in the case of *cis*-isomer, to 4'-hydroxy permethrin. The predominant urinary metabolites are DCVA and 3-phenoxybenzoic acid, which are excreted mainly as polar conjugates.

Ivie and Hunt (1980) drew attention to the following similarities and differences between goats, cows and rats.

1. In each of the 3 species, a greater percentage of an administered *cis*-permethrin dose was eliminated in the feces than was a *trans*-permethrin dose, a pattern that appears to be most pronounced in goats and least in cattle. It was suggested that *trans*-permethrin was absorbed more rapidly than *cis*-permethrin from the gastro-intestinal tract, or alternatively, isomer differences in the rates of biliary excretion of permethrin and/or its metabolites may account for the above observations.
2. Although retention of permethrin by tissues and its excretion into milk of mammals was minimal, *cis*-permethrin and its metabolites in rat, lactating goat and lactating cattle were retained by tissues to a more significant degree than was *trans*-permethrin. In goats, a greater percentage of administered ^{14}C *cis*-permethrin is secreted into milk than in *trans*-permethrin, but these differences are not apparent in cattle.
3. Primary metabolism of permethrin in rats involves attack at 5 major sites, including ester cleavage, hydroxylation at the *cis*- or *trans*-methyl of the germinal dimethyl moiety, and hydroxylation at the 2' or 4'-position of the phenoxybenzyl moiety. Cattle and goats metabolise permethrin similarly, with the exception that 2'-hydroxylation apparently does not occur in these ruminants.
4. Ester metabolites of permethrin were eliminated primarily through the feces of rats, cattle and goats. Cattle eliminate large quantities of ester metabolites of both *cis*-permethrin and *trans*-permethrin in feces but rats and goats eliminate considerably more *cis*-permethrin than *trans*-permethrin esters in feces.
5. In each species, conjugation of permethrin metabolites before urinary excretion was extensive. Rats, cattle and goats eliminated the acid moiety in urine primarily as conjugates of glucuronic acid. The alcohol moiety was excreted, mostly as phenoxybenzoic acid glucuronide or 4'-hydroxyphenoxybenzoic acid-sulphate in rats, but amino acid conjugates of phenoxybenzoic acid comprised most of the excreted products in cattle and goats. Although conjugation of phenoxybenzoic acid with glycine was preferred in goats, conjugation with glutamic acid was favoured in cattle.

21.5.3 Hens

Edwards and Swaine (1977) carried out and reported the analytical work in a hen-feeding study conducted by **Ross *et al.* (1977)**. Groups of 40 laying hens were fed on diets containing 0.4, 3.4 and 33 ppm non-radio-labelled 40:60 *cis:trans* permethrin for 28 days and then returned to a control diet for an additional 14 days. Samples of eggs laid during the study were analysed for permethrin residues. Five hens per group were sacrificed after 21, 28, 35 and 42 days of the study and tissues analysed for permethrin. At the 0.4 ppm dietary rate no residues of permethrin were detected in the albumen and yolks of eggs (limit of detection 0.02 mg/kg) or in the muscle, skin and liver (limit of detection 0.01 mg/kg). At the higher dietary rates no permethrin was detected in egg albumen. In yolks, permethrin residues were up to 0.05 mg/kg and up to 0.64 mg/kg, respectively, at the 3.4 and 33 ppm treatment levels. Residues did not accumulate and declined rapidly when feeding finished, reaching non-detectable levels before the end of the 14-day recovery period in both cases. At the 3.4 ppm dietary rate, permethrin residues in muscl' 'kin and liver were non-detectable (less than 0.01 mg/kg). At the 33 ppm rate permethrin residues in liver were also non-detectable; low residues in muscle and skin of 0.05–0.08 mg/kg fell to 0.02 mg/kg before the end of the recovery period.

Leahey *et al.* (1977b) conducted a study in which hens were dosed for 10 consecutive days with ^{14}C-permethrin (radio-labels in both the acid and alcohol part of the molecule), at a rate equivalent to 10 ppm in the total diet. Radioactive residues in eggs reached a maximum of 0.5 mg/kg in the yolk and 0.04 mg/kg in the albumen within 6–8 days. The hens were sacrificed 4 hours after the final dose and samples of fat, muscle and liver were taken for analysis. The total radioactive residues in these tissues were: fat 0.3–0.7 mg/kg, muscle 0.03–0.13 mg/kg and liver 0.4–1.1 mg/kg of permethrin equivalents. In the fat and the eggs, the residues after dosing with both acid- and alcohol- labelled permethrin were approximately the same, and permethrin was the major compound identified in these tissues. However, in the muscle and liver, higher residues were detected in the hens dosed with acid-labelled permethrin. *Cis/trans-* 3-(-2,2-dichlorovinyl) - 2,2-dimethylcyclopropanecarboxylic acid (*cis:trans* ratio approximately 1:6) was the major residue identified in these tissues.

Radio-carbon from ^{14}C-carbonyl- and ^{14}C-methylene-labelled preparations of [1RS]-*trans*- and [1RS]-*cis*-permethrin administered to laying hens for 3 consecutive days at 10 mg/kg for each dose was largely eliminated from the body within 1 day after the last dose, a portion as $^{14}CO_2$ **(Gaughan *et al.* 1978b)**. The excreta contained all and the eggs most of the following compounds: the unmetabolised pyrethroids; *cis*-permethrin hydroxylated at the 4'-position, at the methyl group *trans* to the carboxyl, and at both of these sites; the dichlorovinyl-phenoxybenzoic acid and the 4'-hydroxy derivatives; sulphate, glucuronide, taurine and other conjugates of these alcohols and acids. Residues of unmetabolised *trans*- and *cis*-permethrin in fat were 0.15 and 0.93 mg/kg respectively, at 7 days after the last dose, and in eggs they reached peak levels of 0.3 and 1.2 mg/kg, respectively, at 3–4 days after the last dose. Almost half of the residues in eggs were unmetabolised *trans*- and *cis*-permethrin in the yolk and the remainder was a great variety of metabolites in the yolk and white including most of those also detected in the excreta.

21.6 Safety to Livestock

Though there is a limited amount of information **(Table 21.10)** from toxicity studies on livestock, this together with the mass of information on the metabolism of permethrin (Section 21.4) and on the fate and distribution of permethrin in livestock (Section 21.5), when combined with the available information on the acute and sub-acute oral toxicity of permethrin to various species **(Table 21.11)** makes it quite clear there is no hazard to livestock from the feeding of grain or milling offals from grain that has been treated with permethrin for protection against stored-product insects.

James (1980) provided the following results which were obtained with permethrin, 25:75 *cis:trans*.

From these results it can be seen that the intravenous toxicity is a hundred times less than for pyrethrins, where doses of 3 to 5 mg/kg are lethal. With sheep and cattle a dose given by the ruminal route is more toxic than by the abomasal route.

Cats are especially sensitive to the synthetic pyrethroids, but this same sensitivity is seen with other classes of compound for example the chlorinated hydrocarbons, DDT or lindane.

Ross *et al.* (1977) studied the effect of incorporation of permethrin in the diet of laying hens on egg production, fertility and hatchability. In 160 laying hens approximately 18 weeks of age, and sixteen 30-week-old cockerels were used in the study. The 28-day treatment period was preceded by an acclimatisation period of 4 weeks for the birds to reach optimum egg production. The birds were allocated to 1 of 4 groups receiving 0, 0.4, 4.0 and 40 ppm permethrin in the ration. All suitable eggs laid on alternate days were incubated. Eggs were examined for fertility by candling on the seventh and eighteenth days of incubation. Chicks which hatched were reared to 10 days of age. Body weights were recorded on days 0 and 10. The health of the birds was not affected by the treatment. Food consumption and body weight changes were considered to be within normal limits. Hens receiving the 2 higher rates of permethrin laid more eggs than the control group but these contained fewer large eggs. There were no adverse effects on fertility and hatchability which could be related to treatment. The chickens hatched from eggs laid by the treated hens were comparable to the untreated controls in viability and there were no significant

Table 21.10. Toxicity of permethrin (25:75) to livestock and domestic animals **(James 1980)**

Species	Trial	Route	Vehicle	Result
Sheep	Acute tox.	Intravenous	Corn oil	300 mg/kg minor effects only
Sheep	Acute tox.	Abomasal	Corn oil	800 mg/kg no toxicity
Sheep	Acute tox.	Ruminal	Corn oil	800 mg/kg minor effects only
Cattle	Acute tox.	Ruminal	Corn oil	400 mg/kg tremors/ataxia 800 mg/kg lethal
Cattle	Acute tox	Abomasal	Corn oil	800 mg/kg no toxicity
Cattle	Acute tox.	Oral	–	1–2 week calves <1g/kg no toxicity
Dog	Acute tox.	Intravenous	Corn oil	400 mg/kg lethal
Dog	Acute tox.	Intravenous	Corn oil	200 mg/kg no effect
Dog	Acute tox.	Oral	Corn oil	1600 mg/kg no effect
Dog	Sub acute	Oral	Neat	14 days 500 mg/kg no effect
Dog	Sub chronic	Oral	Neat	6 months 250 mg/kg no effect
Cat	Acute tox.	Oral	Corn oil	200 mg/kg lethal

Table 21.11. Permethrin: acute oral toxicity to various species

Species	Sex	Vehicle	LD/50(mg/kg)	Author
Rat	M	Water	2 949	**Parkinson 1978**
	F	Water	> 4 000	**Parkinson *et al.* 1976**
	M	DMSO *	1 500	**Clark 1978**
	F	DMSO *	1 000	**Clark 1978**
	M	Corn oil	500	**Jaggers & Parkinson 1979**
	M	Corn oil	430	**Khoda *et al.* 1979**
	F	Corn oil	470	**Khoda *et al.* 1979**
	M & F	Corn oil	1 200	**Braun & Killeen 1975**
	M & F	None	6 000–8 900	**Braun & Killeen 1975**
	F	Nil (neat)	>20 090	**James 1980**
	F	Corn oil	3 185	**James 1980**
	F	Odorless petroleum	> 8 000	**James 1980**
	F	DMSO	> 8 000	**James 1980**
	F	Glycerol formal	> 5 048	**James 1980**
Mouse	F	Water	> 4 000	**Parkinson *et al.* 1976**
	M & F	DMSO *	250–500	**Clark 1978**
	M	Corn oil	650	**Khoda *et al.* 1979a**
	F	Corn oil	540	**Khoda *et al.* 1979a**
Rabbit	F	Water	> 4 000	**Parkinson *et al.* 1976**
Guinea Pig	M	Water	> 4 000	**Parkinson *et al.* 1976**
Hen	F		>15 000	**Milner & Butterworth 1977**
Mallard ducks	M & F	None	>11 275	**Ross *et al.* 1976b**
Japanese quail	M & F	None	>13 510	**Ross *et al.* 1976b**
Starlings	M & F	None	>32 000	**Ross *et al.* 1976b**
Mallard ducklings	M & F	Feed	>23 000 ppm	**Ross *et al.* 1976b**
Japanese quail	M & F	Feed	>23 000 ppm	**Ross *et al.* 1976b**
Pheasants	M & F	Feed	>23 000 ppm	**Ross *et al.* 1976b**
Starlings	M & F	Feed	>23 000 ppm	**Ross *et al.* 1976b**

* DMSO = dimethylsulphoxide

differences between treatments. Chicks from groups treated with permethrin showed greater body weight gains than the control group and this appeared to be a treatment-related effect. The investigators concluded that the inclusion of permethrin in the diet at rates up to 40 ppm had no adverse effects on any of the parameters measured.

Edwards and Iswaren (1977) fed lactating dairy cows for 18–31 days with diets containing 0.2, 1, 10 and 50 ppm of permethrin, applied as a spray to pelleted grassnuts. There was no effect from permethrin throughout the trial upon body weights, milk yields and the general health of the animals. A gross and histopathological examination at slaughter did not reveal any effects which could be attributed to permethrin.

21.7 Toxicological Evaluation

The toxicology of permethrin was evaluated by the Joint FAO/WHO Meeting on Pesticide Residues in 1979, 1981 and 1982 **(FAO/WHO 1980 1982 1983).** Apart from extensive information on biochemistry and metabolism in several species there were special studies for neurotoxicity in rats and hens, special studies on reproduction, teratogenicity, mutagenicity and pharmacology. Data on acute toxicity to numerous species from many investigators clearly showed that the acute toxicity depends on the solvent or vehicle used for administering the insecticide. Vegetable oils and dimethyl sulphoxide obviously potentiate the toxicity by increasing absorption. Acute toxicity studies of varying *cis:trans*-permethrin ratios indicate that *cis*-permethrin has a greater toxicity compared to *trans*-permethrin.

There were no indications of neurotoxicity, no effects on any reproductive parameter, no teratogenicity in mice or rats and all tests for mutagenicity yielded negative findings. Long-term studies in both rats and mice have shown no oncogenic potential. In short-term and long-term studies, permethrin was noted to have an effect on the liver described as an increased liver weight and liver to body weight ratio. This increase, which may be an adaptive response, was accompanied by centrilobular hepatocyte hypertrophy and an increase in the subcellular smooth endoplastic reticulum. The no-effect level was based on the response noted at dosage levels above 100 ppm.

The level causing no toxicological effect in the rat was 100 ppm in the diet equivalent to 5.0 mg/kg body weight. On this basis an estimate was made of the temporary acceptable daily intake for man of 0–0.03 mg/kg-bw which was converted in 1982 to a full ADI of 0.05 mg/kg-bw on the basis of additional data and further evaluation. This ADI is independent of the ratio of *cis:trans* isomers.

21.8 Maximum Residue Limits

The extensive data from supervised residue trials made available to the JMPR enabled the meeting to recommend the following maximum residue limits. These are currently regarded as temporary, pending the submission of additional information.

Commodity	Maximum Residue Limits (mg/kg)
Bran	10
Cereal grains	2
Flour (white)	0.5
Wholemeal	2

Residues are measured and expressed as permethrin.

The methods recommended for the determination of permethrin residues are: **Pesticide Analytical Manual (1979a, 1979); Ambrus *et al.* (1981).** The following methods are recognised as being suitable: **Baker and Bottomley, (1982); Belanger and Hamilton (1979); Chapman and Harris, (1978); Furjie and Fullmer, (1978); Greve and Heusinkveld, (1981); Mestres *et al.* (1979a); Oehler, (1979); Reichel *et al.* (1981); Williams, (1976).**

22. *d*-Phenothrin

d-Phenothrin is a new chrysanthemic acid ester of 3-phenoxybenzyl alcohol synthesised in Japan. Among 4 esters (*d-trans-*, *d-cis-*, *l-trans-*, and *l-cis*-chrysanthemate), the *d-trans*-isomer has proved to be most effective against various insects of medical importance as well as agricultural pests. The *d-cis*-isomer is less effective than the *d-trans*-isomer. Both of the *l*-isomers show very little insecticidal activity.

The racemic mixture known as phenothrin (a 20:80 racemic mixture of (*IRS*), *cis*-and (*IRS*), *trans*-isomers) has been evaluated for several uses including application as a grain protectant but it was later found that it was more economic to standardise commercial production on the mixture known as '*d*-phenothrin' (a mixture of predominantly (*IR*),*cis*- and (*IR*),*trans*-isomers, with a *cis:trans* ratio of 20:80). This product is known commercially as Sumithrin. There is no substantial difference between phenothrin and *d*-phenothrin in the content of impurities. Both materials have been subjected to extensive toxicological investigation and there is clear-cut evidence that, so far as laboratory animals are concerned, both compounds are toxicologically equivalent. There is some confusion in the literature between *d*-phenothrin and phenothrin and in many cases it is not clear which was used in the studies. Some papers refer to the compound as 'fenothrin'. Where it is clear that *d*-phenothrin was used, that term has been used in this review to describe the insecticide. Likewise where there is no doubt that the racemic mixture was employed the term 'phenothrin' has been used. Where the nature of the insecticide is not clear, preference has been given to the term '*d*-phenothrin'.

Among the important uses of *d*-phenothrin are as sprays and aerosols for controlling flying insects including 'disinsection' of aircraft. Its low toxicity, high stability and high potency make it a candidate for use in stored-product pest control and grain protection.

d-Phenothrin has proved particularly useful in combination with organophosphorus insecticides where it is effective in improving the control of *Rhyzopertha dominica*. It is not as potent as bioresmethrin against *Rhyzopertha dominica*, but on a cost-effectiveness basis it shows considerable advantage. Its effectiveness is increased 5-fold by combining with piperonyl-butoxide. *d*-Phenothrin deposits are stable with a half-life of 40 weeks at 25°C. *d*-Phenothrin is readily metabolised by animals. Because of its stability and penetrating effect about 10–15% of the applied dose finds its way from raw wheat into bread. Rice and oats retain most of the deposit on the hulls which are removed before further processing into food commodities.

22.1 Usefulness

Desmarchelier (1977b) put forward a strong argument, supported by extensive data generated from laboratory studies, to show that combinations of pyrethroid and organophosphorus insecticides offered particular advantages for the control of stored-product beetles in grain. He presented evidence to show that such combinations could be used under field conditions as selective treatments against mixed populations of stored product pests and that, by judicious management, residues could be kept to fairly low levels.

Ardley and Halls (1979) noted that *d*-phenothrin was similar in efficiency to bioresmethrin when combined with fenitrothion and synergised with piperonyl butoxide. Wheat treated with *d*-phenothrin (2 mg/kg) plus piperonyl butoxide (10 mg/kg) and fenitrothion (12 mg/kg) gave broad-spectrum insect control for 9 months after application. This combination gave similar protection to that provided by bioresmethrin/piperonyl butoxide/fenitrothion (1/8/12 mg/kg), the standard treatment used since 1975 in eastern States of Australia to control both susceptible and malathion-resistant insects in all classes of cereal grains.

Bengston (1979) explained why it is necessary to use combinations of insecticides and how these should be selected to achieve the most effective control of the broadest possible spectrum of stored-product pests. In order to obtain the highest possible level of freedom from insect pests during a storage period of 9 months, a selected organophosphorus insecticide should be combined with a proven synergised *d*-pyrethroid. *d*-Phenothrin used at the rate of 2 mg/kg, together with piperonyl butoxide at the rate of 10 mg/kg, was regarded as an effective treatment.

Bengston and Desmarchelier (1979) explained how extensive laboratory and field trials conducted in Australia have led to the selection of a range of grain protectants suitable for controlling all species including those which have developed resistance to lindane and malathion. *d*-Phenothrin at 2 mg/kg plus piperonyl butoxide at 10 mg/kg has been selected for use in conjunction with an approved organophosphorus insecticide since synergised *d*-phenothrin, at this rate, has proved effective in controlling *Rhyzopertha dominica* in stored grain for at least 9 months.

Desmarchelier and Bengston (1979a), in a paper summarising results of a continuing program to evaluate grain protectants suitable for use in Australia, described how predictive models are used to determine the fate of insecticides, including phenothrin, on grain in storage. The main use of predictive models is to enable the grain industries to select rates appropriate for given circumstances. They showed how *d*-phenothrin, with a residual half-life of 40 weeks, was selected for use at 2 mg/kg in conjunction with piperonyl butoxide at 10 mg/kg. This approach is further discussed in another paper by the same authors **(Desmarchelier and Bengston 1979b)**.

Fujinami (1980) provided information on the chemistry, physics and biology of *d*-phenothrin and information concerning toxicity to important stored-product pests including strains resistant to other insecticides.

Bengston *et al.* (1980b) reported the outcome of duplicate field trials carried out on bulk wheat in commercial silos in Queensland and New South Wales where a combination of fenitrothion (12 mg/kg) with *d*-phenothrin (2 mg/kg) proved more effective than a pirimiphos-methyl/carbaryl combination against *Sitophilus oryzae* and *Ephestia cautella*. Both treatments were equally effective against *Rhyzopertha dominica, Tribolium castaneum, Tribolium confusum* and *Oryzaephilus surinamensis*, preventing the production of progeny.

Desmarchelier (1983) stressed the importance of temperature control because of its effect upon toxicity of insecticides to stored-product pests. He quoted *d*-phenothrin as an example and showed that at 25°C *d*-phenothrin is vastly superior in reducing the number of progeny of *Tribolium castaneum* over the number produced in the presence of *d*-phenothrin at 35°C. Lowering the temperature not only lowers the reproductive capacity of the insect but also increases the relative potency of the insecticide and prolongs its residual life.

Bengston *et al.* (1983a) carried out duplicate field experiments on the control of insect infestation in stored sorghum under commercial conditions in Queensland. *d*-Phenothrin was applied at the rate of 1 mg/kg along with fenitrothion (12 mg/kg) at each location and bioassays were carried out on samples of treated grain taken at regular intervals during 6 months storage. These established that *d*-phenothrin, used at the rate of 1 mg/kg (without synergist), was only partially effective in controlling *Rhyzopertha dominica* and it was considered that a level of 3 mg/kg would be required for complete control of this species. Natural infestations of *Rhyzopertha dominica* developed in the silos treated with phenothrin. This may have been due to the absence of piperonyl butoxide.

Bengston *et al.* (1983b) reported duplicate field trials carried out on bulk wheat in commercial silos in Queensland and New South Wales. Laboratory bioassays using malathion-resistant strains of insects where carried out on samples of treated grain at intervals over 9 months. These established that treatment with fenitrothion (12 mg/kg) plus *d*-phenothrin (2 mg/kg) plus piperonyl butoxide (8 mg/kg) controlled common field strains of *Sitophilus oryzae* and *Rhyzopertha dominica* and completely prevented progeny production in *Tribolium castaneum, Tribolium confusum* and *Ephestia cautella*.

Bengston *et al.* (1984), who carried out field experiments on bulk sorghum stored for 26 weeks in concrete silos in south Queensland, reported that laboratory bioassays, in which malathion-resistant insects were added to grain samples, indicated that fenitrothion (12 mg/kg) plus phenothrin (2 mg/kg) plus piperonyl butoxide (8 mg/kg) controlled typical malathion-resistant strains of *Sitophilus oryzae, Rhyzopertha dominica, Tribolium castaneum* and *Ephestia cautella*.

Golob *et al.* (1985) determined the susceptiblity of *Prostephanus truncatus* to *d*-phenothrin following topical application and exposure to treated filter papers. *d*-Phenothrin showed similar potency to permethrin following topical application but it was only one fifth as active when the adult beetles were exposed to the filter papers. *d*-Phenothrin was not compared on treated grain or under field-trial conditions.

Desmarchelier *et al.* (1986) reported an extensive series of trials in which 4 separate grain protectant/grain protectant combinations were applied to wheat that was stored at 15 sites throughout 5 Australian States in 63 separate storages. Samples from 12 storages treated with phenothrin were taken at regular intervals for laboratory assays against *Rhyzopertha dominica* and *Tribolium castaneum*. Grain remained free of insects in 60 of 63 storages; partial failure in the other 3 storages was attributed to low or irregular levels of protectants. It was determined that the levels of aged *d*-phenothrin residues required to control *Rhyzopertha dominica* were as follows:

- to control adults within 3 days, 0.8–0.9 mg/kg;
- to control adults during 26 days exposure, less than 0.8 mg/kg;
- to prevent the development of F_1 progeny, less than 0.8 mg/kg;
- to prevent the development of F_2 progeny, less than 0.8 mg/kg.

In view of these observations it was calculated that *d*-phenothrin should be applied at the rate of 1.9 mg/kg in order to leave exactly the required residue level after storage for 9 months.

22.2 Degradation

Information is available from a number of trials on wheat, sorghum and barley grains carried out in Australia in commercial storages of up to 10 000 tonnes treated with phenothrin. The temperature ranged from 18°C to 36°C and the moisture content of the grain from 9% to 13%. Treatments involved several insecticides including *d*-phenothrin in combination with piperonyl butoxide and fenitrothion. Wheat was stored for up to 9 months, barley for up to 5 months and sorghum up to 6 months, and samples were withdrawn for residue analyses at regular intervals. The residues of *d*-phenothrin determined in samples from these trials are set out in **Table 22.1**.

The degradation and fate of *(d)-trans-* and *(d)-cis-* isomers of phenothrin in stored wheat grains were also studied under laboratory conditions in Japan (**Nambu *et al.* 1978**). When *(d)trans* and *(d)cis-* phenothrin labelled with ^{14}C at the methylene group of the alcohol moiety were each applied at a concentration of 4 mg/kg, alone or together with 20 mg/kg piperonyl butoxide and 4 mg/kg fenitrothion to wheat grains with 11–12% moisture content and stored at 15°C or 30°C in the dark for 6 months, both isomers were slightly decomposed, regardless of the presence of piperonyl butoxide and fenitrothion. The distribution of the isomers after storage of treated grain was determined in the germ, endosperm and seed coat and the results are set out in **Table 22.2**.

Desmarchelier and Bengston (1979a) measured the residual half-life of *d*-phenothrin at 30°C and 50% relative humidity to be 38 weeks, when applied alone and 40 weeks in the presence of piperonyl butoxide.

Bengston *et al.* (1980b), reporting the results obtained from duplicate field trials carried out on bulk wheat in commercial silos, recorded that the concentration of *d*-phenothrin declined from approximately 2 mg/kg to 0.8 mg/kg over 37 weeks in a silo where the temperature remained close to 30°C for 31 weeks and then cooled to 20°C over the remaining 6 weeks. In the other silo where the temperature declined steadily from 27°C to 16°C over 25 weeks, the concentration of *d*-phenothrin in the grain remained substantially constant at approximately 2 mg/kg.

Desmarchelier *et al.* (1980a) studied the fate of residues of *d*-phenothrin on barley, unmilled rice, husked rice and polished rice during storage for 6 months. They found phenothrin very persistent on all grains. When the residue data were compared with predictions from models based on a degradation constant and 2 variables, temperature and equilibrium relative humidity, they found that the observed values for residues of *d*-phenothrin agreed precisely with those predicted from the models.

Table 22.1. Residues of d-phenothrin in grain after storage

Grain	Rate* (mg/kg)	Residues after x months storage (mg/kg)									Reference
		1	2	3	4	5	6	7	8	9	
Sorghum	2.0			2.72	1.59			1.56			**Bengston *et al.* (1977)**
Sorghum	1.2	0.4	0.8	0.85	0.75		0.75				**Bengston *et al.* (1978)**
Sorghum	1.0	0.45	0.45	0.5	0.4		0.4				
Wheat	2.0	1.7	0.9	1.1		1.0	0.5		1.1		**Desmarchelier *et al.***
	1.7	2.4	2.1	1.4	1.0	1.1	0.9	0.9	1.0	1.0	**(1986)**
	1.9	1.6	0.8		0.8	1.0	0.5		0.6	0.6	
	2.0	1.5	1.7			1.6					
	2.0	0.7		1.0							
	1.8	1.5	1.3	0.7			0.7				
	1.4	1.5									
	2.0	1.3	1.3		1.1	1.1	0.9	0.8			
	1.9	0.9	0.7	0.8			0.8				
	2.1	1.0		0.7	1.1	0.9					
	2.0	1.0	1.0		0.9	1.5		1.5			
	2.6		1.3	0.6			1.1				
	2.2	1.0		0.8		0.8	1.2				
Barley	2.6	2.0		1.1		1.5					**Desmarchelier *et al.***
	2.9		1.5	1.4	1.1						**(1986)**
Wheat	1.49	1.85	1.35	1.75	1.75						
	2.01	2.25	1.8	2.05	1.6						
	2.01	2.3	1.9	1.6							
Wheat	4	3.8		3.8			3.8				**Nambu *et al.* (1978,**
	4	3.84		3.87			3.69				**1981)**
	4	3.82		3.88			3.69				
	4	3.80		3.80			3.73				
	4	4.03		3.88			3.84				
	4	4.03		3.91			3.76				

* Rate of application calculated from total grams *d*-phenothrin used divided by tonnes of grain treated

Table 22.2. Distribution of *(d)-trans* and *(d)-cis* phenothrin after storage of treated grain **(Nambu *et al.* 1978)**

Compound	Temperature	Months	Residue (mg/kg)			
			Whole grain	Germ	Endosperm	Seed coat
(d)-trans -phenothrin (4 mg/kg)	15°C	0	3.81	0.08	0.62	2.82
		1	3.80	0.1	0.64	2.86
		3	3.78	0.09	0.63	3.09
		6	3.72	0.08	0.51	2.97
	30°C	1	3.90	0.12	0.77	2.95
		3	3.80	0.11	0.69	2.97
		6	3.80	0.12	0.58	3.06
(d)-trans- -phenothrin + PBO + fenitrothion (4+20+4 mg/kg)	15°C	6	3.69	0.10	0.62	2.94
	30°C	6	3.66	0.10	0.60	2.65
(d)-cis- phenothrin (4 mg/kg)	15°C	0	3.85	0.09	0.67	3.14
		1	3.80	0.08	0.70	3.08
		3	3.80	0.08	0.64	3.11
		6	3.73	0.11	0.62	2.63
	30°C	1	4.00	0.08	0.80	3.02
		3	3.93	0.12	0.68	3.28
		6	3.80	0.14	0.58	2.75
(d)-cis- phenothrin + PBO + fenitrothion (4+20+4 mg/kg)	15°C	6	3.76	0.12	0.58	2.92
	30°C	6	3.72	0.10	0.59	2.79

PBO: Technical piperonyl butoxide

Desmarchelier *et al.* (1986) compared the results obtained in extensive pilot studies during 1978–79 (**Table 22.1**) and found there was a high level of agreement between the observed values and the values predicted from the model referred to above (**Desmarchelier *et al.* 1980a**).

Nambu *et al.* (1981) studied the degradation and fate of *d*-phenothrin in stored wheat grains by appling ^{14}C (*d*)-*trans* and (*d*)-*cis* isomers of *d*-phenothrin at the concentration of 4 mg/kg to wheat grains with 11–13% moisture content and stored at 15°C or 30°C in the dark for 12 months. Both isomers of *d*-phenothrin were decomposed very slowly. After 12-month storage at 30°C, 79% of (*d*)*trans*-isomer and 87% of (*d*)*cis*-isomer remained in the grains. At 15°C, 92% of both isomers remained intact.

Noble *et al.* (1982) studied the stability of 4 pyrethroids on wheat in storage. *d*-Phenothrin was applied to wheat, without piperonyl butoxide, at the rate of 2 mg/kg and the wheat was stored for 52 weeks at 25°C or 35°C, and either 12% or 15% moisture content. Rates of loss were calculated from residue analyses of the wheat at 5 intervals during storage. Calculated half-lives for *d*-phenothrin at 25°C and 12% moisture and 35°C and 15% moisture were 72 weeks and 29 weeks respectively. The percentage of *trans*-isomer expressed as a percentage of the total *cis* plus *trans*-isomer content decreased from 80% at the time of application to 63% after 52 weeks at 35° and 15% moisture content. Even at the lower temperature and lower moisture level there was a measurable decrease in the ratio of *trans*-isomer. This work confirms the importance of temperature control wherever possible.

Bengston *et al.* (1983a) found that *d*-phenothrin applied to bulk sorghum stored in commercial silos in south Queensland and central Queensland appeared to degrade quite rapidly during the first 5 weeks after application (at the rate of 1 mg/kg without piperonyl butoxide) but thereafter the residues remained relatively unchanged for the rest of the 24 week storage. The *d*-phenothrin appeared somewhat more stable at the central Queensland site where the moisture content of the grain was almost 1% lower than at the southern site.

Bengston *et al.* (1983b) also studied the fate of phenothrin applied to wheat at the rate of 2 mg/kg in conjunction with piperonyl butoxide (8 mg/kg) and fenitrothion (12 mg/kg) on bulk wheat in commercial silos in Queensland and New South Wales. The residue data gathered on 6 occasions over 9 months indicated a slower rate of degradation than that which occurred on sorghum reported previously. This is no doubt due to the combined effect of piperonyl butoxide and the lower equilibrium moisture content of the wheat. The initial temperature of the grain at the 2 sites was 34°C and 32°C.

Bengston *et al.* (1984), in a further field experiment carried out on bulk sorghum stored for 26 weeks in concrete silos in south Queensland and treated with combinations of organophosphorus and synergised synthetic pyrethroid insecticides, used phenothrin (2 mg/kg) plus piperonyl butoxide (8 mg/kg) with fenitrothion (12 mg/kg). They found very little change in the concentration of phenothrin on the grain at any stage during the 26 weeks storage. The sorghum had a moisture content of 12.8% and at the time of treatment a temperature of 27.3°C. The conditions were obviously more favourable to the persistence of phenothrin.

22.3 Fate in Milling, Processing and Cooking

Nambu *et al.* (1981) applied radio-labelled *d*-phenothrin (both *cis* and *trans*) at the rate of 4 mg/kg to study the fate of the insecticide on wheat grains in storage and their distribution in the various fractions of the grain. At the end of 12 months storage they processed the grain into flour and found that the radioactivity in the wheat grains was localised mainly in the seed coat, resulting in only 0.77 mg/kg of *d*-phenothrin isomers in flour but 11.4 mg/kg in bran. The 2 isomers of *d*-phenothrin were hardly decomposed by the baking process, leaving 0.57 mg/kg of *d*-phenothrin isomers in bread. Thus, about 15% of the original deposit on the grain was carried through to the bread.

A number of studies were carried out in Australia to determine the residues of *d*-phenothrin in cereal products after processing and cooking of wheat, oats, barley and rice grains. Wheat grain containing 0.5–3.9 mg/kg of *d*-phenothrin was used in the studies and the residue results, after processing and cooking, are summarised in **Table 22.3**. Although residues found in bran and pollard after processing were considerably higher than those found in the raw wheat, the residues in flour were much lower (maximum 0.6 mg/kg) – 25% of those found in whole grain. After cooking, the residues in bread did not exceed 0.2 mg/kg in these trials. In gluten, the residue was higher (1.9 mg/kg) than in flour (0.65 mg/kg). However, the pesticide residues in gluten do not cause concern, because normally there would be a 2% maximum addition of gluten in bread. Oats containing 1.2 mg/kg of *d*-phenothrin were processed and cooked **(Desmarchelier 1979e)**. Most of the residues were carried in oat hulls, and low residue levels were found in both groats (0.3 mg/kg) and rolled oats (0.1 mg/kg). In porridge the residue was below the detection limit of 0.05 mg/kg.

Barley containing 3.2 mg/kg and 0.92 mg/kg *d*-phenothrin was processed into malt, wort and cattle feed. Of these products, the highest residues were found in the malt, but these levels, 0.63 mg/kg and 0.15 mg/kg, were less than one-fifth of that in barley. In wort, the residues were less than the detection limit of 0.02 mg/kg. A nominal application of 8 mg/kg *d*-phenothrin was applied to husked (brown) rice and polished (white) rice. Six months after application, husked (6.2 mg/kg of *d*-phenothrin) and polished (5.9 mg/kg) rice were cooked in a minimal amount of boiling water for 25 minutes. Unhusked rice was also treated with *d*-phenothrin at a rate of 8 mg/kg and after 6 months of storage when 4.7 mg/kg of *d*-phenothrin was detected, the rice was milled into husked and polished rice and cooked as described above. As shown in **Table 22.4** the cooking of rice had the effect of reducing the residue by from 34% to 60%. It also

Table 22.3. Residues of phenothrin and d-phenothrin before and after processing and cooking of wheat

	Residue (mg/kg)								
Compound	Wheat	Bran	Pollard	Flour	Wholemeal bread	White bread	Gluten	Starch	Reference
Phenothrin	1.2	5.6	3.6	0.21	0.74	0.16	–	–	**Bengston 1979**
	3.0	10.0	4.1	0.82	–	0.18	–	–	**Desmarchelier 1979**
	1.5	6.0	3.0	0.4	–	0.1	–	–	**Desmarchelier 1979**
	0.9	3.0	2.0	0.15	–	0.05	–	–	**Desmarchelier 1979**
	0.5	1.6	0.9	0.09	–	0.05	–	–	**Desmarchelier 1979**
	3.9	8.1	1.6	0.65	–	–	1.9	0.05	**Desmarchelier 1979**
	1.2	4.0	1.8	0.3	0.4–0.6	0.1–0.2	–	–	**Ardley 1979**
	2.0	–	–	–	–	<0.1	–	–	**Mollard 1979**
	4.3	–	–	–	–	<0.1	–	–	**Mollard 1979**
(*d*)-*trans*-phenothrin	3.78	11.4	–	0.79	–	0.69	–	–	**Nambu *et al.* 1981**
(*d*)-*cis*-phenothrin	3.64	9.24	–	0.91	–	0.66	–	–	

indicated that the milling of rice from husked (paddy) to unhusked (brown) or polished (white) rice has a major effect on residue reduction, that is, 90% and 97% respectively **(Desmarchelier *et al.* 1980a).**

Table 22.4. Residues of d-phenothrin prior to and following processing and cooking of rice **(Demarchelier *et al.* 1980a)**

Residue (mg/kg)			
Unhusked	Husked	Polished	Cooked
–	6.2	–	4.1
–	–	5.9	3.1
4.7	0.50	–	0.2
4.7	–	0.15	<0.1

Bengston *et al.* (1983b), who applied mixtures of organophosphorothioates and synergised synthetic pyrethroids as grain protectants on bulk wheat in commercial silos in Queensland and New South Wales, completed the study by milling some of the wheat from each treatment and processing it into wholemeal bread and white bread. The phenothrin residues were determined in each milling fraction as well as in the bread. The results are given in **Table 22.5**. From these data the investigators calculated the distribution of residues in the milling fractions. The relative proportion of phenothrin in the white flour was higher than that for permethrin, significantly higher than for deltamethrin and almost twice that of fenvalerate. The results indicate that, in this study, phenothrin penetrated the endosperm to a slightly greater extent than the other 3 pyrethroids.

Table 22.5. Fate of phenothrin residues during milling and baking **(Bengston *et al.* 1983b)**

Milling fraction	Residue (mg/kg)	% Distribution
Wheat	1.21	100
Bran	4.4	54.1
Pollard	3.6	31.8
White flour	0.23	14.1
Wholemeal bread	0.81	–
White bread	0.20	–

22.4 Metabolism

Miyamoto *et al.* (1974) studied the metabolism of *d*-phenothrin in mammals. ^{14}C-*d-trans*-phenothrin labelled at the hydroxymethyl group in the alcoholic moiety was administered orally to male Sprague-Dawley rats at the rate of 200 mg/kg. The compound was rapidly absorbed from the gastro-intestinal tract and distributed to various tissues. Radioactivity was rapidly eliminated over 3 days via urine (ca.60%) and faeces (ca.40%). No detectable radioactive carbon dioxide was expired. The urinary and fecal metabolites were separated and identified, the predominant one being 3-(4'-hydroxy)phenoxybenzoic acid and amounting to approximately 55% of the recovered radioactivity. 3-Phenoxybenzoic acid, free and conjugated with glycine, was also identified. *In vitro* studies using liver preparations from rats, mice, guinea pigs, rabbits, and dogs revealed that *d-trans*-phenothrin was hydrolysed to 3-phenoxybenzyl alcohol, which was subsequently oxidised. The *l-trans* isomer was also easily hydrolysed, whereas the *d-cis-* and the *l-cis*-phenothrin were resistant to hydrolytic attack at the ester linkage.

Metabolism of (*d*)-*cis*-phenothrin in male rats was also studied. About 65% of the administered ^{14}C was recovered in feces at 3 days post-treatment. The feces contained 3 ester metabolites which resulted from hydroxylation at the 4'-phenoxy position of the alcohol moiety, oxidation of *trans*-isobutenyl methyl group and hydroxylation of *geminal*-dimethyl group of the acid moiety. A small amount of 3-phenoxybenzoic acid was also found **(Suzuki *et al.* 1976)**.

Thus, metabolism of phenothrin isomers proceeded rapidly in the rat mainly via hydrolysis of the ester linkage and oxidation at several positions on both the alcohol and acid moieties **(Figure 22.1)**. The metabolites derived from the alcohol moiety were almost completely excreted in the urine and feces.

Metabolism studies *in vitro* revealed that liver microsomal enzymes from rats, mice, guinea pigs, rabbits and dogs hydrolysed (*d*)-*trans*-phenothrin much faster than the *cis*-isomers, but the oxidation rate by mouse liver microsomes was somewhat faster with the (*d*)-*cis*-isomer than with the (*d*)-*trans*-isomer **(Suzuki and Miymato 1978; Soderland and Casida 1977a; Casida *et al.* 1979)**.

Soderlund and Casida (1977a, 1977b), reporting studies on the metabolism rates of 44 pyrethroids and 24 model compounds in mouse liver microsomal systems, divided the substrates into 3 groups based on their ease of hydrolysis and oxidation. Primary alcohol esters of *trans*-substituted cyclopropanecarboxylic acids are most rapidly metabolised with hydrolysis generally serving as the major component of the total metabolism rate. Although hydrolysed slowly or not at detectable rates, the primary alcohol *cis*- substituted cyclopropanecarboxylates are rapidly oxidised. The highly insecticidal 2-cyano-3-phenoxybenzyl esters are least susceptible to metabolic attack. *d*-Phenothrin fits the first category.

A carboxycsterase has been isolated and purified from rat liver microsomes. This carboxyesterase was found to hydrolyse *d-trans*-phenothrin faster than *d-cis-* isomers and it accounts in part for the reduced mammalian toxicity of the *d-trans*-isomers. In contrast to the increased hydrolytic rate observed with *d-trans*-phenothrin, the *d-cis*-phenothrin was found to be oxidised at a more rapid rate by mouse liver microsomal preparations. These data may explain the difference in toxicity noted between the

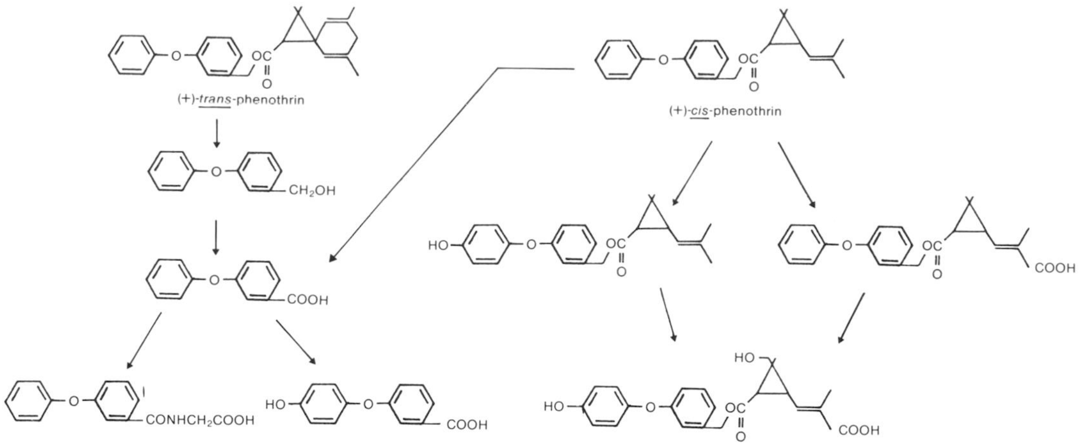

Figure 22.1. Metabolites produced from (+)− *trans*- and (+)− *cis* phenothrin in rats **(Miyamoto *et al.* 1974).**

cis- and *trans* isomers of several of the pyrethroid esters **(Suzuki and Miyamoto 1978; Soderlund and Casida 1977a; Casida *et al.* 1979).**

Casida *et al.* (1979) made a comprehensive review of the comparative metabolism of pyrethroids derived from 3-phenoxybenzyl and *a*-cyano-3-phenoxybenzyl alcohols. Likewise Miyamoto (1981) has dealt extensively with the metabolism of *d*-phenothrin in his review of the chemistry, metabolism and residues analysis of synthetic pyrethroids. Both papers warrant study by those wishing to obtain more detailed information. **Leahey (1985)** reviewed the metabolism and environmental fate of *d*-phenothrin.

Izumi *et al.* (1984) showed that following a single oral administration of each of ^{14}C-[1R, *trans*]-, [1RS, *trans*]-, [1S, *trans*]-, [1R, *cis*]-, [1RS, *cis*]- and [1S, *cis*]-phenothrin [3-phenoxybenzyl(1 RS, *trans/cis*)chrysanthemate] labelled in the alcohol moiety to both rats and mice at 10 mg/kg, the radiocarbon derived from each isomer was almost completely eliminated from the rat and mouse bodies within 6 days after administration and ^{14}C tissue residue levels were generally very low. In both rats and mice, there seemed to be no significant differences in the ^{14}C recovery and ^{14}C tissue residues between the [1R, *trans*]- and [1RS *trans*]-isomers and between the [1R, *cis*]- and [1RS, *cis*]-isomers, whereas the [1S]-isomers in both animals revealed slight differences in the ^{14}C recovery from the other corresponding *optical* isomers. A predominant excretion route was urine with three *trans* isomers, whereas feces was a predominant route for three *cis* isomers in both animals. The major urinary and fecal metabolites were generally common in the nature to both rats and mice, although N-3-phenoxybenzoyltaurine was characteristic to mice. The ester cleaved metabolites in both animals were obtained to a larger extent from *trans* isomer than from *cis* isomer. Cleavage of the ester linkage and of hydroxylation at 4-position of the alcohol moiety occurred to a larger extent in rats than in mice. Both [1S, *trans*]- and [1S, *cis*]-isomers in both animals received cleavage of ester linkage to slightly larger extents than the other *trans* and *cis* *optical* isomers, respectively. It may be concluded that in both animals, there are virtually no differences in metabolic fates between the [1R, *trans*]- and [1RS, *trans*]-isomers and between the [1R, *cis*]- and [1RS, *cis*]-isomers. Metabolic fates of both the [1S, *trans*]- and [1R, *cis*]-isomers were, however, slightly different from the corresponding *optical* isomers to the point of the liability of ester cleavage. Overall, the [1 RS]-isomers of both *stereo* isomers seem to show the metabolic fates close to the corresponding [1R]-isomers rather than the [1S]-isomers.

Nambu *et al.* (1981) applied radio-labelled *d-cis* and *d-trans*-phenothrin to wheat grains and studied the outcome after 6 and 12 months. They found that both isomers were metabolised by hydrolysis of the ester linkage, oxidation of the benzyl alcohol to the benzoic acid and methylation of the benzoic acid as indicated in **Figure 22.2**. The joint application of piperonyl butoxide and feritrothion inhibited the degradation of *d*-phenothrin to some extent.

22.5 Fate in Animals

Haneko *et al.* (1981) studied the absorption and metabolism of dermally applied phenothrin (various isomers) in rats and reported that from 8 to 17% of the applied dose of an EC preparation was absorbed, that the tissue residue levels were very low and that the radiocarbon was completely eliminated within 6 days. Nearly the same metabolites were involved following both oral and dermal application. The *cis*-isomer was excreted preferentially in feces whilst the *trans*-isomer was found mainly in urine. Once absorbed the phenothrin isomers are metabolised in a manner similar to oral administration.

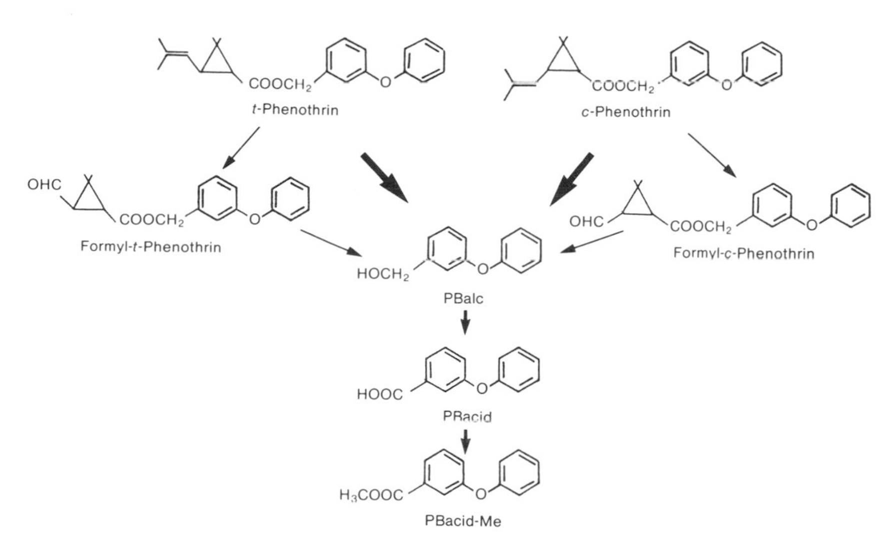

Figure 22.2. Proposed degradation pathways for phenothrin isomers in stored wheat grains (**Mambu *et al.* 1981**).

McColl (1982) reported the results of a study in which dairy cows were fed with grain containing *d*-phenothrin and fenvalerate applied at the rate of 2 mg/kg. Analysis of the grain 48 hours after treatment indicated that only approximately 1 mg/kg of the 2 insecticides was present on the feed. The cows were supplied with this feed for 28 days by which time the residue was determined by analysis to have fallen to approximately 0.5 mg/kg. Milk samples taken from each cow at the end of the feeding period and 3 days after the cows returned to unmedicated food showed no trace of *d*-phenothrin or fenvalerate.

No further studies dealing with the fate of phenothrin in animals other than those mentioned in Section 22.4 have been located. It is clear that *d*-phenothrin is rapidly absorbed, translocated in the body, and excreted predominantly in the urine. From metabolism studies in rats it appears that the majority of administered *d*-phenothrin is excreted within 24 hours.

When *d*-phenothrin is used as a grain-protectant insecticide, the residue level in grain and in milling offals is low (less than 5 mg/kg). Therefore one can be sure that the level of intake of phenothrin by domestic animals will also be low. The chances that *d*-phenothrin residues will be accumulated in meat, milk or eggs is extremely slight.

22.6 Safety to Livestock

No studies dealing specifically with the question of the safety of *d*-phenothrin to livestock have been located.

The acute toxicity in male and female rats and mice is extremely low. The LD/50 was greater than 5000 mg/kg body weight when *d*-phenothrin was administered orally, subcutaneously or by intraperitoneal injection. There were no sex differences noted in acute toxicity studies. Signs of poisoning were noted rapidly following the intravenous administration of *d*-phenothrin. The signs of poisoning include: fibrillation, tremor, slow respiration, salivation, lacrimation, ataxia, and paralysis. The signs of poisoning, evident at one half hour following administration, were rapidly overcome spontaneously to the point where, at 24 hours, there were no signs of toxicity (**Segawa 1976**).

Groups of mice were fed *d*-phenothrin in laboratory chow for 18 months in a standard carcinogenicity study. There were no gross pathological findings (**Murakami *et al.* 1980**).

Groups of rats were fed *d*-phenothrin in the diet at dosage levels of 0, 200, 600, 2000 or 6000 ppm for 2 years. There were no abnormal clinical or behavioural problems associated with this study. The survival rate of all animals of all groups were similar to that of controls and there were no effects of *d*-phenothrin on mortality. Growth, as evidenced by body weight reduction, was significantly affected at 6000 ppm in both males and females. Food consumption was observed to be slightly less in the high-dose male and female animals. All other parameters were not significantly different from controls (**Hiromori *et al.* 1980**).

A teratogenicity study conducted on New Zealand white rabbits receiving *d*-phenothrin at dosage levels ranging from 3 to 30 mg/kg from day 6 to day 18 of gestation showed no apparent teratogenic effect of *d*-phenothrin through at the high dosage level, possible fetal and/or maternal toxicity was observed (**Ladd *et al.* 1976**). In a second study in which pregnant rabbits were administered *d*-phenothrin at dosage levels of 10, 100 or 1000 mg/kg/day from day 6 through day 18 of gestation there were no abnormalities observed in the maternal parameters or fetal data (**Rutter 1974**).

A teratogenic study in which pregnant mice received *d*-phenothrin at 30, 300 or 3000 mg/kg from day 7 to day 12 of gestation and a separate group received dosage levels of 300 or 3000 mg/kg to evaluate post-natal effects revealed no adverse reaction to the administration of *d*-phenothrin. *d*-Phenothrin was neither teratogenic nor embryotoxic in mice at dosage levels up to and including 3000 mg/kg **(Nakamoto *et al.* 1973).** A three-generation reproduction study was conducted **(Takatsuka *et al.* 1980)** with albino rats fed diets containing either 200, 600 or 2000 ppm *d*-phenothrin. Parameters of reproductive performance revealed no consistent reductions which could be attributed to *d*-phenothrin ingestion. No significant changes were observed in the population data pertaining to the numbers of progeny delivered and weaned by test and control dams. Survival data revealed no consistent reductions which could be attributed to *d*-phenothrin ingestion. There were no differences in treated rats which could be attributed to the administration of *d*-phenothrin.

A study in which 14-day-old Mallard ducklings were exposed to appropriate dietary concentrations of *d*-phenothrin for 14 days, and then maintained on a toxicant-free diet for an additional 3 day observation period, revealed the acute LC/50 of *d*-phenothrin in Mallard duck to be greater than 5620 ppm. **(Fink 1978).**

A study in which *d*-phenothrin, dissolved in corn oil, was intubated directly into the crop of 20-week-old Bobwhite quail at concentrations ranging through 398, 631, 1000, 1590, 2510 mg/kg revealed that the LD/50 of *d*-phenothrin in Bobwhite quail is greater than 2510 mg/kg. No mortalities occurred at any dosage level tested, and no overt symptoms of toxicity were noted during the course of the study **(Beavers and Fink 1978).**

An 8-day dietary study conducted with *d*-phenothrin in Bobwhite quail showed the dietary medium lethal concentration (LC/50) of *d*-phenothrin to be in excess of 5000 ppm **(Fletcher *et al.* 1976).**

22.7 Toxicological Evaluation

Having evaluated extensive information on the properties, fate and effects of both phenothrin mixtures the JMPR agreed that the toxicological data on phenothrin was also applicable for the toxicological evaluation of *d*-phenothrin.

The toxicology of *d*-phenothrin was evaluated by the Joint FAO/WHO Meeting on Pesticide Residues in 1980 and 1984 when extensive information on the absorption, distribution, excretion and metabolism revealed that *d*-phenothrin is rapidly absorbed, translocated in the body and excreted predominantly in the urine. It is transformed by hydrolysis and oxidation and the products of metabolism are rapidly excreted. *d*-Phenothrin did not show any pharmacological activity in any of the *in vitro* and/or *in vivo* experiments.

Neither did it induce a neurotoxic effect in rats. Phenothrin was negative in mutagenicity tests. Likewise it was neither teratogenic nor embryotoxic in rabbits or mice and produced no effect in a 3-generation reproduction study in rats.

Acute and short-term studies in several species indicated that *d*-phenothrin has a low toxicity. Long-term studies in mice and rats indicated that *d*-phenothrin is not carcinogenic. The level causing no toxicological effect was:

Mouse: 300 ppm in the diet equivalent to 50 mg/kg bw/day;
Rat: 1000 ppm in the diet equivalent to 50 mg/kg bw/day.
Dog: 30 ppm in the diet equivalent to 7.5 mg/kg bw/day.

On the basis of this information it was possible to estimate a temporary acceptable daily intake for man of 0–0.4 mg/kg bw/day. In order to establish a more adequate basis for the allocation of a full ADI the Meeting required to see the outcome of on-going long-term studies in rats and mice and reproduction studies in rats.

22.8 Maximum Residues Limits

On the basis of the information available to the JMPR in 1980, recommendations were made for the following maximum residue limits.

Commodity	Maximum Residue Limits (mg/kg)
Bran	15
Cereal grains	5

No recommendations have apparently been made for MRLs in other milled cereal products.

Residues are determined and expressed as *d*-phenothrin.

The following method is recognised as being suitable for the determination of *d*-phenothrin residues: **Baker and Bottomley (1982).**

23. Phoxim and Phoxim-Methyl

Phoxim, earlier known as Bayer 77488, Bayer 5621, SRA7502 and now marketed as Baythion, was first evaluated in the mid-1960s. It has a broad spectrum of activity with stomach and contact effects. When applied to plants it has a depth effect but no systemic action. The initial effect is rapid, with a short to moderate duration, depending upon the application. It is used as a foliage and soil-applied insecticide and as a seed dressing. It is also applied to livestock against mites and other ectoparasites. An earlier recommendation for the use of phoxim in stored cereals has been withdrawn by the manufacturer.

Phoxim is diethoxy-thiophosphoryloxyimino-phenylacetonitrile. The corresponding dimethoxy compound known as phoxim-methyl was developed under the code name Bayer SRA-7660. It too has been evaluated against stored-product pests though the literature does not appear to be extensive. Whilst it is recognised that the 2 products are not necessarily similar, there is justification for considering them together since neither appears to be in commercial use for grain protection and the amount of information available on each is limited.

This organophosphorus insecticide has an effect against a broad spectrum of stored-product pests, though it appears to exhibit a degree of cross-resistance to those species which have developed resistance to malathion. Though phoxim was widely tested some years ago, it does not appear to have been registered or adopted as a grain protectant.

23.1 Usefulness

Phoxim was first evaluated in the mid-1960s. **MacDonald and Gillenwater (1967)** found it to be more effective than malathion as a direct-contact toxicant against adult *Tribolium confusum*, *Lasioderma serricorne*, and larvae of *Plodia interpunctella* and *Attagenus unicolor*.

Strong and Sbur (1968), after evaluating the effectiveness of 48 insecticides against 5 major stored-product pests, reported phoxim to be among the most potent. In a further study (Strong 1969) in which the relative susceptibility of 5 stored-product moths to 12 organophosphorus insecticides was determined, phoxim proved to be second only to dichlorvos in potency. It was several times more effective than fenitrothion and malathion and several hundred times more effective than temephos. The author considered it sufficiently promising against stored-product insects to continue research on its use for protective treatment of commodities, space application, surface treatments and structural treatments in storage facilties.

Strong (1969) evaluated 12 organophosphorous insecticides, including phoxim, against 5 stored-product moths and determined their relative susceptibility. In decreasing order of susceptibility the reaction of these insects were as follows: *Cadra cautella*, equal to *Anagasta kuehniella*, greater than *Plodia interpunctella*, greater than *Ephestia elutella*, greater than *Cadra figerulella*. Using a basis of 1 for the most toxic compound, the order and rank in relative effectiveness of insecticides indicated by ratios calculated from the LC/95 values obtained for the least susceptible of the 5 species of moths to each of the insecticides tested were: dichlorvos 1.0, phoxim 2.4, malathion 9.2 and fenitrothion 13.9.

Speirs and Zettler (1969) tested malathion-resistant *Tribolium castaneum* for susceptibility to phoxim by topical application. The malathion-resistant beetles showed a 9.5-fold resistance to phoxim. Only about 1.25 as much phoxim as malathion was required for the LD/50 in the resistant beetles.

In an extensive series of studies of the relative susceptibility of *Tribolium confusum* and *T. castaneum* to 12 organophosphorus insecticides, **Strong (1970)** reported phoxim to be slightly less effective than fenthion and fenitrothion to both of these important beetles.

LaHue and Dicke (1971) made a thorough evaluation of phoxim as a protectant for stored grains by testing it on hard winter wheat, shelled corn and sorghum grain against *Sitophilus oryzae, Tribolium castaneum, T. confusum* and *Rhyzopertha dominica*. Evaluations were made at intervals by determining mortalities after ageing of the deposits, by counting the number of progeny developing after toxicity test exposures, and by an assessment of progeny damage to the treated grain. Phoxim applied at a rate of 5 mg/kg was superior to the standard of 10 mg/kg malathion in all tests. Protection against *S. oryzae* damage was excellent throughout the 12 month storage. A lessening of effectiveness against *Tribolium* spp. and *R. dominica* was noted after 12 months storage but little progeny damage was inflicted.

Wilkin and Hope (1973) tested 20 new pesticides against stored-product mites. They found phoxim effective against all 3 species used in the studies and considered that the compound merited further consideration.

Chawla and Bindra (1973), working in India, screened a number of insecticides against *R. dominica, S. oryzae* and larvae of *Trogoderma granarium* in wheat in a small-scale storage trial lasting 1 year. Phoxim provided protection for a longer period than malathion and was considered comparable with pirimiphos methyl.

Rai and Croal (1973) tested 9 insecticides for their duration of effectiveness on jute sacking. Phoxim was more effective than the others. Phoxim was quite toxic to *Rhyzopertha dominica* and *Tribolium castaneum.*

Golob and Ashman (1974), in the UK, carried out studies on the effect of oil content and insecticides on insects attacking rice bran. Phoxim was the most effective of 4 compounds tested against *Tribolium castaneum* and *Latheticus oryzae*, followed closely by pirimiphos-methyl which did not prevent reproduction of L. *oryzae.*

Carmi (1975), working in Israel, compared the toxicity of phoxim with BHC, malathion and pirimiphos methyl against *Ephestia cautella*. Each insecticide was mixed with wheat at dosages of 2, 4 and 8 mg/kg. Two sizes of larvae were used, up to 3 mm and over 5 mm. The insects were exposed to grain for 24 hours after which mortality counts were taken. The insecticides were ranked in decreasing order of toxicity: phoxim, pirimiphos-methyl, malathion and BHC. The toxicity of pirimiphos-methyl was almost equal to that of phoxim at the 8 mg/kg rate but was less at the lower levels.

Ashman (1975) carried out a trial to compare the effectiveness of 4 insecticides applied to oil-extracted rice bran by admixture against *Tribolium castaneum* and *Latheticus oryzae*. The trial extended over 4 months and included determination of insecticide persistence. Phoxim and pirimiphos-methyl were the most successful insecticides causing high adult mortality and preventing breeding.

Zakladmoi and Bokarev (1976) tested 3 organophosphorus insecticides for protection of stored grain. Phoxim applied at the rate of 5–10 mg/kg was the most effective and most persistent. Only phoxim controlled *Rhyzopertha dominica*. Increase in moisture content (12–17%) decreased the effectiveness.

Tyler and Binns (1977) determined the toxicity of 7 organophosphorus insecticides and lindane to 18 species of stored-product beetles. The results indicated the following overall order of effectiveness: chlorpyriphos-methyl superior to fenitrothion, superior to pirimiphos-methyl, superior to phoxim. Malathion was the least effective of the 7.

Al-Naji *et al.* (1977) tested phoxim-methyl (Bay SRA 7660) the dimethyl analogue of phoxim on soft winter wheat, as a protectant against *Sitophilus oryzae, Tribolium castaneum, T. confusum* and *Rhyzopertha dominica*. Results were evaluated at intervals during 12 months by counting dead insects after 21 days exposure to treated wheat. Phoxim-methyl at 5 mg/kg was effective for 9 months against S. oryzae, for 2 months against *Tribolium* spp. and for 6 months against *R. dominica*. Its effectiveness at a dose of 10 mg/kg against *R. dominica* equalled the protection afforded by malathion; however, the recommended dose of 10 mg/kg malathion gave protection from the other 3 species that was superior to that given by phoxim-methyl.

Loschiavo (1978a) studied the effect of phoxim on the survival and reproduction of 3 species of stored-product insects in laboratory and small-bin experiments. In a laboratory experiment, phoxim was applied in hexane solution to wheat at concentrations of 1, 5, and 10 mg/kg. In a small-bin experiment, it was applied as an emulsion at 5 and 10 mg/kg. Adults of *Cryptolestes ferrugineus, Oryzaephilus surinamensis* and *Tribolium castaneum*, were exposed for 7 days to treated wheat 1 or 2 weeks after initial treatment, and at 10 or 11 other time periods thereafter up to 23 weeks. In both experiments, mortality increased and reproductive capacity decreased with increasing concentration. *Oryzaephilus surinamensis* was the most susceptible, and *Tribolium castaneum* the least susceptible species. The compound was more effective in the small-bin experiment than in the laboratory. In both experiments, progeny production in all species was drastically reduced. In the small-bin experiment, one larva of *Cryptolestes ferrugineus* was recorded in wheat treated at 10 mg/kg. No progeny was produced by *Tribolium castaneum* even though most of the adults survived after the seventh week. In the laboratory experiment the few progeny produced by beetles that survived treatment suffered retarded development.

Al-Naji (1978) reported an extensive study to compare the effectiveness of phoxim-methyl and malathion against stored-grain insects on soft red winter wheat. Various concentrations of phoxim-methyl were compared against malathion applied at 10 mg/kg. Phoxim-methyl applied at 5 mg/kg was effective for 270 days against *Sitophilus oryzae*, for 60 days against *Tribolium castaneum* and *Tribolium confusum* and for 180 days against *Rhyzopertha dominica*. Doses of 10 mg/kg of phoxim-methyl or malathion gave equal protection against *Rhyzopertha dominica* and *Sitophilus oryzae* for up to 365 days storage; however, malathion at 10 mg/kg gave better protection from *Tribolium castaneum* and *Tribolium confusum* than that given by phoxim-methyl after the first 90 days of storage. Phoxim-methyl was more effective than malathion in reducing progeny production of *Rhyzopertha dominica* throughout the 365 day study; however, there was no significant difference between phoxim-methyl and malathion at 10 mg/kg in reducing progeny production of *Sitophilus oryzae* and *Tribolium confusum* throughout the study period, while there was a significant difference between phoxim-methyl and malathion at 10 mg/kg in reducing progeny production of *Tribolium castaneum* after 180 days of the study period.

Pandey *et al.* (1979), working in India, compared the contact and residual toxicity of phoxim at 0.03–0.5% and of malathion at 0.5% on concrete slabs and jute bags. In laboratory tests against *Sitophilus oryzae* and *Rhyzopertha dominica* adults and larvae of *Trogoderma granarium* and adults of *Tribolium castaneum*, contact with 0.3% phoxim on concrete slabs resulted in complete mortality of *Sitophilus oryzae* and *Rhyzopertha dominica*, 85% of *Tribolium* and 28% of *Trogoderma* larvae, as compared with under 60% kill of the first 2 species and 6% of the third with malathion. Phoxim was much superior to malathion when applied to jute bags.

Saba and Matthaei (1979), representing the manufacturers of phoxim, reviewed the use of 3 insecticides that had been developed for protecting stored products. Phoxim is recommended for treating empty warehouses, ships, shops, etc.; dichlorvos is recommended for protecting grain, tobacco, skins, etc., for short periods. For direct treatment and prolonged protection of grain and other stored products, fenitrothion is preferred.

Yadav *et al.* (1980) carried out a laboratory study in India to determine the toxicity of 7 organophosphorus insecticides against 7 species of stored-grain pests. Phoxim and etrimfos were found to be the most toxic compounds against most species.

Kosolapova (1980) lists more than 20 insects that are important pests of stored grain in Kazakhstan (USSR). Insecticides have been used against them in various ways, but continuous use of the same product has led to the development of resistance, so that new preparations are essential. In the tests described, phoxim was tried at various concentrations against 21 species of pests in comparison with trichlorphon. In laboratory tests, amounts of 0.1 g/m^2 had no noticable effect, but higher concentrations of phoxim had greater toxicity. Further tests were carried out in practical conditions and phoxim gave good results at 0.3 g/m^2 proving the most effective treatment.

Al-Naji (1980b), working in Iraq, tested phoxim-methyl on shelled corn as a protectant against *Sitophilus oryzae, Tribolium castaneum, Tribolium confusum* and *Rhyzopertha dominica*. Results were evaluated at intervals during 12 months by counting dead insects after 21 days exposure to treated shelled corn, and by counting the number of progeny developing after toxicity test exposure. Phoxim-methyl applied at 10 mg/kg was superior to the standard dosage of 10 mg/kg malathion in all tests. Protection against *Rhyzopertha dominica* and *Sitophilus oryzae* damage was excellent throughout the 12 months storage.

Viljoen *et al.* (1981a) tested contact insecticides for the control of stored-product pests infesting stacks of bagged maize and groundnuts in the open and in a shed in South Africa. The bag stacks were sprayed after fumigation in October and thereafter at intervals of 4 weeks. A subsample was taken to determine the degree of damage caused by insects to the kernels, and the rest of the sample was kept for 2 months in order to note the numbers of insects that developed in them. Phoxim, though somewhat slow acting, was very effective against *Ephestia elutella* and *Plodia interpunctella* though slightly less effective against *Tribolium castaneum*. It was vastly superior to malathion.

Chahal and Ramzan (1982) conducted laboratory studies in India with pyrethroid and organophosphorus insecticides against third instar larvae of *Trogoderma granarium*. Phoxim, at concentrations as low as 0.025% caused 100% mortality of the larvae but death did not occur for some days.

Kumar *et al.* (1982) tested the effectiveness of phoxim when applied at the rate of 0.15 g/m^2 in godowns in various parts of India, malathion being considered as the standard treatment. Although superior to malathion, phoxim did not perform as well as pirimiphos-methyl.

Tyler and Binns (1982) studied the influence of temperature on the susceptibility of 3 species of stored-product pests to 8 organophosphorus insecticides. Adult *Tribolium castaneum, Oryzaephilus surinamensis* and *Sitophilus granarius* were exposed for 24 hours to a range of deposits of 8 organophosphorus insecticides on filter paper at 10, 17.5 and 25°C. Based on the knock down and kill, the effectiveness of all insecticides was greater at 25°C than at 17.4°C and was markedly lower at 10°C. At 10°C the effectiveness of phoxim, though better than 6 other compounds, was from 10 to 80 times less effective than at 25°C. On the other hand chlorpyrifos-methyl was only slightly less effective at 10°C than at 25°C.

Srivastava and Gopal (1984), working in India, determined the toxicity of phoxim and 6 other organophosphorus insecticides to *Sitophilus oryzae, Trogoderma granarium, Tribolium castaneum, Lasioderma serricorne* and *Callosobruchus maculatus*. On the basis of the LC/50s, etrimfos was most toxic to *Sitophilus oryzae, Tribolium castaneum* and *Trogoderma granarium* and *Lasioderma serricorne*, while phoxim was most toxic to *Callosobruchus maculatus*. It was concluded that phoxim, applied at the LC/99 of 25.7 mg/kg and etrimfos applied at the LC/90 of 10.3 mg/kg against *Trogoderma granarium* and *Callosobruchus maculatus*, respectively, the least susceptible species, would be effective against all the other species.

Zakladnoi (1984) stated that the measures used in the USSR for protecting stored grain against infestation by arthropod pests include spraying of grain and storage premises with phoxim.

23.2 Degradation

Phoxim was tested on hard winter wheat, shelled corn and sorghum as a protectant against 4 major insect species. Lots of 100 g of grain in wide mouth glass jars were treated with measured amounts of a diluted phoxim emulsion in quantities sufficient to apply 5, 10 and 20 mg/kg of actual phoxim. Five replicates of each treatment were prepared. The test jars were fitted with ventilated lids and were held in storage for 12 months (storage temperature not given). Samples were analysed 24 hours after treatment and thereafter 1, 3, 6, 9 and 12 months after treatment. The mean residues determined on the 3 grains from each of the 3 application rates and 6 sampling times are given in **Table 23.1.** Apparently relatively little decomposition of the phoxim occurred during 12 months ageing of the treated grain. This is in contrast with the fate of the malathion applied in a similar manner and at the same time **(LaHue and Dicke 1971).**

Golob and Ashman (1974) studied the effect of insecticides on insects attacking rice bran of varying oil content. Phoxim was applied at the rate of 4 and 8 mg/kg.

Table 23.1. Mean residues of phoxim and malathion on 3 grains at given intervals during storage **(LaHue and Dicke 1971)**

Insecticide applied	Interval after treatment					
	24 hr	1 month	3 months	6 months	9 months	12 months
			Wheat			
Phoxim						
20 ppm	16.6	14.1	12.7	13.5	4.2	8.4
10 ppm	7.5	6.0	7.0	5.1	2.4	3.0
5 ppm	4.6	2.8	4.1	2.8	1.6	2.0
Malathion						
10 ppm	8.4	5.4	4.2	3.5	2.5	2.0
			Corn			
Phoxim						
20 ppm	15.3	43.6	11.8	10.9	9.8	7.3
10 ppm	5.1	6.0	6.4	5.5	6.2	4.7
5 ppm	2.8	2.7	2.6	1.4	2.1	1.9
Malathion						
10 ppm	7.9	5.2	3.2	3.0	1.2	0.6
			Sorghum			
Phoxim						
20 ppm	13.2	16.6	16.8	14.7	6.7	16.1
10 ppm	8.1	8.0	7.5	6.1	9.5	9.1
5 ppm	2.8	4.3	4.4	4.6	4.0	5.7
Malathion						
10 ppm	9.6	4.3	2.4	1.7	0.6	0.6

[a] Means based on 2 samples

Unfortunately, they did not assay the dose actually applied, and their first analytical figures are for 8 day samples, wherein they found 4 mg/kg and 5.6 mg/kg of phoxim. After 6 weeks storage at 27°C the residues had declined to 4.8 mg/kg and 2.6 mg/kg.

Residues data were obtained at 5 intervals over a 30-day period from wheat, corn, and sorghum containing 8 levels of moisture, following 10 mg/kg application of phoxim-methyl emulsion spray **(Kadoum and Al-Naji 1978)**. High-moisture content reduced the effectivenss and persistence of phoxim-methyl in stored grain. Degradation differed significantly (0.05) at each moisture level. After 30 days of storage, 10, 3 and 22.9% of the initial residue deposit remained on the 20% moisture sorghum, wheat and corn respectively. The highest residue deposits remained in the 6% grain moisture levels. After 30 days of storage, 71, 63, and 68% of the initial residue deposit remained on sorghum, wheat and corn respectively. The full data are provided in **Tables 23.2, 23.3 and 23.4.** Apparently higher moisture content does not accelerate residue degradation on corn to the same extent as on sorghum and wheat. This would no doubt be due to differences in hygroscopic equilibria between the various grains.

Al-Naji and Kadoum (1979) reported an extensive study in which phoxim-methyl was applied to separate batches of cleaned, soft red winter wheat as a water emulsion at the rate of 10 mg/kg. Samples of the treated wheat were stored for periods up to 365 days in uncovered drums at 26.5°C and 60% relative humidity. At the end of 1, 7, 14, 21, 30, 60, 90, 180, 270 and 365 days after treatment samples were taken for analysis and for milling in an automatic laboratory mill. The wheat samples were milled after the moisture content was stabilised at 14%. The milling samples were composited to produce bran, shorts and flour fractions. The results of analysis of the whole wheat and milling fractions are given in **Table 23.5.**

Milling yield averaged 16% bran, 10% shorts and 72% flour with 2% weight loss during milling. Phoxim-methyl degraded rapidly so that only 11.2% of the original deposit remained on the whole grain after 365 days of storage.

Al-Naji (1978) studied the effectiveness of phoxim-methyl against stored-grain insects and the fate of the insecticide residues on the grain. Malathion and phoxim-methyl residues were analysed in treated wheat of 12.5% moisture content during the 365 days of storage. Phoxim-methyl degraded rapidly during the first 30 days, but thereafter, loss was gradual; 91.8% of phoxim-methyl and 85% of malathion degraded by the end of the study period.

Table 23.2. Bay SRA 7660 residues covered from corn at different moistures, applied at 10 ppm emulsion spray[a] **(Kadoum and Al-Naji 1978)**

% moisture content	Intervals after application (days) 0	1	7	14	21	30
6	10.3a	9.9a	8.9a	8.0a	7.2a	7.0a
8	9.9a	9.8a	8.8a	7.7b	7.0a	6.5b
10	10.0a	9.0b	8.4b	7.0c	6.1b	6.4a
12	10.0a	8.8bc	8.0c	6.2d	6.0bc	5.9c
14	10.0a	8.6cd	8.0c	5.9e	5.8c	5.4d
16	10.0a	8.4d	7.6d	5.4f	4.9d	4.0e
18	9.9a	8.1e	7.3e	4.1g	3.0e	2.6f
20	10.9a	8.1e	6.6f	3.5h	3.3f	2.5f

a Each number is an average of four replicates. Numbers followed by the same letter in the same column are not significantly different at the 5% level using Duncan's Multiple Range Test.

Table 23.3. Bay SRA 7660 residues covered from sorghum at different moistures, applied at 10 ppm emulsion spray[a] **(Kadoum and Al-Naji 1978)**

% moisture content	Intervals after application (days) 0	1	7	14	21	30
6	10.0a	9.7a	8.8a	8.2a	7.9a	7.1a
8	9.9a	9.2b	8.6a	8.0b	7.5b	6.9b
10	9.9a	9.0bc	7.5b	5.8c	5.5c	4.9c
12	10.0a	8.9bc	6.8c	5.7c	5.9d	4.3d
14	10.0a	8.7c	6.3d	4.9d	4.7e	3.6e
16	10.0a	8.3d	6.0de	4.7e	3.7f	3.5e
18	10.0a	8.1d	6.0de	3.9f	3.0g	2.1f
20	10.0a	7.4e	5.9e	3.8f	1.4h	1.0g

a Each number is an average of four replicates. Numbers followed by the same letter in the same column are not significantly different at the 5% level using Duncan's Multiple Range Test.

Table 23.4. Bay SRA 7660 residues covered from wheat at different moistures, applied at 10 ppm emulsion spray[a] **(Kadoum and Al-Naji 1978)**

% moisture content	Intervals after application (days) 0	1	7	14	21	30
6	10.0a	9.9a	8.5a	7.3a	6.9a	6.3a
8	10.0a	9.7ab	8.2b	7.0b	6.1b	6.0b
10	10.0a	9.5b	8.0c	6.2c	5.9c	5.3c
12	10.0a	8.5c	7.5d	5.5d	5.3d	4.0d
14	9.7a	5.9d	4.9e	3.9e	2.9e	2.0e
16	9.8a	5.7d	4.3f	3.5f	2.4f	2.0f
18	10.1a	5.4e	3.5g	2.0g	1.6g	0.6g
20	10.1a	5.2e	2.6h	1.4h	0.7h	0.3h

a Each number is an average of four replicates. Numbers followed by the same letter in the same column are not significantly different at the 5% level using Duncan's Multiple Range Test.

Table 23.5. Average methyl phoxim residues in parts per million on soft red winter wheat milling fractions after 10-ppm application of methyl phoxim emulsion spray on wheat[a] **(Kadoum and Al-Naji 1979)**

Days of storage	Whole wheat	Milling fractions		
		Bran	Shorts	Flour
1	9.86± 0.24[b]	32.78± 0.46	18.41± 0.36	1.71± 0.08
7	8.00± 0.12	29.06± 0.50	17.73± 0.44	2.07± 0.06
14	7.34± 0.08	24.53± 0.40	16.54± 0.32	1.79± 0.05
21	6.88± 0.09	21.04± 0.26	13.66± 0.12	1.22± 0.08
30	5.80± 0.07	18.21± 0.32	11.80± 0.08	0.84± 0.05
60	4.50± 0.08	13.22± 0.18	9.01± 0.36	0.76± 0.04
90	3.48± 0.03	9.52± 0.22	7.30± 0.22	0.70± 0.05
180	2.55± 0.06	7.72± 0.07	4.60± 0.12	0.54± 0.04
270	1.43± 0.02	5.46± 0.08	3.46± 0.10	0.31± 0.05
365	1.12± 0.04	4.26± 0.06	2.83± 0.08	0.19± 0.02
control	0	0	0	0

a Each number is an average of four replicates.
b Standard deviation.

Pandey *et al.* (1979) evaluated the residual toxicity of phoxim applied to concrete slabs and jute bags against 3 species of stored-product pests. The residual toxicity of phoxim on jute bags persisted for up to 5 months against *Rhyzopertha dominica* and *Tribolium confusum* but for less than 3 months against *Trogoderma granarium*, while that of malathion was about 3 months against the first 2 species and none against the third.

Al-Naji (1980a) reported that residues of phoxim-methyl on hard wheat decreased rapidly during the first month of storage when the insecticide was applied as an aqueous emulsion at the rate of 2–10 mg/kg. Residues decreased more slowly during the subsequent 11 months of storage. In testing the effectiveness of phoxim-methyl against 4 species of stored-grain insects in stored corn, **Al-Naji (1980b)** reported that phoxim-methyl gave excellent protection against *Rhyzopertha dominica* and *Sitophilus oryzae* throughout the 12-month storage but there was a lessening of effectiveness against adult *Tribolium castaneum* and *Tribolium confusum* after 3 months storage, although little progeny damage was inflicted.

Sun *et al.* (1984) who studied the effects of degradation in toxicity of insecticides on the survival and reproduction rate of *Sitophilus zeamais*, reported that phoxim degraded relatively quickly, while malathion and chlorpyrifos degraded slowly. This is contrary to the experience of numerous other investigators.

23.3 Fate in Milling, Processing and Cooking

Golob and Ashman (1974), who studied the effect of oil content and insecticides on insects attacking rice bran, found that phoxim at 8 mg/kg was the most successful treatment in preventing breeding of *Tribolium castaneum* and *Latheticus oryzae*. Although these investigators studied the fate of the insecticide deposit by chemical analysis, unfortunately they did not assay the dose actually applied, and their first analytical figures are for 8 day old samples. At the eighth day assay they had levels in identical bran samples of 5.6 and 4 mg/kg. After 6 weeks storage at 27°C these had decreased to 4.8 and 2.6 mg/kg.

Al-Naji (1978) who applied phoxim-methyl to soft red winter wheat showed that greater residues of phoxim were found in the bran and shorts and very small amounts in the flour. Loss of residues during milling were 8% to 10%. There were no changes in water absorption of flour, dough mixing tolerance, and dough development time between treated and untreated samples. Chlorine bleaching at the rate of 1250 mg/kg of soft wheat flour resulted in 52–55.5% loss in phoxim-methyl residues. Bleached flour was used in cake making. Cake samples were baked at 190°C for 28 minutes. Baking at this temperature resulted in a 91.3–100% loss in phoxim-methyl residues. There were no significant differences in cake volume or internal texture, between cake samples made from treated or untreated wheat.

The only reference located to work on the effect of milling and processing on residues is that of **Al Naji and Kadoum (1979)** relating to phoxim-methyl. This has been referred to under Section 23.2. The data provided there in **Table 23.5** shows that the largest amounts of phoxim-methyl residues appeared in the pericarp and were recovered from the bran and the shorts. Very small amounts of phoxim-methyl were carried through the milling process and into flour. This indicated that the penetration of phoxim-methyl from the pericarp to the endosperm is very slow during the 12 months storage.

In a further study **Al-Naji (1980a)** applied phoxim-methyl to Iraqi hard wheat which was held in store for 12 months. After milling, the bran contained 52.1% of the recovered residues, the shorts 23.1% and the flour 8.8%, with 16.3% lost in the milling process. Cooking destroyed 84.2–100% of the residues.

23.4 Metabolism

Information on the absorption and excretion of phoxim in laboratory animals is provided in Section 23.5.

Urine from white mice after oral treatment with ^{32}P-phoxim at the rate of 114 mg/kg and 955 mg/kg was found to contain 5 metabolites. These were identified as diethyl-phosphoric acid, phoxim, phoxim carboxylic acid, O,O-diethyl phosphorothioic acid and either desethyl phoxim or desethyl-P=O-phoxim (**Vinopal and Fukuto 1971**).

Rowlands (1975) stated that **Mason (1973)** was supposed to have found evidence by means of thin-layer chromatography that phoxim applied to stored wheat was converted mainly to the P=O analogue and to the S-ethyl isomer, but Rowlands had seen only a summary of the work and commented that such metabolic products are more likely to arise on leaf or other surfaces where there is exposure to 'weathering' conditions and light. Mason's work was available only as a PhD thesis and has not been examined by this reviewer.

Studies by **Daniel *et al.* (1978b)** with ring-labelled ^{14}C-phoxim in rats showed that phoxim was largely absorbed and was detected in plasma as a dealkylated compound in the P–S and P–O form. The radioactive metabolites in the urine of rats at 0–24 hours after oral administration of 10 mg/kg consisted of about 90% sulphate and glucuronide conjugates that were hydrolysed enzymatically to *a*-hydroxy-imino-phenylacetonitrile. Hippuric acid was detected as a metabolite; it represented about 5% of the radioactivity eliminated through the kidneys.

23.5 Fate in Animals

Daniel *et al.* (1978a) showed that phoxim was readily absorbed from the gastro-intestinal tract of male rats, with average maximal plasma levels equivalent to 0.35 and 2.44 μg phoxim/mL being achieved within 30 minutes after dosing at 1 and 10 mg/kg, respectively. Nil radioactivity was detected at 24 hours in the plasma of rats dosed with 1 mg/kg, whereas a value of 0.04 μg phoxim equivalents/mL was recorded for rats dosed with 10 mg/kg.

The distribution of radioactivity in the organs and tissues of male rats intubated with ^{14}C-phoxim at 10 mg/kg has been investigated. Apart from the large intestine and its contents, the distribution of radioactivity was essentially similar to that for plasma (**Daniel *et al.* 1978a**). A study has also been made of radioactivity in the gastro-intestinal tract of rats up to 7.5 hours after dosing with phoxim at 1 mg/kg. In these tissues, no accumulation of radioactivity in any organ or tissue was found.

A study was made on excretion of radioactivity in the urine and feces of male and female rats given a single oral dose of ^{14}C-phoxim at 10 mg/kg (**Daniel *et al.* 1978a**). Male rats excreted an average of 92.2% of the radioactivity in the urine and 4.9% in the feces in 10 days, while females excreted 86.1% of the dose in the urine and 6.9% in the feces in the same period. Male rats intubated with ^{14}C-phoxim at a dose of 1 mg/kg excreted 82% of the radioactivity in the urine and 7.9% in the feces in 10 days. The results indicate that most of the radioactivity was eliminated in 24 hours and excretion was virtually complete within 2 days. No evidence was obtained for the presence of $^{14}CO_2$ in expired air. An average of 4.1% of the radioactivity was excreted within 0–24 hours in the bile of male rats intubated with ^{14}C-phoxim at 10 mg/kg (**Daniel *et al.* 1978a**).

Following oral administration of 10.5, 114 and 955 mg/kg ^{32}P-phoxim to mice, the ultimate recovery of administered radioactivity in the urine and feces was in the range of 73–84% (**Vinopal and Fukuto 1971**). However, the radioactivity appeared in the urine and feces at a much slower rate than expected. For example, 24 hours after oral treatment of mice at 10.5 and 114 mg/kg with radioactive phoxim, only 43% and 22%, respectively, of the administered radioactivity was excreted in the urine. At 955 mg/kg, only 17% of the administered radioactivity was excreted in the urine after 30 hours. The studies of **Vinopal and Fukuto (1971)** on white mice were negative for the period 0–48 hours after oral administration of 114 mg/kg ^{32}P-phoxim. The amounts of organosoluble materials (which might include the strong anticholinesterase P=O-phoxim) were essentially insignificant. The reason for the apparent uptake of water-soluble radioactivity in the urinary bladder of the mouse is not clear.

23.6 Safety to Livestock

Though it has not been possible to locate specific studies designed to determine the oral toxicity to livestock, phoxim is generally regarded as having a low mammalian toxicity.

The acute oral toxicity to rats is quoted to the 2170 mg/kg. Likewise the acute dermal toxicity to rats is stated to be greater than 1000 mg/kg. Rats exposed in inhalation chambers withstood maximum concentrations of 3630 mg/m^3 for 4 hours without adverse effect. (**Bayer, undated a**).

The chronic oral toxicity to dogs and rats was tested in 2 year feeding trials in which dogs tolerated 15 ppm and rats 375 ppm in their rations without any indication of somatic damage. No indication of carcinogenic effects were produced (**Bayer, undated a**). In a 3-generation trial in rats, as well as in specifically designed teratogenicity and mutagenicity studies it was ascertained that phoxim does not affect fertility and has no embryotoxic or teratogenic action. Phoxim is not neurotoxic (**Bayer, undated b**).

Phoxim is recommended and widely used for the control of ectoparasites, especially mites, in a wide range of livestock including cattle, goats, horses, pigs and sheep. It is applied by dipping and spraying with concentrations in the range 0.02% to 0.1%. Overdosage trials have shown that livestock species were unaffected by concentrations 4 to 28 times the recommended rate. (**Bayer, undated a**).

In order to study the safety of phoxim in sheep dips, swallowing of the dipwash was simulated by administering dipwash at rates equivalent to 50 and 100 mg/kg-bw by stomach tube. None of the sheep thus treated showed any symptoms. Trials under practical conditions showed that when phoxim was used for the treatment of sheep and pigs there was no adverse effect on mating, reproduction, number and viability of offspring or intolerance symptoms of any kind **(Bayer undated a,b)**.

23.7 Toxicological Evaluation

Phoxim was evaluated in 1982 and 1984 by the Joint FAO/WHO Meeting on Pesticide Residues **(FAO/WHO 1983, 1985)**. It was found to have a mild acute toxicity. In a 2-year dog study, plasma and erythrocyte cholinesterase were more sensitive than brain cholinesterase. The metabolic pathway in mammals follows typical steps, such as hydrolysis, desulphuration and conjunction. The metabolites have a slight to moderate acute oral toxicity. No-effect levels were determined with respect to reproduction and teratogenicity. Mutagenicity and carcinogenicity studies were negative. The available data permitted the determination of no-effect levels in 2 species:

Rat: 5 ppm in the diet, equivalent to 0.56 mg/kg bw;
Dog: 2 ppm in the diet, equivalent to 0.05 mg/kg bw.

The acceptable daily intake for man was estimated to be 0–0.001 mg/kg-bw. The meeting expressed a desire to see observations in humans, particularly effects on cholinesterase together with information on the nature and level of impurities in the technical grade active ingredient.

Phoxim-methyl has not yet been proposed for priority in the Codex system so no independent toxicological evaluation is available.

23.8 Maximum Residue Limits

Although phoxim was evaluated in 1984 by the Joint FAO/WHO Meeting on Pesticide Residues, there was no indication that any country had registered this insecticide for direct application to grain. Neither was there any information concerning the nature and level of residues on grain or the effect of processing and cooking on such residues. The representative of the manufacturers advised that there were no plans to carry out any work to develop the use of phoxim as a grain protectant insecticide.

The status of phoxim-methyl is not clear as no proposals have been made to have this compound evaluated by JMPR.

No recommendations have been made for methods suitable for determining phoxim and phoxim-methyl.

24. Piperonyl Butoxide

Piperonyl butoxide is the accepted common name for 5-[2-(2-butoxyethoxy)ethoxy = methyl]-6-propyl-1,3-benzodioxole. Its development resulted from the work begun in 1934 by Dr L.F. Hederberg who initiated a search for a chemical compound which would increase the efficacy of natural pyrethrins. In 1943 his long and intensive efforts produced a compound called piperonyl cyclonene — a mixture of 3-alkyl-5-(2,4-methylenedioxyphenyl)-6-ethoxy-carbonylcyclohex-2-en-1-one) and 3-alkyl-5-(3,4-methylenedioxyphenyl) = cyclohex-2-en-1-one. This, combined with pyrethrins, showed unusual efficacy against a wide variety of insects. Further synthesis, in cooperation with Dr Herman Wachs, produced a related compound — piperonyl butoxide. These substances have a common toxophoric nucleus considered essential for effectiveness in this group of compounds (**Wachs 1947**). The manufacture and use of piperonyl butoxide were the subject of various patents (**Wachs 1949, 1951**).

The useful activity of piperonyl butoxide was first claimed by **Wachs (1947)** who showed that it was a synergist of pyrethrins against house flies. Piperonyl butoxide itself has virtually no activity against house flies but, when it is mixed with pyrethrins, the activity of pyrethrins is substantially increased, particularly with respect to kill and, to a lesser extent, to rate of knockdown.

Hewlett (1960), in a review of the joint action of insecticides, confined the term synergism to the 'limiting case in which one component is inactive or negligibly active when alone but the mixture with the second component is more active than the second component alone. The first component is described as syngergising the second'. General reviews of the synergism of pyrethrins have been given by **Blackith (1953), Metcalf (1955)** and **Hewlett (1960)**.

It is generally accepted that the syngeristic activity of piperonyl butoxide is due to inhibition, in insects, of oxidative processes which detoxify pyrethrins and a number of other insecticides.

The ratio of pyrethrins to piperonyl butoxide employed is usually between 1:5 and 1:10. Within this range, it is possible to achieve a given effect with one quarter or less of the amount of unsynergised pyrethrins; i.e. the factor of synergism is not less than 4. This finding has allowed the manufacture of domestic insecticides and industrial insecticides at an acceptable price for use in the home and other premises to control a range of insects which are a nuisance, cause damage or discomfort or which are vectors of disease. It has further led to the use of very efficient insecticides in the treatment of large areas against disease vectors and in the widespread use of pyrethrum in bulk food stores.

Because of its interesting pharmacological properties, piperonyl butoxide has served as a tool for investigating many biological and biochemical processes, with the result that there is an enormous literature in which piperonyl butoxide features prominently. However, the major commercial applications of piperonyl butoxide have been in conjunction with the use of natural pyrethrum products and synthetic pyrethroids.

Synergism of pyrethroids is considered in reviews by **Metcalf (1955, 1967), Kearns (1956), Dahm (1957), Hewlett (1960), Wilkinson (1968, 1971), Casida (1970), Hayashi (1972)**), and **Yamamoto (1973)**.

The human toxicity of piperonyl butoxide has been reviewed by **Hatfield (1949), Muirhead-Thomson *et al.* (1952), Coppock (1952), Dove (1953), Enigk (1954), Brown (1970, 1971), Bond *et al.* (1973a,b), Moore (1973)**, and **Haley (1978)**.

Of the numerous reviews of piperonyl butoxide that have been located, none deals with issues directly related to grain or other stored products. An attempt will be made to classify the available information into topics that will be useful for those concerned with this subject.

24.1 Usefulness

The use of piperonyl butoxide in grain protection can conveniently be discussed under 7 broad headings:

1. comparison with other synergists
2. stability of pyrethrins
3. synergism of pyrethrins
4. synergism of synthetic pyrethroids
5. synergism of carbamates
6. synergism of organophosphates
7. effect on other insecticides.

24.1.1 Comparison with Other Synergists

Yamamoto (1973) pointed out that the synergistic action of sesame oil was discovered in 1940 and the active components were found to be sesamin and sesamolin, each having the methylene dioxyphenyl (MDP) moiety in the molecule. Recognition of the importance of the MDP moiety for synergism led to the testing, for synergistic potency, of hundreds of MDP compounds that were synthesised or isolated from natural sources.

Glynne-Jones and Chadwick (1960) undertook a comparison of 4 pyrethrin synergists. Using piperonyl butoxide as a standard, a comparison was made of the relative potencies of 3 other pyrethrin synergists; S.421 (octachlorodipropyl ether), sulfoxide (N-octyl sulfoxide of isosafrole), and bucarpolate (ester of piperonylic acid with the mono-*n*-butyl ether of diethylene glycol). Using house flies and a measured drop technique, the average order of potency relative to piperonyl butoxide was: piperonyl butoxide = 1, sulphoxide = 1.1, bucarpolate = 0.7, S.421 = 0.4. As a space spray against house flies, sulphoxide, piperonyl butoxide and bucarpolate had a very similar effect on the rate of knock-down. S.421 had a reduced effect. Following topical application to *Tribolium castaneum*, the average order of potency relative to piperonyl butoxide was: piperonyl butoxide = 1, sulphoxide = 1.8, S.421 = 0.7, bucarpolate = 0.6. With *Sitophilus granarius* and *Sitophilus oryzae*, sulphoxide and piperonyl butoxide were equal in potency, with bucarpolate and S.421 appreciably less potent.

Matsubara (1961) published the results of an extensive evaluation of the synergistic effect of natural and synthetic synergists on barthrin. The tests used *Culex pipiens* and *Musca domestica* and may not be applicable to stored-product pests.

Chadwick (1961) published a comparison of safroxan and piperonyl butoxide as pyrethrin synergists. Safroxan is 4-(3,4-methylene-dioxyphenyl)-5-methyl-*m*-dioxane. Using techniques similar to those of **Glynne-Jones and Chadwick (1960)** the relative potency of pyrethrins plus safroxan, compared with pyrethrins plus piperonyl butoxide rated at 1.0, was 0.6 with house flies, 1.5 with *Tribolium castaneum*, 1 with *Periplaneta brunnea*, 1.2 with *Sitophilus oryzae* and 0.8 with *Sitophilus granarius*.

Later, **Chadwick (1963a)** published a comparison of MGK 264 and piperonyl butoxide as pyrethrin synergists. MGK 264 is N-(2-ethylhexyl)-bicyclo(2,2,2)-5-heptene-2,3-dicarboximide. The relative potency of pyrethrins plus MGK 264, compared with pyrethrins plus piperonyl butoxide rated as 1.0, was 0.3 with house flies, 0.4 with *Sitophilus oryzae* and *Sitophilus granarius*, 0.5–0.7 with *Tribolium castaneum* and less than 1 with *Periplaneta brunnea*.

Baker (1963) described the use of a combination of piperonyl butoxide and MGK264 for synergising pyrethrins in the formulation of fly sprays. This obviously offers advantages especially against resistant strains of house flies but it is not clear whether the combination has any application against stored-product pests.

Matsubara (1963) published the results of a study of the synergistic effect of various synthetic synergists on carbaryl. A carbaryl:synergist ratio of 1:8 was used to evaluate the knock-down and lethal action on larvae of the mosquito, *Culex pipiens*. The degree of synergism of piperonyl butoxide, sulfoxide, MGK–F 5026, safroxan and *n*-propyl isome was 1.61, 1.61, 1.46, 1.31 and 1.30, respectively. The synergistic effect on lethal effectiveness of carbaryl against the same insect was 3.31, 4.28, 1.69, 3.49, and 4.92, respectively. In general, the combination of carbaryl with various synthetic synergists shows higher orders of synergism on lethal effectiveness than the similar combination with pyrethroids.

Brooke (1967) studied the effect of five methylenedioxyphenyl synergists on the stability of pyrethrins and reported that piperonyl butoxide and sulfoxide considerably increased the length of time for which the pyrethrins are biologically active against larvae of *Aedes aegypti* in subdued light. They pointed out that **Donaldson and Stevenson (1960)**, using *Cadra cautella*, compared irradiated pyrethrins with a mixture of irradiated pyrethrins and piperonyl butoxide. No stabilisation was shown at a 1:5 ratio. This was not surprising when Brooke had shown that even in visible light the degree of stabilisation is not particularly marked at a 1:5 ratio. Intense ultraviolet light would rapidly degrade the pyrethrins, perhaps swamping the effect of the synergist. On the other hand, **Desmarchelier *et al.* (1979a)** reported that piperonyl butoxide enhanced both chemical stability and biological efficacy against *Rhyzopertha dominica*.

Brown *et al.* (1967) pointed out that different synergists provide different factors of synergism against the same insect with the same insecticide. Furthermore, the order of effectiveness changes somewhat when the synergists are applied against different species. Their data illustrating these facts are provided in **Table 24.1**

Singh and Rai (1967) showed that various methylenedioxyphenyl compounds significantly increased the respiration rate of *Tribolium castaneum* adults at least 24 hours after treatment. Sesamex appeared to increase the average oxygen consumption slightly more than did piperonyl butoxide.

Saxena (1967b) compared the rate of knock-down of *Tribolium castaneum* following spraying with pyrethrum synergised with piperonyl butoxide, MGK 264 or a mixture of both. MGK 264 is apparently slightly inferior to piperonyl butoxide under these conditions. A mixture of the 2 synergists is superior to either of them individually.

24.1.2 Stability of Pyrethrins

Over a period of 10 years following the development of piperonyl butoxide in 1947, there was a good deal of confusion and controversy over the effect of piperonyl butoxide on the stability of pyrethrins. **Phipers and Wood (1957)** published the results of an investigation into the reported stabilisation of pyrethrins by piperonyl butoxide. They were able to show that the normal commercial grade of pyrethrum extract, when exposed to intensive artificial irradiation with incandescent and ultraviolet light, degraded almost completely in 4 days. A variety of additives, including another synergist, did not significantly improve the stability; some actually reduced

Table 24.1. Topical application of pyrethrins and synergists at 1:5 ratio to *Tribolium castaneum* and *Musca domestica* **(Brown *et al.* 1967)**

Synergist	Factors of synergism *Tribolium*	*Musca*	Rank with: *Tribolium*	*Musca*
Piperonyl butoxide	2.5	5.0	3	2
Sulfoxide	3.8	5.5	2	1
Bucarpolate	1.3	3.5	6=	3
Safroxan	4.0	3.0	1	4=
Piprotal	2.0	3.0	4	4=
MGK 264	1.3	1.2	6–	7
S 421	1.7	2.0	5	6

the survival quite markedly. A film of the same pyrethrum extract containing piperonyl butoxide at 8 times the concentration of the pyrethrins exhibited exceptional stability showing only minimum loss over 5 days of irradiation.

Brooke (1967) provided a comprehensive review of the available knowledge concerning the stability of pyrethrins to air and light as a prelude to the reporting of an extensive investigation of the stability, as measured by bioassays with mosquito larvae, of various mixtures of pyrethrins and synergists. Piperonyl butoxide certainly improved the stability in the dark and to a lesser extent in diffuse light. Its performance was considerably better than 4 other synergists.

Saxena (1967a) showed that the addition of piperonyl butoxide to pyrethrum in equal amounts increased the persistence of films and toxicity to *Tribolium castaneum*.

Head *et al.* (1968) studied the effect of piperonyl butoxide on the stability of films of crude and refined pyrethrum extracts. The stability of thin films of partly dewaxed pyrethrum extract and decolorised pyrethrum extract was examined by chemical and biological techniques. The results confirmed that the pale extract was more stable than the partly dewaxed extract when exposed to sunlight and that the pyrethrins degrade faster than the cinerins. The addition of piperonyl butoxide improved the stability of partly dewaxed extract in sunlight but hardly affected the rate of degradation of total pyrethrins in pale extract. The nature of the degradation products revealed by chromatography shows that the breakdown of pyrethrins proceeds through degradation of the chrysanthemate moiety, the alcoholic portion remaining unchanged.

Friedman and Epstein (1970) showed that thin films of commercial grade piperonyl butoxide were stable during exposure to intense fluorescent light for periods up to 7 days. In addition, only 40% of the piperonyl butoxide in films produced from insecticide aerosols was lost during 7-day exposure. When commercial grade piperonyl butoxide was heated to 100°C for up to 5 days, there was no appreciable degradation. Piperonyl butoxide was shown to be stable under extreme conditions related to domestic usage.

Desmarchelier *et al.* (1979a) showed in laboratory experiments and field trials that piperonyl butoxide enhanced the stability of pyrethrum on grain in storage (Section 24.1.3).

24.1.3 Synergism of Pyrethrins

Valuable reviews of literature on the usefulness of piperonyl butoxide in the synergism of pyrethrins have been prepared by **Blackith (1953), Metcalf (1955), Dahm (1957), Hewlett (1960), Gillenwater and Burden (1973)**, and **Yamamoto (1973)**.

Dove (1947, 1947a) and **Dove and MacAllister (1947)** published some of the first papers alluding to the use of combinations of piperonyl butoxide with pyrethrins for the control of stored-product pests.

Watts and Berlin (1950) reported a preliminary study designed to show, and hopefully determine, the degree of synergism produced by piperonyl butoxide on pyrethrins formulated as a dust to control *Sitophilus oryzae* in wheat. A remarkable degree of synergism (greater than 10 fold) was shown by a mixture of piperonyl butoxide:pyrethrins in the ratio of 5:1 which produced superior results to those shown by 10 times the amount of pyrethrins alone. These studies showed that the synergism is operative over a wide range of ratios of the two materials.

Dove (1951, 1952, 1952a) described the use and performance of commercial pyrethrum/piperonyl butoxide combinations for admixture with grain and other stored products, and for the treatment of grain storages and mills.

Goodwin-Bailey and Holborn (1952) described laboratory and field experiments conducted in the UK with powder formulations of pyrethrum and piperonyl butoxide for the protection of grain from *Sitophilus granarius* and *Oryzaephilus surinamensis*. They were able to demonstrate degrees of synergism of 9 with a 1:2.5 ratio of pyrethrins:piperonyl butoxide and up to 19 with a 1:20 ratio. Treated wheat was exposed to severe conditions of external infestation by granary insects for 16.5 months. The wheat was examined at intervals for insect numbers, grain injury and weight loss. Concentrations of 1.3 mg/kg pyrethrins with 27 mg/kg piperonyl butoxide in wheat gave good protection from *Sitophilus granarius* and *Oryzaephilus surinamensis* for about a year in comparison with untreated wheat. There was evidence of some repellency in the earlier months following treatment.

Wilbur (1952a) reported the results of extensive field trials to evaluate the effectiveness of a commercial dust containing pyrethrum and piperonyl butoxide. The protectant was applied to the surface of the wheat in the truck at the bin site. Mixing took place during the operation of transferring the wheat from the truck to the bin. The protection afforded by treatment was regarded, in almost every instance, as outstanding. Extensive information is provided on the level and nature of the infestation.

In a futher paper, **Wilbur (1952b)** reported a series of laboratory studies with the same pyrethrum/piperonyl butoxide dusts to determine the effect of moisture on the degree of protection. He showed that on high moisture grain the degree of protection was considerably less favourable than that obtained on wheat of low moisture content.

Dove and Schroder (1955) described the use of water-miscible emulsions of piperonyl butoxide and pyrethrins which could be applied directly to grain to control insects. Studies showed that satisfactory results were obtained from the use of an emulsion containing 2% piperonyl butoxide and 0.2% pyrethrins applied at the rate of from 2 to 12 gallons (U.S.) of emulsion per 1000 bushels of grain (200–1200 mL/m^3).

Blum (1955) presented a thesis on a study of pyrethrum and its activators and showed that piperonyl butoxide was able to synergise pyrethrum when applied to insects after the insecticide had been distributed. In an insect study, large doses of synergists were much more effective than small doses and this investigator considered that since the rate of penetration of compounds through the cuticle depended on concentration, larger doses of synergist permit the synergist to reach critical sites while the pyrethrum is still effective.

Kantack and Laudani (1957) published the results of comparative laboratory tests with emulsion and wettable powder formulations of synergised pyrethrum applied to grain for the control of *Plodia interpunctella*. Adults were most susceptible to residues of pyrethrum plus piperonyl butoxide. The second-instar larvae were much more tolerant and it was found that wettable powder deposits produced quicker action and higher mortalities with adults and larvae than did emulsion formulations.

Kamel (1958) reported that a commercial grain protectant dust containing 0.05% pyrethrins, 0.8% piperonyl butoxide and 99.15% inert ingredients was effective in controlling adults of *Tribolium castaneum, Sitophilus oryzae* and *Rhyzopertha dominica*, though the first species was more tolerant than the other two. Exposure of the preparation to the sun for 39 weeks had no effect on its efficacy but its effectiveness was lost during long periods of storage (presumably of the treated grain). Protection from infestation was superior to the kill of insects already infesting grains.

Quinlan and Miller (1958) evaluated synergised pyrethrum for the control of *Plodia interpunctella* in stored shelled corn. Corn stored in circular metal bins holding approximately 80 tonnes was treated with a synergised pyrethrum spray applied to the top surface of the bulk shelled corn at different dosages and frequencies. The ratio of piperonyl butoxide to pyrethrins was 10:1 and the sprays were applied at 3 concentrations, 3 frequencies and with 3 replicates. None of the treatments gave complete control but the populations in all of the treatments were significantly lower than those in the untreated corn.

Saldarriaga (1958) studied the factors influencing the effectiveness of insecticides used in the protection of stored grain in Bogota, comparing malathion, lindane and synergised pyrethrins. Increases in the moisture content of grain from 12 to 24% caused, in every case, diminution of the insecticide effectiveness. Nontheless, the effectiveness of synergised pyrethrins at 18% moisture content was remarkably good.

Lloyd and Hewlett (1958) studied the relative susceptibility of Coleoptera and Lepidoptera infesting stored products to pyrethrum in oil. Adult moths and beetles of 51 species were investigated, together with the larvae of 16 of these. The insects were treated with 1.3% pyrethrins, and 0.3% pyrethrins plus 3% piperonyl butoxide, in a heavy, highly refined mineral oil. Insects were exposed on filter papers or were directly sprayed. Large differences in susceptibility were encountered. The adult moths and bruchids were susceptible, the adult ptinids were rather tolerant, but otherwise susceptibility showed little correlation with systematic classification. On the whole, 1.3% pyrethrins and 0.3% pyrethrins plus 3% piperonyl butoxide were of about equal toxicity.

Dove (1958) described the development work on pyrethrin/piperonyl butoxide formulations which led to the granting of official approval for the use of these materials in the USA for the protection of stored grains from insect damage.

Brooke (1958) studied the effect of pyrethrins and piperonyl butoxide against *Ephestia elutella* and determined the degree of synergism on several ratios of the 2 components applied topically, as space sprays and as films. The results are reproduced in **Table 24.2**. Synergism was, on the whole, low and was almost nil against larvae. A change in the ratio from 1:1 to 1:10 did not lead to an increase in synergism.

Dove (1959a) described how commercial pyrethrins/piperonyl butoxide emulsifiable concentrates had been developed to have the same order of effectiveness as powder protectants.

Walkden and Nelson (1959) published results of tests with various formulations of pyrethrins and several synergists including piperonyl butoxide. The various formulations were applied at dosages ranging from 0.89 to 2.5 mg/kg pyrethrins with synergist in the ratio of 1:10 or 1:5. Formulations containing 1.5 mg/kg pyrethrins or higher were effective in controlling insect infestation.

Phillips (1959) described the use of pyrethrin/piperonyl butoxide sprays applied to the surface of wheat stored in surplus ships and which had been effective in almost completely suppressing both moth and beetle infestations.

Glynne-Jones and Green (1959) compared the toxicities of pyrethrins and synergised pyrethrins to *Sitophilus oryzae* and *Sitophilus granarius*. They found that the response of each species to piperonyl butoxide as a

Table 24.2. Synergism factor of pyrethrins/piperonyl butoxide against *Ephestia elutella* **(Brook 1958)**

Test method	Larvae Topical	Larvae Space spray	Adults Space spray	Adults Deposit	Adults Topical
Ratio P:PB	1:10	1:10	1:1	1:10	1:10
Syn. factor					
at LD/50	1.3	1.3	1.5	1.6	1.2
at LD/80	1.1	1.3	1.3	1.9	1.3

pyrethrin synergist was markedly different; *Sitophilus granarius*, when exposed to a dust with a pyrethrins:synergist ratio of 1:8, evinced a response exactly 3 times greater than that shown by *Sitophilus oryzae*. The results obtained are summarised in **Table 24.3.**

Table 24.3. Degree of synergism of various ratios of pyrethrum:piperonyl butoxide **(Glynne-Jones and Green 1959)**

Ratio P:PB	Degree of synergism *Sitophilus oryzae*	*Sitophilus granarius*
1:1	2.1	5.5
1:8	5.0	15.0
1:20	6.5	18.7

Walkden and Nelson (1959) provided an extensive report of field trials carried out in the USA over the years 1952–1957 on over 14 000 tonnes of wheat and shelled corn held in small silos. The treatments consisted of pyrethrins alone and mixtures of pyrethrins with piperonyl butoxide, sulfoxide and MGK 264. Products containing pyrethrins/piperonyl butoxide protected wheat and shelled corn against insect infestation for long periods but it would appear that the insect pressure was unusually low. When MGK 264 was mixed with pyrethrins in the ratio of 5:1 it exerted a depressing effect on the insecticidal and repellent properties of the pyrethrins. This work was criticised by **Goodwin-Bailey (1959)** on grounds of low insect infestation, variability of residue data and lack of information about methods of application.

Lloyd and Hewlett (1960) studied the susceptibility to pyrethrins of 3 species of moth infesting stored products. Exposure of adult *Ephestia elutella* to residues on filter paper showed that piperonyl butoxide definitely synergised the action of pyrethrins. However, with piperonyl butoxide and pyrethrins at a ratio of 10:1, the LD/50 of pyrethrins alone was only about 1.6 times that of pyrethrins with synergist. Raising the proportion of synergist:pyrethrins above 5:1 appeared not to increase toxicity. In experiments with last-instar larvae of *Ephestia elutella*, dosing by various techniques failed to show significant synergism of pyrethrins by piperonyl butoxide. These results confirm those of **Brooke (1958)** and **Stevenson (1958).**

Lloyd (1961) studied the effect of piperonyl butoxide on the toxicity of pyrethrins to *Lasioderma serricorne* and reported that for spray treatment the synergism factor was 2.49 and for topical application 1.96.

Strong *et al.* (1961) studied the influence of formulation on the effectiveness of malathion, methoxychlor and synergised pyrethrum protective sprays for stored wheat but unfortunately the concentration of synergised pyrethrum (pyrethrins, 1.5 mg/kg plus piperonyl butoxide, 15 mg/kg) applied to wheat as sprays formulated from solutions, emulsions and wettable-powder suspensions was not toxic to *Tribolium confusum* and insufficiently toxic to *Sitophilus oryzae*. The results therefore do not bring out any differences between formulations; neither do they measure the persistence of the treatment.

Watters (1961) evaluated the effectiveness of pyrethrins/piperonyl butoxide by comparison with lindane, malathion and methoxychlor for the control of *Ptinus villiger* in commercial flour storages in western Canada. As measured by the number of eggs laid in flour sacks, synergised pyrethrins gave less protection to flour stocks at 6 weeks than at 4 weeks after application. The use of an oil-based spray rather than a wettable powder may have had a bearing on the result.

Esin *et al.* (1964) reported that a commercial synergised pyrethrum dust and spray was effective against *Sitophilus granarius* but not against *Tribolium castaneum*. On the other hand malathion and lindane were effective against both species.

Joubert (1965a) published the results of tests with synergised pyrethrum in direct application to maize for the control of grain-infesting insects. The commercial preparation used contained 5% pyrethrins and 50% piperonyl butoxide in an emulsifiable base. The emulsion was sprayed onto the maize in the conveyor system in South Africa and a dosage of 1.25 mg/kg pyrethrins was attained. Total protection of the maize against insect infestation was afforded for a full year. During this period the untreated control was fumigated twice and when final examinations were made, fumigation was again required. An official recommendation was made for treatment with synergised pyrethrum in South Africa.

Kockum (1965) carried out a crib storage experiment in Nairobi in which pyrethrum/piperonyl butoxide powder formulations were assessed against lindane powders for the treatment of maize cobs in wire mesh cribs. Even with a very low concentration of the pyrethrum formulation, a remarkable, almost complete protection was achieved against *Sitotroga cerealella* within the centre of the crib. On the outside, however, no strength of synergised pyrethrum gave a satisfactory protection. The investigator was of the opinion that synergised pyrethrum formulations would be extremely effective in clay covered cribs which would allow the walls to get an insecticidal treatment and at the same time protect the pyrethrum deposit from sunlight.

LaHue (1966) reaffirmed that a water-base spray of synergised pyrethrins applied at the rate of 2.75 mg/kg pyrethrins was an effective protectant for shelled corn. The lengthy protection given by this treatment raises the question as to whether the extended period of protection afforded by synergised pyrethrums is due to toxicity or repellent action.

Brown *et al.* (1967), in discussing the role of synergists in the formulation of insecticides, pointed out that the response of different insect species, even of the same genus, to the same insecticide/synergist system varies. **Figure 24.1** illustrates 4 examples of the response of adult insects. At a 1:8 ratio of pyrethrins to piperonyl butoxide, the factor of synergism against *Sitophilus oryzae* is 5 but against the closely related *Sitophilus granarius* it is 15. Little synergistic effect is seen against *Tribolium castaneum* but a much larger one occurs with *Musca domestica*. The figure also shows the increasing factor of synergism as the proportion of synergist is increased. Piperonyl butoxide does not synergise the activity of pyrethrins against certain stored-product moths.

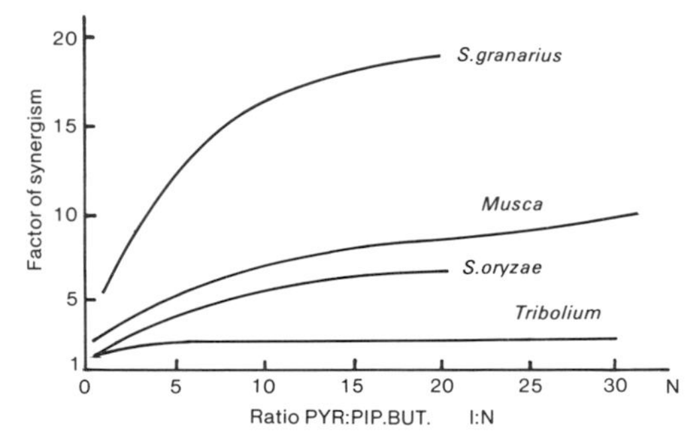

Figure 24.1. Four examples of the response of adult insects to the same insecticide/synergist system **(Brown *et al.* 1981).**

Saxena (1967a), working in India, showed that the addition of piperonyl butoxide to pyrethrum in equal amounts increased the toxicity and persistence of films to *Tribolium castaneum*. **Saxena (1967b)** showed that piperonyl butoxide is superior to MGK 264 in synergising pyrethrum against *Tribolium castaneum*.

Joubert and DeBeer (1968a) reported a field trial carried out in South Africa in a silo consisting of 43 concrete bins containing partially infested yellow maize which was treated with 8 mg/kg premium grade malathion or 1.75 mg/kg of synergised pyrethrum. During the test period 12 species of insects were collected in the silo, *Cadra cautella* being the most numerous, followed closely by *Sitophilus oryzae* and *Tribolium castaneum*. In both treatments the initial infestation was eradicated after several weeks and for 10 months thereafter the treated maize exhibited excellent resistance to reinfestation,. Internal damage to the treated maize remained below 2%, while 2 control masses were totally destroyed.

Joubert and DuToit (1968) reported the outcome of a further study in which mixed yellow dent and flint maize was sprayed with either pyrethrum synergised with piperonyl butoxide, pyrethrum synergised with saffroxan piperonyl butoxide or unsynergised pyrethrum. The insecticides were applied so as to deposit 1.75 mg/kg pyrethrins and the maize was stored in 90 kg jute bags which were stacked for the duration of the experiment in an open store under corrugated iron. The maize was stored for a year and regular sampling and exposure to insects indicated that only the pyrethrins/piperonyl butoxide combination gave results that were entirely satisfactory.

Hewlett (1969) developed a mathematical formula from which it was possible to calculate the relationship between pyrethrin concentrations and piperonyl butoxide concentrations giving fixed levels of mortality. In the case of *Tribolium castaneum* and for infinitely large doses of piperonyl butoxide the mixtures were estimated to be 3 times as toxic as pyrethrins alone.

Weaving (1970a, 1970b) determined the susceptibility of some bruchid beetles of stored pulses to powders containing pyrethrum and piperonyl butoxide. *Acanthoscelides obtectus, Zobrotes subfasciatus* and *Callosobruchus chinensis* were exposed to pyrethrum powders and pyrethrum/piperonyl butoxide powders admixed with wheat. The 3 species had similar susceptibilities to pyrethrins, with LC/50 values of 6.6–8.0 mg/kg but *Callosobruchus chinensis* showed the greatest response to formulations containing the synergist.

Boles (1971) studied the ovipositional response of *Sitophilus oryzae* treated with synergised pyrethrins and was able to show that, notwithstanding the duration of treatment (up to 6 hours), treated females produced up to 50% more progeny than the untreated controls during all oviposition periods. This appears to be the first recorded evidence that production of progeny by *Sitophilus oryzae* can be so affected.

Lloyd (1973) studied the toxicity of pyrethrins, synergised pyrethrins and 5 synthetic pyrethroids to *Tribolium castaneum*, and susceptible and pyrethrin-resistant *Sitophilus granarius*. Piperonyl butoxide synergised all the compounds against the 3 insect strains. Factors of synergism were low against *Tribolium castaneum* (1–4); higher against susceptible *Sitophilus granarius* (4–37); and very high against resistant *Sitophilus granarius* (31–209). Toxicant:synergist ratios higher than 1:10 gave higher factors of synergism with bioresmethrin, but not with pyrethrins.

Wilkin and Hope (1973), who evaluated about 20 pesticides against stored-product mites, found that whereas pyrethrins had only limited activity and piperonyl butoxide no activity against the 3 species of mites, synergised pyrethrins at the same concentration gave 100% mortality of *Acarus siro* and between 50 and 75% mortality of *Tyrophagus putrescentiae* and *Glycyphagus destructor* in 14 days.

Boles (1974) continued the work reported in 1971 on the effect of sub-lethal doses of synergised pyrethrins on the mating efficiency of *Sitophilus oryzae*. Many variables were studied but generally it was found that progeny production was reduced by sub-lethal exposure to the insecticide.

Piperonyl butoxide synergised pyrethrins were recommended for treatment of grain going into storage in New York State **(Matthysse 1974b)**.

Ahmed *et al.* (1976) tested different combinations of selected ingredients for addition to pyrethrins to prolong toxicity against *Tribolium castaneum* adults. The standard for comparison was a pyrethrins/piperonyl butoxide (1:10) formulation. Several of the new formulations were as effective as the standard in maintaining residual toxicity for 8 to 10 weeks.

Desmarchelier *et al.* (1979a) published the results of extensive Australian studies and experience on the stability and efficacy of pyrethrins on grain in storage. In laboratory experiments, piperonyl butoxide enhanced both chemical stability of pyrethrins on grain in storage and their efficiency against *Rhyzopertha dominica*. The persistence of pyrethrins on wheat, barley, oats and milled rice under different grain conditions was determined chemically and the residual efficacy on wheat was also determined biologically. The findings are given in **Tables 24.4 and 24.5**. In commercial applications, synergised pyrethrins plus malathion gave effective control of all species of insects present on wheat and barley in New South Wales.

Bengston and Desmarchelier (1979) in describing the biological efficacy of new grain protectants studied in Australia pointed out that pyrethrins at 3 mg/kg plus piperonyl butoxide at 30 mg/kg was effective against *Rhyzopertha dominica* and protected grain during 9 months storage under Australian conditions. **Greening (1980)** confirmed that 3 mg/kg pyrethrins plus 27 mg/kg piperonyl butoxide controlled malathion-resistant *Rhyzopertha dominica* and *Tribolium castaneum* which infest grain stored on farms in Australia.

Weaving (1980), in small-scale storage trials in Zimbabwe, showed that 2.5 mg/kg pyrethrins synergised 5 times with piperonyl butoxide controlled a range of insects infesting maize in tribal storage.

Bengston *et al.* (1983a) found that when 1.5 mg/kg pyrethrins and 12 mg/kg piperonyl butoxide were combined with 10 mg/kg chlorpyrifos-methyl to control the complex of insects infesting sorghum in Queensland the combination was only partly effective against malathion-resistant *Rhyzopertha dominica*.

Greening (1983) reported that farm-stored grain in New South Wales treated with 1.5 mg/kg pyrethrins and 30 mg/kg piperonyl butoxide remained free of infestation for about 2 months, with the bulk of the grain below the surface remaining uninfested for about 10 months. A low level infestation by *Tribolium castaneum*, however, was found to have started on the surface. The grain remained free of *Sitophilus oryzae, Rhyzopertha dominica* and *Cryptolestes pusillus* which infested untreated grain. In another trial, wheat treated with 2 mg/kg synergised pyrethrins remained free of infestation whereas 12 mg/kg fenitrothion-treated barley became infested with *Rhyzopertha dominica*. In a subsequent trial, however, 2 mg/kg and 1.5 mg/kg synergised pyrethrins failed to give as good a protection to grain as had been achieved with 3 mg/kg synergised pyrethrins in 1976–77.

Desmarchelier (1983) showed that temperature and time of exposure had a pronounced effect on the mortality of adult *Tribolium castaneum* after exposure to natural pyrethrins synergised with piperonyl butoxide (1:8) as illustrated in **Figure 24.2**. In his experiments, 2 groups of insects, each of 50 adults on 140 g of wheat at 55% relative humidity, were removed after different exposure periods and the number of survivors were counted 14 days later. At 35°C, mortality reached a plateau after a few days exposure, whereas at 20°C mortality continued to increase with period of exposure. Thus, mortality after a 30 day exposure to 0.6 mg/kg pyrethrins at 20°C approximately equalled that obtained after a 30 day exposure to 2.5 mg/kg at 35°C.

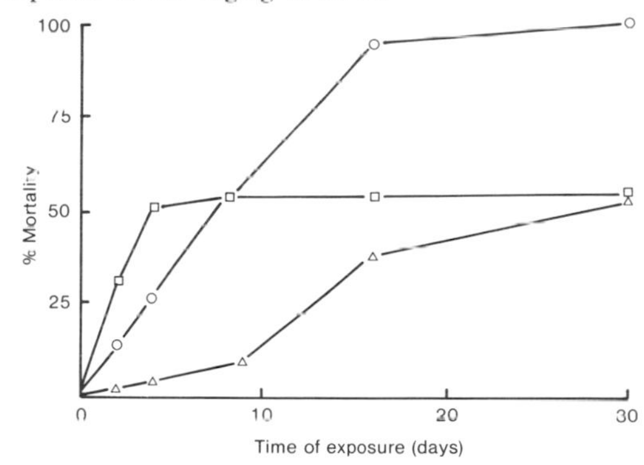

Figure 24.2. Mortality of *Tribolium castaneum* after exposure to synergised pyrethrins: 2.5 mg/kg pyrethrins at 35°C (□); 1.2 mg/kg pyrethrins at 20°C (○); 0.6 mg/kg pyrethrins at 20°C (△) **(Desmarchelier 1983).**

24.1.4 Synergism of Synthetic Pyrethroids

Matsubara (1961) found that piperonyl butoxide synergised the action of barthrin against *Culex pipiens* larvae and adult *Musca domestica* but the degree of synergism was considerably less than that produced with pyrethrins or with allethrin.

Berndt (1963) showed that whereas neither allethrin nor piperonyl butoxide when used alone were significant-

Table 24.4. Period of protection against *R. dominica*, CRD2, given by pyrethrins plus piperonyl butoxide on wheat of 11.5% moisture content and stored at 30°C **(Desmarchelier *et al.* 1979a)**

Nominal deposit (g/t)		Time (t, in weeks) between treatment and assay, % corrected 14-day mortality (m) and % reduction in progeny (r)												Period (weeks) of	
pyrethrins	piperonyl butoxide	t	m	r	t	m	r	t	m	r	t	m	r	complete suppression**	decline of numbers***
2	0	2	44	99.4	6	86	99.4	10	44	95*	14	17	94*	0	6–10
		20	8	74*	26	9	64*	32	3	62*	39	8	58*		
2	5	2	95	100	6	100	100	10	87	99.9	14	89	99.9	6–10	20–26
		20	77	99.2	26	71	92*	32	18	91*	39	13	84*		
2	13	2	99	100	6	100	100	10	97	100	14	100	100	14–20	26–32
		20	95	100	26	87	99.2	32	56	95*	39	44	95*		
2	31	2	99	100	6	100	100	10	99	100	14	100	100	20–26	>39
		20	100	100	26	94	99.8	32	67	98.1	39	93	99.4		
4	0	2	100	100	6	100	100	10	87	100	14	98	100	6–10	26–32
		20	38	99.4	26	28	98.7	32	36	97*	39	34	98.8		
4	10	2	100	100	6	100	100	10	100	100	14	100	100	20–26	>39
		20	100	100	26	94	100	32	84	100	39	98	100		
4	26,62****													>39	>39

* Number of progeny greater than initial number of adults
** 100% mortality plus zero progeny
*** Number of progeny less than initial number of adults
**** m and r were always 100 at weeks 2–39.

Table 24.5. Loss of pyrethrins I on grains treated with pyrethrins plus piperonyl butoxide, 1 to 8 w/w **(Desmarchelier *et al.* 1979a)**

Type of grain	Holding conditions: Temperature (°C)	E.r.h.** (%)	Residue (g/t) at time (weeks) after application*: 3	7	11	16	22	30	Mean (3–30)
wheat	20	47	3.8	2.8	3.7	2.9	3.1	2.6	3.15
wheat	25	47	3.8	3.4	2.7	2.7	3.0	2.5	3.01
wheat	30	47	3.3	2.8	3.2	3.0	2.3	2.3	2.82
wheat	35	47	3.0	2.7	2.9	2.4	2.2	2.0	2.53
barley	30	33	3.1	3.3	3.0	3.1	3.1	2.9	3.08
wheat	30	36	3.4	3.2	2.8	2.6	2.6	2.6	2.85
oats	30	52	–	2.9	2.9	2.1	2.3	2.2	2.48
wheat	30	52	3.0	2.7	3.1	2.7	2.6	2.3	2.73
wheat	30	62	3.0	2.9	2.5	2.4	2.2	1.9	2.48
rice	30	75	2.1	3.0	2.6	2.5	2.5	–	2.54
mean (all grain)	29	50	3.12	3.01	2.95	2.64	2.58	2.41	

* The nominal deposit was 3.5 g/t.
** Grain equilibrium relative humidity.

ly repellent to stored-product pests on wheat. When the 2 were formulated together and tested in the same proportions as when they were tested alone, there was a marked increase in repellency.

Lloyd and Field (1969) found that bioresmethrin was 1.4 times as toxic as resmethrin and 3.6 times as toxic as pyrethrins against *Sitophilus granarius*. When synergised with piperonyl butoxide using an insecticide:synergist ratio of 1:10, a factor of synergism of 5.8 was obtained and this was increased to 13.4 when the ratio was increased to 1:100. Against *Tribolium castaneum*, bioresmethrin was found to be 3.4 times as toxic as resmethrin, and 1.9 times as toxic as pyrethrins. With this species a factor of synergism of 1.5 was obtained with piperonyl butoxide used in the ratio of 1:10. This is about half the factor obtained with resmethrin.

Davies (1972) gave an account of the development of bioallethrin and bioresmethrin wherein he stated that laboratory tests and field trials had demonstrated that bioallethrin and bioresmethrin were synergised by piperonyl butoxide so that a combination of all 3 compounds can be used in fly sprays, aerosols and thermal foggers in place of synergised pyrethrins.

Hayashi (1972) reviewed the work resulting from the search for effective synergists for allethrin and several new synthetic pyrethroids such as tetramethrin, resmethrin, prothrin and proparthrin. Piperonyl butoxide was one of the 4 most effective synergists.

Lloyd (1973) determined the toxicity of pyrethrins and 5 synthetic pyrethroids to *Tribolium castaneum*, and susceptible and pyrethrin–resistant *Sitophilus granarius*. Piperonyl butoxide synergised all the compounds against all 3 strains. Except for tetramethrin, factors of synergism were low against *Tribolium castaneum* (1–4); higher against susceptible *Sitophilus granarius* (4–37); and very high against resistant *Sitophilus granarius* (31–209). Very high factors of synergism were obtained with tetramethrin (43–216) against all 3 strains. Against the weevils, toxicant:synergist ratios higher than 1:10 gave higher factors of synergism with resmethrin and bioresmethrin; but not with pyrethrins.

Wilkin (1975a) showed that bioresmethrin synergised with piperonyl butoxide will control lindane–resistant strains of mites in the fabric of buildings and in stored grain.

Coulon and Barres (1978), working in France, carried out a comprehensive study of the effects of deltamethrin on *Sitophilus granarius*, taking bioresmethrin as the reference product. Based on the concentration of the insecticide in wheat, the LC/50 and LC/90 of deltamethrin were found to be respectively 0.23 and 0.44 mg/kg, while the LC/50 and LC/90 of bioresmethrin was 5.2 and 10.3 mg/kg. Piperonyl butoxide was found to synergise both pyrethroids, but in the case of bioresmethrin the synergistic effect was more pronounced.

Ardley and Desmarchelier (1975) investigated the use of resmethrin and bioresmethrin as potential grain protectants in silo-scale trials in Australia. They found that piperonyl butoxide synergised both compounds quite considerably, allowing synergised bioresmethrin to be recommended as the most cost-effective treatment for long-term storage in Australia.

Ardley (1976) published the results of grain protectant trials comparing 4 mg/kg bioresmethrin plus 20 mg/kg piperonyl butoxide with 12 mg/kg malathion for the long-term protection of Australian grain against the five major species of stored-product pests. The only problem encountered was that *Tribolium* species were somewhat less sensitive to synergised bioresmethrin than the other species.

Desmarchelier (1977b) presented a stong case for the use of selective treatments, including combinations of synergised pyrethroid and organophosphorus insecticides for control of stored-products Coleoptera. Bioresmethrin used in conjunction with piperonyl butoxide was considered an ideal means of supplementing the effectiveness of selected organophosphorus insecticides.

Ardley and Desmarchelier (1978) published the results of field trials with combinations of synergised bioresmethrin and fenitrothion as potential grain protectants. They showed that the potency of such combinations against all species and the residual stability of low concentrations of each of the components made them ideally suited for the control of any and all insect pest complexes.

Ardley and Halls (1979) presented the results of a study carried out jointly in Australia and the United Kingdom to compare *d*-phenothrin, when used alone, when synergised with piperonyl butoxide and when combined with fenitrothion, against similar applications of bioresmethrin. Similar efficiency was demonstrated for synergised *d*-phenothrin when used at approximately twice the rate of synergised bioresmethrin. This corresponds to 2 mg/kg phenothrin used in conjunction with 6–10 mg/kg fenitrothion.

Bengston and Desmarchelier (1979) provided further information on biological efficacy of new grain protectants in which they stressed the value of piperonyl butoxide for use in conjunction with bioresmethrin, fenvalerate, permethrin, and *d*-phenothrin.

Bengston (1979), in discussing the potential of pyrethroids as grain protectants, pointed out that piperonyl butoxide synergises all the synthetic pyrethroids so far tested but that the exact level of synergism varies with each strain of insect and with each insecticide. It is necessary to make appropriate measurements before deciding on the combination. Piperonyl butoxide is important because it considerably reduces the cost of using pyrethroid insecticides.

Desmarchelier (1980), who studied the loss of bioresmethrin, carbaryl and *d*-phenothrin on wheat during storage, was able to show that piperonyl butoxide greatly enhanced the stability of bioresmethrin on wheat. However, the effect on carbaryl was small to negligible. The conclusion drawn from the studies was that *d*-phenothrin was relatively stable in the absence of piperonyl butoxide, and that the effect of piperonyl butoxide had not been quantitatively substantiated.

Bengston *et al.* (1983a) conducted duplicate field experiments on bulk sorghum stored in Queensland and treated with various insecticide combinations including fenitrothion 12 mg/kg plus phenothrin 1 mg/kg. This combination controlled the malathion-resistant strains of *Sitophilus oryzae, Tribolium castaneum* and *Ephestia elutella*, but was only partly effective against malathion-resistant *Rhyzopertha dominica*. In view of other experience by these workers the absence of piperonyl butoxide no doubt contributed to failure to control *Rhyzopertha*.

Bengston *et al.* (1983b) reported the results of duplicate field trials carried out on bulk wheat in commercial silos in Queensland and New South Wales and which were designed to evaluate combinations of organophosphorothioates and synergised synthetic pyrethroids. These trials established that all treatments were generally effective in controlling common field strains of *Sitophilus oryzae* and *Rhyzopertha dominica*. All treatment combinations completely prevented progeny production in *Tribolium castaneum, Tribolium confusum* and *Ephestia cautella*. Pyrethroids included deltamethrin, fenvalerate, *d*-phenothrin and permethrin.

Ishaaya *et al.* (1983), working in Israel, studied the toxicity and synergism of synthetic pyrethroids presented to *Tribolium castaneum* larvae for dietary consumption. Piperonyl butoxide had little or no effect on the toxicity of *trans*-permethrin, but synergised the toxicity of *cis*-cypermethrin about threefold. Piperonyl butoxide also synergised the toxicity of *cis*-permethrin, *trans*-cypermethrin and deltamethrin, but not that of fenvalerate. Oxidases appear to be more important than esterases in pyrethroid detoxification by *Tribolium castaneum* larvae.

Bitran *et al.* (1983a), working in Brazil, showed that deltamethrin synergised with piperonyl butoxide (1:10) applied directly to maize grain was superior to malathion in controlling *Sitophilus zeamais* for 9 months.

Bitran *et al.* (1983b) showed that pyrethroids plus piperonyl butoxide were highly efficient in controlling *Sitophilus zeamais* but this effectiveness was greatly reduced without piperonyl butoxide.

Bodmaryk *et al.* (1984) studied the interaction between synergists and permethrin on adults of *Tribolium castaneum*. No synergism could be demonstrated for the combination of permethrin/piperonyl butoxide (1:2).

Bengston *et al.* (1984) described a field experiment carried out on bulk sorghum stored for 26 weeks in concrete silos in southern Queensland. Vairious combinations of organophosphorus and synergised synthetic pyrethroid insecticides were generally effective but pirimiphos-methyl (6 mg/kg) plus permethrin (1 mg/kg) plus piperonyl butoxide (8 mg/kg) allowed some survival of adults and progeny production by *Sitophilus oryzae* after 12 weeks.

24.1.5 Synergism of Carbamates

Synergism of carbamates has not yet been used on any scale in commercial formulation, although extensive laboratory studies have been made. Many carbamates are strongly synergised by inhibitors of oxidative processes such as piperonyl butoxide and safroxan. However, it would seem that a strongly insecticidal carbamate shows a smaller respose to the synergist, whilst less effective or more easily detoxified carbamates show greater response.

Georghiou and Metcalf (1962) discussed a wide range of carbamate insecticides including their comparative insect toxicity. They stated that all the carbamates tested were markedly synergised by piperonyl butoxide, indicating the importance of detoxication in determining the degree of resistance.

Fukuto *et al.* (1962) studied the synergism of substituted phenyl N-methyl carbamate insecticides by piperonyl butoxide. The toxicity of 40 substituted phenyl N-methyl carbamates to the house fly was re-evaluated with piperonyl butoxide as a synergist. There was no correlation between housefly toxicity and fly brain anticholinesterase activity when the carbamate was used alone. In general, the use of piperonyl butoxide with the carbamate increased the toxicity to levels which parallel cholinesterase activity. An approximate linear relationship between the log of toxicity and cholinesterase inhibition was obtained. The data support previous suggestions that differences in the toxicity of carbamates used alone on different insects is due to different detoxication rates in the insects.

Matsubara (1963) studied the action of synthetic synergists on carbaryl in its knock-down and lethal effect on larvae of *Culex pipiens*. All synergists tested were synergistic on knock-down of the insect and the degree of synergism of piperonyl butoxide was 1.61. All synergists tested enhanced the lethal effectiveness of carbaryl and the factor of synergism of piperonyl butoxide was 3.31.

Metcalf *et al.* (1966) showed that the insecticidal carbamates are synergised by a wide variety of methylene dioxyphenyl compounds. These act as inhibitors of phenolase enzymes which detoxify the carbamates largely by ring hydroxylation. The active inhibitors appear to require a 3-point attachment to the phenolase enzyme to orient the methylene carbon atom so that interaction with a nucleophilic group at the enzyme-active site takes place. The susceptibility of individual carbamates to enzymic detoxication is greatly influenced by the nature of the aryl ring. The methylene dioxynaphthalenes and piperonyl carbamates are exceptionally active carbamate synergists.

Brattsten and Metcalf (1970) studied the synergistic ratio of carbaryl and piperonyl butoxide as an indicator of the distribution of multifunctional oxidase in the Insecta. The synergistic ratio of carbaryl with piperonyl butoxide was evaluated with 54 species of insects from 8 orders and 37 families. The topical LD/50 for carbaryl ranged from 0.3 to 4000 μg/g and the synergistic ratio from 1.5 to 400. The wide ranges of susceptibility to carbaryl and of synergistic ratio are interpreted as responses to differences in the activities of multifunction oxidase detoxication which is inhibited by piperonyl butoxide. These factors are highly variable in the Insecta. This work was extended so that **Brattsten and Metcalf (1973a)** presented reports of experiments on an additional 20 insects. Similarly wide variations in the results were found.

Desmarchelier (1977c and 1980), who studied the loss of carbaryl on grains in storages, included piperonyl butoxide in his trial design with no thought to Australian conditions or intended usage, but because of the possible use of carbaryl plus piperonyl butoxide under tropical conditions, since it is known that piperonyl butoxide greatly enhances the biological efficacy of carbaryl. The effect of piperonyl butoxide on the loss of carbaryl ranged from zero to slight and was considered of little practical significance.

24.1.6 Effect on Organophosphorus Insecticides

Eddy *et al.* (1954) observed that several pyrethrum synergists, including piperonyl butoxide, increased initial kill and residual effectiveness of certain phosphorus compounds against the body louse.

Hoffman *et al.* (1954) evaluated 7 phosphorus compounds with 19 synergists, including piperonyl butoxide, against DDT-resistant house flies. The synergists were effective with all of the phosphorus compounds except malathion. **Rai *et al.* (1956)** found that piperonyl butoxide was antagonistic to malathion against susceptible and DDT-resistant house flies. **Craig (1956)**, using topical applications, found that piperonyl butoxide was synergistic in combination with malathion to the German cockroach, *Blattella germanica*, and antagonistic to the Maderia roach, *Leucopheae maderae*.

Ware and Roan (1957) studied the interaction of piperonyl butoxide with malathion and 5 analogues applied topically to male house flies. They found that whereas malathion was antagonised, some malathion analogues were synergised and some were antagonised by piperonyl butoxide used in the ratio of 1:10. The reasons for this were not clear.

Rai and Roan (1959) studied the nature of the antagonistic effect of piperonyl butoxide to the toxic action of malathion on house flies. Malathion was applied topically to various sites on the house fly body with varying degrees of toxicity. The same phenomenon occurred for the relative toxicity of malathion in combination with piperonyl butoxide. The antagonistic effects of piperonyl butoxide were demonstrated even when it was applied on a separate area of the integument, separated from the malathion treatment site, in nearly simultaneous topical applications to individual flies. The antagonistic effect of piperonyl butoxide decreased slowly over a 30 hour period, when it was applied first, followed at varying time intervals by the toxicant. In reverse order of treatment such antagonistic effects decreased rapidly over a range of 30 minutes between the applications of malathion and piperonyl butoxide. The application of piperonyl butoxide 1 to 2 hours before malathion markedly increased the magnitude of the antagonism observed in simultaneous applications.

Sun and Johnson (1960) showed that several methylene dioxyphenyl compounds, including piperonyl butoxide, commonly used with pyrethroids, antagonised the action of phosphorothionate insecticides in house flies, presumably by preventing oxidative activation.

Sudershan and Naidu (1961) carried out experiments by an isolated cockroach heart technique to study the antagonistic action of piperonyl butoxide on malathion. They found malathion induced an immediate increase in heartbeat rate followed by a steady decline. Piperonyl butoxide worked similarly. When the two were used simultaneously, the initial stimulatory action of malathion was reduced and was followed by a rapid decline, indicating an antagonistic action of piperonyl butoxide towards malathion. If malathion was added after the piperonyl butoxide, the antagonistic action was enhanced. It appeared that piperonyl butoxide interfered with malathion in the formation of its isomer, malaoxon, which is a potent cholinesterase inhibiter and a more active insecticide. Since the laboratory evidence indicates that piperonyl butoxide does not interfere with malaoxon, but inhibits the oxidation of malathion, it is suggested that pure malathion may have little action on the heartbeat frequency.

Rowlands (1966c) studied the activation and detoxication of 3 organic phosphorothionate insecticides applied to stored wheat and showed that piperonyl butoxide SKF 525–A, and butylated hydroxytoluene prevented oxidases present in the seed coats of wheat from oxidising malathion and bromophos to their thiolate and phosphate analogues, respectively. In the case of piperonyl butoxide and SKF 525–A, the enzymes slowly recovered presumably as these compounds were degraded. Thus, with the degradation of insecticides in stored grain being hydrolytic or oxidative or both, there is reason to believe that synergists of the methylenedioxyphenyl type will increase their persistence **(Rowlands, 1967a)**.

Dyte and Rowlands (1970) studied the effects of some insecticide synergists on the potency and metabolism of bromophos and fenitrothion on *Tribolium castaneum*. They found that sesamex did not affect the potency of fenitrothion to *Tribolium castaneum* but it antagonised bromophos and more effectively antagonised malathion. SKF 525–A antagonised all 3 insecticides, and was a more effective antagonist than sesamex for bromophos and malathion.

It was shown that sesamex or SKF 525–A strongly inhibited the production of the O-desmethyl derivative of both compounds, and also inhibited the formation of the oxygen analogue from fenitrothion. The effects of the synergists on the potency of fenitrothion is probably due to the inhibition of both toxication and detoxication processes. The stronger antagonism of malathion is attributed to the known importance of the O-desmethylation as a detoxication pathway for this insecticide. The ability of the synergists to inhibit O-desmethylation of some organophosphorus compounds may help to explain the very different effects of these synergists on the potency of different organophosphorus insecticides.

Attia *et al.* (1980) showed that larvae of the last instar of *Plodia interpunctella* from susceptible and multiple organophosphorus-resistant strains, pretreated with piperonyl butoxide, were more susceptible to the synergised insecticides, except malathion. The toxicity of fenitro-oxon was found to have been increased. This result indicates that mixed function oxidases play a significant role in resistance to dichlorvos, pirimiphos-methyl, diazinon and fenitrothion.

24.1.7 Effect on other Insecticides

Piperonyl butoxide blocked the conversion of aldrin to dieldrin in treated soils and increased the persistence of the residues by factors of 1.5 to 1.8 within 4 months **(Lichtenstein *et al.* 1963)**. It also significantly inhibited soybean root phorate sulphoxidation **(Krueger 1975)**. Piperonyl butoxide inhibited the metabolism of cyclodiene insecticides and increased their toxicity **(Rueckert and Ballschmitter 1973)**.

Udeaan and Kalra (1983) studied the effect of piperonyl butoxide on the toxicity and metabolism of lindane in susceptible and resistant strains of *Tribolium castaneum*. Piperonyl butoxide enhanced the toxicity of lindane against both strains. The synergistic ratio was 5.0 and more than 40.5 against the susceptible and resistant strains, respectively. A significant proportion of lindane resistance in *Tribolium castaneum* was shown to be overcome by the use of the synergist. The resistant strain was found to degrade more lindane than the susceptible strain. The amount of lindane recovered from the resistant strain corresponded to about 2% and 26% when applied without and with piperonyl butoxide, respectively. In the susceptible strain the recovery was 16% and 35%, respectively.

Quistad *et al.* (1975d) were able to show that piperonyl butoxide increased the morphogenetic effect of methoprene on larval mosquitoes and house flies.

Solomon and Metcalf (1974) studied the effect of piperonyl butoxide and triorthocresyl phosphate on the activity and metabolism of methoprene in *Tenebrio molitor* and *Oncopeltus fasciatus*. They found that the juvenilising activity of methoprene in *Tenebrio molitor* was increased by these synergists but the activity in *Oncopeltus fasciatus* was considerably decreased indicating the blocking of an activation reaction. This was confirmed by the fact that 4 methoprene metabolites were considerably more active than the parent compound against *O. fasciatus*.

Rowlands and Dyte (1979) found that piperonyl butoxide and a juvenile hormone mimic were metabolised more quickly in resistant than in susceptible *Tribolium castaneum* adults. A main site of attack on both compounds in the resistant strain was the methylene group of the methylenedioxyphenyl moiety, but in the susceptible strain other sites on the molecule were of greater importance. Developing larvae of the resistant strain were resistant to the effects of the hormone mimic and the toxic effects of high doses of piperonyl butoxide.

24.2 Degradation

Notwithstanding the length of time that piperonyl butoxide has been available for use in grain protection and its importance in increasing the usefulness of pyrethrum, the amount of information available on its degradation in stored grain in remarkably small.

Quinlan and Miller (1958) applied synergised pyrethrum spray to the top surface of bulk shelled corn of 12 to 13% moisture content at different dosages and frequencies to control infestation by *Plodia interpunctella*. The residues of piperonyl butoxide found in composited samples of the surface 2.5 cm of grain taken from each bin, 3 and 6 months after the treatments were begun, are given in **Table 24.6**. The data show that greater amounts of piperonyl butoxide were recovered following the use of greater concentrations and more frequent applications. Much of the piperonyl butoxide had disappeared by the time the samples were taken; about 75% at the end of 3 months and nearly 90% at the end of 6 months.

Dove (1958) reported several trials in which the fate of piperonyl butoxide residues had been determined. Three corn storages in Georgia and Maryland were treated with pyrethrum/piperonyl butoxide powder so as to deposit 14.3 mg/kg piperonyl butoxide. These showed 7 mg/kg after 4 months, 6.5 mg/kg after 9 months, 6 mg/kg after 10 months, and 2 mg/kg after 15 months of storage. In Missouri, two storages of corn, 3300 bushels each, were treated with pyrethrum/piperonyl butoxide powder to give a theoretical initial deposit of 29.5 mg/kg of technical piperonyl butoxide. Eighteen samples from each of these bins, taken 18 months after treatment, averaged 5.6 mg/kg and 5.2 mg/kg, respectively.

Blinn *et al.* (1959) sprayed wheat of 14% moisture content with 51.1 g of a wettable-powder formulation containing 2% pyrethrins and 20% piperonyl butoxide per 3.9 L of water at the rate of 260 ml of spray/45 kg of grain. The treated grain was stored in tightly sealed metal cans at 32°C. From the residues of pyrethrins and piperonyl butoxide found at intervals over 90 days of storage **(Table 24.7)** these workers concluded that piperonyl butoxide is much more persistent than pyrethrins under these conditions. As a result of this work they questioned the practice, then widespread, of determining the piperonyl butoxide residue level and from the knowledge of the ratio of pyrethrin:piperonyl butoxide in the original treatment, of calculating the pyrethrin residue. These investigators calculated that pyrethrins have an apparent residue half-life of 5.8 weeks in contrast with that for piperonyl butoxide of 9.9 weeks.

Table 24.6. Residues of piperonyl butoxide in the surface inch of shelled corn at intervals following periodic application of three concentrations of pyrethrum-piperonyl butoxide spray, 1956 **(Quinlan and Miller 1968)**

Frequency of treatment and conc. of piperonyl butoxide[a]	Piperonyl butoxide applied and recovered in shelled corn (ppm)			
	After 3 months[b]		After 6 months[c]	
	Amount applied July 9 to Sept. 27	Amount recovered	Amount applied July 9 to Oct. 25	Amount recovered
Semi-weekly				
1.25%	786	203	1,048	127
2.50%	1,572	386	2,096	247
3.75%	2,358	964	3,144	356
Weekly				
1.25%	393	107	524	56
2.50%	786	256	1,048	119
3.75%	1,179	359	1,572	206
Biweekly				
1.25%	197	54	262	31
2.50%	393	149	524	69
3.75%	590	182	786	95

a The proportion of pyrethrins to piperonyl butoxide was 1:10.

b These samples were drawn on Oct. 1, 1956, and represent the amount of insecticide accumulated during the first three months of the study.

c These samples were drawn on Jan. 1, 1957, and represent the amount present six months after the first and three months after the last application.

Table 24.7. Residues of pyrethrins and piperonyl butoxide found on wheat of 13% moisture content, stored at 30°C **(Blinn *et al.* 1959)**

Days after treatment	Parts per million	
	Pyrethrins[a]	Piperonyl butoxide[b]
1	0.8	7.7
14	0.6	6.2
30	0.4	5.0
90	0.2	3.5

a Average of two values. Corrected for background (0.1 ppm) and for recovery (89%).
b Average of two values. Corrected for background (0.1 ppm) and for recovery (98%).

Walkden and Nelson (1959), in evaluating synergised pyrethrum for the protection of stored wheat from insect attack, determined the residues of pyrethrins and piperonyl butoxide. They found that, 2 months after the application of the synergised pyrethrins, 83% of the piperonyl butoxide could be recovered from the treated grain. The same grain at the end of 12 months yielded 61% of the piperonyl butoxide originally applied, whilst at the end of 24 months this residue had declined only to 58% of the original application rate.

Strong *et al.* (1961) reported the results of laboratory tests made to determine the influence of formulation on the effectiveness and fate of synergised pyrethrum protective sprays on stored wheat. Synergised pyrethrum (pyrethrins 1.5 mg/kg, plus piperonyl butoxide 15 mg/kg) was applied to wheat in sprays formulated as solutions, emulsions, and wettable powder suspensions. Relative to each formulation, the residues remaining on wheat at various intervals after application of sprays, the proportional amount of residue on the wheat that is removed by the usual handling and cleaning practices followed at milling, the residues finally in the milled fractions, and the effect on quality of bread as evaluated by taste and odour panels, were determined. Two temperature levels and two levels of moisture content were incorporated in the series of tests composed of 36 primary lots of treated wheat and an equivalent amount of unsprayed wheat for appropriate controls. Residues persisting on the whole grain at various intervals after application of spray decreased with time. Results from chemical analysis for piperonyl butoxide **(Table 24.8)** indicated that cleaning operations used in this study were not highly effective in removing chemical residues from treated wheat, probably because most of the piperonyl butoxide retained by wheat was in the pericarp and endosperm.

Table 24.8. Residue of piperonyl butoxide found in wheat during milling and in the milling fractions of wheat three months after applying initial deposits of synergised pyrethrum (1.5 ppm pyrethrins and 15 ppm piperonyl butoxide in various spray formulations **(Strong et al. 1961)**

Formulation	Storage temperature (°F)	Moisture content of wheat (%)	Recovery of piperonyl butoxide (ppm)[a] from: Whole grain			Milling fractions from Buhler mill			
			Before cleaning	After cleaning	Clean-out (dockage)	Flour (composite)	Bran	Shorts	Low grade (middlings)
Emulsion	90	13	6.0	4.8	38.3	1.7	12.2	0.15	2.5
		10	6.2	5.5	29.3	3.3	13.0	7.2	9.1
	60	13	6.1	3.9	18.8	0.1	11.0	2.3	4.2
		10	7.3	8.6	20.5	1.9	10.9	4.8	2.6
Wettable powder suspensions	90	13	11.6	3.6	31.2	<0.5	10.6	<0.5	2.3
		10	5.4	–	29.2	3.2	14.6	7.5	1.0
	60	13	9.0	7.6	26.9	2.9	12.2	0.1	5.3
		10	10.9	8.8	25.3	4.3	7.6	8.4	4.2
Tetrachloroethylene solution	90	13	4.1	4.4	5.7	3.2	9.5	4.3	3.3
		10	6.7	4.5	13.7	4.6	13.8	5.0	–
	60	13	7.9	8.6	34.7	2.4	7.2	1.4	1.6
		10	5.4	4.2	8.3	4.5	11.4	1.9	1.1

a For the purpose of this study it was assumed that synergised pyrethrum maintained its original ratio of pyrethrins to piperonyl butoxide (1 to 10) and that recoveries of piperonyl butoxide by chemical analysis could be used to estimate residues of pyrethrins.

Joubert (1965a) described a field trial set up in South Africa to study the toxicity of contact insecticides to seed-infesting insects. 250 tonnes of yellow maize was treated with synergised pyrethrum (1:10) which was applied to the maize to give an initial deposit of 20 mg/kg piperonyl butoxide. Chemical analysis of the maize directly after treatment showed the piperonyl butoxide deposit ranged from 17.2 to 20.0 mg/kg with a mean just below 20 mg/kg. Temperatures taken in the grain during the test period ranged from 21.5°C to 28°C. At the end of 12 months storage the maize was again sampled and was found to contain 4.7 mg/kg piperonyl butoxide.

LaHue (1965) treated wheat of 11% moisture content with pyrethrins plus piperonyl butoxide (1:10) and determined the piperonyl butoxide residue. His calculated application was 22.4 mg/kg of piperonyl butoxide; however, immediately after treatment he recovered only 6.6 mg/kg. After 1, 3, 6, and 12 months his recoveries were 10.7, 12.9, 8.4, and 9.9 mg/kg, respectively. He concluded that the piperonyl butoxide residues degraded in an erratic pattern during the first few samplings, but apparently stablized after 6 months. The moisture content varied from 9.6 to 12.6% during the storage period.

LaHue (1966) treated shelled corn with pyrethrum plus piperonyl butoxide (1:10) and determined the piperonyl butoxide residue. His calculated application was 27.5 mg/kg of piperonyl butoxide; however, immediately after treatment he recovered 12.3 mg/kg, which was lower than expected. After 1, 3, 6 and 12 months his recoveries were 10.2, 7.7, 5.7 and 6.1 mg/kg, respectively. He concluded that the residue degraded in a uniform pattern for the first 6 months, apparently reaching a level of stability at that time.

Friedman and Epstein (1970) showed that thin films of commercial grade piperonyl butoxide were stable during exposure to intense fluorescent light for periods up to 7 days. In addition, only 40% of the piperonyl butoxide in films produced from insecticide aerosols was lost during 7-day exposure. When commercial grade piperonyl butoxide was heated to 100°C for up to 5 days, there was no appreciable degradation.

Rowlands (1975), in his comprehensive review of available information on metabolism of insecticides in grain, noted that nothing further seemed to have been done to elucidate the fate of piperonyl butoxide in stored grains.

Haley (1978), in his extensive review of the literature on piperonyl butoxide, makes no mention of any publications dealing with the degradation of piperonyl butoxide on grain or related commodities.

Halls (1981a), as part of the evaluation of permethrin as a grain protectant in Australia (see Chapter 21.2) determined the piperonyl butoxide content of wheat over 8 months. Initial treatment at 10 mg/kg was made when the temperature of the wheat was 22°C–25°C. The temperature declined gradually until it reached 10°C–13°C at the end of 5 months and thereafter gradually rose to the original temperature at the end of 9 months. The results of the determination of piperonyl butoxide are presented in **Table 24.9.**

24.3 Fate in Milling, Processing and Cooking

Watters (1956) treated elevator boot stocks in a flour mill with 2.5% pyrethrins and 25% piperonyl butoxide at the rate of 1 ml of insecticide per 50 g of stock. Flour passed over a treated boot showed a faint odour of the insecticide in loaves baked from the first sample, but the slight odour on subsequent samples was not noticeable after fermentation and baking.

Dove (1958) provided some information on the fate of piperonyl butoxide in milled products from wheat and maize. He stated that wheat, maize and other grains milled after 6 to 7 months storage following treatment with synergised pyrethrum were found to have some residues in the germ fractions but the majority appeared in the initial screenings and scourings. No piperonyl butoxide was found in the germ or the flour milled from grain treated 12 months previously. In Kansas, two storages of 3300 bushels each were treated with emulsifiable spray of pyrethrum/piperonyl butoxide to give a theoretical initial deposit of 12.2 mg/kg piperonyl butoxide (technical). Milled lots at 8 months after treatment showed no piperonyl butoxide in grits, flour or hominy feed. Each milled lot showed 1 mg/kg of piperonyl butoxide in meal, a grain fraction that accounted for 23% and 24% of the weight of whole grain. In one of these, the germ

Table 24.9. Residues of piperonyl butoxide on wheat after indicated periods of storage **(Halls 1981a)**

Target application (mg/kg)	Residue analysis (mg/kg)					
	Initial	1 month	2 months	3 months	6 months	8 months
10	3.9	4.5	4.4	5.0	5.7	3.8
10	7.9	4.6	4.1	5.5	6.7	4.0
10	4.2	3.4	3.0	4.0	4.0	3.5
10	5.6	7.3	6.9	6.6	6.4	5.9

represented 15% of the whole maize and showed 16 mg/kg piperonyl butoxide. In the other, the germ represented 17.9% of the whole grain and showed 11.5 mg/kg of piperonyl butoxide.

Feuersenger (1960b) conducted large-scale experiments with pyrethrum/piperonyl butoxide against weevils in grain and found that the flour made from the grain contained up to 6 mg/kg piperonyl butoxide.

Strong *et al.* (1961), in a comprehensive study of the influence of formulation on the effectiveness and fate of synergised pyrethrum sprays for stored wheat, concluded that cleaning operations used in their study were not highly efficient in removing chemical residues from treated wheat, probably because most of the insecticide retained by wheat after cleaning was in the pericarp and endosperm. In milled samples, residues of piperonyl butoxide recovered were greater in bran than in the other fractions. Residues found in flour were considerably higher than expected **(Table 24.10)**.

Isshiki *et al.* (1978) made an extensive survey in Japan of residual piperonyl butoxide in agricultural products. They examined 531 samples of 10 species of grain, both domestic and imported. Piperonyl butoxide was detected in 3 samples of barley and 3 samples of wheat harvested in USA and Australia. The concentration of piperonyl butoxide residues ranged from 0.2 to 1.4 mg/kg. It was not detected in 98.87% of the samples analysed. The limit of determination of their method was 0.1 mg/kg.

Simonaitis (1983) published details of a method for the determination of piperonyl butoxide residues in bread made from cornmeal and wheat flour, based on chromatographic clean-up of the extract in methyl acetate/pentane and colorimetric quantitation. The recoveries recorded were greater than 91%. Part of the study involved the determination of piperonyl butoxide in baked bread made from flour or corn meal fortified with piperonyl butoxide. Since the amount of piperonyl butoxide found by analysis represented 94–100% of the calculated weight of piperonyl butoxide in each sample it seems unlikely that a significant proportion of the piperonyl butoxide was lost or destroyed in the bread-making process. The bread samples were analysed on a dry weight basis after drying them in a microwave oven at 2450 mHz for 10 minutes or until no additional weight loss was observed.

24.4. Metabolism

Piperonyl butoxide exerts a synergistic action, among others, by inhibiting the oxidative metabolism of insecticides catalysed by the mixed-function oxidase system of microsomes. Because of the importance of this biochemical process, piperonyl butoxide has frequently been used to study biotransformation of many chemicals, including drugs and environmental contaminants. There is therefore an enormous literature in which piperonyl butoxide features but much of this has neither direct nor indirect bearing on grain protectant insecticides. No attempt has been made to review the literature on the

Table 24.10. Residue of piperonyl butoxide found during milling and in the milling fractions of wheat 3 months after applying initial deposits of synergised pyrethrum (1.5 ppm pyrethrins and 15 ppm piperonyl butoxide) in various spray formulations **(Strong *et al.* 1961)**

			Recovery of insecticide (ppm) from						
Formulation	Storage temperature (°F)	Moisture content of wheat (%)	Whole grain			Milling fractions from Buhler Mill			
			Before cleaning	After cleaning	Clean-out (dockage)	Flour (composite)	Bran	Shorts	Low grade (middlings)
Emulsion	90	13	6.94	4.84	38.80	1.65	12.17	0.149	2.53
		10	6.24	5.53	29.34	3.30	13.00	7.20	9.10
	60	13	6.08	3.90	18.76	0.11	10.99	2.31	4.18
		10	7.3	8.63	20.53	1.92	10.88	4.78	2.64
Wettable powder suspension	90	13	11.64	3.58	31.20	<0.45	10.62	<0.45	2.25
		10	5.4	–	29.20	3.20	14.64	7.5	4.00
	60	13	9.04	7.61	26.93	2.90	12.17	0.11	5.34
		10	10.94	8.75	25.26	4.26	7.56	8.36	4.24
Tetrachloroethylene	90	13	4.07	4.35	5.65	3.20	9.51	4.30	3.26
		10	6.74	4.51	13.70	4.57	13.76	5.00	–
	60	13	7.87	8.59	34.66	2.40	7.17	1.41	1.63
		10	5.43	4.19	8.32	4.53	11.43	1.90	1.12

biotransformation of piperonyl butoxide or the effect of piperonyl butoxide on the biotransformation of a wide range of other chemicals.

Haley (1978) has provided a very fruitful source of such information and his paper is recommended for those desiring more comprehensive data.

Casida *et al.* (1966) showed that the methylene group of the ^{14}C-labelled piperonyl butoxide was hydroxylated, yielding formate in the microsomal NADPH system *in vitro* and CO_2 in living houseflies and mice. These workers deduced that methylene dioxyphenyl compounds apparently serve as alternate substrates for the enzymic hydroxylation system of microsomes, and thus reduce the rate of metabolism and thereby prolong the action of certain insecticides.

Rowlands (1966c) showed that piperonyl butoxide prevented the seed coat oxidases from converting malathion and bromophos to their P= =O analogues.

Metcalf *et al.* (1966) showed that piperonyl butoxide, in the process of synergising insecticidal carbamates, acts as an inhibiter of phenolase enzymes which detoxify the carbamates largely by ring hydroxylation. This appears to require a 3-point attachment to the phenolase enzyme to orient the methylene-C so that interaction with a nucleophilic group at the enzyme-active site takes place.

Fishbein *et al.* (1967) used thin-layer chromotography of rat bile and urine to study the metabolites of piperonyl butoxide following intravenous administration. The elimination of piperonyl butoxide after single intravenous injections in rats occurred in the bile and urine. Noticeable chemical alterations were apparent but the rate of elimination of the compounds, although it was high, did not reach a rapid peak with a rapid decline, indicating prolonged elimination of the metabolites and suggesting to the investigators hazards from repeated and prolonged contact with the compound.

Falk and Kotin (1969), in a review of potential hazards of pesticide synergists and their metabolites, noted that these compounds are essentially non-toxic when used alone. The action of piperonyl butoxide which permits a reduction in the amount of insecticide without loss in effectiveness is attributed to the inhibition of enzyme action or the preventing of enzyme induction. The writers commented that this induction which prevents detoxification may prolong the action of hormones, pharmaceutical agents, pesticides and agricultural chemicals.

Esaac and Casida (1969) used piperonyl butoxide with the carbon labelled in the methylenedioxy system and another sample with the label on the methylene group of the carbon atom adjacent to the ring. Following injections into house flies, about 80% had been metabolised in 24 hours with corresponding elimination of labelled material either in the excreta or as expired carbon dioxide, the latter being a small proportion (about 10%) of the total. There are 2 routes of metabolism; removal of the carbon atom from the methylenedioxy group occurs and the excretion products resemble those of injected formic acid. Alternatively, the side chain derived from butyl carbitol is attacked and fission may occur at any of the 3 oxygen atoms present in this chain; there is no evidence of attack on the propyl group. The metabolism of the oxygen-containing chain warrants comment as this route is not possible with precursors of piperonyl butoxide, a substantial proportion of which is excreted unchanged.

The results of a comprehensive study of the metabolism of piperonyl butoxide in the rat have been reported by **Fishbein *et al.* (1969)** who used material labelled in the same positions as that used by **Esaac and Casida (1969)**. A total of 26 metabolites was detected but none of them was characterised. However, it was shown that there was a high initial rate of excretion after intravenous injection of labelled piperonyl butoxide followed by a slower rate, observations being made up in a period of 7 hours. Within this period, unchanged piperonyl butoxide was found in the lungs, but the high rate of metabolism indicates that there is unlikely to be any cumulative effect.

Kuwatsuka (1970) discussed the biochemical aspects of methylenedioxyphenyl compounds in relation to their synergistic action and pointed out that while they inhibit the microsomal hydroxylation of various insecticides, they are themselves hydroxylated by the same microsomal system, releasing formate from the methylene moiety. By using rat liver microsomes, the biochemical nature of demethylation of the methylenedioxyphenyl group and the methoxyphenyl group was investigated. Neither demethyleneation nor demethylation was biochemically different from aniline ring hydroxylation, and the inhibition of hydroxylation by methylenedioxyphenyl and methoxyphenyl compounds had no relation to either enzyme affinity or reaction velocity. Even though the enzymes for substrate hydroxylation and inhibitor hydroxylation are identical, inhibition was shown to occur not only by competitive reaction but also by other means.

Albro and Fishbein (1970) showed that when piperonyl butoxide was given orally to rats at the rate of 1 g/kg, 5 minutes before the administration of hydrocarbons, significantly increased deposition of hydrocarbons in the heart, lung, pancreas, and spleen occurred. The effect of piperonyl butoxide on octadecane deposition was greater than that on monacosane deposition.

Jaffe and Neumeyer (1970) compared the effects of piperonyl butoxide and N-(4-pentynyl)phthalimide on mammalian microsomal enzyme functions. Piperonyl butoxide markedly synergises the insecticidal effects of pyrethrins, but is usually less effective with synthetic pyrethrin analogues such as allethrin; on the other hand N-(4-pentynyl)phthalimide synergises allethrin strongly and pyrethrins weakly. These studies showed that both synergists produced marked and similar inhibition of hepatic microsomal enzyme functions in mice.

Brattsten and Metcalf (1970) interpreted the wide range of susceptibility to carbaryl and of the synergistic ratio as a response to differences in the activities of multi-function oxidase detoxication which is inhibited by piperonyl butoxide. These factors are highly variable in the Insecta.

Although they are usually considered inhibitors of hepatic microsomal enzymes, **Wagstaff and Short (1971)** showed that technical piperonyl butoxide and some of its analogues increased the activity of these enyzmes when fed to female adolescent rats for 15 days. Maximum activity was obtained from 10 000 ppm piperonyl butoxide in the feed. Having examined a number of analogues, the authors concluded that there was greater response to the methylenedioxyphenyl structures than to dimethoxyphenyl compounds. The conclusions drawn from these studies were: (1) inductive ability following repeated exposure is directly proportional to inhibitary ability following acute exposure, (2) these compounds apparently act as alternative substrates of microsomal enzymes, and (3) there is a possibility of induction in animals and man due to the ubiquity and potential additivity of inductive effects for compounds of this class.

Friedman *et al.* (1971) reported preliminary studies designed to test the possibility that piperonyl butoxide and related MDP derivatives may interact with naturally occurring MDP derivatives in foods. The synergists were tested on mice by injection of the compounds dissolved in corn oil. The data presented raised the possibility that piperonyl butoxide and other related pesticidal synergists may interact with low levels of MDP derivatives in food to produce interactive inhibition of microsomal enzyme functions.

Fishbein *et al.* (1972) showed that after intravenous administration of labelled piperonyl butoxide to rats, the radioactivity was widely distributed to various tissues, but after oral administration the synergist was rather poorly absorbed from the gastro-intestinal tract and rapidly excreted in the urine and feces. Intratracheal administration leads to initially high levels of biliary excretion followed by a prolonged period of elimination. Lower lung tissue residues of unmetabolised piperonyl butoxide were found after both oral and intratracheal administration as compared with that following intravenous dosage.

Conney *et al.* (1972), who studied the effects of piperonyl butoxide on drug metabolism in rodents and man, showed that rats require a more than 100-fold higher dose of piperonyl butoxide than mice for inhibiting antipyrine metabolism. Studies in man revealed that oral administration of 0.71 mg of piperonyl butoxide per kilogram of body weight did not influence antipyrine metabolism. Since this dose is considerably greater than the daily exposure of individuals using sprays extensively in enclosed areas, it is unlikely that environmental exposure to piperonyl butoxide inhibits human microsomal enzyme function.

Bock and Fishbein (1972) showed that there was a dose-dependent increase in rat hepatic microsomal demethylase activity associated with an increase in relative liver weight after repeated intraperitoneal injections of piperonyl butoxide. A similar increase was not found in other organs or tissues. The degree of induction was similar to that seen after treatment with DDT.

Franklin (1972, 1972a) showed that the oxidative metabolism of piperonyl butoxide produces a compound which forms a complex with cytochrome P-450. The complex is formed rapidly but is destroyed only slowly. This could contribute to the relatively slow elimination reported by **Fishbein *et al.* (1972).**

Yamamoto (1973) reviewed the mode of action of synergists, including piperonyl butoxide, in enhancing the insecticidal activity of pyrethrum and pyrethroids and he provided considerable information on the metabolism of piperonyl butoxide in insects. He concluded that piperonyl butoxide and related compounds appear to act by inhibiting the mixed-function oxidase system and minimising the rate of formation of less toxic oxidation products. The optimal synergist for each pyrethroid depends on the relative rate of entry of the insecticide and synergist into the insect, the importance of various sites of metabolic attack in detoxifying the pyrethroid, and the potency of the synergists for inhibiting the enzyme(s) involved in attacking the biodegradable sites of the pyrethroid molecule.

Goldstein *et al.* (1973) showed that oral administration of high concentrations (5000 and 10 000 ppm in diet) of piperonyl butoxide to rats increased liver weight and microsomal protein, causing enlargement and proliferation of the smooth endoplastic reticulum and increasing the activities of the drug-metabolising enzymes. Lower doses of piperonyl butoxide are known to inhibit the microsomal detoxifying enzymes. Active levels of piperonyl butoxide are quite high and are unlikely to be achieved with environmental contamination.

Hodgson *et al.* (1973) showed that *in vivo* long-term experiments involving feeding piperonyl butoxide to mice at concentrations up to 1000 ppm in the diet gave results which indicated that binding of piperonyl butoxide to cytochrome P-450 increased with duration of feeding. However, increase in liver weight occurred only at 1000 ppm.

Solomon and Metcalf (1974) showed that the activity of methoprene to *Tenebrio molitor* was increased by simultaneous treatment with piperonyl butoxide, indicating the blocking of a degradation reaction.

Quistad *et al.* (1975d) showed that piperonyl butoxide increased the morphogenetic effect of methoprene on larval mosquitoes and house flies.

Kulkarni and Hodgson (1976) have provided an extensive paper on the interactions of piperonyl butoxide with microsomal cytochrome P-450.

Levine and Murphy (1977) made a valuable contribution to the knowledge of the complexity of interaction between piperonyl butoxide and phosphorothionate insecticides by showing this varied from antagonism to potentiation depending upon minor variations in the structure of the phosphorus compounds.

Rowlands and Dyte (1979) showed that piperonyl butoxide and a juvenile hormone mimic (compound 2b of

WS Bowers) were both metabolised more rapidly in resistant than in susceptible *Tribolium castaneum* adults. The main site of attack on both compounds in the resistant strain was the methylene group of the methylenedioxyphenyl moiety.

Udeaan and Kalra (1983) showed that a significant portion of lindane resistance in *Tribolium castaneum* was overcome by the use of piperonyl butoxide. The resistant strain of *Tribolium castaneum* was found to degrade more lindane than the susceptible strain but this could be prevented by the use of piperonyl butoxide.

Casida (1970) and **Fishbein *et al.* (1970)** have both provided extensive reviews of the metabolism of piperonyl butoxide and other insecticide synergists.

24.5 Fate in Animals

Notwithstanding extensive use of piperonyl butoxide and the many metabolism studies which have appeared, the available information on the fate of piperonyl butoxide in animals is extremely sparse.

Fishbein *et al.* (1972) studied the metabolic fate of piperonyl butoxide in the rat. They were able to show that many sites on piperonyl butoxide were attacked, as evidenced by the large number of metabolites found following all routes of administration. The rather poor absorption of the synergist from the gastro-intestinal tract following oral administration, as well as its comparatively rapid elimination from the urine and feces, were noteworthy.

Moore (1971, 1972) published the results of a series of trials to determine the level, distribution and fate of residues of pyrethrum and piperonyl butoxide in poultry following repeated dipping of the hens in baths containing 2 concentrations of synergised pyrethrum. The hens were dipped twice weekly for 7 weeks and were then held for a further 4 weeks without further treatment. One treatment consisted of an emulsion containing 0.085% pyrethrin and 0.85% piperonyl butoxide and the other contained 0.85% pyrethrins and 8.5% piperonyl butoxide. The concentration of material in the first preparation would be about the amount recommended as a spray, but the dosage obtained from dipping is extremely high by comparison. The number of applications is also out of all proportion to that which would be experienced in actual use. The second preparation, being 10 times more concentrated, is therefore still more exaggerated. The severity of the test was intentional so as to reach residue levels that could be measured. The concentration of piperonyl butoxide in eggs reached a peak after about 10 dippings when approximately 1.8 mg/kg was found following treatment with the less concentrated emulsion and 4.2 mg/kg following treatment with the higher concentration. After 4 weeks without treatment the concentration of piperonyl butoxide in eggs fell to 0.5 and 2.25 mg/kg, respectively. Residues in muscle, liver and gizzard taken from birds following 7 weeks of the dip-treatments ranged between 1.4 and 2.0 mg/kg in the case of the lower concentration dip and between 3.5 and 5.6 mg/kg in the case of the higher concentration. These residue levels fell over the succeeding 4 weeks to 0.18–2.5 mg/kg and 1.2–2.6 mg/kg, respectively. Much higher residues were found in the fat (16 and 42 mg/kg) and in the skin (66 and 575 mg/kg). These declined to 3 and 21 mg/kg and 8 and 117 mg/kg respectively, after 4 weeks of no dipping. Very high levels in the skin included the material still on the outside of the skin from the residual dip. From these data it is not possible to draw conclusions which could be applied to the feeding of poultry with rations containing anticipated residues of piperonyl butoxide other than to recognise that piperonyl butoxide is readily eliminated from poultry tissues following withdrawal of intake.

24.6 Safety to Livestock

No papers dealing specifically with the safety of piperonyl butoxide to livestock have been located.

Dove and MacAllister (1947), in a general article on piperonyl butoxide, stated that rats appear to tolerate a diet containing about 2.5% piperonyl butoxide without ill effects. **Dove (1947a)** indicated that piperonyl butoxide is well tolerated by warm-blooded animals, the LD/50 for rabbits, rats and dogs being 5, 7.5 and 10 mL/kg, respectively.

Dove (1949) referred to chronic toxicity experiments conducted in the USA where undiluted piperonyl butoxide had been fed at different concentrations in all the food consumed by growing rats and other animals. There were no ill effects on the parent rats or their offspring. Tissues from about 300 animals used in the feeding experiments had been sectioned and studied by histopathologists. These showed that the extremely high dosage level of 1% piperonyl butoxide in the feed has no toxic effect upon the organs of the animals. He advanced the opinion that this lack of toxicity undoubtedly lies in the experimental demonstration that at least 85% of the chemical passes through the intestine of the animal within 48 hours of being ingested.

Sarles (1949) referred to toxicity studies carried out during the previous 4 years, including extensive chronic toxicity tests on dogs and rats. Piperonyl butoxide had been found to be one of the least toxic insecticides and therefore of special value for the control of arthropod pests in situations where safety was of prime importance. **Lehman (1948)** reported piperonyl butoxide to have a particularly low toxicity.

Sarles *et al.* (1949) recorded that the oral LD/50 dose of undiluted piperonyl butoxide had been found to lie between 7.5 and 10 mL/kg for rats, between 2.5 and 5 mL/kg for rabbits and above 7.5 mL/kg for cats and dogs.

Hatfield (1949) reviewed the toxicological information available at that time.

Wachs *et al.* (1950) referred to a 2–6 year study of chronic toxicity then nearing completion which was reported to show a high degree of safety of this substance when used alone or mixed with pyrethrum.

Sarles and Van de Grift (1952) discussed the chronic oral toxicity and related studies on animals with piperonyl butoxide indicating that acute toxicity tests on rats, subacute toxicity tests on rats, graded dosage chronic toxicity tests on rats and dogs and low-dosage chronic tests on a goat and a short-term experiment of low and moderate dosages on monkeys had all shown piperonyl butoxide to be of a low order of toxicity.

Dove (1958) reported that, in a 1-year feeding study in dogs which were fed 6 times a week by capsule, the administration of 3000 ppm of piperonyl butoxide was tolerated with only a moderate toxic effect. From these studied it was concluded that technical piperonyl butoxide was not carcinogenic.

Fishbein *et al.* (1967) showed that following intravenous injections, the rate of elimination of piperonyl butoxide and its metabolites, although high, did not reach a rapid peak with a rapid decline, indicating prolonged elimination of the metabolites and suggesting hazards from repeated and prolonged exposure.

Epstein *et al.* (1967) conducted experiments to demonstrate the enhancement by piperonyl butoxide of acute toxicity due to Freons, benzo(*a*)pyrene and griseofulvin in infant mice. Infant mice were injected subcutaneously with these compounds, alone and with piperonyl butoxide. In all instances the synergist enhanced acute toxicity. This increased toxicity was accompanied by anomalous weight increase in surviving mice, generally becoming pronounced by 21 days. The authors recommended that the possibility of a toxic hazard, whether synergistic or additive in nature, due to piperonyl butoxide in conjunction with other drugs or environmental pollutants should be further investigated.

Bond *et al.* (1973a, 1973b), in summarising the results of extensive investigations into the interaction of piperonyl butoxide and other chemicals, stated that piperonyl butoxide can be considered to have an extremely low toxic potential when used in recommended amounts.

Kennedy *et al.* (1977) carried out classical studies to evaluate the teratogenic potential of piperonyl butoxide in the rat. Using doses of either 300 or 1000 mg/kg administered by gavage, these studies revealed no indication of potential for the production of teratra.

These indirect indications suggest that occasional or continuous consumption of traces of piperonyl butoxide in animal feeds should have no deleterious effect upon livestock.

24.7 Toxicological Evaluation

Piperonyl butoxide has been evaluated by the Joint FAO/WHO Meeting on Pesticide Residues on 3 occasions in 1965, 1966 and 1972 **(FAO/WHO 1965, 1967, 1973)**. Studies on the metabolism in the rat indicated that its breakdown is rapid, although clearance from the body is relatively slow. Long-term studies in rats showed no toxicological effect at 100 ppm in the diet. A short-term study in dogs showed no toxicological effect at 3 mg/kg/day. Carcinogenic studies in mice showed no increase in tumours at levels of 890 ppm. Administration of extremely high doses of piperonyl butoxide together with Freon propellant administered parenterally to neonatal mice resulted in an increase in hepatomas. Acute studies in man showed no effects of piperonyl butoxide at a level of 0.71 mg/kg body weight (bw). The no-effect levels demonstrated in the rat (5 mg/kg-bw per day), and the dog (3 mg/kg-bw per day) served as a basis for the establishment of an ADI for man of 0.03 mg/kg-bw. Indicated areas for further research comprise studies on the effect of this compound on a litter of dogs and reproduction studies in at least one more species.

24.8 Maximum Residue Limits

On the basis of the above ADI, and a limited amount of residue data, together with the results of official decisions taken in the USA, the JMPR **(FAO/WHO 1967, 1968, 1970, 1973)** recommended the following maximum residue limit for piperonyl butoxide in all cereal grains and milled cereal products.

Commodity	Maximum Residue Limit (mg/kg)
Cereals and cereal products	20

The residue is determined and expressed as piperonyl butoxide.

In view of decisions taken more recently with respect to other grain protectant insecticides and the current knowledge of the distribution of piperonyl butoxide within milling fractions, a review of these recommendations appears to be justified.

The methods recommended for the determination of piperonyl butoxide are: **AOAC (1980c); Methodensammlung (1982 — XII-6; S19)**. The following methods are recognised as being suitable: **Pesticide Analytical Manual (1979d): Methodensammlung (1982 — 163); Specht and Tillkes (1980).**

25. Pirimiphos-Methyl

Pirimiphos-methyl is a fast-acting broad-spectrum organophosphorus insecticide with both contact and fumigant action. It shows activity against a wide spectrum of insect pests, and possesses only limited biological persistence on leaf surfaces but gives long lasting control of insect pests on inert surfaces such as wood, sacking and masonry. It retains its biological activity when applied to stored agricultural commodities including raw grain, nuts, pulses, dates and cheese.

Its acute oral and dermal toxicity to animals is low but the minimum effective dose against a wide range of stored-product pests is lower than most other organophosphorus insecticides in use or under development as grain protectants **(McCallum-Deighton 1978)** thus providing a desirable degree of additional safety.

Since pirimiphos-methyl was first developed in the UK in the late 1960s many publications have appeared. This reviewer has collected over 220 references to the use of pirimiphos-methyl in grain storage. There is also a wide literature on the use of pirimiphos-methyl against pests of fruit, vegetables, ornamentals and field crops as well as against insects of public health concern.

Pirimiphos-methyl has been widely evaluated and used in many situations against stored-product pests. It is potent against beetles, weevils, moths and mites, but not sufficiently effective against some strains of *Rhyzopertha dominica*. It is useful against immature stages within the individual grains and it appears quite effective against many malathion-resistant strains. It is considerably more potent than malathion, bromophos, dichlorvos, iodophenphos and approximately equal in effect to fenitrothion and chlorpyrifos-methyl but somewhat less effective than methacrifos at low temperatures. Pirimiphos-methyl deposits are extremely stable on grain but eventually degrade on grain and rapidly in animals to non-toxic metabolites. The half-life on wheat at 30°C and 50% relative humidity is 45 weeks. Most of the pirimiphos-methyl applied to wheat is removed with the bran and almost all of the deposit on rice in husk is removed when the rice is milled. Between 85% and 90% of the residue in wheat is lost in the preparation of white bread. Deposits on barley are destroyed in the malting process. Likewise residues in bran are destroyed in the processing of breakfast cereals. About 1.5% of the deposit on paddy rice remains on the cooked rice.

25.1 Usefulness

The **Ministry of Agriculture Fisheries and Food of the UK (1969)**, in its advisory leaflet on insects and mites in farm-stored grain, recommended pirimiphos-methyl for the spraying of storage structures and for admixture with grain at the rate of 200 g of 2% dust per tonne. **Chawla and Bindra (1971)** evaluated the residual stability and effectiveness of pirimiphos-methyl in comparison with malathion, bromophos and iodofenphos. They treated wheat with a moisture content of 8.6% at dosages of 30 and 50 mg/kg and stored the grain in gunny sacks under normal storage conditions. Analysis at the end of six months revealed that pirimiphos-methyl was the most persistent compound. The residues of bromophos and iodofenphos were observed to dissipate at a significantly faster rate, similar to those of malathion.

Coulon *et al.* (1972) compared the activity of pirimiphos-methyl against *Sitophilus granarius* with that of malathion, dichlorvos, bromophos and iodofenphos. The residual effect of pirimiphos-methyl was greatly superior to the other compounds. **Coulon (1972)** reported studies with 10 new insecticides, including both organophosphorus compounds and synthetic pyrethroids against *Sitophilus granarius* at rates between 5 and 20 mg/kg. Pirimiphos-methyl, dichlorvos and several synthetic pyrethroids showed a remarkable effect against juvenile forms of the insect secreted within the grain.

In the Philippines, **Morallo-Rejesus and Santhoy (1972)** evaluated the toxicity of 6 organophosphorus insecticides to field collected strains of 2 stored-product pests that were known to be resistant to DDT. These studies involved the determination of LD/50 values based on 24 hour mortality. Pirimiphos-methyl appeared particularly effective.

In Ecuador **(INIAP 1972)**, trials were made with a number of insecticides against *Sitophilus oryzae* in maize. Admixture of pirimiphos-methyl at the rate of 2, 4 and 6 mg/kg controlled this pest for 143, 199 and 339 days, respectively. These studies showed that pirimiphos-methyl did not affect the reproduction of adult weevils, but did control recently emerged adults.

In Cyprus, the **Cyprus Agricultural Research Institute (1972)** evaluated pirimiphos-methyl and a number of other organophosphorus insecticides against selected species of stored-grain insects. Most of the chemicals had lost their effectiveness by the third month but pirimiphos-methyl and fenitrothion was still effective 4 months after applications.

In Malawi, 17 strains of *Tribolium castaneum* collected from warehouses and stores throughout Malawi were found to be resistant to lindane and 11 of the 17 were resistant to malathion. Pirimiphos-methyl proved very effective against these resistant strains **(Pieterse and Schulten 1972)**.

In the Philippines, studies were carried out on the comparative toxicity and duration of effectiveness of 5 insecticides when admixed with corn grain. Observations made over a 6-month storage period showed pirimiphos-methyl, whether applied as dust or emulsion, was effective as a protectant of corn grain against a variety of pests, especially *Sitophilus oryzae* **(Morallo-Rejesus 1973)**.

Morallo-Rejesus and Nerona (1973) showed pirimiphos-methyl to be effective also for impregnating sacks used to store grain.

In Nicaragua, **Giles (1973)** reported that for protecting maize in cribs, pirimiphos-methyl and malathion dusts were superior to other insecticides.

Nine insecticides were tested for their residual effectiveness on jute sacking. Pirimiphos-methyl proved among the most effective giving a quick knock-down of several species **(Rai and Croal 1973)**.

In the United Kingdom, **Wilkin and Hope (1973)** found pirimiphos-methyl particularly effective against a number of stored-product mites.

Cogburn (1973) reported from the USA that pirimiphos-methyl appeared, from preliminary trials, to offer the best protection of rice at acceptably low dosage levels.

McCallum-Deighton (1974) reviewed information on the properties, performance and use of pirimiphos-methyl and stated that pirimiphos-methyl was approximately as potent as fenitrothion but more potent than malathion, bromophos, iodofenphos, tetrachlorvinphos or dichlorvos. Pirimiphos-methyl was considered to be eminently suitable as a stored-product insecticide.

In France, the activity of pirimiphos-methyl was compared with dichlorvos and malathion against *Sitophilus granarius* where it was found to destroy adults of this species within 48 hours when applied to grain at the rate of 8 mg/kg **(Graciet *et al.* 1974)**.

Trials in India showed that, of 9 insecticides screened for effectiveness against the adults of 3 stored-product pests, pirimiphos-methyl proved promising and was tested in a small-scale storage trial lasting 1 year. At comparable doses, pirimiphos-methyl provided protection for a longer period than malathion **(Bindra 1974)**. These studies were reported in more detail by **Chawla and Bindra (1973)**. The value of pirimiphos-methyl for treating structural surfaces was confirmed by Indian work, carried out by **Girish *et al.* (1974)**. Likewise from India, **Verma and Ram (1974)** found pirimiphos-methyl effective against malathion-resistant strains of *Tribolium castaneum* and these workers considered that pirimiphos-methyl could safely replace malathion if malathion resistance became a problem.

In the Philippines, **Morallo-Rejesus and Carino (1974)** studied the residual toxicity of pirimiphos-methyl in comparison with 4 other insecticides against *Sitophilus* spp. and *Rhyzopertha dominica* infesting corn. Though pirimiphos-methyl performed well, it was not the most potent nor the most stable under prevailing conditions. The studies were extended to evaluate the effectiveness of combinations of grain protectant and insecticide-impregnated sacks for the control of storage pests of shelled corn. Pirimiphos-methyl proved more effective when the treated corn was packed in pirimiphos-methyl impregnated sacks than when stored in malathion impregnated sacks. Such combinations were satisfactory over a 9 month storage period but were not sufficiently effective for 12 months **(Morallo-Rejesus and Eroles 1974)**.

Seth (1974) reported on the use of pirimiphos-methyl for control of pests of stored rice in South East Asia. An admixture treatment at 4 mg/kg controlled all pests present and protected against insects entering rice sacks for up to 6 months after treatment. Pirimiphos-methyl was reported to give effective and prolonged control of *Tribolium castaneum, Oryzaephilus surinamensis, Sitophilus oryzae, Ephestia cautella, Rhyzopertha dominica* and *Sitotroga cerealella*. The application of pirimiphos-methyl to the surface of rice sacks at the rate of 250–500 mg/m^2 controlled all insect pests entering the sacks for up to 6 months after treatment.

Golob and Ashman (1974) studied the effect of oil content and insecticides on insects attacking rice bran. Pirimiphos-methyl applied at the rate of 8 mg/kg was equal to phoxim applied at the same rate in causing the highest mortality of *Latheticus oryzae* but did not prevent reproduction of this species after 31 days. This work was further extended and reported by **Ashman (1975)** when pirimiphos-methyl was considered the most successful of the insecticides evaluated.

Carmi (1975), working in Israel, evaluated the toxicity of pirimiphos-methyl to *Ephestia cautella* in comparison with malathion, lindane and phoxim. The toxicity of pirimiphos-methyl was almost equal to that of phoxim at the 8 mg/kg level, but less at lower levels. Large larvae of *Ephestia cautella* were more difficult to kill.

Pinniger (1975) assessed pirimiphos-methyl along with 7 other insecticides for the control of *Oryzaephilus surinamensis* as a residual spray on wood, metal and concrete. The persistence of pirimiphos-methyl was regarded as satisfactory and it gave good control of this pest.

Wilkin (1975b) studied the effects of mechanical handling and admixture of acaricides on mites in farm-stored barley. Pirimiphos-methyl dust was the more persistent of the treatments evaluated and despite the barley having a moisture content of more than 18%, virtually no breakdown of the insecticide was detected.

McCallum-Deighton (1975a), describing the properties and uses of pirimiphos-methyl in a Russian journal, indicated that treatment of wheat grain with 6 mg/kg pirimiphos-methyl 41 days after oviposition by *Sitophilus granarius* prevented the development of adults and formation of a second generation. In a later paper **McCallum-Deighton (1975b)**, pointed out that a number of the species which have developed strains highly resistant to malathion have been found to be fully susceptible to pirimiphos-methyl. He went on to provide a detailed comparison between pirimiphos-methyl, malathion, fenitrothion and dichlorvos against the major stored-product pests but little mention was made of *Rhyzopertha dominica* against which pirimiphos-methyl is not particularly effective. It was noted in the annual report of the **Queensland Department of Primary Industries (1975)** that pirimiphos-methyl was not particularly active against *Rhyzopertha dominica*. It was stated that combinations with bioresmethrin or other synthetic pyrethroid insecticides appear necessary.

McCallum-Deighton (1975b) presented a resume on pirimiphos-methyl in which one of the outstanding features claimed was its ability to eradicate infestations of the larvae of primary grain insects.

In Australia, **Bengston *et al.* (1975)** compared pirimiphos-methyl, bioresmethrin and chlorpyrifos-methyl as grain protectants against malathion-resistant insects in wheat. Resistance factors for malathion, as measured by impregnated paper assays, were particularly high in the case of *T. castaneum* and *S. oryzae*. Pirimiphos-methyl was ranked less potent than chlorpyrifos-methyl. All organophosphorus compounds were considered relatively less potent against *R. dominica*, against which bioresmethrin proved outstandingly effective. Stability trials under conditions prevailing in bulk grain stores led to a recommendation that a combination of pirimiphos-methyl with bioresmethrin should be used as a grain protectant.

LaHue (1975c) tested pirimiphos-methyl as a short-term protectant of grain against stored-product insects and reported that some insects survived and reproduced after 3 months. At the higher rates of application, the residual protection lengthened considerably. Later, the same investigator **(LaHue 1975a)** tested pirimiphos-methyl alongside a number of other grain protectants against the moth *Sitotroga cerealella*. It proved superior to the standard malathion treatment.

Grain protectants for use under tribal storage conditions in Rhodesia were evaluated **(Weaving 1975)**. Pirimiphos-methyl, applied at a dosage of 5 mg/kg or fenitrothion (8 mg/kg) performed satisfactorily and showed good persistence for 12 months.

Wilkin (1975a) and **Wilkin and Haward (1975)** used pirimiphos-methyl in experiments designed to control stored-product mites. The studies showed that pirimiphos methyl controlled mites in the fabric of buildings and in stored grain and that this insecticide also controlled lindane-resistant strains.

Studies in the USA designed to find an alternative for malathion, to which many stored-product insects were developing resistance, included pirimiphos-methyl **(Cogburn 1975, 1976)**. Rice in small bins, subjected to very heavy infestation pressure from a multi-species insect population, and treated with pirimiphos-methyl at dosages of 10 and 15 mg/kg, was protected for 9 months, even from some malathion-resistant insects.

The toxicity of pirimiphos-methyl to *Ephestia cautella* was compared with malathion and several other insecticides in Israel. Pirimiphos-methyl was one the most effective materials. Trials to control mites in stored grain showed that when pirimiphos-methyl was applied at the rate of 4 mg/kg, mite control was rapid and reinfestation was prevented for at least 3 months **(McCallum-Deighton and Pascoe 1976)**.

MacDonald and Gillenwater (1976), who tested pirimiphos-methyl against 6 species of stored-product insects, considered it to be one of the most promising grain protectant insecticides then available.

LaHue (1976) found pirimiphos-methyl, applied to seed corn at a concentration of 8 mg/kg, effective against stored-product insects over a 21 month period. The number of exit holes in the grain and the number of progeny produced at the end of 21 months was only about 4% of that found in untreated controls and only about 10% of the number in the standard malathion treatment. Similar studies involving high-moisture sorghum grain showed pirimiphos-methyl gave excellent protection during a 12 month storage period. About 50% of the initial deposit remained on the sorghum at the end of 12 months **(LaHue and Dicke 1976a)**.

Pirimiphos-methyl proved to be one of the most effective of 9 insecticides tested in India against 3 of the most important stored-product species **(Chawla and Bindra 1976)**.

Davies (1976) and **Davies and Desmarchelier (1981)** showed that carbaryl could be effectively combined with pirimiphos-methyl to supplement the control of *Rhyzopertha dominica*.

LaHue and Dicke (1976b) evaluated 4 insecticides as protectants for shelled corn against stored-grain insects and reported pirimiphos-methyl superior to the standard dosage of malathion. Pirimiphos-methyl residues degraded relatively slowly and provided a high standard of protection after 12 months storage.

Morallo-Rejesus and Eroles (1976) evaluated pirimiphos-methyl for the impregnation of polypropylene sacks as well as for admixture with corn to control storage pests in the Philippines. Greatest reduction in insect population and kernel damage was obtained with pirimiphos-methyl dust admixed with corn stored in pirimiphos-methyl treated sacks. When applied at the rate of 10 mg/kg, combined with pirimiphos-methyl impregnated sacks, corn was effectively protected against damage by *Rhyzopertha dominica*, *Tribolium castaneum* and *Sitophilus* spp. for 12 months. When evaluating 5 contact insecticides as protectants against insects in stored corn, **Morallo-Rejesus and Carino (1976a)** reported that at 10–30 mg/kg pirimiphos-methyl dust was the most effective; the higher the concentration the less damaged were the grains.

This work was extended **(Morallo-Rejesus and Carino 1976b)** to evaluate the residual toxicity of the 5 insecticides on 3 varieties of corn and sorghum under laboratory conditions in the Philippines. The residual toxicity increased over the range 10, 20, 30 and 50 mg/kg. Pronounced differences in the susceptibility of the test insects to the insecticides were noted. *Sitophilus zeamais* was more susceptible than *Rhyzopertha dominica* to pirimiphos-methyl. The type of grain influenced the effectiveness of the insecticides. Pirimiphos-methyl was more persistent on maize than on sorghum.

McCallum-Deighton (1976) reviewed trials with pirimiphos-methyl against all major pests of stored grain and other stored food at a conference covering South East Asia and the Pacific region.

Tyler and Binns (1977) evaluated the toxicity of 7 organophosphorus insecticides and lindane against 18 species of stored-product beetles and ranked pirimiphos-methyl third following chlorpyrifos-methyl and fenitrothion. The knock down effect of pirimiphos-methyl was good at the higher dosages except against *Rhyzopertha dominica, Ptinus tectus* and *Dermestes maculatus*. Kill was also poor against these species.

LaHue(1977b) studied a gradient of effective doses of pirimiphos-methyl on hard winter wheat against 4 species of adult stored-product insects. The decreasing order of tolerance was *Tribolium confusum, Rhyzopertha dominica, Tribolium castaneum, Sitophilus oryzae*. Although a dose of 3 mg/kg was sufficient to give 100% kill of *Sitophilus oryzae* at the end of 12 months, 4 mg/kg was required to do the same against *Tribolium castaneum*, 10 mg/kg against *Tribolium confusum* but 10 mg/kg was not sufficient to kill all *Rhyzopertha dominica* under the same circumstances.

LaHue and Dicke (1977) evaluated candidate protectants for wheat against stored-grain insects in small bins. Pirimiphos-methyl applied at a calculated dose of 7.8 mg/kg gave excellent protection for 12 months against continuous infestation of mixed populations of insects. Once again *Rhyzopertha dominica* was the most difficult species to control.

Desmarchelier (1977b) demonstrated how selective treatments, appropriate to simulated storage conditions, reinfestation pressures and species or strains of insects present in wheat, required less insecticide than any 'blanket' or all-purpose treatment. He showed how bioresmethrin, synergised bioresmethrin or synergised pyrethrins which were especially effective against *Rhyzopertha dominica*, combined readily with organophosphorus insecticides such as pirimiphos-methyl which were especially effective against *Tribolium* and *Sitophilus* species. There was no antagonism between any of the pyrethroids and any of the organophosphorus insecticides when they were applied in combinations, in that the period of protection given by combinations was the period of protection given by each of the components against the species for which they were effective. It was found that the net effect of lower temperature in reducing toxicity but increasing persistence was to increase the period of protection given by insecticides in cooled systems. It was suggested that the amounts of insecticides applied to grain could be considerably reduced if more consideration were given to grain storage conditions.

Ofasu (1977) reported laboratory studies carried out in Ghana on the persistence of pirimiphos-methyl applied in water-based emulsions to shelled maize against *Sitophilus zeamais* and *Tribolium castaneum*. Pirimiphos-methyl was effective against *Sitophilus zeamais* and *Tribolium castaneum* after 24 weeks storage when applied at rates of 8 and 12 mg/kg, respectively.

Rai (1977) reported laboratory studies on admixture of pirimiphos-methyl with paddy rice which indicated that whereas 3–7.5 mg/kg pirimiphos-methyl was 100% effective against adult *Tribolium castaneum* for 12 months, 18.8 mg/kg was insufficient to control *Rhyzopertha dominica* 4 months after treatment. However, in a field trial in which paddy rice was treated with 3.13 mg/kg pirimiphos-methyl and stored in jute bags in a stack containing 150 tonnes, rice was protected against *Sitotroga cerealella, Rhyzopertha dominica* and *Tribolium castaneum* for 6 months. This reviewer has some doubts about the figure of 3.13 mg/kg pirimiphos-methyl applied to the grain since the intended rate of application was 13.7 mg/kg.

Zettler and Jones (1977), working in the USA, tested the toxicity of 7 insecticides to malathion-resistant *Tribolium castaneum*. Cross-resistance was not detected to any of the insecticides at the LD/50 level. Pirimiphos-methyl was the most toxic insecticide and was more than twice as toxic as malathion to the susceptible strain.

Bengston *et al.* (1977) reported the results of extensive field trials to compare pirimiphos-methyl with a range of other insecticides for the control of malathion-resistant insects infesting wheat in Australia. They found that the relative potency of each material varied depending upon which insect pest was used to challenge it in the bioassay. Pirimiphos-methyl was generally the most effective, except against *Rhyzopertha dominica* where methacrifos was undoubtedly more potent.

Pirimiphos-methyl applied at the rate of 8 mg/kg proved more effective than malathion, applied at a rate of 10.5 mg/kg as a protectant on hard winter wheat against *Tribolium* spp. in a 12 month storage study using small bins **(LaHue 1977a)**. In a separate, contiguous study to determine the potency of pirimiphos-methyl against 4 common species of stored-grain insects, *Tribolium confusum* proved to be the most tolerant and *Sitophilus oryzae* the most susceptible. Related studies using seed corn, showed pirimiphos-methyl to be significantly more effective than chlorpyrifos-methyl, fenitrothion and malathion **(LaHue 1977c)**.

Bengston *et al.* (1978a), reporting field experiments with grain protectants for the control of malathion-resistant insects in stored sorghum under typical conditions in Queensland, found that pirimiphos-methyl, applied at the 4 mg/kg rate, controlled prevalent strains of *S. oryzae, T. castaneum* and *E. cautella* but was not entirely satisfactory against *R. dominica*. From these and related studies came recommendations for combinations of pirimiphos-methyl and bioresmethrin or carbaryl.

Advisory leaflets from the Nigerian Stored-Products Research Institute **(Adesuyi and Cornes 1978)** recommended pirimiphos-methyl for the protection of stored maize.

LaHue (1978) studied the effectiveness of diatomaceous earth dusts impregnated with pirimiphos-methyl and malathion in order to find simple methods for controlling insects in small lots of low-moisture wheat stored at relatively high temperatures; conditions which apply throughout most of the tropical regions. The dusts were formulated by adding pirimiphos-methyl in sufficient quantity to yield a deposit of 4.5, 6 and 7.5 mg/kg when the dust was incorporated in the grain at the rate of 0.5 kg dust/tonne. Dusts delivering a concentration of 6 to 7.5 mg/kg were 100% effective in destroying *Sitophilus oryzae* and *Sitophilus zeamais* weevils for the duration of the trial (8 months). The author noted that these dust formulations of pirimiphos-methyl looked exceedingly promising for use on small farms in under-developed agricultural areas that experience high temperatures and low humidities following the harvest season. He visualised that the insecticidal protectant dusts could be pre-packaged for distribution according to the capacity of the normal storage stuctures in a geographical area and then used for application to the relatively small lots of grain placed in farm storages. No special equipment would be needed, unlike spray applications, and the grain could be easily washed free of the dust before use.

Kadoum *et al.* (1978) studied the efficacy of pirimiphos-methyl applied at 2 rates to wheat for milling. They found that the lower rate (7.3 mg/kg) was fully effective against *Sitophilus oryzae, Tribolium castaneum* and *Tribolium confusum* for 12 months whereas 14.6 mg/kg was fully effective against *Rhyzopertha dominica* for 6 months but less than 85% effective after 12 months. However, no *Rhyzopertha dominica* progeny developed in the wheat treated with the 14.6 mg/kg dose even after 12 months.

Taylor *et al.* (1978) observed that in Nigeria maize treated with pirimiphos-methyl became infested with *Araecerus fasciculatus* which is considered a major pest on maize and yam products and only occasionally occurs on other produce. In subsequent investigations it was shown that the increase in numbers of *Araecerus fasciculatus* was due principally to wide differences in the effectiveness of pirimiphos-methyl for the control of *Sitophilus zeamais* and *A. fasciculatus*. The tolerance factor compared with *Sitophilus zeamais* is of the order of over 500 times and these investigators expressed the concern that the use of low doses of pirimiphos-methyl adequate for *Sitophilus* control would result in *A. fasciculatus* replacing *Sitophilus* as the principal pest of treated stored maize.

McCallum-Deighton (1978), in a review of pirimiphos-methyl and other insecticides used in the control of stored-product insects, pointed to the value of dichlorvos and pirimiphos-methyl in eradicating a population of primary grain insects from infested grain. It was concluded that this eradicant effect is probably due to vapour phase activity. He described a test in which *Sitophilus granarius* adults were allowed to lay eggs in wheat of 14% moisture content for 1 week. The adults were removed and at intervals after oviposition, portions of the grain were either treated with several concentrations of pirimiphos-methyl or dichlorvos or left untreated. The numbers of adults emerging from the treated and untreated samples were counted through 2 successive generations. When treatments were applied 9 days after oviposition both pirimiphos-methyl and dichlorvos had considerable effect.

While the higher rates of dichlorvos killed almost all the larvae, those that did survive the treatment developed into normal viable adults which, in turn, gave rise to a succeeding generation. The lowest rate of pirimiphos-methyl did not reduce the number of larvae maturing to adults to the same extent as did the higher rates of dichlorvos; nevertheless, insects which did emerge died rapidly and did not give rise to a following generation. Higher rates of pirimiphos-methyl both reduced the number of larvae maturing to adults and also killed such adults as did emerge. Similar results were obtained when the treatments were applied 25, 33, 41 and 46 days after egg laying. This eradicant effect was probably due to a combination of the vapour phase activity of pirimiphos-methyl on pre-adult stages and adults prior to emergence, and to the persistence of pirimiphos-methyl in the pericarp of the grain, resulting in the death of adults as they tried to emerge, or soon after emergence and before oviposition.

Adesuyi (1979) described the requirements for selection of a new insecticide for use in maize storage at the farmer's level in Nigeria to replace lindane which was beginning to fail through the development of resistant strains. Pirimiphos-methyl and iodofenphos were considered equally effective and acceptable but it was necessary to use them at a rate equivalent to 10 mg/kg to obtain 7 months protection. It was considered important that the insecticide recommended should be readily available, especially during the storage season, that the formulation should be of very low active ingredient content and that it should be packaged in quantities that are practicable for the farmer to handle and exhaust in one operation.

Mensah and Watters (1979b) applied aqueous formulations of malathion, bromophos, iodofenphos and pirimiphos-methyl to dry (12% moisture content) and tough (16% moisture content) wheat at 2 dosage levels to compare their persistence and effectiveness on susceptible and malathion-resistant strains of *Tribolium castaneum*. Pirimiphos-methyl at 4 and 6 mg/kg was the most effective compound during the 24 weeks against both strains of *Tribolium castaneum* and prevented reproduction.

Bitran *et al.* (1979b), working in Brazil, was able to show that infestation by *Sitophilus zeamais* and *Sitotroga cerealella* of maize stored in farm bins increased weight losses. The best protection afforded against these pests was obtained by fumigation with hydrogen phosphide followed by treatment with pirimiphos-methyl at the rate of 4 mg/kg. Treatment with malathion at 8 mg/kg without fumigation was not satisfactory.

Bengston and Desmarchelier (1979a), in discussing the biological efficacy of new grain protectants, showed from LC/50 values determined by impregnated-paper assays of candidate grain protectants in non-volatile solvents that pirimiphos-methyl was relatively less toxic to the more important insect species than was malathion (and considerably less toxic than chlorpyrifos-methyl). However, the chemical and biological stability of pirimiphos-methyl in deposits renders it more effective in controlling insect pest species in practice. For example, LC/99.9 values, based on 3-day mortality of adults exposed to newly sprayed grain, showed pirimiphos-methyl comparable with chlorpyrifos-methyl and fenitrothion against a range of malathion-susceptible strains. Thus, pirimiphos-methyl was more potent when measured by its ability to kill all exposed insects on grain, than when measured on impregnated paper. On this basis, pirimiphos-methyl is relatively much more potent than malathion and comparable with fenitrothion, being only a little less potent than chlorpyrifos-methyl and methacrifos. All of these materials, however, prove relatively weak by comparison with bioresmethrin against *Rhyzopertha dominica* and the authors drew the conclusion that combinations of currently approved insecticides are required to provide effective protection against the entire pest complex.

Desmarchelier and Bengston (1979a) discussed the residual behaviour of chemicals in stored grain and the fact that the residual life can be predicted from the water activity (interstitial relative humidity) and temperature and a rate constant for each insecticide. **Desmarchelier and Bengston (1979b)** extended this concept to show how the model could assist in calculating the rate of application for a given set of storage conditions and thus enable the storage conditions to be manipulated to gain the maximum effect from the grain protectant insecticides. Pirimiphos-methyl was used as one of the examples.

O'Donnell (1980) studied the toxicities of 4 insecticides to *Tribolium confusum* under 2 sets of conditions of temperature and humidity, and showed that all the products under test were less potent at low temperatures. However, pirimiphos-methyl showed the lowest difference under the 2 sets of conditions. This observation further assists us to understand why pirimiphos-methyl consistently out-performs malathion and iodofenphos, either under low temperatures or under fluctuating temperatures such as occur under practical conditions in temperate climates.

Taylor and Evans (1980) evaluated dilute dusts containing pirimiphos-methyl and permethrin for control of bruchid beetles attacking stored pulses. Damage to the pulses was almost entirely prevented by the application of pirimiphos-methyl at the rate of 2.5 mg/kg.

Bengston *et al.* (1980a) reported the results of extensive field trials with various insecticide combinations carried out on bulk wheat in commercial silos in Queensland, South Australia and Western Australia. Laboratory bioassays on samples of treated wheat at intervals over 8 months, using malathion-susceptible and malathion-resistant strains, enabled the investigators to rank the insecticides and insecticide combinations in order of effectiveness against each of 6 species and many strains of these insects. Pirimiphos-methyl at 4 or 6 mg/kg was rated highly, except against *Rhyzopertha dominica* where it rated well only in combination with bioresmethrin.

Quinlan *et al.* (1980) reported field tests conducted in 1.5 tonne lots of hard winter wheat of 14.6% moisture content held at 25°C and artificially infested with *Sitophilus oryzae, Tribolium castaneum, Rhyzopertha dominica* and *Oryzaephilus surinamensis*. Pirimiphos-methyl applied at a rate of 7.8 mg/kg was more effective than malathion (10.4 mg/kg) throughout a 9 month period. Pirimiphos-methyl persisted longer on the wheat than did malathion.

Bengston *et al.* (1980b) described duplicate field trials carried out on bulk wheat in commercial silos in Queensland and New South Wales where pirimiphos-methyl (6 mg/kg) plus carbaryl (10 mg/kg) was compared against a mixture of fenitrothion (12 mg/kg) plus phenothrin (2 mg/kg). The second combination was more effective against *Sitophilus oryzae* and *Ephestia cautella* but the order of protectiveness was reversed for *Sitophilus granarius*. Against *Rhyzopertha dominica, Tribolium castaneum, Tribolium confusum* and *Oryzaephilus surinamensis*, both treatments effectively prevented the production of progeny for 9 months.

Hindmarsh and MacDonald (1980) conducted field trials to control insect pests in farm-stored maize in Zambia. *Sitophilus zeamais* and *Sitotroga cerealella* are the main pests causing up to 92% damage to grain 8 months after harvest. Pirimiphos-methyl applied at the rate of 4 mg/kg kept damaged grains below 10% for up to 8 months after harvest but there was evidence of continued *Sitotroga cerealella* infestation.

Bansode *et al.* (1981) tested a field-collected strain of *Plodia interpunctella* and found it to be more than 227-fold resistant to malathion. However, its tolerance to pirimphos-methyl was only 1.9-fold. It was determined that increased carboxyesterase was the cause of the detoxification of malathion in this strain.

Al-Saffar and Al-Iraqi (1981) carried out studies in which pirimiphos-methyl was applied at 5, 10 and 15 mg/kg as an emulsion to wheat, barley and maize, which were then stored under warehouse conditions, and the residual effectiveness of the insecticides was tested with fourth-instar larvae of *Trogoderma granarium* and 14 day old adults of *Tribolium confusum*. Pirimiphos-methyl residues appeared to be more persistent than those of malathion on all treated grains. The persistence was roughly in proportion to the concentration applied and the effectiveness against the insects was greater on wheat than on maize, and greater on maize than on barley. *Tribolium* adults proved more susceptible than *Trogoderma* larvae. It was concluded that pirimiphos-methyl would be a suitable substitute for malathion to protect grains against malathion-resistant strains of stored-product insects.

Hampson (1981) published a review of the properties and uses of pirimiphos-methyl for the protection of stored products in Indonesia.

Newton (1981) presented the results of studies conducted in the United Kingdom to determine the susceptibility of numerous strains on *Oryzaephilus mercator* to pirimiphos-methyl incorporated into oatmeal. On the basis of these studies it was recommended that pirimiphos-methyl be incorporated in oatmeal rodent baits at a concentration of 40 mg/kg which would be effective against the most tolerant strains of this pest.

The **Tropical Products Institute (1981)** reported on the first outbreak of *Prostephanus truncatus* on the African continent on farms in Tanzania. To prevent the heavy losses which this beetle causes, farmers were advised to treat shelled maize at the rate of 10 mg/kg with pirimiphos-methyl applied as a dust.

Desmarchelier *et al.* (1981b) reported extensive pilot usage of grain-protectant combinations, including pirimiphos-methyl (6 mg/kg), together with bioresmethrin (1 mg/kg), applied commercially in 1976 to 21 storages. None of the 21 storages became infested with any species although a very low level of infestation with *Rhyzopertha dominica* occurred in 3 of 21 similar storages treated with a mixture of fenitrothion (12mg/kg) and bioresmethrin (1 mg/kg). These infestations were apparently caused by failure of the application equipment at several stages during the receival and treatment of the grain.

A similar series of trials involving a combination of pirimiphos-methyl and carbaryl was carried out the following year **(Desmarchelier *et al.* 1986)**. Again all storages containing wheat treated with pirimiphos-methyl and carbaryl remained free of infestation and bioassay of samples taken at regular intervals indicated that the grain would withstand invasion by insect pests during the storage period of more than 9 months.

Following extensive trials to evaluate grain protectants for use under tribal storage conditions in Zimbabwe, **Weaving (1981)** considered that pirimiphos-methyl dust applied to maize at the rate of 4 mg/kg would give at least 12 months protection against the major pest, *Sitophilus zeamais*, but there was reason to believe that some damage might be caused by *Sitotroga cerealella*. Both of these pests infest the maize cobs prior to harvest.

Viljoen *et al.* (1981b) reported that tests with small bins of maize treated in South Africa with pirimiphos-methyl at various rates showed that even the lowest rate of 2 mg/kg afforded full protection against *Sitophilus granarius* and *Sitophilus zeamais* for a test period of 55 weeks. Treatments at 2 and 4 mg/kg, however, failed to give protection against *Tribolium confusum* but 8 mg/kg was considered effective. *Rhyzopertha dominica* was not controlled by rates of 16 mg/kg or higher. In a field trial with bulk-stored maize, pirimiphos-methyl at 6.5 mg/kg was compared with a standard malathion treatment at 13 mg/kg and the results confirmed the small-bin tests. It was considered, however, that the application rate should be raised to 8 mg/kg to reduce the likelihood of resistance developing in *Tribolium confusum*.

Tyler and Binns (1982) studied the influence of temperature on the susceptibility to 8 organophosphorus insecticides of susceptible and resistant strains of *Tribolium castaneum, Oryzaephilus surinamensis* and *Sitophilus granarius*. Based upon knock down and kill, the effectiveness of all insecticides, including pirimiphos-methyl, was greater at 25°C then at 17.4°C and was markedly lower at 10°C. Pirimiphos-methyl proved effective against the malathion-resistant strains of all species, though at the lower temperatures death took several days.

Gelosi (1982), in a comprehensive article on the biology, importance and control of *Sitophilus oryzae* in Italy, recommended the spraying with pirimiphos-methyl of storages and vehicles used for transporation. **Gelosi and Arcozzi (1982)**, following laboratory and warehouse tests in Italy, with liquid and dust formulations of pirimiphos-methyl applied at 4–15 mg/kg for the protection of stored wheat against *Rhyzopertha dominica, Sitophilus oryzae, Sitophilus granarius* and *Oryzaephilus surinamensis*, found that the insecticide had good immediate and residual effects although it did not give complete control of *Rhyzopertha dominica* even at 8mg/kg. It was stated that pirimiphos-methyl had been approved in Italy for application to cereals. An almost identical article was published by **Cavaler and Frigato (1982)** following several years of experimentation with pirimiphos-methyl for the protection of stored cereals.

Ferguson and Waller (1982) compared the effectiveness of malathion and pirimiphos-methyl when applied to wheat in the laboratory in New Zealand. They reported that the LC/95 against adults of *Sitophilus granarius* was 0.75 mg/kg for pirimiphos-methyl and 3.4 mg/kg for malathion. The concentrations that gave 95% reduction in the progeny were 0.5 mg/kg for pirimiphos-methyl and 2.0 mg/kg for malathion.

Kumar *et al.* (1982) conducted field trials with some newer organophosphorus insecticides against insect pests of stored food grains in godowns in various parts of India, malathion being considered as the standard treatment. Although all the other compounds were superior to malathion, pirimiphos-methyl gave the best protection to grains stored either loose or in bags. It controlled *Tribolium castaneum, Rhyzopertha dominica, Oryzaephilus surinamensis, Cryptolestes pusillus* and *Ephestia cautella*.

Chahal and Ramzan (1982) evaluated synthetic pyrethroids and organophosphorus insecticides, including pirimiphos-methyl, against third-instar larvae of *Trogoderma granarium*. Although pirimiphos-methyl produced only a low level of mortality in the first 24 hours, a progressive increase in mortality was observed and after 7 days 100% mortality was achieved even with sprays of 0.025%.

Golob *et al.* (1982) set up trials under ambient conditions in southern Malawi to determine the effectiveness of admixing locally available powders with maize to protect the grain during storage. The protection afforded by wood ash admixed at 30% by weight was of the same order as that provided by admixing pirimiphos-methyl at 8.8 mg/kg.

Zhang *et al.* (1982), in laboratory tests in Zhejiang, China, showed that 30 mg/kg pirimiphos-methyl had no significant effects on the germination of rice, wheat, maize, sorghum and barley. There was an adverse effect on the germination of maize treated with 10 000 mg/kg of each of several pesticides.

Barson (1983) studied the effects of temperature and humidity on the toxicity of 3 organophosphorus insecticides, including pirimiphos-methyl, to adult *Oryzaephilus surinamensis*. All insecticides were more effective at 25°C than at 12.5–20°C, the difference in the LD/50 value for pirimiphos-methyl being 4.4-fold.

Bitran *et al.* (1983b) evaluated the residual action of some pyrethroid and organophosphorus insecticides for the control of *Sitophilus zeamais* infestation on stored maize in Brazil. Pirimiphos-methyl was more effective than malathion.

Bengston *et al.* (1983a) carried out duplicate experiments on bulk sorghum stored in concrete silos in south Queensland and in central Queensland. One of 4 treatments, consisting of pirimiphos-methyl (4 mg/kg) plus carbaryl (8 mg/kg), controlled typical malathion-resistant strains of *Sitophilus oryzae, Rhyzopertha dominica, Tribolium castaneum* and *Ephestia cautella*, as demonstrated by bioassay at frequent intervals over the 24 weeks storage.

Bengston *et al.* (1983b) reported duplicate field trials carried out on bulk wheat in commercial silos in Queensland and New South Wales in which pirimiphos-methyl (4 mg/kg) plus permethrin (1 mg/kg) plus piperonyl butoxide (8 mg/kg) controlled common field strains of *Sitophilus oryzae* and *Rhyzopertha dominica* and completely prevented progeny production in *Tribolium castaneum, Tribolium confusum* and *Ephestia cautella*.

Longstaff and Desmarchelier (1983) studied the effects of the temperature/toxicity relationship of pirimiphos-methyl and deltamethrin upon the population growth of *Sitophilus oryzae*. Pirimiphos-methyl and deltamethrin were shown to have opposite relationships with temperature, the toxicity of pirimiphos-methyl increasing with temperature over the range 21–32.3°C. The authors proposed several strategies to take advantage of this relationship. These strategies present considerable savings in the cost of insecticide and a reduction in the level of residues.

Golob *et al.* (1983) reported the results of a preliminary field trial to control *Prostephanus truncatus* attacking maize in Tanzania. Pirimiphos-methyl at 10 mg/kg gave adequate protection to grain during 27 weeks storage. Other materials tested were considerably less effective.

Measures recommended for the control of insect pests of stored grain in Bulgaria include treatment with pirimiphos-methyl at 8–20 mg/kg **(Tsvetkov 1983)**.

Wohlgemuth (1984) conducted comparative laboratory trials with insecticides under tropical conditions. He evaluated 18 commercial insecticides involving 14 different compounds applied to sorghum and stored at 36°C and 50% r.h. for 24 months. Samples were infested with 5 species of stored-product beetles at 1, 3, 6, 9, 12, 18 and 24 months pirimiphos-methyl was applied at the rate of 2 mg/kg instead of the intended 4 mg/kg. Although this rate was effective against 3 of the 5 species immediately after its application it failed to control *Rhyzopertha dominica* or *Tribolium castaneum*. It likewise failed to prevent reproduction of *Tribolium castaneum* but was effective in preventing reproduction of 3 other species for 1 to 6 months.

The response of larvae of *Tribolium castaneum* to pirimiphos-methyl was studied by exposing them to lethal concentrations in the flour/yeast rearing medium or to pirimiphos-methyl vapour **(Mondal 1984a)**. Larvae of all instars except the first were repelled by contact with the treated medium, those in the fifth and sixth instars showed the strongest response. There was no significant difference in the response to the different concentrations. No significant response was observed in larvae of any instar to pirimiphos-methyl vapour. In a separate study, **Mondal (1984b)** determined the dose/mortality response of *Tribolium castaneum* larvae to pirimiphos-methyl. Following 24 hours exposure to a treated flour/yeast medium, the LD/50 values were calculated. He found that first instar larvae were greater than 80 times more tolerant than second instar larvae which were the most susceptible stage. Tolerance increased gradually throughout the succeeding instars. It was concluded that the high tolerance of first instar larvae was due to their inability to consume significant quantities of the treated feeding medium.

White (1984) studied the residual activity of several organophosphorus and pyrethroid insecticides applied to wheat stored under simulated western Canadian conditions. Wheat treated with insecticides at 2, 4 or 8 mg/kg was bioassayed with adults of *Tribolium castaneum* and *Cryptolestes ferrugineus* at 30°C and 70% R.H. after 4 days, 2 weeks, then at 11 one-month intervals. Pirimiphos-methyl was more toxic to *Tribolium castaneum* than to *Cryptolestes ferrugineus* whereas the reverse was true for malathion. Pirimiphos-methyl gave nearly 100% control of both species for 9 months at 4 mg/kg and 11 months at 8 mg/kg, and of *Acarus siro* for at least 4 weeks at 8 mg/kg.

Bengston *et al.* (1984) reported a field experiment carried out on bulk sorghum stored for 26 weeks in concrete silos in south Queensland. Pirimiphos-methyl (6 mg/kg) plus permethrin (1 mg/kg) plus piperonyl butoxide (8 mg/kg), which was one of 4 combinations tested, allowed some survival of adults and progeny production by *Sitophilus oryzae* after 12 weeks, and by 1 strain of *Rhyzopertha dominica* throughout. All of the other combinations controlled typical malathion-resistant strains of *Sitophilus oryzae*, *Rhyzopertha dominica*, *Tribolium castaneum* and *Ephestia cautella*.

Golob and Muwalo (1984) described how, in Malawi, maize is stored on the cob, with the husk leaves intact, in cylindrical woven baskets. Until recently, lindane dust was recommended to protect the cobs from damage by *Sitophilus* spp. and *Sitotroga cerealella* but the use of this insecticide had been prohibited. Studies carried out in Malawi in 1978–80 evaluated pirimiphos-methyl as an alternative. The best results were obtained with a 2% dust applied as the cobs were being placed in the cylindrical store at a rate to give 18 mg/kg pirimiphos-methyl on the total cob weight (ie, 23 mg/kg on the shelled-out grain weight). When the stores were subsequently plastered with mud to minimise moisture uptake during the rainy season, this treatment resulted in less than 6% loss of grain weight over the 10 months of the experiments, as compared to 19–28% where there was no treatment.

Raju (1984), in an extensive discussion of various types of preventative measures adopted by farmers in India for preserving stored grains, emphasised the use of grain protectants, special attention being paid to *Rhyzopertha dominica*. Admixture of insecticides with grains may be extended to include pulses but not to flour or milled rice. Malathion was still the most widely used compound followed by pirimiphos-methyl.

Sun *et al.* (1984) carried out laboratory experiments in Taiwan to study the decrease in toxicity of insecticides and the survival and reproductive rate of *Sitophilus zeamais*. Adults were treated with sub-lethal doses of 4 insecticides, including pirimiphos-methyl. Whilst one increased the reproductive rate and another reduced the reproductive rate, a third had no effect at any of the doses tested. The reproductive rate of adults treated with the LD/50 of pirimiphos-methyl was 27.7% higher than that of untreated insects. Pirimiphos-methyl was reported to degrade 'relatively quickly'.

Carmi *et al.* (1984) reported that 900 tonnes of Israeli wheat at 32°C and 10–12% moisture content were treated with pirimiphos-methyl at 15 mg/kg using a gravity feed method with the insecticide being dripped onto the grain stream entering the silo. This treatment gave total protection from insects for 5 months, whereas an adjacent control bin and all other bins in the elevator were heavily infested with *Oryzaephilus surinamensis*.

Zakladnoi (1984), in describing the measures used in the USSR for protecting stored grain against infestation by arthropod pests, stated that grain and storage premises are treated with insecticides including pirimiphos-methyl.

Evans (1985) evaluated the effectiveness of pirimiphos-methyl and four other insecticides against 10 specimens of 5 species of stored-product pests collected in Uganda and which were found to be substantially resistant to lindane or/and malathion. Pirimiphos-methyl at 10 mg/kg was fully effective against *Tribolium castaneum, Tribolium confusum* and *Sitophilus zeamais* throughout 24 weeks when the culture medium was held at 27°C and 27% r.h.. However, it failed to prevent reproduction of *Sitophilus zeamais*, *Zabrotes subfasciatus* and *Callosobruchus maculatus* 42 days after adults were subjected to insecticide-treated substrate.

Golob *et al.* (1985) evaluated the susceptibility of *Prostephanus truncatus* to 11 insecticides including pirimiphos-methyl. Pirimiphos-methyl was relatively ineffective following topical application or exposure to treated filter papers but when applied at the rate of 10 mg/kg to maize grain it produced a high level of kill for 3 months and inhibited reproduction for the same period. When applied to maize cobs and shelled maize in the form of a 2% dust at the rate of 20 and 10 mg/kg, pirimiphos-methyl substantially reduced the damage to maize grain.

25.2 Degradation

In India, **Chawla and Bindra (1971, 1973)** reported that the rate of decline of pirimiphos-methyl residues from grain was significantly slower than that of malathion, bromophos or iodofenphos. The **Cyprus Agricultural Research Institute (1972)** observed, when testing a range of insecticides as grain treatments against a variety of stored-grain insects that, by the third month, most of the chemicals had lost their effectiveness, whereas pirimiphos-methyl was still effective 4 months after application. In the USA, **LaHue (1974)** studied the patterns of degradation of a number of potential grain protectants compared with the standard treatment, malathion. It was observed that pirimiphos-methyl residues were not as greatly influenced by high-moisture content of the grain at time of treatment as were those of malathion and fenitrothion.

Bowker (1973) studied the degradation of radio-labelled pirimiphos-methyl during 8 months storage, when applied as a 2% dust to wheat at 4 mg/kg. In grain of moisture content below 14%, only 20% degradation was observed during the storage time, whereas in grain with about 18% moisture, about 70% to 80% degradation was recorded. He found no significant levels of degradation products, other than the hydrolysis product 2-diethylamino-6-methyl pyrimidin-4-ol. In particular, amounts of the very unstable 'oxon' (III) were less than 0.01 mg/kg and of the N-desethylated thionate (II), approximately 0.05 mg/kg **(Plant Protection Division 1974)**.

Residue data were obtained during a 12 month period from hard winter wheat, shelled corn and sorghum grain treated with pirimiphos-methyl stored in bins in a laboratory **(LaHue 1974). Table 25.1** shows the average residues of pirimiphos-methyl in the stored grain over the 12 month period at 27°C. It is noticeable that the degradation appears to be independent of the moisture content of the grain under these conditions. Malathion, included in the same trial, degraded to about one-tenth of the initial dosage within the same period. In the case of the high moisture sorghum, it was almost entirely destroyed within 1 month.

Table 25.1. Average residues of pirimiphos-methyl (mg/kg) in stored grain over 12 months at 27°C **(LaHue 1974)**

Grain	Moisture content (%)	Intended dosage (mg/kg)	Post-treatment periods					
				Months				
			24 hrs	1	3	6	9	12
Wheat	12.5	7.8	6.5	6.1	5.6	6.2	4.9	5.4
Corn	12.5	8.4	7.9	6.3	4.5	4.1	3.4	3.0
Sorghum	17.6	8.4	7.5	6.5	4.3	3.8	3.8	3.7

Seth (1974) reported trials to evaluate pirimiphos-methyl for the control of stored-rice pests in South East Asia. Rice in husk and polished rice were treated with EC and dust formulations at rates equivalent to 2, 4 and 8 mg/kg. The rice was analysed at monthly intervals throughout the 4 month trial. In the case of the polished grains, the residue at the end of 4 months was approximately 25% of the residue found on the first day of the trial. In the case of rice in husk, it is not possible to deduce the degradation because the residue data are expressed on hull and milled grain separately. However, there appeared to be relatively little loss of pirimiphos-methyl over the 5 month period. Relatively little of the insecticide (generally less than 0.5 mg/kg) transferred to milled grain. There was no significant difference between the dust and EC formulations.

Udeaan *et al.* (1974), using pirimiphos-methyl in an 'insecticide treating machine study', found a grain-to-grain variation of 0.45 to 1.90 μg/wheat grain. They went on to discuss the variations in insecticidal deposit on the treated wheat but as the treatment involved soaking grains in acetone solution of pirimiphos-methyl for 24 hours it is of little value to attempt any assessment of uptake and distribution from their work.

Rowlands (1975) reported that L. Horvath studied the breakdown and recovery of radio-labelled pirimiphos-methyl in wheat treated at 4 mg/kg with a solution in hexane and stored under laboratory conditions at 21°C. He found that, after 8 months, 78% of the radioactivity was still recoverable as intact parent compound and 9–10% was bound to lipid and proteinaceous matter in the grain. This could only be recovered by digestion and was thought to be associated chiefly with the protein. The only metabolite detected was the hydroxy-pyrimidine (IV).

Rowlands and Wilkin (1975) applied solutions of radio-labelled pirimiphos-methyl and also a 2% dust to wheat of 14% and 18% moisture to give 4 mg/kg final treatment. The samples were then stored in a laboratory at 20°C for 6 months. They, too, found that approximately 10% of the radioactivity was bound to lipoprotein material and was unextractable except by digesting the aleurone regions of the grain. Such labelled material, as was recovered by this means, appeared to be unchanged pirimiphos-methyl, but degradation during this extraction and subsequent purification hampered identification. During the storage period, the breakdown rates were approximately the same for both solvent and dust treatments. After 6 months, 15% and 50% degradation had occurred at the 14% and 18% moisture levels, respectively. The main degradation products were the free hydroxy-pyrimidine (IV) and the N-desethyl-pyrimidinol (V).

Bengston *et al.* (1975), in studying the effect of pirimiphos-methyl against malathion-resistant insects and its fate in bulk wheat under typical semi-tropical conditions, found the deposit degraded rather slowly. Deposits of 2 and 4 mg/kg degraded to 1.62 and 2.94 mg/kg after 25 weeks respectively. The authors estimated the half-life of such deposits to be of the order of 45 weeks (temperature and moisture conditions not stated).

Weaving (1975) found pirimiphos-methyl, applied to maize and sorghum so as to give a deposit of 5 and 10 mg/kg, gave complete kill of *Sitophilus zeamais* for 12 months under tribal storage conditions in Rhodesia.

LaHue (1975a) applied pirimiphos-methyl at the rate of 8.4 mg/kg to shelled maize in a trial to control *Sitotroga cerealella*. The moisture content of the corn averaged 13.4% and it was held in a room maintained at 26.4°C and 60% r.h. Residues analysis carried out at monthly intervals showed that the concentration of pirimiphos-methyl had declined from 7.9 mg/kg, 24 hours after application, to 4.0 mg/kg, 8 months later.

Markem (1975b) reported that farm-stored barley of 18% moisture content, treated with pirimiphos-methyl to control mite infestations, caused virtually no breakdown of pirimiphos-methyl over 3 months.

LaHue and Dicke (1976a) found that pirimiphos-methyl, applied at the rate of 8.4 mg/kg to high-moisture sorghum, gave excellent protection against damage by 6 different insect pests for a period of 12 months. About 55% of the initial deposit remained after 12 months.

Morallo-Rejesus and Carino (1976b) reported pirimiphos-methyl was more persistent on maize than on sorghum but apparently little allowance had been made for any difference in temperature or moisture content.

Desmarchelier (1977b) reported that the degradation of pirimiphos-methyl over a 35 week period was too slow to allow accurate measurements of temperature effects (experiments conducted over the range 20°C to 30°C and 11.5% to 12.5% moisture content).

LaHue (1977a) noted that pirimiphos-methyl degraded much more slowly than did malathion under similar conditions because 83% of the original deposit of pirimiphos-methyl remained on wheat 12 months after treatment, whereas only 16% of the malathion deposit remained under identical conditions.

Cerna and Benes (1977) provided results of a study carried out in Czechoslovakia where 400 tonnes of wheat, containing 12.6–14.2% moisture at a temperature of 8–10°C, was treated with pirimiphos-methyl at a rate to provide 4 mg/kg. The wheat was analysed at the time of treatment and at intervals thereafter until discharged for milling at the end of 286 days. Over this period, the residue levels declined from 3.7 to 1.62 mg/kg.

Banks and Desmarchelier (1978) discussed the finding that the loss of insecticide residues from stored grain follows pseudo first-order reaction kinetics and that this model applies equally well to pirimiphos-methyl **(Desmarchelier 1977b; Desmarchelier 1978)**. They pointed to the influence of water vapour and temperature on changes in pesticide residue levels with time and drew attention to the errors introduced in calculating the rate of degradation from a 'linear' or a semi-logarithmic model.

Desmarchelier and Bengston (1979a) further developed this concept and explained how the mathematical models are developed and used. They compared the rate of degradation of 12 grain protectants. The half-life of pirimiphos-methyl at 30°C and 50% relative humidity was given as 70 weeks. This compared with 12 weeks for malathion under similar conditions. The interesting point is that the co-efficient of variation with respect to temperature is much smaller for pirimiphos-methyl than all other compounds considered.

Cerna *et al.* (1978) determined the fate of pirimiphos-methyl residues in wheat which had been held in storage for 9.5 months after treatment with pirimiphos-methyl at the rate of 4 mg/kg. They found a decrease in residue concentration of 56.2% in this time.

Mensah *et al.* (1979b) conducted an extensive experiment in which the fate of 4 insecticides and their distribution throughout the grain kernel was determined following the application of 2 dosage rates on wheat of 2 different moisture contents. The results **(Table 25.2)** indicate that the degradation is not significantly affected by the dosage or the moisture content of the wheat. This contrasts sharply with similar data for malathion, bromophos and iodofenphos which show a much more rapid rate of degradation with a relatively greater rate on the wheat of higher moisture content.

Hindmarsh and MacDonald (1980), who conducted extensive field trials to control insect pests of farm-stored maize in Zambia, found that pirimiphos-methyl, applied as a dust at the rate of 4 mg/kg, degraded by less than 30% over the storage period of 7 months when the fluctuating moisture level in the shelled maize ranged from 9.8% to 12.2%.

Quinlan *et al.* (1980), conducted an experiment to evaluate pirimiphos-methyl as a protectant for high-moisture stored wheat. The field tests were conducted in 1.5 tonne plywood bins and the wheat of 14.5% moisture content was treated with pirimiphos methyl at a calculated dosage of 7.8 mg/kg. The recovery of pirimiphos-methyl immediately after treatment averaged 41% of the amount applied; 9 months later, about 38% was recovered; in contrast, the immediate recovery of malathion averaged 82% of the amount applied; 9 months later 6% was recovered. In a laboratory study carried out simultaneously, pirimiphos methyl was applied to wheat of 10, 12, 14, and 16% moisture content at a dosage rate of 7.5 mg/kg. The samples of grain, held in sealed jars at 26.6°C, were analysed at various intervals over 9 months and the results are presented in **Table 25.3**. This shows a different result to that obtained in the field trial and contrasts with the results of **Mensah *et al.* (1979b) (Table 25.2)**.

Table 25.2. Fate of pirimiphos-methyl residues on stored wheat **(Mensah *et al.* 1979b)**

Storage period months	Dosage (mg/kg)	12% Moisture		16% Moisture	
		Residues	% Remaining	Residues	% Remaining
1	4	3.29	82	3.24	81
3	4	2.98	75	3.04	76
6	4	2.54	64	2.26	57
1	6	4.93	82	4.92	82
3	6	4.54	75	4.44	74
6	6	3.66	61	3.86	64

Table 25.3. Fate of pirimiphos-methyl residues on wheat at 4 moisture levels after application at 7.5 mg/kg. Wheat was stored in sealed jars at 26.6°C **(Quinlan *et al.* 1980)**

Interval after treatment	Pirimiphos-methyl residues (mg/kg) at indicated moisture content (% of original dose applied)			
	10%	12%	14%	16%
24 hours	5.2 (69)	4.9 (65)	4.7 (63)	5.1 (68)
6 weeks	4.8 (64)	4.6 (61)	4.7 (63)	2.9 (39)
3 months	4.3 (57)	4.2 (56)	4.3 (57)	1.8 (24)
6 months	4.1 (55)	3.9 (52)	4.0 (53)	1.2 (16)
9 months	3.9 (52)	3.9 (52)	3.5 (47)	0.9 (12)

Sherjugjit Singh and Chawla (1980) conducted periodic analysis by GLC of stored wheat, maize and paddy, treated with pirimiphos-methyl at 4 and 8 mg/kg, and showed that the insecticide degraded slowly by about 50% in 6 months. The persistence pattern was similar on all 3 varieties of grain.

Desmarchelier *et al.* (1981b) provided extensive information from 21 commercial wheat storages, which were treated with pirimiphos-methyl at the rate of 6 mg/kg. Grain condition and protectant residue levels were regularly monitored and it was found that the residue level declined from 6 to 4 mg/kg over ten months on grain that remained at 30°C for seven months and then cooled gradually to 20°C. The moisture content of this wheat was between 11 and 12%. The mean observed and predicted residue levels of pirimiphos-methyl were plotted at intervals after application and **Figure 25.1** shows the high level of agreement with the model based on pseudo first-order kinetics.

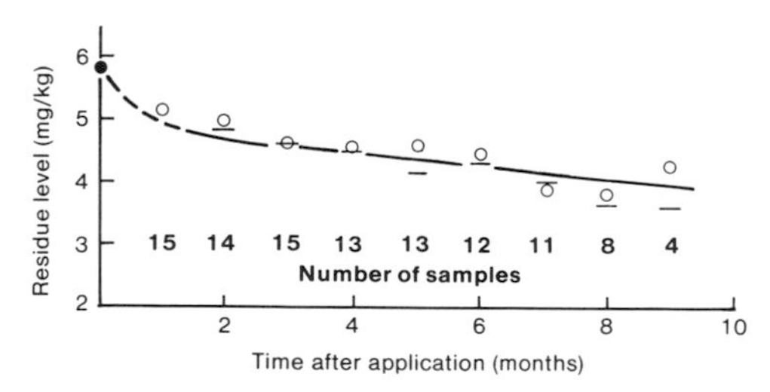

Figure 25.1. Mean observed (○) and predicted (−) residues level of pirimiphos-methyl at intervals after application **(Desmarchelier *et al.* 1981b).**

Bengston *et al.* (1980b) reported duplicate field trials carried out on bulk wheat in commercial silos, where pirimiphos-methyl was applied at the rate of 6 mg/kg in conjunction with phenothrin (2 mg/kg). The effectiveness against a range of stored-product pests and the residue levels were monitored throughout a period of 9 months during which the pirimiphos-methyl residue level did not change appreciably. The moisture content of the grain ranged from 10.7% to 11.1% and the temperature from 26°C to 29°C.

Bengston *et al.* (1980a) provided data from field trials with various pesticide combinations carried out on bulk wheat in commercial silos in Queensland, South Australia and Western Australia. Once again, the concentration of pirimiphos-methyl declined only slightly during the 8 months storage period.

Desmarchelier *et al.* (1980b) published the results of an extensive collaborative study of residues on wheat of methacriphos, chlorpyrifos-methyl, fenitrothion, malathion and pirimiphos-methyl. Pirimiphos-methyl degraded more slowly and to a lesser extent than the other compounds. The measured values of residues of pirimiphos-methyl on wheat at the experimental sites agreed completely with the values predicted from the calculation using first-order kinetics, grain temperature and interstitial relative humidity. The rate of degradation was slow.

Desmarchelier *et al.* (1980a) reported studies on the fate of pirimiphos-methyl and 5 other grain protectants on rice and barley after storage and during processing. The level of residues were determined on unhusked rice, husked rice, polished rice and barley over a storage period of 6 months. The observed levels were close to levels predicted from use of a model which relates rate of loss of residue levels to a rate constant and only 2 variables, temperature and equilibrium relative humidity. The difference between predicted and observed values of residues of pirimiphos-methyl on the 4 commodities was 8%.

Rowlands (1980) showed about 40% loss of ^{14}C-pirimiphos-methyl from wheat held at 20°C for 6 months. The major product detected was the free 2-diethylamino-6-methyl-pyrimidin-4-ol with lesser amounts of the free N-desethyl pyrimidinol. Some 15% of the applied radioactivity was not recoverable by solvent extraction but was released by sequential enzyme digestion (amylase, lipase, protease and cellulase). The nature of the bound material was not determined.

Weaving (1981) applied pirimiphos-methyl at the rate of 8 mg/kg to shelled maize under tribal storage conditions in Zimbabwe though only 37.5% of this (2.97 mg/kg) was recovered by analysis at the end of 4 months. There was apparently no further degradation by 12 months after treatment. After 24 months, half of the amount present at 4 months (1.4 mg/kg) was still recoverable by analysis.

Ivbijaro (1981), working in Nigeria, concluded from the nature and level of infestation of crib-stored maize with insects, especially *Sitophilus spp.*, that the application of pirimiphos-methyl sprays to dehusked maize at the rate of 10 and 15 mg/kg was effective for only 3 months. At 5 mg/kg the protectant action did not last beyond 1 month. He believed retreatment would be necessary to protect maize seed during long storage because of insect damage to the grain embryo.

Leahy and Curl (1982) studied the degradation of pirimiphos-methyl on wheat, paddy rice and brown rice grains stored at moisture contents of 13% and 20%. The grains were sprayed with ^{14}C-pirimiphos-methyl and stored at 20°C under controlled humidity conditions for 12 and 20 weeks. At a moisture content of 13%, degradation was slow, so that at least 70% of the radioactive residue was unchanged pirimiphos-methyl after 24 weeks. Faster degradation occurred at 20% moisture content, but the major component of the radioactive residue was still unchanged pirimiphos-methyl after 24 weeks. Two major degradation products were found on the grains; one was characterised as the pyrimidin-4-ol, and the other a conjugate of the same compound. They also identified trace amounts of the N-monoethyl parent compound and the corresponding pyrimidinol, and an unknown metabolite. Radioactivity not extractable by solvents, as determined by combustion, accounted for 3–7% of the applied activity after 3 months storage and 2–13% after 6 months. However, there was virtually no 'bound' material in the dehusked rice and this was matched by the almost negligible breakdown of pirimiphos-methyl on dehusked rice; only 5% loss after 6 months storage as against 25% in paddy and in wheat where the seed coats were present. This work seems to confirm the importance of the seed coat/husk in the degradative process.

Bengston *et al.* (1983a) carried out duplicate field experiments on the control of insect infestation in stored sorghum. Residues of pirimiphos-methyl declined from 3.9 to 3.0 mg/kg over a period of 24 weeks, the average temperature being 26°C and the average moisture content 12.4%.

Bengston *et al.* (1983b) applied pirimiphos-methyl (4 mg/kg) plus permethrin (1 mg/kg) plus piperonyl butoxide (8 mg/kg) to compare with other organophosphorothioates and synergised synthetic pyrethroids as grain protectants on bulk wheat in commercial silos in Queensland and New South Wales. Although the amount of residue recovered by analysis, 1 week after treatment, was only 57% and 70% of the calculated application rate, the amount remaining at the end of 9 months storage was 77% and 78% of the amount present after 1 week **(Table 25.4)**.

Bengston *et al.* (1984) treated bulk sorghum stored in concrete silos in south Queensland for 26 weeks with pirimiphos-methyl (6 mg/kg) plus permethrin (1 mg/kg) plus piperonyl butoxide (8 mg/kg). The amount of residue determined analytically 1 week after treatment was approximately 30% less than the calculated application rates but thereafter there was practically no change over the remaining 26 weeks. The mean grain temperature was 21.1°C and the mean moisture content 11.6%.

Desmarchelier *et al.* (1986) applied pirimiphos-methyl (4 mg/kg) plus carbaryl (8 mg/kg) to bulk grain in commercial silos at 15 sites throughout Australia. They monitored the temperature and moisture conditions and the residue levels frequently during 9 months storage. The results obtained **(Table 25.5)** were compared with those predicted from the mathematical models described by **Desmarchelier and Bengston (1979a).** Notwithstanding all of the difficulties that can arise in such a study, there was a remarkably high level of agreement.

Table 25.4. Calculated pirimiphos-methyl application rate and analytically determined residues on treated wheat stored in Queensland and New South Wales at varying times following spray application **(Bengston *et al.* 1983b).**

Site	Calculated application (mg/kg)	Pirimiphos-methyl residues (mg/kg) at various times after application (months)						
		0.25	1.5	3	4.5	6	7	9
Qld	4.0	2.3	2.3	2.1	1.8	1.9	–	1.8
NSW	3.7	2.6	–	2.8	2.9	2.3	2.2	2.0

Table 25.5. Residues of pirimiphos-methyl on wheat in commercial storages (**Desmarchelier *et al.* 1986**)

Amount applied		Residues (mg/kg) at time (months) after application								
		1	2	3	4	5	6	7	8	9
Mean	4.0	2.8	2.6	2.6	2.9	2.9	2.4	2.5	1.5	1.8
Predicted	4.0	3.1	3.0	2.9	2.7	2.6	2.5	2.4	2.4	2.3
Number*	(16)	(13)	(12)	(6)	(8)	(13)	(11)	(5)	(5)	(4)

* Number of analytical results on which mean is calculated.

25.3 Fate in Milling, Processing and Cooking

Bullock (1973, 1974) reported many separate experiments which demonstrated that residue levels of pirimiphos-methyl are signficantly reduced during the milling and baking processes. **Table 25.6** summarises the results of residue trials carried out in the UK on wheat that had been treated to contain nominally 4 mg/kg pirimiphos-methyl.

Table 25.7 summarises results of a residue trial reported by **Bullock (1973)** to have been carried out in the

Table 25.6. Effect of milling and baking on residues in wheat admixed with pirimiphos-methyl at 4 mg/kg—UK (**Bullock 1973, 1974**)

Grain fraction	Interval between treatment and sampling (months)	Residues (mg/kg)[1] Highest	Lowest	Mean	
Whole grain	0	4.2	1.9	2.9	(9)
		3.0*	–*	3.0*	
	1	4.1	1.5	2.8	(9)
	2	4.1	1.6	2.6	(8)
		2.5*	2.4*	2.5*	(3)
	3	3.6	1.3	2.3	(7)
Wholemeal flour	0	1.3	0.94	1.1	(3)
	1	2.1	1.1	1.7	(4)
	2	2.2	1.2	1.7	(3)
		1.5*	1.4*	1.5*	(3)
	3	2.1	1.0	1.5	(4)
White flour	0	0.88	0.30	0.52	(6)
		0.56*	0.53*	0.55*	(3)
	1	0.77	0.44	0.59	(5)
	2	0.64	0.24	0.56	(6)
		0.29*	0.24*	0.27*	(3)
	3	0.67	0.38	0.56	(3)
Wholemeal bread	0	0.72	0.53	0.64	(4)
	1	0.91	0.55	0.79	(4)
	2	1.1	0.65	0.93	(4)
		0.97*	0.82*	0.88*	(3)
	3	0.54	0.21	0.49	(3)
White bread	0	0.28	0.19	0.23	(6)
		0.26*	0.24*	0.25*	(3)
	1	0.36	0.22	0.30	(5)
	2	0.45	0.31	0.36	(8)
		0.15*	0.13*	0.14*	(3)
	3	0.54	0.21	0.43	(3)

1 All results are from field trials except those marked* which are from a small-scale trial. Figures in parentheses are the numbers of results upon which the means are based.

Table 25.7. Residues of pirimiphosmethyl in whole grains and in milling and baking fractions of whole grains treated in a laboratory trial in UK at 8 mg/kg. **(Bullock 1963)**

Interval between treatment and sampling	Residues, mg/kg[1]				
	Whole grain	White flour	White bread	Wholemeal flour	Wholemeal bread
0 days	6.0	0.86	0.52	–	–
	6.0	0.91	0.56	–	–
	6.0	1.0	0.57	–	–
9 weeks	4.8	0.47	0.28	3.2	1.7
	5.2	0.59	0.33	3.0	1.6
	5.2	0.60	0.30	3.2	1.5

1 In all cases, no residues of the phosphorus-containing compounds (II) or (III) were detected. (Limit of detection: 0.01 ppm in each case.)

UK with wheat nominally treated to contain 8 mg/kg pirimiphos-methyl. These data are substantially in agreement with those of **Bengston *et al.* (1975)** and show that there is relatively little penetration beyond the seed coat, even throughout a storage period of nine weeks.

Seth (1974) discussed experiments concerned with the admixture of pirimiphos-methyl with paddy rice (rice in husk) in South East Asia. Both emulsion and dust formulations were used, being applied at the rate of 2, 4 and 8 mg/kg. Portion of the rice was milled immediately after treatment and further portions at the end of 1, 2, 3 and 5 months, to produce milled (husked or brown) rice and rice husks. It was found that the deposit of pirimiphos-methyl on both the husk and the milled grain was proportional to the amount applied, but the concentration on the husk was always about 25 times that on the milled grain. The residue on the husk and milled grain declined steadily over the 5 month storage period. Though initially the concentration on the husk was as high as 12 mg/kg, the residue on the milled rice rarely exceeded 0.5 mg/kg and even at the highest rate of application was always below 1 mg/kg. These experiments clearly showed that there is minimum transfer of pirimiphos-methyl from the husk to the kernel of rice, even during prolonged storage. The degree of penetration is comparable to that reported by **Kadoum and La Hue (1974)** for malathion on wheat, corn and sorghum.

Residues of pirimiphos-methyl in flour are relatively stable to the conditions found during baking to bread and biscuits. However, because of the dilution which the flour undergoes during these processes, flour initially containing 1 mg/kg pirimiphos-methyl is likely to yield bread/biscuits containing residues of the order of 0.5 mg/kg. **Bullock *et al.* (1976)** studied the fate of pirimiphos-methyl during processing of flour into bread and biscuits. In studies with radio-labelled compound, flour was dosed with 2-^{14}C-labelled pirimiphos-methyl and baked to produce white bread, wholemeal bread and biscuits. Although pirimiphos-methyl is known to be a relatively volatile compound, there was no significant loss of radioactivity by volatilisation. Distribution of radioactivity throughout the bread was fairly uniform. Unchanged pirimiphos-methyl accounted for 75–90% of the radioactivity in the baked product. The major degradation product formed during the baking was hydroxypyrimidine, which accounted for 3–10% of the radioactivity in the final product.

Similar results were obtained by residue analysis in a second set of studies. After correcting for the weight increase when flour is converted to bread, residues of pirimiphos-methyl fell by 11–18%. Likewise, residue analysis of biscuits showed average losses of 8%. However, owing to the dilution of the flour, which occurs during baking, the residue of pirimiphos-methyl in bread will be lower than that in the corresponding flour. **Bullock *et al.* (1976)** found levels of pirimiphos-methyl in bread and biscuits to be about 50% of those in the flour from which they were derived. These findings agreed well with the earlier work of **Bullock (1973, 1974)**. No residues of the oxon or the N-desethyl pirimiphos-methyl (phosphorus-containing compounds II and III), **(Figure 25.2)** were detected in bread baked from flour treated with pirimiphos-methyl at 1 mg/kg or in biscuits baked from flour containing up to 5 mg/kg **(Bullock *et al.* 1976)**.

The hydroxypyrimidine (IV) **(Figure 25.2)** also undergoes little degradation during the baking process. This compound constitutes only a minor part (generally less than 0.5 mg/kg) of the residue in stored grains **(FAO/WHO 1975)**. **Bullock *et al.* (1976)** used radio-labelled compound IV and found that it degraded by less than 10% during the baking of bread. It can therefore be concluded that residues of compound IV in baked products will never exceed 0.2 mg/kg and will normally be considerably lower.

Bullock and May (1976) studied the fate of residues in wheat during processing to semolina and pasta. White semolina, prepared from Durum wheat treated at 10 mg/kg, contained only 1.6 mg/kg pirimiphos-methyl. Pirimiphos-methyl levels in both white and wholemeal pasta were approximately 85–90% of those in the corresponding semolina. Of the pirimiphos-methyl residue in semolina 70% was transferred unchanged to cooked pasta. However, the weight of pasta increases by 100% on cooking so that the concentration of pirimiphos-methyl in cooked pasta is likely to be approximately 35% of that in the corresponding semolina.

Results of experiments in Czechoslovakia **(Cerna and Benes 1977; Cerna *et al.* 1978)** indicated that the residues of pirimiphos-methyl in grain, which had been in store for 9 months, were substantially removed by the milling process. The bulk of the residue was removed in the bran, there being no substantial difference in the concentration in the different bran fractions. When white flour was made into white bread, there was a further loss of approximately 50% so that the concentration of the residue in the bread was only 10–15% of its concentration in raw grain. Some residue was destroyed during milling.

Magallona and Celino (1977), wishing to extend some of the information available on the fate of pirimiphos-methyl residues on rice subjected to preparation and cooking **(FAO/WHO 1975; Bullock 1973)**, carried out a series of experiments in which rice was washed and cooked according to traditional Asian practices. Milled rice was first treated with pirimiphos-methyl at 10, 20 and 30 mg/kg by application of a diluted pirimiphos-methyl emulsion concentrate in the conventional manner. After storing for 1 week the rice was washed and cooked and the amount of pirimiphos-methyl in the uncooked rice, washings and cooked rice was determined. Washing and cooking destroyed or removed approximately 50% of the initial deposit (45.0%–52.4%) with from 9–14% being found in the washings. When the rice was cooked, in an unwashed condition, only an average of 18% (10.9–28.5) of the residues were removed. It would perhaps be unsafe to extrapolate from these data to the practical situation which occurs in commercial practice where (a) milled rice is not likely to be treated with pirimiphos-methyl and (b) the amount of residue on milled rice derived from the treatment of paddy rice would be in the range of 1 mg/kg, 10–30 times less than used in these experiments.

Rai (1977) carried out field studies with paddy rice treated with pirimiphos-methyl at an intended dosage of 13.7 mg/kg and stored in jute bags in a warehouse in Guyana. The rice was sampled by a probe at frequent intervals and subjected to mechanical milling or parboiling and milling through a standard laboratory rice mill. Samples were sent to the United Kingdom for analysis. The treated paddy rice was reported to contain 3.13 mg/kg pirimiphos-methyl and the residues 0, 1, 2, 3 and 4 months after treatment where 9.8, 9.0, 9.9, 7.0 and 9.4 mg/kg respectively for husk and 0.64, 0.66, 0.92, 0.70 and 0.72 mg/kg respectively for dehusked grain indicating that the bulk of the deposit was on the husk and that the residues did not decrease significantly with time, at least over a 4 month period. The dehusked grain obtained after parboiling had a higher residue than was present in the mechanically dehusked grain, whilst the residue level in the husk after parboiling was lower than that found in the mechanically removed husk. Thus, it would appear that parboiling caused some movement of the insecticide from husk to grain. A considerable reduction of residue level occurs during rice whitening (milling to remove the bran — sometimes referred to as polishing). The reduction occurs whether or not paddy is parboiled, although the reduction is smaller after parboiling. This could be due to the penetration of insecticide into the grain during the parboiling process.

Mensah *et al.* (1979b) studied the distribution of pirimiphos-methyl residues in milled fractions of dry (12% moisture) or tough (16% moisture) wheat held in storage for 1, 3 or 6 months after the application of 10 and 15 mg/kg pirimiphos-methyl. The results obtained are given in **Table 25.8.**

It is clear that these data indicate a relatively slow decomposition with time and a limited penetration into the endosperm. However, it is difficult to know whether the differences shown represent a true difference in distribution due to moisture content or whether they reflect the difficulties involved in obtaining a uniform separation of the milling fractions with every sample processed under laboratory conditions.

Tempone (1979) reported a series of studies on the effects of insecticides on barley malting and the resulting residues. In the first of these trials, barley treated with pirimiphos-methyl at the rate of 6 mg/kg or 18 mg/kg was converted into malt but no residues were detected in the wort (unfermented extracts from the malted barley), the limit of determination being 0.004 mg/kg. Wort, prepared from barley which had been malted after being in store for 3, 6 and 9 months, likewise showed no residues. In the second series of similar trials the residue was determined in the malt (germinated grain which had been calcined). Residues could be found in the malt but at a level of from 10–20% of that in the barley before malting. Barley which had been kept in storage for 3 months after being treated with pirimiphos-methyl transferred significantly less residue to the malt than did the barley malted immediately after treatment, no doubt because there was significantly less residue on the outside of the kernel where it would be protected from the enzymatic activity within the barley grain during germination.

Bengston *et al.* (1980b) arranged for wheat from bulk grain treated with pirimiphos-methyl and held in commercial silos to be processed through to wholemeal bread and white bread. The results obtained in these trials are set out in **Tables 25.9** and **25.10.** During processing from wheat to white bread, residues were reduced by 85–91%.

Desmarchelier *et al.* (1980b), having treated bulk wheat in commercial silos with pirimiphos-methyl, arranged for a portion to be milled and for the white flour to be converted into white bread. The results are given in **Table 25.11**. It is of interest to note that the residues in flour and bread were higher in the hard wheat, held for 22

Table 25.8. Mean[a] ± SE pirimiphos-methyl residues (ppm) on dry (12% moisture content) and tough (16% moisture content) wheat, and milled fractions after 1, 3, and 6 months of storage on wheat treated at 4 or 6 ppm (**Mensah *et al.* 1979b**)

Dosage (ppm)	Storage period (months)	12% moisture content				16% moisture content			
		Whole wheat[b]	Bran	Middlings[c]	Flour	Whole wheat[b]	Bran	Middlings[c]	Flour
4	1	3.29 ± 0.07	15.66 ± 0.79	21.44 ± 1.96	1.94 ± 0.25	3.24 ± 0.03	19.81 ± 1.63	15.35 ± 0.58	1.57 ± 0.03
	3	2.98 ± 0.10	13.93 ± 0.26	15.10 ± 1.00	1.93 ± 0.12	3.04 ± 0.01	13.45 ± 0.80	13.51 ± 0.36	1.16 ± 0.06
	6	2.54 ± 0.07	13.16 ± 0.51	13.80 ± 1.36	1.56 ± 0.06	2.26 ± 0.18	11.98 ± 0.67	10.68 ± 0.60	0.81 ± 0.23
6	1	4.93 ± 0.13	20.87 ± 0.82	23.15 ± 1.47	2.86 ± 0.14	4.92 ± 0.04	27.33 ± 1.41	21.83 ± 1.73	1.91 ± 0.07
	3	4.54 ± 0.10	20.54 ± 1.38	19.34 ± 0.74	2.42 ± 0.22	4.44 ± 0.04	20.90 ± 2.92	17.89 ± 0.59	1.50 ± 0.08
	6	3.66 ± 0.23	19.32 ± 0.73	16.05 ± 1.35	2.23 ± 0.13	3.86 ± 0.26	18.60 ± 0.73	16.89 ± 0.56	1.33 ± 0.14

a Mean of 3 replicates
b Ground wheat
c Shorts, wheat germ, and coarse particles of flour

Table 25.9. Mean residues of pirimiphos-methyl following milling and baking of wheat stored at 2 sites (**Bengston *et al.* 1980b**)

Site	Storage (weeks)	Pirimiphos-methyl residue mg/kg						
		Wheat	Bran	Pollard	Wholemeal flour	White flour	Wholemeal bread	White bread
Site B	13	6.2	20.6	10.7	5.9	2.0	3.1	0.92
Site D	19	3.2	12.6	5.6	3.1	0.72	1.4	0.30

Table 25.10. Reduction in pirimiphos-methyl residue contents following milling and baking of wheat stored at 2 sites (**Bengston *et al.* 1980b**)

Site	Reduction in residue %					
	Wheat to Wholemeal Flour	Wheat to White Flour	Wholemeal flour to Wholemeal Bread	White Flour to White Bread	Wheat to Wholemeal Bread	Wheat to White Bread
Site B	5	68	47	54	50	85
Site D	0	78	55	58	56	91

Table 25.11. Residues of pirimiphos-methyl in wheat, its milling products and bread **(Demarchelier *et al.* 1980b)**

Type of Wheat	Storage (weeks)	Pirimiphos-methyl residues (mg/kg)					
		Initial deposit	Whole grain	bran	pollard	flour	bread
Soft	11	5.2	4.2	9.1	8.3	0.2	0.1
Hard	22	6.1	4.5	12.4	10.4	1.3	0.35

weeks before milling, than in the soft wheat, held for 11 weeks before milling.

Desmarchelier *et al.* (1980a) studied the fate of pirimiphos-methyl and a number of other grain-protectant insecticides, applied to unhusked rice, husked rice, polished rice and barley over a storage period of 6 months and subsequently during the processing and cooking of these grains. The results of these trials are given in **Table 25.12**. These show that only about 10–15% of the residue present on husked rice or polished rice is destroyed in the cooking process. However, if the treatment is applied to unhusked rice, approximately 70% of the residue is removed with the husk and a great deal more when the husked rice is milled for the removal of the bran. Subsequent cooking of the husked rice or polished rice brings about a further substantial reduction in the residue level.

These same workers found that pirimiphos-methyl, applied to barley destined for malting, was substantially lost during the malting process. Barley treated at the rate of 6 mg/kg and held in storage for 6 months was found to contain 4.9 mg/kg pirimiphos-methyl. When this grain was malted, the pirimiphos-methyl residue in the prepared malt was only 0.9 mg/kg.

Sherjugjit Singh and Chawla (1980) compared the persistence of residues of pirimiphos-methyl on wheat, maize and paddy rice in storage. They found that 97.2–98.3% of the residue on paddy rice was in the husk. In maize 77.1–81.9% of the pirimiphos-methyl was lost during processing into popcorn.

Wilkin and Fishwick (1981) treated wheat with the recommended dose of 6 grain-protectant insecticides and after keeping the wheat in storage for varying periods ranging from 4 to 36 weeks converted it into wholemeal flour by passing it through a hammer mill. The wholemeal flour was processed into wholemeal bread and each of the commodities was analysed for residues. The data on pirimiphos-methyl are given in Table 25.13. These show that in converting the aged wheat to bread approximately 50% of the primiphos-methyl was lost (based on calculations made on a 'moisture free' basis). The authors were concerned that the limits recommended by JMPR for residues were uncomfortably tight.

Table 25.12. Residues of pirimiphos-methyl (mg/kg) on husked rice, polished rice and unhusked rice after storage for 3 and 6 months and subsequent processing/cooking **(Desmarchelier *et al.* 1980a)**

Grain	Application level	Residues (mg/kg)			
		After 3 months	After 6 months	After cooking at 3 months	After cooking at 6 months
Husked rice	6	6.5	4.9	4.9	4.0
Polished rice	6	6.0	6.0	4.9	4.9

Grain	Application level	After 3 months	After 6 months	Residues (mg/kg) after 6 months storage and			
				Milling		Milling & cooking	
				Husked rice	Polished rice	Husked rice	Polished rice
Unhusked rice	6.0	6.0	4.6	1.4	0.3	0.6	0.1

Table 25.13. Pirimiphos-methyl residues in wholemeal flour and bread produced from treated wheat (stored 36 weeks) **(Wilkin and Fishwick 1981)**

	'As is' basis (mg/kg)	'Moisture free' basis (mg/kg)	Loss in process (%)	
Recommended dose	4.0	–		
Actual dose	3.4	3.97		
Aged wheat	3.1	3.62	storage –	8.8
Wholemeal flour	2.3	2.73	milling –	24.6
Bread	1.2	1.85	baking –	32.2
	–		from wheat to bread	48.9

25.4 Metabolism

Residues of pirimiphos-methyl on wheat grains are degraded and detoxified by hydrolysis of the phosphorus-ester side chain to give principally the hydroxypyrimidine (IV) **(Figure 25.2)** and also the related compounds (V and VI). At a given temperature, the rate of breakdown increases with increasing moisture content of the grains. Levels of the N-desethyl phosphorus compound (II) were always extremely low (approximately 0.05 mg/kg over a period of 32 weeks in wheat grain treated at 4 mg/kg). No residue of the chemically-unstable oxygen analogue (III) was detected at the limit of detection of 0.01 mg/kg **(Bowker 1973)**.

Autoradiograms of grain sectioned after 4 months showed that the insecticide and its degradation products were concentrated in the seed coat so that residues in white flour and bread are likely to be lower than in bran and wholemeal products. The general pattern of breakdown on stored rice was similar to that found for wheat grain. The insecticide and its degradation products were concentrated in the husk in which the rate of breakdown appeared to be unaffected by the moisture content of the rice **(Bowker 1973; Bullock 1973)**.

Degradation was marginally more rapid in contact with the grain than in the isolated formulation but whether this additional breakdown was caused by factors within the grain, or by the associated microflora, was not known. At higher-moisture content (approximately 18%) less pirimiphos-methyl was recovered on analysis but increased levels of the hydrolysis product (IV) were obtained, suggesting that a more rapid degradation of the insecticide occurred. **Bowker (1973)** concluded that grain may lack the enzymatic activity, believed to be present in plants and soil, to cleave the pyrimidine N-ethyl bonds. It is likely therefore, that following the treatment of wheat and brown rice with pirimiphos-methyl, the major residues during storage will be the insecticide itself and the simple hydrolysis product (IV). Under optimum conditions, the maximum level of compound (IV) following treatment at 4 mg/kg was found to be 0.17 mg/kg; under poor storage conditions with high-moisture content grain, 0.62 mg/kg. The effect of freshly harvested wheat was apparently not considered.

Solutions and also a 2% dust of radio-labelled pirimiphos-methyl were applied in the laboratory to wheat grains of 14% and 18% moisture to give levels of 4 mg/kg. Throughout a storage period of 6 months, the residues of pirimiphos-methyl were found almost entirely in the seed coat and aleurone layer, with only traces present in the germ or endosperm. It was also clear that, as with malathion but to a greater extent, there was transfer of insecticide between grains, possibly in the vapour phase. About 10% of the total residual radioactivity was bound to lipoprotein material in the aleurone region of the grain and was not extractable, except by digestion of the aleurone protein. By contrast with other organophosphates studied, the bound material appeared to be unchanged pesticide, rather than a metabolite. Certainly it was a P=S compound, but degradation during the liberation of the bound material complicated identification **(Rowlands *et al.* 1974)**.

The metabolism of pirimiphos-methyl in rats and in dogs was studied. Twelve metabolites were detected. None of these possessed anti-cholinesterase activity. No parent pirimiphos-methyl was detected in the urine. In both rats and dogs, 2 ethylamino 6-methyl-pyrimidin-4-ol(V) was the major urinary metabolite **(Bratt and Jones 1973)**. The principal metabolites are shown in **Figure 25.2.**

When rats were given a single oral dose of radio-labelled pirimiphos-methyl at 7.5 mg/kg, both the uptake of radioactivity into blood and its subsequent disappearance from the blood stream were rapid. More than 50% of the radioactivity present in the blood 30 minutes after dosing had disappeared at one hour after dosing. Unchanged pirimiphos-methyl normally represented less than 10% of the total residue in the blood 24 hours after dosing. When radio-labelled pirimiphos-methyl was administered orally to rats at 7.5 mg/kg/day for 4 days, total radioactive residues in the liver, kidney and fat, did not normally exceed 2 mg/kg pirimiphos-methyl equivalents. There was no evidence to show that either pirimiphos-methyl or its metabolites accumulated in the liver, kidneys or fat of rats, following daily dosing with the insecticide over 4 days **(Mills 1976)**.

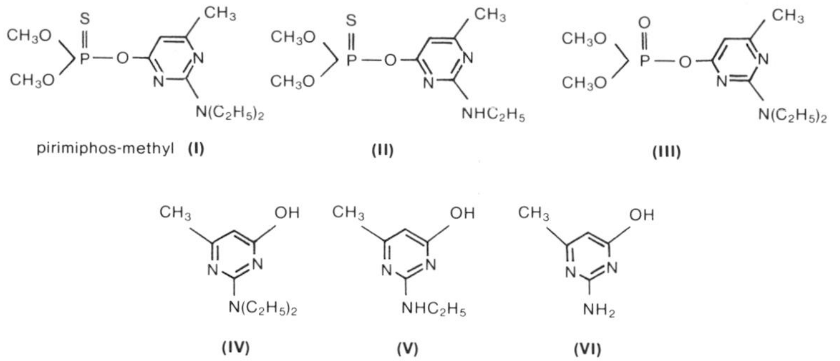

Figure 25.2. The structure of pirimphos-methyl and its metabolites **(Bratt and Jones 1973).**

Rowlands and Wilkin (1975) applied solutions of ^{14}C-pirimiphos-methyl to wheat of 14% and 18% moisture, and also as a 2% dust, to give 4 mg/kg final treatment. The samples were then stored in the laboratory at 20°C for 6 months. They found that approximately 10% of the radioactivity was bound to lipoprotein material and was unextractable except after digesting the aleurone regions of the grain. Such labelled material as was recovered by this means appeared to be unchanged pirimiphos-methyl, but degradation during this extraction and subsequent purification hampered identification. The P=S and hydroxy-pyrimidine moieties were intact but the investigators could not detect whether both N-ethyl or P-O-methyl groups were present. During the storage period, the breakdown rates were appoximately the same for both solvent and dust treatments. After 6 months, 15% and 50% degradation had occurred at the 14% and 18% moisture levels, respectively. The only metabolic products positively identified were the 3-pyrimidinol, a starch-pyrimidinol complex, and the free N-desethyl pyrimidinol.

Morallo-Rejesus *et al.* (1975) noted, but did not identify, breakdown products in shelled corn stored in pirimiphos-methyl impregnated bags.

Rowlands (1980) studied the metabolism of pirimiphos-methyl by stored wheat grains under laboratory conditions on wheat containing 12% and 18% moisture. Only 2 metabolites other than pirimiphos-methyl were found and these were identified as the pyrimidinol, the main product, and barely significant quantities of the N-desethyl pyrimidinol, found in the damper wheat only.

Leahey and Curl (1982) studied the fate of ^{14}C-pirimiphos-methyl during 6 months storage at 20°C on wheat grains and on both paddy and brown rice. They found that the breakdown was qualitatively and quantitatively similar on both wheat and rice, the main pathway being to the free 2-diethylamino-6-methyl-pyrimidin-4-ol (2–23%) and to a more polar compound – probably a conjugate – that yielded the same pyrimidinol on acid hydrolysis (1.1%). They also identified trace amounts of the N-monoethyl parent compound and the corresponding pyrimidinol, the bis-N-dealkylated pyrimidinol and an unknown metabolite. Radioactivity not extractable by solvents, as determined by combustion, accounted for 3–7% of the applied activity after 3 months storage and for 2–13% after 6 months. However, there was virtually no 'bound' material in the dehusked rice (only 1–2% after 6 months) and this was matched by the almost negligible breakdown of pirimiphos-methyl in dehusked rice; only 5% loss after 6 months storage as against 25% loss in paddy and in wheat where the seedcoats were present. This work seems to confirm the importance of the seedcoat/husk in the degradative process.

25.5 Fate in Animals

When 0.6 mg/kg of 2-^{14}C-ring-labelled pirimiphos-methyl was given orally to adult male rats, 73 to 81% of dose was excreted in the urine during the first 24 hours. This indicates rapid absorption. At the end of 120hrs after oral dosing, the entire dose could be accounted for by excretion of radioactive products in the urine (86% of dose) or feces (15.2% of dose). The concentration of the label in the abdominal fat of female rats given four daily doses of 5 mg/kg of ^{14}C-pirimiphos-methyl revealed that the metabolites were stored to some extent in fat: 0.46, 1.87 and 1.00 mg/kg pirimphos-methyl equivalents in fat after 1, 2, 3 and 4 doses, respectively **(Bratt and Jones 1973)**.

Two male dogs given 17–18 mg/kg orally also excreted most of the dose in the urine during the first 24–48 hours. Chromatographic and autoradiographic studies of the urine of dosed rats and dogs revealed extensive metabolism by the presence of 9 radioactive chromatogram spots **(Bratt and Dudley, 1970)**.

25.5.1 Goat

When 2-^{14}C-labelled pirimiphos-methyl was given orally to a lactating goat at a rate equivalent to approximately 6 ppm in the daily diet, 91% of the label was excreted in the following 8 days – 87% in the urine and 4% feces. Only 0.4% of the label was secreted in the milk, primarily during the first 24 hours. The maximum residue in the milk was 0.026 mg/kg pirimiphos-methyl equivalents, of which pirimiphos-methyl itself represented 0.003 mg/kg (**Bowker *et al.* 1973**).

The radioactive residues in the liver and kidneys of a goat dosed daily for 7 consecutive days with pirimiphos-methyl at the rate of 30 ppm in its diet were examined by **Curl and Leahey, (1980)**. A radioactive residue of 0.25–0.30 mg/kg of pirimiphos-methyl equivalents was detected in the liver and of this 0.015–0.018 mg/kg represented unchanged pirimiphos-methyl. A radioactive residue of 0.6–0.7 mg/kg of pirimiphos-methyl equivalent was found in the kidneys but none of this was parent compound. Most of the residues were pyrimidinol derivatives.

25.5.2 Cow

The pattern of excretion in the cow is very similar to that in the goat. When a single dose of 2-^{14}C-labelled pirimiphos-methyl was given orally to a cow at 0.5 mg/kg bw (equivalent to approximately 17 ppm in the daily diet), the label was quantitatively recovered during the following 7 days – 85% in the feces and 14% in the urine. Only 0.37% of the labelled material was secreted in the milk, nearly all during the first 3 days. The highest levels of radioactivity in milk, urine and feces were found during the first day after dosing. The milk contained 0.036 mg/kg pirimiphos-methyl equivalents of which about 0.004 mg/kg was pirimiphos-methyl (**Bullock *et al.* 1974a**).

Groups of 3 cows were maintained for 30 days on diets containing 0,5,15 and 50 ppm pirimiphos-methyl. Residues of pirimiphos-methyl in milk samples taken every 2 days and in samples of kidney, liver, heart, fat and muscle, taken at the end of the study, did not exceed 0.02 mg/kg. Butter prepared from the milk of cows dosed at the 50 ppm rate only contained about 0.025 mg/kg pirimiphos-methyl (**Bullock *et al.* 1974b**).

25.5.3 Pigs

Davis *et al.* (1976) maintained groups of pigs for 21 and 29 days on diets containing 3, 10 and 34 ppm pirimiphos-methyl. No residues of pirimiphos-methyl or of its phosphorus containing metabolites were detected in the kidneys, liver, lung, heart or muscle. Small residues of pirimiphos-methyl and N-desethyl pirimiphos-methyl were found in the fat of pigs fed 10 and 34 ppm.

25.5.4 Chickens

From hens, maintained for 18 days on diets containing 1, 4 and 8 ppm of pirimiphos-methyl, 90 eggs were found to contain no detectable residues of pirimiphos-methyl and 13 eggs contained residues up to 0.008 mg/kg. In the yolks and whites of eggs from hens fed radio-labelled pirimiphos-methyl for 28 days, levels of radioactivity increased steadily to 0.038 mg/kg of pirimiphos-methyl equivalents over 15 days, and then remained constant. At no time did the concentration of pirimiphos-methyl itself in yolks and whites exceed 0.001 mg/kg. Of the radioactivity in eggs 90% is present as water-soluble metabolites. Hens, fed at the artificially high rate of 32 ppm of radio-labelled insecticide for 7 days, gave a maximum level of 0.007 mg/kg of pirimiphos-methyl in the whites of the eggs and 0.012 mg/kg in the yolks. Neither pirimiphos-methyl nor its O-analogue were detected in the muscle of hens taken at the termination of the 32 ppm study (**Green *et al.* 1973**).

Broiler chickens were maintained for 6 weeks from the age of 2 weeks on rations containing pirimiphos-methyl at graded rates ranging from 4 to 48 ppm. At the end of this time they were slaughtered and samples of muscle, skin and fat were analysed. No residue was detected in any sample except the fat from those receiving 48 ppm. Even so the highest concentration found was only 0.017 mg/kg (**Graham and Jenkins 1974**).

25.6 Safety to Livestock

Though only a limited number of studies have been carried out to determine the possible effect of feeding pirimiphos-methyl to livestock over a long period, these studies make it clear that pirimiphos-methyl is well tolerated by cows. Because pirimiphos-methyl, like many other organophosphorus insecticides, is more toxic to birds, including poultry, than it is to mammals, the manufacturers sponsored a number of trials in which chickens and ducks received substantial quantities of pirimiphos-methyl continuously in their feed. Whilst these revealed that there is a definite limit to how much can be tolerated without adverse effects, there appears to be an adequate margin above the level which is likely to occur in grain or milling offals incorporated into poultry feeds.

A reliable indication of the margin of safety is given by the acute oral toxicity values to a range of species set out in **Table 25.14.**

Table 25.14. Acute oral toxicity of pirimiphos-methyl

Animal	Sex	mg/kg-bw	Reference
Rat	F	2050 (1840–2260)	**Clark 1970**
	F	1415	**Barnes 1971**
Mouse	M	1180 (1030–1360)	**Clark 1970**
Guinea pig	F	1000–2000	**Clark 1970**
Rabbit	M	1150–2300	**Clark 1970**
Cat	F	575–1150	**Clark 1970**
Dog	M	>1500	**Gage 1972**
Hen	F	30–60	**Clark 1970**
Quail	F	≈140	**Gage 1971**
Greenfinch		200–400	**Gage 1972**
Pigeon		<800	**Gage 1971**

25.6.1 Cows

Four groups of 3 cows were fed for 30 days on diets containing 0, 5, 15 and 50 ppm of pirimiphos-methyl.

The animals accepted the diet and were in good health throughout the trial. Blood samples taken from the animals during the trial showed no significant reduction in cholinesterase level. At the end of the trial the animals were slaughtered. There were no visual pathological effects attributable to pirimiphos-methyl. Neither were any histological effects discernible **(Bullock *et al.* 1974b)**.

25.6.2 Chickens

Pirimiphos-methyl was fed daily to laying hens at dietary levels up to 32 ppm for periods of up to 28 days. There were no adverse effects. Food consumption and egg production were unaltered irrespective of the dose administered **(Green *et al.* 1973)**.

Pirimiphos-methyl, fed at 4, 12 and 40 ppm in the diet of laying hens for 28 days, produced no adverse effects on egg production, egg quality, feed consumption, body weight gain or upon the general health of the birds. There was a statistically significant increase in 'dead in shell' eggs found at the 12 and 40 ppm rates, which was not considered to be toxicologically significant, and there were no significant effects on either viability or body weight of the chicks which hatched **(Ross *et al.* 1974)**. This trial was considered invalid due to a faulty incubator and was later repeated.

Pirimiphos-methyl was fed at 0, 4, 12 and 40 ppm in the diet to groups of 4 cockerels and 35 laying hens for 28 days. There were no mortalities in the treated groups. The birds remained in good health throughout and showed no signs of stress or abnormal behaviour. There were no significant effects on body weight, food consumption, egg production, egg quality or total egg weights. There were no effects on egg fertility and hatchability, chick mortality was within normal limits and there were no significant effects on body weights. From the foregoing results it was concluded that, at levels up to 40 ppm in the diet, pirimiphos-methyl has no ill effects on the health of chickens, on egg production or on chicken reproduction **(Ross *et al.* 1976a)**.

The inclusion of varying levels of pirimiphos-methyl in the diet of broiler birds from 4 ppm to 48 ppm temporarily depressed feed intake and growth rate during the first 3 weeks of another trial **(Graham and Jenkins 1974)**. After week 3, growth rate, feed intake and feed efficiency were comparable to the non-treated control group. All birds on the pirimiphos-methyl treated diet grew normally with no apparent effect on health or wellbeing.

Pym *et al.* (1984) reported a series of multi-factorial experiments in which layer strain hens were fed with feed containing dichlorvos, malathion and pirimiphos-methyl either singly or in combinations. The hens receiving feed containing 200 ppm of pirimiphos-methyl showed a measurable depression of egg production. Feed containing 50 ppm pirimiphos-methyl together with 100 ppm malathion and 15 ppm dichlorvos very significantly depressed food consumption and egg production. The level and degree of suppression of egg production was consistent with the level and degree of depression of food intake. Hens deprived of feed to the same extent produced similar numbers and weights of eggs and the authors concluded that the effect on egg production was mediated via a reduced food intake.

25.6.3 Ducks

Fink determined the acute LC/50 of technical pirimiphos-methyl to Mallard ducklings to be 633 ppm in their diet. He reported that pirimiphos-methyl had a marked effect on body weight and food consumption at all dosage levels. Symptoms of toxicity included lethargy, laboured respiration and loss of co-ordination. Depression of cholinesterase was observed at the lowest dosage **(Fink 1974)**.

25.7 Toxicological Evaluation

The toxicology of pirimiphos-methyl was evaluated by JMPR in 1974 and 1976 **(FAO/WHO 1975, 1977).** Acute toxicity studies were available on the following species: rat, mouse, guinea pig, rabbit, cat, dog, hen, quail and finch. Its moderate acute toxicity is due to its inhibition of cholinesterase. After single toxic doses, the onset of inhibition of cholilnesterase and appearance of toxic signs were delayed for several hours and persisted for several days. Sub-acute and 2-year feeding studies in rats and dogs showed that cholinesterase inhibition did not reach an equilibrium for several weeks. The compound was not teratogenic in rats or rabbits, although hydronephrosis was noted in rats. However, this was within the normal limits for the rats tested. A decrease in pregnancy rates was noted in a 3-generation reproduction study and cholinesterase activity was depressed at all dosage levels. Other parameters of reproduction were not affected. No compound-related histopathological effects were detected at dosage rates considerably above those that inhibited cholinesterase, except that, in a 90-day study in dogs, liver injury was observed in dogs receiving 10 mg/kg/day or above.

Results of more recent studies on the effect of pirimiphos-methyl on rat reproduction in a 3-generation test, on avian-egg production and hatchability and on special tests for mutagenicity were all negative. Hydronephrosis incidence noted in a previous test has been shown to be well within the limits of control values pertaining to the same species tested. In an 80-week mouse feeding study, the tumour incidence (including that of liver) was comparable to test control groups. In a 3-month dog study, slight bile duct proliferation was observed in animals at a dosage level of 25 mg/kg body weight per day. However, the effect was not noted at 50 mg/kg body weight per day. Focal inflammation occurred equally in the livers of control and dosed animals. In a 56-day human study, 0.25 mg/kg/day did not induce any changes in liver function tests, blood count or erythrocyte and plasma cholinesterase activity. No-effect levels have been demonstrated in the mouse (0.5 mg/kg body weight per day), rat (0.5 mg/kg body weight per day) and man (0.25 mg/kg body weight in a 56-day period). An ADI for man of 0.01 mg/kg body weight was allocated.

25.8 Maximum Residue Limits

The JMPR has evaluated information on residues of pirimiphos-methyl in food on 5 occasions, 1974, 1976, 1977, 1979 and 1983, **(FAO/WHO 1975, 1977, 1978, 1980, 1984).** On the basis of the information available, maximum residue limits have been recommended for pirimiphos-methyl in a number of raw cereal grains and milling products as follows:

Commodity	Maximum Residue Limits (mg/kg)
Bran	20
Bread (wholemeal)	1
Bread (white)	0.5
Cereal grains (except rice, hulled and polished)	10
Flour (white)	2
Rice (hulled)	2
Rice (polished)	1

The residue is determined and expressed as pirimiphos-methyl.

The methods recommended for the determination of pirimiphos-methyl residues are: **Pesticide Analytical Manual (1979a; 1979b); Methodensammlung, (1982 — S8); Ambrus *et al.* (1981); Desmarchelier *et al.* (1977b); Panel (1980).** The following methods are recognised as being suitable: **Methodensammlung, (1982 — 476); Zweig, (1976); Mestres *et al.* (1979a, 1979b).**

26. Pyrethrum

Pyrethrum is the oily extract from the pyrethrum flower (*Chrysanthemum cinerariaefolium*). Technical pyrethrum extract contains many different and related substances. The major components are pyrethrin I and pyrethrin II, which are highly insecticidal, and cinerin I and cinerin II, which are very much less active but contribute to the efficacy of the pyrethrins. It should be noted that the term 'pyrethrum' refers to the plant, the flower or the crude, concentrated or refined extract. The term 'pyrethrin' is reserved for describing the active constituent(s) of pyrethrum.

Pyrethrum has a very low acute oral toxicity. The highly refined grades now used appear to have no untoward effect on people using them or on laboratory and domestic animals. Technical pyrethrum is a dark-brown oily liquid with a characteristic odour. It is insoluble in water but is miscible with oils such as kerosene or industrial white oil. These oil solutions have a bright yellow-green colour.

Although it is possible to produce pyrethrum dusts and wettable powders, practically all commercial preparations consist of oil solutions, solution concentrates and to a lesser extent emulsifiable solutions. Oil solutions are used as the basis of many personal, domestic and industrial aerosols.

The outstanding feature of pyrethrum is the rapid action on a wide variety of insects. Pyrethrum preparations form the basis of most knockdown sprays and aerosols. Pyrethrum sprays have the additional effect of stimulating insects to move rapidly and to come out of hiding. Pyrethrum is often added to other sprays as a 'flushing' agent.

Unfortunately, many insects which are rapidly immobilised by pyrethrum recover after several hours. They have the ability to detoxify the small amount of pyrethrum which they have received. To overcome this tendency, synergists, particularly piperonyl butoxide, are added to pyrethrum sprays to enhance the effect and to reduce the cost.

Pyrethrum is not a very stable substance and its insecticidal activity is rapidly lost from treated surfaces, especially in the presence of direct or indirect sunlight. Pyrethrum is satisfactory for the treatment of grain to destroy an active infestation and good residual protection is obtained especially in the presence of piperonyl butoxide. Little or no residual activity results from the treatment of structures or grain surfaces. Such treatment must therefore be repeated frequently. Pyrethrum is effective against all types of stored-product insects including moths and their larvae. The action of pyrethrum against moths is not improved by the addition of synergists. Where moth control is the prime requirement, unsynergised pyrethrum preparations should be used.

Pyrethrum has been used for over 40 years for the control of stored-product pests and as a grain protectant. It is effective on its own or, more so, in combination with organophosphorus insecticides. Pyrethrum extracts are not stable on grain unless they are properly formulated and combined with a suitable anti-oxidant, such as piperonyl butoxide. Suitable commercial products with the requisite stability are available. The addition of piperonyl butoxide potentiates the efficacy of natural pyrethrum extracts considerably. Pyrethrum is not stable in high-moisture grain but when used on grain of normal or low-moisture content, and when applied correctly, it is capable of giving protection for 12 months. Notwithstanding the wide experience with its use, there is little or no information about the fate of pyrethrins following milling and cooking. It is assumed, from the knowledge of the chemistry of the compounds, that residues are probably destroyed in cooking.

26.1 Usefulness

Pyrethrum, especially in the form of a powder, was used for the protection of grain and for the destruction of grain insects well before World War II and there is every reason to believe that there was localised use for this purpose going back many generations.

The use of unsynergised pyrethrum, especially on damp grain, has seldom given satisfactory protection for long, due to the inherent instability of the various components. However, **Beckley (1948)** reported that satisfactory protection of wheat could be obtained by the admixture of 25 to 50 parts per million of pyrethrins as a powder, presumably containing approximately 1% total pyrethrins.

Since that time, a considerable world literature on the use of pyrethrum preparations for the control of stored-product pests has developed. Of the early publications, only those that were immediately accessible have been reviewed.

Watts and Berlin (1950) further developed the work initiated by **Beckley (1948)** by investigating the synergistic effect of piperonyl butoxide on crude pyrethrum powders. Small-scale laboratory tests against *Sitophilus oryzae* in wheat were conducted with piperonyl butoxide alone, pyrethrum alone, and with combinations of these 2 materials in dusts. Piperonyl butoxide used alone was almost completely ineffective against *Sitophilus oryzae*. Pyrethrum alone at dosages from 10 to 20 times that used in the combinations gave from moderate to good control. Combinations of the 2 materials in 4 different ratios gave from good to complete control over the 30 day test period. The ratios of piperonyl butoxide to pyrethrum ranged from 5:1 to 20:1. No significant difference was noticed over this range.

Goodwin-Bailey and Holborn (1952) described a laboratory method for comparing the effect of pyrethrins and pyrethrins/piperonyl butoxide mixtures on *Sitophilus granarius*. Applied as a powder mixed with wheat, the toxicity of pyrethrin formulations was greatly increased by the addition of piperonyl butoxide. The toxicity and degree of synergism is dependent on the ratio of piperonyl butoxide to pyrethrins and the type of carrier. Fine millers wheat offal known as 'bees wing' proved to be the best carrier and a ratio of 1:20 for pyrethrins:piperonyl butoxide gave a degree of synergism of ×19.

This information was used in formulating a grain protectant for use in a field experiment with stored bagged wheat. The treatment resulted in the application of 1.3 mg/kg pyrethrins and 27 mg/kg piperonyl butoxide. Treated wheat was exposed to severe conditions of external infestation by stored-product insects for 16 months. Good protection was obtained in comparison with untreated wheat for about a year. There was evidence of some repellency in the early months following treatment.

Wilbur (1952a) reported 2 series of field trials carried out in Kansas, USA, in 1950 and 1951 to evaluate the effectiveness of several dust formulations containing piperonyl butoxide and pyrethrum for protecting wheat against insects. The paper is worthy of detailed study because of the efforts taken to monitor the effectiveness of the treatments and to evaluate their performance on each farm notwithstanding the complications brought about by variations in infestation level and cross-infestation. One of the formulations showed outstanding protectant value when applied to newly-harvested wheat of high-moisture content and held in storage in farm bins of wooden construction. The other experimental formulae showed varying degrees of protectant value.

Wilbur (1952b) and **Chao Yung-Chang and Delong (1953)** considered the role of moisture content of grains as a single factor in determining the efficacy of pyrethrins, which was known to vary considerably depending upon climate and storage conditions.

Hewlett and Clayton (1952) recommended 0.4% lindane as the upper limit for use in admixture with pyrethrins for control of *Ptinus tectus, Tribolium confusum* and *Tribolium castaneum* but the mixture was not considered satisfactory for control of *Oryzaephilus surinamensis*.

LePelley and Kochum (1954), in a 15 page report described 5 experiments in the use of insecticides against *Sitophilus oryzae* in stored maize in Kenya. They used several pyrethrum/piperonyl butoxide mixtures and pyrethrum powders at concentrations ranging from 6.25 to 25 mg/kg and reported that these gave good protection for up to 10 months. The authors concluded that at least 25mg/kg was required to provide long-term protection against *Sitophilus oryzae*. However, this might be a measure of the rather poor formulations available at that time.

Warner (1954) described the properties and use of Pybuthrin, a proprietary mixture of pyrethrins and piperonyl butoxide available in a variety of formulations. The results of trials carried out in the United Kingdom with insect-free grain stored in heavily infested premises had shown that the protection obtained with Pybuthrin grain protectant is good up to a year on the criteria of numbers of live insects, grain injury and weight loss in comparison with untreated grain stored alongside. Even after 16 months the treated grain was found in better condition than the untreated grain. Protection was due to the repellent, as well as the insecticidal, property of the formulation, which is effective against a wide range of grain pests.

Schroeder (1955) found that dehusked maize, stored at 27°C, required more pyrethrum dust for effective control of infestation when the moisture content was 15% than when it was 10%.

Dove and Schroeder (1955) showed, in laboratory tests, the value of water-miscible emulsions of synergised pyrethins for the protection of stored grains.

Lloyd and Hewlett (1958) determined the relative susceptibility to pyrethrum in oil of 47 species of stored-product insects. They tested pyrethrum and pyrethrum plus piperonyl butoxide in highly refined mineral oil by direct application and by exposure of the insects to sprayed filter papers. Large differences in susceptibility were encountered. General susceptibility showed little correlation with systematic classification. The larva of a given species was usually more resistant than the adult. Among the adults of the different species, susceptibility to pyrethrum appeared to be correlated with high activity of the normal insect. On the whole, 1.3% pyrethrins and 0.3% pyrethrins plus 3% piperonyl butoxide were of about equal toxicity.

Walkden and Nelson (1959) demonstrated that both dust and spray formulations of synergised pyrethrins, applied at rates ranging from 1.5 to 2.5 mg/kg, were effective protectants for wheat and corn in bulk storage.

Glynne-Jones and Green (1959) compared the toxicities of pyrethrins and synergised pyrethrins to *Sitophilus oryzae* and *Sitophilus granarius*. They were able to show that *Sitophilus granarius* was approximately twice the weight of *Sitophilus oryzae* when both species were reared on wheat; both species, when transferred to maize, reproduced to give weevils with a 25–40% increase in mean weight. *Sitophilus oryzae* raised on maize was almost equal in weight to *Sitophilus granarius* fed on wheat. These differences were related to the tolerance of each species to pyrethrins. There was a significant positive correlation between bodyweight and tolerance to pyrethrins which was independent of the species. A parallel result was also obtained using lindane. The response of each species to the addition of piperonyl butoxide as a pyrethrum synergist was markedly different; *Sitophilus granarius* when exposed to a dust with a synergist/pyrethrins ratio of 1:8 evinced a response exactly 3 times greater than that shown by *Sitophilus oryzae*.

Phillips (1959) reported the use of pyrethrum sprays, applied to the surface of wheat stored in surplus ships at periodical intervals during 1957 in lieu of the previous method of fumigating the wheat, was effective in supressing stored-grain insect infestations at the James River and Hudson River Reserve Fleets, USA. At the James River Fleet, a spray containing 0.3% pyrethrins and 3% piperonyl butoxide, applied at the rate of approximately 1 L/10 m^2 of grain, ship skin, bulkhead, and underside tween-deck surfaces. 5 or more times from April through July, suppressed both moth and beetle infestation to a very low population level. At the Hudson River Fleet 4 periodical applications of the same spray almost completely suppressed both moths and beetle infestations.

The potentialities of pyrethrum for the protection of stored grain were assessed by **Parkin (1961)** but only the spraying of liquid emulsion directly onto the grain was seen to be really economic. A concentrate of 0.2% pyrethrins, with 3.2% piperonyl butoxide, showed considerable promise. In practice, the pyrethrins may be effective over longer periods than their half-lives would suggest since sub-lethal concentrations usually exert some repellent action on insects; this is not true of the synthetic pyrethroids.

Calderon and Shaaya (1961) noted the adverse effect of grain moisture during their attempt to control *Sitophilus oryzae* in stored seeds by admixing pyrethrum dust. For protecting stored grain from insect infestation, allethrin is much less effective initially than an equivalent amount of pyrethrum; 12 mg/kg allethrin is equivalent to 3 mg/kg total pyrethrins **(Chadwick 1962)** and it does not display the repellent effects towards some insects at low dosages. However, allethrin is considerably more stable on wheat grain than synergised or unsynergised pyrethrum and therefore exerts a protective effect for longer periods.

Watters (1961) compared the effectiveness of lindane, malathion, methoxychlor and pyrethrins/piperonyl butoxide against *Ptinus villiger* in field trials in 32 flour and feed storage warehouses in Manitoba. 0.1% pyrethrins/ 1.0% piperonyl butoxide in deodorised base oil was much superior to malathion but not significantly different to methoxychlor and lindane in suppressing the number of eggs laid in flour sacks.

Strong *et al.* (1961) studied the influence of formulation on the effectiveness of malathion, methoxychlor and synergised pyrethrin protective sprays for stored wheat. They used *Sitophilus oryzae* and *Tribolium confusum* as the test insects. Erratic results, but recognisable toxic effects, were recorded in tests with *Sitophilus oryzae*. Mortalities of *Sitophilus oryzae* declined rapidly in tests made after storage of synergised pyrethrum-treated wheat for more than 1 month after application of sprays. Results from repellency tests conclusively illustrated repellency of the synergised pyrethrum sprays to both species of test insects. Differences between formulations were not clearly defined. *Tribolium castaneum* was vastly more tolerant than *Sitophilus oryzae*.

King *et al.* (1962) reported a series of bioassays of 7 insecticides applied to sorghum challenged by 7 coleopterous pests. Pyrethrum/piperonyl butoxide alone and with oil did not provide adequate control. The moisture content of the sorghum ranged from 13% to 17%.

Coulon (1963) showed that the efficacy and persistence of the insecticidal action of pyrethrins was greatly improved by the addition of piperonyl butoxide.

McFarlane (1963) examined the prospects for pyrethrum with particular reference to its use in the control of pests of stored foodstuffs. He suggested that the apparent increase in the persistence of pyrethrum, observed by bioassay under laboratory conditions, was probably due to the effect of the synergist on the very low levels of pyrethrum, which otherwise would not be effective.

Joubert (1965a) tested a pyrethrum formulation containing 5% pyrethrins and 50% piperonyl butoxide in an emulsifiable base for direct application to maize for the control of grain-infesting insects. The emulsion was sprayed on to the maize in the conveyor system and a dosage of 1.25 mg/kg was attained. Total protection of the maize against insect infestation was obtained for a full year. During this period, untreated maize held under similar conditions for comparison was fumigated twice and, when final examinations were made, fumigation was again required. The moisture content of the maize at the onset of the test ranged from 12.6 to 13.6% and at the termination of the test, 11.1–11.2%. The author considered that the test had not been carried out under severe moisture conditions but nonetheless felt that the high moisture content of the grain apparently had little, if any, effect. Temperatures taken in the maize during the test period ranged from 21.5°C to 28°C and the author stated that temperature, as experienced, did not exert a measurable influence. The paper suggested that the repellent action of pyrethrum played an important role in the protective action of this insecticide against grain pests. There were no insects in the treated maize and well over 10 000 in corresponding samples drawn from the unsprayed maize, which had been fumigated several times during the period. The effect was therefore considered to be 'significant'.

Joubert (1965b) strongly recommended the use of pyrethrum preparations for disinfesting flour mills troubled with moths and beetles.

Kockum (1965) carried out an extensive field trial to compare pyrethrum/piperonyl butoxide powders with lindane powder for the control of *Sitophilus zeamais* and *Sitotroga cerealella* in maize stored in cribs in Nairobi. He found that even at a very low concentration of the pyrethrum formulation, a remarkable, almost complete protection was achieved inside the bulk of the cribs. On the outside, however, no strength of pyrethrum gave a satisfactory protection. This was no doubt due to the sensitivity of pyrethrum to light. This investigator concluded that pyrethrum would be more effective in clay-covered cribs than in the wire mesh cribs used in these trials.

LaHue (1966) reaffirmed that a water-based spray of synergised pyrethrins, applied at the rate of 2.75 mg/kg, was an effective protectant for shelled corn.

LaHue (1967) compared synergised pyrethrum against malathion for protecting sorghum grain against insects in small bins. Insect infestations were not adequately controlled by any of the treatments as damaging populations became established in all bins during the 12-month period. The synergised pyrethrum killed most of the *Sitophilus oryzae* during the first week, but not after that, and did not kill many of the *Tribolium confusum* at any time.

Partially infested yellow maize in South Africa was treated with 8 mg/kg of premium-grade malathion. A similar mass was treated with 1.75 mg/kg synergised pyrethrum. In both treatments the initial infestation was eradicated after several weeks and thereafter the treated maize exhibited excellent resistance to reinfestation, for 10 months. Internal damage in the treated maize remained below 2%, while 2 control masses were totally destroyed **(Joubert and de Beer 1968a).**

Joubert and de Beer (1968b) used synergised pyrethrum at the rate of 1.75 mg/kg pyrethrins as a standard to compare bromophos and malathion in a series of trials in commercial silos each containing approximately 800 tonnes of maize. The pyrethrum treatment gave the result the authors had learned to expect – the treated maize remaining free of living insects for the full period of the experiment (50 weeks). Both bromophos and malathion were inferior, even at 10 mg/kg.

Piperonyl butoxide synergised pyrethrum, saffroxan/piperonyl butoxide synergised pyrethrum and unsynergised pyrethrum formulations were applied to yellow maize at a dosage of 1.75 mg/kg of actual pyrethrins **(Joubert and du Toit 1968)**. The maize was stored for 1 year and regular sampling and exposure to insects indicated that only the 3% pyrethrins-30% piperonyl butoxide combination gave results that were entirely satisfactory. This formulation was recommended for increased use, especially when alternated annually with the cheaper synthetic insecticides as a precaution against the selection of insecticide-resistant strains.

McDonald (1968) compared the relative effectiveness of piperotol (Tropital) and piperonyl butoxide as synergists for pyrethrins against stored product insects. Results of these studies showed that piperotol does synergise pyrethrins, but in general, the synergistic action of piperotol is less than that of piperonyl butoxide for pyrethrins against 3 species of stored-products insects. The ratio of pyrethrins:piperonyl butoxide of 1:10 appears optimum against *Tribolium confusum, Lasioderma serricorne* and *Attagenus megatoma*.

Coulon (1968) showed that synergised pyrethrins were superior to malathion and lindane at equivalent rates of application against *Sitophilus granarius, Tribolium castaneum* and *Oryzaephilus surinamensis* but unfortunately at these rates the pyrethrin preparations were significantly more expensive.

Speirs and Zettler (1969) reported a 13.3-fold resistance to pyrethrins in a field-collected strain of rust red flour beetle, *Tribolium castaneum*.

Talekar and Mookherjee (1969) studied the effect of temperature and grain moisture on the deterioration of pyrethrum-based grain protectants. The pyrethrum was used as a dust with and without piperonyl butoxide. Unfortunately, there appeared to be an error in the information since it is stated that the dosages of pyrethrins used were 5, 10 and 20 mg/kg with or without synergists. This was almost 10 times as high as the standard rate of treatment. It was observed that even a dose as high as 20 mg/kg of pyrethrins had little toxic effect right from the beginning when used singly. In conjuction with piperonyl butoxide, 20 mg/kg of pyrethrins was effective for 1 to 2 months under constant temperature and humidity conditions after which there was a decline in toxicity. Under fluctuating conditions of temperature and humidity, 20 mg/kg of synergised pyrethrins was effective for 3 months after which there was a decline in the toxicity.

Hewlett (1969) developed a mathematical model for relating the mortality of *Tribolium castaneum* to the concentration of pyrethrins and piperonyl butoxide in a combination. For large ratios of piperonyl butoxide, the mixtures were estimated to be 3 times as toxic as pyrethrins alone.

Cichy (1969) investigated the influence of temperature, atmospheric relative humidity, age, sex and stages of development of *Tribolium castaneum* on the susceptibility to pyrethrum/piperonyl butoxide. It was found that all of these factors modified the susceptibility of the insects.

Lloyd (1973) compared the toxicities of 5 synthetic pyrethroids with that of natural pyrethrins against susceptible *Tribolium castaneum* and susceptible and pyrethrum-resistant *Sitophilus granarius*, using topical application techniques. When the synthetic toxicants were used alone against 3 strains of insect, bioresmethrin was by far the most toxic compound. Depending upon the strain tested, bioresmethrin had up to 16 times more toxicity and tetramethrin as little as one-hundredth the toxicity of pyrethrins. Piperonyl butoxide synergised all the compounds against the 3 strains. The 'non-insecticidal' fraction of pyrethrum extract was found to be toxic, at half concentration, to resistant *Sitophilus granarius*.

Gillenwater and Burden (1973) provided a brief summary of the usefulness of synergised pyrethrins for incorporation into stored commodities.

Wilkin and Hope (1973), who evaluated 22 pesticide compounds against 3 species of stored product mites, found that, whereas pyrethrins applied at the rate of 2 mg/kg were relatively ineffective, pyrethrins (2 mg/kg) plus piperonyl butoxide (20 mg/kg) were fully effective against *Acarus siro*, and moderately effective against *Tyrophagus putrescentiae* and *Glycyphagus destructor*.

Harein and De Las Casas (1974), in an extensive review of chemical control of stored-grain insects, devoted one and a half pages to pyrethrins explaining that the toxic effect of pyrethrins is lost rapidly when applied to stored grain, whereas the repellent action remains the primary factor for insect protection for 6 to 12 months thereafter. It is suggested that synergised pyrethrins should be applied to the grain surface immediately after the grain is levelled off upon reaching its storage site. This appears to ignore much fine work that was available at that time and which indicated the value of incorporating synergised pyrethrum in grain, particularly grain of low moisture content.

Median lethal doses of five insecticides were measured for *Sitophilus zeamais* on maize and sorghum using laboratory-formulated dusts **(Weaving 1975)**. Fenitrothion was the most toxic and pyrethrins the least toxic. The optimum ratio of pyrethrins:piperonyl butoxide was 1:15, smaller ratios demanding a higher deposit of pyrethrins than is normally recommended. The 1:15 ratio gave a factor of synergism of 24. Nonetheless, the mortality of the test species was as low as 20% immediately after admixture; thereafter the mortality declined sharply.

Ahmed *et al.* (1976) tested different combinations of selected ingredients for prolonging the toxicity of pyrethrins against *Tribolium castaneum* adults. The emulsifiable formulations of pyrethrins which resulted were tested for their efficacy and stability in comparison with the standard pyrethrins/piperonyl butoxide (1:10) formulation. It was confirmed through bioassay and chemical analysis that the new formulations were as effective as the standard in maintaining residual toxicity for 8 to 10 weeks. The aqueous extract of de-oiled neem seeds incorporated in one of the formulations showed an added effect on the residues of pyrethrins.

Morallo-Rejesus and Carino (1976a) evaluated 5 contact insecticides as protectants against insects on stored corn and rated synergised pyrethrum applied at the rate of 10 mg/kg equivalent to similar rates of pirimiphos-methyl and chlorpyrifos-methyl for protecting maize against *Sitophilus zeamais, Tribolium castaneum* and *Rhyzopertha dominica* for up to 12 months.

Desmarchelier (1977b) showed how combinations of pyrethrum and organophosphorus insecticides were especially effective against mixed populations of stored-product pests, particularly where *Rhyzopertha dominica* was a problem. There was no antagonism between pyrethrum and any of the organophosphorus insecticides when they were applied in combination, in that the period of protection given by combinations was the same as the period of protection given by the organophorphorus component against *Tribolium* and the pyrethrins against *Rhyzopertha*. The advantage of reducing the temperature of the stored grain, in order to increase the residual life of the insecticide deposit, was demonstrated.

Desmarchelier and Bengston (1979a) supplied further information in support of the argument that predictive models, from which one can calculate the persistence of a chemical on any grain at any constant or varying conditions of temperature and moisture content, provide an excellent way of improving the residual protection whilst reducing the cost and lowering the residues present at the time the grain is taken out of storage. The paper gave examples of the advantage of cooling treated grain. The half-life of pyrethrins at 30°C and 50% relative humidity was stated to be 34 weeks, 3 times that of malathion. Further evidence in support of these proposals is contained in the paper by **Desmarchelier and Bengston (1979b)**.

Bengston and Desmarchelier (1979) reviewed extensive laboratory and field trials conducted in Australia to control a wide range of species. They stated that pyrethrins at 3 mg/kg plus piperonyl butoxide at 30 mg/kg are effective against *Rhyzopertha dominica*.

Being aware of contradictory accounts of the stability and efficacy of synergised pyrethrins as grain protectants, **Desmarchelier *et al.* (1979a)** undertook a series of experiments in the laboratory and under commercial conditions in Australia to study the stability of pyrethrum. The effect of piperonyl butoxide on the residual biological efficacy of pyrethrins was also investigated. The experience gained from treating 2 million tonnes of wheat and barley in southern New South Wales in 1977 with mixtures of malathion and synergised pyrethrins was included in the report. In commercial applications, synergised pyrethrins were applied at the rate of 3 mg/kg, together with 28 mg/kg piperonyl butoxide. In laboratory experiments, pyrethrins, applied at the rate of 2 mg/kg, synergised with piperonyl butoxide in the ratio of 1:15, gave up to 26 weeks complete suppression of *Rhyzopertha dominica*. This was compared with unsynergised pyrethrins applied at the same rate which failed to suppress the insect pests even at the end of 2 weeks. The dose of pyrethrins on wheat that was lethal to *Rhyzopertha dominica* after an exposure of 3 days decreased as the amount of piperonyl butoxide on wheat increased over the range 1 to 30 mg/kg. The LD/99.9 fell from 2.3 mg/kg, for wheat containing 1mg/kg piperonyl butoxide to 1.24 mg/kg, for wheat containing 30 mg/kg piperonyl butoxide. In providing protection against reinfestation by *Rhyzopertha dominica*, pyrethrins (2 mg/kg) plus piperonyl butoxide (5 mg/kg) was almost as effective as pyrethrins (4 mg/kg); pyrethrins (2 mg/kg) plus piperonyl butoxide (13 mg/kg) was slightly more effective than pyrethrins (4 mg/kg). Thus, the residual efficacy of pyrethrins was doubled by addition of between 5 and 13 mg/kg piperonyl butoxide. While much of this increased efficacy was due to the synergistic effect of piperonyl butoxide, chemical determination of pyrethrin I revealed an enhanced stability in the presence of piperonyl butoxide. Under commercial conditions, the combination of malathion (to control *T. castaneum*) with synergised pyrethrins (to control *R. dominica*) greatly lowered the incidence and severity of infestation with *R. dominica* and *T. castaneum* notwithstanding high infestation pressures.

Lloyd and Ruczkowski (1980) showed that a strain of *Tribolium castaneum*, with a specific resistance to malathion and its carboxylic ester analogues, had no cross-resistance to topical applications of natural pyrethrins. Another strain of *Tribolium castaneum*, showing resistance to many organophosphorus insecticides, was cross-resistant (x 34) to pyrethrins.

Weaving (1980) evaluated synergised pyrethrins alongside 5 organosphosphorus insecticides as possible grain protectants for use under tribal storage conditions in Zimbabwe. When used at the rate of 1.85 mg/kg, synergised with 5 times the quantity of piperonyl butoxide, pyrethrins were the least effective of the insecticides tested. The level of protection conferred on maize and sorghum was significantly better than untreated control grain only during the first 1 to 2 months. Thereafter, the level of infestation was no better than untreated controls.

Weaving (1981) reported further results of his studies of grain protectants for use under tribal storage conditions in Zimbabwe. He tested 7 insecticide admixtures under field conditions in traditional tribal grain-storage bins. Synergised pyrethrins were less effective and less persistent than 5 other insectides against the dominant insect pest, *Sitophilus zeamais*. The author concluded that synergised pyrethrins appear to be unsuitable for use in tribal storage.

Viljoen *et al.* (1981a) studied the protection of maize and groundnuts in bag stacks against reinfestation. Small bag stacks of maize and groundnuts in the open and in a shed in South Africa were sprayed with various contact insecticides after fumigation and thereafter at intervals of 4 weeks. A mixture of diazinon with pyrethrins was effective against *Ephestia cautella* and *Plodia interpunctella* but was less so against *Tribolium castaneum*.

Harein (1982), in a further review entitled 'Chemical Control Alternatives for Stored-grain Insects', repeated the assertion made by **Harein and De Las Casas (1974)** that it is the repellent properties of synergised pyrethrins that are most important in protecting grain from 5 major pest species.

Gelosi (1982), in a comprehensive treatise on the morphology, biology, importance and control of *Sitophilus oryzae* in Italy, stated that control consists of thorough cleaning of cereal storehouses and of containers and vehicles used for transportation followed by treatment with contact insecticides, including pyrethrins.

Bengston *et al.* (1983a) reported duplicate experiments carried out on bulk sorghum stored in south Queensland and central Queensland. Bioassays of treated grain, conducted during 6 months storage, established that chlorpyrifos-methyl at 10 mg/kg plus pyrethrins at 1.5 mg/kg plus piperonyl butoxide at 12 mg/kg controlled typical malathion-resistant strains of *Sitophilus oryzae*, *Rhyzopertha dominica* and *Tribolium castaneum* but were not fully effective against *Ephestia cautella*.

Greening (1983) reported that farm-stored wheat treated with 1.5 mg/kg of pyrethrins and 30 mg/kg piperonyl butoxide remained free of infestations for about 2 months, with the bulk of the grain, sub-surface, remaining uninfested for at least 10 months. A low-level infestation by *Tribolium castaneum*, however, was found to have started on the warmer side of the bin and spread peripherally to the cooler side. Pyrethrum-treated grain remained free of *Sitophilus oryzae*, *Rhyzopertha dominica* and *Cryptolestes ferrugineus*, which infested untreated grain. In another trial, wheat treated with 2 mg/kg pyrethrins remained free of infestation whereas barley treated with 12 mg/kg fenitrothion was infested with *Rhyzopertha dominica*. In a subsequent trial, however, 2 and 1.5 mg/kg pyrethrins failed to give as good a protection to grain as had been achieved with 3 mg/kg pyrethrins previously.

Wohlgemuth (1984) conducted comparative laboratory trials with 18 commercial insecticides, including pyrethrum/piperonyl butoxide dust under tropical conditions against 5 stored-product beetles. When applied to sorghum at the rate of 1.65 mg/kg pyrethrins plus 26.6 mg/kg piperonyl butoxide, the treated sorghum failed to control any species immediately after application and allowed all species to reproduce. These results leave some doubts about the quality of that commercial formulation.

26.2 Degradation

The available information on the degradation of pyrethrins on grain is neither adequate nor convincing. This is in no small measure due to the fact that, even today, the methods of analysis for pyrethrin residues are anything but simple. Until comparatively recently, most chemical estimatations depended upon the determination of piperonyl butoxide, which was used in a fixed ratio along with the pyrethrins in the formulation. It was assumed that the distribution and rate of degradation of piperonyl butoxide was identical with that of pyrethrins but the fallacy of this assumption will be brought out by some of the evidence presented in this section.

Goodwin-Bailey and Holborn (1952) reported that a synergised pyrethrum dust, applied to wheat so as to deposit 1.3 mg/kg of pyrethrins, together with 27 mg/kg piperonyl butoxide, gave good protection against several insect species for about 1 year.

Chao Yung-Chang and Delong (1953) pointed to the role of grain moisture as the single most important factor in influencing the rate of degradation of synergised pyrethrins, applied to grain.

Quinlan and Miller (1958) applied synergised pyrethrum spray to the top surface of bulk shelled corn of 12–13% moisture content at different dosages and frequencies to control infestation by Indian meal moth. The residues of piperonyl butoxide found in composite samples of the surface 2.5 cm of grain taken from each bin, 3 and 6

Table 26.1. Residues of piperonyl butoxide in the surface inch of shelled corn at intervals following periodic application of three concentrations of pyrethrum-piperonyl butoxide spray, 1956 **(Quinlan and Miller 1968)**

Frequency of treatment and conc. of piperonyl butoxide[a]	Piperonyl butoxide applied and recovered in shelled corn (ppm)			
	After 3 months[b]		After 6 months[c]	
	Amount applied July 9 to Sept. 27	Amount recovered	Amount applied July 9 to Oct. 25	Amount recovered
Semi-weekly				
1.25%	786	203	1048	127
2.50%	1572	386	2096	247
3.75%	2358	964	3144	356
Weekly				
1.25%	393	107	524	56
2.50%	786	256	1048	119
3.75%	1179	359	1572	206
Biweekly				
1.25%	197	54	262	31
2.50%	393	149	524	69
3.75%	590	182	786	95

a The proportion of pyrethrins to piperonyl butoxide was 1:10.
b These samples were drawn on Oct. 1, 1956, and represent the amount of insecticide accumulated during the first three months of the study.
c These samples were drawn on Jan. 4, 1957, and represent the amount present six months after the first and three months after the last application.

months after the treatments were begun, are given in **Table 26.1.** The data showed that greater amounts of piperonyl butoxide were recovered from treatments at higher concentrations and with more frequent applications. Much of the piperonyl butoxide had disappeared by the time the samples were taken; about 75% at the end of 3 months, and nearly 90% at the end of 6 months.

Walkden and Nelson (1959) determined the residues of pyrethrins and piperonyl butoxide applied to stored wheat. They found that 2 months after application of pyrethrins, an average of about 50% of the dosage applied was recovered. Twelve months after application, this was reduced to about 25% and remained at about this level for 2 years.

Blinn *et al.* (1959) sprayed wheat of 13% moisture content with a wettable powder formulation containing 2% pyrethrins and 20% piperonyl butoxide, so as to deposit approximately 1 mg/kg pyrethrins. The treated grain was stored in tightly sealed cans at 30°C. They concluded that under these conditions, the half-lives were 5.8 weeks for pyrethrins and 9.9 weeks for piperonyl butoxide thus demonstrating the unsatisfactory nature of the practice of correlating an un-assayed pyrethrin level with that of the piperonyl butoxide found by chemical assay **(See Table 26.2)**. However, the authors stated that this practice can be condoned for purposes of defining toxicological hazard if the actual residual relationship of these two materials is considered.

Table 26.2. Residues of pyrethrins and piperonyl butoxide found on wheat of 13% moisture content, stored at 30°C **(Blinn et al. 1959)**

Days after treatment	Parts per million	
	Pyrethrins[a]	Piperonyl butoxide[b]
1	0.8	7.7
14	0.6	6.2
30	0.4	5.0
90	0.2	3.5

a Average of two values. Corrected for background (0.1 ppm) and for recovery (89%).
b Average of two values. Corrected for background (0.1 ppm) and for recovery (98%).

Joubert (1965a) tested a commercial formulation containing 5% pyrethrins and 50% piperonyl butoxide in an emulsifiable base under large-scale commercial conditions in South Africa. He applied the emulsion to maize so as to deposit 1.25 mg/kg. Total protection of the maize against insect infestation was afforded for a full year.

LaHue (1965) treated wheat at 11% moisture content with pyrethrins plus piperonyl butoxide (1:10) and determined the piperonyl butoxide residue. His calculated application was 22.4 mg/kg of piperonyl butoxide but immediately after treatment he recovered only 6.6 mg/kg. After 1, 3, 6 and 12 months, his recoveries were 10.7, 12.9, 8.4 and 9.9 mg/kg respectively. He concluded that piperonyl butoxide residues degraded in an erratic pattern during the first few samplings, but apparently stabilised after 6 months. The residual pyrethrins were estimated by assuming that the residue of the pyrethrins and those of the piperonyl butoxide were in the same proportion as the original formulation. The moisture content varied from 9.6 to 12.6% during the storage tests.

LaHue (1966) treated shelled corn with pyrethrins plus piperonyl butoxide (1:10) and determined the piperonyl butoxide residue. This calculated application was 27.5 mg/kg of piperonyl butoxide. However, immediately after treatment, he recovered 12.3 mg/kg, which was lower than expected. After 1, 3, 6 and 12 months his recoveries were 10.2, 7.7, 5.7 and 6.1 mg/kg respectively. He concluded that the residue degraded in a uniform pattern for the first 6 months, apparently reaching a level of stability in that time. He assumed that the recovery of piperonyl butoxide can be used to estimate the pyrethrins' residues at a 10:1 ratio, the proportion in the emulsifiable concentrate.

LaHue (1967) applied synergised pyrethrin to sorghum grain held in small bins. The commercial pyrethrin formulation was applied at the recommended rate and was reported to have killed most of the *Sitophilus oryzae* during the first week, but not after that and not to kill many of the *Tribolium confusum* at any time. The calculated rate of application of pyrethrins was 2.4 mg/kg. The sorghum grain was sampled immediately after treatment and at the end of 1, 3, 6, 9 and 12 months. The residues of piperonyl butoxide recovered by chemical analysis were 13, 12, 8.4, 7.1, 4.4 mg respectively. It was again assumed that recoveries of piperonyl butoxide could be used to estimate the residues of pyrethrins at a 10:1 ratio.

Joubert and de Beer (1968a) reported that maize, treated with synergised pyrethrum at the rate of 1.75 mg/kg, exhibited excellent resistance to reinfestation for 10 months. **Joubert and de Beer (1978b)** evaluated several synergists and several ratios of pyrethrin:piperonyl butoxide and confirmed that a dosage of 1.75 mg/kg pyrethrins, prepared from a formulation containing 5% pyrethrins and 50% piperonyl butoxide, protected yellow maize, when challenged by insect pests for a period exceeding 12 months. Similar results were reported in a separate study **(Joubert and du Toit 1968).**

Lindgren *et al.* (1968) reviewed the literature on residues in raw and processed foods resulting from post-harvest insecticidal treatments. Though several pages were devoted to pyrethrins plus piperonyl butoxide, virtually all of the information available for consideration at that time was dependent upon the determination of the piperonyl butoxide residue since there were no specific methods available for measuring pyrethrin residues. The review of the metabolism of contact insecticides in stored grains by **Rowlands (1971)**, likewise, does not throw much light on the fate of pyrethrum in stored grains. Mention is made of hydrolytic activity at the chrysanthemic acid linkage in permethrin I and some oxidative activity to unidentified products in older wheat.

Talekar and Mookherjee (1969) reported that unsynergised pyrethrins lost their effect almost immediately they were added to stored grain, whereas high concentrations of synergised pyrethrins were effective for 1 to 2 months under constant temperature and humidity conditions. There was, however, some doubt about the quality of the pyrethrum extract used for these experiments and the rate of pyrethrins applied to the grain.

Moore (1970) referred to the work of **Head *et al.* (1968)** which showed that piperonyl butoxide stabilises pyrethrins exposed to sunlight and air. These studies were made by exposing a thin film of pyrethrins and piperonyl butoxide on glass plates to sunlight and air. They used electron capture gas chromatography to determine the level of pyrethrins. This method was described by **Head (1966)**. Head, in unpublished work, has investigated the stability of pyrethrins on wheat exposed to sunlight and air. **Table 26.3** shows the residue of pyrethrins (cinerins

Table 26.3. Cinerins and pyrethrins residues on wheat **(Head, unpublished data)**

Sample	Residues (ppm)							Loss total pyrethrins (%)
	Cin.I	Py.I	Cin.II	Py.II	'Py.I'	'Py.II'	Tot.	
Formulation	0.71	1.88	1.08	1.42	2.59	2.50	5.09	–
Control (unexposed)	0.66	1.81	1.08	1.35	2.47	2.43	4.90	–
10-hr. exp.	–	–	0.006	0.002	–	0.008	0.008	99.8
15-hr. exp.	–	–	0.004	0.001	–	0.005	0.005	99.9
Wheat control	0.76	1.76	1.05	1.39	2.52	2.43	4.95	–
10-hr. exp.	0.24	0.41	0.51	0.57	0.65	1.08	1.73	65
15 hr. exp.	0.28	0.41	0.51	0.77	0.69	1.28	1.97	60
24-hr. exp.	0.19	0.20	0.23	0.28	0.39	0.51	0.90	81

and pyrethrins) remaining on the wheat. The 'pyrethrins' on wheat showed much greater stability than in the thin film exposure of the untreated controls. The pyrethrin II components tended to be more stable than the pyrethrin I components and the cinerins were more stable than the pyrethrins. Although the extract from the wheat was typical of degraded material, both pyrethrin I and pyrethrin II were still clearly present. It is possible therefore, that when wheat is sprayed with pyrethrum, part of the active constituents are absorbed and protected while the material at the surface suffers rapid degradation. The electron capture gas chromatographic tracings showed that the degradation of pyrethrins on grain is typical of the degradation of pyrethrin films on glass.

Weaving (1975), in a laboratory evaluation of grain protectants for use under tribal storage conditions in Zimbabwe (Rhodesia), showed that the addition of piperonyl butoxide to pyrethrins greatly increased the effectiveness of a dosage in the 2–4mg/kg range apparently by increasing the persistence and stability of the deposit. Greater than 90% mortality of *Sitophilus zeamais* and complete prevention of breeding resulted from the application of 4 mg/kg pyrethrins synergised with piperonyl butoxide in the ratio of 1:5. The same concentration of unsynergised pyrethrins failed to produce any mortality and reduced breeding by only 25–50%.

Ahmed *et al.* (1976) evaluated a range of additives in the hope of prolonging the residual activity of pyrethrins applied to grain. It was confirmed through bioassay and chemical analysis that several of the new formulations were as effective as the standard pyrethrum/piperonyl butoxide product for maintaining residual toxicity for 8 to 10 weeks.

Desmarchelier (1977b) applied combinations of pyrethroid and organophosphorus insecticides for selective control of stored product Coleoptera at 2 temperatures. He showed that 4 mg/kg synergised pyrethrins gave more than 30 weeks protection, defined as 28 day adult mortality at 20°C, but only 20 weeks at 30°C. If the criterion for effectiveness was suppression of reproduction, 4 mg/kg synergised pyrethrins remained effective for 9 months.

Desmarchelier and Bengston (1979a, 1979b), discussing the residual behaviour of chemicals in stored grain, recorded the half-life of pyrethrum at 30°C and 50% relative humidity as 34 weeks.

One of the most valuable studies on the stability and degradation of pyrethrins on grain in storage was that of **Desmarchelier *et al.* (1979a)**. These workers showed, in laboratory experiments, that piperonyl butoxide enhanced both the chemical stability of pyrethrins on grain in storage and their efficacy against *Rhyzopertha dominica*. The persistence of pyrethrins on wheat, barley, oats and milled rice under different conditions was determined chemically and the residual efficacy on wheat was also determined biologically. **Table 26.4** shows the loss of pyrethrins on grains treated with a commercial formulation of pyrethrum and piperonyl butoxide (1:8). In all instances the half-life exceeded 30 weeks which bears out the previous estimate of the half-life being 34 weeks **(Desmarchelier and Bengston 1979b).** On average, 77% of the amount of pyrethrins on grain 3 weeks after application was still present on grains after a holding period of 30 weeks at average temperature 29°C and average equilibrium relative humidity 50% **(Table 26.4). (Desmarchelier *et al.* 1979a)**. Loss is seen to be dependent on temperature and equilibrium relative humidity by comparing the mean from 3 to 30 weeks for the different sets of conditions **(Table 26.4, last column)**. For example, on wheat in equilibrium with 47% relative

Table 26.4. Loss of pyrethrins I on grains treated with pyrethrins plus piperonyl butoxide, 1 to 8 w/w **(Desmarchelier and Bengston 1979b)**

Holding conditions			Residue (g/t) at time (weeks) after application*						
Type of grain	Temperature (°C)	E.r.h.** (%)	3	7	11	16	22	30	Mean (3–30)
wheat	20	47	3.8	2.8	3.7	2.9	3.1	2.6	3.15
wheat	25	47	3.8	3.4	2.7	2.7	3.0	2.5	3.01
wheat	30	47	3.3	2.8	3.2	3.0	2.3	2.3	2.82
wheat	35	47	3.0	2.7	2.9	2.4	2.2	2.0	2.53
barley	30	33	3.1	3.3	3.0	3.1	3.1	2.9	3.08
wheat	30	36	3.4	3.2	2.8	2.6	2.6	2.6	2.85
oats	30	52	–	2.9	2.9	2.1	2.3	2.2	2.48
wheat	30	52	3.0	2.7	3.1	2.7	2.6	2.3	2.73
wheat	30	62	3.0	2.9	2.5	2.4	2.2	1.9	2.48
rice	30	75	2.1	3.0	2.6	2.5	2.5	–	2.54
mean (all grain)	29	50	3.12	3.01	2.95	2.64	2.58	2.41	

* The nominal deposit was 3.5 g/t.
** Grain equilibrium relative humidity.

humidity, the average residue value was 3.15 mg/kg at 20°C, but only 2.53 mg/kg at 35°C. At 30°C, the average residue value was 3.08 mg/kg on wheat in equilibrium with 36% relative humidity but only 2.48 mg/kg on wheat in equilibrium with 62% relative humidity. Clearly, rate of loss increased with increasing temperature and equilibrium relative humidity.

Otieno and Pattenden (1980) summarised the then-available knowledge about the degradation of natural pyrethrins under photo-chemical, thermal and acid or basic conditions. They pointed out that structure/activity data showed that the potency of natural pyrethrins is critically sensitive to chemical and stereochemical changes in the structure. Changes in the structural constitution and stereochemical details of the pyrethrins occur rapidly when they are exposed to light and heat, acid or base, and microbial activity. These conditions are met when the insecticides are used in the control of insects and also during the refining process leading to the production of the commercial products. In spite of their superior environmental qualities, the general instability of the natural pyrethrins has considerably restricted their development as all-purpose crop protection agents. The structure of the pyrethrins and the processes leading to the formation of about 40 derivatives were provided.

26.3 Fate in Milling, Processing and Cooking

There appears to be practically no information concerning the fate of pyrethrum residues during milling, processing and cooking. This is no doubt understandable in view of the lack, until recently, of suitable analytical methods, the low concentration of the residue initially in the whole grain and the general acceptance that pyrethrum residues are without hazard to consumers.

Wheat of 10 and 13% moisture content was treated with synergised pyrethrum (1 mg/kg pyrethrins and 15 mg/kg piperonyl butoxide) in various spray formulations and piperonyl butoxide residues were determined after storage for 3 months at 15° and 30°C (**Strong *et al.* 1961**). They concluded that cleaning operations used in this study were not highly efficient in removing chemical residues from treated wheat, probably because most of the insecticide retained by wheat after cleaning was in the pericarp and endosperm. In milled samples, residues of piperonyl butoxide recovered were greater in bran than in other fractions. Residues found in flour were considerably higher than expected. The detailed results obtained in this study are given in **Table 26.5.**

Schoeggl *et al.* (1983) showed that, in laboratory studies with wheat sprayed in the normal commercial manner so that the residues ranged from 1.5–2.0 mg/kg, it was possible to reduce the residues by about 30% by thorough cleaning of the grain prior to milling. During milling, a variable distribution resulted in individual fractions, bran and pollard containing the bulk of the residues. During storage in a closed system, pyrethrins remained stable for a relatively long time, whereas in open storage rapid decomposition occurred. Residues in

Table 26.5. Residue of piperonyl butoxide found in wheat during milling and in the milling fractions of wheat three months after applying initial deposits of synergised pyrethrum (1.5 ppm pyrethrins and 15 ppm piperonyl butoxide in various spray formulations **(Strong et al. 1961)**

Formulation	Storage temperature (°F)	Moisture content of wheat (%)	Recovery of piperonyl butoxide (ppm)[a] from: Whole grain: Before cleaning	Whole grain: After cleaning	Whole grain: Clean-out (dockage)	Milling fractions from Buhler mill: Flour (composite)	Bran	Shorts	Low grade (middlings)
Emulsion	90	13	6.0	4.8	38.3	1.7	12.2	0.15	2.5
		10	6.2	5.5	29.3	3.3	13.0	7.2	9.1
	60	13	6.1	3.9	18.8	0.1	11.0	2.3	4.2
		10	7.3	8.6	20.5	1.9	10.9	4.8	2.6
Wettable-powder suspensions	90	13	11.6	3.6	31.2	<0.5	10.6	<0.5	2.3
		10	5.4	–	29.2	3.2	14.6	7.5	1.0
	60	13	9.0	7.6	26.9	2.9	12.2	0.1	5.3
		10	10.9	8.8	25.3	4.3	7.6	8.4	4.2
Tetrachloro-ethylene solution	90	13	4.1	4.4	5.7	3.2	9.5	4.3	3.3
		10	6.7	4.5	13.7	4.6	13.8	5.0	–
	60	13	7.9	8.6	34.7	2.4	7.2	1.4	1.6
		10	5.4	4.2	8.3	4.5	11.4	1.9	1.1

a For the purpose of this study it was assumed that synergised pyrethrum maintained its original ratio of pyrethrins to piperonyl butoxide (1 to 10) and that recoveries of piperonyl butoxide by chemical analysis could be used to estimate residues of pyrethrins.

breakfast rolls were found to be only about 50% of the level in the flour used for their production. Decomposition of pyrethrins during baking of rye and whole-wheat bread averaged about 75%.

26.4 Metabolism

Perry (1960) reviewed the then available information on the metabolism of insecticides by various insect species. He pointed out that the reversal of paralytic symptoms and knockdown induced by pyrethrum suggest that insects may possess a detoxifying mechanism capable of attacking the compound at certain reactive sites. It was first suggested **(Acree *et al.* 1936)** that hydrolytic enzymes such as esterases might be involved in decomposing pyrethrins to non-toxic products. Detoxication of pyrethrins was demonstrated first in the southern army worm by bioassaying tissue extracts from treated insects against mosquito larvae **(Woke 1939)**. Greatest detoxication *in vitro* was brought about by fat body, followed by skin and muscle, digestive tract, and blood. Lipase extracts of roaches and houseflies readily hydrolysed pyrethrin esters to non-toxic derivatives **(Chamberlin 1950).** Piperonyl butoxide inhibited lipase activity to some extent, and consequently, detoxication of pyrethrins was diminished. The American roach has been shown to hydrolyse pyrethrins and cinerins to the corresponding keto alcohols and chrysanthemum monocarboxylic acids, plus unchanged esters and several unidentified metabolites. Of the radioactivity 8 to 12% was excreted as CO_2.

House flies were shown to metabolise significant amounts of pyrethrins to non-toxic substances (non-pyrethroid derivatives) within 24 hours after application **(Winteringham *et al.* 1959).**

Chang and Kearns (1964) studied the metabolism of ^{14}C-pyrethrin I and cinerin I by houseflies with special reference to the synergistic mechanism. They found that the metabolism of pyrethrin and of cinerin follow a similar pattern and the differences are quantitative rather than qualitative. The difference in potency of pyrethrin and cinerin was due to the rate of detoxification. There is no difference in the absorption rate. More than 96% of the absorbed dose of pyrethrin I or cinerin I was detoxified 4 hours after topical application. The mechanism of synergism was explained by the fact that the synergist prevents the detoxification of the absorbed pyrethroids although the absorption of pyrethroids is less in the presence of sesamex than in its absence. Five metabolites plus chrysanthemic acid each from pyrethrin and cinerin had been isolated and partially characterised. The amount of free chrysanthemic acid never exceeds 2.6% of the applied dose. The direct hydrolysis of the ester linkage in either molecule is not a major detoxification mechanism. Of 5 metabolites, 3 showed an intact chrysanthemic acid moiety and ester linkage. The detoxification process is apparently initiated on the keto-alcohol moiety while the acid moiety and the ester linkage are still intact. Further breakdown of the altered molecule into more polar compounds is indicated.

Verschoyle and Barnes (1972) studied the toxicity of natural and synthetic pyrethrins to rats and were able to show that the intravenous toxicity of natural pyrethrins (1–5 mg/kg) was greatly higher than the acute oral toxicity (>1400 mg/kg). It is clear that the nervous systems of the insect and mammal are almost equally sensitive to the natural pyrethrins and that they respond in a similar manner. This is not inconsistent with the findings of **Narahashi (1971)** and **Narahashi and Haas (1968)** that their effects on nerve membranes are not dissimilar. By their speed of action after intravenous injection it seems likely that the pyrethrins act *per se* and no metabolic conversion to a more toxic molecule is required for their action.

The natural pyrethrins are highly unstable to light, moisture, alkali, acid, oxidizing agents and air. The general stability is of the order: cinerin I > pyrethrin I > jasmolin I > cinerin II > jasmolin II > pyrethrin II, and various stabilisers have been incorporated to preserve the active constitutents from oxidation or hydrolysis during normal storage of the formulations **(Rowlands 1967)**.

There are no published accounts of the nature of metabolism of pyrethrum in stored grain but it is reasonable to suggest that initial oxidation of unsynergised pyrethrins may occur as they penetrate the seed coat, especially in freshly harvested grain, followed by subsequent hydrolysis as they reach the aleurone layer or the germ. The presence of a synergist may prevent the initial oxidation taking place during penetration and the subsequent or initial hydrolysis may also be hindered until the synergist itself has been degraded **(Rowlands 1967a).**

Utilising an *in vitro* enzyme system from insects in the presence of $NADPH_2$, **Yamamoto and Casida (1966)** showed that pyrethrin I was converted to at least 10 metabolites. A major metabolite was characterised as a product which had undergone oxidation of a methyl group in the iso-butenyl moiety to the carboxylic acid. In a more comprehensive study, these authors concluded that oxidation rather than hydrolysis in insects might be the mode of metabolism of pyrethroid chemicals **(Yamamoto *et al.* 1969)**.

The first steps in metabolism of pyrethrins in insects and mammals are now understood. The metabolic systems that have been examined include esterase and oxidase enzyme systems, living houseflies, mice and rats, and various micro-organisms. The detoxication reactions involved are given in **Figure 26.1**.

Much of the present knowledge of the metabolism of pyrethrins in mammals comes from the studies by Elliott, Casida, and co-workers **(Elliott *et al.* 1969a, 1969b, 1972a, 1972b; Casida *et al.* 1971; Yamamato *et al.* 1971).**

The cyclopropane carboxylicester group is cleaved to a small extent **(a, Figure 26.1)** in living rats in the case of allethrin but almost no cleavage occurs with the pyrethrins. This differentiates these cyclopentononyl esters from

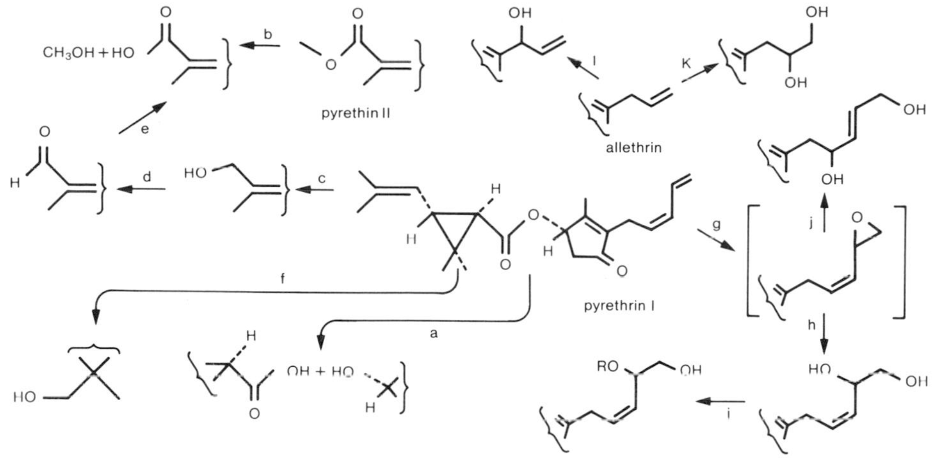

Figure 26.1. Detoxication reactions of pyrethrin I and II, and allethrin, based on studies with living rats, mice and houseflies, and various oxidase and esterase enzyme systems from these organisms.

the synthetic pyrethroids that are chrysanthemates of primary alcohols; the primary alcohol esters of the *trans*-chrysanthemates (resmethrin, tetramethrin, and others) are extensively hydrolysed in rats and an esterase in mouse liver microsomes readily cleaves the *trans*- but not the *cis*-esters.

The pyrethrate methoxycarbonyl group is readily cleaved **(b, Figure 26.1)** by esterases present in rat and mouse liver, resulting in rapid detoxication of the pyrethrates. The majority of the metabolites of pyrethrin I and allethrin arise by various sites of oxidative attack **(c & e, Figure 26.1)**.

Pyrethrins I and II have also been shown to be oxidatively metabolised in rats. Oxidation was found to occur at the *trans*-methyl group of pyrethrin I as well as on the pentadienyl sidechain to produce 2 diols. These metabolites were also found in conjugate form. The oral administration of radio-labelled pyrethrin I, or pyrethrin II to rats produced several urinary metabolites. Each contained a *trans*-2-carboxyprop-1-enyl sidechain resulting from the oxidation of the chrysanthemate carbonyl group. Also, the *cis*-2',4'-pentadienyl sidechain of pyrethrin I and pyrethrin II was modified to give a *cis*-4',5'-dihydroxypent-2'-enyl group, a 4' conjugate of this diol, or a *trans*-2',5'-dihydroxypent-3'-enyl group **(Elliott *et al.* 1972b, 1972c)**. Pyrethrins, in addition to metabolism by oxidation, were hydrolysed, as evidenced by $^{14}CO_2$ in expired air of rats following treatment with pyrethroid labelled in the carboxyl group attached to the C_1 of the cycloproprane ring.

A comprehensive review of the metabolism and biochemistry of the pyrethrins is provided by **Casida (1973)**. The metabolism and environmental degradation of pyrethrum has been reviewed by **Leahey (1985)**.

The microsomal mixed function oxidase systems of houseflies or rat liver **(Elliott *et al.* 1972b)**, as well as living houseflies, quickly oxidise the *trans*-methyl group of the isobutenyl moiety to the alcohol **(c, Figure 26.1)**, which is then oxidised by other enzymes to the aldehyde **(d, Figure 26.1)** and the acid **(e, Figure 26.1)**, ultimately producing the same acid moiety resulting from pyrethrin II by simple esteratic hydrolysis. Rats fed chrysanthemic acid excrete chrysanthemum dicarboxylic acid, among other products, so a portion of any chrysanthemic acid liberated undergoes oxidative steps **c, d, e** or **Figure 26.1.** The acid moiety also may be oxidatively modified as noted with allethrin, by converting one of the geminal dimethyl groups to the alcohol **(f, Figure 26.1)**. The final metabolic modification of the acid moiety, liberating CO_2 from chrysanthemates, including allethrin and pyrethrin I labelled with ^{14}C in the 1 position, must involve cleavage of both the cyclopropane ring and the ester grouping, but the products of the reaction, other than CO_2, are not identified.

The alcohol side chain of the pyrethrins is rapidly modified in mammals, resulting in a series of mono- and dihydroxylated derivatives **(g-l Figure 26.1)**. With pyrethrins administered orally to rats, the initial attack probably involves epoxidation of the terminal double bond of the pentadienyl group **(g, Figure 26.1)** and the epoxide rearranges to the 4',4',5'-diol **(h, Figure 26.1)** which undergoes conjugation with an unidentified moiety **(R, i, Figure 26.1)**, which possibly is a phenylacetic acid derivative. Alternatively, the epoxide can open and rearrange to the 2',5'-diol **(j, Figure 26.1)**.

There are many unidentified metabolites that possibly arise from metabolic attack at 2 or more of the sites already identified and from conjugation of the metabolites. For example, it is likely that house flies conjugate the alcohol derivative **(c, Figure 26.1)** to form a glucoside.

The sites of metabolic alteration on the acid moieties of cinerins I and II and jasmolins I and II undoubtedly are the same as with pyrethrins I and II, with a small degree of cyclopropanecarboxylic ester hydrolysis.

The enzymes primarily involved in limiting the toxicity of pyrethrin II to rats appear to be esterases, whereas those involved with pyrethrin I are microsomal mixed function oxidases. The bio-degradability of the pyrethrins probably accounts for their low toxicity to mammals and the failure of these compounds to yield persisting residues in environmental organisms.

Otieno and Pattenden (1979) summarised the then available knowledge about the degradation of the natural pyrethrins under photo-chemical, thermal and acid or base conditions. They illustrated by means of molecular-structure diagrams the nature of pyrethrins and about 40 degradation products.

26.5 Fate in Animals

Following ingestion, the pyrethrins are hydrolysed by various digestive systems in their gastro-intestinal tract. However, a small proportion of the insecticidally active compounds or their derivatives are absorbed as shown by their toxicity and their effect on the liver. The pyrethrins or their metabolites are not known to be stored in the body or to be excreted in milk.

Pyrethrins are absorbed from the gastro-intestinal tract following oral administration. Studies in male rats receiving 3 mg/kg orally resulted in almost complete absorption and metabolism within 100 hours. No pyrethrin was excreted in urine, although substantial quantities of metabolites were present. In feces, small quantities of the parent pyrethrin were observed, again accompanied by metabolites.

It is quite clear that the feeding of grain or milling offals from grain that had been treated with pyrethrum will not result in residues in foods of animal origin. Further reference should be made to the monograph by **Casida (1973).**

26.6 Safety to Livestock

No information has been located on studies designed to show the safety of administering pyrethrum to livestock by the oral route. It is known that many years ago crude pyrethrum preparations were used for the control of intestinal parasites in various livestock and domestic animals as well as humans.

McLellan (1964) published a literature review containing 38 references to the use of pyrethrum powder and extracts for the treatment of a wide variety of helminth parasites of humans, dogs, horses, poultry and sheep. The author drew attention to the absence of any ill effects on any of the hosts submitted to pyrethrins therapy.

The moderate to low acute toxicity of pyrethrum to laboratory animals as indicated in **Table 26.6** provides a high degree of reassurance that the residues encountered in animal feedstuffs are not likely to produce an acute reaction in livestock. Long-term studies involving the administration of pyrethrum to rats in their diet has shown that continuous administration of concentrations as high as 5000 ppm in the ration for 2 years had no significant effect on their growth or survival. Slight, though definite, liver damage characterised by bile duct proliferation and focal necrosis was found in rats receiving 1000 and 5000 ppm pyrethrum for 2 years **(Lehman 1965).**

Whilst the acute oral toxicity of pyrethrum is relatively low, studies have shown that the intravenous lethal dose to a dog is of the order of 6–8 mg/kg body weight. The relatively high inherent toxicity of pyrethrum should thus be noted. The very marked difference in the oral and intravenous toxicities indicates a low rate of absorption from the gastro-intestinal tract, very efficient destruction by the liver, or a combination of the two.

The acute effects resemble veratrine intoxication, proceeding from excitation to convulsions and tetanic convulsions, except that pyrethrum also causes muscle fibrillation. Death is caused by respiratory failure. Persistent tremor is occasionally seen in animals that recover from a single large dose.

There appear to be no apparent teratogenic effects elicited by pyrethrum in rabbits receiving 90 mg/kg body weight/day, orally from day 8–16 of gestation **(Weir 1966a)**.

The acute signs of poisoning in rats include: depression, rapid and/or laboured respiration, ataxia, incoordination, convulsions and muscular tremors. Necropsy findings include: congestion of the lungs, liver, kidneys, adrenals and pancreas and slight gastric inflammation **(Weir 1966b; Malone and Brown 1968).**

26.7 Toxicological Evaluation

The pyrethrins have an acute single dose oral LD/50 against white rats variously reported at 200 to 1500 mg/kg, with 200 to 400 probably being the range for less pure pyrethrum preparations and 400 to 800 being the range for the more purified preparations. Dosages for guinea pigs and mice fall within these same ranges.

The pyrethrins show little evidence of causing chronic problems of toxicity. Experimental animals can be fed a large portion of an acute LD/50 every day without harmful effects.

Lehman (1954) gave a chronic LD/50 for rats of 250 mg/kg/day. His acute LD/50 was 200 mg/kg for single doses.

Dermal toxicity of pyrethrins is negligible, because they seem to be poorly absorbed through the intact skin. The consensus of reported values for the dermal LD/50 of the pyrethrins runs from 1350 mg/kg to about 5000 mg/kg. There is little evidence of human toxicity problems from the pyrethrins. Long years of experience have produced no clear-cut cases of human poisoning except allergic reactions to impurities in crude pyrethrum extracts and these were eliminated by commercial processing many years ago..

A valuable review of the toxicology and pharmacology of pyrethrum to mammals is provided by **Bartel (1973).**

Table 26.6. Pyrethrum: acute oral toxicity to various species

Animal	Sex	Grade of pyrethrum	Pyrethrin content (%)	Vehicle	LD/50 (mg/kg)	Conf. limits	Ref.
rat	–	PD	20.3	Neat	794.3	604–1045	1
	–	Pale	20.9	Neat	584.3	481–710	1
	–	OR	27.2	Neat	634.0	495–812	1
	–	OR	20.0	OPD	584.3	452–755	1
	–	NP	77.8	Neat	900.0	733–1106	1
	M	Pyrethrin I	–	DMSO		260–420	2
	M	Pyrethrin II	–	DMSO		>600	2
	–	Pale	–	Neat	710	–	3
	–	OR	–	OPD	820	–	4
	–	PD	–	PO	1870	–	4
	–	OR	–	PO	>1500	–	8
	–	OR	–	PO	>2600	–	6
	F	NP	–	GF	>1400	–	2
	–	OR	–	OPD	820	–	4
	–	PD	–	PO	1870	–	4
	–	OR	–	PO	200	–	5
	–	OR	–	PO	>2600	–	6
	M	–	–	–	710	–	7
	–	OR	–	Neat	634		1
	–	PD	–	Neat	794		1
	–	Pale	–	Neat	584		1
	–	NP	–	Neat	900		1
mouse	–	OR	–	OPD	273	–	1
	–	NP	–	OPD	786	–	1
guinea pig	–	OR	–	PO	>1500	–	8

Sources:

1. **Malone and Brown (1968)**
2. **Verschoyle and Barnes (1972)**
3. **Griffin (1973**
4. **Carpenter *et al.* (1950)**
5. **Lehman (1951)**
6. **Ambrose and Robbins (1951)**
7. **Weir (1966b)**
8. **Shimkin and Anderson (1936)**

GF Glycerol formal
PO Petroleum oil
OR Pyrethrum oleoresin
NP Nitromethane concentrate of pyrethrins
OPD Odourless petroleum distillate
PD Partially dewaxed oleoresin
Pale Pale extract
DMSO Dimethyl sulphoxide

There are several early chronic toxicity studies reported with pyrethrins, but by modern standards some of these are not adequate. In the 2-year rate study there were effects at 5000 ppm but the no-observed-effect level (NOEL) has been quoted as either 200 or 1000 ppm.

Griffin (1973) reported that **Hunter and Newman (1972)** administered pyrethrum in the diet of rats for 5 weeks. Those receiving the 8000 ppm rate showed a few prominent Peyers patches in the ileum of 2 rats.

Bond *et al.* (1973) reported a study in which rats receiving up to 360 mg/kg by way of the diet for 13 weeks showed marked reduction in bodyweight, liver eosinophilia with early cell necrosis and kidney tubular degeneration.

The JMPR evaluation **(Vettorazzi 1975)** concluded that the no-observed-effect level in a 2 year rat feeding study was 200 ppm. Bile duct proliferation and focal necrosis of the liver were observed at high doses.

Griffin (1973) reported that dogs receiving 500 ppm in the diet for 13 weeks displayed ataxia, laboured respiration, salivation mainly in the first month and vacuolation of the liver.

The toxicology of pyrethrins was evaluated by JMPR on 4 occasions, 1965, 1966, 1970 and 1972. The data available included numerous acute studies on rats, mice, guinea pigs, dogs and chicks, together with short-term toxicity studies in man, rats, and rabbits. There was also a long-term feeding study involving rats and special studies of reproduction and skin sensitisation.

On the basis of these studies, particularly the long-term feeding study in rats which showed that the no-effect level was 200 ppm in the diet, equivalent to 10 mg/kg body weight/day, the JMPR allocated an ADI for man of 0.4 mg/kg body weight.

26.8 Maximum Residue Limits

The JMPR has evaluated a variety of information on residues of pyrethrins on nine occasions between 1965–75. As a result of these evaluations, maximum residue limits have been recommended as follows:

Commodity	Maximum Residue Limits (mg/kg)
Cereal grains	3
Cereal products	1

These recommendations are not altogether consistent with the data on pyrethrum residues on milling products, especially bran.

Residues are determined and expressed as 'total pyrethrins'.

The methods recommended for the determination of pyrethrum residues are: **Methodensammlung, (1982 — XII-6; S19); Pesticide Analytical Manual, (1979d); Zweig (1976); Mestres *et al.* (1979a); Sprecht and Tillkes (1980).**

27. Tetrachlorvinphos

Tetrachlorvinphos was developed in the mid 1960s under the code number SD8447. It has been marketed for more than 15 years under the trade name Gardona. Although its potency is relatively low it has a broad-spectrum activity and has found a minor place in agriculture, home gardens, animal care and personal protection.

Tetrachlorvinphos has been shown to have good effect against many species of stored-product pests and good stability in dry grain. It appears to be comparable with bromophos in its spectrum of effect and persistency. Tetrachlorvinphos deposits do not penetrate the individual grains to any extent and appear to be removed on hulls and bran. Residues are readily degraded in cooking and even the simple procedures followed under tribal conditions appear to degrade the residues completely. No action has been taken to have tetrachlorvinphos fully developed and registered for admixture with grain.

27.1 Usefulness

Lemon (1966) was among the first to evaluate tetrachlorvinphos against a range of stored-product pests. He showed that it was superior to malathion against *Tribolium confusum* and *T. castaneum*. In view of the promising results obtained with tetrachlorvinphos and its extremely low mammalian toxicity, **Lemon (1967a, 1967b)** extended the investigation to include a further 8 species. The compound proved more effective than malathion against 6 of the species. It was, however, relatively ineffective against *Oryzaephilus spp*. although its toxicity to these species exceeded that of bromophos. Against *Rhyzopertha dominica* it equalled fenithrothion in toxicity.

Lemon (1967b) determined the relative susceptibilities of *Tribolium confusum* and *Tribolium castaneum* to 7 organophosphorus insecticides by applying the compounds topically at 4 dosage rates. Tetrachlorvinphos was more effective than malathion against both species and was evaluated against 8 more species of stored-product beetles. It was more toxic than malathion to *Rhyzopertha dominica, Lasioderma serricorne, Stegobium paniceum* and *Sitophilus zeamais* but less so against *Oryzaephilus surinamensis, Oryzaephilus mercator, Ptinus tectus* and *Sitophilus granarius*. *Sitophilus zeamais* was about 11 times more susceptible to tetrachlorvinphos than *Sitophilus granarius*.

Strong and Sbur (1968) tested the effectiveness of 48 insecticides against 5 stored-product insects by spraying adult insects and larvae with a range of 5 concentrations of each insecticide from 0.01% to 1%. Tetrachlorvinphos was one of the more effective materials, being somewhat more effective than malathion.

Strong (1970) reported the comparative susceptibility of *Tribolium confusum* and *T. castaneum* to 12 organophosphorus insecticides applied directly to adult insects in the form of a spray solution. Tetrachlorvinphos was second only to fenthion in its effectiveness against *T. confusum* though it ranked midway and equal to malathion in effectiveness against *T. castaneum*.

Coulon (1972), reporting laboratory studies with 10 new insecticides (organophosphorus or pyrethroid) against *Sitophilus granarius* stated that tetrachlorvinphos was found to be less effective than other substances tested.

McGauhey (1972a) tested tetrachlorvinphos, dichlorvos and a mixture of tetrachlorvinphos and dichlorvos as protectants for rice in small bins. Tetrachlorvinphos at 15 and 20 mg/kg, dichlorvos at 7.5 and 10 mg/kg and the combination at 20 and 10 mg/kg eliminated pretreatment infestations, greatly reduced reinfestations for 3, 1 and 9 months respectively, and did not affect germination of the rice. Tetrachlorvinphos was highly toxic to *S. oryzae* for 6 months and to *R. dominica* and *T. confusum* for 1 month. The 20:10 mg/kg combination was highly toxic to *S. oryzae* for 12 months, to *R. dominca* for 1 month and *T. confusum* for 3 months.

Morallo-Rejesus and Santhoy (1972) tested the efficacy of 6 organophosphorus insecticides applied topically to adults of DDT-resistant strains of *Sitophilus oryzae* and *Tribolium castaneum* collected in the Philippines. When tested against *Sitophilus oryzae* on maize, tetrachlorvinphos was one of the most effective materials and when tested on sorghum it was significantly better than all the others. It was also the best against *T. casteneum* when tested on wheat flour and yeast.

Harein and Rao (1972) reported mortalities of 95 to 100% in *Sitophilus granarius* after 20-day exposure to wheat with a residue of 8 mg/kg tetrachlorvinphos.

Morallo-Rejesus (1973) studied the toxicity and duration of protection of maize when mixed with 5 insecticides. Tetrachlorvinphos applied as a wettable powder was effective against a range of insect pests particularly *Sitophilus oryzae* in 6 months storage.

LaHue (1973) made a thorough evaluation of tetrachlorvinphos as a protectant against insects in stored wheat. Tetrachlorvinphos was applied at dosages of 5, 10 and 20 mg/kg to hard winter wheat containing 10, 12 and 13.5% moisture. The effectiveness was tested after ageing the treated wheat by counting the progeny developing after the wheat had been exposed to *Sitophilus oryzae, Tribolium casteneum*, T. *confusum* and *Rhyzopertha dominica*. Malathion applied at a rate intended to give a deposit of 10 mg/kg was used as the basis for comparison. Tetrachlorvinphos at 10 mg/kg was generally not so effective as malathion, but considerable protection was afforded wheat treated with this insecticide. High-moisture content of the wheat lowered the efficacy of tetrachlorvinphos, but the adverse effect of high moisture was compensated for to some extent when the dosage was increased.

Chawla and Bindra (1973), working in India, tested a range of 9 organophosphorus insecticides against the 3 principal pests of stored wheat *R. dominica, S. oryzae* and *T. granarium*. Tetrachlorvinphos was not amongst the 5 most promising compounds tested.

Giles (1973), studying the control of insect pests in maize cribs in Nicaragua, reported that tetrachlorvinphos dust was always inferior to other materials for the first 16 weeks but at 24 weeks it had a relatively better performance.

McGaughey (1973) tested the usefulness of a mixture of tetrachlorvinphos/dichlorvos as a protection for rice in husk against *Sitotroga cerealella*. The mixture (2:1, tetrachlorvinphos:dichlorvos) when applied to the surface of unhusked rice gave a high degree of control and prevented reinfestation by *Sitotroga*.

Morallo-Rejesus and Nerona (1973) evaluated the effectiveness of sacks impregnated with 2 and 4% malathion, pirimiphos-methyl, dichlorvos and tetrachlorvinphos. Tetrachlorvinphos was inferior to pirimiphos-methyl and malathion but superior to dichlorvos.

Morallo-Rejesus and Carino (1974), studying the residual toxicity of 5 insecticides in corn against *Sitophilus spp.* and *Rhyzopertha dominica*, found that tetrachlorvinphos was more effective against *Sitophilus spp.* and that its residual effectiveness against *Rhyzopertha* was greater than the other 4 compounds.

Harein (1974) drew attention to the fact that notwithstanding widespread use of malathion, it frequently failed to control stored-product moths. By comparison, tetrachlorvinphos was less than one-quarter as effective as malathion against the least susceptible of the 5 important moths species.

Verma and Ram (1974) reported studies carried out in India on a strain of *Tribolium castaneum* which was resistant to malathion. The development of resistance to malathion did not result in any serious cross-resistance to tetrachlorvinphos and the authors concluded that this insecticide may safely replace malathion if malathion resistance in *Tribolium castaneum* became a problem.

Morallo-Rejesus and Eroles (1974) found that both pirimiphos-methyl and tetrachlorvinphos were more effective as protectants when the treated grain was stored in pirimiphos-methyl impregnated sacks. When shelled corn was treated with tetrachlorvinphos emulsion concentrate at the rate of 10 mg/kg and the treated maize was stored in pirimiphos-methyl impregnated sacks, comparatively little damage to the grain occurred during 12 months storage. The main species involved were *Sitophilus zeamais, Rhyzopertha dominica, Tribolium castaneum, Lophacoteres pusillus* and *Palorus ratzeburgii*.

Weaving (1975) determined the median lethal doses of 5 insecticides to *Sitophilus zeamais* on maize and sorghum using laboratory formulated dusts. Tetrachlorvinphos was rated fourth after fenitrothion, fenthion and iodofenphos.

Chawla and Bindra (1976) determined the relative toxicity of 7 organophosphorus insecticides and 2 synthetic pyrethroids against *Rhyzopertha dominica, Sitophilus oryzae* adults and larvae of *Trogoderma granarium* under controlled laboratory conditions. Malathion was employed as standard. The bioassay procedure consisted of exposing the test species to wheat treated with graded doses of the insecticides. Pirimiphos-methyl and phoxim were found to be the most promising. Tetrachlorvinphos was appreciably less toxic than malathion except against adults of *Rhyzopertha dominica*.

Morallo-Rejesus and Eroles (1976) continued the studies of grain treatment combined with sack impregnation to control pests of stored corn. Eight treatment combinations were screened over 12 months storage. The insecticides were applied to the grain at the rate of 10, 15 and 30 mg/kg. Grains admixed with tetrachlorvinphos, when stored in pirimiphos-methyl-impregnated sacks had a lower level of infestation than those stored in malathion-impregnated sacks. Powder formulations of tetrachlorvinphos and pirimiphos-methyl gave superior results than treatments with the emulsifiable concentrates.

McCallum-Deighton (1978) tabulated the minimum effective doses of 8 insecticides against 7 important coleopterous pests of stored grain. Tetrachlorvinphos was shown to be significantly less potent than all other insecticides against all species except *Rhyzopertha dominica* where it was shown to be more effective than malathion, bromophos and iodofenphos.

DeLima (1978b) presented a thesis on the bionomics and control of *Sitophilus zeamais* and *Sitotroga cerealella* in stored maize under laboratory and field conditions in Kenya. Tetrachlorvinphos was one of the insecticides evaluated in this study.

Adesuyi (1979) reported trials carried out to assess the efficacy of iodofenphos, pirimiphos-methyl and tetrachlorvinphos for protecting maize cobs stored in cribs at the village level in Nigeria. Tetrachlorvinphos applied as a dust at the rate of 5, 10 and 15 mg/kg resulted in the maize having a lower percentage insect damage right throughout an 11 month storage than corresponding treatments with either iodofenphos or pirimiphos-methyl. This investigator concluded that the results of these trials showed that any of these insecticides applied at 10 mg/kg could be used instead of lindane by Nigerian farmers to treat their maize stored in cribs, especially as they do not generally store for longer than 8 months in such a structure. Even so the percentage insect damage at the end of 8 months would reach 10%.

Bitran *et al.* (1980c) showed that the fumigation of

corn cobs with phosphine followed by treatment with 8 mg/kg tetrachlorvinphos protected the corn against *Sitophilus zeamais* and *Sitotroga cerealella* better than treatment with malathion or tetrachlorvinphos alone. Therefore, phosphine fumigation followed by spraying with tetrachlorvinphos was recommended for the reduction of infestation of maize in Brazil.

Hindmarsh and MacDonald (1980) carried out field trials to control insect pests of farm-stored maize in Zambia by applying insecticidal dusts to shelled maize stored in mud-walled, timber and thatch cribs. They found that malathion, tetrachlorvinphos and iodofenphos applied at 12 mg/kg all kept damaged grains below 10% up to 10 months after harvest.

Weaving (1980) studied grain protectants for use under tribal storage conditions in Zimbabwe. Tetrachlorvinphos applied at 8 mg/kg, though effective initially, was significantly less persistent than fenitrothion, fenthion, malathion and iodofenphos.

Weaving (1981) reported further evaluations of grain protectants for admixture with maize stored in traditional grain bins under tribal storage conditions in Zimbabwe. Seven insecticide admixtures were tested under field conditions. Pyrethrins and tetrachlorvinphos were less effective and less persistent than fenitrothion, fenthion, iodofenphos or pirimiphos-methyl. Tetrachlorvinphos was regarded as satisfactory for controlling *S. zeamais* for up to 4 months whereas several of the other materials were equally effective for at least 12 to 15 months.

Morallo-Rejesus (1981) carried out laboratory toxicity tests in the Philippines with tetrachlorvinphos and several other insecticides against DDT- and lindane-resistant strains of *Sitophilus zeamais* and *Tribolium castaneum*. Tetrachlorvinphos was more toxic than malathion but inferior to chlorpyrifos-methyl and pirimiphos-methyl.

Viljoen *et al.* (1981a) used tetrachlorvinphos and a number of other insecticides to protect maize and groundnuts in bag stacks to prevent re-infestation after fumigation. Small bag stacks of maize and groundnuts in the open and in a shed in South Africa were sprayed with various contact insecticides directly after fumigation. Tetrachlorvinphos was very effective against Lepidoptera but less so against *Tribolium castaneum*.

Tyler and Binns (1982) studied the influence of temperature on the susceptibility to 8 organophosphorus insectides of susceptible and resistant strains of *Tribolium castaneum*, *Oryzaephilus surinamensis* and *Sitophilus granarius*. Based upon knock down and kill, the effectiveness of all insecticides was greater at 25°C than at 17.5°C and was markedly lower at 10°C. At 10°C, tetrachlorvinphos was virtually ineffective against *Sitophilus granarius*.

27.2 Degradation

Very little information appears to have been published on the degradation of tetrachlorvinphos in stored grain. **Hall *et al* (1973)** published the results of studies of the persistence and distribution of tetrachlorvinphos on wheat and maize. The grain was held in metal farm-storage bins holding approximately 3 tonnes each. Two such bins were treated with tetrachlorvinphos at a rate equivalent to 26 mg/kg and 2 bins were filled with maize treated at the rate of 22 mg/kg. The wheat was sampled immediately and 98, 194 and 260 days after treatment. The maize was sampled immediately at 92, 133 and 223 days after treatment. The residue was extracted with acetonitrile and determined by GLC equipped with a flame photometric detector. The results of analysis are given in Table 27.1. These indicate a relatively slow rate of degradation; less than 25% in 260 days in the case of wheat and 60% in 223 days in the case of maize. Unfortunately, no information is given concerning the moisture content of the grain or the temperature during storage.

Weaving (1975) evaluated 10 grain protectants on maize and sorghum for potential use under tribal storage conditions in Zimbabwe. Though tetrachlorvinphos produced a high level of mortality of *Sitophilus zeamais* initially, the effect fell away rapidly so that by the end of 4 months comparatively little activity was apparent.

Table 27.1. Tetrachlorvinphos residues on wheat and maize held in farm storage **(Hall *et al.* 1973)**

Grain	Replicate	Sample date (time in storage, days)			
		8 Oct 1970 (0 days)	14 Jan 1971 (98 days)	20 April 1971 (194 days)	25 June 1971 (260 days)
Wheat	1	25.75	25.16 ± 2.74	22.76 ± 2.50	18.13 ± 3.03
	2	24.00	28.83 ± 4.49	32.51 ± 2.51	24.71 ± 3.99
		11 Nov 1970 (0 days)	18 Feb 1971 (92 days)	16 Mar 1971 (133 days)	29 June 1971 (223 days)
Maize	1	12.28	15.17 ± 1.73	12.43 ± 1.16	5.28 ± 0.54
	2	15.50	18.43 ± 0.83	13.55 ± 1.77	6.36 ± 0.82

Mian and Kawar (1979) studied the persistence of dichlorvos and tetrachlorvinphos on wheat grains in Pakistan. Wheat grain treated with 10, 15 and 20 mg/kg of insecticide was analysed after 8 months of storage. The residues of tetrachlorvinphos were considerably higher than those of dichlorvos.

Hindmarsh and Macdonald (1980), following field trials to control insect pests of farm-stored maize in Zambia, reported the fate of the deposit of tetrachlorvinphos on the grain over a 10 month storage. Though the treatment was aimed at depositing 12 mg/kg tetrachlorvinphos on the shelled maize, the concentration found by analysis was only 7.0 mg/kg. This concentration fell slowly and consistently through 5.6 mg/kg after 4 months, 3.2 mg/kg after 8 months to 2.5 mg/kg after 10 months. The mean moisture content of the grain fluctuated during the 10 month storage due to prevailing atmospheric conditions.

Weaving (1980), in further studies of grain protectants for use under tribal storage conditions in Zimbabwe, found that, though tetrachlorvinphos was as effective as the best treatment initially, at the end of 2 months the effectiveness began to fall away and by the end of 4 months it was inferior to fenitrothion, fenthion and iodofenphos and equal to malathion in the degree of protection provided.

In the third of a series of studies into grains protectants for use under tribal storage conditions in Zimbabwe, **Weaving (1981)** found that the effect of tetrachlorvinphos fell away sharply after 4 months in its action upon *Sitophilus zeamais*. No indication is given of the actual residue levels involved.

27.3 Fate in Milling, Processing and Cooking

Hall *et al.* (1973), who studied the persistence and distribution of tetrachlorvinphos in wheat and maize, reported that when the treated grains which had been in storage for approximately 8 months were milled, the bulk of the residues was found on the bran, fine feed and germ portion of the maize and on the bran and shorts of the wheat. Though the bran, fine feed and germ constituted only 55% of the milled product from maize, they contained 94% of the total tetrachlorvinphos residues. The bran, shorts and flour contained 64%, 3% and 30%, respectively, of the tetrachlorvinphos residue. In an separate experiment the milling products from the maize were oven dried (temperature not stated). There was a consistent, but relatively small, loss of tetrachlorvinphos indicating that the insecticide is relatively non-volatile. The results are shown in **Tables 27.2 and 27.3.**

The above authors used the wheat and maize flour from the above experiments to produce sweet-milk corn bread and yeast bread. The corn bread was baked at 230°C for 30 minutes and the wheat bread at 200°C for 35 minutes. The results **(Table 27.4)** show that 73% of the tetrachlorvinphos remained in the corn bread and 60% in the wheat bread. This is considered remarkable.

Table 27.2. Tetrachlorvinphos residues in milling products from maize after 223 days in storage **(Hall *et al.* 1973).** (Unmilled maize contained 5.8 mg/kg residues.)

Milling fraction	Yield (%)	Residues (mg/kg)
Bran	17.13	4.90
Fine feed	21.35	7.96
4-grits	5.06	<0.25
Coarse grits	7.07	0.51
Medium grits	5.48	0.44
Fine grits	7.59	0.63
Coarse meal	7.07	0.84
Fine meal	6.56	0.45
Germ	16.14	2.50

Table 27.3. Tetrachlorvinphos residues in milling products from wheat after 194 days in storage **(Hall *et al.* 1973).**

Milling fraction	Replicate 1		Replicate 2	
	(%)	(mg/kg)	(%)	(mg/kg)
Unmilled grain	100.00	22.76	100.00	32.51
Bran	21.79	20.08	22.10	26.13
Shorts	2.52	10.30	2.37	13.05
Flour	72.20	2.93	71.00	3.41

Table 27.4. Effects of baking on tetrachlorvinphos residues in bread **(Hall *et al.* 1973).**

Product	Replicate	Residue before baking (mg/kg)	Residue after baking (mg/kg)	% Residue remaining
Wheat bread	1	1.57	0.90	57.3
	2	1.61	1.03	63.9
Corn bread	1	0.46	0.34	73.9
	2	0.46	0.33	71.7

Lockwood *et al.* 1974 studied the degradation of organophosphorus pesticides in cereal grains during milling and cooking in India. They applied tetrachlorvinphos to rice in husk in amounts equivalent to 10, 40 and 80 mg/kg on the whole grain. They were able to recover a substantial portion of this amount by analysis of samples taken immediately after treatment. Portion of the rice was milled to produce polished rice immediately following treatment and after 1 and 2 months storage. Only in the rice milled from the grain treated at 40 and 80 mg/kg were residues of tetrachlorvinphos detected and then only at levels ranging from 1 to 2.3 mg/kg. This indicates that very little of the insecticide deposited on the husk of the rice migrates into the grain itself. Boiled rice and steamed, fermented rice (iddli) were prepared from the polished rice. Using methods capable of detecting less than 0.1 mg/kg, these investigators were unable to obtain any peaks on the gas-liquid chromatograms of samples extracts from the cooked rice. Bulk quantities of wheat and sorghum were also treated with tetrachlorvinphos at rates designed to deposit 10, 40, and 80 mg/kg and the grain was stored for 1 and 2 months in India though the storage conditions are not described. The sorghum and wheat samples taken immediately after treatment and at the end of 1 and 2 months were milled using an electrically powered disc mill, a type common in cities and villages throughout India. The grain sample (500g) was ground until 90% passed through a 32-mesh sieve. The results are reported in **Tables 27.5 and 27.6.**

Chapaties, thin unleavened cakes cooked on a grill, were prepared from the wheat flour and the sorghum flour was made into a dough which was formed into balls and cooked by boiling. All trace of the tetrachlorvinphos residue was lost from the sorghum as a result of the wet-cooking. The average loss of tetrachlorvinphos from the wheat flour during the preparation of chapaties was 75% **(Table 27.7) (Lockwood *et al.* 1974).**

27.4 Metabolism

Glutathione-dependent demethylation of dichlorvos has been demonstrated in experiments *in vitro* investigating the influence of different alkyl groups on the rate of the O-dealkylation reaction **(Hollingsworth 1969)**. Tetrachlorvinphos undergoes similar dealkylation, which can also be performed by liver microsomes. The highest

Table 27.5. Losses of indicated insecticide residues during machine grinding of wheat

Insecticide	Treatment amount (ppm)	Insecticide residue recovered from: Whole grain (ppm)	Machine-ground grain (ppm)	Loss of residue: ppm	%	Avg. %	Std. dev. %
Malathion	10	5.8	2.2	3.6	62		
	40	31.9	9.3	22.6	71	71	8.5
	80	77.1	16.5	60.6	79		
Gardona	10	18	13.5	5.5	31		
	40	39	27.5	11.5	29	30	1.2
	80	65	45	20	31		
Sumithion	2	2	1	1	50		
	8	8.3	3	5.25	64	62	11.6
	16	15.5	4.2	11.3	73		

Table 27.6. Losses of indicated insecticide residues during machine grinding of sorghum

Insecticide	Treatment amount (ppm)	Insecticide residue recovered from: Whole grain (ppm)	Machine-ground grain (ppm)	Loss of residue: ppm	%	Avg. %	Std. dev. %
Malathion	10	5.76	3.4	2.23	41		
	40	22.4	13.5	8.9	40	40	1.5
	80	53.3	33	20.3	38		
Gardona	10	6.5	2.75	3.75	58		
	40	32.5	25.5	7	22	39	17.5
	80	85	53	32	38		
Sumithion	2	2.5	1.8	0.7	28		
	8	9.25	6.5	2.75	30	28	2.0
	16	15.5	11.5	4	26		

Table 27.7. Loss of indicated insecticide residues during the preparation of chepaties from wheat flour (**Lockwood *et al.* 1974**)

Insecticide	Treatment amount (ppm)	Months after treatment	Insecticide residue recovered from: Machine-ground grain (ppm)	Chapati (ppm)	Loss during cooking: ppm	%	Avg. %	Std. Dev. %
Malathion	10	0	2.2	1.5	0.7	32		
		1	0.8	0.23	0.57	71		
		2	0.53	0.23	0.3	57		
	40	0	9.3	5.8	3.5	38		
		1	3.5	1.3	2.2	63	51	24.0
		2	1.85	0.95	0.9	49		
	80	0	16.5	16	0.5	3		
		1	12	3.5	8.5	71		
		2	7.5	1.6	5.9	79		
Gardona	10	0	13.5	b.d[a]	13.5	100		
		1	5	2.85	2.15	43		
		2	2.35	0.65	1.70	72		
	40	0	27.5	1.7	25.8	94		
		1	27.5	7.5	20	73	75	18.3
		2	8	3	5	63		
	80	0	45	3	42	93		
		1	50	12.5	37.5	75		
		2	18	6.5	11.5	64		
Sumithion	2	0	1	0.53	0.47	47		
		1	0.2	b.d.	0.2	100		
		2	0.13	0.13	0	0		
	8	0	3	1.5	1.5	50		
		1	1.1	0.09	1.01	92	52	35.1
		2	0.58	0.38	0.2	34		
	16	0	4.2	3.6	0.6	14		
		1	1.9	0.23	1.67	88		
		2	1.08	0.65	0.43	40		

a b.d. = below detectability.

activity has been located in the supernatant fraction and requires reduced glutathione as described for other demethylations. Methylated glutathione has, in this case, also been identified as the primary reaction product (**Hutson *et al.* 1968**).

The leaving groups of tetrachlorvinphos have been found to give rise to a well defined pattern of metabolites, but remarkable differences have been observed between rat and dog metabolism. Whereas in the dog O-dealkylation dominated (46% occurred as dealkylated derivatives in the urine), esteratic cleavage and conjugation have been found to be the main pathways used by the rat. Both reductive and hydrolytic dechlorination had taken place as di- and tri-chlorophenyl ethan-1-ol and the corresponding ethanediol have been identified, mostly in the form of their *b*-D-glucuronides, (35% in the rat, 0% in the dog) as metabolites in urine.

Data on the residual behaviour of ^{32}P-labelled tetrachlorvinphos on its general metabolism (**Whetstone *et al.* 1966**) and its residue behaviour in cattle (**Ivey *et al.* 1968**), poultry (**Ivey *et al.* 1969; Yadav and Shaw 1970**) and bovine milk and excreta (**Oehler *et al.* 1969; Miller *et al.* 1970**) are available.

27.5 Fate in Animals

Miller and Gordon (1973) studied the effect of feeding tetrachlorvinphos to dairy cows over extended periods. In tests in Maryland in 1967–70, a wettable powder containing 75% tetrachlorvinphos was fed at 36–252 ppm in the ration to 4 dairy cows for 1 complete gestation period and to 1 cow for 2. The rates corresponded to averages of about 1 mg/kg body weight daily for 2 cows, about 2.5 mg/kg for a third and about 7 mg/kg for the remaining 2, but varied during the gestation periods. Residues of tetrachlorvinphos in the milk and feces were determined continually, and the toxicity of the feces to larvae of house flies was measured. Shortly after calving, the cows were slaughtered and the residues in tissues were determined. The residues in tissues were undetectable (less than 0.02 mg/kg), except in one sample of fat. The residues in feces never reached 1 mg/kg, and the highest recorded was 0.91 mg/kg for a cow that was receiving tetrachlorvinphos at 8.2 mg/kg during the sampling period concerned. The residues in milk were poorly correlated with the tetrachlorvinphos intake in mg/kg, but on the whole exceeeded 0.1 mg/kg only when the intake exceeded 2 mg/kg. The toxicity of the feces to larvae of house flies indicated that well-tolerated levels of tetrachlorvinphos in the ration of dairy cows should give good control of susceptible but not of organophosphorus-resistant house fly larvae in the feces.

Kokhtyuk and Sukhomlinova (1977) studied the fate of tetrachlorvinphos in poultry. They found that when birds were treated at a dosage of 1000 mg/kg, a dose which produces no mortality, the pesticide first accumulated in the muscles, liver, kidneys, brain and ovaries, but disappeared 15 days after administration. Poisoning of birds with tetrachlorvinphos is diagnosed on the basis of the following symptoms: disturbance of the functioning of the central nervous system, lack of coordination of movement, general depression, paresis and paralysis, depression of blood-cholinesterase activity, accumulation of blood in the internal organs, and catarrhal haemorrhagic enteritis.

27.6 Safety to Livestock

The studies reported by **Miller and Gordon (1973) (Section 27.5)** showed that the feeding of tetrachlorvinphos at levels as high as 7 mg/kg during 1 or 2 complete gestation periods did not adversely affect calving or blood-cholinesterase levels. This corresponds to up to 250 ppm in the ration.

The paper by **Kokhtyuk and Sukhomlinova (1977)** (Section 27.5) indicates that 1000 mg/kg tetrachlorvinphos is not lethal when fed to poultry but that poisoning does occur when an unstated higher level is administered.

27.7 Toxicological Evaluation

Tetrachlorvinphos has not been recommended for priority for evaluation by the Joint FAO/WHO Meeting on Pesticide Residues. On several occasions over recent years the Codex Committee on Pesticide Residues has discussed the desirability of including tetrachlorvinphos in its list of priorities but on each occasion it has been recognised that the overall usage of tetrachlorvinphos was relatively small and that those uses which had been approved generally did not give rise to significant residues in food. From this, it must be concluded that tetrachlorvinphos has not been widely accepted for use as a grain protectant insecticide.

27.8 Maximum Residue Limits

Tetrachlorvinphos had not been evaluated by the Joint FAO/WHO Meeting on Pesticide Residues and no maximum residue limits have been recommended for any food commodity.

No methods have been recommended for the determination of tetrachlorvinphos residues.

28. Effect of Temperature

When one studies the extensive literature on various aspects of grain protection one cannot help being concerned over major differences in results reported by different investigators. Undoubtedly these differences derive from many sources, including differences in scientific techniques and experimental conditions as well as differences in the quality and formulation of insecticides available and in the biology of the test insects.

Lack of attention to the importance of temperature and moisture has undoubtedly contributed greatly to creating or exaggerating these differences. Many papers failed to even record the temperature or moisture content (humidity) that prevailed during the experiments. On the other hand, there are many excellent studies which have been devoted to obtaining a better understanding of the function which temperature plays on the outcome of pest-control techniques. Recognition of temperature as one of the major factors in determining the outcome of pest-control practices would go a long way towards the development of practical and economical pest-control techniques.

Whilst many of these studies have been referred to in the review of effectiveness of individual insecticides and their degradation within the grain storage system, an attempt will be made to review some of these papers under 3 subheadings

1. Effect of Temperature on Insect Biology
2. Effect of Temperature on Insecticide Toxicity
3. Effect of Temperature on Insecticide Persistence.

28.1 Effect of Temperature on Insect Biology

As a group, grain-damaging insects are mostly of subtropical origin and do not hibernate. They have not developed tolerance to low temperatures, so that in many of the grain-growing areas of North America, USSR and China they are rarely abundant enough to cause serious damage to grain in storage. A thorough knowledge of the limiting effect of low temperatures is invaluable in formulating management programs for the safe storage of grain. Records of the relative susceptibilities of various stored-grain insects to low temperatures are given in **Table 28.1.**

Temperatures that are not quickly lethal indirectly cause the death of many insect pests of stored grain by rendering them inactive and by preventing them from feeding. The life processes of insects that do not hibernate are not sufficiently retarded by low temperatures to enable the food reserves in their bodies to sustain them through an extended period of dormancy. As a result, they die from starvation. A few insects such as *Tenebroides mauritanicus* are true hibernators and can survive exposure to low temperatures for a long time **(Cotton and Wilbur 1974)**.

According to **Robinson (1926),** *Sitophilus oryzae* is dormant at temperatures of 7°C or below, and *Sitophilus granarius* at 2°C or below. **Anderson (1938)** noted that neither species mated when the temperature fell below 12–13°C. Although **Richards (1947)** has placed the lower limit of oviposition at 10°C, most writers agree that few eggs are laid at temperatures below 15°C. Hatching and development of larvae at temperatures between 13 and 15°C are very slow. *Tribolium confusum* and *Oryzaephilus surinamensis* do not lay eggs at 15°C, hence reproduction ceases at that temperature. Tyroglyphid mites are able to breed in stored wheat at temperatures between 4°C and 10°C if moisture conditions are favourable.

Tables 28.1. Temperature tolerance of insects that attack stored-grain products **(Cotton 1950)**

	Days exposure required to kill all stages at						
	−18°to −15°C	−15° to −12°C	−12° to −10°	−10° to −6.5°C	−6.5° to −4°C	−4° to −1°C	−1° to +2°C
Sitophilus oryzae	1	1	1	3	6	8	16
Sitophilus granarius	1	3		14	33	46	73
Oryzaephilus surinamensis	1	1	1	3	7	23	26
Tribolium confusum	1	1	1	1	5	12	17
Tribolium castaneum	1	1	1	1	5	8	17
Plodia interpunctella	1	3	5	8	28	90	
Anagasta kuehniella	1	3	4	7	24	116	

Subject to certain upper limits, the rate of development and reproduction of all grain-infesting insects increases with rising temperature **(Table 28.2)**. A grain temperature of 21°C is considered to be favourable for insects and constitutes the dangerline. At 21°C or higher temperatures, severe damage to stored grain from insects may be expected, whereas below this temperature serious damage is not likely to occur. With few exceptions, temperatures above 35°C are unfavourable for the reproduction of most grain-infesting insects; oviposition ceases and the adults are short-lived. *Rhyzopertha dominica* is an exception, however, since reproduction by this species has been recorded by **Gay and Ratcliffe (1941)** at temperatures of 37.8°C.

Table 28.2. Reproduction of *Tribolium castaneum* (25 pairs) in clean wheat as affected by temperature and grain moisture, indicated by number of progeny after 5 months **(Cotton *et al* 1960)**

Moisture (%)	Temperature 15°C	24°C	27°C	30°C	32°C	35°C
9	0	0	0	0	33	0
12	20	17	28	112	257	44
15	90	24	91	144	370	165

Temperature directly influences a weevil's ability to reproduce in grain. Even when provided with a favourable 14% moisture wheat, few *Sitophilus oryzae* and *Sitophilus granarius* progeny were produced at 15°C; but the progeny greatly increased at 21°C, peaked at 27°C, and fell off appreciably at 32°C **(Figure 28.1)**.

Howe (1965b) has provided information giving the minimum temperature at which 43 species of beetle, 9 species of moth and 1 mite can multiply sufficiently to become pests, and the range of temperature most favourable to each. An estimate of the maximum rate of increase for each species is also given. Howe pointed out that a successful species must be able to multiply rapidly when conditions are favourable and also be able to withstand unfavourable conditions. Data on both of these aspects for a number of species have been provided. For most species there is a range of temperatures covering some 3–4°C at which the rate of increase is greatest, the maximum temperature for each species being usually less than 5°C above the optimum.

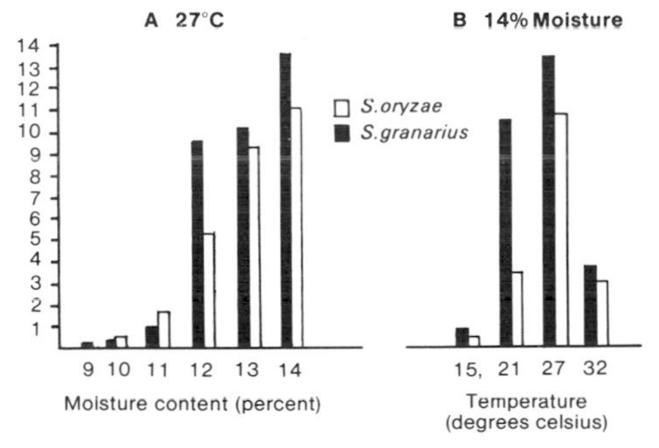

Figure 28.1. Reproduction of *Sitophilus oryzae* and *Sitophilus granarius* (50 pairs each) in wheat as affected by temperature and grain moisture, indicated by number of progeny after 5 months **(Cotton *et al.* 1960)**.

Burges and Burrell (1964) reported that the optima of most of the important species lie between 28° and 38°C and that lightly infested grain cooled below 17°C is safe from damage by insect infestation over a long period since the rate of increase of the insects is low at these temperatures. Burrell (1967) showed in laboratory and field tests that cooling can be used to prevent insect infestations from building up in bulk grain. However, when a heavy infestation is already present, cooling is unlikely to destroy all insects unless the grain can be kept cool for a period of many months.

Srivastava and Perti (1971) reviewed the literature on the effect of temperature and humidity on the susceptibility to insecticides of insects that affect human health, with particular reference to flies, mosquitoes and cockroaches. Susceptibility was shown to be markedly influenced by the temperature and humidity of the environment, and it was concluded that to achieve effective control of insect pests, any measures taken, whether chemical, physical or genetic, must be integrated with the environment.

Evans (1977a, b, c) carried out a complex series of studies to determine the capacity for increase at low temperature of several Australian populations of *Sitophilus oryzae* and *Sitophilus granarius*. He showed that there were significant differences between these populations in their capacity for increase at 15°C, which is considered a marginal temperature in wheat of 14% moisture content. The capacity for increase at 15°C of a given population was found to be correlated with its fertility at 27°C and with its body weight, rather than with its cold-tolerance, as evidenced by chill-coma temperature, or its previous temperature-history. Later, **Evans (1979)** made a similar comparison of Australian populations of these weevils with 3 populations of each from Great Britain and Canada. It was concluded that the observed differences were related to differences in the vigour of the populations rather than to physiological differences in cold-tolerance.

Mills (1978) reviewed the potential and limitations of the use of low temperatures to prevent insect damage in stored grain. He pointed out that whereas the optimum temperature for most stored-grain insects is between 25 and 39°C, any lowering of the temperature below the optimum for a species will have some adverse effect, but to cool insects enough to prevent feeding and reproduction, the temperature must be lowered considerably; below 5–17°C for most species. It is known that insects acclimate if exposed to gradually decreasing temperatures, and thus are able to survive cold exposure and/ or lower temperatures longer than those not acclimated.

Evans (1982) showed that temperature and grain moisture content had a very pronounced influence on the intrinsic rate of increase of *Sitophilus oryzae* in wheat grain, especially at temperatures above 15°C. Later **Evans (1983)** showed that relative humidity and thermal acclimation had a pronounced influence on the survival of adult *Sitophilus oryzae* in cooled wheat. Using five major beetle species, he showed that survival at low temperatures differed between species and there was considerable interaction between the effects of temperature and humidity.

Longstaff (1981, 1984) demonstrated how insect-population growth rate declines if temperature or moisture content drop below or go above the optimum levels. This can be easily visualised from **Figure 28.2**. The population growth rate is derived from a combination of 3 demographic features: (a) the duration of the developmental period and its effect upon a mean generation time; (b) the age-specific fecundity; (c) age-specific mortality. For *Sitophilus oryzae*, a drop in temperature from 27°C to 15°C increases the development period from 4 weeks to 30 weeks. The total number of eggs laid by females appears to be relatively unaffected by such temperature drops, at least in *Sitophilus*, but changes in grain moisture content can have quite pronounced effects **(Table 28.3)**. The rate of oviposition is progressively reduced as temperature drops, and the eggs are laid over a longer period.

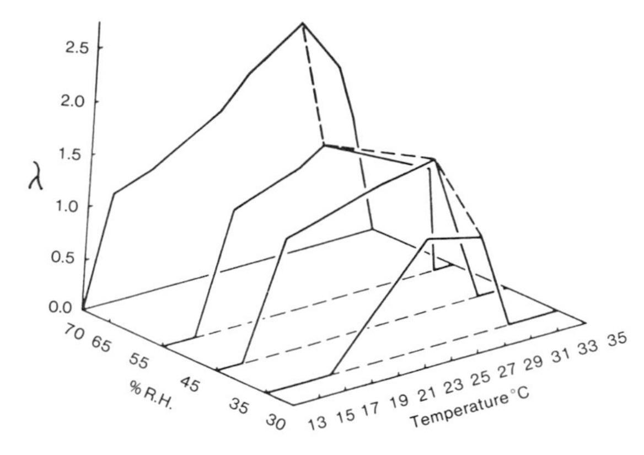

Figure 28.2. Relationship between insect-poulation growth rate (λ), temperature and grain moisture content **(Longstaff 1984).**

In most species, declining temperature increases the rate of immature mortality, although for *Sitophilus oryzae* it is not until the temperature drops below about 18°C that this increase becomes significant in wheat of 14% moisture content. However, in drier grain, the developing insect is much more sensitive to temperature changes **(Figure 28.3).**

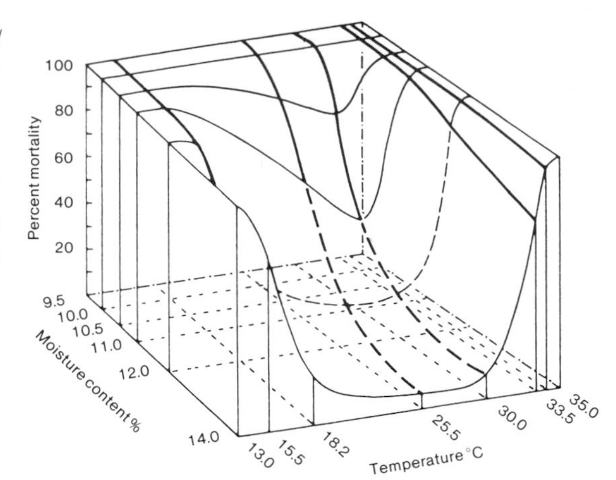

Figure 28.3. The influence of temperature and grain moisture content upon the immature mortality of *Sitophilus oryzae* **(Birch 1945).**

Table 28.3. Total number of eggs laid per female *Sitophilus granarius* **(from Eastham and McCully 1943)**

Relative humidity (%)	Temperature (°C) 20	22.5	25	27.5
40	65	60	55	–
50	80	90	80	75
60	123	100	95	90
70	125	120	135	115
80	–	80	90	120

28.2 Effect of Temperature on Insecticide Toxicity

Organophosphorus insecticides seem to have a positive temperature coefficient, i.e. toxicity is greater when insects are held at higher temperatures after dosing **(Rai *et al.* 1956; Hadaway and Barlow 1957; Norment and Chambers 1970)**. Pyrethrins and the synthetic pyrethroids usually have negative temperature coefficients, i.e. toxicity is greater when insects are held at low temperatures after dosing **(Hadaway and Barlow 1957; Harris and Kinoshita 1977; Harris, Svec and Chapman 1978; de Vries and Georghiou 1979; Sparks *et al.* 1983; Scott and Georghiou 1984)**. Inconsistent effects of temperature on toxicity have been observed for lindane **(Hadaway and Barlow 1957; Busvine 1971; de Vries and Georghiou 1979)** and carbamates such as carbaryl **(Busvine 1971; Harris and Kinoshita 1977; Reichenbach and Collins 1984).**

Tyler *et al.* (1967) showed that low temperature at the time of treatment was responsible for the slow rate of action of bromophos against *Oryzaephilus surinamensis*. Laboratory experiments showed that 50% mortality of adult *Oryzaephilus surinamensis* exposed on grain freshly treated with bromophos at 8 mg/kg occurred after 6.5 hours at 25°C and after 32 hours at 10°C.

Teotia and Pandey (1967) studied the influence of temperature and humidity on the contact toxicity of some insecticide deposits to *Tribolium castaneum* in India. The contact toxicity of dichlorvos, malathion, dieldrin and carbaryl to adults of *Tribolium castaneum* was studied on the basis of their LC/50 at 2 temperatures (27° and 35°C), and about 75% humidity. The mortality rates obtained after exposure to residual films of the insecticides for 24 hours showed that the first 3 insecticides gave better control at the higher temperature and the last one at the lower temperature.

Iordanou and Watters (1969) used filter papers impregnated with DDT, methoxychlor, lindane, malathion and bromophos to compare their toxicities at 26.7°, 15.5° and 10°C against *Tribolium castaneum, Tribolium confusum, Oryzaephilus surinamensis, Oryzaephilus mercator* and *Cryptolestes ferrugineus*. Beetles were exposed to treated filter papers for 24 hours, and mortality was assessed 72 hours after the beetles were returned to flour at the same temperature. Methoxychlor was ineffective against all species except *Tribolium castaneum*. DDT was more toxic to all species at 10° and 15° than at 26.7°C but was slightly more toxic at 15.5° than at 10.0°C. Lindane was more toxic at 10° than at 15°C to *Tribolium castaneum* and *Oryzaephilus surinamensis*, but was more toxic at 15° than at 10°C to *Oryzaephilus mercator* and *Cryptolestes ferrugineus*. Malathion and bromophos showed positive temperature coefficients against all species. Bromophos was much less effective at 15° and 10° than at 26.7°C against both *Tribolium* species and *Oryzaephilus surinamensis*.

Baker (1968) found that the time to obtain 50% kill of *Tyrophagus putrescentiae* using malathion varied inversely with temperature over the range 18.9–30.0°C. **Baker (1969)** found lindane to be more effective against *Tyrophagus putrescentiae* at 20.6°C than at 15.6°C.

Wilkin and Haward (1975) found that the time required to obtain complete kill of *Acarus siro* with a mixture of lindane and malathion increased from 4 days at 25°C to 21 days at 10°C and complete kill was not obtained at 5°C within 21 days. These differences were even more marked with *Tyrophagus putrescentiae* and *Glycyphagus destructor*. Similar results were obtained with pirimiphos-methyl at temperatures at 10°C or more but its effectiveness was severely restricted at 5°C. Fenitrothion against *Tyrophagus putrescentiae* and *Glycyphagus destructor* and menazon against *Glycyphagus destructor* also had little effect at 5°C. Although temperature had a marked effect on the speed of action of all the compounds tested it did not affect their specificity. For example, fenitrothion was effective against *Tyrophagus putrescentiae* and *Glycyphagus destructor* but even at 25°C failed to kill a significant number of *Acarus siro*.

Hall (1979) studied the effect of a range of temperatures on the knockdown speed of etrimfos and pirimiphos-methyl against pests of stored grain. He found that temperature did affect knockdown speed and the results showed that the knockdown speed was always slower at 2–5°C than at 20–25°C. This effect could be due to the reduced activity of the insects. The insect takes up chemical either by feeding or through the cuticle. At low temperatures, insects enter a state of dormancy as observed at 2–5°C. Whilst dormant, they do not feed or move around and their metabolic rate is very slow so that there is very little exchange of materials through the cuticle. Hence, at low temperatures knockdown speed is slower, due to reduced uptake of chemical.

O'Donnell (1980) studied the toxicities of 4 insecticides to *Tribolium confusum* in 2 sets of conditions of temperature and humidity and found that the effective dose differed by factors of less than 10 in the cases of malathion and pirimiphos-methyl and more than 20 in the cases of fenitrothion and iodofenphos **(Table 28.4).**

Table 28.4. The toxicities of four insecticides to adult *T. confusum* in two sets of conditions of temperature and humidity **(O'Donnell 1980)**

Conditions	Insecticide	ED_{50} and fiducial limits (mg/m²)	ED_{95} and fiducial limits (mg/m²)	Slope b ± S.E.
	pirimiphos-methyl	9.3 (8.1 , 10.7)	15.3 (12.8 , 21.2)	7.7±1.3
25°C	fenitrothion	4.5 (4.0 , 5.1)	6.9 (6.0 , 8.5)	8.8±1.0
70% r.h.	iodofenphos	4.3 (4.0 , 4.6)	9.3 (8.3 , 10.6)	4.9±0.3
	malathion	69.9 (65.9 , 74.2)	128 (116 , 114)	6.3±0.5
5°C	pirimiphos-methyl	80.0 (75.6 , 84.6)	139 (127 , 156)	6.9±0.5
82% r.h.	fenitrothion	92.6 (87.0 , 98.4)	180 (163 , 204)	5.7±0.4
to 10°C	iodofenphos	101 (86.8 , 118)	363 (286 , 502)	3.0±0.3
79% r.h.	malathion	522 (454 , 598)	974 (815 , 1300)	6.1±0.8

Note: The insects were exposed to the insecticides for 24 hr, and the knock down response was assessed after a 24 hr post-exposure period at 25°C and 70% r.h.

Tyler and Binns (1982) evaluated the influence of temperature on the susceptibility to 8 organophosphorus insecticides of susceptible and resistant strains of *Tribolium castaneum, Oryzaephilus surinamensis* and *Sitophilus granarius*. Based upon knockdown and kill, the effectiveness of all insecticides was greater at 25°C than at 17.5°C and was markedly lower at 10°C. At 10°C, tetrochlorvinphos, bromophos and iodofenphos were virtually ineffective against *Sitophilus granarius* even at 5000 mg/m^2. By contrast, at 25°C, 100 mg/m^2 was adequate to give complete knockdown of all species with most insecticides, the exceptions being malathion and tetrachlorvinphos. Knockdown was invariably followed by mortality although at the lower temperatures this took several days.

Longstaff and Desmarchelier (1983) studied the effects of the temperature/toxicity relationships of deltamethrin and pirimiphos-methyl on the population growth of *Sitophilus oryzae*. In the case of pirimiphos-methyl, they found that for each time period and each level of exposure, mortality increased over the range 21–32.3°C **(Table 28.5)**. Deltamethrin was more toxic at 21°C than at either 27 or 32.3°C but there was no significant difference in its toxicity between the last 2 temperatures. It was shown that only approximately 0.4 mg/kg of pirimiphos-methyl was required to prevent population growth of *Sitophilus oryzae* at 32.3°C and only approximately 0.1 mg/kg of deltamethrin was required to prevent population growth at 21°C. **(Table 28.6)**. These rates of application are only a small fraction of the rates of either pirimiphos-methyl (4 mg/kg) or deltamethrin (2 mg/kg) recommended for control of stored-product insects over the range of temperatures found in Australian wheat **(Bengston and Desmarchelier 1979)**. Thus, if one has or provides a temperature that is less than optimal for the insect, and if one chooses an insecticide that is particularly effective at that temperature, a high level of control can be obtained.

Table 28.6. The intrinsic rate of increase per week of *Sitophilus oryzae* exposed to deltamethrin (0.1 mg/kg) or pirimiphos-methyl (0.4 mg/kg) **(Longstaff and Desmarchelier 1983)**

Temperature (°C)	Control	Deltamethrin	Pirimiphos-methyl
21	0.336	0.083	0.285
27	0.679	0.452	0.241
32.3	0.465	0.247	–0.073

Barson (1983) reported on the effects of temperature and humidity on the toxicity of 3 organophosphorus insecticides to adult *Oryzaephilus surinamensis* showing that all the insecticides were more effective at 25°C than when the insects were exposed to a diurnal cycle of 12.5–20–12.5°C to simulate grain-store conditions in the UK during spring and autumn.

Herve (1985) presented unpublished data from P.R. Carle *et al.* showing the lethal doses and thermodependence of deltamethrin and bioresmethrin on 5 species of grain beetles. These demonstrate that whilst deltamethrin and synergised deltamethrin are generally about twice as toxic at 30°C as they are at 10°C, the toxicity of synergised bioresmethrin in relatively unaffected by temperature over this range.

Samson (1985), in his discussion of biological efficacy of residual pesticides at high humidities and moisture contents, pointed out that there is unlikely to be a single

Table 28.5. The cumulative survival of mature *S. oryzae* exposed to pirimiphos-methyl and deltamethrin over a period of 4 weeks. Results are presented as mean cumulative survival **(Longstaff and Desmarchelier 1983)**

Treatment	Temperature (°C)	Weeks 1	2	3	4
Control	21	1.00	1.00	1.00	
	27	1.00	0.99 ± 0.02	0.99 ± 0.02	0.99 ± 0.02
	32.3	1.00	1.00	0.99 ± 0.02	0.96 ± 0.05
Deltamethrin and piperonyl butoxide (0.05 mg/kg)	21	0.71 ± 0.14	0.50 ± 0.11	0.42 ± 0.11	0.32 ± 0.12
	27	0.88 ± 0.07	0.80 ± 0.09	0.76 ± 0.11	0.70 ± 0.09
	32.3	0.95 ± 0.08	0.89 ± 0.07	0.86 ± 0.07	0.75 ± 0.13
Deltamethrin and piperonyl butoxide (0.1 mg/kg)	21	0.42 ± 0.10	0.30 ± 0.13	0.22 ± 0.12	0.14 ± 0.09
	27	0.69 ± 0.16	0.58 ± 0.17	0.55 ± 0.15	0.50 ± 0.13
	32.3	0.84 ± 0.14	0.72 ± 0.09	0.61 ± 0.10	0.50 ± 0.12
Pirimiphos-methyl (0.25 mg/kg)	21	0.97 ± 0.05	0.96 ± 0.06	0.96 ± 0.05	0.91 ± 0.06
	27	0.95 ± 0.04	0.73 ± 0.13	0.47 ± 0.16	0.38 ± 0.20
	32.3	0.56 ± 0.17	0.22 ± 0.14	0.13 ± 0.10	0.11 ± 0.11
Pirimiphos-methyl (0.4 mg/kg)	21	0.95 ± 0.07	0.89 ± 0.13	0.87 ± 0.12	0.77 ± 0.16
	27	0.47 ± 0.09	0.24 ± 0.11	0.03 ± 0.06	0.01 ± 0.02
	32.3	0.17 ± 0.12	0.01 ± 0.02	0.01 ± 0.02	0.0

relationship between post-treatment temperature and insect responsiveness, as there are several component processes leading up to toxic action; penetration of insecticide through the cuticle, metabolism (toxification or detoxification), storage or excretion, penetration to the target site of toxic action, attack upon the target, and consequent insect mortality or recovery. Each of these may have its own relationship to temperature, and the relative importance of each in determining the overall temperature/toxicity relationship may depend on the insecticide, the species of insect, and even the resistance status of particular strains of a given species. Also, the apparent relationship between temperature and toxicity may vary depending upon the time when insect response is assessed: whether it be an arbitrary time after dosing or after end-point mortalities are reached.

28.3 Effect of Temperature on Persistence of Insecticides

The many papers devoted to the study of the degradation of insecticides on grain in storage have been reviewed in Section 2 of each Chapter devoted to individual insecticides (Section 2 of Chapters 7 through 27). It is noticeable that the major proportion of these papers have paid little or no attention to the importance of temperature and some have merely observed that the rate of degradation was increased at higher temperatures.

Although a few good systematic studies were carried out in the early days, the first real breakthrough seems to have come with the work of Desmarchelier and co-workers. **Desmarchelier (1976b)** showed, in a study of the kinetic constants for breakdown of fenitrothion on grains in storage, that the breakdown on wheat, paddy rice, oats and sorghum in storage at 4 temperatures, was first-order with respect to grain moisture. At a given moisture content, water activities were dependent upon, but rate constants were independent of, grain type. The effect of temperature on breakdown was in the form of the Arrhenius equation. A simple expression was derived from these basic concepts of kinetics which accurately predicted residues under 24 different conditions set by altering grain type, moisture content and temperature **(FAO/WHO 1977)**.

This concept was further developed, evaluated under a wide range of field conditions, and discussed in relation to a variety of management options. Full details may be found in the following papers: **Desmarchelier (1977a, 1978a, 1980); Banks and Desmarchelier (1978); Desmarchelier and Bengston (1979a, 1979b); Desmarchelier *et al.* (1979b)**.

Whilst it is now known that temperature and moisture are the critical factors determining the fate of insecticides deposited on stored grains, many empirical studies have been carried out since the mid-1950s and it is important to recognise how this knowledge has provided a valuable basis for the initiatives taken approximately 20 years later. All of these papers have been reviewed in the chapters dealing with the insecticides to which they directly relate and therefore the details will not be repeated here.

Strong and Sbur (1960) showed that the effectiveness of malathion applied at 10 mg/kg to wheat of 10% moisture content and stored in temperature-controlled cabinets at graded ranges of 15°, 21°, 27°, 32°, 43°, and 49°C decreased with increases in temperature, but maximum safe temperature ranges which may determine the persistence of malathion at effective levels on grain in storage were not clearly defined.

In a study of the influence of moisture content and storage temperature on the residual persistence and effectiveness of selected dosages of 5 insecticides, **Strong and Sbur (1964a)** demonstrated the influence of storage temperature on insecticides applied to wheat by reduced mortalities of *Sitophilus oryzae* and increased numbers of progeny with increases in storage temperature. They used wheat of 10% moisture content stored at graded ranges similar to those used previously.

Strong and Sbur (1965b), in a complex study of the inter-relation of moisture content, storage temperature, and dosage on the effectiveness of diazinon as a grain protectant against *Sitophilus oryzae*, showed the influence of temperature on degrading the effectiveness of diazinon deposits by the reduced mortalities of *Sitophilus oryzae*, and increased number of progeny with increases in the storage temperature of wheat. These investigators recommended the adjustment of dosages of diazinon to compensate for adverse effects of temperature that may be found in actual storage of grain.

Tyler and Green (1968) observed that both fenitrothion and malathion became ineffective after 2 weeks on moist, heating grain near the centre of farm storage bins but residues remained effective for at least 12 weeks at the cooler, outer edges of the bins and in those bins remaining cool after the grain was turned for treatment.

Kadoum and LaHue (1969), who studied the effect of variety, moisture content, foreign material and storage temperature, on the degradation of malathion residues in grain sorghum found that storage temperature significantly affected residue retention.

Minett and Belcher (1970), investigating the loss of dichlorvos residues in stored wheat at 21° and 35°C, showed that the disappearance of dichlorvos residues is dependent on storage temperature as well as moisture content.

McDougall (1973) used wheat of 5 moisture levels between 9.5% and 13% and 2 temperatures (22°C and 35°C) to study the effect of moisture and temperature on the persistence of methacrifos. He showed that under these conditions the half-life of methacrifos in wheat varied from 238 days to 19 days, the half-life decreasing as moisture and temperature increased **(Tables 19.1 and 19.2)**.

Eichler (1974) and **Eichler and Knoll (1974)** carried out a series of large-scale laboratory trials to determine the degradation of bromophos at different temperatures. They showed that the rate of degradation increased -with increasing temperature. At 26°C about 60% of the deposit was degraded in 12 months, whereas at 15°C only about 40% was lost.

Moore and McDougall (1974a) provided the results of a valuable study of factors affecting the dissipation of methacrifos in wheat. Increases in temperature over the range 28°-42°C greatly increased the degradative effect of enzymes on the insecticide applied to wheat.

Desmarchelier (1978a) studied the loss of fenitrothion on grains in storage and determined the activation energy for fenitrothion breakdown on grain using the well-known Arrhenius equation. This is the first activation energy to be published for a pesticide on a stored product since Winteringham and Harrison's determination with methyl bromide on glutenin in 1946, but the general absence of such studies is directly the consequence of a lack of understanding of the kinetics of breakdown at a fixed temperature. If one does not calculate constants — and cannot because of the manner of the presentation of data — one obviously cannot calculate activation energies. Combining the quantitative measurement of the effect of temperature with the quantitative measurement of the effect of water activity, it is possible to describe the breakdown of fenitrothion or any other insecticide in a single equation or in graphical form as in **Figure 14.2 (Banks and Desmarchelier 1978).**

Takimoto *et al.* (1978) studied the degradation of fenitrothion applied to harvested rice grains using ^{14}C-fenitrothion. They showed the decomposition proceeded more rapidly at 30°C than at 15°C with the respective half-lives of about 4 months and more than 12 months being independent of the applied concentrations.

Nambu *et al.* (1978) found that ^{14}C-phenothrin applied to wheat of 11–12% moisture content degraded so slowly over 6 months that the effect of temperature (15° and 30°C) could not be measured.

Desmarchelier (1979d) developed a mathmatical model for the loss of carbaryl on grains in storage based on relative humidity and temperatures and demonstrated that this model held true under widely varying commercial conditions in various parts of Australia.

Watters and Mensah (1979), in studies in Canada, showed that the stability of malathion applied to dry or high-moisture wheat was significantly affected by storage temperature both in the laboratory and in farm granaries. The breakdown rate was higher at 20° than at 10° and highest at 30°C.

Desmarchelier and Bengston (1979a) described the development and use of predictive models for loss of residues of 12 grain protectants on grains under any constant or varying condition of temperature and moisture content.

Desmarchelier *et al.* (1979a) demonstrated that temperature over the range 20–35°C had a commercially significant influence in determining the rate of loss of pyrethrins on grains treated with synergised pyrethrum.

Desmarchelier *et al.* (1979b) showed how temperature and moisture manipulation could be combined with the use of grain protectants to provide greatly superior and more economical pest control.

Thorpe and Elder (1980) discussed the use of mechanical refrigeration to improve the storage of pesticide-treated grain. They cited models to show that the loss of methacrifos on treated wheat initially at 30°C and 11% moisture content could be reduced during 6 months storage from 90% to about 30% by prompt cooling to 20°C. The cost of such cooling was more than offset by the saving of insecticide.

Desmarchelier (1980) explained how the loss of bioresmethrin, carbaryl and phenothrin on wheat during storage was related quantitatively to wheat temperature and equilibrium relative humidity.

Abdel-Kader (1981), Abdel-Kader and Webster (1980) and **Abdel-Kader** *et al.* (1980, 1982) studied the effect of storage temperatures on the rate of degradation of malathion and fenitrothion in stored wheat. They showed that very little breakdown of the insecticide residue occurred on wheat that had been stored at -35° and -20°C for 72 weeks. As the temperature increased, both malathion and fenitrothion residues decreased, the rate depending upon the age of the deposit. At the end of 72 weeks storage, 26%, 41%, 74%, 95%, and 96%, of the initial deposit of malathion and 18%, 35%, 56%, 90% and 96% of the initial deposit of fenitrothion had degraded from wheat stored at -5°, 5°, 10°, 20°, and 27°C, respectively.

Thorpe and Elder (1982) showed how aeration is able to reduce the rate of degradation of insecticides applied to stored grain, and to render the rate of decay relatively insensitive to initial grain conditions. In the temperate and sub-tropical wheat growing regions of Australia, aeration can reduce usage of methacrifos by factors of 7 and 4, respectively.

Noble *et al.* (1982), who carried out a highly refined and systematic study of the stability of pyrethroids on wheat in storage, found that the half-lives for the pyrethroids at 25°C, 12% moisture and 35°C, 15% moisture were: permethrin 252 and 44 weeks, phenothrin 72 and 29 weeks, fenvalerate 210 and 74 weeks, and deltamethrin 114 and 35 weeks, respectively.

Joia (1983) showed that the half-life of fenvalerate on wheat of 13.3% moisture content and -5°C was 385 weeks and that the corresponding half-life for cypermethrin on grain was 169 weeks.

Desmarchelier (1983) discussed several ways of maximising the benefit to be obtained from cooling grain soon after it is put in storage. Such cooling not only reduces the degradation of the insecticide but reduces the reproductive rate of insect pests. This enables the amount of insecticide to be reduced considerably. Rapid cooling of insecticide-treated grain has the additional advantage that the grain is available for immediate shipment, if that is required. On the other hand, should the grain have to be held over the residual insecticide deposit will remain effective during the extended storage, thus obviating the need for retreatment or the risk of infestation.

Table 28.7. Theoretical effects of thermal lability on half-life **(Desmarchelier 1985)**

Compound	Thermal lability* °C	Half-life at (°C)				
		20° (weeks)	30° (weeks)	40° (weeks)	100° (minutes)	230° (minutes)
A	5	32	8	2	10	«1
B	8	32	14	6	200	9

* Temperature rise required to halve the half-life.

Desmarchelier (1985) pointed to the practical importance of temperature-lability of insecticides used for the protection of stored grain. Regulating the storage temperature will preserve the effectiveness of the insecticide deposit whilst keeping the final residue level under control. Temperature-labile insecticides can be most useful if the storage temperature can be lowered so as to extend their persistence at an effective level right throughout the storage period. By the end of the storage period the residue level will be comparatively low and, because of its labile nature, be degradable during processing and cooking. This has been illustrated in **Table 28.7** for 2 chemicals, A and B, with different thermal labilities of 5° and 8°C, but each with a half-life of 32 weeks at 20°C. It can be seen that A, and to a lesser extent B, become insufficiently stable as storage temperatures increase to 30–40°C. However, A is extensively degraded during cooking of rice, as the half-life at 100°C (10 minutes) is much less than the cooking time. Compound B, however, will not be extensively degraded, as the half-life at 100°C of 200 minutes exceeds most cooking times. Compound A will be completely degraded during baking of bread, whereas compound B will be only partly degraded. Data for compounds A and B correspond to experimental data for methacrifos and fenitrothion, respectively,

29. Effect of Moisture

29.1 Effect of Moisture on Insect Biology

Grain moisture is an important factor in the life economy of insect pests of stored grain because they depend on their food supply for the moisture needed to carry on their life processes. Up to a certain point, increasing the grain moisture favours a rapid increase in the numbers of insects. Beyond that point, microorganisms take over and destroy the insects, except the fungus feeders, along with the grain. If, on the other hand, the moisture content of the grain is low, the water required for carrying on vital life processes must be obtained by breaking down the food supply or the food reserves in the fatty tissues of the insect body.

The moisture requirements natually differ with different species of insects, as does the ability of the insects to produce the water they need. *Sitophilus oryzae*, *Sitophilus zeamais*, and *Sitophilus granarius* are unable to reproduce in grain with a moisture content below 9%, and the adults soon die in dry grain. *Sitophilus oryzae* adults survived for only one week in 8% moisture wheat at 30°C; at 9% moisture, about 70% were dead by 3 weeks, though a few lived for 7 weeks; at 12% moisture they lived for over 15 weeks. The critical moisture content in wheat for both *Sitophilus oryzae* and *Sitophilus granarius* development is between 11 and 12% **(Cotton *et al.* 1960).**

Tribolium castaneum and *Tribolium confusum*, on the other hand, produce progeny in flour or grain dust from which practically all moisture has been removed. Nevertheless, in whole-kernel grain, high moisture is essential for reproduction and survival. The presence of grain dockage or dust is also of vital importance; without it, dry grain is unfavourable for reproduction **(Figures 29.1 and 29.2). McGregor (1964)** found that dockage greatly enhances the ability of *Tribolium castaneum* to develop in wheat, the number of insects present being directly proportional to the percentage of available dockage. The effect of grain moisture of 8%, 9% and 12% on the survival of *Tribolium confusum* in clean wheat is shown in **Figure 29.1**. The longevity of the insect is reduced as the grain moisture is lowered, although low moisture has less effect on *Tribolium confusum* than on *Sitophilus oryzae*. However, when dockage was added to the 8% moisture wheat, nearly 100% of the beetles were alive after 16 weeks **(Figure 29.2)**.

The grain moisture content can have quite pronounced effects on the total number of eggs laid by females, at least in the case of *Sitophilus granarius* **(Eastham and McCully 1943) (Table 29.1).**

Birch (1945) demonstrated that grain moisture content greatly influenced the mortality of immature *Sitophilus oryzae* at both low and high temperatures.

Cotton *et al.* (1960) showed that the reproduction of *Tribolium castaneum* in clean wheat is affected to a considerable degree by grain moisture, as indicated by the number of progeny after 5 months. This was clearly illustrated in **Figure 28.2.**

Joffe (1958) reported that moisture migration in horizontally stored bulk maize intensified the development of grain-infesting insects under South African conditions. This is typical of the additional problems of protecting stored grains from insects under tropical conditions.

Above 65% or 70% relative humidity, fungi developed,

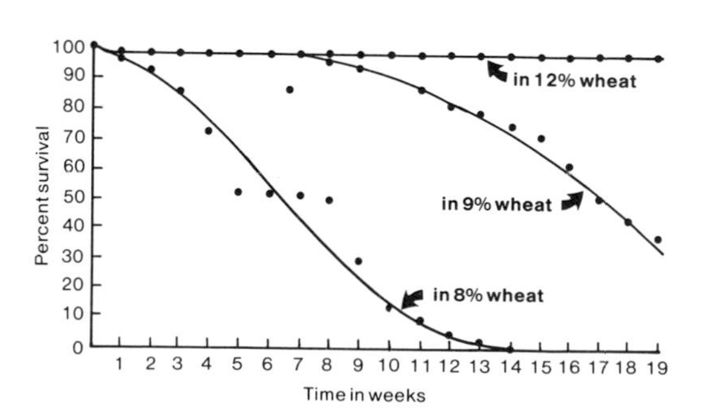

Figure 29.1. Survival of *Tribolium confusum* adults at 27°C in clean wheat of different moisture contents **(Cotton *et al.* 1960).**

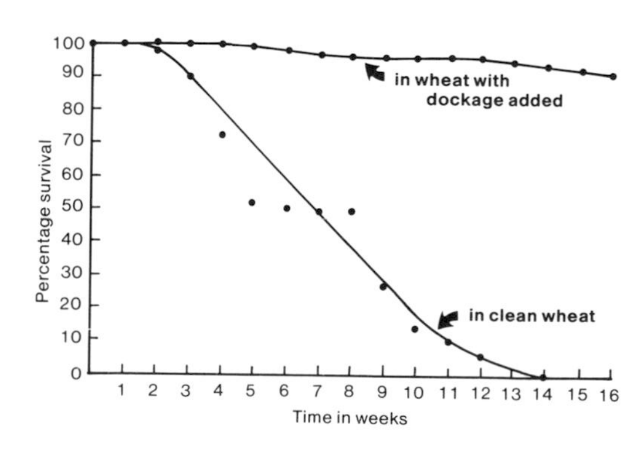

Figure 29.2. Survival of *Tribolium confusum* adults in 8% moisture wheat as affected by dockage **(Cotton *et al.* 1960).**

producing heat and an increase in temperature which may reach 63°C given what is termed 'damp grain heating' **(Sinha and Wallace 1965)**. At relative humidities below 70%, insect infestation may develop and an increase in temperature, resulting from the respiration of the insects, may occur. Temperatures may reach 42°C given what is termed 'dry grain heating' **(Sinha 1961)**. At a temperature of about 27°C maize, wheat and sorghum with a moisture content of 13.5% is in equilibrium with interstitial air of 70% relative humidity. For paddy rice the equilibrium is reached at 15% and for milled rice at 13%.

Howe (1965b) presented a table giving the minimum humidity and temperature at which 43 species of beetle, 9 species of moth and 1 mite can multiply sufficiently to become pests.

Srivastava and Perti (1971) reviewed the literature on the effect of temperature and humidity on the susceptibility to insecticides of insects, particularly flies, mosquitos and cockroaches. Susceptibility was shown to be markedly influenced by the temperature and humidity of the environment, and it was concluded that to achieve effective control of insect pests any measures taken whether chemical, physical or genetic, must be integrated with the environment.

Evans (1982) showed that grain moisture content as well as temperature had an important influence on the intrinsic rate of increase of *Sitophilus oryzae*, the rate at 14% moisture content being more than double that in wheat of 11.2% moisture content.

Evans (1983) showed that relative humidity had a positive influence on determining the survival of adult grain beetles in cooled grain, the survival being shorter in 45% than in 70% relative humidity.

29.2 Effect of Moisture on the Toxicity and Availability of Insecticides to Insects

Although it is logical to expect that moisture and humidity during the post-treatment period could affect the reaction of insects to insecticides, an extensive search of the literature has produced relatively little useful information. Many titles incorporate such phrases as 'effectiveness of insecticides on high-moisture grain', 'effect of moisture on the toxicity of insecticies to stored-product insects', but a careful examination has revealed that all of these refer to studies involving the degradation of the insecticide deposit under varying moisture regimes.

Measurement of the effect of humidity on insect responsiveness requires that test insects initially receive a standard dose of insecticide. This criterion is not fulfilled by studies in which the insects are allowed to dose themselves under different conditions, such as when they are continually exposed to treated surfaces at different humidities. Responsiveness can be measured by exposing insects to insecticide deposits, or to topical application, but only if they are exposed under a single condition of humidity and then transferred to the different experimental conditions. However, this procedure has the disadvantage that the exposure time may occupy a significant proportion of the time required for reaction to the insecticide so insects may react before they reach the experimental regimes. Experiments in which humidity was varied only during the post-treatment period have produced conflicting results.

Pradhan (1949) found that the toxicity of DDT to *Tribolium castaneum* increased with higher relative humidity in the range 0–84%. **Hadaway and Barlow (1957)**, however, found that relative humidity in the range 20–95% had no effect on the toxicity of several chemicals, including DDT, to mosquitoes and house flies, and a similar lack of effect of post-treatment humidity was found by **Crauford-Benson (1938)** and **Elmosa and King (1964)**. Several more papers have shown increased toxicity of some chemicals at lower relative humidity **(Harries *et al.* 1945; Sales 1979; Reichenbach and Collins 1984)**, but the effect seems to have been slight, e.g. the increase in toxicity of propoxur to German cockroaches reported by Reichenbach and Collins was only 1.6 times at 20% relative humidity compared with 100%.

The procedure for measuring effects of post-treatment humidity include sources of error. It is sometimes difficult to establish different humidity regimes quickly after dosing, because the chambers in which such experiments are usually performed may take a significant time to re-achieve the desired humidity after opening. Also, the method of applying a known topical dose is not foolproof, as a considerable proportion of the applied dose can be lost from the insect before it is absorbed through the cuticle.

Samson (1985) reported some experiments to measure the effect of humidity on the response of *Tribolium castaneum* to fenitrothion applied topically in cyclohexanone. Three different post-treatment holding procedures were tried: (i) treated insects were placed in maize conditioned to 9% or 20% m.c. in desiccators containing appropriate salt solutions to give either low or high relative humidity, (ii) treated insects were placed into desiccators without grain but with a paper foothold, and (iii) treated insects were placed into holding jars conditioned to appropriate relative humidity by forced ventilation, and later transferred to appropriate relative humidities in controlled environment rooms. Toxicity was reduced at the higher humidity when treated insects were held in grain but a residue of fenitrothion was found in the grain after the insects were removed, suggesting some of the applied dose had rubbed off. Lack of effect of humidity using method (ii), could have been due to delay in achieving the required relative humidities each time the desiccators were opened. No such proviso could be applied to method (iii), however, as the test insects were exposed to the desired relative humidities immediately after treatment, but again there was no effect of humidity. The authors referred to the conclusion of **Hadaway and Barlow (1957),** that 'atmospheric humidity may affect the insecticide and its availability rather than the insects'.

Barson (1983) studied the effects of temperature and humidity on the toxicity of 3 organophosphorus insecticides to adult *Oryzaephilus surinamensis* and although the results were not clear-cut they did indicate higher toxicities at the higher relative humidity. In a second experiment, toxicity studies were made with chlorpyrifos-methyl under various constant conditions of temperature and humidity ranging from 5°-30°C and 30%, 50%, 70% and 90% relative humidity. Chlorpyrifos was more toxic to both susceptible and resistant strains of *Oryzaephilus surinamensis* at the highest humidity (90%) throughout the whole temperature range. The LD/50 value for each strain decreased at each temperature as the water vapour concentration was increased.

Teotia and Pandey (1967) studied the influence of temperature and humidity on the contact toxicity of some insecticide deposits to *Tribolium castaneum*. They used 2 levels of humidity, 70% and 90%, with a temperature of about 27°C. Dichlorvos, dieldrin and carbaryl were more effective at the higher than at the lower humidity, whereas the reverse was true for malathion. The differences were statistically significant only in the case of malathion.

29.3 Effect of Moisture on the Fate of Insecticide Residues

The importance of moisture content of grains as a single factor determining the efficacy of insecticide treatments was considered by many investigators right from the early days of the use of pyrethrins. **Wilbur (1952b)** showed that a high moisture content of grain resulted in a quick breakdown of pyrethrins and that piperonyl butoxide could prolong the insecticidal action even in the presence of moisture.

Schroeder (1955) found that de-husked maize stored at 27°C required more pyrethrum dust for effective control of infestation when the moisture content was 15% (i.e. 'critical') than when it was 13%, which would suggest that the pyrethrins in dust were more rapidly degraded in grain of more than critical moisture content.

Watters (1959a) showed that malathion applied at 2 mg/kg to wheat of 13.5% moisture content and stored in airtight containers at 9°C caused 99% mortality of *Cryptolestes ferrugineus* at 8 months after treatment. A dosage of 16 mg/kg was required for comparable control in wheat of 15.5% moisture content but this dosage was ineffective in wheat of 18% moisture content 5 months after treatment.

Strong and Sbur (1960) studied the influence of grain moisture on the effectiveness of malathion as a grain protectant and found that when malathion was applied at 10 mg/kg to wheat with a graded moisture range of 10%, 12%. 14%, 16%, 18% and 20% and stored at 15°C there was obvious reduction in mortalities of *Sitophilus granarius, Sitophilus oryzae* and *Tribolium confusum* with increases in moisture content. Results indicated that a moisture content of 12% was about the maximum safe level and 14% appeared to be the critical level of moisture in wheat when considered in regard to the persistence of biologically effective malathion deposits.

Strong and Sbur (1964a) also studied the influence of moisture content on the residual persistence and effectiveness of selected doses of 5 insecticides in protecting stored grain against insects. When the insecticides were applied to wheat with a graded moisture range of 10%, 13% and 16%, and stored at 15°C, the influence of moisture was shown by reduced mortalities of *Sitophilus oryzae* and increased numbers of progeny with increases in moisture content of grain. Later, the same authors **(1965b)** showed a similar influence of moisture on the effectiveness of diazinon deposits where less diazinon was required for protection of dry grain than grain stored at higher safe levels of moisture. The authors recommended adjusting dosages of diazinon to compensate for adverse effects of moisture found in actual storage of grain.

Tyler and Green (1968) found that when fenitrothion (2 mg/kg) and malathion (10 mg/kg) were applied to warm, moist grain stored in bins in a British farm granary there was a relationship between the persistence of the insecticides and the physical condition of the grain. Both fenitrothion and malathion became ineffective after 2 weeks on moist, heating grain near the centre of the bins. At the doses used, the persistence of fenitrothion was similar to that of malathion on warm, damp grain but somewhat superior on cooler, drier grain.

Minett *et al.* (1968) showed that when malathion was applied to wheat there was a critical moisture level of approximately 11.8% for a storage temperature of 21°C and 11.6% for a temperature of 32°C.

LaHue (1973) showed that high moisture of wheat lowered the efficacy of tetrachlorvinphos when applied to wheat, but the adverse effect of high moisture was compensated for to some extent when the dosage was increased.

Telekar and Mookherjee (1969) showed that the residual toxicity of synergised pyrethrins (20 mg/kg) was lost within 4 months at 25°-35°C whereas unsynergised pyrethrins at 20 mg/kg lose their effectiveness in a little more than one month under similar conditions.

Kadoum and LaHue (1969), who studied the effect of variety, moisture content, foreign materials and storage temperature on the degradation of malathion residues in grain sorghum found that hybrid had little effect on malathion degradation but grain moisture content and storage temperature significantly affected residue retention.

McDougall (1973) studied the effect of moisture and temperature on the persistence of methacrifos in wheat. For the moisture range of 9.5%–13.0% and temperatures of 22°C and 35°C the half-life of methacrifos in wheat varied from 238 days to 19 days, half-life decreasing as moisture and temperature increased.

LaHue (1974) presented residue data obtained during a 12-month period from hard winter wheat, shelled corn and sorghum treated with 4 insecticides. He found that pirimiphos-methyl and chlorpyrifos-methyl residues were not as greatly influenced as malathion by a high moisture content of the grain at the time of treatment.

Moore and McDougall (1974a) studied the factors affecting the degree of dissipation of methacrifos in wheat, finding that the half-life of the compound decreased as moisture and temperature increased. They derived a mathematical formula for the expression of the half-life of methacrifos as influenced by moisture and temperature.

Rowlands and Wilkin (1975) applied ^{14}C-labelled pirimiphos-methyl to wheat of 14% and 18% moisture which they stored in the laboratory at 20°C for 6 months. After 6 months, 15% and 50% degradation had occurred at the 14% and 18% moisture levels, respectively. Under similar conditions the breakdown of chlorpyrifos-methyl was 15% and 35%, respectively.

Desmarchelier (1978a) showed that the loss of fenitrothion from post-harvest application to wheat, oats, paddy rice and sorghum followed a second-order rate process, with the rate of loss being proportional, at a fixed temperature, to the amount of fenitrothion and the activity of water, which was obtained from the equilibrium partial pressure of water vapour. A chart relating half-life to temperature and relative humidity was presented in **Figure 14.2.** It should be noted that the relative humidity range employed in these studies had an upper limit of about 65% to avoid the complication that would arise if mould developed due to excessive humidity.

One major factor obscuring the correct chemical interpretation of insecticide breakdown on grain is the widespread use of moisture content as a measure of the water present in grain. Moisture content, arbitrarily defined in terms of the weight loss on heating under specified conditions, is not linearly related to water activity for a particular grain and differs substantially between commodities of the same water activity. Water activity is a term derived from the laws of chemical thermodynamics, which state that the activity of a chemical that exists in equilibrum in two phases, e.g. water on grain and interstitial relative humidity, is the same in each phase. This concept enables an easy estimation to be made of the activity, i.e. 'real' concentration in the chemical sense, of water on grain from the measurement of relative humidity. Thus, activity equals 0.01% equilibrum relative humidity.

Moisture content is, however, easy to measure and compilations **(Hall 1963; Gough and Bateman 1977)** are available to convert it into water activity for the material under test. A comprehensive set of references to the study of equilibrium relative humidity and moisture content of a range of stored cereals is to be found in Section 4.2.7.

The conclusion arising from widespread use of moisture content may be illustrated by a statement from a recent authoritative review which said 'the persistence of fenitrothion residue was greater in wheat than in barley; however, the moisture content of barley was greater than that of wheat'. This and other statements have given the impression that major distinctions can be made between different cereals. However, if water activities were used as a measure of water present, the breakdown rate of fenitrothion and other organophosphorus insecticides is found to be independent of grain type **(see Figure 14.4)** and first-order with respect to water activity.

Combining the quantitative measurement of the effect of water activity with the quantitative effect of temperature, it is possible to describe the breakdown of fenitrothion on grains under the conditions that prevail across Australia in a single equation or in graphical form as **Figure 14.2**. The chemical model has proved accurate in extensive commercial applications on wheat and barley. Similar models are available for other pesticides **(Desmarchelier *et al.* 1979b; Desmarchelier and Bengston 1979a; Desmarchelier 1980).**

Kadoum and Al-Naji (1978) investigated the effect of grain moisture content on the degradation rate of phoxim-methyl in corn, sorghum and wheat. Residue data were obtained at 5 intervals over a 30-day period from wheat, corn and sorghum containing 8 levels of moisture, following 10 mg/kg applications of phoxin-methyl emulsion sprays. High-moisture content reduced the effectiveness and persistence of phoxim-methyl in stored grain. Degradations differed significantly at each moisture level. After 30 days, 10%, 3% and 23% of the initial residue deposit remained on the 20% moisture sorghum, wheat and corn, respectively. At the 6% grain moisture levels, 71%, 63%, and 68% of the initial residue deposit remained on sorghum, wheat and corn, respectively, after 30 days.

Desmarchelier and Bengston (1979a), Desmarchelier *et al.* (1979b) and **Desmarchelier (1980)** have provided extensive information on the residual behaviour of insecticides on stored grain and the effect of moisture on degradation. They describe the predictive models for loss of residues of 12 protectants on grains and from these models one can calculate the persistence of a chemical on any grain at any constant or varying conditions of moisture content and temperature.

Kandoum and LaHue (1979a, 1979b) obtained residue data during a 12-month period from corn and wheat with various levels of moisture following 10 mg/kg application of malathion emulsion spray. The sequence of malathion residue losses did not vary greatly among the different moisture levels 24 hours after the initial deposits; however, the ratios of degradation were distinct for each moisture level. After 12 months storage, 41%, 21%, 7% and 3% of the initial residue deposit remained on wheat containing 10%, 12%, 14% and 16% moisture, respectively; for corn the deposits were 34%, 24%, 8% and 5% remaining, respectively.

Quinlan *et al.* (1979) found that chlorpyrifos-methyl was much more effective than malathion in controlling insects on 14.6% moisture red winter wheat stored in plywood bins throughout a 9-month period. This was because chlorpyrifos-methyl persisted longer on the wheat than malathion did.

Thorpe (1979) and **Thorpe and Elder (1982)** developed a mathematical model to describe the effects of moisture transfer and heat on the degradation of insecticides applied to bulks of aerated wheat.

Quinlan *et al.* (1980) showed that the higher the moisture content of stored wheat, the greater the breakdown of both malathion and pirimiphos-methyl. Pirimiphos-methyl was significantly more persistent.

Noble *et al.* (1982), who studied the stability of 4 pyrethroids on wheat in storage, found that an increase in the moisture content from 12% to 15% resulted in a three- to six-fold reduction in the half-lives of these relatively stable insecticides.

Samson (1985) provided a comprehensive discussion of the biological efficacy of residual pesticides at high humidities and high moisture contents. He not only drew attention to the chemical degradation that is promoted by high moisture contents but discussed the disproportionate fall-off in biological efficacy that can be readily demonstrated by careful bioassay and by a simultaneous measurement of the 'available' and 'unavailable' fractions of the residue on the grain. Experimental evidence presented appears to confirm that the declining biological activity of insecticides on high moisture grain is due to the fact that the insecticide is somehow not readily available to be taken up by wandering insects. This concept merits further investigations.

30. Use Under Small-Farmer/Tribal Systems

There is considerable literature on the use of grain protectant insecticides under small-farmer conditions in Africa, the Indian Sub-continent and South East Asia. This has been reviewed in the separate chapters dealing with individual insecticides but there are many important logistical, social and economic considerations and recommendations which can be drawn from this literature and which need special consideration.

In this chapter an attempt is made to draw together the observations and recommendations of scientists who have devoted their attention to the special needs of subsistence farmers and others who store relatively small quantities of grain mainly for personal needs but possibly for some sale between one harvest and the next.

Kockum (1958) pointed out that the climatic conditions in Kenya normally prevent maize from drying out in the fields sufficiently to allow immediate shelling. At harvest time the husk is removed and the maize on the cobs is stored in cribs until dry enough for shelling and delivery. For this and various other reasons the harvested crop is often kept in this manner for 6 to 8 months or even longer. Before any control methods were introduced, ideal conditions existed during the storage time on the farm for many pests to increase to large numbers causing considerable losses in weight and quality. While excellent insect control over a period of 6 months is achieved by dusting maize on the cob with insecticide dusts as the maize is placed in cribs with ordinary wire-netting walls, this control becomes unsatisfactory when the maize is to be stored over a period of 13 months. Virtually complete protection for 13 months can be obtained from the same treatment provided the crib is surrounded with a wall of hessian cloth treated with a persistent insecticide. Mud or other solid material may be used in place of the hessian but it too must be treated with insecticide. The surface treatment needs to be repeated at regular intervals during storage. A further publication **(Kockum 1965)** confirmed the importance of protecting the treated maize cobs from the degrading effects of weather.

Weaving (1975), who made extensive studies of grain protectants for use under tribal conditions in Zimbabwe (Rhodesia), pointed out that the choice of insecticide under traditional tribal storage conditions is governed more by availability than suitability.

Golob (1977) produced a guide describing ways of mixing insecticidal powders with grains and pulses, just before they are stored, in order to prevent damage by insects. This guide was designed for use by extension workers to help them introduce the methods to farmers. It was illustrated with pictures as well as simple descriptions of the steps involved. With the exception of the insecticidal powders which can be purchased in ready-to-use packets for mixing with cereal grains, the equipment and facilities recommended are available in all traditional farm storages. The importance of complete and thorough mixing of the powder with the grain is stressed. The guide was easily understood even by illiterate persons.

Taylor and Webley (1979) wrote about constraints on the use of pesticides to protect stored grain under rural conditions, quoting results of a limited survey conducted through the Tropical Products Institute. They found that, given appropriate guidance, most farmers would use insecticides if the need for them was made clear. The traditional use of ash or minerals leads naturally to the use of dilute insecticide dusts, which the farmer may regard as a form of mineral treatment. Pesticides for control of storage insects are either in the form of dilute insecticide dusts or insecticide sprays. In the survey, 75% of the replies indicated that malathion was the principal insecticide recommended for storage use. It may sometimes be necessary to avoid malathion either because insects may be resistant to it or because the dust formulation may be poor and the product unstable. Several alternative insecticides are approved internationally for application to stored cereals. Many of these compounds are rather more expensive than malathion but they may often be more effective. Not all of these insecticides are available as a dilute dust, which is probably the safest and most effective formulation that farmers can use for mixing with properly dried gain. Large drums of insecticide may be available in urban centres but the provision, by manufacturers, of properly labelled, small packs is one of the problems associated with the supply of insecticide for small farm use. These packs generally increase the costs and the necessity to accurately determine the requirements for each area. The supply of too much insecticide can create problems because it may have a limited shelf-life and deteriorate if kept, or if the product is entirely used up, over-dosing may result. Dilute dusts for use on stored grains should be clearly recognisable and distinguishable from other insecticides and the use of symbols has proved beneficial. Full instruction for use, in the appropriate language, must be included. Where the usage of insecticides is low it may not be economic for suppliers to compete by offering alternative products. A strong lead by a government department or agricultural institute making definite recommendations may provide the necessary impetus to get a product established and used. In some countries it may also be necessary for government departments responsible for the control of pesticides to extend approval to insecticides that have received authoritative approval elsewhere.

Taylor and Webley (1979) also pointed out that failure of the distribution network is a major constraint. There appear to be four commonly employed means of distributing pesticides to rural communities for protection of stored crops. These are retail suppliers, the agricultural extension service, co-operative unions, and the agricultural marketing organisations. The survey indicated that retail suppliers and the agricultural extension service are the most widely used as major distributors of pesticides but that co-operative unions and agricultural marketing organisations may also be important sources of supply. In some countries, several organisations distribute pesticides for stored-crop protection, while in others the agricultural extension service alone may provide this service. Pesticides may be distributed free by government or they may be subsidised by marketing boards. The use of insecticides to protect crops stored under rural conditions does not appear to be linked directly with the use of other agricultural chemical inputs by farmers. There are, however, situations in which farmers regularly employ fertilisers and insecticides on growing crops but do not use insecticides on their stored crops. Here, the further training of farmers may be necessary and the appropriate insecticides should be made available.

Boxall *et al.* (1979) reported a study of farm-level food grain losses carried out in Andhra Pradesh, India, from 1976 to 1978. The object of the project was to provide a social cost/benefit analysis of farm-level storage improvements. The study concentrated on the storage of paddy rice as the staple crop. Various improved storage practices were tested, including improvements to traditional stores and use of metal bins. Insecticides were not included in the program. The social cost/benefit analysis revealed that efficiency gains by preventing losses are still sufficiently large to justify an extensive public investment in a program to reduce farm-level storage losses. The use of improved stores and metal bins can reduce losses by 3.2 %, representing a significant benefit. Because of the physical characteristics of paddy rice, the loss estimates are likely to be lower than for other food grains which are more susceptible to insect infestation, and consequently higher social cost/benefit ratios can be expected for other main dietary staples in India. Metal bins have been successfully introduced in the wheat-growing areas of northern India where the rate of return is attractive to the private sector. These workers recommended that the on-going extension service be expanded; that more attention be paid to the distribution of input through liaison with the pesticide industry; and that further research be conducted in other areas of India.

Adesuyi (1979) summarising extensive experience in the selection of new insecticides for use in maize storage at the farmer's level in Nigeria, discussed the factors affecting choice and use of insecticides. The formulation should be of very low active ingredient content. This has the advantage that large quantities of the dust will have to be applied and therefore better coverage of the produce will be achieved. The insecticide should be cheap. It should be readily available, especially during the storage season. Experience has shown that there is often an acute shortage of recommended insecticide at the time that it is needed. This is discouraging to farmers, who have taken to official recommendations and seen the beneficial effects. As a result, it is not possible to restrict recommendations to one insecticide at a time. It has been observed that recommending more than one insecticide at a time results in the sales agents complementing one another rather than competing. Most farmers who use insecticides for storage require small quantities. It is therefore desirable that insecticide manufacturers pack them in quantities small enough for it to be practicable for the farmer to exhaust his supplies in one operation. This will reduce the dangers of careless handling, poisoning, wastage and under-dosing or over-dosing.

A recommendation for or against use of pesticides at the small-farm level is an important one as it affects many people. It is estimated that 100 million farms are less than 5 ha in size and of these over 50 million are less than 1 ha **(Brader 1979).**

Hindmarsh and Macdonald (1980), from experience gained in Zambia, pointed out that the local varieties of maize have a higher proportion of small cobs with tight husks, and this feature together with the low moisture content of the grain for 6 or 7 months of the year contributes to a delay in the build up of insects in the storage crib. The introduction of high-yielding hybrids resulted in an increased storage problem when the farmer attempted to store new varieties in the traditional way. Poor husk cover, and the inherent susceptibility of the grain to attack by *Sitophilus zeamais* are contributing factors. However, minor and inexpensive modifications to traditional cribs significantly reduced the damage caused by insect pests. Nevertheless, without the use of insecticides insect damage levels were still unacceptably high.

Schulten (1981) pointed out that a complete change from the traditional storage system means that a small-farmer has to invest a considerable part of his yearly income in better storage. Even if it can be shown that the rate of return over a number of years is high, farmers are most reluctant to make such a large investment, as is demonstrated for example by the failure of many projects to introduce small silos of various designs. In general, the farmer will try to reduce losses by a small change in his traditional system, which frequently means the use of insecticides. There are also social costs/benefits to be considered. These are the costs a country has to make to improve the storage system and the benefits it may obtain as a consequence. Among the social costs are subsidies for equipment or insecticides and the cost of the extension effort to induce farmers to modify their traditional systems. Possible benefits could be an increase of available maize, an increase in the income of the farming community and so on. Farmers will use any available insecticide if they consider that insect damage is serious. This practice may not only be uneconomic but also dangerous. A change from storage of ears to shelled maize often means a large investment which farmers are not always willing to make. To reduce losses in such situations there are a number of options, such as the growing of maize varieties with a good husk cover and low inherent susceptibility, the improvement of marketing and introduction of co-operative storage, the subsidising of small silos, more effective insecticide and so forth. All these systems have their costs and benefits. A detailed cost/benefit analysis from the point of view of the farmer and from the social point of view, as well as an understanding of the motivations of a farmer to accept or reject a change, is the only way to find the right answer for the prevention of losses not only in stored maize ears but in stored produce in general.

Schulten (1981) considered that many farmers are in practice willing to use insecticides but government planners and local authorities can be reluctant to take the necessary steps to make insecticides available to the small-farmer. It should be acknowledged that insecticides are a necessary component in the strategy of integrated pest control, a strategy which is as important in post-harvest as in pre-harvest control. He considered that insecticides intended for use at the small-farm level should have the following properties:

- very safe in terms of toxicity and residues
- long shelf life to facilitate distribution
- easy to apply
- effective at least against primary pests
- rapid kill of adults and if possible of immature stages
- effective on shelled grains at dosages not exceeding the residue limits recommended by the Joint FAO/WHO Meeting of Experts on Pesticide Residues
- effective on shelled grains for a period of 6–12 months
- effective on unshelled produce for a period of 4–8 months.

McCallum-Deighton (1981) pointed out that approximately half of the world-wide production of cereals is grown in developing countries, and probably more than 75% of cereals from these countries is stored on the farm or in the villages. Damage by insects to this produce stored under rural conditions is frequently high. Although effective pesticides are available, it is thought that their use at the farmer level is minimal. He examined possible reasons for this failure and suggested remedial measures. Although the market in totality is large enough to be commercially attractive, the fragmentation of the market into many countries, and in these many countries broken into many farmers each producing relatively small amounts of cereals, reduces the manufacturer's interest in the market. The cost of local distribution is extraordinarily high. In developing countries these costs may be ten times those encountered in industrialised countries. The small farmer has difficulty in finding money for insecticides, and the cost of the insecticides as a proportion of his income is very high when compared with farmers in developed countries. Pesticide formulations and application techniques should be designed to suit the traditional storage techniques. Occasionally, it might be possible to introduce new systems as a result of modern techniques, but generally speaking these will be too expensive, will not use local materials and will be incompatible with the peasant farmer's way of life.

The economics of the situation inhibit manufacturers from doing more by way of extension and education because of the costs incurred from the point of entry into the country. Therefore, only in those developing countries which are relatively wealthy can the farmer afford to pay for the education and distribution costs to be recovered in the price of the insecticide, and so the poorer developing countries lose out. It is the responsibility of the government to provide an infrastructure which will allow the farmer to be educated. Some countries are trying to recognise the needs of the subsistence farmer, but inefficient government agencies either do not package in time, where there is local packaging, or they do not distribute in time for the farmer to treat his produce. Availability of the pesticide at the right time is all-important. If the farmer applies insecticide at the wrong time and in consequence obtains an effect less than anticipated, he will not use it again. Pesticide manufacturers must ensure that formulations available at the farmer level are stable and effective. Unless the farmer has cash crops which he can sell immediately before his grain harvest, he will have very little money available to buy the pesticide for his stored grain. Central government purchase, with a subsidy sufficient to cover the distribution costs, is probably most appropriate.

Webley (1981a) drew attention to the problem of instability and consequently poor shelf-life of many commercial dust formulations due to the use of unsuitable materials as fillers. He pointed out that higher standards of stability are needed in developing countries because of the higher temperatures encountered, and the long delays in distribution. He also pointed out that the consequences of supplying poor quality dusts are very serious indeed and recommended regular checks and analysis followed by the removal of unsatisfactory products from the market place.

Taylor (1981) reported the outcome of the survey initiated by **Taylor and Webley (1979)**. A summary of major constraints to the use by farmers of insecticides in grain storage in less developed countries is given below:

Constraint	Number of countries reporting = 32
Non-availability	24
High costs	16
Small packs unavailable	13
Farmers prefer other grain protection methods	11
Ignorance of modern insecticides among farmers	7

Pinniger and Halls (1981), in summarising the

discussion on the use of pesticides at farmer and village level at the GASGA Seminar in 1981, pointed out that the value of insecticides at the farm level is determined by economics of costs versus loss. If a farmer at subsistence level experiences major losses then he will use anything to hand to reduce these losses. In a less desperate situation he will need proof that treatments work and bring more profit. Living insects may be seen as a treatment failure and differences in pest response may not be recognised by the farmer. He needs to reduce damage but not necessarily achieve complete control. If his capacity to grow more is increased he needs to store more. He may have to store higher yielding varieties which have greater susceptibility to insect attack. He then needs assistance to kill insects and prevent losses by use of pesticides so that he can sell his surplus on the market. Treatment of bags and the store structure in addition to admixture may be useful but it is necessary, where possible, to have the same insecticide for all treatments so as to avoid or reduce confusion. It was suggested that pesticides should be supplied to the farmer as part of an economic package deal of seed, fertiliser and storage insecticide. The success of this would depend upon central government input and adequate extension.

Golob (1981c) observed that it is difficult for the extension officer in a developing country to comprehend the storage loss and the need for marketing advice. Virtually all the advice provided by the agricultural extension officer is related to pre-harvest husbandry, with none to post-harvest matters. Only rigorous training and demonstration will persuade the extension officer of the need the farmer has for advice on storage. Such training must be part of the standard tuition of the extension officer before he goes out into the field — all too often it is not.

Calverley (1981) pointed out that in developing countries a farmer may not use pesticides for the grain his family will consume partly because he has no cash and partly because he can compensate for the loss by retaining and storing more grain at less cost to himself than buying insecticides.

Golob (1984) and **Golob and Muwalo (1984)** described experience and experimental work over 3 years in Malawi which led them to the conclusion that small farmers growing traditional varieties of maize could reduce loss and damage by plastering their cribs with mud as protection against rain during the wet season but they would not gain additional benefit from the use of insecticides. However, national policies encourage farmers to grow improved grain varieties. As the use of these becomes more widespread, the local, farm-stored varieties will change in character due to cross-breeding and will become more susceptible to storage pests. To prevent the high weight-loss that would otherwise occur, the application of 2% pirimiphos-methyl dust allows these varieties to be stored as successfully as local maize. The insect attack on stored maize becomes critical at the end of 4 months storage. If, at this stage, the cobs are shelled and the shelled maize is treated with pirimiphos-methyl or other suitable insecticide, hybrid or improved maize varieties can be stored for 9 months or more with relatively little loss. However, this technique requires a major change in farm practice and a substantial additional cost which will not readily be accommodated.

A summary report of discussions and recommendations of the GASGA Seminar held in 1981 **(Anon 1984)** indicated that it was agreed that, for the control of stored-products pests in developing countries, there is a present and continuing need for the use of pesticides in the conservation of grains and some other agricultural products; pesticides should only be used against storage pests when they were needed to complement other good storage practices; and insufficient consideration is given to the economics and technological limitations of using pesticides in relation to the pest complex, type of structure, the climate and the total post- harvest system. The Seminar recommended that national governments should undertake, as a matter of urgency, a review of current post-harvest pest-control problems in relation to materials and techniques currently used in the country; the level of efficiency and success being achieved; constraints affecting the use of pesticides, such as availability, suitability of formulations currently available and problems of registration; the potential for improvements in existing pest control procedures; and the ability of locally available manpower to use pesticides properly.

31. Regulation and Registration

31.1 Introduction

One of the prerequisites of a pesticide is that it should be toxic to the target organism when applied in a convenient manner at a predetermined rate. Since few pesticides possess a high degree of specificity most present at least a potential hazard to non-target organisms, including man. It has been accepted that the availability and use of pesticides should be controlled in the public interest.

The goal in regulating pesticides is to provide society with adequate protection from adverse effects while not denying it access to benefits.

The principal method of establishing the manner in which a pesticide may be marketed and used is through the registration requirements. The term 'registration' used in this context should not be confused with the registration of a motor vehicle, a trade mark or a dog. In these cases the procedure simply involves the recording in a register of a few salient details which establish ownership, evidence of which is then provided by a document for which the registrant pays a designated fee. Such operation entails the minimum of time, expense or documentation. In the case of pesticides, registration implies the acceptance, by statutory authority, of extensive documented proof submitted in support of all claims for efficacy and safety made for the proposed product. Registration implies a number of different controls among which evaluation is the most important. For a pesticide to be adequately assessed for registration purposes extensive scientific information must be developed by the manufacturer on all aspects of the product, its properties and performance.

Evaluation involves the mature judgement of experienced professionals using a multi-disciplinary approach. In the evaluation of pesticides, as in other fields of human endeavour, some degree of risk must be considered acceptable to society. The alternative would be needless prohibition of important benefits.

There are potential problems with pesticide usage but the purpose of the large amount of research going into the generation of data for registration is to tackle the issues before they become problems.

Registration enables authorities to exercise control over use levels, claims, labelling, packaging and advertising and thus to ensure that the interests of end-users are well protected. The registration legislation provides a system under which the public's interest and the manufacturer's rights are protected.

Most nations are committed by law, policy and traditions to assure their constituents that their food supply is adequate, safe, clean and wholesome. In order to give effect to such laws and policies it is necessary to develop criteria and protocols that are effective, workable, and enforceable. It should be the objective to achieve these goals with minimum dislocation of production or trade, but under no circumstances should adverse effects on people or the environment be countenanced to serve economic goals. While pesticides are intended to effectively control organisms that destroy or endanger man's food, health or environment they, like virtually every chemical, may have physiologial effects on other organisms living in the environment, including man himself. Whether the effects occur or not is simply a question of the dosage and of proper use.

How best to reduce the hazards of pesticides to man and animals is a problem that has occupied many individuals and organisations the world over. In electing to control the introduction of pesticides through some type of registration scheme, national authorities have been mindful of the needs of the many inter-related and inter-dependent segments of the community.

31.2 Responsibility

There are four levels of responsibility associated with the registration of pesticides:

31.2.1 Manufacturer

The prime responsibility rests with the manufacturer who must first be satisfied that the product fulfils the many requirements demanded by the public and the government authorities charged to watch the public interest. The manufacturer must ensure that there is adequate scientific evidence to support all claims for efficacy and safety. It is not generally recognised that registration authorities do not usually ask more difficult or different questions to those demanded by corporate management of those charged with research and development responsibilities for new pesticides.

The manufacturer of a pesticide must be satisfied that he has generated sufficient scientific information to effectively and positively answer at least the following questions about it:

Is it effective?
Is it efficient?
Is it reliable?
Is it safe to users?
Is it safe to bystanders?
Is it safe to consumers?

Is it safe to livestock?
Is it safe to wildlife?
Is it acceptable in the environment?

Implicit in these questions are many issues and aspects which the manufacturer must consider and on which appropriate scientific data must be forthcoming. If and when all this information is available the manufacturer may approach regulatory authorities in confident expectation that they will judge the data adequate and acceptable.

31.2.2 Government

In most countries it is recognised that we have entered a period characterised both by a fuller understanding of the risks and advantages of pesticides and a desire to provide adequate controls, either voluntary or mandatory, to ensure that the use of pesticides does not affect public health, the environment or trade.

Public policy must be aimed at protecting the public and the environment from excessive exposure to harmful substances while also preserving and increasing the great variety and utility of those products that have contributed so much to the improvement of our food supply, protection of our health, the increase in trade and the standard of life.

Governments must establish legislation to regulate the manufacture, sale and use of pesticides. Such legislation must be based on regulations that establish a permissible safe use pattern for each chemical. This use pattern must be described on the labelling for each product and the labels need government approval. In addition, safe legal limits must be established for residues in food and feed.

Some countries exercise control over both safety in use and efficacy while others control one or the other. In some countries, the protection of the operator stops with the label directions, but in others, the law imposes responsibility on employers in respect of their employees. Many countries make use of the idea of an experimental permit, temporary clearance or licensing to allow new pesticides to be field tested and some registration authorities undertake a critical laboratory and field examination of new products.

In summary, the responsibility of government is to:

- protect the unwary from the unscrupulous;
- prevent unsubstantiated claims;
- ensure adequate directions for use;
- highlight precautions and limitations in use;
- protect the uninitiated from their own ignorance;
- safeguard reputable manufacturers from spurious claims by disgruntled users;
- engender confidence in the system by the general public.

Pesticides legislation requires manufacturers and distributors of products classified as pesticides to obtain registration of their products and product labels before offering them for sale. The registration requirements are most exacting. They provide protection for the general public from fraud or misrepresentation but, in addition, are designed to ensure that the registered labels contain adequate directions for safe, effective and proper use in the interests of all concerned.

31.2.3 Vendors

Those engaged in the distribution and sale of pesticide products carry a heavy responsibility to ensure that they do not offer for sale products which are not registered and that they do not promote uses which are not recommended on approved labels. Users rely heavily upon their suppliers for guidance in the safe and effective use of pesticides and it is recognised that such sales outlets provide the major source of information reaching users. Because of this, the role of supplier carries with it both privilege and responsibility.

31.2.4 Users

Users must recognise the responsiblity to themselves, their families, their neighbours, the community, the environment and those who might ultimately consume the produce protected with the aid of pesticides.

The directions on registered labels have been developed at great cost in time, money and scientific manpower, have been evaluated by experienced scientists and have been approved by government authorities. The claims and directions are made in the knowledge that if they are followed the result will be entirely satisfactory and there will be no untoward hazard. Unless users accept their responsibility the efforts of manufacturers and government will have been to no avail.

31.3 National Requirements for Insecticides to be Used in Grain Storage

Only those insecticides that have been specifically approved should be used on and around grain. The choice of insecticides that may be used is limited by the very strict requirements that must be enforced to ensure absolute safety for consumers of these important basic food commodities. To qualify for selection as a possible candidate material for use on or around grain, the insecticide must fulfil the following ten requirements **(FAO 1982a):**

1. it must have a wide spectrum of high insecticidal activity
2. it must present no hazard to consumers of grain and grain products
3. it must be acceptable to health authorities
4. it must be acceptable to the international grain trade
5. legal limits must be established for the resulting residues under the laws of the country where the grain is stored
6. it must not affect the quality, flavour, smell, or handling of grain
7. it must be capable of being used without undue hazard to operators

8. it must be effective at economic rates of use
9. it must not be flammable, explosive, or corrosive
10. its method of use must be compatible with established grain handling procedures.

The requirements for insecticides used on seed are similar but, under circumstances where there is no possiblilty of seed being used as food or feed grain, materials of higher mammalian toxicity can be used. Additional requirements are:

11. no detrimental effect on germination of seed and seedling growth
12. compatibility with fungicides used for pre-emergent and seedling diseases.

31.4 Efficacy

Because of the wide variety of stored-product pests that can occur in a particular type of grain, region or country, detailed information is required concerning the effectiveness against each important species of stored-product pest. This can include information concerning the biological activity against several life stages and, where appropriate, information concerning the susceptibility of species which have already been selected for resistance to other pesticides.

Such data are generally developed under controlled laboratory conditions using cultures of stored-product pests the history of which is known and which are exposed to a range of concentrations of the insecticide applied to a substrate upon which the insects will feed, reproduce and live successfully. In most instances this may be whole raw grain.

These studies are generally designed to determine the lowest concentration of insecticide required to kill adult insects and the concentration that will prevent reproduction and development of immature stages. It is essential that the studies should be carried out under known and controlled conditions of temperature and humidity. These conditions should preferably coincide with those found in stored grain in the region.

Since the insecticide is intended to protect grain from insect attack rather than to destroy existing heavy infestations it is usual to design some of the experiments to measure the susceptibility of treated grain to infestation by the most important species encountered in the region. Samples of grain which have been treated uniformly and accurately with insecticide at a graded range of concentrations and which have been held under controlled conditions of storage for varying periods should be challenged with known numbers of insects. The mortality rate should be determined after an exposure period (generally 3 and 26 days) and the number of progeny should be determined after a period sufficient to allow for their development.

Such data should be used to decide the optimal rate of application of insecticide that would be most effective in providing the degree of protection required. It is generally necessary to carry out pilot studies in which small bulks of grain (100–200 kg) are treated with the insecticide at a predetermined rate prior to storage under conditions typical of those encountered in the region. Samples of this bulk grain should be taken at regular intervals for bioassay with selected stored-products insects. The object of such studies is to determine the length and degree of protection provided by the insecticidal treatment and to establish a reliable indication of the minimum effective concentration of insecticide that should be applied to grain.

It is absolutely vital that the rate of application should be no higher than the concentration that will confer an adequate degree of protection for a reasonable period when the commodity is stored under conditions which minimise insect attack. Insecticides are to be regarded as a supplement to, not a substitute for, good storage practices.

Because of the many pitfalls inherent in scaling up from small-scale laboratory conditions to commercial-scale grain storage and handling, it is generally necessary to take the results of laboratory and pilot-scale studies and verify them in typical commercial practice. Such practical trials should be overseen by scientists and technical personnel who should be responsible for monitoring treatment, collecting data on temperature, humidity, rate of treatment, etc. Samples of treated grain should be collected for chemical analysis and bioassay immediately after treatment and at intervals during storage.

Such a regime of experimentation and investigation should lead to the development of practical directions for use of the insecticide. In order that the information can be evaluated by relevant authorities it is essential that all details of experiments and their results should be systematically recorded and reported.

Recognition that efficacy studies conducted in the field in accordance with internationally accepted guidelines can produce data supportive of the results of similar field studies carried out under different climatic, meteorological and agricultural conditions in some other part of the world has greatly reduced the cost of generating adequate data on efficacy but it does not do away with the need for adequate field trials in the region. Efficacy studies should be designed to determine the optimum method and rate of use.

The amount of grain protectant insecticide required depends on the type of insect species present, the temperature and moisture content of the stored commodity, type of storage and the duration of protection required. For example, moths can usually be controlled in bulk grain storage by treatment of the space above the grain and by application of a suitable protectant to the grain surface, rather than by admixture with all of the grain. Other examples of optimum use include:

(1) selection of the insecticide or combination of insecticides most effective against the species likely to occur;
(2) selection of rates providing adequate protection under local storage conditions for the anticipated period of storage, but which give rise to minimum

residues at the time the grain is taken for processing;

(3) reduced rates of application when grain is cool, being cooled or aerated;

(4) careful supervision of application and a program of worker training to ensure that the application is as uniform and complete as possible, thus avoiding pockets of grain containing either too little or unnecessarily high deposits of insecticide.

31.5 Fate

Comprehesive information concerning the fate of the insecticidal deposit on the grain is essential for the proper understanding of the biological activity under prolonged storage as well as the knowledge of the level and nature of residues in the treated commodity when it is removed from storage and is passed into trade channels.

For these reasons it is essential that the pilot studies and supervised field trials should be monitored by chemical analysis of samples of the stored commodity. The frequency of sampling should be such as will enable the rate of degradation to be determind with a fair degree of accuracy. It is now possible to predict the fate of the insecticidal deposit from a knowledge of the storage temperature and relative humidity of the interstitial space within the grain **(Desmarchelier 1978)** and this calculation should be made to anticipate and confirm the results of the residue trials.

The climatic conditions surrounding stored grain, expecially bulk grain, are much more regular than those to which field crops are exposed. For example, temperature and moisture content of stored grain are relatively stable and stored grain is sheltered from wind, rain and light. Under such conditions, it is logical to expect that the rate of disappearance of the insecticide deposit would be predictable. **Desmarchelier (1978)** showed that the loss of fenitrothion from post-harvest application to wheat, oats, rice in husk and sorghum, followed a second-order rate process, with rate of loss being proportional, at a fixed temperature, to the amount of fenitrothion and the activity of water, which was obtained from the partial pressure of water vapour in the interstitial spaces in equilibrium with the moisture absorbed on the grain. The effect of temperature was in the form of an Arrhenius equation.

A chart relating half-life to temperature and relative humidity was presented in a form suitable for field use **(Figure 31.1),** and a mechanism was proposed for loss of fenitrothion. The proposed mechanism is that an absorbed molecule of fenitrothion is desorbed by replacement by a water molecule. The desorbed molecule is more likley to be degraded than an absorbed molecule because it has a greater chance of collision with enzymes, metal ions and other active molecules.

This general model, developed for fenitrothion, has been extended to other insecticides, including bioresmethrin, phenothrin and carbaryl **(Desmarchelier 1980a, b)**, pyrethrum **(Desmarchelier *et al.* 1979a), pirimiphos-methyl, chlorpyrifos-methyl and methacriphos (Desmarchelier *et al.* 1980)**, and several photostable pyrethroids **(Desmarchelier and Bengston 1979)**.

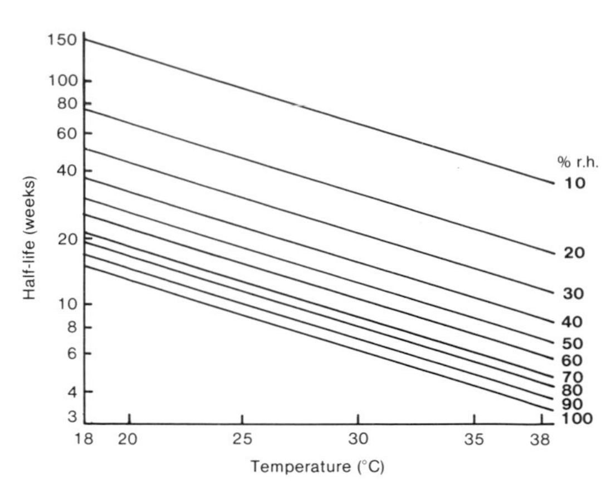

Figure 31.1. Half-life of fenitrothion on grain at different relative humidities and different temperatures **(Desmarchelier 1978).**

Agreement is good between data predicted by the models and results obtained by careful monitoring of extensive field use involving tens of thousands of tonnes of various grains **(Desmarchelier *et al.* 1980a, 1981a, in press)**.

The studies by **Desmarchelier (1978)** and **Desmarchelier and Bengston (1979)** enable a direct comparison to be made of the 'reference half-lives' of different insecticides, i.e. time required to degrade to half the original concentration at a fixed temperature and relative humidity within the stored commodity (reference point — 50% r.h.) The half-lives of most of the insecticides under consideration are given in **Table 31.1.**

Moisture content of stored products, arbitrarily defined in terms of the weight loss on heating under specified conditions, is not linearly related to water activity for a particular grain and differs substantially between different commodities of the same water activity. Moisture content is, however, easy to measure and compilations are available **(Hall 1963; Gough and Bateman 1977)** to convert it into water activity in equilibrium with the grain under test **(Banks and Desmarchelier 1978)**. According to these workers, if water activity is used as a measure of water present, the breakdown rate of various insecticides is found to be independent of grain type **(see Figure 31.2)** and is a first-order reaction with respect to water activity.

31.6 Residues

Residues in food are not a novelty of the twentieth century and their occurrence is not only associated with the use of pesticides. Food legislation in most countries has evolved as a result of the need to protect consumers from the risks of adulteration and contamination. Limits for chemical contaminants in food appeared in food

Table 31.1. Important features of insecticides currently used or under development as grain protectants

Insecticide	In use since	Under development	Rate of application (mg/kg)	Synergist used[1]	Half-life at 30°C and 50% r.h. (wweks)	Temperature coefficient (K/°C)
Bioresmethrin	1975	–	1	+	38	0.031
Bromophos	1968	–	10	–	/	/
Carbaryl	1979	–	5	–	21	0.031
Chlorpyrifos-methyl	1978	–	5–10	–	19	0.040
Deltamethrin	–	+	1	+ –	>50	/
Dichlorvos	1966	–	4–10	–	2	/
Etrimphos	–	+	10–15	–	/	/
Fenitrothion	1977	–	6–12	–	14	0.036
Fenvalerate	–	+	2	+ –	>50	/
Malathion	1960	–	8–20	–	12	0.050
Methacriphos	–	+	5–15	–	8	0.055
Permethrin	–	+	2	+ –	>50	/
d-Phenothrin	–	+	2	+ –	40	0.029
Pirimiphos-methyl	1969	–	4–8	+ –	70	Small
Pyrethrins	1935	–	2–3	+ +	55	0.022

1 + = yes; – = no/not; + – = yes and no; + + definitely; / = no information yet available.

legislation in the United Kingdom and the United States in the early 1900s. Pure Food Acts and Food and Drug Acts were introduced in Australia well back in its history. The proliferation of standards (tolerances) for pesticide residues in food commenced in 1952 when, as a result of public hearings, limits were fixed for DDT and other pesticide residues in many raw agricultural commodities and foods in the USA.

The science and practice of evaluating residues and establishing legal limits has spread beyond national boundaries and has become part of the Food Programme of the United Nations conducted by the Food and Agriculture Organization and the World Health Organization working in close collaboration to protect public health and to facilitate trade in foodstuffs.

There have been a few instances where people have been injured by gross misuse of pesticides. The most notable examples were where HCB and methyl-mercury treated seed was used directly for human food and where people have been injured as a result of pesticides leaking into food transported or stored in close proximity. There are, however, no known instances of injury to consumers resulting from the consumption of food containing residues derived from the proper use of chemicals. The modern attitude is, however, that food should be as free as possible of man-made contaminants.

No legal recognition has been provided to cover accidental contamination or the misuse of pesticides. Neither is it anticipated that such residues will be accorded legitimate status. The deliberate application of insecticides or fumigants for the destruction of insects in

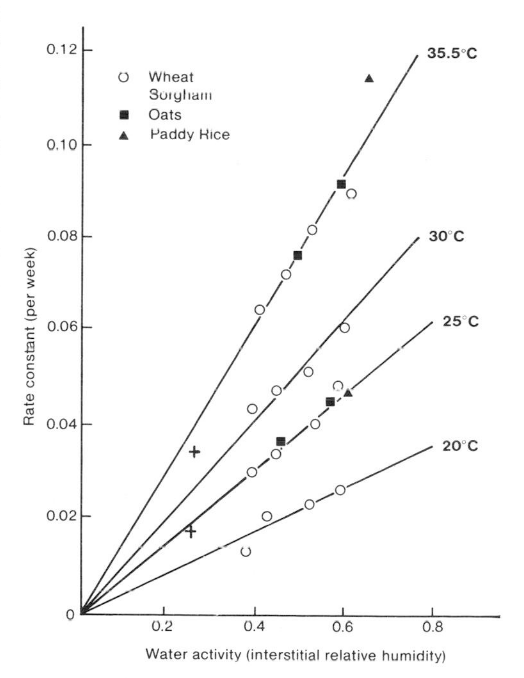

Figure 31.2. Variation in breakdown rate of fenitrothion on whole grains as a function of water activity and temperature **(Banks and Desmarchelier 1978).**

stored grain or for protection against insect attack presents quite different problems when it comes to consideration of residues.

Residues resulting from use of pesticides during the growth of the crop occur only occasionally and then only at relatively low levels so that the intake of residues in the diet from this source is relatively insignificant. When chemicals are deliberately added to stored grain the chances are that all or most of the grain will be treated and that the residues will be at a relatively high level. The intake in the diet could, theoretically, therefore be highly significant. Toxicologists and health authorities require greater assurance and more extensive evidence of safety before authorising the deliberate addition of potentially toxic substances to food.

These authorities are conservative, and unless the scientific data which are presented are conclusive and leave no room for doubt, no recommendation for use or for a maximum residue limit will be made. The number and variety of insecticides which have been cleared for application to stored grain and for which maximum residue limits have been established is strictly limited. These are dealt with in Chapters 6–27 which provide information on the current status of each insecticide.

The ability of a pesticide to persist for a certain length of time can be desirable and has been recognised as important in some situations for successful control of pests. Thus, a knowledge of residues of a pesticide, or arising from the use of a pesticide, is useful in establishing its efficacy. However, the assessment of the potential human hazards arising from very small quantities of a pesticide in food and the environment has become an important part of the overall risk/benefit evaluation and is essential before a pesticide can be introduced into commercial use.

One of the basic requirements of such assessments is the availability of reliable data on pesticide residues in food, feed and the environment so that a realistic estimate can be made of the human exposure. The increasing demands of national registration and health authorities include residue data on treated crops and commodities and additionally in water, soil, air and wildlife. These authorities will only reach conclusions and make decisions if they are satisfied that the data are reliable.

However, variations in methods and techniques used in obtaining these data, including the selection, preparation and analysis of samples, have made it difficult to compare results and decide if the results are valid. Secondly, the validity of a set of results depends primarily on an adequate design of the trial. These variations have made it difficult to compare information from different sources and have contributed to differences in the regulations adopted in different countries.

These difficulties are most apparent when considering the conclusions reached by national authorities during the registration of pesticides and the use of residues data to set and enforce legal maximum residue limits for pesticides in food and feed. These limits have become important in the movement of food and feed commodities in international trade. The harmonisation of the methods used in the production of residue data and a more uniform approach to evaluating the data are urgently needed.

Guidance on the many aspects of producing and evaluating residue data is desirable. It will be of particular value to those countries still in the process of initiating procedures for the official control of pesticides. The need for guidance has been recognised by a number of national and international organisations and committees and several are already making contributions.

Prior to registration, data have to be developed to allow a reasonable judgment to be made of the residues left in a commodity when the insecticide has been applied according to the recommendation for use. Such data are essentially predictive and enable a registration authority to estimate the maximum residue level which might be expected. This estimate is normally based on data from supervised trials and may be used as a guide to what level may be expected when the pesticide is used. Subsequently, after considering the potential toxicity of such a residue to man and using appropriate safety factors, legal maximum residue limits (MRLs) may be established.

After a pesticide has been registered and used, it is desirable for a competent authority to be able to confirm that the estimate of expected residues made at the time of registration is a valid one. If doubts arise about the validity of the estimate, surveillance and monitoring studies may have to be carried out to ascertain if any revision of the estimated maximum residue level is required. Enforcement programs of MRLs also produce information relevant to the need to reconsider maximum residue limits.

The estimation of a maximum residue level is based mainly on a knowledge of the residues which occur following the use of a pesticide in accordance with good agricultural practice, normally obtained by the analysis of samples from supervised trials. This may be supplemented by selective surveys of commodities where there is detailed information available on the use of the pesticide.

Data obtained from trials and studies are limited by practical considerations and the estimation of a maximum residue level must be part *assessment* and part *prediction*. It is obviously impossible to carry out sufficient trials to cover all the various conditions under which a pesticide may be used. Therefore, although well planned trials demonstrate a range of residues, emphasis should be directed towards the identification of conditions and factors which lead to the highest residue levels following recommended use patterns.

Well planned trials take all factors into account so that the residue data represent the widest range of treatment conditions possible. Although the number of variables can be reduced in a supervised trial it is rarely possible to isolate the influence of an individual parameter and subsequently use the information accurately in predictions.

Insecticides are available in a number of different types

of formulation, e.g. liquid, emulsifiable concentrate, suspension, wettable powder, and dust. They may be applied as sprays or dusts, by methods ranging from relatively simple techniques, such as those used for maize in cribs or stacked commodities in sacks, to automated systems, such as those used in large central storages. In none of these will the application be completely uniform, and representative sampling presents considerable difficulties, particularly from bulk transports and bulk storages. The difficulty is further aggravated by segregation, which inevitably occurs when the commodity is moved, turned, or transported. The presence or absence of grain dust and dockage influences the level of residues found in non-representative samples and numerous studies have drawn attention to the need for care in taking samples and interpreting the results of analysis (**Snelson 1971, 1974**).

For this reason it has been considered necessary to establish maximum residue limits for grain protectant insecticides somewhat above the maximum rate needed in good storage practice to allow for variations that cannot be avoided in sampling and analysis. Usually a factor of about 2 is regarded as appropriate to cover these contingencies but, nonetheless, those responsible for the application of insecticides must take extreme care to keep the variation within such limits.

The cleaning of grain preparatory is milling removes dust and dockage containing disproportionately high concentrations of insecticide and the blending that occurs during milling and processing make it unnecessary to provide a significant margin to cover variations in the residue levels due to sampling difficulties in milled products.

After application, the pesticide, depending on its chemical constitution and the nature of the commodity, may move from the surface of the individual kernels to internal tissues. The extent of penetration can range from complete retention of the residue on the surface, to near equilibrium throughout the whole grain. The processing of grain, e.g. removal of hull, husk and bran, usually results in the insecticide becoming concentrated in the hull, husk or bran, making it important to consider the uses to which such fractions might be put. **Table 31.2** records information collated from numerous studies designed to determine the effect of milling, processing and cooking on insecticide residues in a variety of stored grains. The data have been expressed in terms of the percentage reduction in residues in converting various treated commodities to processed grain fractions or prepared food. Although many useful data have been published, more are needed to reflect the fate of various insecticide residues after milling, processing and cooking under various conditions typical of different regions of the world.

31.7 Toxicological Requirements

The assessment of safety basically depends upon toxicological studies, most of which are conducted on laboratory animals. The World Health Organization has published a review of the principles and methods of

Table 31.2. Percentage reduction of residues brought about by various steps in processing raw grain for human consumption.

Insecticide	Wheat to wholemeal	Wheat to white flour	Wheat to wholemeal bread	Wheat to white bread	Rice in husk to husked rice	Rice in husk to polished rice	Rice in husk to cooked rice	Barley to malt	Barley to wort
Bioresmethrin	0	35	100	100	85	93	97	90	99
Bromophos	0	63	72	98	/	/	/	/	/
Carbaryl	57	98	75	99	93	98	99	97	100
Chlorpyrifos-methyl	67	(94)	(83)	98	90	86	/	93	>99
Deltamethrin	0	80	30	80	/	/	/	/	/
Dichlorvos	50	93	95	100	95	96	100	/	/
Etrimphos	0	70	55	90	/	/	/	/	/
Fenitrothion	40	92	80	99	92	97	99	80	>99
Fenvalerate	0	85	30	90	/	/	/	/	/
Malathion	20	75	80	95	90	97	99	99	>99
Methacriphos	50	87	100	100	93	97	99	99	>99
Methoprene	/	85	85	95	/	/	/	/	/
Permethrin	0	75	68	85	/	/	/	/	/
Phenothrin	0	78	46	87	90	97	98	90	>99
Pirimiphos-methyl	5–16	40–79	76	90	70	93	98	90	100
Pyrethrum	/	/	100[a]	100[a]	/	/	/	/	/

/ No information available. a Assumed to be destroyed by cooking but no information available.

evaluating the toxicity of chemicals **(WHO 1978)** and this supplies details which can help the investigator to select the most suitable technique for a specific study. It must be noted that the toxicological issues relevant to biologically-active chemicals used as pesticides may be considerably different from those for conventional toxic chemicals.

Acute toxic hazards to operators, bystanders and those exposed during transport or storage are determined by the short-term toxicological properties of the formulated product and may not necessarily be reflected by tests done on the technical active ingredient. An elaborate review of toxicological investigations appropriate for pesticides has been published by the **Council of Europe (1981, 1984).** WHO, through its International Program on Chemical Safety (IPCS), convened a Scientific Working Group during 1983 to establish the principles and methodology for evaluating environmental epidemiology **(WHO 1983)**. In order to promote mutual acceptance of toxicological test data, the Organization for Economic Co-operation and Development (OECD) has issued guidelines for individual test parameters **(OECD 1981a)**.

The aim of test guidelines for toxicity testing is to produce a framework for each toxicity test which is sufficiently well defined to enable it to be carried out in a similar manner in different countries and to produce results that will be fully acceptable to various regulatory bodies. The growing demands for testing and evaluating the toxicity of chemical substances will place an increasing pressure on personnel and laboratory resources. A harmonised approach, promoting the scientific aspects of toxicity testing and ensuring a wide acceptability of test data for regulatory purposes will avoid wasteful duplication or repetition and contribute to the efficient use of laboratory facilities and skilled personnel.

The objective of all safety testing is to ensure attainment of the desired benefits of use without incurring needless risks. There must of course be some balance between the benefit and the cost of assessment, just as there needs to be a balance between the benefit and acceptable risk. Thus, to subject all pesticides to a single rigid routine of study would be gravely off the mark and self-defeating. The big questions should be asked first and the more detailed ones broached sequentially as the need for more detail is demonstrated.

Our understanding of the effects of chemicals is increasing very rapidly. Hence, it would be unwise to establish a rigid evaluation scheme at this time. Any testing procedure should be flexible enough to permit updating as scientific understanding advances and as new procedures become available.

To demand too much testing would prevent the development of some socially and technologically beneficial chemicals; to demand too little would permit the development of certain products whose net impact on society could be harmful.

It is unrealistic to expect that any system of pre-market evaluation will ensure absolute safety. With our present incomplete knowledge, we cannot expect to predict all the potential hazards of each new chemical. Even with a reasonably elaborate evaluation scheme, potential hazards associated with some chemicals could well go unrecognised. A more reasonable goal is to minimise the hazard within the limitations imposed by our knowledge and resources, with periodic review.

No test procedure provides an exact measure of all the potential effects that need to be identified. Toxicological tests on laboratory animals must be extrapolated to predict potential effects on man at much lower doses, with considerable resulting uncertainty. Even after a pesticide has been released into the environment in quantity, only a limited number of its effects, on possibly non-representative species, can be measured. All tests are thus models and, as predictive tools, are subject to error.

31.8 Labelling

The best insecticides will be found wanting if used incorrectly and the presentation to the users of the product (that is, the label), must therefore be as clear and concise as possible. A great deal of time and effort is put into labelling both by the manufacturer and the registration authority. Agreement of the claims and the directions for use are the final stage in the granting of registration. The aim is to ensure that the registered label of each product carries an adequate amount of well authenticated information.

It is well recognised that failure to understand and follow the directions on labels is one of the main causes of disappointment, misadventure and injury following the handling and use of pesticides.

The topic of pesticide labelling is currently being discussed in several national and international arenas in an endeavour to find ways of communicating effectively with illiterate and semi-literate users.

Several national authorities have issued guidelines on the labelling of pesticides. A similar guideline suitable for international use is currently being developed by FAO. This became available during 1986 **(FAO 1986)**.

Many factors influence the amount, nature, and distribution of the residue. The most important of these factors are the chemical, its formulation, the rate of application, method of application, time of treatment, the number of treatments, temperature and equilibrium relative humidity of the grain, use of adjuvants, and the interval between the last application and the release of the commodity into trade channels.

In order to reduce the incidence and level of residues of chemicals occurring in raw agricultural commodities (and hence in foodstuffs), it is essential to adopt good agricultural practices in the use of chemicals.

The concept of good agricultural practice in the use of chemicals in the realm of residues embraces all interrelated and essential factors and functions which ensure that the desired effect will be achieved without leaving behind more than the minimum of residues necessary for effective performance.

Good agricultural practice in the use of chemicals is

therefore the officially approved usage of a chemical which is essential for the control of pests under all practical conditions, bearing in mind all the difficulties and hazards involved.

It is absolutely vital that the concept of good agricultural practice in the use of pesticides should be appreciated and applied so as to control the pest but to leave the minimum amount of residue that is practicable.

The directions on labels of registered pesticide products are designed to produce the required effect without giving rise to residues in excess of legal limits. The legal limits for residues in raw agricultural commodities are based on residue trials, and users of pesticides may rest assured that their produce will not contain residues in excess of approved limits if they follow the directions on the registered label.

In the case of specific chemicals offered for sale to the general public, all the above factors except the pesticide and its formulation are under the direct control of the user. Directions for use are designed to guide users to apply the product correctly and in a manner which ensures, not only that the desired effect will be obtained, but also that, should residues occur, they will be within legally acceptable limits. Too little stress is placed on the value and importance of label directions. The message which should be brought before users of pesticides regularly and repeatedly is *'READ THE LABEL — FOLLOW THE LABEL'*.

31.9 Surveillance

Whilst it is very important to have legislation and to try and educate people to do the right thing, it is none the less essential that there should be continuous monitoring to ensure that everything is as it should be.

Most industrially developed countries have introduced some form of monitoring of food for residues. Some such systems are highly sophisticated and continuous whilst others depend on regular or ad hoc surveys of critical food commodities. Whichever system is considered appropriate for the particular country it should be capable of determing whether the bulk of food produced, imported, consumed or exported conforms to acceptable standards so far as residues are concerned.

In the event that a result is found to be above the permitted level or in conflict with national or international limits, action should be taken to investigate the cause and to modify practices accordingly. Grain handling authorities should initiate quality control analysis to check the effectiveness of operator training and supervision. In this way they can maintain the effectiveness and efficiency of their pest control practices whilst gauging compliance with government standards and trade requirements.

However, it is well recognised that such surveys of residues in raw commodities do not provide a measure of the amount of pesticide residues ingested by consumers since much or most of the residue is lost during the preparation, processing and cooking prior to consumption. In order to accurately gauge the intake of pesticide residues by consumers, total diet studies, otherwise known as market basket surveys, are conducted. In these surveys a typical diet for a young adult consuming more than the average amount of food is chosen and appropriate quantities of food are purchased in retail shops. The surveys are generally repeated four times throughout the year to represent food available in the four separate seasons. The food is then cooked (where appropriate) or otherwise prepared for eating, and samples of the ready-to-eat food are forwarded for analysis. The results reflect the intake of residues by consumers and may be compared with the Acceptable Daily Intake (ADI) to determine the relative hazard posed by the residues.

Governments, representing the interest of the public as consumers, have attempted to minimise any hazard from pesticide residues in one or other of two basic ways.

- By controlling the use of pesticides, legally or by means of advice, so that good agricultural practice is carefully followed. Such control, with cooperation of users, should ensure that residues in food do not exceed the acceptable maximum residue levels estimated from data from supervised trails.
- By the establishment and enforcement of legal maximum residue limits.

Residue levels on grain in storage do not, except in the case of immediate consumption, indicate in any way the amount of pesticide which may be consumed. Residues of most pesticides continue to degrade and information on the further disappearance on storage and transport enables an estimate to be made of the residue level in the commodity when it is normally offered for sale. These levels are usually appreciably lower than the maximum residue limit.

Residues are often reduced even further during food preparation, processing or cooking and a realistic *prediction* of consumer hazard is possible only when all these factors are taken into account. The only realistic way to *assess* consumer hazard is by carrying out actual intake studies.

When the legal limit is based on the maximum residue level and has been arrived at from the consideration of reliable data then a residue determined during enforcement to be in excess of the MRL can be regarded as a clear indication that (a) good agricultural practice has not been followed, (b) there has been a deliberate misuse or (c) there has been some accidental contamination of the food.

A residue in excess of the MRL does not in itself imply a health risk, although an enforcing authority could take appropriate action on the basis of a 'substandard' food produced as a result of one of the three indications above. A legal limit does not have any real effect unless it is enforceable and a clearly 'substandard' food ought to be rejected for trade or consumption.

The chance of a food produced by good agricultural practice being rejected in this way is very small since the recommended sampling method is aimed at determining the *average* pesticide residue content of a lot of goods.

This average would then be compared with the *maximum* residue limit and there should be an ample safety margin for the producer against a false rejection.

The real risk to a commodity lot lies in the situation where a country has based its legal maximum residue limits on either limited data or on average data from supervised trials or both. This will result in a falsely low legal MRL which can be exceeded by many samples especially if the samples are drawn from commodities not covered by the supervised trials.

Some food control activities are necessary, both for the direct protection of the consumer and in relation to the acceptability of commodities in trade. However, both commodity monitoring and dietary studies should be undertaken only after a careful study of the real need for such activities. These of course may be justifiable on the basis of administrative 'reassurance' of the consumer but it is difficult to justify massive monitoring programs for pesticides in food on the basis of current scientific evidence.

The scientific arguments for initiating or continuing monitoring programs are weak but there is a political and administrative need to continually reassure consumers that their food is not contaminated. The decision on how much reassurance can be afforded will vary from country to country but where analytical resources are at a premium a very close examination should be made of the real benefits of monitoring. The position of minimal scientific return from routine monitoring has probably been reached.

31.10 Detection and Determination of Residues

The development of complex, new, and sensitive (and expensive) electronic equipment has revolutionised analytical chemistry and has been largely responsible for the current insight into the question of residues. It has brought about a new era of analytical methodology much of which does not any longer depend upon chemical reaction but rather on the measurement of physical and electronic responses to a series of carefully standardised physical stimuli. The responses from purified extracts made from the sample are compared with those given by standard samples of known composition and quality, and the concentration is determined by comparing the magnitude of the separate responses.

Over the past ten years, methods for the detection and determination of minute traces of pesticide residues have become highly sophisticated, specific, and sensitive. It is now possible to measure very small amounts of many substances. The determination of 0.01 mg/kg lindane is considered quite straightforward and commonplace. Determination of 0.0001 mg/kg of lindane (1 gram of lindane in 10 000 tonnes of grain) is possible.

The operation of detection equipment requires skilled and highly trained operators, well equipped laboratories, and funds for the purchase and operation of expensive equipment. Some grain handling authorities and some commercial organisations have well equipped residue laboratories which are now considered essential for the maintenance of proper standards, for the protection of health, and to safeguard valuable markets.

Methods of residue analysis have been worked out in official and industrial laboratories and these methods have been examined by such international bodies as the Food and Agricultural Organization, the Association of Official Agricultural Chemists, and the International Union of Pure and Applied Chemistry. There is, as yet, no international agreement on methods of residue analysis — largely because residue analysis methodology is constantly changing, becoming more sensitive, more accurate, and more reproducible. The Codex Committee on Pesticide Residues, however, has recently issued a list of 'Recommended Methods' for determining a wide range of residues in many food commodities **(FAO 1983a)** and a Code of Good Analytical Practice **(Bates 1982).** These methods are referred to at the end of each chapter on specific insecticides (Chapters 6–27).

Modern methods and equipment have made it possible to carry out complex analyses on as little as a few grams of sample containing very small traces of complex substances. The speed with which these determinations can now be executed is such that it has been possible to carry out a substantial surveillance of food moving in commerce including food moving in international trade. As a result, there are extensive data on the level of residues in many commodities and it has become possible for administering authorities to take regulatory action as a result of their examination of a significant sample of the foods moving in commerce.

Generally, as little as 10–25 grams of grain is required to carry out a determination of the various residues which may be present. Enormous problems are encountered in obtaining a truly representative sample from a bulk of grain. Infinite care and effort are required to be sure that the sample drawn from any bulk is truly representative of the whole.

31.11 Maximum Residue Limits and How They are Established

In order to limit the contamination of food with chemical residues, it has been customary to fix administrative action levels to gauge whether chemicals have been used in accordance with registered directions and good agricultural practices. Governments of several countries established limits which they referred to as 'tolerances'. This was an unfortunate choice of terms because it conjures in the mind of most people, the idea of biochemical or toxicological tolerance, that is, a safe limit beyond which danger would ensue. However, the term means legal limit — literally the amount which is tolerated within the law. For these reasons, the word 'tolerance' is gradually being abandoned and preference is shown for the phrase 'maximum residue limit' (MRL).

Fundamentally, the maximum residue limit reflects the

maximum residue that could result when the chemical is used according to approved directions and the crop is harvested, the grain stored, or the cereal product processed as the case may be. Residues in excess of the maximum residue limit are tantamount to evidence that the chemical has been misused or 'good agricultual practice' has not been followed.

Maximum residue limits are established on the results of extensive supervised trails designed to determine the nature and level of residues resulting from the approved use of the chemical. These trials are conducted in a number of different regions or situations in order to determine the maximum concentration of residue likely to occur in or on the food. In addition to experiments carried out at the normal rate of application, it is usual to also conduct parallel experiments at double the approved rate and to sample the produce at varying intervals thereafter up to and beyond the normal date of harvest, storage, shipment, and processing, etc.

Such trials are the responsibility of the manufacturer of the chemical and normally the trials are conducted in a manner simulating the most extreme conditions likely to be encountered in commercial practice. Generally, such studies are supplemented by additional studies to show the effect of storage, processing, preparation, and cooking on the level and nature of residues reaching consumers. Further studies are carried out to determine the effect of plants and animals on the chemical and its conversion into metabolites. If the metabolites in plants and domestic animals are not identical in nature and similar in magnitude to those formed in laboratory animals used for toxicological studies, additional toxicological studies will be carried out on the metabolites themselves.

In order to gauge the safety of such residues to consumers, it is necessary to carry out extensive long-term feeding studies on laboratory animals. Such studies usually involve two distinct species for periods approaching their life-span, during which time a complete veterinary record is kept of each animal in the trial, and a complete histo-pathological study is carried out on all important organs of all animals which die, as well as those which are sacrificed at the end of the trials. In addition, studies of reproduction, teratology [(1)], mutagenesis [(2)], carcinogenesis [(3)] and other features appropriate to the chemical in question must be carried out and all data submitted to the authority.

From these studies, the level of intake which causes no discernible effect on the most susceptible species is ascertained and this is used to calculate the level of intake which could be considered safe for humans if consumed daily for a whole lifetime. A large safety factor (usually 100) is incorporated as an additional safeguard. This acceptable daily intake (ADI) is used to gauge the acceptability of the maximum residue limit needed to cover residues arising from use in 'good agricultural practice'. Some agricultural commodities will require higher limits than others. Some chemicals likewise require limits higher than others. The legal limit is, however, not an indication of the relative risk (or hazard) associated with a particular chemical.

On the basis of the evaluation of the data, a maximum residue limit is established. There is thus a large margin of safety built into the legal limit fixed for the residue in the specific raw agricultural commodity. The knowledge that only some of the food contains the residue, that only some of this fraction contains residues at levels approaching the limit, and that much or all of the residue is removed in preparation or processing for eating gives further reassurance for the safety of the consumer. The numerical value of all such residue limits is generally rather small.

Limits known as 'tolerances' are established in the USA by the Environmental Protection Agency, and in Canada by the Food and Drug Directorate. Similar limits are established in many countries. Each authority examines similar though not necessarily identical data and applies generally similar criteria in reaching its decisions. Although there may be minor differences in the numerical values and in the foods in which the residues may occur, both philosophy and practice in all countries are basically similar. Some variation in numerical values is sometimes necessitated by variations in the use pattern from one country to another, and efforts are being made to reach international agreement on residue limits to reduce the effect of such variations on international trade.

The basis for such international agreement is provided under the Food Programme of the United Nations by the recommendations of the Food and Agriculture Organization's Panel of Experts on Pesticide Residues and the World Health Committee of Experts on Pesticide Residues. Working in joint session (known as the Joint FAO/WHO Meeting of Experts on Pesticide Residues), these bodies examine all available scientific information on the properties, use, and residues of selected pesticides and evaluate their effects on laboratory animals and man. On the basis of this evaluation, recommendations on acceptable daily intake (ADI), MRLs, methods of analysis, metabolism, fate and effect of residues are published for the information and guidance of governments.

The recommendations become the basis for agreement between member governments of the Codex Committee on Pesticide Residues which provides a forum for member governments to discuss and agree upon maximum residue limits for pesticides in food commodities which can be

---

(1) [teratology] the study of the effect of chemicals on the structure and viability of developing embryos.

(2) [mutagenesis] the process of causing mutations, the process of heritable variations in plants and animals resulting from a new combination of genes and chromosomes.

(3) [carcinogenesis] the process by which cancer is induced.

adopted into national legislation and thus provide a means to protect public health and facilitate trade in needed foodstuffs.

The basic principle which is followed in arriving at the maximum residue limit is that it should reflect the residue resulting from good agricultural practice.

The functioning of the registration procedures is such that good agricultural practice becomes the registered and recommended usage of a chemical which is necessary for the control of pests, etc. under practical conditions. Recommendations made by manufacturers of pesticides, resellers, and government officials must be such that the maximum residue limits are not exceeded when the commodity goes first into commercial channels.

31.12 International Harmonisation

The complexity of the pesticide residue problem and its international implications were recognised by the Food and Agriculture Organization of the United Nations as early as 1959, when the FAO Panel of Experts on the Use of Pesticides in Agriculture made the recommendation that FAO, jointly with the World Health Organization, should study:

(a) the hazard to consumers arising from pesticides residues in and on food and feedstuffs;
(b) the establishment of principles governing the setting of pesticide maximum residue limits;
(c) the feasibility of preparing an international code for the toxicological and residue data required in achieving the safe use of a pesticide.

As a result of this recommendation, a joint meeting between the FAO Panel of Experts on the Use of Pesticides in Agriculture and the WHO Expert Committee on Pesticide Residues was held in 1961. The purpose of the meeting was to consider the establishment of maximum residue limits for pesticide residues in food from the aspect of consumer safety. The first regular Working Session of the FAO and WHO Expert Groups took place in 1963 and meetings have been held on an annual basis since 1965. These regular sessions have since become familiar as the Joint Meeting on Pesticide Residues (JMPR).

31.12.1 Joint FAO/WHO Meeting of Experts on Pesticide Residues (JMPR)

The JMPR consists of experts in their individual capacity (i.e. not representing governments), invited by the Directors-General of FAO/WHO. Their task is to establish the Acceptable Daily Intake (ADI) figures for individual pesticides on the basis of toxicological evidence, to recommend maximum residue limits for pesticide residues in food, and to recommend acceptable methods for chemical analysis to be used by food inspection authorities for regulatory purposes.

WHO assembles a group of experts with special competence in matters related to toxicology of pesticides, whilst FAO experts are chosen for their knowledge and experience in the use, fate and analysis of pesticides.

Firstly, the WHO part of the JMPR is responsible for proposals with respect to Acceptable Daily Intake (ADI) for each individual pesticide under consideration. The ADI of a chemical is defined as 'the daily intake, which during an entire lifetime, appears to be without appreciable risk on the basis of all the known facts at the time'. It is expressed in milligrams of the chemical per kilogram of body weight. It is therefore a purely toxicological concept.

Secondly, the FAO part of the JMPR is responsible for recommending maximum residue limits for each individual pesticide under consideration and on each separate food commodity or group of food commodities on which the pesticide is being used. These recommendations take into account the worldwide use pattern. A maximum residue limit is defined as 'the maximum concentration of a pesticide residue resulting from the use of a pesticide according to good agricultural practice directly or indirectly for the production and/or protection of the commodity for which the limit is recommended'. The maximum residue limit should be legally recognised. It is expressed in milligrams of the residue per kilogram of the commodity.

Thirdly, the FAO part of the JMPR makes recommendations for methods of chemical analysis, suitable for regulatory actions by those responsible for enforcement of maximum residue limits.

Fourth, the joint session of both FAO and WHO experts critically examines the compatibility of recommended maximum residue limits with ADI figures.

Maximum residue limits are based on, among other things, good agricultural practice. The concept of good agricultural practice in the use of pesticides is defined as 'the officially recommended or authorised usage of pesticides under practical conditions at any stage of production, storage, transport, distribution and processing of food and other agricultural commodities, bearing in mind the variations in requirements within and between regions and taking into account the maximum quantities necessary to achieve adequate control, the pesticides being applied in such a manner as to leave residues that are the smallest amounts practicable and that are toxicologically acceptable'. The definition implies that a maximum residue limit should be based on two main considerations. On the one hand, the limit should be low enough that the total amount of residues reaching the consumer does not exceed the ADI; on the other hand the limits should be high enough to give an adequate degree of pest control.

The JMPR depends on information and background data on toxicological, agricultural and chemical aspects provided by industry and member countries so that it can properly evaluate the pesticide under consideration.

31.12.2 The Codex Committee on Pesticide Residues

Parallel with the establishment of the JMPR another development took place — the establishment of the Codex

Alimentarius Commission. Based on initiatives taken by the Government of Austria in the early 1960s, the Codex Alimentarius Commission was established as part of the Joint FAO/WHO Food Standards Program and an initial meeting was held in Rome in 1963. The Codex Alimentarius Commission is charged with the establishment of food standards and it comprises a great number of committees dealing with standards for individual food groups and for more general subjects related to food. In order to make the Codex machinery operative, member countries were asked to take responsibility for the organisation and accommodation of regular sessions. Thus, The Netherlands was asked to take the responsibility for the two Codex committees on general subjects namely, the Codex Committee on Food Additives and the Codex Committee on Pesticide Residues (CCPR). The Codex committees consist of delegates from member countries in their capacity as government representatives, but sessions are also attended by observers from other international organisations and from the agrochemical industry.

The prime objective of the CCPR is to reach agreement on internationally acceptable maximum limits for pesticide residues in food commodities moving in international trade.

From the beginning of the work of the CCPR, it was stipulated that a close collaboration with the JMPR should be the basis on which a world-wide program of harmonisation of pesticide residue limits should be developed.

On completion of its evaluation the JMPR publishes a report and monographs setting out its evaluation of each pesticide and these are submitted to the CCPR for formal consideration at the government level. In dealing with these proposals, the CCPR follows the procedure laid down in the Procedural Manual of the Codex Alimentarious Commission. In theory, 11 steps are involved, but in practice some of these steps are combined. The Codex Step Procedure is illustrated in **Figure 31.3**.

Although the procedure is long, it has the advantage that member countries are given ample opportunity to comment on the proposals between and during the CCPR sessions, and this opportunity is given at several stages of the procedure. After each CCPR session progress is formally reported and submitted to the Codex Alimentarius Commission for approval. Thus, countries not present at the CCPR session but attending the meeting of the Codex Alimentarius Commission (comprising 117 member countries) also have an opportunity to comment. Proposals for maximum residue limits which have reached Step 9 of the procedure are published and are formally submitted to governments for acceptance.

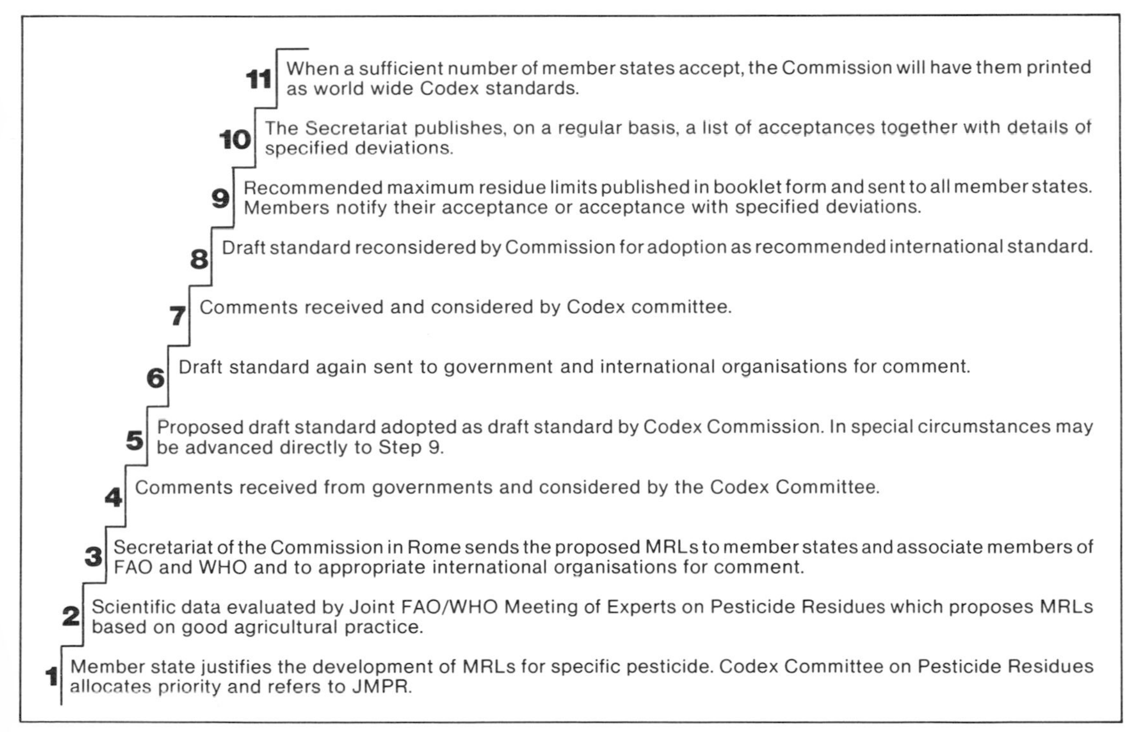

Figure 31.3. The eleven steps in the procedure for the elaboration of Codex Maximum Residue Limits (MRLs).

31.12.3 Acceptance of Codex International Maximum Limits for Pesticides Residues

The legal implications of the acceptance procedure pertaining to international food standards, including the obligation to incorporate in national legislation any such standards when accepted, for a long time hampered progress in the field of maximum residue limits for pesticides. Acceptance with minor or specified deviations, as provided for in the Codex Procedural Manual, was not applicable to a maximum residue limit, as this involved a single figure. It became increasingly clear that pesticide residues presented a special problem which required adjustment in the acceptance procedure. It was also recognised that the requirements for maximum residue limits were greatly dependent on regional, climatic and/or pest control conditions, and that it was hardly possible to cover all requirements in one single figure applicable world-wide, particularly when this was coupled with an obligation to adopt this figure in the legislation of individual countries. It was a fundamental step forward when the CCPR was able to agree on a modified acceptance procedure which provides, among other things, for limited acceptance. This implies that a country would not hinder the importation of food complying with the Codex maximum residue limit, and that it would not impose a Codex maximum residue limit which would be more stringent than it applied domestically. This new procedure has enabled member countries to accept CCPR proposals more readily.

The CCPR has recently initiated a review of legal problems inhibiting the acceptance of Codex maximum residue limits as a further step in the harmonisation procedure.

31.12.4 Factors Inhibiting Acceptance

During the years that I served as a delegate at the Codex Committee on Pesticide Residues I noticed the steady evolution of an organisation that serves not only as a forum for the exchange of views between governments but as a valuable piece of machinery for decision making. The democratic processes that are followed are slow and somewhat cumbersome but they do provide reassurance that the maximum residue limit, when adopted, is not only scientifically sound but also politically acceptable.

Unfortunately, many people, including myself, have been somewhat frustrated by the slowness with which the process has evolved and the apparent reluctance of many food importing (industrialised) countries to adopt the Codex MRLs into their legislation. When one makes allowance for the many complexities involved, 20 years is perhaps not such a long time after all. Let us look at some of the reasons which have delayed or slowed down the adoption of international limits for pesticide residues. These include:-

1. Failure by many people and national authorities to recognise the need to use chemicals to protect valuable food, ensure the availability of staple commodities as a buffer against famine, maintain economy and meet the demand for food to feed the increasing population.
2. Lack of knowledge about the limitations of available non-chemical measures to control pests.
3. Lack of understanding of the needs and agricultural practices of trading partners.
4. Lack of sympathy for those who live under tropical and semi-tropical conditions.
5. Belief that man-made chemicals are somehow different to chemicals that occur in nature.
6. Tradition that foods, particularly staples, should be 'pure' and that nothing should be deliberately added to food.
7. A political attitude opposed to the concept of residues.
8. The development of the 'natural food' cult and the attendant rackets in 'health foods'.
9. Political pressure by merchants, domestic producers and other self-interest groups to create misgivings in order to produce non-tariff barriers to trade.
10. The sensation-seeking news media.
11. Fear of the unknown. What cannot be seen could well be dangerous!
12. Inability to understand the significance of toxicology studies on laboratory animals, the dose-related effect and the concept of no-toxic-effect level.
13. Failure to understand and accept the concept of acceptable daily intake.
14. Mathematical calculations of intake of residues based on the assumption that every lot of each commodity contains residues and that residues always occur at the level of the MRL.
15. Laws that lay down rigid procedures for establishing MRLs in national legislation.
16. Existing MRLs that are lower than those being recommended for international acceptance.
17. Legislative procedures that make amendments difficult.

You might well ask what science has done to break down these barriers to the acceptance of residues of chemicals used for protecting world food supplies. I believe that science has produced adequate data to convince informed scientists of the safety and acceptability of these chemicals. Whether it has done sufficient to convince the sceptics and the non-scientific segment is open to question.

The majority of delegations that attend the annual session of the Codex Committee on Pesticide Residues come from countries which import food so they naturally take the consumers' point of view. In its simplest form this point of view is that they would prefer to have no residues in food. Unfortunately, many delegations are not familiar with the problems facing agriculturalists generally and food producers in the semi-tropics and tropics in particular. It is therefore understandable why they often appear unsympathetic to the needs of countries producing and exporting from other regions. However, in the process of exchanging comments at the CCPR, a better understanding has developed and in recent years there has been noticeable softening of attitudes towards the presence of residues.

Unfortunately, some of the officials in developing countries, who do not have access to comprehensive technical information and advice, are seriously disturbed by the alarmist publicity in the news media about the alleged danger of chemicals, particularly pesticides. Since they take their responsibilities seriously they are reluctant to accept the use of insecticides lest there should be adverse effects upon consumers, particularly in countries where staple foods, such as raw grain, are converted for consumption with a minimum of preparation and cooking.

We must therefore accept that the process of achieving an extensive set of international maximum residue limits will be slow, the more so because the resources available in FAO and WHO to provide the requisite amount of technical information and educational material are sorely limited. Even these are being whittled down by inflation and the escalating costs of the increasingly complex information which is being generated.

31.12.5 Harmonisation of Registration Requirements

The idea of achieving a high level of harmony between the requirements of different countries was often discussed privately but remained little more than a dream until 1975 when, at the FAO Ad Hoc Government Consultation on Pesticides in Agriculture and Public Health, it was proposed that the Director-General of FAO convene a consultation between government and industry to discuss the possibility of harmonising registration requirements for pesticides **(FAO 1975)**. Among the many resolutions made at the Consultation, this received the highest priority and FAO convened a further Consultation in October 1977. This Consultation was attended by almost 50 governments, and many international agencies and chemical industry. The level of agreement achieved and the spirit of co-operation, which was so evident, surprised everyone.

The report of the 1977 Consultation **(FAO 1977a),** of which 7500 copies were distributed, is a blueprint for the guidance of government and industry alike. Whilst drawing attention to all of those aspects and requirements, which could be harmonised, or even standardised, it drew attention to those issues where national, international and collaborative effort was required in order to develop standards, guidelines, test procedures, codes of practice and other information which could serve as a basis for harmonised requirements. Many governments, agencies, organisations and local committees responded to the challenge and most of the missing information was developed, co-ordinated and published in the next few years.

In order to consolidate the achievements of the 1977 Consultation, to draw attention to the subsequent developments and to seek a commitment from governments and industry, FAO convened a second Consultation on International Harmonisation of Pesticide Registration Requirements in Rome in October 1982. This was attended by over 60 governments, 9 international organisations and chemical industry under the aegis of the International Group of National Associations of Agrochemical Manufacturers (GIFAP). The initiative and level of agreement, once again, astounded even the most enthusiastic supporters. The report on the Consultation **(FAO 1983b)** is proof of what can be achieved when people of goodwill forget their political, economic and cultural differences and agree to work together in the interests of international understanding.

The objective of the Consultation was to agree upon test procedures, practices and presentation, which would adequately delineate the properties, effect and fate of biologically active chemicals in a manner which would adequately demonstrate the suitability, efficacy and safety of each compound under conditions of use representative of the practices that would be followed by farmers and other users. The Consultation accepted the concept that scientific data, which has been generated under standardised laboratory conditions by competent people using good test methods and well-defined procedures of 'good laboratory practice', should be transportable and acceptable anywhere in the world **(OECD 1981b)**.

Recognition that efficacy studies conducted in the field in accordance with internationally accepted guidelines can produce data supportive of the results of similar field studies carried out under different climatic, meteorological and agricultural conditions in some other part of the world has greatly reduced the cost of generating adequate data on efficacy.

31.12.6 Residues Trials Methodology

Variations in methodologies in conducting trials to determine residues (including the selection, preparation and analysis of samples) have created difficulties in evaluating the significance of residues on commodities during their production, storage, preparation for market and processing. These variations have also made it difficult to compare information from different sources and have contributed to differences in the MRLs adopted in different countries.

In response to an invitation from the Ad Hoc Government Consultation in 1977 **(FAO 1977b)**, the Codex Committee on Pesticide Residues (CCPR), through its Working Groups, has developed 'Guidelines on Residue Trials Methodology' and these have been published **(Department of Primary Industry 1981; FAO 1981, 1986)**. Proposals to harmonise procedures for reporting laboratory results and for developing data for foods of animal origin are also being published by FAO **(FAO 1986)**.

Further guidance on methods of sampling, the portion

of the agricultural commodity to be analysed, recommended methods of analysis and on good analytical practice in residue analysis has also been prepared by CCPR **(FAO 1979, 1982, 1983a, 1984, 1986)** and this has also been published by the International Union of Pure and Applied Chemistry (IUPAC) **(Bates 1982).**

31.13 Code of Conduct in the Distribution and Use of Pesticides

The action by FAO to develop, in conjunction with a number of UN agencies and other organisations, an International Code of Conduct on the Distribution and Use of Pesticides, has not occurred in isolation to many other events, some going back 25 years, all designed to benefit the international community and to serve to increase international confidence in the availability, regulation, marketing and use of pesticides for the improvement of agriculture, public health and personal comfort.

The Director-General of FAO, in addressing a meeting in 1981, suggested that such a code could help to overcome a number of difficulties associated with pesticides. The FAO Panel of Experts on Pesticide Specifications, Registration Requirements and Application Standards, at its meeting in 1982, agreed that the control of export and import of pesticides, and thereby their safe use, might be best dealt with through the adoption of a code of conduct and to that end prepared a working paper for the 1982 Inter-Governmental Consultation.

The formal decision to produce the Code was taken at that Consultation, which was the Second Government Consultation on the International Harmonisation of Pesticide Registration Requirements convened by FAO, in Rome, in October 1982, and which recommended that the Director-General, in consultation with appropriate UN organisations and bodies and international organisations, draft such a code. Because of its wide interests and responsibilities in the use of pesticides in agriculture, FAO has given high priority to the preparation of the Code.

A number of organisations and countries have expressed concern about the propriety of supplying pesticides to countries which do not have infrastructures to register pesticides or to ensure that these materials are used safely and effectively. There has also been concern over the possibility that residues of pesticides, not needed or not permitted to be used in some countries, are present in imported agricultural commodities produced in countries where such restrictions do not apply. While recognising that it is impossible to eliminate such incidents because of diverging pest control needs, it is essential that every effort should be made to apply pesticides only in accordance with good and recognised practices. It is therefore important for industrially developed countries to recognise the pest control needs of developing countries, particularly those situated in the tropics.

In the absence of an effective pesticide registration process and infrastructure for controlling the availability of pesticides, countries importing pesticides must depend heavily on the pesticide industry to promote the safe and proper distribution and use of pesticides.

The export to developing countries of pesticides which have been banned in one or more other countries or whose use has been severely restricted in some industrialised countries has been a subject of discussions on whether the exporting country can assume responsibility for the marketing and use of such products in the importing country. In this respect it is essential to note that when pesticides are banned it is generally for toxicological, environmental or political reasons. Valid and adequate toxicological reasons justifying banning a product are of concern, though not necessarily of equal importance, to most countries. Consequently, such products should not be exported or imported without careful consideration of the toxicological implications for those likely to be exposed.

Whilst a Code of Conduct may not solve all the problems, nevertheless it should go a long way towards defining and clarifying the responsibilities of the various parties involved in the development, distribution and use of pesticides, and should be of value in countries which do not yet have control procedures.

The aim of the Code is to establish standards of conduct for all those engaged in the regulation, production, distribution and use of pesticides of all types and for all purposes, in order to ensure that adverse effects on people and the environment are restricted to the maximum extent possible and that pesticides are used properly and effectively for the improvement of agricultural production and human, animal and plant health.

The code which was approved by the FAO Committee on Agriculture in March 1985, **(FAO 1985b)**, by the FAO Council in June 1985 and by the FAO Conference in November 1985, has been recommended for adoption by all member governments, non-government organisations and chemical industry. It is accompanied by a series of comprehensive guidelines on regulation and registration, evaluation of efficacy, labelling, packaging, disposal of containers and unwanted pesticides and control of hazards. It is anticipated that the Code and guidelines will go a long way towards promoting safe, efficient and effective use of pesticides.

31.14 Conclusion

The regulatory requirements for pesticides used in grain storage systems have become strict and demanding but it is accepted that they are not inconsistent with the responsibilities of manufacturers, governments, vendors and users. These requirements have been embodied in legislation in most countries and international efforts to harmonise the legislation and requirements have been outstandingly successful.

Whilst it is essential that information on effectiveness should be generated, or at least confirmed, under conditions typical of those encountered in practice in each country or region, recognition that scientific data generated by field trials carried out by qualified scientists in accordance with accepted guidelines should be accepted in support of applications for registration irrespective of where such studies were conducted, has reduced the cost and extent of such testing.

Procedures for evaluating the toxicological implications of such uses of pesticides have been accepted by national and international authorities as have the procedures for determining maximum residue limits in raw agricultural commodities and food. Guidance on such matters is available from meetings of experts convened by FAO and WHO which organisations also provide the forum for discussion and adoption of such limits into national legislation. This serves to provide assurance for the safety of consumers and to facilitate trade in essential foodstuffs.

There is a need to encourage and support these efforts in order that the full value of pesticides in contributing to the improvement in grain storage practices and in reducing loss and damage of valuable food stocks can be realised with minimum delay.

References

Abbott, D.C., Crisp, S., Tarrant, K.R. and Tatton, J. O'G. 1970. Pesticide residues in the total diet in England and Wales, 1966–1977. III. Organophosphorus residues in the total diet. Pestic. Sci. 1: 10–13.

Abdel-Aal, Y.A.I. and Hussein, M.H. 1983. Adult emergence and survival of the cowpea seed beetle exposed as eggs to the hormonoid Altosid. Z. Angew. Entomol. 95(1): 30–33.

Abdel-Kader, M.H.K. 1981. Malathion and fenitrothion in stored wheat: low temperature degradation, distribution, and quantification of metabolites. Diss. Abstr. Int. B 1982, 42(8).

Abdel-Kader, M.H.K. and Webster, G.R.B. 1980. Low temperature degradation and translocation of malathion and (the insecticide) fenitrothion in stored wheat. In: Cessna, A.J., Ed., Proc. 15th Ann. Workshop for Pesticide Residue Analysts (Western Canada). Regina, April-May 1980. pp. 143–153.

Abdel-Kader, M.H.K. and Webster, G.R.B. 1981. Analytical methods development and quantitative determination of malathion and fenitrothion and metabolites in stored wheat. In: Muir, D.C.G., Ed., Proc. 16th Ann. Workshop for Pesticide Residue Analysts (Western Canada) pp. 93–97.

Abdel-Kader, M.H. and Webster, G.R. 1982. Analysis of fenitrothion and metabolites in stored wheat. Int. J. Environ. Anal. Chem. 11(2): 153–65.

Abdel-Kader, M.H.K., Webster, G.R.B., Loschiavo, S.R. and Watters, F.L. 1980. Low temperature degradation of malathion in stored wheat. J. Econ. Entomol. 73: 654–657.

Abdel-Kader, M.H.K., Webster, G.R.B. and Loschiavo, S.R. 1982. Effects of storage temperatures on rate of degradation of fenitrothion in stored wheat. J. Econ. Entomol. 75(3): 422–424.

Abernathy, C.L. and Casida, J.E. 1973. Pyrethroid insecticides: esterase cleavage in relation to selective toxicity. Science 179: 1235–1236.

Abernathy, C.L., Ueda, K., Engel, J.L., Gaughan, L.C. and Casida, J.E. 1973. Substrate specificity and toxicological significance of pyrethroid hydrolyzing esterases in mouse liver microsomes. Pestic. Biochem. Physiol. 3: 300–311.

Aboul-Nasr, S., Salama, H.S., Ismail, I.J. and Salem, S.A. 1973. Ecological studies on insects infesting wheat grains in Egypt. Z. Angew. Entomol. 73(2): 203–212.

Acree, F., Shaffer, P. and Haller, H.L. 1936. Constituents of pyrethrum flowers, III. The pyrethrin content of fresh flowers. J. Econ. Entomol. 29: 601.

Acton, F.E. and Parouchais, C. 1966. Malathion levels in wheat and wheat products. Food Technol. Aust. 18: 77–81.

Adams, J.M. 1976a. A guide to the objective and reliable estimation of food losses in small scale farmer storage. Trop. Stored Prod. Inf. 32: 5–12.

Adams, J.M. 1976b. Weight loss caused by development of *Sitophilus zeamais* in maize. J. Stored Prod. Res. 12: 269–272.

Adams, J.M. 1977a. The evaluation of losses in maize stored on a selection of small farms in Zambia, with particular reference to methodology. Trop. Stored Prod. Inf. 33: 19–24.

Adams, J.M. 1977b. A review of the literature concerning losses in stored cereals and pulses published since 1964. Trop. Sci. 19: 1.

Adams, J.M. 1977c. Post-harvest losses of cereals and pulses. The results of a questionnaire survey, June 1976. Trop. Stored Prod. Inf. 34: 23–48.

Adams, J.M. 1977d. A bibliography on post-harvest losses in cereals and pulses with particular references to tropical and semi-tropical countries. Trop. Prod. Inst. Rep. G 110: 23 pp.

Adams, J.M. and Harman, G.W. 1977. The evaluation of losses in maize stored in a selection of small farms in Zambia with particular reference to the development of methodology. Trop. Prod. Inst. Rep. G 109: 160 pp.

Adams, J.M. et al. 1975. Farmers act to reduce losses. Trop. Stored Prod. Inf. 30: 6.

Adams, P.H. 1985. The fate of chlorpyrifos-methyl on single wheat kernels. Ministry of Agriculture, Slough Laboratory, Report. 20 pp.

Adesuyi, S.A. 1978. Comparative effectiveness of different doses of iodofenphos (Nuvanol N) in controlling insect infestation on maize in cribs. Nigerian Stored Prod. Res. Inst. Rep. 31–34.

Adesuyi, S.A. 1979. Selection of new insecticides for use in maize storage at the farmer's level in Nigeria. Proc. 2nd Int. Working Conf. Stored-Product Entomol. Ibadan, Nigeria, 10–16 September, 1978, 406–411.

Adesuyi, S.A. 1982. Field trials with permethrin dust for control of insect infestation on stored maize in southern Nigeria. J. Stored Prod. Res. 18(3): 125–130.

Adesuyi, S.A. and Cornes, M.A. 1978. Operation feed the nation. Storing your produce. Advisory Leaflet No. 1. Maize. Nigerian Stored Products Research Institute, Lagos.

Agarwal, H.C. and Pillai, M.K.K. 1967. Laboratory evaluation of certain organophosphorus insecticides against *Trogoderma granarium*. Indian J. Entomol. 29(4): 346.

Ahmad, M. 1973. Radiation disinfestation of wheat. Sci. Chron. (Karachi) 11: 258–261.

Ahmed, S.M., Gupta, M.R. and Bhavanagary, H.M. 1976. Stabilization of pyrethrins for prolonged residual toxicity. Part II. Development of new formulations. Pyrethrum Post 13(4): 119–121.

Ahmadu Bello University. 1963. Institute for Agricultural Research. Annual Report 1962/63, Samaru, Nigeria.

Akhtar, M.H. 1984. Metabolism of deltamethrin by cow and chicken liver enzyme preparations. J. Agric. Food Chem. 32: 258–262.

Akhtar, M.H., Hamilton, R.M.G. and Trenholm, H.L. 1985. Metabolism, distribution, and excretion of deltamethrin, by Leghorn hens. J. Agric. Food Chem. 33: 610–617.

Akintonwa, D.A.A. and Hutson, D.H. 1967. Metabolism of 2-chloro-1-(2,4,5-trichlorophenyl) vinyl dimethyl phosphate in the dog and rat. J. Agric. Food Chem. 15: 632.

Akram, M., Ahmad, S. and Forgash, A.J. 1978. Metabolism of etrimfos in bean and corn plants. J. Agric. Food Chem. 26: 925–31.

Albro, P.W. and Fishbein, L. 1970. Short-term effects of piperonyl butoxide on the deposition of dietary hydrocarbon in rat tissues. Life Sci., Part 2, 9(13): 729–739.

Alessandrini, M.E. 1965. Determination of the persistence and fate of various insecticides in or on wheat during storage, milling and during the baking or cooking of the products made from the treated wheat. Final Technical Report USDA Project No. E15-AMS-8(a), Grant No. FG-1t-124, Mar 1, 1963 through Feb. 28, 1965. Institute Superiore di Sanita, Rome, Italy.

Alessandrini, M.E., Leoni, V., Disimone, L., Imbroglini, G. and Angelelli, L. 1968. Andamento dii residui di malathion nei grain trattati in scala semiindustriale e nei relativi prodotti di molitura. Rass. Chim. No. 5: 201.

Al-Naji, L.K. 1978. Effectiveness of methyl phoxim and malathion on stored-grain insects and the fate of the insecticide residues on the properties of the grain products. Diss. Abstr. Int. B 1978 39(5): 2117.

Al-Naji, L.K. 1980a. Distribution of methyl phoxim residues on Iraqi hard wheat milling fractions and in bread during storage. Zanco, Ser. A 6(2): 43–9.

Al-Naji, L.M. 1980b. The effectiveness of methyl phoxim and malathion residues on four species of stored grain insects in stored corn. Zanco, Ser. A 6(3): 69–80.

Al-Naji, L.K. and Kadoum, A.M. 1979. Residues of methyl phoxim in wheat and milling products. J. Agric. Food Chem. 27: 583–584.

Al-Naji, L., Kadoum, A.M. and La Hue. 1977. Phoxim methyl (Bay 7660): a candidate grain protectant for wheat. J. Econ. Entomol. 70(1): 98–101.

Al-Saffar, Z.Y. and Al-Iraqi, R.A. 1981. Pirimiphos-methyl and malathion as grain protectants against some stored- product insects in Iraq. Anz. Schaedlingsk. 54(4): 54–55.

Al-Saffar, Z.Y. and Kansouh, A.S.H. 1979. The survey of stored grain insects in the bins of Mosul silo and the protection measurements. Mesopotamia J. Agric. 14(2): 131–150.

Ambika, B. and Abraham, C.C. 1982. Effect of juvenile hormone analogue methoprene (ZR-515) on development of eggs and larvae of *Corcyra cephalonica* Stainton (Lepidoptera: Pyralidae). Agric. Res. J. Kerala 20(1): 60–62.

Ambrose, A.M. and Robbins, D.J. 1951. Fed. Proc. 10: 276.

Ambrus, A., Lantos, J., Visi, E., Csatlos, I. and Sarvari, L. 1981a. General method for determination of pesticide residues in samples of plant origin, soil and water. I. Extraction and cleanup. J. Assoc. Off. Anal. Chem. 64: 733–742.

Ambrus, A., Hargatai, E., Karoly, G., Fulop, A. and Lantos, J. 1981b. General method for the determination of pesticide residues in samples of plant origin, soil and water. II. Thin layer chromatographic determination. J. Assoc. Off. Anal. Chem. 64: 743–748.

Ambrus, A., Visi, E., Zakar, F., Hargitai, E., Szabo, L. and Papa, A. 1981c. General method for determination of pesticide residues in samples of plant origin, soil and water. III. Gas chromatographic analysis, and confirmation. J. Assoc. Off. Anal. Chem. 64(3): 749–768.

American Chemical Society. 1978. A Chemical Perspective (2nd Ed.): pp 41 ff, 320 ff. American Chemical Society, Washington, D.C.

American Cyanamid Co. 1977. Toxicity data report on malathion (technical). Report No A77–153, 16 Sept. 1977.

Amos, T.G. and Williams, P. 1975. Insect growth regulators: Some effects of Altosid on *Tribolium castaneum*. Proc. 1st Int. Working Conf. Stored-Product Entomology, Savannah, Georgia, 7–11 October 1974. pp. 507–510.

Amos, T.G. and Williams, P. 1977. Insect growth regulators: some effects of methoprene and hydroprene on productivity of several stored grain insects. Aust. J. Zool. 25: 201–206.

Amos, T.G., Williams, P. and Semple, R.L. 1978. Sterilising activity of methoprene and hydroprene in *Tribolium castaneum*. Experientia 34: 469–470.

Amos, T.G., Williams, P. and Minett, W. 1979. Non-uniform application of grain protectants in commercial storages. Proc. 2nd Int. Working Conf. Stored-Product Entomol., Ibadan, Nigeria, 10–16 September 1978. pp. 344–349

Anderegg, B.N. and Madisen, L.J. 1983a. Effect of insecticide distribution and storage time on the degradation of [14C]malathion in stored wheat. J. Econ. Entomol. 76(5): 1009–1013.

Anderegg, B.N. and Madisen, L.J. 1983b. Effect of dockage on the degradation of [14C]malathion in stored wheat. J. Agric. Food Chem. 31(4): 700–4.

Anderegg, B.N. and Madisen, L.J. 1983c. Degradation of 14C (carbon isotope)-malathion in stored corn and wheat inoculated with *Aspergillus glaucus*. J. Econ. Entomol. 76(4): 733–736.

Anderson, J.T. 1938. Der Kornkafer (*Calandra granaria*) Biologic und Bekamfung. Monograph — Angew. Entomol. 13. P. Pavey, Berlin.

Andrawes, N.R., Chancey, E.L., Crabtree, R.J., Herret, R.A. and Weiden, M.H.J. 1972. Fate of naphthyl carbaryl in laying chickens. J. Agric. Food Chem. 20(3): 608–617.

Annis, P.C., Banks, H.J. and Sukardi 1984. Insect control in stacks of bagged rice using carbon dioxide treatment and an experimental PVC-membrane enclosure. CSIRO Aust. Div. Entomol. Tech. Pap. No. 22.

Anon. 1960. The determination of malathion residues in cereals and oilseeds. Report of the Malathion Panel. Analyst 85: 915–921.

Anon. 1979a. Effectiveness of liquid fenitrothion against pests in stored grain. Milling Feed and Fertiliser 162(12): 30–32.

Anon. 1979b. Etrimfos. Acute oral, subcutaneous and i.p. LD/50 in male and female mice. Sandoz Ltd Reports Agro. Dok. CBK 3909/79, 3910/79 and 3911/79. (Unpublished).

Anon. 1979c. Etrimfos. Acute oral, subcutaneous, i.p. and dermal LD/50 in male and female rats. Sandoz Ltd Reports Agro. Dok. CBK 3912/79, 3913/79, 3914/75 and 3915/79. (Unpublished).

Anon. 1979e. Disaster in dribbles. Ceres May-June 1979. FAO, Rome.

Anon. 1981. Report of discussions and recommendations. Proc. GASGA Seminar on the Appropriate Use of Pesticides for the Control of Stored Products Pests in Developing Countries, Slough, 17–20 February 1981. pp. 3–11, 63–66.

Anon. 1982. Tanzanian pest outbreak seen as major threat to stored produce. Ceres Nov.-Dec. 1982, p. 10. FAO, Rome.

Anon. 1984. Appropriate use of pesticides for the control of stored products pests in developing countries. Trop. Stored Prod. Inf. 47: 4–9.

Anon. 1985. Southern African Development Coordination Conference. Report SADCC Agriculture Towards 2000. FAO, Rome.

A.O.A.C. 1980a. Multiresidue methods for chlorinated and certain organophosphorus pesticides. Sections 29.001- 29.018. Official Methods of Analysis (14 Ed.) Association of Official Analytical Chemists. Washington, D.C.

A.O.A.C. 1980b. Carbamates "Holden" multiresidue method. Section 20.058–29.063. Official Methods of Analysis, (14th Ed.) Association Official Analytical Chemists. Washington, D.C.

A.O.A.C. 1980c. Section 29.082–29.090. Official Methods of Analysis (14th Ed.). Association of Official Analytical Chemists, Washington, D.C.

A.O.A.C. 1980d. Sections 29.161–29.164. Official Method of Analysis (13th Ed.). Association of Official Analytical Chemists, Washington, D.C.

Arbogast, R. 1984. Biological control of stored-product insects: status and prospects. In: Baur, F.J., Ed., Insect Pest Management for Food Storage and Processing. American Association of Cereal Chemists, St Paul, Minnesota. pp. 225–235.

Arcozzi, L. and Contessi, A. 1984. Synthetic pyrethroids against *Sitophilus granarius* L. in mini-silos. Camera di Commercio Industria Artigianato e Agricoltura. pp. 127–135.

Ardley, J.H. 1973. Further field evaluation of bioresmethrin as a potential replacement for malathion in grain protection. Wellcome Australia Ltd. Report TD/5/73/K1-36.

Ardley, J.H. 1975. Use and application of resmethrin and bioresmethrin as potential grain protectants. Report of Wellcome Australia Ltd., Concord, N.S.W. (Unpublished).

Ardley, J.H. 1976. Synergised bioresmethrin as a potential grain protectant. J. Stored Prod. Res. 12: 253–259.

Ardley, J.H. and Desmarchelier, J.M. 1975. Investigations into the use of resmethrin and bioresmethrin as potential grain protectants. Proc. 1st Int. Working Conf. Stored-Product Entomology, Savannah, Georgia, 7–11 October 1974. pp. 511–516.

Ardley, J.H. and Desmarchelier, J.M. 1978. Field trials of bioresmethrin and fenitrothion combinations as potential grain protectants. J. Stored Prod. Res. 14: 65–67.

Ardley, J.H. and Halls, G.R.H. 1979. Phenothrin: a potential grain protectant. "Preprints", ASC/CJS Chemical Congress, Honolulu, Hawaii, 1–6 April, 1979. pp. 38–54.

Ardley, J.H. and Sticka, R. 1977. The effectiveness of fenitrothion and malathion as grain protectants under bulk storage conditions in New South Wales, Australia. J. Stored Prod. Res. 13: 159–168.

Armstrong, M.T. 1961. Comparison between malathion spraying and lindane-whitewash mixture for controlling *Ephestia elutella* in warehouses. Ann. Appl. Biol. 49: 578–587.

Armstrong, M.T. and Hill, B.G. 1959. The use of lindane and whitewash mixtures to control infestations of *Ephestia elutella* in warehouses. J. Sci. Food Agric. 10: 79.

Arthur, G.N. 1981. A new sorptive dust insecticide. Proc. 1st Aust. Stored Grain Pest Control Conf., Melbourne, May 1981. pp. 7/33–35.

Arthur, V.W. and Casida, J.E. 1957. Metabolism and selectivity of dichlorvos and its acetyl and vinyl derivatives. J. Agric. Food Chem. 5(3): 186–191.

Ash, J.S. and Taylor, A. 1964. Trials to measure the effects of a normal mercury/*gamma*-BHC seed-dressing on breeding pheasants (interim report). Rep. Game Research Assoc. 1963.

Ash, J.S. and Taylor, A. 1965. Further trials on the effect of BHC seed-dressings on breeding pheasants. Rep. Game Research Assoc. 1964.

Ashman, F. 1975. Rice bran storage trial with pirimiphos-methyl. Trop. Prod. Inst. Rep. 1972–1975, 57.

van Asperen, K. and Oppenoorth, F.J. 1954. Metabolism of *gamma*-benzene hexachloride in the animal body. Nature 173: 1000.

Atallah, Y.H. 1981. Pitfalls in trace analysis of chlordane in environmental samples. Report dated 12 March, 1981, Velsicol Chemical Corporation, Chicago, Ill.

Atkins, W.G. and Greer, E.N. 1953. The storage of flour in jute bags treated with insecticide. J. Sci. Food Agric. 4: 155.

Attfield, J.G. and Webster, D.A. 1966. Dichlorvos. Chem. Ind. 12 Feb. 1966, 272.

Attia, F.I. 1976. Insecticide resistance in *Cadra cautella* in New South Wales, Australia. J. Econ. Entomol. 69(6): 773–4.

Attia, F.I. Shanahan, G.J. and Shipp, E. 1980. Synergism studies with organophosphorus resistant strains of the Indian meal moth (*Plodia interpunctella*). J. Econ. Entomol. 73(2): 184–185.

Ayerst, G. 1965. Determination of water activity of some hygroscopic food materials by a dew-point method. J. Sci. Food Agric. 16(2): 71–78.

Azab, A.K., el Nahal, K.M. and el Hafaway, M.A. 1971. Studies on the susceptibility of *Tenebroides mauritanicus* to certain fumigants, contact insecticides and inert materials. Bull. Entomol. Soc. Egypt Econ. Ser. 5: 181–196.

Babbitt, J.D. 1949. Observations on the adsorption of water vapour by wheat. Can. J. Res. 27F: 55–72.

Bailey, S.W. 1962. The effects of percussion on insect pests of grain. J. Econ. Entomol. 55: 301–4.

Bailey, S.W. 1969. The effects of physical stress on the grain weevil, *Sitophilus granarius*. J. Stored Prod. Res. 5: 311–324.

Bailey, S.W. 1979. The irradiation of grain : an Australian viewpoint. In: Evans, D.E., Ed., Australian Contributions to the Symposium on the Protection of Grain against Insect Damage During Storage, Moscow 1978. CSIRO Division of Entomology, Canberra. pp. 136–138.

Bailey, S.W. and Banks, H.J. 1974. Use of controlled atmospheres for the storage of grain. Proc. 1st Int. Working Conf. Stored-Product Entomol., Savannah, Georgia, 7–11 October 1974. pp. 362–374.

Bailey, S.W. and Banks, H.J. 1980. A review of recent studies of the effects of controlled atmospheres on stored product pests. In: Shejbal, J. Ed., Controlled Atmosphere Storage of Grains. Elsevier, Amsterdam. pp. 101–118.

Bakal, C. 1963. The mathematics of hunger. Saturday Review, 27 April 1963, 18.

Baker, G.J. 1963. The "dual synergist system" of piperonyl butoxide and MGK 264. Pyrethrum Post 7(1): 16–18.

Baker, P.S. 1968. Effectiveness of malathion against four species of mite that inhabit stored grain. J. Econ. Entomol. 51: 944–946.

Baker, P.G. and Bottomley, P. 1982. Determination of synthetic pyrethroids in fruit and vegetables by gas-liquid and HPLC chromatography. Analyst 107: 206–212.

Baker, J.E. and Lum, P.T.M. 1976. Comparative effects of dietary methoprene on symbiotic and aposymbiotic rice weevils and asymbiotic granary weevils. J. Ga. Entomol. Soc. 11: 213–216.

Baker, V.H., Taboada, O. and Wiant, D.E. 1953. Lethal effects of electrons on insects infesting wheat and flour. Part I. Agric. Eng. St. Joseph, Michigan 34: 755–758.

Baker, V.H., Taboada, O. and Wiant, D.E. 1954. Lethal effects of electrons on insects infesting wheat, flour and beans. Part II. Agric. Eng. St. Joseph, Michigan 35: 407–410.

Bakke, J.E. and Price, C.E. 1975. Metabolism of chlorpyrifos-methyl in sheep and rats. Personal communication for H.J. Dishburger, Dow Chemical, USA. (Unpublished).

Bandyopadhyay, B. and Ghosh, M.R. 1984. Efficacy of etrimfos to control *Sitophilus oryzae* infesting wheat in storage. Pestology 8(1): 5–6.

Bang, Y.H. and Floyd, E.H. 1962. Effectiveness of malathion in protecting stored polished rice from damage by several species of stored grain insects. J. Econ. Entomol. 55: 188.

Banham, E.J. and Crook, L.J. 1966. Susceptibility of the confused flour beetle and the rust-red flour beetle to gamma radiation. In: Cornwell, P.B., Ed., The Entomology of Radiation Disinfestation of Grain. Pergamon Press, Oxford. Paper VIII.

Banks, H.J. 1972. A review of insecticide residues in storage cereals in Australia. Proc. 22nd Ann. Conf.

RACI Cereal Chem. Div., September 1972, Melbourne. pp. 187–196.

Banks, H.J. 1976. Physical control of insects. J. Aust. Entomol. Soc. 15: 89–100.

Banks, H.J. 1978. Recent advances in the use of modified atmospheres for stored-product pest control. Proc. 2nd Int. Working Conf. Stored-Product Entomol., Ibadan, Nigeria, 10–16 September 1978. pp. 198–217.

Banks, H.J. 1981a. Alternatives to contact insecticides. Proc. GASGA Seminar on the Appropriate Use of Pesticides for the Control of Stored-Products Pests in Developing Countries, Slough, 17–20 February 1981. pp. 195–200.

Banks, H.J. 1981b. Effects of controlled atmosphere storage on grain quality: a review. Food Technol. Aust. 33(7): 335–340.

Banks, H.J. 1984a. Modified atmosphere and hermetic storage — effects on insect pests and the commodity. In: Champ, B.R., and Highley, E., Ed., Proc. Aust. Dev. Asst. Course on Preservation of Stored Cereals. CSIRO Division of Entmology, Canberra. pp. 521–532.

Banks, H.J. 1984b. Modified atmospheres — testing of storage structures for gastightness. In: Champ, B.R., and Highley, E., Ed., Proc. Aust. Dev. Asst. Course on Preservation of Stored Cereals. CSIRO Division of Entomology, Canberra. pp. 533–543.

Banks, H.J. 1984c. Modified atmospheres — generation using externally supplied gases. In: Champ, B.R., and Highley, E., Ed., Proc. Aust. Dev. Asst. Course on preservation of Stored Cereals. CSIRO Division of Entomology, Canberra. pp. 544–557.

Banks, H.J. 1984d. Modified atmospheres — hermetic storage. In: Champ, B.R., and Highley, E., Ed., Proc. Aust. Dev. Asst. Course on Preservation of Stored Cereals. CSIRO Division of Entomology, Canberra. pp. 558–573.

Banks, H.J. and Annis, P.C. 1977. Suggested procedures for controlled atmosphere storage of dry grain. CSIRO Aust. Div. Entomol. Tech. Pap. No. 13. 23pp.

Banks, H.J. and Desmarchelier, J.M. 1978. New chemical approaches to pest control in stored grain. Chem. Aust. 45(6): 276–281.

Banks, H.J., Annis, P.C., Henning, R.C. and Wilson, A.D. 1980. Experimental and commercial modified atmosphere treatments of stored grain in Australia. In: Shejbal, J., Ed., Controlled Atmosphere Storage of Grains. Elsevier, Amsterdam. pp. 207–224.

Bansch, R., Holtmann, H. and Knoll 1974. Biological evaluation of bromophos for the control of storage pests. Proc. 1st Int. Working Conf. Stored-Products Entomol., Savannah, Georgia, 7–11 October 1974. pp. 533–541.

Bansode, P.C., Campbell, W.V. and Nelson, L.A. 1981. Toxicity of 4 organophosphorus insecticides to a malathion resistant strain of the Indian meal moth in North Carolina USA. J. Econ. Entomol. 74(4): 382–384.

Bare, C.O. 1942. Some natural enemies of stored-tobacco insects, with biological notes. J. Econ. Entomol. 35: 185–9.

Barlow, F. 1947. Determination of BHC in the blood of cattle. Nature 160: 719.

Barnes, J.M. and Verschoyle, R.D. 1974. Toxicity of new pyrethroid insecticide. Nature 248: 711.

Baron, R.C., Sphon, J.A., Chen, J.T., Lustig, E., Doherty, J.D., Hansen, E.A. and Kolbye, S.M. 1969. Confirmatory isolation and identification of a metabolite of carbaryl in urine and milk. J. Agric. Food Chem. 17: 883–887.

Barrett, J.R., Deay, H.O. and Hartstock, J.G. 1971. Striped and spotted cucumber beetle response to electric light traps. J. Econ. Entomol. 64: 413–416.

Barson, G. 1983. The effects of temperature and humidity on the toxicity of three organophosphorus insecticides to adult *Oryzaephilus surinamensis* (L.). Pestic. Sci. 14(2): 145–152.

Bartel, W.J. 1973. Toxicity of pyrethrum and its constituents to mammals. In: Casida, J.E., Ed., Pyrethrum, the Natural Insecticide. Academic Press, New York. pp. 123–142.

Bates, A.N. and Rowlands, D.G. 1964. Interference by extractives in the determination of malathion residues in rice bran. Analyst 89: 286–287.

Bates, A.N. and Rowlands, D.G. 1964a. The recovery of malathion from a range of stored products. Analyst 1964, 89(1057): 288.

Bates, A.N., Rowlands, D.G. and Harris, A.H. 1962. The removal of plant extractives interfering in the determination of malathion residues in barley and rice bran. Analyst 87(1037): 643–649.

Bates, J.A.R. 1974. Analytical requirements of pesticide regulatory authorities. In: Environmental Quality and Safety (Suppl. Vol. III). Georg Thieme, Stuttgart. pp. 75–79.

Bates, J.A.R. 1982. Recommended approaches to the production and evaluation of data on pesticides residues in food. Pure Appl. Chem. 54(7): 1361–1450.

Batte, E.G., Moncol, D.J., Todd, A.C. and Isenstein, R.S. 1965. Critical evaluation of an anthelmintic for swine. Vet. Med. Small Anim. Clin. 60: 539.

Batte, E.G., Robinson, O.W. and Moncol, D.J. 1969. Effects of dichlorvos on sow reproduction. J. Anim. Sci. 28: 148.

Baur, F.J. 1984. Insect Management for Food Storage and Processing. American Association of Cereal Chemists, St Paul, Minnesota. 384 pp.

Baur, F.J. 1984a. Important behavioural aspects of selected insects. In: Baur, F.J., Ed., Insect Pest

Management for Food Storage and Processing. American Association of Cereal Chemists, St Paul, Minnesota.

Bazer, F.W., Robinson, O.W. and Ulberg, U.C. 1969. Effect of dichlorvos and PMS on reproduction in swine. J. Anim. Sci. 28: 145.

Beadles, M.L., Miller, J.A., Chamberlain, W.F., Eschle, J.L. and Harris, R.L. 1975. The horn fly: methoprene in drinking water of cattle for control. J. Econ. Entomol. 68: 781–785.

Beard, R.L. 1972. Lethal action of UV irradiation on insects. J. Econ. Entomol. 65(3): 650.

Beaument, J.W.L. 1961. The water relations of insect cuticle. Biol. Rev. 36: 281–320.

Beavers, J.B. and Fink, R. 1977. Acute oral LD/50 — Mallard duck — Technical Decis. Final report W1–77.06. 06/A, Wildlife International, USA Submitted to WHO by Roussel Uclaf. (Unpublished).

Beavers, J.B. and Fink, R. 1978. Acute oral LD/50 Bobwhite quail — Sumethrin. Report to Sumitomo Chemical Co. Ltd. from Wildlife International Ltd 29 December 1978 (Unpublished).

Becker, H.A. 1960. On the absorption of liquid water by the wheat kernel. Cereal Chem. 37: 309–323.

Beckley, V.A. 1948. Pyrethrum as a protectant against weevil attack. Pyrethrum Post 1: 3.

Belanger, A. and Hamilton, H.A. 1979. Determination of disulfoton and permethrin residues in an organic soil and their translocation into lettuce, onion and carrot. J. Envir. Sci. Health B14(2): 213–226.

Beliles, R.P., Makris, S.L. and Weir, R.J. 1978. Three- generation reproduction study in rats. Report from Litton Bionetics, Inc. under contract to Sumitomo Chemical Company Ltd. (Unpublished).

Benes, V. and Cerna, V. 1970. Contributions to the toxicological evaluation of fenitrothion and its residues. In: Halas, et al. Ed., Proc. Pesticide Symposium Inter-American Conferences on Toxicology and Occupational Medicine, Miami, Florida.

Bengston, M. 1976. Timed daily emission of dichlorvos for control of *Ephestia cautella* infesting stored wheat. J. Stored Prod. Res. 12: 157–164.

Bengston, M. 1979. Potential of pyrethroids as grain protectants. In: Evans, D.E., Ed., Australian Contributions to the Symposium on the Protection of Grain Against Insect Damage During Storage, Moscow, 1978 CSIRO Division of Entomology, Canberra. pp. 88–98.

Bengston, M. and Desmarchelier, J.M. 1979. Biological efficacy of new grain protectants. In: Evans, D.E., Ed., Australian Contributions to the Symposium on the Protection of Grain Against Insect Damage During Storage, Moscow, 1978 CSIRO Division of Entomology, Canberra. pp. 81–87.

Bengston, M., Cooper, L.M. and Grant-Taylor, F.J. 1975. A comparison of bioresmethrin, chlorpyrifos-methyl and pirimiphos-methyl as grain protectants against malathion resistant insects in wheat. Queensl. J. Agric. Anim. Sci. 32: 51–78.

Bengston, M., Connell, M., Crook, I., Desmarchelier, J., Hart, R., Phillips, M., Snelson, J. and Sticka, R. 1977. Field trials to compare chlorpyrifos-methyl, fenitrothion, pirimiphos-methyl, malathion and methacrifos for control of malathion-resistant insects infesting wheat in Australia. Submitted to J. Stored Prod. Res. and later published in Queensl. J. Agric. Anim. Sci. (1986).

Bengston, M., Cooper, L.M., Davies, R., Desmarchelier, J., Hart, R. and Phillips, M. 1978a. Grain protectants for the control of malathion-resistant insects in stored sorghum. (Was submitted for publication and eventually appeared in same form in 1983. See same authors 1983a).

Bengston, M., Davies, R.A.H., Desmarchelier, J.M., Phillips, M. and Simpson, B.J. 1978b. Additional grain protectants for the control of malathion-resistant insects in stored sorghum. Report of Australian Working Party on Grain Protectants.

Bengston, M., Davies, D., Desmarchelier, J.M., Elder, W.B., Henning, R., Minett, W., Murray, W., Ridley, E.G., Ripp, B.E., Sieraikowski, C.J., Sticka, R., Snelson, J.T., Thomas, D., Wallbank, B.E. and Wilson, A. 1979a. Final Report on Pilot Usage Trials, 1978–79 Working Party on Grain Protectants. Department of Primary Industries, Queensland, Australia. (Unpublished).

Bengston, M., Davies, D., Desmarchelier, J.M., Elder, W.B., Henning, R., Minett, W., Murray, W., Ridley, E.G., Ripp, B.E., Sieraikowski, C.J., Sticka, R., Snelson, J.T., Thomas, D., Wallbank, B.E. and Wilson, A. 1979b. Final Report on Silo Scale Experiments 1977–78 Working Party on Grain Protectants. Department of Primary Industries, Queensland, Australia. (Unpublished).

Bengston, M., Davies, R.A.H., Desmarchelier, J.M., Phillips, M. and Simpson, B.J. 1979c. Additional Grain Protectants for the Control of Malathion-Resistant Insects in Stored Sorghum. Department of Primary Industries, Entomology Branch, Queensland, Australia. (Unpublished).

Bengston, M., Connell, M., Davies, R., Desmarchelier, J., Elder, B., Hart, R., Phillips, M., Ridley, E., Ripp, E., Snelson, J. and Sticka, R. 1980a. Chlorpyrifos-methyl plus bioresmethrin; methacrifos; pirimiphos-methyl plus bioresmethrin; and synergised bioresmethrin as grain protectants for wheat. Pestic. Sci. 11: 61–76.

Bengston, M., Connell, M., Davies, R., Desmarchelier, J., Phillips, M., Snelson, J. and Sticka, R. 1980b. Fenitrothion plus phenothrin and pirimiphos-methyl plus carbaryl, as grain protectant combinations for wheat. Pestic. Sci. 11: 471–482.

Bengston, M., Cooper, L.M., Davies, R.A., Desmarchelier, J.M., Hart, R.J. and Phillips, M.P. 1983a. Grain protectants for the control of malathion-resistant insects in stored sorghum. Pestic. Sci. 14, 385–398.

Bengston, M., Davies, R.A.H., Desmarchelier, J.M., Henning, R., Murray, W., Simpson, B.W., Snelson, J.T., Sticka, R. and Wallbank, B.E. 1983b. Organophosphorothioates and synergised synthetic pyrethroids as gain protectants on bulk wheat. Pestic. Sci. 14(4): 373–384.

Bengston, M., Davies, R.A.H., Desmarchelier, J.M., Phillips, M.P. and Simpson, B.W. 1984a. Organophosphorus and synergised synthetic pyrethroid insecticides as grain protectants for stored sorghum. Pestic. Sci. 15: 500–508.

Bengston, M. et al. 1984b. Final report on silo-scale experiments 1982–83. Report of Australian Wheat Board Working Party on Grain Protectants, Queensland Department of Primary Industries, Brisbane, 30 May 1984. (To be published.)

Bengston, M. et al. 1986a. Final report on commercial scale experiments 1984–85. Report of Australian Wheat Board Working Party on Grain Protectants, Queensland Department of Primary Industries, Brisbane, 15 May 1986. (To be published.)

Bengston, M. et al. 1986b. Final report on silo-scale experiments with methoprene 1984–85 and preliminary report on pilot usage trials with methoprene plus fenitrothion, 1985–86. Report of Australian Wheat Board Working Party on Grain Protectants, Queensland Department of Primary Industries, Brisbane, 15 May 1986. (To be published.)

Berck, B. 1980. Surface residues and insecticidal effectiveness of permethrin and cypermethrin (NRDC 143 and 149) as bin sprays and grain protectants (against *Tribolium castaneum, Cryptolestes ferrugineus*). In: Cessna, A.J., Ed., Proc. 15th Ann. Workshop for Pesticide Residue Analysts (Western Canada), Regina, April-May 1980. pp. 77–86.

Bergot, B.J. 1972. Metabolism of ^{14}C-Altosid by the dwarf lima bean plant. Report Zoecon Corporation, Palo Alto, California. (Unpublished).

Berndt, W.L. 1963. Synergism in the repellent action of combinations of piperonyl butoxide and allethrin. (Ph.D. Thesis, Kansas St. Univ., 1963, 72 pp.) (Diss. Abstr. 24(4), 1753–1754.)

Berry, M.R. Jr and Dickerson, R.W. Jr 1973. Moisture absorption isotherms for selected feeds and ingredients. Trans. Amer. Soc. Agric. Eng. 16(1): 137–139.

Bewick, D.W. and Leahey, J.P. 1976. Permethrin absorption in cows. Report No.TMJ 1357B. ICI Plant Protection Division. (Unpublished).

Bewick, D.W. and Leahey, J.P. 1978. The analysis of the permethrin metabolite 3-(2,2-dichlorovinyl)-2-methylcyclopropane-1,2-dicarboxylic acid in the excreta of rats given a single oral does of ^{14}C-permethrin. Report from ICI Plant Protection Ltd. (Unpublished).

Bhatnagar-Thomas, P.L. 1972. Laboratory evaluation of a synthetic juvenile hormone analogue for the control of *Trogoderma granarium*. Indian J. Entomol. 34: 87–93.

Bhatnagar-Thomas, P.L. 1973. Control of insect pests of stored grain using juvenile hormone analogue. J. Econ. Entomol. 66: 277–278.

Bindra, O.S. 1974. Grain protection and pesticide residues tolerances in India. Proc. 1st Int. Working Conf. Stored-Product Entomology, Savannah, Georgia, 7–11 October 1974. pp. 499–503.

Bindra, O.S. and Sidhu, T.S. 1972. Dissipation of malathion residues on maize grain in relation to dosage, storage conditions and baking. J. Food Sci. Technol. 9: 29.

Bindra, O.S. and Udeaan, A.S. 1975. Determination of dosage of malathion grain protectant for treatment of wheat in September for rural storage in the Punjab. Indian J. Entomol. 35(4): 351–352.

Bioresearch 1982. Acute oral toxicity study in albino rats administered fenitrothion formulation with: (1) Cyclosol 63 and Triton X114; (2) Cyclosol 63 and Triton X100; (3) Triton X114; and (4) Triton X100. Project numbers 50326, 50325, 50281, 50280. Bioresearch Laboratories Ltd. Report to Forest Protection Ltd, New Brunswick, Canada.

Birch, L.C. 1945. The mortality of the immature stages of *Calandra oryzae* and *Rhyzopertha dominica* in wheat of different moisture contents. Aust. J. Exp. Biol. Med. Sci. 23: 141–45.

Bitran, E.A. 1974. Ensaios biologicos para avaliacao da acao residual do malathion e tetrachlorvinphos (Gardona), no controle do caruncho de cafe *Araecerus fasciculaus* (De Geer, 1775) (Coleoptera, Anthribidae). Biologico 40: 111–6.

Bitran, E.A. and Campos, T.B. 1978. Resultados preliminares na avaliacao da acao residual do piretroide Decis (RU-22974) no controle de pragas de graos armazenados. In: Congresso Latinoamericano de Entomologia, 3., and Congresso Brasileiro de Entomologia, 5., Ilheus, Itabuna, 1978. Resumos.

Bitran, E.A. and Oliveira, D.A. 1977. Phytosanitary treatment of stored coffee with malathion under conditions of high relative humidity. Biologico 43 (7/8): 157–166.

Bitran, E.A., Campos, T.B. and Oliveira, D.A. 1975. Ensaios de avaliacao da acao residual de dois piretroides sinergizados e do inseticidas experimental DOWCO-214 em cafe armazenado. In: Congresso Brasileiro de pesquisas cafeeiras, 3., Curitiba. Resumos. Rio de Janeiro, IBC/GERCA. pp. 46–8.

Bitran, E.A., Campos, T.B. and Oliveira, D.A. 1979a. Avaliacao da persistencia residual de inseticidas na protecas de milho e cafe durante o armazenamento. 1 — productos organoforados. Biologico 45(11/12): 255–262.

Bitran, E.A., Campos, T.B., Oliveira, D.A. and Araujo, J.B.M. 1979b. Experiment on protection of stored maize in farm bins by using malathion and pirimiphos-methyl, with or without fumigation. An. Soc. Entomol. Bras. 8 (1), pp. 29–38.

Bitran, E.A., Campos, T.B., Oliveira, D.A. and Araujo, J.M.B. 1980a. Evaluation of the action of experimental product CGA-20168 (methacrifos) on the protection of corn in farm storage. Biologico 46(5/6): 85–96.

Bitran, E.A., Campos, T.B. and Oliveira, D.A. 1980b. Evaluation of the residual persistence of insecticides on corn and coffee protection during storage — II — pyrethroids (in Portuguese). Biologico 46(3/4): 45–57.

Bitran, E.A., Campos, T.B., Oliveira, D.A. and Araujo, J.B.M. 1980c. Experimental evaluation of the effect of phosphine, malathion and tetrachlorvinphos on the preservation of maize ears with husks in farm bin conditions (infestation of *Sitophilus zeamais* and *Sitotroga cerealella*). Cienc. Cult. 32(2): 209–214.

Bitran, E.A., Campos, T.B., Oliveira, D.A. and Araujo, J.B.M. 1981. Experimental evaluation of the action of the pyrethroid decamethrin in the treatment and conservation of unhusked maize in cribs. An. Soc. Entomol. Bras. 10(1): 105–117.

Bitran, E.A., Campos, T.B., Oliveira, D.A. and Araujo, J.B.M. 1982a. Protection of farm-stored maize cobs after treatments based on malathion and dichlorvos, preceded or not by fumigation. Biologico 48(11): 281–287.

Bitran, E.A., Campos, T.B., Kastrup, L.F.C., Ishizaki, T. and Oliveira, D.A. 1982b. Evaluation of the efficiency of malathion/dichlorvos mixtures for the protection of sacked corn. Biologico 48(10): 239–49.

Bitran, E.A., Campos, T.B., Oliveira, D.A. and Chiba, Soyako. 1983d. Evaluation of residual effectiveness of the pyrethroid deltamethrin in stored grains. Biologico 49(9–10): 237–46.

Bitran, E.A., Campos, T.B., Oliveira, D.A. and Chiba, Soyako. 1983b. Evaluation of the residual action of some pyrethroid and organophosphorus insecticides in the control of *Sitophilus zeamais* Motschulsky, 1855 infestation on stored maize. Biologico 49(11–12): 265–73.

Black, B.J.G., Getty, J., Jameson, H.R. and Pirie, H. 1950. The effect of continuous ingestion by poultry of benzine hexachloride. Br. Vet. J. 106: 386.

Blackith, R.E. 1953. Pyrethrum synergists: a review. Pyrethrum Post 3(2): 20–29.

Blackman, D.G. 1969. Tests of newer insecticides. Pest Infest. Res. 1968: 46.

Blattmann, P. 1978. Degradation of CGA-20168 in stored wheat grains, including their processing. Unpublished report PR 42/78 from Ciba-Geigy Limited, Basle, Switzerland. 29 pp.

Blinn, R.C., Dorner, R.W. and Gunther, F.A. 1959. Comparative residual behaviour of pyrethrins and piperonyl butoxide on wheat. J. Econ. Entomol. 52: 703–4.

Block, S.S. 1948a. Residual toxicity tests on insecticidal protective coatings. Soap Sanit. Chem. 24: 155.

Block, S.S. 1948b. Insecticidal surface coatings Parts I and II. Soap Sanit. Chem. 24: 138, 151.

Blum, M.S. 1955. A study of pyrethrum and its activators. (Ph.D. Thesis, Univ. Illinois, 96 pp, Doc. Diss. No. 11,489.) (Diss. Abstr. *15*(5): 711).

Board on Science and Technology for International Development 1978. Post-Harvest Food Losses in Developing Countries. National Academy of Sciences, Washington, D.C. 206 pp.

Bock, M. and Fishbein, L. 1972. Inductive effect of piperonyl butoxide on microsomal demethylase activity. Sci. Total Environ. 1(2): 197.

Bodnaryk, R.P., Barker, P.S. and Kudryk, L. 1984. Interaction between synergists and permethrin in adults of the red flour beetle, *Tribolium castaneum*. Pestic. Sci. 25: 481–486.

Boehringer, C.H. 1965. Critical review on the use of bromophos for the treatment of cereals. C.H. Boehringer Sohn, Ingelheim, West Germany.

Boles, H.P. 1971. Ovipositional responses of the rice weevil, *Sitophilus oryzae* treated with synergised pyrethrins. J. Kans. Entomol. Soc. 44: 70–75.

Boles, H.P. 1974. Effect of sublethal doses of pyrethrins on the mating efficiency of the rice weevil, *Sitophilus oryzae*. J. Kans. Entomol. Soc. 47: 444–451.

Bond, E.J. 1974. Future needs and developments for control of stored product insects. Proc. 1st Int. Working Conf. Stored-Products Entomol., Savannah, Georgia, 7–11 October 1974. pp. 317–322.

Bond, E.J. 1975. Introduction to symposium on pesticides, toxicity and insect resistance. Proc. 1st Working Conf. Stored-Products Entomol., Savannah, Georgia, 7–11 October 1974. p. 236.

Bond, E.J. 1984. Manual of Fumigation for Insect Control. FAO, Rome.

Bond, E.J. and Buckland, C.T. 1979. Development of resistance to carbon dioxide in the granary weevil. J. Econ. Entomol. 72: 770–771.

Bond, H., Mauger, K. and De Feo, J.J. 1972. The oral toxicity of pyrethrum, alone and combined with synergisers and common drugs. Report, Department of Pharmacology and Toxicology, Uni. of Rhode Island. (Unpublished).

Bond, E.J., Monro, H.A.U., Dumas, T., Benazet, J. and Turtle, E.E. 1972a. Control of insects in empty cargo ships with dichlorvos. J. Stored Prod. Res. 8: 11–18.

Bond, H., Mauger, K. and De Feo, J.J. 1973a. Interaction with the toxicity of other chemicals to mammals. In: Casida, J.E., Ed., Pyrethrum, the Natural Insecticide. Academic Press, New York. pp. 177–194.

Bond, H., Mauger, K. and De Feo, J.J. 1973b. The oral toxicity of pyrethrum, alone and combined with synergists and common drugs, and pathological effects produced. Pyrethrum Post 12(2): 59–63.

Borah, B. and Mohan, B. 1981. Efficacy of aluminum phosphide and malathion formulations against *Sitotroga cerealella* Oliv. and *Sitophilus oryzae* Linn. infesting stored paddy seeds. J. Res. Assam Agric. Univ. 2(1): 74–79.

Boulanger, R.J., Boerner, W.M. and Hamid, M.A.K. 1971. Microwave and dielectric heating systems. Milling 153(2): 18–21, 24–28.

Bourne, J.R. and Arthur, B.W. 1967. Diazinon residues in the milk of dairy cows. J. Econ. Entomol. 60: 402–405.

Bourne, M.C. 1977. Post-harvest food losses — the neglected dimension in increasing the world food supply. Cornell International Agriculture Mimeograph 53. New York State College of Agriculture and Life Sciences, April 1977.

Bowers, W.S. 1969. Juvenile hormone: activity of aromatic terpenoid esters. Science 164: 323–5.

Bowers, W.S. 1971. Insect hormones and their derivatives as insecticides. Bull. WHO 44: 381–9.

Bowker, D.M. 1973. Pirimiphos-methyl: fate in stored wheat and rice grain in the laboratory. ICI Plant Protection Ltd Report No. AR2457 AR. (Unpublished).

Bowker, D.M., Griggs, B.F. and Harper, P. 1973. Pirimiphos- methyl excretion by a goat. ICI Plant Protection Ltd. Report No. AR2458B. (Unpublished).

Bowman, M.C. and Beroza, M. 1967. Temperature-programmed gas chromatography of 20 phosphorus containing insecticides on 4 different columns and its application to the analysis of milk and corn silage. J. Assoc. Off. Anal. Chem. 50: 1228–1236.

Bowman, M.C., Holder, C.I. and Rushing, L.G. 1978. Trace analysis of etrimfos and two degradation products in corn and alfalfa. J. Agric. Food Chem. 26: 35–42.

Boxall, R.A. and Gillett, R. 1982. Farmer level storage losses in eastern Nepal. Trop. Prod. Inst. Rep. G157. 29 pp.

Boxall, R.A., Tyler, P.S. and Prevett, P.F. 1978. Loss assessment methodology — the current situation. Proc. 2nd Int. Working Conf. Stored-Products Entomol., Ibadan, Nigeria, 10–16 September 1978. pp. 29–36.

Boxall, R.A., Greeley, M. and Tyagi, D.S. with Lipton, M. and Neelakanta, J. 1979. The prevention of farm level food grain losses in India. A social cost-benefit analysis. Trop. Stored Prod. Inf. 37: 11–17.

Boyd, E.M. 1969. Dietary protein and pesticides toxicity in male weanling rats. Bull. WHO, 40: 801–895.

Boyd, E.M. and Carsky, E. 1969. Kwashiokorigenic diet and diazinon toxicity. Acta Pharmacol. Toxicol. 27: 284–294.

Boyer, A.C. 1976. Metabolism of WL 43775 by rat liver enzymes. Report of Shell Development Co. (Unpublished).

Boyer, A.C. 1977a. Residues in rat tissues from rats fed ^{14}C-SD43775. Report of Shell Development Co. (Unpublished).

Boyer, A.C. 1977b. Identification of metabolites found in the faeces of rats fed ^{14}C-SD43775. Report of Shell Development Co. (Unpublished).

Boyer, A.C. 1977c. Identification of metabolites in the livers of rats fed ^{14}C-SD43775. Report of Shell Development Co. (Unpublished).

Bradbury, F.R. 1957. Absorption and metabolism of BHC in susceptible and resistant houseflies. J. Sci. Food Agric. 8: 90.

Bradbury, F.R. and Standen, H. 1959. Metabolism of benzene hexachloride by resistant houseflies (*Musca domestica*). Nature 183: 983.

Bradbury, F.R., Nield, P. and Newman, J.F. 1953. Amount of *gamma*-benzene hexachloride picked up by resistant houseflies bred in a medium containing benzene hexachloride. Nature 172: 1052.

Brader, L. 1979. Integrated pest control in the developing world. Ann. Rev. Entomol. 24: 225–254.

Branson, D.R. and Litchfield, N.H. 1971. Absorption, excretion and distribution of O,O-dimethyl-0–3,5–6-trichloro-2–6,^{14}-2-pyridylphosphorothiate (C^{14}-Dowco 214) in rats. The Dow Chemical Company, Midland, Michigan. (Unpublished report NBA-8).

Bratt, H. and Dudley, L.A. 1970. Pirimiphos-methyl: excretion by rats and dogs. Report from ICI Industrial Hygiene Research Laboratories. (Unpublished)

Bratt, H. and Jones, L.A. 1973. Pirimiphos-methyl: metabolism in rats and dogs. Report from ICI Industrial Hygiene Research Laboratories. (Unpublished).

Brattsten, L.B. and Metcalf, R.L. 1970. Synergistic ratio of carbaryl and piperonyl butoxide as an indicator of the distribution of multifunctional oxidase in the Insecta. J. Econ. Entomol. 63: 101–104.

Brattsten, L.B. and Metcalf, R.L. 1973. Synergism of carbaryl toxicity in natural insect populations. J. Econ. Entomol. 66(6): 1347–1348.

Braun, W.G. and Killeen, J.C. 1975. Acute oral toxicity in the rat: compound No.FMC33297. Report from Bio-Dynamics Inc. to F.M.C. Corporation. (Unpublished).

Breese, M.H. 1955. Hysteresis in the hygroscopic equilibrium of rough rice at 25°C. Cereal Chem. 32(6): 481–487.

Bressani, R., Medrano, J.F., Elias, L.G., Gomez-Brenes, R., Gonzalez, J.M., Navarrete, D. and Klein, R.E. 1982. Studies on the control of insects to preserve stored Opaque-2 maize, and effects on its nutritive value. Turrialba 32(1): 51–58.

Bressau, G. 1966. Residues of storage protection agents - especially malathion — in grains. Dtsch. Lebensu. Rundsch. 62: 390–395.

Bridges, R.G. 1958a. Fate of labelled insecticide residues in food products. VI. Retention of *gamma*-benzene hexachloride by wheat and cheese. J. Sci. Food Agric. 9: 431.

Bridges, R.G. and Trotman, C.N. 1957. Loss of *gamma*-benzene hexachloride from whole wheat during storage and milling. Pest Infest. Res. 1956, 42.

Bris, E.J. 1968. Some physiological responses of ruminants to 2,2-dichlorovinyl dimethyl phosphate. Diss. Abst. B. 29 (2): 421.

Bris, E.J., Dyer, I.A., Howes, A.D., Schooley, M.A. and Todd, A.C. 1968. Anthelmintic activity of 2,2-dichlorovinyl dimethyl phosphate in cattle. J. Am. Vet. Med. Assoc. 152: 175.

Brodeur, J. and DuBois, K.P. 1963. Comparison of acute toxicity of anticholinesterase insecticides to weanling and adult male rats. Proc. Soc. Exp. Med. 114(2): 509–511.

Bronisz, H, Bidzinski, Z. and Lenicka, J. 1962. Metabolism of *gamma*-hexachlorocyclohexane I. Urinary excretion of — hexachlorocyclohexane in rats. Med. Pr. 13: 449–458.

Brooke, J.P. 1958. Effect of pyrethrins and piperonyl butoxide against the cacao moth *Ephestia elutella*. Chem. Ind. Lond. 1958: 387.

Brooke, J.P. 1961. Protection of grain in storage. World Crops 13: 27–30.

Brooke, J.P. 1967. The effect of five methylene dioxyphenyl synergists on the stability of pyrethrins. Pyrethrum Post 9(1): 18–30.

Brower, J.H. 1974. Radiosensitivity of *Gnathocerus maxillons* Fla. Entomol. 57: 91–95.

Brower, J.H. 1975. Potential for genetic control of stored- product insect populations. Proc. 1st Int. Working Conf. Stored-Products Entomol., Savannah, Georgia, 7–11 October 1974. pp. 167–180.

Brower, J.H. 1980a. Inheritance of partial sterility in progeny of irradiated males of *Ephestia cautella* (Lepidoptera: Pyralidae) and its effect on theoretical population suppression. Can. Entomol. 112(2): 131–140.

Brower, J.H. 1980b. Reduction of *Ephestia cautella* population in simulated storages by the release of genetically incompatible males. J. Econ. Entomol. 73: 415–418.

Brower, J.H. and Tilton, E.W. 1972. Insect disinfestation of shelled pecans, almonds and walnuts by *gamma*-irradiation. J. Econ. Entomol. 65: 222–224.

Brown, G.A. and Davis, R. 1972. Sensitivity of grain mite eggs to *gamma* radiation as influenced by dose rate and treatment age. J. Econ. Entomol. 65(6): 1619–21.

Brown, G.A. and Davis, R. 1973. Sensitivity of red flour beetle to *gamma* radiation as influenced by dose treatment age and dose rate. J. Ga. Entomol. Soc. 8: 153–157.

Brown, N.C. 1970. A review of the toxicology of piperonyl butoxide. Rep. Cooper Tech. Bur. (Wellcome Foundation, Res. & Dev.) A28/52. 46 pp.

Brown, N.C. 1971. A review of the toxicology of piperonyl butoxide. Pyrethrum Post 11(2): 66–68.

Brown, N.C., Chadwick, P.R. and Wickham, J.C. 1967. The role of synergists in the formulation of insecticides. Int. Pest Control 9(6): 10–13.

Bruce, W.A. 1975. Effect of UV radiation on egg hatch of *Plodia interpunctella*. J. Stored Prod. Res. 11: 243.

Bruce, W.A. and Lum, P.T.M. 1976. UV radiation: effect on synthetic sex pheromone of the Indian meal moth. Fla. Entomol. 59(1): 53.

Bruce, W.A. and Lum, P.T.M. 1978. The effects of UV radiation on stored-product insects. Proc. 2nd Int. Working Conf. Stored-Products Entomol, Ibadan, Nigeria, 10–16 September 1978. pp. 271–277.

Bruce, R.B., Howard, J.W. and Elsea, J.R. 1955. Toxicity of 0,0-diethyl 0-(2-isopropyl-6-methyl-4-pyrimidinyl) phosphorothioate (diazinon). J. Agric. Food Chem. 3: 1017.

Bull, D.L. 1972. Metabolism of organophosphorus insecticides in animals and plants. Residue Rev. 43(1).

Bull, M.S. 1983a. The distribution of methacrifos and its dealkylated metabolite CGA-90953 in gluten and milling fractions following treatment of wheat with DAMFIN. Ciba-Geigy Australia Ltd. Technical Report No.83/12/975.

Bull, M.S. 1983b. The uptake of residues of methacrifos and its dealkylated metabolite CGA-90953 in the fat and tissues of beef cattle following consumption of wheat treated with methacrifos. Ciba-Geigy Australia Ltd. Technical Report No.83/12/976.

Bulla, L.A. and LaHue, D.W. 1975. Progress report - studies on the development of new grain protectant treatment — chlorpyrifos-methyl. Report to Dow Chemical Co., Michigan from USDA, ARS, Grain Marketing Research Centre, Manhattan, Kansas. 12 pp.

Bullock, D.J.W. 1973. Pirimiphos-methyl: residues in stored grain. ICI Plant Protection Ltd Report No. AR 2472. (Unpublished).

Bullock, D.J.W. 1974. Pirimiphos-methyl: residues in stored grain, bread, flour and milled products. ICI Plant Protection Ltd Report. (Unpublished).

Bullock, D.J.W. and May, M.S. 1976. Pirimiphos-methyl: residue transfer from durum wheat to semolina and pasta. ICI Plant Protection Ltd Report No. TMJ1345A. (Unpublished).

Bullock, D.J.W., Day, S.R. and Griggs, B.F. 1974a. Pirimiphos-methyl: metabolism and residue transfer into the milk and meat of a lactating cow. ICI Plant Protection Ltd. Report AR2552A. (Unpublished).

Bullock, D.J.W., Day, S., Hemingway, R.J. and Jekatheeswaren, T. 1974b. Pirimiphos-methyl residue transfer study with cows. ICI Plant Protection Ltd. Report AR2551A. (Unpublished).

Bullock, D.J.W., Harrison, P.J. and Day, S.R. 1976. Degradation of residues in flour during baking. ICI Plant Protection Ltd. Report No. AR2666A. (Unpublished).

Burges, H.D. and Burrell, N.J. 1964. Cooling bulk grain in the British climate to control storage insects and to improve keeping quality. J. Sci. Food Agric. 15: 32–50.

Burke, F. 1946. Crickets in bakeries. Their habits and some suggestions for their control. Food 15: 37.

Burke, J.A. and Corneliussen, P.E. 1975. Quality assurance in the Food and Drug Administration pesticide residue analytical laboratories. In: Environmental Quality and Safety (Suppl. Vol. III) George Thieme, Stuttgart. pp. 28–31.

Burkholder, W.E. 1979. Application of pheromones and behaviour-modifying techniques in detection and control of stored product insects. Proc. 2nd Int. Working. Conf. Stored-Products Entomol., Ibadan, Nigeria, 10–16 September 1978. pp. 56–65.

Burkholder, W.E. 1984. Use of pheromones and food attractants for monitoring and trapping stored-product insects. In: Baur, F.J., Ed., Insect Pest Management for Food Storage and Processing. American Association of Cereal Chemists, St Paul, Minnesota. pp 69–82.

Burrell, N.J. 1974. Chilling and aeration. In: Christensen, C.M., Ed., Storage of Cereal Grains and their Products. American Association of Cereal Chemists, St Paul, Minnesota. Vol. 5, pp. 420–478.

Burrell, N.J. 1967. Grain cooling studies — II — effect of aeration on infested grain bulks. J. Stored Prod. Res. 3(2): 145–154.

Busvine, J.R. 1971. A Critical Review of the Technique for Testing Insecticides. Commonwealth Agricultural Bureaux, Slough. 345 pp.

Calandra, J.C. 1972. Acute oral toxicity study with Altosid technical in albino rats. Report 1BT No.A2240 (Oct. 5, 1972) submitted by Zoecon Corporation. (Unpublished).

Calderon, M. 1981. The ecosystem approach for apprehending the extent of postharvest grain losses. Phytoparasitica 9(1): 157–167.

Calderon, M. and Desmarchelier, J.M. 1979. Susceptibility of *Tribolium castaneum* adults to some organophosphorus compounds after pre-exposure to carbon monoxide. J. Stored Prod. Res. 15: 33–36.

Calderon, M. and Gonen, M. 1971. Effect of *gamma* radiation on *Ephestia cautella* 1. Effects on adults. J. Stored Prod. Res. 7: 85–90.

Calderon, M. and Shaaya, E. 1961. Grain storage study — I - control of *Sitophilus oryzae* in stored seeds by admixed insecticidal dusts. Agric. Pub. Div. Israel Min. Agric. Pub., 34.

Caliboso, F.M. 1977. Studies on the losses of stored grains due to pest infestation. Mimeo Report to the Action Oriented Workshop, Alor Setar, Kedah, Malaysia.

Caliboso, F.M. 1982. Warehouse loss assessment. In: Rottger, U., and SEARCA Team, Ed., Proc. GASGA Seminar on Paddy Deterioration in the Humid Tropics, Bagiuo, Philippines, 11–18 October 1981. GTZ, Eschborn. pp. 81–85.

Caliboso, F.M., Sayaboc, P.D. and Amoranto, M.R. 1985. Pest problems and use of pesticides — the current situation in the Philippines. NAPHIRE News 3(3): 4–6.

Caliboso, F.M. and Teter, N.C. 1985. Warehouse stock inventory and loss assessment. In: Teter, N.C. et al., Ed., Maintaining Good Grain Quality: Proc. 6th Ann. Workshop on Grain Post-Harvest Technology (Southeast Asia Cooperative Post-Harvest Research and Development Programme), Bogor, Indonesia, 3–6 May 1983. pp. 181–189.

Calverley, D.J.B. 1981. The appropriate use of pesticides for the control of stored products in developing countries. Proc. GASGA Seminar on the Appropriate Use of Insecticides for the Control of Stored-Products Pests in Developing Countries, Slough, 17–20 February 1981. pp. 25–30.

Campbell, A. and Sinha, R.N. 1976. Damage of wheat by feeding of some stored product beetles. J. Econ. Entomol. 69(1): 11–13.

Carl, M. 1979. Internal laboratory quality control in the routine determination of chlorinated pesticide residues. In: Geissbuhler, H., Ed., Advances in Pesticide Science. Pergamon Press, Oxford. Part 3.

Carle, P.R. 1979. CRBA Roussel-Uclaf Procida, Marseille. La deltamethrin dans la lutte contres les insects des denrees stockees. Organisation Europeenne et Mediterranaenne pour la Protection des Plantes — Rabat 1979.

Carlson, S.D. and Ball, H.J. 1962. Mode of action and insecticidal value of diatomaceous earth as a grain protectant. J. Econ. Entomol. 55: 964–70.

Carmi, Y. 1975. The toxicity of pirimiphos-methyl to *Ephestia cautella*. Trop. Storage Abstr. 1975(1): 3.

Carmi, Y., Bulbul, O. and Kashanchi, J. 1984. Non-uniform distribution of insecticide in wheat. Progress report for the year 1982/83 of the Stored Products Division. Bet Dagan, Israel, Ministry of Agriculture VII(7): 35–44.

Carpenter, S.J. 1947. Control of stored food insects with benzene hexachloride. J. Econ. Entomol. 40: 136.

Carpenter, C.P., Weil, C.S., Pozzani, U.C. and Smyth, H.F. 1950. Comparative acute and subacute toxicities of allethrin and pyrethrins. Arch. Ind. Hyg. Occup. Med. 2: 420–432.

Carpenter, C.P., Weil, C.S., Palm, P.E., Woodside, M.W., Nair, J.H. and Smythe, H.F. Jr 1961. Mammalian toxicity of 1-naphthyl-N-methylcarbamate (Sevin Insecticide). J. Agric. Food Chem. 9: 30.

Carpy, S. and Klotzsche, C. 1975a. Etrimfos. 4-week feeding study in rats. Sandoz Ltd. Report Agro. Dok. CBK 754b/73 (18 March 1975). (Unpublished).

Carpy, S. and Klotzsche, C. 1975b. Etrimfos, 3-month feeding study in rats. Sandoz Ltd. Report Agro. Dok. CBK 1720/75 (18 March 1975) . (Unpublished).

Carpy, S. and Klotzsche, C. 1977. Etrimfos. 2-Year feeding study in dogs. Sandoz Ltd. Rep. Agro. Dok. CBK2568/77 (3 August 1977). (Unpublished).

Carshalton 1964. OMS 43. Summary. Mammalian toxicology. Report of the Toxicology Research Unit, Carshalton to Farbenfabriken Bayer AG. (Unpublished).

Carson, L.J. 1981. Modified Storrher method for determination of organophosphorus pesticides in non-fatty food total diet composites. J. Assoc. Off. Anal. Chem. 64: 714–719.

Carter, S.W. 1975. Laboratory evaluation of three novel insecticides inhibiting cuticle formation against some susceptible and resistant stored product beetles. J. Stored Prod. Res. 11: 187–193.

Carter, S.W., Chadwick, P.R. and Wickham, J.C. 1975. Comparative observations on the activity of pyrethroids against some susceptible and resistant stored products beetles. J. Stored Prod. Res. 11: 135–142.

Casida, J.E. 1970. Mixed-function oxidase involvement in the biochemistry of insecticide synergists. J. Agric. Food Chem. 18: 753–772.

Casida, J.E. 1973. Biochemistry of the pyrethrins. In: Casida, J.E., Ed., Pyrethrum, the Natural Insecticide. Academic Press, New York. pp. 101–120.

Casida, J.E., McBride, L. and Niedermeier, R.P. 1962. Metabolism of dichlorvos in relation to residues in milk and mammalian tissue. J. Agric. Food Chem. 10: 370.

Casida, J.E., Engel, J.L., Esaac, E.G., Kamienski, F.X. and Kuwatsuka, S. 1966. Methylene-^{14}C-dioxyphenyl compounds: metabolism in relation to their synergistic action. Science 153: 1130–1133.

Casida, J.E., Kimmel, E.C., Elliott, M. and Janes, N.F. 1971. Oxidative metabolism of pyrethrins in mammals. Nature 230: 326–327; Pyrethrum Post 11(2): 58–59, 71.

Casida, J.E., Gaughan, L.C. and Ruzo, L.O. 1979. Comparative metabolism of pyrethroids derived from 3-phenoxybenzyl and α-cyano-3-phenoxy-benzyl alcohols. In: Geissbuhler, H., Ed., Advances in Pesticide Science. Pergamon Press, Oxford. Part 2, pp. 182–189.

Cavaler, L.C. and Frigato, V. 1982. Several years of experimentation with pirimiphos-methyl for the protection of stored cereals. Camera di Commercio Industria Artigianato c Agricoltura 137–154.

Central Food Technological Research Institute, Mysore 1965. Rice in India: a technico-economic review (Abstract). Rice J. 69(2): 34–35.

Cerna, V. and Benes, V. 1977. Residues of pirimiphos-methyl in wheat, mill products and white bread. Report by Czechoslavakian Institute of Hygiene in Epidemiology, Prague — 30 June 1977.

Cerna, V., Benes, V. and Horak, E. 1978. Dynamics of pirimiphos-methyl residues in crops. II. Residues after application to stored grain. Cesk. Hyg. 23(8): 321–5.

Chadwick, P.R. 1961. A comparison of safroxan and piperonyl butoxide as pyrethrum synergists. Pyrethrum Post 6(2): 30–37.

Chadwick, P.R. 1962. Studies on the sub-lethal effects of pyrethrins on the grain weevil *Calandra oryzae* L. Pyrethrum Post 6: 20.

Chadwick, P.R. 1963a. A comparison of MGK 264 and piperonyl butoxide as pyrethrum synergists. Pyrethrum Post 7(1): 11–15, 48.

Chahal, B.S. and Ramzan, M. 1982. Relative efficacy of synthetic pyrethroids and some organophosphate insecticides against the larvae of khapra beetle (*Trogoderma granarium* Everts). J. Res. Punjab Agric. Univ. 19(2): 123–126.

Chamberlain, R.W. 1950. An investigation of the action of piperonyl butoxide with pyrethrum. Amer. J. Hyg. 52: 153–183.

Chamberlain, S.J. 1981. Etrimfos residues in rapeseed oil during laboratory scale refining. J. Stored Prod. Res. 17: 183–185.

Chamberlain, W.F., Hunt, L.W.M., Hopkins, D.E., Gingrich, A.R., Miller, J.A. and Gilbert, B.N. 1975a. Absorption, excretion and metabolism of methoprene by a guinea pig, a steer and a cow. J. Agric. Food Chem. 23(4): 736–742.

Chamberlain, W.F., Hunt, L.M., Hopkins, D.E., Miller, J.A., Gingrich, A.R. and Gilbert, B.N. 1975b. A balance study with ^{14}C-labelled methoprene in a dairy cow. Official Report from US Department of Agriculture, Kerrville (apparently incorporated in previous reference).

Champ, B.R. 1965. Lindane resistance in *Sitophilus oryzae* and *S. zeamais* in Queensland. J. Stored Prod. Res. 1, 9.

Champ, B.R. 1968. A test method for detecting insecticide resistance in *Sitophilus oryzae*. J. Stored Prod. Res. 4, 175.

Champ, B.R. 1976. World production of cereals and their movement in international trade — the infestation problem. International Training Programme in Grain Storage and Handling, Australian Development Assistance Bureau. pp. 3–14.

Champ, B.R. 1977. FAO global survey of pesticide susceptibility of stored grain pests. FAO Plant Prot. Bull. 25(2): 49–67.

Champ, B.R. 1984. Distribution, importance and economic significance of storage insect pests. In: Champ, B.R., and Highley, E., Ed., Proc. Aust. Dev. Asst. Course on Preservation of Stored Cereals. CSIRO Division of Entomology, Canberra. pp.28–40.

Champ, B.R. and Cribb, J.N. 1965. Lindane resistance in *Sitophilus oryzae* (L.) and *Sitophilus zeamais* Motsch. (Coleoptera, Curculionidae) in Queensland. J. Stored Prod. Res., 1: 9–24.

Champ, B.R. and Dyte, C.E. 1976. Report of the FAO Global Survey of Pesticide Susceptibility of Stored Grain Pests. FAO Plant Production and Protection Series No. 5. FAO, Rome. 297 pp.

Champ, B.R. and McCabe, J.B. 1984. Grain storage in earth covered bunkers. Proc. 3rd Int. Working Conf. Stored-Products Entomol., Manhattan, Kansas, October 1983. pp. 398–415.

Champ, B.R., Steele, R.W., Genn, B.G. and Elms, K.D. 1969. A comparison of malathion, diazinon, fenitrothion and dichlorvos for the control of *S. oryzae* and *R. dominica* in wheat. J. Stored Prod. Res. 5: 21–48.

Chang, S.C. and Kearns, C.W. 1964. Metabolism in vivo of ^{14}C-labelled pyrethrin II and cinerin I by houseflies with special reference to synergistic mechanisms. J. Econ. Entomol. 57: 397.

Chao Yung-Chang and Delong, D.N. 1953. The stability of lindane and pyreone-impregnated dusts and the relative toxicity to granary weevils. J. Econ. Entomol. 46(5): 908–950.

Chapman, R.A. and Harris, C.R. 1978. Extraction and liquid- solid chromatography cleanup procedures for the direct analysis of four pyrethroid insecticides in crops by gas- liquid chromatography. J. Chromatog. 166: 513–518.

Chawla, R.P. and Bindra, O.S. 1971. Residues of some promising grain protectants on stored wheat. In: Bindra, O.S., and Kalra, R.L., Ed., Proc. Symp. Progress and Problems in Pesticide Residue Analysis Punjab Agric. Univ., Ludhiana and Indian Counc. Agric. Res., New Delhi. pp. 68–74.

Chawla, R.P. and Bindra, O.S. 1973. Toxicity of some safe insecticides to stored grain pests and residues of potential grain protectants. Indian J. Plant Protection 1(1): 57–59.

Chawla, R.P. and Bindra, O.S. 1976. Laboratory screening of some safe insecticides as grain protectants. Pesticides India, 10(2): 29–31.

Chen, T-N. and Liang, C-J. 1956. Oral toxicity of lindane and its tolerance in poultry and mice. J. Agric. Assoc. China 15: 78.

Cherian, M.C. and Rao, P.R.N. 1945. Trials with DDT and 666 against pests of stored grain. Indian Farming 6: 572.

Cherkovskaya, A. Ya., Zakladnoi, G.A. and Abdullaev, M.M. 1980. Carbophos for the protection of flour from pests. Zashch. Rast. (Moscow) (12): 15–16.

Christensen, C.M. 1971. Mycotoxins. Crit. Rev. Environ. Control 2: 57–80.

Christensen, C.M. and Kaufmann, H.H. 1969. Grain Storage: the Role of Fungi in Quality Loss. University of Minnesota Press, Minneapolis. 153 pp.

Chung, D.S. and Pfost, H.B. 1967a. Absorption and desorption of water vapour by cereal grains and their products. Part I: Heat and free energy changes of adsorption and desorption. Trans. ASAE 10: 549–551.

Chung, D.S. and Pfost, H.B. 1967b. Absorption and desorption of water vapour by cereal grains and their products. Part II: Development of general isotherm equation. Trans. ASAE 10: 552–555.

Chuwit Sukprakarn 1984. Control of stored grain insect pests, in maize. In: Semple, R.L., and Frio, A.S., Ed., Health and Ecology in Grain Post-Harvest Technology: Proc. 7th Technical Seminar on Grain Post-Harvest Technology, Kuala Lumpur, Malaysia, 21–24 August 1984. Manila, ASEAN Crops Post-Harvest Programme. pp. 95–99.

Ciba-Geigy 1971. Nogos/Nuvan. Technical bulletin published by Ciba-Geigy, Basle, Switzerland. 227 pp.

Cichy, D. 1969. The influence of some ecological factors on the susceptibility of *Tribolium castaneum* to Pybuthrin. Ekol. Pol. Ser. A 17(9): 159–166. (Polish).

Ciegler, A. and Lillehoj, E.B. 1968. Mycotoxins. Adv. Appl. Microbiol. 10: 155 219.

Claborn, H.V. 1956. Insecticide residues in meat and milk. US Department of Agriculture ARS 33–25.

Claborn, H.V. 1963. Residues in body tissues of livestock treated with Sevin or Sevin in the diet. J. Agric. Food Chem. 11: 74–76.

Claborn, H.V., Bowers, J.W., Wells, R.W., Radeleff, R.D. and Nickerson, W.J. 1953. Meat contamination from pesticides. Agric. Chem. 8: 37

Clark, D.G. 1970. The toxicity of PP511 (pirimiphos-methyl). Report from ICI Industrial Hygiene Research Laboratories. (Unpublished).

Clark, D.G. 1978. Toxicology of WL43479 (permethrin): Acute toxicity of WL43479. Report of Shell Research Limited. (Unpublished).

Coaker, T.H. 1959. "In-sack" treatment of maize with insecticide against stored products pests in Uganda. East Afr. Agric. J. 24: 244.

Cochrane, W.P. and Whitney, W. 1979. The Canadian check sample program on pesticide residues analysis; reliability and performance. In: Geissbuhler, H. Ed., Advances in Pesticide Science. Pergamon Press, Oxford Part 3, pp. 664–667.

Cochrane, W.P., et al. 1979. J. Envir. Sci. Health B14: 197–212.

CODEX 1983. Codex recommendations for methods of residue analysis and guidelines on good analytical practice in pesticide residue analysis. Codex Committee on Pesticide Residues, 1983. Report Appendix II, 30 pp. FAO, Rome. (To be published as Section 8 of Codex Guide.)

CODEX Committee on Pesticide Residues 1979. A 79/24, App IV Annex 1. FAO, Rome.

Cogburn, R.R. 1967. Laboratory tests of five new insecticides as protectants for rough rice. J. Econ. Entomol. 60: 1286.

Cogburn, R.R. 1973. Insect resistance — stored rice studies. Rice J. 76(7): 72–73.

Cogburn, R.R. 1975. Stored rice insects research. Rice J. 78(7): 78.

Cogburn, R.R. 1976. Pirimiphos-methyl as a protectant for stored rough rice: small bin tests. J. Econ. Entomol. 69(3): 369–373.

Cogburn, R.R. 1977. Susceptibility of varieties of stored rough rice to losses caused by storage insects. J. Stored Prod. Res. 13: 29–34.

Cogburn, R.R. 1981. Comparison of malathion and three candidate protectants against insect pests of stored rice and advantages of encapsulation. Southwest Entomol. 6(1): 38–43.

Cogburn, R.R., Brower, J.H. and Tilton, E.W. 1971. Combination of *gamma* and infrared radiation for control of Angoumois grain moth in wheat. J. Econ. Entomol. 64: 923–925.

Cogburn, R.R., Tilton, E.W. and Brower, J.H. 1972. Bulk-grain *gamma* irradiation for the control of insects infesting wheat. J. Econ. Entomol. 65: 818–821.

Cogburn, R.R., Tilton, E.W. and Brower, J.H. 1973. Almond moth: *gamma* radiation effects on the life stages. J. Econ. Entomol. 66: 745–751.

Cogburn, R.R., Calderwood, D.L., Webb, B.D. and Marchetti, M.A. 1983. Protecting rough rice stored in metal farm bins from insect attack. J. Econ. Entomol. 76(6): 1377–1383.

Cohen, I.C., Norcup, J., Ruzicka, J.H.A. and Wheals, B.B. 1970. An electron-capture gas chromatographic method for the determination of some carbamate insecticides as 2,4-dinitrophenyl derivatives of their phenol moieties. J. Chromatog. 49: 215–221.

Coleman, D.A. and Fellow, H.C. 1925. Hygroscopic moisture of cereal grains and flaxseed exposed to atmospheres of different relative humidities. Cereal Chem. 12: 275–287.

Collett, J.N. and Harrison, D.L. 1968. Lindane residues on pastures and in the fat of sheep grazing pasture treated with lindane prills. N.Z. J. Agric. Res. 11: 589.

Collins, J.A., Schooley, M.A. and Singh, V.K. 1971. The effect of dietary dichlorvos on swine reproduction and viability of their offspring. Toxicol. Appl. Pharmacol. 19: 377.

Connell, P.J. and Johnston, B.H. 1981. Costs of alternative methods of grain insect control. Occasional Paper, Bureau of Agricultural Economics, Australia 61: 77pp.

Conney, A.H., Chang, R., Levin, W.M., Garbut, A., Munro-Faure, A.D., Peck, A.W. and Bye, A. 1972. Effects of piperonyl butoxide on drug metabolism in rodents and man. Arch. Environ. Health 24(2): 97–106.

Coper, H. et al. 1951. Pharmacology and toxicology of chlorinated cyclohexanes. Arch. Exp. Pathol. Pharmakol. 212: 463–471.

Coppock, J.B.M. 1952. Analytical and pharmacological problems arising from the use of organic chemicals as processing aids and hygiene aids in the food industry. J. Sci. Food Agric. 3: 115–122.

Cornwell, P.B. 1964. Insect Control. In: Jefferson, S., Ed., Massive Radiation Techniques. Newnes, London. Chapter 5.

Cornwell, P.B. 1966. The Entomology of Radiation Disinfestation of Grain. Pergamon Press, London.

Cornwell, P.B., Crook, L.J. and Bull, J.O. 1957. Lethal and sterilizing effects of *gamma* radiation on insects infesting cereal commodities. Nature 179: 670–672.

Cotton, R.T. 1950. Insect Pests of Stored Grain and Grain Products (Rev. Edition). Burgess Pub. Co., Minneapolis, Minnesota.

Cotton, R.T. and Frankenfeld, J.C. 1949. Silica aerogel for protecting stored seed or milled cereal products from insects. J. Econ. Entomol. 42: 553.

Cotton, R.T. and Good, N.E. 1937. Annotated list of the insects and mites associated with stored grain and cereal products, and of their arthropod parasites and predators. U.S. Dep. Agric. Misc. Publ. 258. 81 pp.

Cotton, R.T. and Wilbur, D.A. 1974. Insects. In: Christensen, E.M., Ed., Storage of Cereal Grains and their Products. American Association of Cereal Chemists, St Paul, Minnesota. pp. 193–231.

Cotton, R.T., Walkden, H.H, White, G.D. and Wilbur, D.A. 1960. Causes of outbreaks of stored grain insects. Kansas Agric. Exp. Stn. Bull. 416. 35 pp.

Coulon, J. 1963. Comparative efficiency of pyrethrins with synergist and of lindane on *Sitophilus granarius*. Phytiatr.-Phytopharm. 12(2): 67–9.

Coulon, J. 1968. Efficacite de quelques products insecticides pour le traitement des cereales. Pyrethrum Post 9(3): 45–53.

Coulon, J. 1972. Centre National Agronomic Research. Laboratory studies on the activity of some substances applied to grain contaminated by grain weevils *Sitophilus granarius*. CILDA Inf. Sheet No. 4.

Coulon, J. and Barres, P. 1968. Study of some treatment procedures (other than spraying) applicable to the clearance of insects from empty barges. Phytiatr.-Phytopharm. 17(3): 249.

Coulon, J. and Barres, P. 1975. Laboratory studies on the persistance of activity of some insecticide toxicants used against *Sitophilus granarius* for protecting stored grain. Phytiatr.-Phytopharm. 24: 245–54.

Coulon, J. and Barres, P. 1978. INRA laboratoire de phytopharmacie: resultats obtenus avec quelques pyrethrinoides appliques an charancon du ble. Bulletin CILDA No.9 (Avril 1978).

Coulon, J., Barres, P. and Poutier, J.-C. 1970. Results from the application of dichlorvos on grain contaminated with *Sitophilus* and with various species of insects. Phytiatr.-Phytopharm. *19*: 93–100.

Coulon, J., Bares, P. and Delorme, R. 1972. Comparing the activity of malathion, dichlorvos, bromophos, iodofenphos and pirimiphos-methyl on *Sitophilus granarius*. Phytiatr.-Phytopharm. 21: 15–26.

Coulon, J., Barres, P. and Delorme, R. 1972a. Activities of different formulations of dichlorvos applied to grain contaminated internally by *Sitophilus granarius*. Influence of the amount and of the homogeneity of the dispersion on emergences. Phytiatr.-Phytopharm. 21: 27. (CA 78, 120211).

Coulon, J., Barres, P. and Daurade, M.H. 1979a. Comparative activities of methacrifos and methyl chlorpyrifos used to control the grain weevils (*Sitophilus granarius*). Phytiatr.-Phytopharm. 28(3): 169–176.

Coulon, J., Barres, P. and Daurade, M.H. 1979b. Residues and residue activities of methacrifos and methyl chlorpyrifos applied on wheat stored in silos (Includes effects on *Sitophilus granarius*). Phytiatri.-Phytopharm. 28(3): 177–184.

Council of Europe 1973. Importation of cereals treated with pesticides. Coun. Eur. Resolution AP 73(3), Appendix. With Minist. Agric. Fish Fd., Tech. Inf. Circ. 1973 50: 1–12.

Council of Europe 1981. Pesticides (5th Edition). Council of Europe, Strasbourg.

Council of Europe 1984. Pesticides (6th Edition). Council of Europe, Strasbourg.

Craig, J.G.O. 1956. A comparative study of the toxicity of malathion and malathion plus piperonyl butoxide by topical application on the German cockroach, *Blattella germanica* and Madera cockroach, *Leucopheae maderae*. Masters Thesis, Kansas State College, Manhattan, Kansas.

Crauford-Benson, H.J. 1938. An improved method for testing liquid contact insecticides in the laboratory. Bull. Entomol. Res. 29: 41–56.

Cristobal, U.L. 1953. The preservation of stored grains. A new technique with the synthetic insecticides. Rev. Fac. Agron. Eva Peron Univ. 3: 29, 85. (Through Rev. Applied Entomol. A42: 344, 1954).

Cummings, J.G. Zee, K.T., Turner, V., Quinn, F. and Cook, R.E. 1966. Residues in eggs from low level feeding of five chlorinated hydrocarbon insecticides to hens. J. Assoc. Off. Anal. Chem., 49: 35–4.

Cummings, J.G., Eidelman, M., Turner, V., Reed, D., Zee, K.T. and Cook, R.E. 1967. Residues in poultry tissue from low level feeding of five chlorinated hydrocarbon insecticides to hens. J. Assoc. Off. Anal. Chem. 50: 418.

Curl, E.A. and Leahey, J.P. 1980. Pirimiphos-methyl: the radioactive residues found in the liver and kidney of a goat after administration of a multiple dose. ICI Plant Protection Division Report RJ0144B 9 July 1980. (Unpublished).

Cyprus Agricultural Research Institute 1972. Chemical control of stored grain insects 1972. Cyprus Agric. Res. Inst. Ann. Rep. 1972. p. 69.

Cyprus Agricultural Research Institute 1976. Chemical control of insects which attacked stored grain. Cyprus Agric. Res. Inst. Ann. Rep. 1975, 1976, p. 40.

Dahlen, J.D. and Haugen, A.D. 1954. Acute toxicity of certain insecticides to the bobwhite quail and morning dove. J. Wild. Manage. 18 : 477.

Dahm, P.A. 1957. The mode of action of insecticides exclusive of organic phosphorus compounds. Ann. Rev. Entomol. 2: 247–260.

Dale, W.E., Miles, J.W., and Weathers, A. 1973. Measurement of residues of dichlorvos absorbed by food exposed during disinsection of aircraft. J. Agric. Food Chem. 21 : 858–860.

Daniel, J.W., Swanson, S. and McLean, J. 1978. Phoxim - pharmacokinetics and biotransformation in the rat. Life Science Research U.K. Report 78/BAG5/194. (Unpublished).

Daniel, J.M., McLean, J. & Pringeur, M. 1978b. Phoxim - biotransformation in the rat. Life Science Research, U.K. Report. 78/BAG6/317. (Unpublished).

Davey, P.M. and Elcoate, S. 1965. Moisture content relative humidity equilibria of tropical stored produce (Part 1, Cereals; Part 2, Oilseeds). Trop. Stored Prod. Inf., 11: 439–467; 12: 495–512.

Davey, P.M. and Elcoate, S. 1966. Moisture content humidity equilibria of tropical stored produce (Part II). Trop. Stored Prod. Inf. 12: 495–512.

Davey, P.M. and Elcoate, S. 1967. Moisture content/humidity equilibria of tropical stored produce (Part III). Trop. Stored Prod. Inf. 13: 15–34.

Davidow, B. and Frawley, J.P. 1951. Tissue distribution, accumulation and elimination of the isomers of benzene hexachloride. Proc. Soc. Exp. Biol. Med. 76: 780.

Davies, J.C. 1959. A note on the control of bean pests in Uganda. East Afr. Agric. J. 24: 174.

Davies, J.C. 1960. Experiments in crib storage of maize in Uganda. E. Afr. Agr. J. 26: 71; through Rev. Appl. Entomol. A49: 569.

Davies, M. 1972. An account of the development of bioallethrin and bioresmethrin synergised by piperonyl butoxide for control of *Musca domestica* L. Proc. 3rd Br. Pest Control Conf., St Helier 1971, 35–41.

Davies, R.A.H. 1976. Carbaryl plus pirimiphos-methyl for stored grain protection. Special report for J.T. S-nelson, Pesticides Co-ordinator. ICI Australia Ltd, Rural Division, Melbourne, August.

Davies, R.A.H 1981. Permethrin: an alternative insecticide for control of *Rhyzopertha dominica*. Proc. 1st Aust. Stored Grain Pest Control Conf., Melbourne, May 1981. pp. 7/15–21.

Davies, R.A.H. and Desmarchelier, J.M. 1981. Combinations of pirimiphos-methyl and carbaryl for stored grain protection. Pestic. Sci. 12: 669–677.

Davis, J.A., Day, S.R., Hemmingway, R.J., Jegatheeswaren, T., and Bullock, D.J.M. 1976. Pirimiphos-methyl: residue transfer study in pigs. ICI Plant Protection Ltd Report No. AR 2665A. (Unpublished).

Davison, K.L. 1976. Carbon-14 distribution and elimination in chickens given methoprene-14C. Agric. Food Chem. 24(3) 641–643.

Deahl, H.Z. and Tucker, K.E.B. 1974. Residue studies on wheat stored in vertical concrete silos in Queensland and N.S.W. (Australia). Dow Chemical (Australia) Ltd. Altona, Victoria, Australia. 29 pp. (Unpublished report GHF-P-024).

Deahl, H.Z. and Tucker, K.E.B. 1975. Determination of chlorpyrifos-methyl residues in wheat grain stored in a vertical concrete silo in Queensland. Residue (Melbourne) 3(1): 14–34.

De Breve, M., Raboud, G., Sieber, J., Perdomo, J.A. and Valasquez, J.E. 1982. Projecto Post-Cosecha Informe Sobre los primeros resultados. Ministerio de Recursos Naturales; Co-operation Suiza Al Desarrollo, Tegucigalpa, Honduras.

Dedek, W. and Schwartz, H. 1969. Z. Naturforsch. B. 24: 744.

De Las Casas, E., 1984. FAO's programme on prevention of food losses. Proc. 3rd Int. Working Conf. Stored-Products Entomology, Manhattan, Kansas, October 1983. pp. 621–628.

De Lima, C.P.F. 1976. An ecological study of traditional on-farm maize storage in Kenya and the effects of a control action. Proc. 15th Int. Conr. Entomol., Washington, D.C.

De Lima, C.P.F. 1978a. A review of the use of physical storage procedures in East Africa: aspects for improvement and extension. Proc. 2nd Int. Working Conf. Stored-Products Entomol., Ibadan, Nigeria, 10–16 September 1978. pp. 237–243.

De Lima, C.P.F. 1978b. A study of the bionomics and control of *Sitophilus zeamais* (Motschulsky) and *Sitotroga cerealella* (Olivier), and associated fauna in stored maize, under laboratory and field conditions in Kenya. Thesis, London University, U.K. 1978. 240 pp.

De Lima, C.P.F. 1979a. Appropriate techniques for use in the assessment of country loss in stored produce in the tropics. Trop. Stored Prod. Inf. (38): 15–19.

De Lima, C.P.F. 1979b. The assessment of losses due to insects and rodents in maize stored for subsistence in Kenya. Trop. Stored Prod. Inf. (38): 21–26.

De Lima, C.P.F. 1982. Strengthening of the Food Conservation and Crop Storage Section (Ministry of Agriculture and Co-operatives). Project SWA/002/ PFL Final Report. FAO, Rome.

Delouche, J.C. 1975. Seed processing and storage. In: Proc. Int. Symp. on Role of Seed Science and Technology in Agricultural Development (FAO and Government of Austria), Vienna, 1–6 October 1973. pp. 108–124.

Department of Primary Industry 1981. Guidelines on residue trials. Document PB 412, Department of Primary Industry, Canberra.

Derbyshire, J.C. and Murphy, R.T. 1962. Diazinon residues in treated silage and milk of cows fed with powdered diazinon. J. Agric. Food Chem. 10(5): 384–386.

Dermott, T. and Evans, D.E. 1978. An evaluation of fluidized-bed heating as a means of disinfesting wheat. J. Stored Prod. Res. 14: 1–12.

Desmarchelier, J.M., 1975a. Results of studies to determine fate of bioresmethrin on stored grain. CSIRO Stored Grain Research Laboratory, Canberra, Australia. Submission to FAO for consideration by JMPR.

Desmarchelier, J.M. 1975b. The development of new grain protectants. Bulk Wheat 9:36–37.

Desmarchelier, J.M. 1976b. Field trials with fenitrothion and phenothrin and pirimiphos-methyl plus carbaryl. Part 2. Residue studies. Preliminary report of ongoing study. CSIRO Division of Entomology, Canberra.

Desmarchelier, J.M. 1976c. Residues from chlorphyrifos-methyl and fenitrothion in baked and milled wheat products (Manuscript). CSIRO Division of Entomology, Canberra, Australia.

Desmarchelier, J.M. 1976d. Five tables showing the effect that storage and processing has on residues of CGA 20168 and other insecticides on different grains. Unpublished report from CSIRO Division of Entomology, Canberra, Australia.

Desmarchelier, J.M. 1977a. A model of the breakdown of dichlorvos in grain. J. Exp. Agric. Anim. Husb. 17: 818–825.

Desmarchelier, J.M. 1977b. Selective treatments including combinations of pyrethroid and organophosphorus insecticides for control of stored product Coleoptera at two temperatures. J. Stored Prod. Res. 13: 129–137.

Desmarchelier, J.M. 1977c. Carbaryl data obtained since the JMPR submission by Snelson in 1976. Report to J.T. Snelson and Australian Wheat Board Working Party on Grain Protectants, October 1977.

Desmarchelier, J.M. 1978a. Loss of fenitrothion on grains in storage. Pestic. Sci. 9: 33–38.

Desmarchelier, J.M. 1978b. Application of protectants to bulk grain in Australia — some practical aspects. (A summary of results of a field survey, Oct. 1977-Jan. 1978.) CSIRO Division of Entomology, Canberra.

Desmarchelier, J.M. 1978c. Mathematical examination of availability to insects of aged insecticide deposits on wheat. J. Stored Prod. Res. 14: 213–222.

Desmarchelier, J.M. 1979a. Analysis of formulations and residues — some current considerations. Proc. 2nd Int. Working Conf. Stored-Products Entomol., Ibadan, Nigeria, 10–16 September 1978. pp 193–197.

Desmarchelier, J.M. 1979c. Transfer of grain protectant insecticide residues from grain to wheat gluten. Unpublished report, CSIRO Division of Entomology, Canberra, Australia.

Desmarchelier, J.M. 1979d. Loss of carbaryl on grains in storage. Unpublished report, CSIRO Division of Entomology, Canberra, Australia.

Desmarchelier, J.M. 1979e. The fate of phenothrin residues during processing of wheat, barley and oats. Unpublished report, CSIRO Division of Entomology, Canberra, Australia.

Desmarchelier, J.M. 1980. Loss of bioresmethrin, carbaryl and d-phenothrin on wheat during storage. J. Pesticide Sci. 5: 533–537.

Desmarchelier, J.M., 1980a. Comparative study of analytical methods for bioresmethrin, phenothrin, pyrethrum, carbaryl, fenitrothion, methacrifos, pirimiphos-methyl and dichlorvos on various grains. J. Pestic. Sci, 5; 521–532.

Desmarchelier, J.M. 1981a. The use of insecticide for the control of infestation of exported grain and pulses. Proc. GASGA Seminar on the Appropriate Use of Pesticides for the Control of Stored Product Pests in Developing Countries, Slough, 17–20 February 1981. pp.96–109.

Desmarchelier, J.M. 1983. Maximising benefit: risk ratios from insecticide. Proc. 3rd Int. Working Conf. Stored-Product Entomol., Manhattan, Kansas, 23–28 October 1983. pp. 172–182

Desmarchelier, J.M. 1985. Behaviour of pesticide residues on stored grain. In: Champ, B.R., and Highley, E., Ed., Pesticides and Humid Tropical Grain Storage Systems: Proc. Int. Seminar, Manila, Philippines, 27–30 May 1985. ACIAR Proceedings No. 14. pp. 151–156.

Desmarchelier, J.M. and Banks, H.J. 1984. Applying protectants to grain and improving their performance in use. Proc. Aust. Dev. Asst. Course on Preservation of Stored Cereals CSIRO Division of Entomology, Canberra. Vol. 2, pp. 824–832.

Desmarchelier, J.M. and Bengston, M. 1979a. The residual behaviour of chemicals in stored grain. Proc. 2nd Int. Working Conf. Stored-Products Entomol., Ibadan, Nigeria, 10–16 September 1978. pp. 138–151.

Desmarchelier, J.M. and Bengston, M. 1979b. Chemical residues of newer grain protectants. In: Evans, D.E., Ed., Australian Contributions to the Symposium on the Protection of Grain against Insect Damage During Storage, Moscow 1978. CSIRO Division of Entomology, Canberra. pp. 108–115.

Desmarchelier, J.M. and Elek, J. 1978. Sampling of bulk grain for pesticide residues. CSIRO Aust. Div. Entomol. Rep. No. 7. 9 pp.

Desmarchelier, J.M. and Hogan, J.P. 1978. Reduction of insecticide residues in grain dust by treatment with alkali. Aust. J. Exp. Agric. Anim. Husb. 18: 453–458.

Desmarchelier, J.M. and Marsden, P. 1979. Effect of baking on fenitrothion residues in wheat gluten. Unpublised report, CSIRO Stored Grain Research Laboratory, Division of Entomology, Canberra, Australia.

Desmarchelier, J.M. and Wilson, A.D. 1981. Distribution of fenitrothion applied to wheat by gravity feeding and spraying. Aust. J. Exp. Agric Anim. Husb. 21: 432–434.

Desmarchelier, J.M., Banks, H.J., Williams, P. and Minett, W. 1977a. Toxicity of dichlorvos vapour to insects in aerated and non-aerated wheat and comparison of the vapour action of dichlorvos and malathion. J. Stored Prod. Res. 13: 1–12.

Desmarchelier, J.M., Bengston, M., Connell, M., Minett, W., Moore, B., Phillips, M., Snelson, J., Sticka, R. and Tucker, K. 1977b. A collaborative study of residues on wheat of methacriphos, chlorpyrifos-methyl, fenitrothion, malathion and pirimiphos-methyl. Part I. Method Development. Pestic. Sci. 8: 473–483.

Desmarchelier, J.M., Bengston, M. and Sticka, R. 1979a. Stability and efficacy of pyrethrins on grain storage. Pyrethrum Post 15: 3–8.

Desmarchelier, J.M., Bengston, M., Evans, D.E., Heather, N.W. and Whyte, G. 1979. Combining temperature and moisture manipulation with the use of grain protectants. In: Evans, D.E., Ed., Australian Contributions to the Symposium on the Protection of Grain against Insect Damage During Storage, Moscow 1978. CSIRO Division of Entomology, Canberra. pp 61–73.

Desmarchelier, J.M., Goldring, M. and Horan, R. 1980a. Predicted and observed residues of bioresmethrin, carbaryl, fenitrothion, d-phenothrin, methacriphos and pirimiphos- methyl on rice and barley after storage, and losses of these insecticides during processing. J. Pestic. Sci. 5: 539–545.

Desmarchelier, J.M., Bengston, M., Connell, M., Minett, W., Moore, B., Phillips, M., Snelson, J., Sticka, R. and Tucker K. 1980b. A collaborative study of residues on wheat of methacrifos, chlorpyrifos-methyl, fenitrothion, malathion and pirimiphos-methyl. II. Rates of decay. CSIRO Aust. Div. Entomol. Rep. No. 20. 21 pp.

Desmarchelier, J.M., Bengston, M., Connell, M., Henning, R., Ridley, E., Ripp, E., Sierakowski, C., Sticka, R., Snelson, J. and Wilson, A. 1981b. Extensive pilot usage of the grain protectant combinations, fenitrothion plus bioresmethrin and pirimiphos-methyl plus bioresmethrin. J. Pestic. Sci. 12: 365–74.

Desmarchelier, J.M., Bengston, M. and Sticka, R. 1981c. Stability and efficacy of pyrethrins on grain in storage. Pyrethrum Post 15(1): 3–8.

Desmarchelier, J.M., Bengston, M., Connell, M., Minett, W., Moore, B., Phillips, M., Snelson, J., Sticka, R. and Tucker, K. 1981d. Field trials with chlorpyrifos-methyl and methacrifos for control of malathion-resistant insects infesting wheat in Australia. Unpublished report, CSIRO Division of Entomology, Canberra, Australia.

Desmarchelier, J., Bengston, M., Davies, R., Elder, B., Hart, R., Henning, R., Murray, W., Ridley, C., Ripp, E., Sierakowski, C., Sticka, R., Snelson, J., Wallbank, B. and Wilson, A. 1986. Pilot usage of the grain protectants chlorpyrifos-methyl plus bioresmethrin, fenitrothion plus d-phenothrin, methacrifos and pirimiphos-methyl plus carbaryl. (Submitted to J. Pestic. Sci., April 1986).

De Vos, R.H., Bosma, M.P.M.M. and Brouwer, A.E. 1974. Rapid analysis of dieldrin, lindane, PCNB and TCNB residues in lettuce by automated gas-liquid chromatography. J. Chromatog. 93: 91–98.

De Vries, D.H. 1976. Residues of SD43775 in milk from cows fed radiolabelled SD43775. Shell Development Co. Report. (Unpublished).

De Vries, D.H. and Georghiou, G.P. 1979. Influence of temperature on the toxicity of insecticides to susceptible and resistant horseflies. J. Econ. Entomol. 72: 48–50.

De Witt, J.B. and Menzie, C.H. 1972. Effect of Sevin on quail and pheasants. Cited in Toxicol. Appl. Pharmacol. 21: 401.

Dhari, K., Dass, N. and Dass, R. 1978. Persistent toxicity of some insecticides to the adults of pulse beetle, *Callosobruchus chinensis* Linn. Indian J. Entomol. 39(4): 361–4.

Dichter, D. 1976. The stealthy thief. Ceres (FAO Review) 9(4): 51–53, 55.

Dobie, P. 1975. The susceptibility of different types of maize to post-harvest infestation by *Sitophilus zeamais* and *Sitotroga cerealella* and the importance of this factor at the small-scale farm level. Proc. 1st Int. Working Conf. Stored-Products Entomol., Savannah, Georgia, 7–11 October 1974. pp. 98–118.

Dobie, P. 1984a. Biological methods for integrated control of insects and mites in tropical stored products. I: The use of resistant varieties. Trop. Stored Prod. Inf. 48: 4–8.

Dobie, P. 1984b. Biological methods for integrated control of insects and mites in tropical stored products VI: Integrated control: The role of biological methods. Trop. Stored Prod. Inf. 48: 37–45.

Donaldson, J.M and Stevenson, J.H. 1960. The stabilising effect of piperonyl butoxide on pyrethrum exposed to ultra-violet light. J. Sci. Food Agric. 11: 370–373.

Donat, H.J. 1979a. Determination of acute toxic dose of etrimfos to *Sitophilus granarius*. Report No. 721/79 (60, 979) 6.7.79 Sandoz Ltd Basle, Switzerland. (Unpublished).

Donat, H.J. 1979b. Etrimfos: Effect upon various stages of development of *Sitophilus granarius* in wheat grain. Report No. 744/79 (60, 982) 19.10.79, Sandoz Ltd Basle Switzerland. (Unpublished).

Donat, H.J. 1979c. Persistence of etrimfos on wheat under various climatic conditions. Report No. 750/79 (60, 989) 26.11.79 — Sandoz Ltd, Basle, Switzerland. (Unpublished).

Donat, H.J. 1979d. Persistent effect of dust formulations of etrimfos. Report No. 747/79 (60, 986) 5.11.79. Sandoz Ltd, Basle, Switzerland. (Unpublished).

Donninger, C., Hutson, D.H. and Pickering, B.A. 1967. The metabolism of chlorfenvinphos in the dog and rat. Biochem. J. 102: 133.

Dorough, H.W. 1967. Carbaryl-14C metabolism in a lactating cow. J. Agric. Food Chem. 15: 261–266.

Dorough, H.W. 1970. Metabolism of insecticidal methyl-carbamates in animals. J. Agric. Food Chem. 18: 1015.

Dorough, H.W., 1971. Paper presented at the Int. Symp. on Pesticide Terminal Residues, Tel Aviv, Israel, 17–19 February 1971.

Dorough, H.W. and Casida, J.E. 1964. Nature of certain carbamate metabolites of the insecticide Sevin. J. Agric. Food Chem. 12: 294.

Dove, W.E. 1947. Piperonyl butoxide, a new and safe insecticide for the household and field. Pests 15(9): 30.

Dove, W.E. 1947a. Piperonyl butoxide, a new and safe insecticide for household and field. Am. J. Trop. Med. 27(3): 339–345.

Dove, W.E. 1949. Progress of non-hazardous methods of insect control. (Paper read at UNESCO Int. Tech. Conf. on Protection of Nature, U.S.A., Aug.-Sept., 1949) UNESCO. Pyrethrum Post 1951, 2(2): 9–11.

Dove, W.E. 1951. Piperonyl butoxide and pyrethrins in the protection of grain and similar products from insect damage. (Paper read at 9th Int. Congr. Entomol., Amsterdam 1951) Rev. Appl. Entomol. 1953, 41(8): 234.

Dove, W.E. 1952. An evaluation of residual insect sprays for mill insect control. Milling Prod. 17(6): 20.

Dove, W.E. 1952a. Piperonyl butoxide and pyrethrins for the protection of grains and similar products from insect damage. Trans. 9th Int. Cong. Entomol., Amsterdam Vol. 1, pp. 875–879.

Dove, W.E. 1953. Piperonyl butoxide. Proc. 39th Mid-Year Mtg. Chem. Spec. Mfrs Assoc. May 1953, 184–105.

Dove, W.E. 1957. Piperonyl butoxide and pyrethins officially endorsed for protection of stored products in the U.S.A. World Crops 1958 10(2): 56.

Dove, W.E., 1958. Protection of stored grains with pyrethrins and piperonyl butoxide. Proc. 10th Int. Congr. Entomol., Montreal 1956. Vol. 4, pp. 65–71.

Dove, W.E. 1958a. Concerning official tolerances and clearances for pyrethrins and piperonyl butoxide in the U.S.A. Pyrethrum Post 4(4): 20–24.

Dove, W.E. 1959a. Powder and spray stored grain protectants. Pest Control 27(7): 32–33, 35, 38.

Dove, W.E. 1960. Piperonyl butoxide and pyrethrins officially endorsed for protection of stored products in the U.S.A. Proc. 4th Int. Congr. Crop Protect., Hamburg 1957. Vol. 2, pp. 1755–1759.

Dove, W.E. and MacAllister, L.C. 1947. Piperonyl butoxide for control of stored grain insects. Paper read at 59th Ann. Mtg Amer. Assoc. Econ. Entomologists, 1947.

Dove, W.E. and Schroeder, H.O. 1955. Protection of stored grain with sprays of pyrethrins-piperonyl butoxide emulsion. J. Agric. Food Chem. 3: 932–936.

Dow Chemical Pacific Ltd. 1981. Evaluation of Reldan insecticide (chlorpyrifos-methyl) as a protectant for stored grains. Pests of stored products. Proceedings of BIOTROP Symposium on Pests of Stored Products, Bogor, Indonesia, 24–26 April 1978. pp. 217–229.

Draeger, G. 1968. Gas-chromatographic method for determining dichlorvos residues in plants and milk. Pflanzenschutz-Nachr. (Engl. Ed.) 2: 373–380.

DuBois, K.P. and Puchala, F. 1960. The acute toxicity and anticholinesterase action of Bayer 41831 (fenitrothion). Report from the Department of Pharmacology, University of Chicago to Farbenfabriken Bayer AG. (Unpublished).

Duguet, J.S. 1985a. (Protection of maize in village granaries and in cribs in a hot climate with deltamethrin). Document published in France in the Proceedings of the First Symposium of CILDA No. 16/1985. pp. 83–92. (French).

Duguet, J.S., 1985b. (Contribution to the evaluation of the comparative effectiveness of deltamethrin on the 3 species: *Sitophilus granarius*, *S. oryzae* and *S. zeamais*). Report of Roussel Uclaf — Division Scientifique — Agrovet prepared for IV Congress on the Protection of Human Health and Crops in the Tropical Environment — Marseille.

Duguet, J.S. and Wu Gin Xin 1984. (Evaluation of the effectiveness of deltamethrin on *Callosobruchus chinensis* and *Callosobruchus maculatus*). Bulletin du CILDA No. 15: 15–27. (Reviewed in English in Int. Pest Contr.)

Du Monceau, D. 1753. Traite de la conservation des grains, Paris, Hippolyte-Louis Guerin et Louis-Francois Delatour. 294 pp. (quoted by Burrell, 1974.)

Du Monceau, D. 1765. Supplement au traite de la conservation des grains, Paris, Hippolyte-Louis Guerin et Louis-Francois Delatour. 144 pp. (quoted by Burrell, 1974.)

Dunachie, J.F. and Fletcher, W.W. 1966. Effects of some insecticides on the hatching rate of hens eggs. Nature (London) 212: 1062–1063.

Dyte, C.E. 1972. Resistance to synthetic juvenille hormone in a strain of the flour beetle, *Tribolium castaneum*. Nature (London) 238: 48–49.

Dyte, C.E. and Blackman, D.G. 1972. Laboratory evaluation of organophosphorus insecticides against susceptible and malathion-resistant strains of *Tribolium castaneum*. J. Stored Prod. Res. 8: 103.

Dyte, C.E. and Rowlands, D.O. 1970. The effects of some insecticide synergists on the potency and metabolism of bromophos and fenitrothion in *Tribolium castaneum*. J. Stored Prod. Res. 6: 1.

Earl, F.L., Melveger, B.E., Rainwall, J.E., Bierbower, G.W. and Curtis, J.M. 1971. Diazinon toxicity — comparative studies in dogs and miniature swine. Toxicol. Appl. Pharmacol. 18(2) 285–95.

Eastham, L.E.S. and McCully, S.B. 1943. The oviposition response in *Calandria granaria*. J. Exp. Biol. 20:35 42.

Ebeling, W. 1969. Use of mineral dusts for protection against insect pests with special reference to cereal grains. In: Majunder, S.K., and Vanugopal, J.S., Ed., Grain Sanitation. Pest Control Science, Mysore, India. p. 103.

Ebeling, W. 1971. Sorptive dusts for pest control. Ann. Rev. Entomol. 16: 123–158.

Ebeling, W. 1973. Dust desiccants: effect of prolonged exposure of films on insecticidal efficacy. J. Econ. Entomol. 66: 280–281.

Eddy, G.W., Cole, M.M. and Marulli, A.S. 1954. Tests of synergists with phosphorus compounds against the body louse increased the initial activity 10 times. Soap Chem. Spec. 30: 121–23, 143.

Edson, E.F. and Noakes, D.N. 1960. The comparative toxicity of six organophosphorus insecticides in the rat. Toxicol. Appl. Pharmacol. 2: 523–539.

Edwards, J.P. 1981. Alternative chemicals. Proc. GASGA Seminar on the Appropriate Use of Pesticides for the Control of Stored Products pests in Developing Countries, Slough, 17–20 February 1981. pp 190–194.

Edwards, M.J. and Iswaren, T.J. 1977. Permethrin: residue transfer and toxicology study with cows fed treated grass nuts. ICI Plant Protection Division Report No. TMJ1519/B. (Unpublished).

Edwards, J.P. and Menn, J.J. 1981. The use of juvenoids in insect pest management. In: Wengler, R., Ed., Chemie der Pflanzenschutz und Shadlingsbekampfungsmittel Vol. 6, Springer-Verlag, Berlin. pp. 185–214.

Edwards, J.P. and Rowlands, D.G. 1978. Metabolism of synthetic juvenile hormone (JHI) in two strains of the grain weevil *(Sitophilus granarius)*. Insect. Biochem. 8: 23–28.

Edwards, J.P. and Short, J.E. 1984. Evaluation of three compounds with insect juvenile hormone activity as grain protectants against insecticide-susceptible and resistant strains of *Sitophilus* spp. J. Stored Prod. Res. 20(1): 11–15.

Edwards, M.J. and Swaine, H. 1977. Permethrin: incorporation of permethrin in the diet of laying hens: residues in eggs and tissues. ICI Plant Protection Division Report No. TM1520/B. (Unpublished.)

Eichler, D. 1972. Bromophos and bromophos-ethyl residues. Residue Rev. 41: 65–112.

Eichler, D. 1974. Degradation of bromophos in stored wheat. Report to CELAMERCK, Ingelheim, Germany.

Eichler, D. and Knoll, H.A. 1974. Degradation of bromophos in stored wheat. Proc. 1st Int. Working Conf. Stored-Products Entomol., Savannah, Georgia, 7–11 October 1974. pp. 582–590.

Eichner, M. 1978. Rapid pesticide-residue control of plants and animal food, tobacco and tobacco-products by sweep-co-distillation analysis. Part 1. Z. Lebensm. Unters. Forsch. 167(4): 245–249.

Elder, W.B. 1972. New aeration techniques for the preservation of stored cereals. Proc. 22nd Annual Conf., The Royal Australian Chemical Institute, Cereal Chemistry Div., Bronte, N.S.W. pp. 205–212.

Elder, W.B. 1984a. Aeration with naturally occurring and refrigerated air — introduction to aeration. In Champ, B.R., and Highley, E., Ed., Proc. Aust. Dev. Asst. Course on Preservation of Stored Cereals. CSIRO Division of Entomology, Canberra. pp. 419–425.

Elder, W.B. 1984b. Control of aeration operation. In: Champ, B.R., and Highley, E., Ed., Proc. Aust. Dev. Asst. Course on Preservation of Stored Cereals. CSIRO Division of Entomology, Canberra. pp. 443–449.

Elder, W.B., Hunter, A.J. and Griffiths, H.J. 1975. Refrigeration of bulk wheat in a thermally insulated silo for control of insect infestation. Federal Conf. Aust. Inst. Refrig. Air Condit. Heat., Hobart, Australia.

Eldumiati, I.I. and Levengood, W.C. 1972. Summary of attractive responses in Lepidoptera to electromagnetic radiation and other stimuli. J. Econ. Entomol. 65: 291–293.

Elgar, K.E., Marlow, R.G. and Mathews, B.L. 1970. The determination of residues of dichlorvos in crops and tissues. Analyst 95: 875–878.

Elliott, N.T., Farnham, A.W., Janes, N.F., Needham, P.H. and Pearson, B.C. 1967a. Benzyl-3-furylmethyl chrysanthemate: a new potent insecticide. Nature (London) 213: 493–494.

Elliott, M. Farnham, A.W. Janes, N.F. Needham, P.H., Pearson, B.C. and Stevenson, J.A. 1967b. New synthetic insecticidal compounds related to pyrethrins. Proc. 4th Br. Insectic Fungic. Conf., Brighton, 1967. pp. 437–443.

Elliott, M, Needham, P.H. and Potter, C. 1969. Insecticidal activity of pyrethrins and related compounds — II. Relative toxicity of esters from optical and geometrical isomers of chrysanthemic, pyrethric and related acids and optical isomers of cinerolone and allethrolone. J. Sci. Food. Agric. 20: 561–565.

Elliott, M., Kimmel, E.C. and Casida, J.E. 1969a. ^{3}H-pyrethrin I and — pyrethrin II: preparation and use in metabolism studies. Pyrethrum Post 10(2): 3–8.

Elliott, M., Janes, N.F., Kimmel, E.C. and Casida, J.E. 1972a. Mammaliam metabolites of pyrethroids. In: Tahori, A.S., Ed., Insecticides. (Proc. KUPAC Congress Pest. Chem., Israel, Feb 1970) Gordon and Breach, New York. Vol. 1, pp. 141–162. Also in Pyrethrum Post 11(3): 94–103.

Elliott, M., Janes, N.F. Kimmel, E.C. and Casida, J.E. 1972c. Metabolic fate of pyrethrin I, pyrethrin II and allethrin administered orally to rats. J. Agric. Food Chem. 20: 300–313.

Elliott, M., Farnham, A.W., Janes, N.F., Needham, P.H., Pulman, D.A. and Stevenson, J.H. 1973. A photostable pyrethroid. Nature (London) 246: 169–170.

Elliott, M., Farnham, A.W. Janes, N.F., Needham, P.H. and Pulman, D.A. 1974. Synthetic insecticide with a new order of activity. Nature (London) 248: 710–711.

Elliott, M., Janes, N.F., Pulmans, D.A., Gaughan, L.C., Unai, T. and Casida, J.E. 1976. Radiosynthesis and metabolism in rats of the 1R isomers of the insecticide permethrin. J. Agric. Food Chem. 24(2): 270–276.

Elliott, M., Janes, N.F., Stevenson, J.H. and Walters, J.H.H. 1983. Insecticidal activity of the pyrethrins and related compounds. Part XIV: selectivity of the pyrethroid insecticides between *Ephestia kuehniella* and its parasite *Venturia canescens*. Pestic. Sci. 14: 423–426.

Elmosa, H.M. and King, H.L. 1964. Some effects of temperature, humidity, age, and sex on the toxicity of dieldrin and ethion to resistant onion maggots, *Hylemya antiqua*. J. Econ. Entomol. 57: 649–650.

Elms, K.D., Kerr, J.D. and Champ, B.R. 1972. Breakdown of malathion and dichlorvos mixtures applied to wheat. J. Stored Prod. Res. 8: 55–63.

Ely, R.E., Moore, H.D. and Carter, R.H. 1952. The effect of various dosage levels of crystalline lindane on the concentration of lindane in cow's milk. J. Dairy Sci. 35: 733.

Ely, R.E., Underwood, B., Moore, A. Mann, H.D. and Carter, R.A. 1953. Observations on lindane poisoning in dairy animals. J. Am. Vet. Med. Assoc. 123: 448.

Enigk, K. 1954. Biological principles in the control of ectoparasites. Monatsh. Tierheilkd. 6: 61–70.

England, D.C. and Day, P.E. 1971. Natal traits of litters from dams fed dichlorvos. J. Anim. Sci. 33 208.

Epstein, S.S. Andrea, J., Clapp. P., Mackintosh, D. and Mantel, N. 1967. Enhancement by piperonyl butoxide of acute toxicity due to farens, benzo(*a*)pyrene and griseofulvin in infant mice. Toxicol. Appl. Pharmacol. 11: 442–448.

Erdman, H.E. 1973. Effect of chronic *gamma* radiation dose rate, temperature and cohabitation on population performance of flour beetles. Environ. Entomol. 2: 41–48.

Esaki, K., Tanioka, Y., Tsukada, M., Izumiyama, K. and Oshio, K. 1973. Effects of DOWCO 214 against fetuses of experimental animals. Exper. Animal Central Laboratories Kawasaki, Japan. (English Translation, 41 pp.)

Esin, T., Guvener, A., Gunay, Y. and Sevintuna, C. 1964. Biological activity and residual amounts of insecticides admixed to stored grain. Bitki Koruma Bul. 4: 160–72.

Essac, E.C. and Casida, J.E. 1969. Metabolism in relation to mode of action of methylendroxyphenyl synergists in houseflies. J. Agric. Food Chem. 17: 539.

Esser, H.O. 1972. Terminal residues of organophosphorus insecticides in animals. In: Tahore, G., Ed., Pesticide Terminal Residues (Proc. IUPAC Int. Symp., Tel Aviv, 1971). Butterworths, London. pp. 33–56.

Evans, D.E. 1977a. The capacity for increase at a low temperature of several Australian populations of *Sitophilus oryzae* (L.). Aust. J. Ecol. 2: 55–67.

Evans, D.E. 1977b. The capacity for increase at a low temperature of some Australian populations of granary weevil, *Sitophilus granarius* (L.). Aust. J. Ecol. 2: 69–79.

Evans, D.E. 1977c. Some aspects of acclimation to low temperature in grain weevils *Sitophilus oryzae* (L.). and *S. granarius* (L.). Aust. J. Ecol. 2: 309–18.

Evans, D.E. 1979a. A comparison of the capacity for increase at a low temperature of foreign and Australian populations of *Sitophilus oryzae* (L.) and *S. granarius* (L.). Aust. J. Ecol. 4: 105–114.

Evans, D.E. 1979b. The response of grain weevils to low temperatures. In: Evans, D.E., Ed., Australian Contributions to the Symposium on the Protection of Grain Against Insect Damage During Storage, Moscow, 1978. CSIRO Division of Entomology, Canberra. pp. 39–46.

Evans, D.E. 1981c. The influence of some biological and physical factors on the heat tolerance relationships for *Rhyzopertha dominica* and *Sitophilus oryzae*. J. Stored Prod. Res. 17: 65–72.

Evans, D.E. 1982. The influence of temperature and grain moisture content on the intrinsic rate of increase of *Sitophilus oryzae* (L.) (Coleoptera: Curculionidae). J. Stored Prod. Res. 18(2): 55–66.

Evans, D.E., 1983. The influence of relative humidity and thermal acclimation on the survival of adult grain beetles in cooled grain. J. Stored Prod. Res. (1983) 19(4): 173–180.

Evans, D.E. 1984a. Miscellaneous physical methods of pest control. In: Champ, B.R., and Highley, E., Ed., Proc. Aust. Dev. Asst. Course on Preservation of Stored Cereals. CSIRO Division of Entomology, Canberra. pp. 515–520.

Evans, D.E. 1984b. Biological control of stored grain pests. In: Champ, B.R., and Highley, E., Ed., Proc. Aust. Dev. Asst. Course on Preservation of Stored Cereals. CSIRO Division of Entomology, Canberra. pp. 574–592.

Evans, D.E. and Dermott, D. 1979. The potential of a fluidized-bed heating system for the disinfestation of grain. In: Evans, D.E., Ed., Australian Contributions to the Symposium on the Protection of Grain against Insect Damage during Storage, Moscow, 1978. CSIRO Division of Entomology, Canberra. pp. 32–38.

Evans, D.E. and Dermott, D. 1981. Dosage/mortality relationship for *Rhyzopertha dominica* exposed to heat in a fluidised-bed. J. Stored Prod. Res. 17: 53–64.

Evans, D.E., Thorpe G.R. and Dermott, T. 1983. The disinfestation of wheat in a continuous-flow fluidized bed. J. Stored Prod. Res. 19(3): 125–137.

Evans, D.E., Thorpe, G.R. and Sutherland, J.W. 1984. Large scale evaluation of fluid-bed heating as a means of disinfesting grain. Proc. 3rd Int. Working. Conf. Stored-Product Entomol., Manhattan, Kansas, 7–11 October 1985. pp. 523–530.

Evans, N. 1985. The effectiveness of various insecticides on some resistant beetle pests of stored products from Uganda. J. Stored Prod. Res. 21(2): 105–109.

Falk, H.L. and Kotin, P. 1969. Pesticide synergists and their metabolites; potential hazards. Ann. N.Y. Acad. Sci. 160(1): 299–313.

Fall, M., Hernandez, S. and Ly, M. 1979. Control trial in storage of traditional millet in a country area of Senegal. Report of the proceedings. Part I. Tropical crops. Part II. Humans and animal health. Congres sur la lutte contre les insectes en milieu tropical. Chambre de Commerce et d'industrie de Marseille. pp. 611–620.

FAO 1975. Resolution XII, Report of the Ad Hoc Government Consulation on Pesticides and Public Health, Rome, 7–11 April 1975, to the Council of the Food and Agriculture Organisation of the United National, Rome. Document CL 66/21, May 1975, FAO, Rome.

FAO 1977a. Report of the Ad Hoc Government Consultation on International Standardisation of Pesticide Registration Requirements, Rome, 24–28 October 1977. Document AGP: 1977/M/9, FAO, Rome.

FAO 1977b. Resolution IV, Report of Ad Hoc Government Consultation on International Standardisation of Pesticide Registration Requirements, Document AGP: 1977/M/9, FAO, Rome.

FAO 1979. Recommended Method of Sampling for Determination of Pesticide Residues. Codex ALINORM 79/24, Appendix IV, Annex 1. FAO, Rome, 1979.

FAO 1981. Guidelines on Pesticide Residue Trials to Provide Data for Registration. Plant Protection Bulletin 29(1–2) 1981. FAO, Rome.

FAO 1982a. Important features of insecticides currently used or under development for the protection of stored grain. FAO Plant Production and Protection Paper No. 42. pp. 548–561. FAO, Rome.

FAO 1982b. Recommendations for Methods of Analysis for Pesticide Residues. Codex ALINORM 83/24 ADD. I. FAO, Rome 1982.

FAO 1983a. Recommendations for methods of analysis. CX/PR Report Appendix II Annex I ALINORM 85/24A. FAO, Rome.

FAO 1983b. Report of the Second Government Consultation on International Harmonisation of Pesticide Registration Requirements Rome, 11–15 October 1982. FAO, Rome.

FAO 1984. Recommendations for methods of analysis. Part 8 of the Guide to the Codex Alimentarius. (CAC/PR 8–1984) FAO, Rome.

FAO 1985a. FAO Guidelines on the Labelling of Pesticides. FAO, Rome.

FAO 1985b. International Code of Conduct on the Distribution and Use of Pesticides. Document CL 87/9 and Supplement, May 1985. FAO, Rome.

FAO 1986. FAO Guidelines for the Regulation and Management of Pesticides. A series of 10 guidelines issued as supplements to the International Code of Conduct on the Distribution and Use of Pesticides. Plant Protection Service, F.A.O., Rome.

FAO/WHO 1965. Evaluation of the toxicity of pesticide residues in food. FAO Meeting Report, No.PL:1965/10/1: WHO Food Add./27.65.

FAO/WHO 1967. Evaluation of some pesticide residues in food. FAO:PL/CP/15: WHO Food Add./67.32.

FAO/WHO 1968. 1967 evaluation of some pesticide residues in food. FAO PL:1967/M/11/1: WHO Food Add./68.30.

FAO/WHO 1969. 1968 evaluation of some pesticide residues in food. FAO PL:1968/M/9/1: WHO Food Add./69.35.

FAO/WHO 1970. 1969 evaluations of some pesticide residues in food. FAO PL:1969/M/17: WHO Food Add./70.38.

FAO/WHO 1971. 1970 evaluations of some pesticide residues in food. AGP:1970/M/12/1: WHO Food Add./71.42.

FAO/WHO 1972. 1971 evaluations of some pesticide residues in food. AGP:1971/M/9/1: WHO Pesticide Residues Series, No.1.

FAO/WHO 1973. 1972 evaluations of some pesticide residues in food. AGP:1972/M/9/1: WHO Pesticide Residues Series, No.2.

FAO/WHO 1974. 1973 evaluations of some pesticide residues in food. AGP:1973/M/9/1: WHO Pesticide Residues Series, No.3.

FAO/WHO 1975. 1974 evaluations of some pesticide residues in food. AGP:1974/M/11: WHO Pesticide Residues Series, No.4.

FAO/WHO 1976. 1975 evaluation of some pesticide residues in food. AGP:1975/M/12: WHO Pesticide Residue Series, No.5.

FAO/WHO 1977. 1976 evaluations of some pesticide residues in food. AGP:1976/M/14.

FAO/WHO 1978. 1977 evaluations of some pesticide residues in food. FAO Plant Production and Protection Paper 10 Sup.

FAO/WHO 1979. 1978 evaluations of some pesticide residues in food. FAO Plant Production and Protection Paper 15 Sup.

FAO/WHO 1980. 1979 evaluations of some pesticide residues in food. FAO Plant Production and Protection Paper 20 Sup.

FAO/WHO 1981. 1980 evaluations of some pesticide residues in food. FAO Plant Production and Protection Paper 26 Sup.

FAO/WHO 1982. 1981 evaluations of some pesticide residues in food. FAO Plant Production and Protection Paper 42.

FAO/WHO 1983. 1982 evaluations of some pesticide residues in food. FAO Plant Production and Protection Paper 49.

FAO/WHO 1984. 1983 evaluations of some pesticide residues in food. FAO Plant Production and Protection Paper 61.

FAO/WHO 1985. 1984 evaluations of some pesticide residues in food. FAO Plant Production and Protection Paper 68.

Farebrother, D.A. 1973. NRDC 107. Whole body radioautographal study in male albino rats. Wellcome Foundation Report HEFH 73–1.

Faulkner, M.D. and Wratten, F.T. 1969. The Louisiana State University infra-red preheat rice dryer. 61st Annu. Exp. Stn. Progr. Rep. p. 101.

Fazlul Huq, A.K. 1980. Some estimates of farm-level storage losses of rice in Bangladesh. Trop. Stored Prod. Inf. No. 39: 5–12.

Federal Working Group on Pest Management 1974. Guidelines on Analytical Methodology. Washington D.C. 20460.

Federal Working Group on Pest Management 1974a. Guidelines on Sampling. Environmental Protection Authority, Washington D.C. 20460 USA.

Ferguson, A.J. and Waller, J.B. 1982. A laboratory evaluation of pirimiphos-methyl and difluron admixed with wheat against the granary weevil. N.Z. J. Agric. Res. 25(1): 113–117.

Feuersenger, M. 1960a. Uber die Bestimmung von Schadlingsbekampfungs-mitteln in Lebensmitteln. Dtsch. Lebensm. Rundsch. 59: 69.

Feuersenger, M. 1960b. Uber die Bestimmung von Schadlingsbekamfunsmitteln in Lebensmitteln, Bundesgesundheitsblatt (10): 149.

Feuersenger, M. 1962. Ruckstandfragen bei de Anwendung von Kontaktininsektizieden in Getreidenschutz. Nachrichtenbl. Dtsch. Pflanzenschutzdienst. 14: 189.

Fielder, O'G. H. 1954. Lindane dust for grain weevil control. J. Ent. Soc. South Africa 17: 115.

Fink, R. 1972a. Eight-day dietary LC/50 to Mallard ducks — ZR-515. Report 777–100 of Hazleton Laboratories to Zoecon Corporation, 31 July 1972. (Unpublished).

Fink, R. 1972b. Eight-day dietary LC/50 — Bobwhite quail — ZR-515. Report 777–101 of Hazleton Laboratories to Zoecon Corporation, 31 July 1972. (Unpublished).

Fink, R. 1973. Eight-day dietary LC/50 — Chickens: Methoprene — Report to Zoecon Corporation from Truslow Farms Inc, 28 August 1973. Project 102–101. (Unpublished).

Fink, R. 1974. Eight-day dietary LC/50 — Mallard ducks, Technical pirimiphos-methyl Report to ICI United States Inc. by Truslow Farms Inc. Regulatory Affairs Office Project 123–102, 20 August 1974. (Unpublished).

Fink, R. 1978. Eight-day dietary LC/50 — Mallard ducks: Sumethrin (phenothrin). Report to McLaughlin Gormley King Company from Wildlife International Ltd, 22 May 1978. Report 163–105 Document ET-81–0017. (Unpublished).

Fink, R. and Beavers, J.B. 1978. Acute oral LD50 Bobwhite quail — Sumethrin (phenothrin). Report to Sumitomo Chemical America Inc. by Wildlife International Ltd dated 29 December 1978. Project 166–102 Doc. No. ET-81–0025. (Unpublished.)

Fink, R. and Rero, F.E. 1973a. Methoprene technical : one-generation reproduction study — Mallard ducks, Project 777–109. Report to Zoecon Corporation from Hazleton Laboratories Inc. 21 June 1973. (Unpublished).

Fink, R. and Rero, F.E. 1973b. Methoprene technical: one-generation reproduction study — Bobwhite quail — Project 777–108. Report to Zoecon Corporation from Hazleton Laboratories Inc. June 21 1973. (Unpublished).

Fishbein, J., Fawkes, H.L., Falk, and Thompson S. 1967. Thin layer chromatography of rat bile and urine following intravenous administration of pesticidal synergists. J. Chromatog. 97(1): 153–66.

Fishbein, L., Falk, H.L., Fawkes, J., Jordan, S. and Corbett, B. 1969. Metabolism of piperonyl butoxide in the rat with C-14 in the methylenedioxy or α-methylene group. J. Chromatog. 41(1): 61.

Fishbein, L., Falk, H.L., Fawkes, J. and Jordan, S. 1970. Metabolism of pesticidal synergists. In: Deichmann, W.B., Ed., Collect. Pap. 7th Inter-Am. Conf. Toxicol. Occup. Med., 6th Pestic. Symp. 1968/70. Miami: Halos Ass. Inc. pp. 83–97.

Fishbein, L., Falk, H.L., Fawkes, J. and Jordan, S. 1972. Metabolism of carbon-14-labeled piperonyl butoxide in rat. In: Tahori, A.S., Ed., Proc. 2nd Int. IUPAC Congr. Pestic. Chem. Tel Aviv, 1971. Gordon and Breach, NY. Vol. 6, pp. 503–519.

Fletcher, D. 1971. Oral toxicity study with Sumithion technical in Mallard ducks. Report from Industrial Bio-Test Inc. to Sumitomo Chemical Co. Ltd, Osaka, Japan. (Unpublished).

Fletcher, D. 1974. Toxicity and reproductrion study with Sumithion (fenitrothion) in Mallard ducks. Report to Sumitomo Chemical Co Ltd from Industrial Biotest Laboratories, 1BT-No 651–04052, 26 June 1974. (Unpublished).

Fletcher, D., Jenkins, D.H., Keplinger, M.L. and Francher, O.E. 1971. Acute oral toxicity study with Sumithion (fenitrothion) in ringneck pheasants. Report to Sumitomo Chemical Company Ltd from Industrial Biotest Laboratories Inc. No. J9993. (Unpublished).

Fletcher, D., Jenkins, D.H. and Keplinger, M.L. 1974. Toxicity and reproduction study with Sumithion (fenitrothion) in Bobwhite quail. Report to Sumitomo Chemical Company Ltd from Industrial Biotest Laboratories Inc. IBT No. 651–04052, July 1974. (Unpublished).

Fletcher, D., Jenkins, D.H. Kinoshita, F.K. and Keplinger, M.L. 1976. 8 day dietary LC/50 study with S-2539 Forte (phenothrin) in Bobwhite quail. Report to Sumitomo Chemical Co Ltd from Industrial Biotest Laboratories, Inc. IBT No. 65–07917, 12 Jan 1976. (Unpublished).

Fletcher, D., Jenkins, D.H., Kinoshita, F.K. and Keplinger, M.L. 1978. 8 day dietary LC/50 study with S-2539 Forte (phenothrin) in Bobwhite quail. Report to Sumitomo Chemical Co. from Industrial Biotest Laboratories Inc. IBT No. 651–07917. (Unpublished).

Floyd, E.H. 1961. Effectiveness of malathion dust as a protectant for farm-stored corn in Louisiana. J. Econ. Entomol. 54: 900–904.

Floyd, E.H. and Newson, L.D. 1956. Protection of stored corn with lindane impregnated sawdust. J. Econ. Entomol. 49: 753.

Floyd, E.H. and Smith, C.E. 1953. Pyrethrum and lindane in the protection of corn and rough rice from stored grain insects. J. Econ. Entomol. 46: 771.

Fogleman, R.W. 1954. Parathion and malathion — poultry study — acute oral administration, cholinesterase study, sub-acute feeding, dust and spray studies. Report to American Cyanide Company from Hazleton Laboratories, Falls Church, Virginia, 12 February 1954. (Unpublished).

Foote, C.S., Wuesthoff, M.T., Wexler, S., Burnstain, I.G., Denny, R., Schenck, G.O. and Schulte-Elte, K.H. 1967. Tetrahedron 23: 2583.

Formica, G. 1973. CGA 20168. Gas chromatographic residue determination in wheat grain, wheat meal, bran, groats and bread. Report REM 20/73 from Ciba-Geigy Limited, Basle, Switzerland. (Unpublished).

Formica, F. 1974. CGA 20168. Residues in chicken tissues. Report RVA 205/74 from Ciba-Geigy Ltd, Basle, Switzerland.

Formica, G. 1975. Degradation of CGA 20168 in wheat and maize. Report RVA 314/75 from Ciba-Geigy Limited, Basle, Switzerland. (Unpublished).

Formica, G. 1977a. Methacrifos (CGA 20168/dichlorvos). Determination of CGA 20168 and dichlorvos in feedgrain treated with different formulation and stored in drums. Report RVA 615/77 from Ciba-Geigy Limited, Basle, Switzerland. (Unpublished).

Formica, G. 1977b. CGA 20168 Residues in cow's milk after oral administration. Report RYA612/77 from Ciba-Geigy Ltd Basle, Switzerland. (Unpublished).

Formica, G. 1978a. Methacrifos (CGA 20168). Determination in French barley after direct spray treatment with formulation EC 950. Report RVA 620/78 from Ciba-Geigy Limited, Basle, Switzerland. (Unpublished).

Formica, G. 1978b. Methacrifos (CGA 20168). Determination in wheat and barley after direct spray treatment with the formulation EC 950. Carinena, Spain. Report RVA 626/78 from Ciba-Geigy Limited, Basle, Switzerland. (Unpublished).

Formica, G. 1978c. Methacrifos (CGA 20168). Determination in barley and oats after direct spray treatment with the formulation EC 950. Report RVA 623/78 from Ciba-Geigy Limited, Basle, Switzerland. (Unpublished).

Formica, G. 1978d. Methacrifos (CGA 20168). Determination of residues in stored grain and its products 7 months after direct spray treatment with the formulation SO 050. Report RVA 610/78 from Ciba-Geigy Limited, Basle, Switzerland. (Unpublished).

Formica, G. 1978e. Methacrifos (CGA 20168). Determination of residues in stored grain and its products 21 months after direct spray treatment with the formulation EC 950. Report RVA 611/78 from Ciba-Geigy Limited, Basle, Switzerland. (Unpublished).

Formica, G. 1978f. Methacrifos (CGA 20168). Determination of residues in stored feedgrain and their products after direct spray treatment with the formulation EC 1000. Report RVA 618/78 from Ciba-Geigy Limited, Basle, Switzerland. (Unpublished).

Formica, G. 1978g. Methacrifos (CGA 20168). Dissipation of methacrifos residues in maize grain after a 'sandwich treatment' with the dust formulation P 2. Report RVA 624/78 from Ciba-Geigy Limited, Basle, Switzerland. (Unpublished).

Foster, J.R. 1968. Effect of DDVP on reproductive performance of swine. J. Anim. Sci. 27: 1774.

Foster, T.S. 1974. Physiological and biological effects of pesticide residues in poultry. Residue Rev. 51: 69–121.

Franklin, M.R. 1972. Inhibition of hepatic oxidative xenobiotic metabolism of piperonyl butoxide. Biochem. Pharmacol. 21(24): 3287–3299.

Franklin, M.R. 1972a. Piperonyl butoxide metabolism by cytochrome P-450 affecting the formation and disappearance of the metabolite-cytochrose P-450 complex. Xenobiotica 2(6): 517–527.

Frawley, J.P., Fyat, N.N., Hagan, E.C., Blake, J.R. and Fitzhugh, O.G. 1957. Marked potentiation in mammalian toxicity from simulataneous administration of 2 anticholinesterase compounds. J. Pharmacol. Exp. Ther. 121: 96–106.

Freeman, J.A. 1957. Infestation of grain in international trade. A review of problems and control measures. J. Sci. Food Agric. 11: 623–629.

Freeman, J.A. 1974a. A review of changes in the pattern of infestation in international trade. EPPO Bulletin 4: 251.

Freeman, J.A. 1974b. Infestation of stored food in temperate countries with special reference to Great Britain. Outl. Agric. 8(1): 34–41.

Freeman, J.A. 1976. Food in Store. In: Gunn, D.L., and Stevens, J.G.R., Ed., Pesticides and Human Welfare. Oxford University Press. pp. 169–178.

Freeman, J.A. and Turtle, E.E. 1947. Insect pests of food; the control of insects in flour mills. Her Majesty's Stationery Office: London.

Frehse, H. and Timme, G. 1980. Quantitative residue analytical reliability: beatitude through application of latitude. Residue Rev. 73: 27–47.

Friedman, M.A. and Epstein, S.S. 1970. Stability of piperonyl butoxide. Toxicol. Appl. Pharmacol. 17(3): 810–812.

Friedman, M.A. and Saunders, V. 1976. Effects of piperonyl butoxide on dimethylnitrosamine metabolism and toxicity in Swiss mice. J. Toxicol. Environ. Health 2(1): 67–75.

Friedman, M.A., Arnold, E., Bishop, T. and Epstein, S.S. 1971. Additive and synergistic inhibition of mammalian microsomal enzyme functions by piperonyl butoxide, safrols and other methylenedioxyphenyl derivatives. Experientia 27(9): 1052–1054.

Friestad, H.O. 1961. Restkonsentrasjoner av lindan i mel. Nord. Hyg. Tidskr. 42: 64.

Frolov, B.A. and Berlin, A.A. 1969. Toxicity of some insectoacaricides for chicks. Tr. Vses. Nauchno-Issled. Inst. Vet. Sanit. 32: 333 (CA 78, 119924).

Fujie, G.H. and Fullmer, O.H. 1978. Determination of FMC 33297 residues in plant, animal and soil matrices by gas-chromatography. J. Agric. Food Chem. 26: 395–398.

Fujinami, A., 1980. Phenothrin (Wellcide) insecticide mainly for stored grain control, toxicity, residues. Jpn. Pestic. Inf. (37): 30–36.

Fukuto, T.R., Metcalf, R.L., Winton, M.Y. and Roberts, P.A. 1962. The synergism of substituted phenyl N-methylacarbamates by piperonyl butoxide. J. Econ. Entomol. 55(3): 341–345.

Funch, F.H. 1981. Analysis of residues of 7 pesticides in some fruits and vegetables by high pressure liquid chromatography. Z. Lebensm. Unters. Forsch. 173: 95–98.

Gaafar, S.M. and Turk, R.D. 1957. The toxicity of malathion to chickens. Amer. J. Vet. Res. 18: 180–182.

Gage, J.C. 1971. Pirimiphos-methyl: avian toxicity. Report No HO/1H/R/329 from ICI Industrial Hygeine Laboratories. (Unpublished).

Gage, J.C. 1972. Pirimiphos-methyl: oral toxicity in the dog and in a passerine bird species. Report from ICI Industrial Hygeine Research Laboratory. (Unpublished).

Gaines, T.B. 1960. The acute toxicity of pesticides to rats. Toxicol. Appl. Pharmacol. 2: 88–99.

Gaines, T.B. 1966. Detoxification of dichlorvos by rat liver. Nature (London) 209: 88.

Gaines, T.B. 1969. Acute toxicity of pesticides. Toxicol. Appl. Pharmacol. 14: 515–534.

Gaines, T.B., Hayes, W.J. and Linder, R.E. 1966. Liver metabolism of anticholinesterase compounds in live rats: Relation to toxicity. Nature (London) 209: 88.

Gane, R. 1941. The water content of wheats as a function of temperature and humidity. J. Soc. Chem. Ind., London 60: 40–46.

Gasser, R. 1953. A new insecticide with a wide range of activity. Z. Naturforsch. 8b: 225.

Gast, M. and Sigaut, F., Eds 1979. 'Les techniques de conservation des grains a long terme'. Centre National de la Recherche Scientifique: Paris. 232 pp.

Gaughan, L.C., Unai, T. and Casida, J.E. 1976. Permethrin metabolism in rats and cows, and in bean and cotton plants. Paper delivered at 172nd ACS National Meeting, San Francisco, (August 1976).

Gaughan, L.C., Unai, T. and Casida, J.E. 1977. Permethrin metabolism in rats. J. Agric. Food Chem. 25(1): 9–17.

Gaughan, L.C., Ackerman, M.E., Unai, T. and Casida, J.E. 1978a. Distribution and metabolism of *trans-* and *cis*-permethrin in lactating jersey cows. J. Agric. Food Chem. 26(3): 613–618.

Gaughan, L.C., Robinson, R.A. and Casida, J.E. 1978b. Distribution and metabolic fate of *trans-* and *cis*-permethrin in laying hens. J. Agric. Food Chem. 26(6): 1374–1380.

Gay, F.J. 1946. The effect of temperature on the moisture content — relative humidity equilibria of wheat. Counc. Sci. Ind. Res. J. 19: 187–9.

Gay, F.J. and Ratcliff, F.N. 1941. The importance of *Rhyzopertha dominica* as a pest of wheat under wartime conditions. J. Counc. Sci. Ind. Res. 14: 173–180.

Geigy, 1963. Scientific data on diazinon submitted to WHO J.R. Geigy S.A., Basle, Switzerland. (Unpublished).

Gelosi, A. 1982. Rice weevil (*Sitophilus oryzae* Linneus). Inf. Fitopatol. 32(12): 31–34.

Gelosi, A. and Arcozzi, L. 1982. Protection of stored wheat with pirimiphos-methyl. Inf. Fitopatol. 32(7/8): 45–50.

Georghiou, G.P. and Metcalf, R.L. 1962. Carbamate insecticides: comparative insect toxicity of Sevin, Zectran, and other new materials. J. Econ. Entomol. 55(1): 125–127.

Gfeller, W. 1974. CGA20168 Oral toxicity in pigs. Daily administration for 28 days. Report of Ciba-Geigy Ltd., Basle, Switzerland, submitted to WHO. (Unpublished).

Ghadiri, M. 1968. Malformation and toxic effects of two organic phosphates and a carbamate on the chicken embryo and the adult chicken. Diss. Abstr. B. 29: 1314.

Ghadiri, M., Greenwood, D.A. and Binns, W. 1967. Feeding of malathion and carbaryl to laying hens and roosters. Presented 6th Ann. Meeting Soc. Toxicol. Atlanta, Ga. March 1967.

Ghaly, T.F. 1984. Aeration trial of farm-stored wheat for the control of insect infestation and quality loss. J. Stored Prod. Res. 20(3): 125–131.

Ghosh, A.K. 1947. The use of "Gammexane" for the control of insect pests of stored rice. Indian Farming 8: 129.

Giannone, C. and Formica, G. 1979a. Methacrifos (CGA 20168). Residue in maize cobs after surface treatment with Damfin® 950 EC. Report RVA 4010/79 from Ciba-Geigy Limited, Basle, Switzerland. (Unpublished).

Giannone, C. and Formica, G. 1979b. Methacrifos (CGA 20168). Gas chromatographic determination of total residues in maize cobs after surface treatment with Damfin® 950 EC. Report RVA 4010/79A from Ciba-Geigy Limited, Basle, Switzerland. (Unpublished).

Giannone, C. and Formica, G. 1979c. CGA 20168. Gas chromatographic determination of total residues in beer after processing of malt treated with CGA 20168 EC 950. Report RVA 4013/79A from Ciba-Geigy Limited, Basle, Switzerland. (Unpublished).

Giannone, C. and Formica, G. 1980a. Gas chromatographic determination on methacrifos (CGA 20168) and its de-alkylated metabolites in barley and maize after treatment with Damfin® SO 050 by means of a Vobamatic B Sprayer. Report RVA 4001/79A from Ciba-Geigy Limited, Basle, Switzerland. (Unpublished).

Giannone, C. and Formica, G. 1980b. Gas chromatographic determination of methacrifos (CGA 20168) and its de-alkylated metabolites in wheat and barley after treatment with Damfin® SO 050 by means of a Vobamatic B Sprayer. Report RVA 626/78A from Ciba-Geigy Limited, Basle, Switzerland. (Unpublished).

Giannone, C. and Formica, G. 1980c. Determination of methacrifos (CGA 20168) and its de-alkylated metabolites in wheat grain after treatment with Damfin® SO 050. Report RVA 622/78A from Ciba-Geigy Limited, Basle, Switzerland. (Unpublished).

Giannone, C. and Formica, G. 1980d. Gas chromatographic determination of methacrifos and its de-alkylated metabolites in sorghum after treatment with DAMFIN® SO 050 or Damfin® P 2. Report RVA 4045/79A from Ciba-Geigy Limited, Basle, Switzerland. (Unpublished).

Giannone, C. and Formica, G. 1980e. Methacrifos (CGA 20168). Dissipation in barley and maize grain after treatment with the formulation SO 050 by means of a Vabomatic B. Sprayer. Report RVA 4001/79 from Ciba-Geigy Limited, Basle, Switzerland. (Unpublished).

Giese, A.C. 1964. Photophysiology. Vol. 1, General Principles; Action of light on plants. Academic Press, New York.

Gilbert, D. 1984. Insect electrocutor light traps. In: Baur, F.J., Ed., Insect pest management for food storage and processing. American Association of Cereal Chemists, St Paul, Minnesota. pp. 87–107.

Giles, P.H. 1964. Lindane contamination in stored sorghum and millet in northern Nigeria. Trop. Sci. 6: 113.

Giles, P.H. 1965. Control of insects infesting stored sorghum in Northern Nigeria. J. Stored Prod. Res. 1: 145.

Giles, P.H. 1973. Stored product maize crib trial. Sepral Quart. Rep. Jul-Sept 2–3.

Gillenwater, H.B. and Burden, G.S. 1973. Pyrethrum for control of household and stored-product insects. In: Casida, J.E., Ed., Pyrethrum, the Natural Insecticide. Academic Press, N.Y. pp 243–257.

Gillenwater, H.B., McDonald, L.L. and Simonaitis, R.A. 1978. Potential of permethrin as a protectant for wheat and corn against rice weevils. J. Ga. Entomol. Soc. 13(2): 109–113.

Gilzin, V.M. 1984. Grain pests eliminated by radiation disinsectization. Soviet Export 2(149): 48–49.

Gingrich, J.B., Pratt, J.J. Jr. and Mandels, G.R. 1977. Potential of UV radiation for control of American cockroach populations. Ent. Exp. Appl. 21: 112.

Girish, G.K., Goyal, R.K. and Krishnamurthy, K. 1970. Studies on stored grain pests and their control I. Efficacy and residual toxicity of iodofenphos and malathion. Bull. Grain Technol. 8: 103–6.

Girish, G.K., Gopal, K. and Krishnamurthy, K. 1970a. Efficacy of DDVP against various developing stages of *Sitophilus oryzae*. Bull. Grain Technol. 8: 166.

Girish, G.K., Goyal, R.K. and Krishnamurthy, K. 1974. Pirimiphos-methyl (50 per cent e.c.) versus malathion (50 per cent e.c.) — its efficacy and residual toxicity against some stored grain insect pests. Bull. Grain Technol. 11(2): 106–112.

Glenn, M.S. and Sharpf, W.G. 1977. Synthetic route to the acid portion of permethrin. A.C.S. Symposium Series 42: 116.

Glomot, R. 1979. Acute toxicity study by oral route in the rat. Report RU-79 803–54/A2 from Roussel Uclaf Submitted to WHO by Roussel Uclaf. (Unpublished).

Glomot, R. and Chevalier, B. 1969. Etude de la toxicite argue du RU11484 (NRDC 107) Report from Roussel Uclaf submitted to WHO by The Wellcome Foundation Ltd. (Unpublished).

Glomot, R. and Chevalier, B. 1976. Deltamethrin — Acute toxicity study, mouse and rat by oral route. Report Toxicol 76810/A from Roussel Uclaf. (Unpublished).

Glomot, R., Chevalier, B., Collas, E. and Audegond, L. 1977. Deltamethrin. Acute toxicity study by oral route in male and female beagle dogs. Report Toxicol. 7804/JL-5 from Roussel Uclaf. (Unpublished).

Glomot, R., Audegond, L. and Collas, E. 1979. Acute oral toxicity study in the rat — Decis 25 g/l Mixofluid. Report RU 79824/A from Roussel Uclaf. (Unpublished).

Glomot, R., Audegond, L. and Collas, E. 1980. Deltamethrin, single administration study by oral route in the dog. (Decis Wettable Powder 2.5%). Report RU8O: 194/A from Roussel Uclaf. (Unpublished).

Glomot, R., Audegond, L. and Collas, E. 1981. Deltamethrin Single administration study by oral route in the rat. Decis Wettable Powder, 2.5%. Report RU 80.194/A from Roussel Uclaf. (Unpublished).

Glynne-Jones, G.D. and Chadwick, P.R. 1960. A comparison of four pyrethrum synergists. Pyrethrum Post 5(3): 22.

Glynne-Jones, G.D. and Green, E.H. 1959. A comparison of the toxicity of pyrethrins and synergised pyrethrins to *Sitophilus oryzae* and *Sitophilus granarius*. Pyrethrum Post 5(2): 3–7.

Godavari Bai S. 1965. Persistence of malathion on treated food grains under different conditions of storage and processing. Abstr. Symp. Pestic. Mysore, 21–22 December 1964. p. 39.

Godavari Bai S. 1968. Persistence of malathion residues on treated food grains under different conditions of storage and processing. In: Majumder, S.K., Ed., Pesticides. Academy of Pest Control Sciences, India.

Godavari Bai S., Krishnamurthy, K. and Mujumder, S.K. 1960. Studies on malathion for good grain product insect control. Pest. Technol. 2(7): 12. Also in Publ. Hlth. Sanit. 2: 12.

Godavari Bai S., Krishnamurthy, K. and Majumder, S.K. 1964. Malathion for stored product insect control. Int. Pest Contr. 6(6): 9–10.

Goldblatt, L.A. (Ed.) 1969. Aflatoxin — Scientific Background, Control and Implications. Academic Press, New York. 472 pp.

Goldstein, J.A., Hickman, P. and Kimbrough, R.D. 1973. Effects of purified and technical piperonyl butoxide on drug-metabolizing enzymes and ultrastructure of rat liver. Toxicol. Appl. Pharmacol. 26(3): 444–458.

Golebiowska, Z. 1969. The feeding and fecundity of *Sitophilus granarius, S. oryzae* and *Rhyzopertha dominica* in wheat grain. J. Stored Prod. Res. 5: 143–155.

Golob, P. 1977. Mixing insecticides with grain for storage. Rural Technology Guide, Tropical Products Institute, (3): 14 pp.

Golob, P. 1981a. A practical appraisal of on-farm storage losses and loss assessment methods in Malawi 1: the Shire Valley agricultural development area. Trop. Stored Prod. Inf. (40): 5–13.

Golob, P. 1981b. A practical appraisal of on-farm storage losses and loss assessment methods in Malawi 2: the Lilongwe land development programme area. Trop. Stored Prod. Inf. (41): 5–12.

Golob, P. 1981c. The use of pesticides at farmer and village level storage. Proc. GASGA Seminar on the Appropriate Use of Pesticides for the Control of Stored Products Pests in Developing Countries, Slough, 17th-20th February 1981. pp. 60–61.

Golob, P. 1984. Improvements in maize storage for the smallholder farmer. Trop. Stored Prod. Inf. (50): 14–19.

Golob, P. and Ashman, F. 1974. The effect of oil content and insecticides on insects attacking rice bran. J. Stored Prod. Res. 10(2): 91–103.

Golob, P. and Hodges, R. 1982. Study of an outbreak of *Prostephanus truncatus* (Horn) in Tanzania. Report, Tropical Products Institute, (G 164) 23 pp.

Golob, P. and Muwalo, E. 1984. Pirimiphos-methyl as a protectant of stored maize cobs in Malawi (Storage insect pests, *Sitophilus* spp., *Sitotroga cerealella*). Int. Pest Contr. 26(4): 94–96.

Golob, P., Mwambula, J., Mhango, V. and Ngulube, F. 1982. The use of locally available materials as protectants of maize grain against insect infestation during storage in Malawi. J. Stored Prod. Res. 18(2): 67–74.

Golob, P., Dunstan, W.R., Evans, N., Meik, J., Rees, D. and Magazini, I. 1983. Preliminary field trials to control *Prostephanus truncatus* in Tanzania. Trop. Stored Prod. Inf. (45): 15–17.

Golob, P., Changjaroen, P., Ahmed, A. and Cox, J. 1985. Susceptability of *Prostephanus truncatus* to insecticides. J. Stored Prod. Res. 21(3): 141–150.

Golz, H.H. and Shaffer, C.B. 1956. Malathion. Summary of Pharmacology and toxicology. American Cyanamid Company, New York. 20 N.Y. USA.

Gonen, M. and Schwartz 1978. A controlling effect of a juvenile hormone analogue on *Ephestia cautella* by non-direct application. Proc. 2nd Int. Working Conf. on Stored-Product Entomol., Ibadan, Nigeria, 10–16 September 1978. pp.106–115.

Goodwin-Bailey, K.F. 1959. Field trials of wheat and shelled corn protection. Pyrethrum Post 9(1): 18–19.

Goodwin-Bailey, K.F. and Holborn, J.M. 1952. Laboratory and field experiments with pyrethrins/piperonyl butoxide powders for the protection of grain. Pyrethum Post 2(4): 7–17.

Gore, K.S. 1958. Laboratory experiments with some organophosphorus insecticides as wheat protectants. PhD Thesis. Cornell Univ., Ithica NY.

Goresline, H.E. 1973. The need for action on disinfestation of grain and grain products by irradiation. Aspects of the Introduction of Food Irradiation in Developing Countries. Proceedings of a panel organised by the Joint FAO/IAEA — Bombay, November 1972. Pub by International Atomic Energy Agency, Vienna.

Goto, G., Masuoka, Y. and Hiraga, K. 1974. Photooxidation of the sex pheromone (Z,E)-9,12-tetraceceadienyl-1-acetate. Chem. Lett. 11: 1275.

Goto, S., Nishikawa, N. and Yamada, T. 1980. Transfer of fenitrothion to chicken eggs analyzed by gas chromatography. Keibyo Kenkyukai Ho. 16(3): 139–42.

Gough, M.C. 1975. A simple technique for the determination of humidity equilibria in particulate foods. J. Stored Prod. Res. 11: 161–166.

Gough, M.C. and Bateman, G.A. 1977. Moisture/humidity equilibria of tropical stored products — Part 1, Cereals. Trop. Stored Prod. Inf. (33): 25.

Gough, M.C. and King, P.E. 1980. Moisture content/relative humidity equilibria of some tropical cereal grains. Trop. Stored Prod. Inf. (39): 13–17.

Gough, M.C. and Lippiatt, G.A. 1978. Moisture/humidity equibria of tropical stored produce Part III. Legumes, spices and beverages. Trop. Stored Prod. Inf. (35): 15–29.

Graciet, B., Cangardel, H. and De Mallman, R.J. 1974. Activity of pirimiphos-methyl against *Sitophilus granarius* compared with dichlorvos and malathion. Quatrilmes Journees de Phytiatrie et de Phytopharmacie Circum Mediterraneenes, 15–18 September 1974.

Graham, C.A. and Jenkins, S.L. 1974. The effect of pirimiphos-methyl on the growth rate of broiler chicks. Report ICI Australia Ltd. Merrindale Research Station. (Unpublished).

Grant, A.E. and Imperial Chemical Industries Ltd 1939. Benzene hexachloride. British patent 504,569 (April 26 1939).

Gray, A.J. and Rickard, J. 1981. Distribution of radiolabel in rats after intra-venous injection of toxic doses of ^{14}C-acid-, ^{14}C-cyano-labelled deltamethrin. Pestic. Biochem. and Physiol. 16: 79–85.

Green, A.A. 1969. Control of insects in stores of home-grown grain. IV. The use of insecticides. Chem. Ind. 41: 1452–1454.

Green, A.A. and Kane, J. 1960. A comparison of lindane, DDT and malathion for the protection of bagged groundnuts from infestations of *Tribolium castaneum*. Trop. Sci. 1: 290; through Rev. Appl. Entomol. A49: 502.

Green, A.A. and Tyler, P.S. 1966. A field comparison of malathion, dichlorvos and fenitrothion for the control of *Oryzaephilus* infesting stored barley. J. Stored Prod. Res. 1: 273–285.

Green, A.A. and Wilken, D.R. 1969. The control of insects in bagged grain by the injection of dichlorvos. J. Stored Prod. Res. 5: 11–19.

Green, A.A., Kane, M.J., Tyler, P.S., Wilkin, D.R., Bristow, S.G., Horler, D.F. and Clements, J.E. 1965. Field trials with dichlorvos. Pest Infest. Res. 1965: 27.

Green, A.A., Kane, M.J. and Gradidge, J.M.G. 1966a. Experiments in the control of *Ephestia elutella* using dichlorvos vapour. J. Stored Prod. Res. 2: 147–157.

Green, A.A., Kane, M.J. and Mahon, P.A. 1966b. Protective treatment of grain — laboratory experiments. Pest Infest. Res. 1965 24.

Green, A.A., Kane, M.J., King, P.A. and Horler, D.F. 1966c. Protective treatment of grain — laboratory experiments. Pest Infest. Res. 1966: 25–26.

Green, A.A., Wilkin, D.R., Bristow, S.G. and Rowlands, D.G. 1967. Treatment of infested grain — Injection of dichlorvos. Pest Infest. Res. 1967: 26.

Green, A.A., Kane, J., Heuser, S.G. and Scudamore, K.A. 1968. Control of *Ephestia elutella* (Hb.) (Lepidoptera, Phycitidae) using dichlorvos in oil. J. Stored Prod. Res. 4: 69–76.

Green, A.A., Tyler, P.S., Kane, M.J. and Rowlands, D.G. 1970. An assessment of bromophos for the protection of wheat and barley. J. Stored Prod. Res. 6: 217–228.

Green, T., Monks, I.H. and Phillips, P.J. 1973. Pirimiphos-methyl (PP 511): sub-acute oral toxicity and residue studies in hens. ICI Industrial Hygiene Research Laboratories Report No. HO/IH/P/65B. (Unpublished).

Greenberg, R.S. 1981. Determination of fenvalerate, a synthetic pyrethroid, in grapes, peppers, apples and cottonseeds by gas-liquid chromatography. J. Agric. Food Chem. 29: 856–860.

Greening, H.G. 1976. Wheat protectant dust trials 1965–67, NSW, Department of Agriculture, Biological and Chemical Research Institute, Rydalmere, NSW. Report prepared for publication September 1976.

Greening, H.G. 1979a. Observations on the occurrence of insect pests of stored grain in New South Wales. In: Evans, D.E., Ed., Australian Contributions to the Symposium on the Protection of Grain Against Insect Damage During Storage, Moscow, 1978. CSIRO Division of Entomology, Canberra. pp. 15–22.

Greening, H.G. 1979b. Chemical control of insects infesting farm-stored grain and harvesting machinery. In: Evans, D.E., Ed., Australian Contributions to the Symposium on the Protection of Grain Against Insect Damage During Storage, Moscow, 1978. CSIRO Division of Entomology, Canberra. pp. 74–80.

Greening, H.G. 1980. Chemical methods of controlling insects which infest grain stored on farms and harvest technology. Tr., Vses Nauchno-Issled. Inst. Zerna Prod. Ego Pererab. 93: 99–102.

Greening, H.G. 1983. An investigation of protectant treatments for farm-stored grain in New South Wales. Pyrethrum Post 15(3): 78–84.

Greve, P.A. and Grevenstuk, W.B.F. 1975. A convenient small-scale cleanup method for extracts of fatty samples with basic alumina before GLC-analysis on organochlorine pesticide residues. Meded. Fac. Landb. Gent. 40: 1115–1124.

Greve, P.A. and Heusinkveld, H.A.G. 1981. Comparison of various silver nitrate/alumina columns for the cleanup of plant extracts prior to gas chromatographic analysis with electron capture detection. Meded. Fac. Landb. Gent 46: 317–324.

Griffin, C.S. 1973. Mammalian toxicology of pyrethrum. Pyrethrum Post 12(2): 50–58.

Grover, P.L. and Sims, P. 1965. The metabolism of γ-2,3,4,5, 6-pentachlorocyclohex-1-ene and γ-hexachlorocyclohexane in rats. Biochem. J. 96: 521.

Grue, C.E. 1982. Response of common grackles to dietary concentrations of four organophosphate pesticides. Arch. Environ. Contam. Toxicol. 11: 617–626.

Guerra, A.A., Ouye, M.T. and Bullock, H.R. 1968. Effect of UV irradiation of egg hatch, subsequent larval development, and adult longevity of the tobacco budworm and the bollworm. J. Econ. Entomol. 61(2): 541.

Gunther, F.A. 1970. Pesticide residues in the total environment; reliable detection and determination, mitigation and legislative control and surveillance programs. Pure Appl. Chem. 21: 355–376.

Gunther, F.A. 1980. Interpreting residue data. Residue Rev. 76: 155–171.

Gunther, F.A., Lindgren, D.L. and Blinn, R.C. 1958. Biological effectiveness and persistence of malathion and lindane used for protection of stored wheat. J. Econ. Entomol. 51: 843.

Gupta, P.K. and Paul, B.A. 1977. Biological fate of ^{32}P-malathion in *Gallus domesticus* (desi poultry birds). Toxicology 7: 169–177.

Gupta, H.C.L., Pareek, B.L. and Kavidia, V.S. 1978. Residues of lindane in stored cereals. Bull. Grain Technol. 16(2): 132–134.

Gyrisco, C.L.B., Norton, G.W., Trimberger, R.F., Holland, P.J., McEnerney and Muka, A.A. 1959. Effects of feeding low levels of insecticide residues on hay to dairy cattle on flavour and residues in milk. J. Agric. Food Chem. 7: 707.

Gyrisco, G.C., Lisk, D.J., Fertig, S.N., Huddleston, E.W., Fox, F.H., Holland, R.F. and Trimberger, G.W. 1960. The effects of feeding high levels of Sevin on residues, flavour and odour of the milk of dairy cattle. J. Agric. Food Chem. 8: 409–410.

Gysin, H. and Margot, A. 1958. Chemistry and toxicological properties of 0–0-diethyl-0-(2-isopropyl-4-methyl-6- pyrimidenyl) phosphorothioate (diazinon). J. Agric. Food Chem. 6: 900.

Hadaway, A.B. and Barlow, F. 1957. The influence of temperature and humidity upon the action of insecticides I During the post-treatment period. Ann. Trop. Med. Parisitol. 51: 187–193.

Haines, C.P. 1982a. Pest management in stored products. Protection Ecology 4: 321–330.

Haines, C.P. 1982b. Paddy loss to insects. In: Rottger, U., and SEARCA Team, Ed., Proc. GASGA Seminar on Paddy Deterioration in the Humid Tropics, Baguio, Philippines, 11–18 October 1981. GTZ, Eschborn. pp. 67–76.

Haines, C.P. 1984. Biological methods for integrated control of insects and mites in tropical stored products. III: the use of predators and parasites. Trop. Stored Prod. Inf. (48) 17–25.

Hais, K. and Franz, J. 1965. On the problem of Metathione (fenitrothion) residues in milk after disinfestation. Cesk. Hyg. 10: 205–8.

Haley, T.J. 1978. Piperonyl butoxide: a review of the literature. Ecotoxicol. Environ. Safety 2: 9–31.

Hall, D.W. 1963. Some essential considerations on the storage of food grains (cereals, legumes and oilseeds) in tropical Africa. Informal Working Bulletin No. 24, Agricultural Engineering Branch, FAO, Rome.

Hall, D.W. 1970. Handling and storage of food grains in tropical and sub-tropical areas. FAO Agricultural Development Paper (90), FAO, Rome.

Hall, R. 1979. Effect of a range of temperatures on the knockdown speed of etrimfos and primiphos-methyl against pests of stored grain. Report of Sandoz Products Ltd. Agrochemicals, Kings Langley, UK, 5 September 1979. (Unpublished).

Hall, C.W. and Rodriquez-Arias, J.H. 1958. Equilibrium moisture content of shelled corn. Agric. Eng. 39: 466–470.

Hall, R.C., Ballee, D.L., Bennett, G.W. and Fahey, J.E. 1973. Persistence and distribution of Gardona and dichlorvos in grain and grain products. J. Econ. Entomol. 66(2): 315–318.

Halls, G.R.H. 1981a. The fate of permethrin residues on wheat during nine months storage in Australia. Wellcome Foundation Ltd Report HEFH 81–1.

Halls, G.R.H. 1981b. An assessment of the biological activity of permethrin as a grain protectant using wheat treated and stored in Australia. Wellcome Foundation Ltd Report HEFH 81–4.

Halls, G.R.H. and Periam, A.W. 1980a. The fate of residues of deltamethrin on wheat during storage and after milling and baking Report after 9 months storage. Wellcome Research Laboratories Report HEFH 80–4, November 1980. (Unpublished).

Halls, G.R.H. and Periam, A.W. 1980b. The fate of permethrin residues on wheat during storage and after milling and baking — Report after 9 months storage. Wellcome Foundation Ltd Report HEFH 80–3.

Halls, G.R.H. and Periam, A.W. 1981. The fate of permethrin residues on wheat after storage for 12 and 15 months : Results of more recent chemical analyses on the wheat described in HEFH 80–3. Wellcome Foundation Ltd Report HEFH 81–2.

Hambock, H. 1978. Metabolism of CGA 20168 in the rat. Report from Ciba-Geigy Limited, Basle, Switzerland, submitted to the World Health Organisation. (Unpublished).

Hamburger, F. and Klotzsche, C. 1975. Etrimfos. toxicity tests. Sandoz Ltd. Report. Agro. Dok. EBK 754c/73, 10 March 1975. (Unpublished).

Hamid, M.A.K. and Boulanger, R.J. 1969. A new method for the control of moisture and insect infestations of grain by microwave power. J. Microwave Power 4(1): 11.

Hammock, B.D. and Quistad, G.B. 1976. The degradative metabolism of juvenoids by insects. In: Gilbert, L.I., Ed., The Juvenile Hormones. Plenum Press, New York. pp. 374–393.

Hammock, B.D. and Quistad, G.B. 1981. The metabolism of insect growth regulators. In: Hutson, D.H., and Roberts, T.R., Ed., Progress in Pesticide Biochemistry. John Wiley and Sons, N.Y. Vol. 1, pp. 1–84.

Hampson, S.J. 1981. Significance of the insecticide 'Actellic' and rodenticide 'Klerat' for the protection of stored products. In: Proc. BIOTROP Symp. on Pests of Stored Products, Bogor, Indonesia, 24–26 April 1978. pp. 197–216.

Hansen, L.B., Castillo, G.D. and Biehl, E.R. 1981. Gas chromatographic-mass spectrometric characterization of an alteration product of malathion detected in stored rice. J. Assoc. Off. Anal. Chem. 64(5): 1232–1237.

Harada, T. 1967. DDVP fumigant. 1. Efficiency tests as grain protectant and insecticide. Shokuryo Kenkyujo Kenkyu Hokoku (22): 14. (CA 66, 114894.)

Hardie, T. 1940. Manufacture of benzene hexachloride. U.S. patent 2,218,148 (Oct. 15, 1940).

Harein, P.K. 1974. Evaluation of pesticides for control of stored-product insects. Proc. 1st Int. Working Conf. Stored-Product Entomology, Savannah, Georgia, 7–11 October 1974. pp. 237–245.

Harein, P.K. 1982. Chemical control alternatives for stored- grain insects. In: Christensen, C.M., Ed., Storage of Cereal Grains and their Products (3rd edition). American Association of Cereal Chemists Inc., St Paul, Minnesota. Vol. 5, Chapter 10, pp. 319–362.

Harein, P.K. and De Las Casas, E. 1974. Chemical control of stored-grain insects and associated organisms. In: Christensen, C.M., Ed., Storage of Cereal Grains and their Products (2nd edition). American Association of Cereal Chemists Inc., St Paul, Minnesota. pp. 232–291.

Harein, P.K. and Rao, H.R.G. 1972. Dichlorvos and Gardona as protectants for stored wheat against granary weevil infestations in laboratory studies. J. Econ. Entomol. 65(3): 1402.

Hargreaves, P., Bengston, M. and Alder, J. 1982. Inactivation of deltamethrin on stored wheat. Pestic. Sci. 13: 639–646.

Harries, F.H., de Coursey, J.D. and Hofmaster, R.N. 1945. Some factors affecting the insecticidal action of pyrethrum extracts on the beet leafhopper. J. Agric. Res. 71: 553–565.

Harris, C.R. and Kinoshita, G.B. 1977. Influence of post-treatment temperature on the toxicity of pyrethroid insecticides. J. Econ. Entomol. 70: 215–218.

Harris, K.L. and Lindbad, C.J. 1978. Post-harvest grain loss assessment methods. A manual of methods for the evaluation of post-harvest losses. American Association of Cereal Chemists Inc., St. Paul, Minnesota. 193 pp.

Harris, R.L., W.F. Chamberlain and Frazar, E.D. 1974. Horn flies and stable flies: free choice feeding of methoprene mineral blocks to cattle for control. J. Econ. Entomol. 67: 384–386.

Harris, C.R., Svec, H.J. and Chapman, R.A. 1978. Laboratory and field studies on the effectiveness and persistence of pyrethroid insecticides used for cabbage looper control. J. Econ. Entomol. 71: 642–644.

Harrison, D.L. and Hastie, B.A. 1965. Diazinon residues in the milk of cows and fat of sheep after feeding on pasture treated with diazinon. N.Z. J. Agric. Res. 9: 1–7.

Harrison, D.L., Poole, W.S.H. and Nol, J.C.M. 1963. Observations on feeding lindane-fortified mash to chickens. N.Z. Vet. J. 11: 137.

Hart, R.J. and Moore, B. 1975. The use of CGA 20168 for the protection of rice stored in bags. Technical Report No. 75/10/553 from Ciba-Geigy Australia Limited. (Unpublished).

Hart, R.J. and Moore, B. 1976a. The protection of stored grain sorghum from insect attack with CGA 20168. Technical Report No. 76/6/564 from Ciba-Geigy Australia Limited. (Unpublished).

Hart, R.J. and Moore, B. 1976b. The protection of stored barley from insect attack with CGA 20168. Technical Report No. 76/10/595 from Ciba-Geigy Australia Ltd. (Unpublished).

Hartstack, A.W., Hollingsworth, J.P., Ridgeway, R.L. and Hunt, H.H. 1971. Determination of trap spacings required to control an insect population. J. Econ. Entomol. 64: 1090–1100.

Hasan, S.B., Deo, P.G. and Majumder, S.K. 1983. Synergistic effect of some organophosphorus pesticides on the toxicity of Nexion to four stored grain insect pests. Pesticides 17(6): 9–11.

Hasegawa, M., Ono, T., Karashimada, T., Sato, H., Kameyama, K., Kinoshita, Y., Ogawa, E. and Kitagaki, T. 1973. Studies on the safety of the newly introduced organophosphorus insecticide, chloropyrifos-methyl Part 1. Acute toxicity of chlorpyrifos-methyl to rats and mice. (Report Hokkaido Inst. Pub. Health 23: 57–61. Japanese. English summary in Pesticide Abstracts 7(4): 74–0891.

Hastie, A. 1963. The metabolism and elimination of diazinon from animals, animal tissues and foodstuffs. Geigy Agricultural Chemicals, Australia. (Bulletin).

Hatfield, I. 1949. Fly-control chemicals and toxicities. Pests 17: 9–14, 26.

Hattori, K. 1974. Toxicological studies on the effects of chemicals on birds. II. Oral acute toxicity of three organophosphate insecticides to the common pigeon. Hokkaidoritsu Eisei Kenkyushoho 24: 154.

Hattori, K., Sato, H., Tsuchiya, K., Yamamoto, N. and Ogawa, E. 1974. Toxicological studies on influences of chemical to birds (Part 1). Acute oral toxicity and cholinesterase inhibition of three organophosphate insecticides in Japanese quail. Hokkaidoritsu Eisei Kenkyushoho 24: 35–38.

Hayashi, A. 1972. Pyrethroid synergists — effect and mode of action. Kagasku (Tokyo) 42: 376–385.

Hayes, D.K., Sullivan, W.N., Oliver, M.Z. and Schecter, M.S. 1970. Photoperiod manipulation of insect diapause. A method of pest control? Science 169: 382–383.

Hazleton, L.W. and Holland, E. 1953. Long-term feeding of malathion to rats. AMA Arch. Industr. Hyg. Occup. Med. 8: 399–405.

Hazleton Laboratories 1954. Acute toxicity studies on diazinon. Report to Geigy Chemical Corporation. (Unpublished).

Head, S.W. 1966. The quantitative determination of pyrethrins by gas/liquid chromatography. Part I. Detection by electron capture. Pyrethrum Post 8(4): 3.

Head, S.W., Sylvester, N.K. and Challinor, S.K. 1968. The effects of piperonyl butoxide on the stability of crude and refined pyrethrum extracts. Pyrethrum Post 9(3): 14–22.

Heather, N.W. and Wilson, D. 1983. Resistance to fenitrothion in *Oryzaephilus surinamensis* in Queensland. J. Aust. Entomol. Soc. 22: 210–211.

Hend, R.W. and Butterworth, S.T.C. 1976. Toxicity studies of the insecticide WL43775. A short-term feeding study in rats. Report from Shell Development Co. (Unpublished).

Henrick, C.A. 1982. Juvenile hormone analogs: Structure/activity relationships. In: Insecticide Mode of Action. Academic Press, Inc. N.Y. pp. 315–402.

Henrick, C.A., Staal, G.B. and Siddal, J.B. 1973. Alkyl 3,7, 11-trimethyl-2,4-dodeca-dienoate, a new class of potent insect growth regulators with juvenile hormone activity. J. Agric. Food Chem. 21: 354–359.

Henrick, C.A., Willy, W.E. and Staal, G.B. 1976. Insect juvenile hormone activity of alkyl (2E, 4E)-3,7,11- trimethyl-2,4-dodecadienoates. Variations in the ester function and in the carbon chain. J. Agric. Food Chem. 24: 207–218.

Herrick, G.M., Fry, J.L., Fong, W.G. and Golden, D.C. 1969. Pesticide residues in eggs resulting from the dusting and short-time feeding of low levels of chlorinated hydrocarbon insecticides to hens. J. Agric. Food Chem. 17: 291.

Herve, J.J. 1985. Agricultural, public health and animal health usage of pyrethroids. In: Leahey, J.P., Ed., The Pyrethroid Insecticides. Taylor and Francis, London. pp. 343 417.

Hesseltine, C.W. 1969. Mycotoxins. Mycopathol. Mycol. Appl. 39: 371–383.

Hesseltine, C.W. 1974. Conditions leading to mycotoxin contamination of foods and feeds. In: Rodricke, J.V., Ed., Symp. 168th Meeting Amer. Chem. Soc. on Mycotoxins and Other Fungal Related Food Problems, Atlantic City, N.J., 11–13 September 1974. ACS, Washington. Advances in Chemistry Series No. 149, pp. 1–22.

Hewlett, P.S. 1960. Joint action in insecticides. In: Metcalfe, R.L., Ed., Advances in Pest Control Research. Interscience Publishers Inc., New York and London. Vol. III, pp. 27–74.

Hewlett, P.S. 1969. The toxicity to *Tribolium castaneum* of mixtures of pyrethrins and piperonyl butoxide: fitting a mathematical model. J. Stored Prod. Res. 5(1): 1–9.

Hewlett, P.S. and Clayton, B. 1952. Pyrethrum/BHC sprays for stored product insects. 1950 Pest. Infest. Res. Report p. 14.

Hibbs, A.N. 1968. Proc. Sanitation Workshop, Grain and Cereal Products, Kansas State University, Manhattan, 8–11 September. NPCA, Inc., Elizabeth, N.J. 1967.

Hill, E.G. and Border, B.S.J. 1967. Breakdown of malathion during malting processes. Chem. Ind. 1967: 363–364.

Hill, E.G. and Thompson, R.H. 1968. Pesticide residues in foodstuffs in Great Britain. V. Malathion in imported cereals. J. Sci. Food Agric. 19: 119.

Hill, R. 1972. Acute oral toxicity of ZR 515 for dogs. Syntex Research, Department of Toxicology Report No. 99-D-72-ZR515-PO-TX (26 October 1972) submitted by Zoecon Corporation. (Unpublished).

Hill, E.F., Heath, R.G., Spann, J.W. and Williams, J.D. 1975. Lethal dietary toxicities of environmental pollutants to birds. U.S. Fish. Wildl. Serv. Spec. Sci. Rep. 191: 61 pp.

Hindmarsh, P.S. and Macdonald, I.A. 1980. Field trials to control insect pests of farm stored maize in Zambia. J. Stored Prod. Res. 16(1): 9–18.

Hiromori, T., Koyama, Y., Okuno, Y., Arai, M., Ito, N. and Miyamoto, J. 1980. Two-year chronic toxicity study of S-2539 in rats. Report from Sumitomo Chemical Co. of a study performed by Industrial Bio-Test Laboratories, Inc. validated by Sumitomo Chemical Co. and submitted to the World Health Organization by Sumitomo Chemical Co. (Unpublished).

Hladka, A. and Nosal, M. 1967. The determination of the exposition of Metathion (fenitrothion) on the basis of excreting its metabolite p-nitro-m-cresol through urine in rats. Int. Arch. Gewerbepath. Gewerbehyg. 23: 209–14.

Hodges, R.J. 1984a. Biological methods for integrated control of insects and mites in tropical stored products. II: the use of pheromones. Trop. Stored Prod. Inf. (48): 9–15.

Hodges, R.J. 1984b. Biological methods for integrated control of insects and mites in tropical stored products. IV: the use of insect diseases. Trop. Stored Prod. Inf. (48): 27–31.

Hodges, R.J. 1984c. Biological methods for integrated control of insects and mites in tropical stored products. V: the use of sterile insects. Trop. Stored Prod. Inf. (48): 33–36.

Hodges, R.J. and Meik, J. 1984. Infestation of maize cobs by *Prostephanus truncatus* (Horn) (Coleoptera: Bostrichidae) — aspects of biology and control. J. Stored Prod. Res. 20(4): 205–213.

Hodges, R.J., Dunstan, W.R., Magazini, I. and Golob, P. 1983. An outbreak of *Prostephanus truncatus* in East Africa. Protection Ecology 5: 183–194.

Hodgson, E. and Casida, J.E. 1962. Mammalian enzymes involved in the degradation of dichlorvos. J. Agric. Food Chem. 10: 208.

Hodgson, E., Philpot, R.M., Baker, R.C. and Mailman, R.B. 1973. Effect of (pesticide) synergists on drug metabolism. Drug Metab. Dispos. 1(1): 391–401.

Hoffman, R.A., Hopkins, T.L. and Lindquist, A.W. 1954. Tests of synergists with pyrethrum synergists combined with some organic phosphorus compounds against DDT-resistant flies. J. Econ. Entomol. 47: 72–76.

Hollingsworth, R.M. 1969. Dealkylation of organophosphorus esters by mouse liver enzymes *in vitro* and *in vivo*. J. Agric. Food chem. 17: 987.

Hollingworth, R.M., Metcalf, R.L. and Fukuto, T.R. 1967a. The selectivity of Sumithion compared with methyl parathion. Metabolism in the white mouse. J. Agric. Food Chem. 15: 242–249.

Hollingworth, R.M., Fukuto, T.R. and Metcalf, R.L. 1967b. Selectivity of Sumithion compared with methyl parathion. Influence of structure on anticholinesterase activity. J. Agric. Food Chem. 15: 235–241.

Hooper, G.H.S. 1971. Sterilization and competitiveness of the Mediterranean fruit fly after irradiation of the pupae with fast neutrons. J. Econ. Entomol. 64: 1369–1372.

Hoppe, T. 1974. Effect of a juvenile hormone analogue on Mediterranean flour moth in stored grains. J. Econ. Entomol. 67: 789.

Hoppe, T. 1976. Microplot trial with an epoxy phenylether insect growth regulator against several pests of stored wheat grain. J. Stored Prod. Res. 12: 205–209.

Hoppe, T. 1981. Testing of methoprene in resistant strains of *Tribolium castaneum* (Herbst) (Col.; Tenebrionidae). Z. Angew. Entomol. 91(3): 241–51.

Hoppe, T. and Sucky, M. 1975. Present status of research on insect growth regulators for the protection of stored grain. EPPO Bulletin 5(2): 193–6.

Horler, D.F. 1966. Protective treatment of grain. Pest Infest. Res. 1965: 25.

Horler, D.F. 1972. Note on recovery of carbon-14 labelled malathion from aged samples of wheat grain. In: Proceedings of a Combined Panel and Research Coordination Meeting, Vienna, 25–29 October 1971. IAEA Vienna. p 143.

Horler, D.F. and Clark, J.S. 1974. Effect of time of storage on the recovery of malathion from treated wheat. In: Proceedings and Report of Research Coordination Meeting on Isotope Tracer Studies — Chemical Residues in Food and the Agricultural Environment, 1972. IAEA Vienna. pp. 25–27.

Horler, D.F. and Rowlands, D.G. 1967. Penetration of grains by insecticides. Pest Infest. Res. 1967: 30.

Horrigan, W. 1984. Export inspection in the future. Proc. 1st Aust. Stored Grain Pest Control Conf., Melbourne, May 1981. pp. 9, 21–25.

Horton, P.M. 1984. Evaluation of South Carolina field strains of certain stored-product Coleoptera for malathion resistance and pirimiphos-methyl susceptibility. J. Agric. Entomol. 1(1): 1–5.

Horwitz, W. 1982. Evaluation of analytical methods used for regulation. J. Assoc. Off. Anal. Chem. 65: 525–30.

Horwitz, W., Kamps, L.R. and Boyer, K.W. 1980. Quality assurance in the analysis of foods for trace constituents. J. Assoc. Off. Anal. Chem. 63: 1344–1354.

Houston, D.F. 1952. Hygroscopic equilibrium of brown rice. Cereal Chem. 29(1): 71–75.

Howe, R.W. 1962. A study of the heating of stored grain caused by insects. Ann. Appl. Biol. 50: 137–158.

Howe, R.W. 1965a. Losses caused by insects and mites in stored foods and feeding stuffs. Nutr. Abstr. Rev. 35: 285–293.

Howe, R.W. 1965b. A summary of estimates of optimal and minimal conditions for population increase of some stored products insects. J. Stored Prod. Res. 1: 177–84.

Howes, A.D. 1966. Dichlorvos effects on the bovine. M.S. Thesis, Washington State University, Pullman, Washington.

Hsieh, F.K., Hsu, S.L., Wu, I.S. and Hsieh, G.C. 1983. Toxicity of commonly used insecticides to the maize weevil and lesser grain borer. Plant Protection Bulletin, Taiwan 25,(4): 285–289.

Hubbard, J.E., Earle, F.R. and Senti, E.R. 1957. Moisture relations in wheat and corn. Cereal Chem. 34: 422–433.

Hudson, R.H. 1972. Acute oral toxicity of ZR-515 to mallard drakes. Denver Wildlife Research Centre Internal Report Series in Pharmacology ZR-515 (19 December 1972). Submitted by Zoecon Corporation. (Unpublished).

Hudson, R.H., Haegele, M.A. and Tucker, R.K. 1979. Acute oral and percutaneons toxicity of pesticides to mallards; correlations with mammalian toxicity data. Toxicol. Appl. Pharmacol. 47: 451.

Hunt, L.M. and Gilbert, B.N. 1977. Distribution and excretion rates of ^{14}C-labelled permethrin isomers administered orally to four lactating goats for 10 days. J. Agric. Food Chem. 25(3): 673.

Hunter, B. and Newman, A.J. 1972. Toxicity of pyrethrins and piperonyl butoxide in dietary administration to rats. Huntington Research Centre. (Preliminary survey).

Hunter, A.J. and Taylor, P.A. 1980. Refrigerated aeration for the preservation of bulk grain. J. Stored Prod. Res. 16: 123–131.

Huq, A.K.F. 1980. Some estimates of farm-level storage losses of rice in Bangladesh. Trop. Stored Prod. Inf. (39): 5–12.

Hurlock, E.T. 1965. Some observations on the loss in weight caused by *Sitophilus granarius* to wheat under constant experimental conditions. J. Stored Prod. Res. 1: 193–195.

Hurlock, E.T. 1967. Some obervations on the amount of damage caused by *Oryzaephilus surinamensis* on wheat. J. Stored Prod. Res. 3: 75–78.

Hurlock, E.T., Llewelling, B.E. and Stables, L.M. 1979. Microwaves can kill insect pests. Food Manuf. 54(8): 37, 39.

Hussain, S.I. and Qayyum, H.A. 1972. Effectiveness of some insecticides for the control of *Trogoderma granarium* at different temperatures and relative humidities. Pak. J. Sci. Res. 24(3–4): 227–233.

Husted, S.R., Mills, R.B., Foltz, V.D. and Crumrine, M.H. 1969. Transmission of *Salmonella montevideo* from contaminated to clean wheat by the rice weevil. J. Econ. Entomol. 62(6): 1489–1491.

Hutson, D.H. 1979. Metabolism of pyrethroid insecticides. In: Bridges, J.W., and Chausseaud, L.F., Ed., Progress in Drug Metabolism. John Wiley. Vol. 3, p. 215.

Hutson, D.H., Pickering, B.A. and Donniger, C. 1968. Phosphoric acid triester: glutathione alkyltransferase. Biochem. J. 106: 20.

Hutt, R.B. and White, L.D. 1973. Recovery of fertility in male codling moths treated with 38 krad of *gamma* irradiation. J. Econ. Entomol. 66: 388–389.

Hyari, S., Kadoum, A.M. and LaHue, D.W. 1977. Laboratory evaluations of emulsifiable and encapsulated formulations of malathion and fenitrothion on soft red winter wheat against attack by adults of four species of stored-product insects. J. Econ. Entomol. 70(1): 480–482.

Hyde, M.B. 1969. Hazards of storing high moisture grain in airtight silos in tropical countries. Trop. Stored Prod. Inf. (18): 9–11.

Hyde, M.B. 1974. Airtight storage. In: Christensen, C.M., Ed., Storage of Cereal Grains and their Products. American Assosiation of Cereal Chemists, Inc., St Paul, Minnesota. Vol. 5, pp. 383–419.

Hyde, M.B., Baker, A.A., Ross, A.C. and Lopez, C.O. 1973. Airtight grain storage. Agricultural Services Bull. No. 17, FAO, Rome. 71 pp.

Ibvijaro, M.F. 1981. The efficacy of pirimiphos methyl in crib storage of maize in south western Nigeria (mainly against *Sitophilus* spp.) Insect Sci. Appl. 1(3): 295–296.

Ida, M. and Katsuya, S. 1956. Studies on the effect of lindane fumigation on the optimum period to control insect injury to stored cereals. Botyu Kagaku 21: 92.

Ifflaender, U. and Mucke, W. 1975. Distribution, degradation and excretion of CGA 20168 in the rat. Report from Ciba-Geigy Limited, Basle, Switzerland, submitted to the World Health Organisation. (Unpublished).

Immel, R. and Geisthardt, G. 1964. Meded. Landbouhogesch. Wageningen 29: 1242.

INIAP (Instituto Nacional de Investigaciones Agropecuaries) 1972. Control of *Sitophilus oryzae* in maize. INIAP Rep., Ecuador 1972. 168.

Innes, J.R.M. et al. 1969. Bioassay of pesticides and industrial chemicals for tumorigenicity in mice: a preliminary note. J. Natl. Cancer Inst. 42(6): 1101–1114.

INRA 1976a. Toxicity of deltamethrin or DECIS in single ingestion in game duck; *Anas platyrhynchos* L. Report INRA — 76.21.1Z/A, Institut National de la Recherche Agronomique, Jony-en-Jossas, France. Submitted to WHO by Roussel Uclaf. (Unpublished).

INRA 1976b. Toxicity of deltamethrin or DECIS by single ingestion in grey partridge, *Perdix perdix* L. and red partridge, *Alectons rufa* L. Report INRA — 76.28.09/A Institut National de la Recherche Agronomique, Jony-en- Jossas, France. Submitted to WHO by Roussel Uclaf. (Unpublished).

Ioannou, U.M. and Dauterman, W.C. 1978. *In vitro* metabolism of etrimfos by rat and mouse liver. Pestic. Biochem. Physiol. 9: 190–195.

Iordanou, N.T. and Watters, F.L. 1969. Temperature effects on the toxicity of five insecticides against five species of stored-product insects. J. Econ. Entomol. 62: 130–135.

Ishaaya, I., Elsner, A., Ascher, K.R.S. and Casida, J.E. 1983. Synthetic pyrethroids: toxicity and synergism on dietary exposure of *Tribolium castaneum* (Herbst) larvae. Pestic. Sci. 14(4): 367–72.

Isshiki, K., Tsumura, S. and Watanabe, T. 1978. Residual piperonyl butoxide in agricultural products. Bull. Environ. Contam. Toxicol. 19: 518–523.

Ito, T., Kageyama, Y. and Hirose, G. 1976. Fate of fenitrothion residues in rice grains. Report of Research, Department, Sumitomo Chemical Co. Ltd., Takarazuka City, Japan. (Submitted to J. Stored Prod. Res. 1975.)

Ivie, G.W. and Hunt, L.M. 1980. Metabolites of *cis*- and *trans*-permethrin in lactating goats. J. Agric. Food Chem. 28: 1131–1138.

Ivey, M.C., Hoffman, R.A. and Claborn, H.V. 1968. Residues of Gardona (tetrachlorvinphos) in the body tissues of cattle sprayed to control *Hypoderma* spp. J. Econ. Entomol. 61: 1647.

Ivey, M.C., Hoffman, R.A., Claborn, H.V. and Hogan, B.F. 1969. Residues of Gardona in the body tissues and eggs of laying hens exposed to treated litter and dust boxes for control of external arthropod parasites. J. Econ. Entomol. 62: 1003.

Ivey, M.C., Miller, J.A. and Ivie, G.W. 1982. Methoprene: residues in fat of cattle treated with methoprene boluses. J. Econ. Entomol. 75(2): 254–6.

Izumi, T., Kaneko, H., Matsuo, M. and Miyamoto, J. 1984. Comparative metabolism of the six stereoisomers of phenothrin in rats and mice. J. Pestic. Sci. 9: 259–267.

Jackson, J.B., Drummond, R.O., Buck, W.B. and Hunt, L.M. 1960. Toxicity of organic phosphorus insecticides to horses. J. Econ. Entomol. 53: 602–604.

Jacobson, R. and Pinniger, D.B. 1982. Eradication of *Oryzaephilus surinamensis* from a grain store. Int. Pest Contr. 24(3): 68–74.

Jaffe, H. and Neumeyer, J.L. 1970. Comparative effects of piperonyl butoxide and N-(4-pentynyl)phthalimide on mammalian microsomal enzyme functions. J. Med. Chem. 13(5): 901–903.

Jaggers, S.E. and Parkinson, G.R. 1979. Permethrin. Summary and review of acute toxicities in laboratory species. Report ICI Central Toxicology Laboratory. (Unpublished).

James, J.A. 1980. The toxicity of synthetic pyrethroids to mammals. In: van Miert, A., Frans, J., and van der Kreek, F., Ed., Proc. 1st European Cong. Vet. Pharmacol. Toxicol., Zeist, Netherlands, 25–28 September 1979. Elsevier, Amsterdam. pp. 249–255.

Jao, L.T. and Casida, J.E. 1974. Esterase inhibitors as synergists for (+)-*trans*-chrysanthemate insecticide chemicals. Pestic. Biochem. Physiol. 4(4): 456–464.

Jay, E. 1984. Control of *Rhyzopertha dominica* with modified atmospheres at low temperatures. J. Agric. Entomol. 1(2): 155–160.

Jay, E. 1984a. Recent advances in the use of modified atmospheres for the control of stored-product insects. In: Baur, F.J., Ed., Insect Pest Management for Food Storage and Processing. American Association of Cereal Chemists, St. Paul, Minnesota. p.239–252.

Jefferies, M.G. 1979. Quality maintenance in grains and allied foods and feedstuffs for export and dometic markets. In: Champ, B.R., and Highley, E., Ed., Grain Storage Research and its Application in Australia. CSIRO Division of Entomology, Canberra. pp. 85–90.

Joffe, A. 1958. Moisture migration in horizontally-stored bulk maize: the influence of grain-infesting insects under South African conditions. S. African J. Agric. Sci. 1: 175–190.

Johnson, J.C. Jr. and Bowman, M.C. 1972. Responses from cows fed diets containing fenthion or fenitrothion. J. Dairy Sci. 55: 777–781.

Johnson, J.C. Jr., Jones, R.L., Leuck, D.B. and Bowman, M.C. 1974. Persistence of chlorpyrifos-methyl in corn silage and effect of feeding dairy cows the treated silage. J. Dairy Sci. 57: 1467–1473.

Joiner, R.L. 1975. Safety evaluation of Sumithion technical by a five-day feeding study in Mallard ducks. Report by Stauffer Chemical Co. Richmond Research Centre to Sumitomo Chemical Co. Ltd. T-5606 (HW-51–0195), 6 October 1975. (Unpublished).

Joia, B.S. 1983. Insecticidal efficacy and residues of cypermethrin and fenvalerate in stored wheat. Diss. Abstr. Int. B. 44(1): 37–8.

Joia, B.S., Webster, G.R.B. and Loschiavo, S.R. 1985a. Cypermethrin and fenvalerate residues in stored wheat and milled fractions. J. Agric. Food Chem. 33: 618–622.

Joia, B.S., Loschiavo, S.R. and Webster, G.R.B. 1985b. Cypermethrin and fenvalerate as grain protectants against *Tribolium castaneum* and *Cryptolestes ferrugineus* at different moisture levels and temperatures. J. Econ. Entomol. 78(3): 637–641.

Joint FAO/WHO Food and Animal Feed Contamination Monitoring Programme 1981. Analytical Quality Assurance. FAO-ESN/MON/AQA/81/8 FAO, Rome.

Jones, K.H., Sanderson, D.M. and Noakes, D.M. 1968. Acute toxicity data for pesticides. World Rev. Pest Control (London) 7: 138.

Jorgensen, T.A. and Sasmore, D.R. 1972. Toxicity studies of ZR-515 (Altosid Tech) Stanford Research Institute Report No. 1/LSC-1833/Sept. 6, 1972. Submitted by Zoecon Corporation. (Unpublished).

Joubert, P.C. 1965a. The toxicity of contact insecticides to seed-infesting insects. Pyrethrum Post 8(2): 6–15.

Joubert, P.C. 1965b. Disinfect mills with pyrethrum. (Reprinted from "Farming in South Africa".) Pyrethrum Post 8(2): 16–20.

Joubert, P.C. and De Beer, P.R. 1968a. The toxicity of contact insecticides to seed-infesting insects. Tests with pyrethrum and malathion on infested maize. Technical Communication No. 73, Department of Agriculture, Pretoria, South Africa. South African Government Printer, Pretoria.

Joubert, P.C. and De Beer, P.R. 1968b. The toxicity of contact insecticides to seed-infesting insects. Tests with bromophos on maize. Technical Communication No. 84, Department of Agriculture, Pretoria, South-Africa. South African Government Printer, Pretoria.

Joubert, P.C. and du Toit, D.M. 1968. The toxicity of contact insecticides to seed-infesting insects. Tests with various pyrethrum formulations. Technical Communication No. 83, Department of Agriculture, Pretoria, South Africa. South African Government Printer, Pretoria.

Juliano, J.O. 1964. Hygroscopic equilibria of rough rice. Cereal Chem. 41(3): 191–197.

Kadota, T., Okuno, Y. and Miyamoto, J. 1975. Acute oral toxicity and delayed neurotoxicity of 5 organophosphorus compounds, Salithion, Cyanox, Surecide, Sumithion and Sumioxon in adult hens. Botyu-Kagaku (Sci. Pest. Control.) 40: 49–53.

Kadota, T. and Kagoshima, M. 1976. Acute oral toxicity and delayed neurotoxicity of Sumithion in hens. Report of Pesticides Division, Sumitomo Chemical Co. Ltd., 16 October 1976 HT-60–0026. 22 pp. (Unpublished).

Kadota, T. and Miyamoto, J. 1975. Acute and sub-acute toxicity of Sumithion in Japanese quails. Botyu-Kagaku (Sci. Pest. Contr.) 40: 54–58.

Kadota, T., Kagoshima, M. and Miyamoto, J. 1974. Acute and sub-acute toxicity of Sumithion (fenitrothion) in Japanese quail. Report of Pesticides Division, Sumitomo Chemical Co. Ltd., July 1974 HT-40–0050. (Unpublished).

Kadota, T., Okuno, Y. and Miyamoto, J. 1975. Acute oral toxicity and delayed neurotoxicity of 5 organophosphorus compounds including (fenitrothion) in adult hens. Botyu Kagaku (Sci. Pest. Contr.) 40(2): 49–53.

Kadoum, A.M. 1969. Extraction and cleanup methods to determine malathion and its hydrolytic products in stored grains by gas-liquid chromatography. J. Agric. Food Chem. 17(6): 1178–1180.

Kadoum, A.M. and Al-Naji, L. 1978. Effect of grain moisture content on the degradation rate of methyl phoxim in corn, sorghum and wheat. J. Agric. Food Chem. 26(2). 507–9.

Kadoum, A.M. and LaHue, D.W. 1969. Effect of hybrid, moisture content, foreign materials and storage temperature on the degradation of malathion residues in grain sorghum. J. Econ. Entomol. 62: 1161.

Kadoum, A.M. and LaHue, D.W. 1972. Degradation of malathion on viable and sterilised sorghum grain. J. Econ. Entomol. 65(2): 497.

Kadoum, A.M. and LaHue, D.W. 1974. Penetration of malathion in stored corn, wheat and sorghum grain. J. Econ. Entomol. 67(4): 477–478.

Kadoum, A.M. and LaHue, D.W. 1977. Degradation of malathion in wheat and milling fractions. J. Econ. Entomol. 70: 109–110.

Kadoum, A.M. and LaHue, D.W. 1979a. Degradation of malathion on wheat and corn of various moisture contents. J. Econ. Entomol. 72: 228–229.

Kadoum, A.M. and LaHue, D. 1979b. Effect of grain moisture content on malathion residue degradation. (Abstract). Proceedings of the North Central Branch of the Entomological Society of America 33: 47.

Kadoum, A.M. and Sae, S.W. 1970. Effect of some organophosphorus compounds and their metabolites on sorghum grain esterase and certain insects attacking sorghum. Bull. Environ. Contam. Toxicol. 5: 213.

Kadoum, A.M., LaHue, D.W. and Al-Naji, L. 1978. Efficacy and fate of pirimiphos-methyl residue applied at two dosage rates to wheat for milling. J. Econ. Entomol. 71(1): 50–52.

Kalonji, M. Mbombo No date. Faculte des sciences agronomiques de l'Etat, Gembloux, Belgique. Etude comparative de quelques matieres actives dans la lutte contre le charancon du ble. Thesis.

Kamel Abd El-Hakim 1958. Pyrenone for the protection of stored grains against insects. Agr. Res. Rev. Cairo 36: 96–109.

Kanazawa, J. 1973. Survey of pesticide residue in agricultural commodities in Japan. Japan Pestic. Inf. No. 11: 5–16.

Kane, J. and Aggarwal, S.L. 1970. Protective treatment of bagged barley. Pest Infest. Res. 1969, p. 33.

Kane, J. and Green, A.A. 1968. The protection of bagged grain from insect infestation using fenitrothion. J. Stored Prod. Res. 4: 59–68.

Kane, M.J. 1965. Treatment of infested grain. Pest Infest. Res. 1965 p. 27.

Kane, M.J., Green, A.A. and Horler, D.F. 1967. Protective treatment of grain. Pest Infest. Res. 1967 pp. 24–25.

Kane, M.J. Clark, J.S., Green, A.A. and Horler, D.F. 1971. The distribution of insecticide among grains. Rept. of Director, Pest Infestation Control Lab., Slough, 1968. 70: p. 112.

Kaneko, H., Ohkawa, H. and Miyamoto, J. 1979. Comparative metabolism of fenvalerate in rats and mice. Report of Sumitomo Chemical Co Ltd. (Unpublished).

Kaneko, H., Ohkawa, H. and Miyamoto, J. 1981. Absorption and metabolism of dermally applied phenothrin in rats. J. Pestic. Sci. 6: 169–182.

Kaneko, H., Ohkawa, H. and Miyamoto, J. 1981a. Comparative metabolism of fenvalerate and the [2S, S]-isomer in rats and mice. J. Pestic. Sci. 6: 317–326.

Kaneko, H., Izumi, T., Matsuo, M. and Miyamoto, J. 1984. Metabolism of fenvalerate in dogs. J. Pestic. Sci. 9: 269–274.

Kansouh, A.S.H. 1975. Diazinon metabolism in stored wheat grains. Bull. Entomol. Soc. Egypt 8: 47–51. (Seen through Chemical Abstracts 1976, 85, 14955g.)

Kantack, B.H. and Laudani, H. 1957. Comparative laboratory tests with emulsion and wettable-powder residues against the Indian-meal moth. J. Econ. Entomol. 50: 513–14.

Karapally, J.C. 1975. Absorption, blood levels, distribution and excretion of etrimfos in the rat following a single oral dose. Sandoz Ltd. Report Agro. Dok. CBK 1829/75/21 April 1975. (Unpublished).

Karapally, J.C. 1977. Metabolism of ^{14}C-etrimfos in the rat. Sandoz Ltd. Report Agro. Dok. CBK 3005/10 May 1977. (Unpublished).

Karapally, J.C. 1982. Personal communication to WHO. Sandoz Ltd., Basle. Quoted in JMPR monograph 1982 p. 235.

Karnavar, G.K. 1973. Effects of synthetic juvenile hormone on diapause and metamorphosis of stored-grain pest, *Trogoderma granarium*. Indian J. Exp. Biol. 11: 138–140.

Karon, M.L. and Adams, M.E. 1949. Hygroscopic equilibrium of rice and rice grains. Cereal Chem. 26(1): 1–12.

Kashi, K.P. 1972. An appraisal of fenitrothion as a promising grain protectant. Int. Pest Contr. 14: 20–22.

Kashi, K.P. 1981. Controlling pests in stored grain with carbon dioxide. Span 24.2.1981: 69–71.

Kateman, G. and Pijpers, F.W. 1981. Quality control in analytical chemistry. In: Elving, P.C., and Winefordner, J.D., Ed., Chemical Analysis. Wiley. Vol. 60.

Kavlock, R., Chernoff, N., Baron, R., Linder, R., Rogers, E. and Carver, B. 1979. Toxicity studies with deltamethrin, a synthetic pyrethroid insecticide. J. Environ. Pathol. Toxicol. 2: 751–765.

Kawamura, Y., Takeda, M., Uchiyama, M., Sakai, K. and Ishikawa, H. 1980. Organophosphorus residues in flour. Shokuei-shi 21(1): 71–76.

Kearns, C.W. 1956. The mode of action of insecticides. Ann. Rev. Entomol. 1: 123–147.

Keller, J.G. 1957. Malathion Technical 95%. poultry study — final report. Report from Hazleton Laboratories, Falls Church, Virginia to American Cyanamid, 13 December 1957. (Unpublished).

Kennedy, G.L., Smith, S.H., Kinoshita, F.K., Keplinger, M.L. and Calandra, J.C. 1977. Teratogenic evaluation of piperonyl butoxide in the rat. Food Cosmet. Toxicol. 15(4): 337–339.

Kent-Jones, D.W. and Amos, A.J. 1957. Modern Cereal Chemistry (5th Ed.). Northern Publishing Co. Liverpool.

Kettering Laboratory 1964. Neurological effects of Vapona insecticide on chickens. Report to Ciba Ltd from Kettering Research Laboratory (18 February).

Khare, B.P. 1972. Final technical report. Insect pests of stored grain and their control in Uttar Pradesh. Res. Bull. G.B. Pant Univ. Agric. Technol., Patnagar, U.P., Coll. Agric. Exp. Stn., Dep. Entomol., No. 5, 153 pp.

Khare, B.P. 1973. Insect pests of stored grain and their control in Uttar Pradesh. G.P. Pant, Univ. of Agric. and Technol., Pantnagar, Naintal.

Khera, K.S. and Lyon, D.A. 1968. Chick and duck embryos in the evaluation of pesticide toxicity. Toxicol. Appl. Pharmacol. 13: 1.

Khmelevskii, B.N. 1968. The action of Sevin on the organism of chickens during prolonged feeding. Veterinariya 9: 59.

Khoda, H., Kadota, T. and Miyamoto, J. 1976. Teratogenic study on S5602 in mice. Report from Sumitomo Chemical Co Ltd. (Unpublished).

Khoda, H., Kadota, T. and Miyamoto, J. 1979. Acute oral, dermal and subcutaneous toxicities of permethrin in rats and mice. Report Sumitomo Chemical Co. Ltd. (Unpublished).

Kimmerle, G. and D. Lorke, 1968. Toxicology of insecticidal organo-phosphates. Pflanz.-Nachr. Bayer 21: 111–142.

King, D.R., Morrison, E.O. and Sundman, J.A. 1962. Bioassay of chemical protectants and surface treatments for the control of insects in stored sorghum grain. J. Econ. Entomol. 55(4): 506–510.

Kinkel, H.J. 1964. Ermittlung der akuten peroralen Toxizitat und Untersuchungen uber die Neurotoxizitat des Preparates Bromophos (CELA S 1942) an Huhnern. Report. Battelle Institut. (Unpublished).

Kinkel, H.J., Muacevic, G., Sehring, R. and Bodenstein, G. 1966. Arch. Toxikol. 22: 36.

Kiritani, K. 1964. Insect infestation of stored rice in Japan. In: Proc. 12th Int. Congr. Entomol., London, 1964 p. 630.

Kirkpatrick, R.L. 1975a. Infra-red radiation for control of lesser grain borers and rice weevils in bulk wheat. J. Kans. Entomol. Soc. 48: 100–104.

Kirkpatrick, R.L. 1975b. The use of infra-red and microwave radiation for control of stored-product insects. In: Proc. 1st Int. Conf. Stored- Products Entomology, Savannah, Georgia, 7–11 October 1974. pp. 431–437.

Kirkpatrick, R.L., Harein, P.K. and Cooper, C.V. 1968. Laboratory tests with dichlorvos applied as a wheat protectant against rice weevils. J. Econ. Entomol. 61: 356–358.

Kirkpatrick, R.L. and Tilton, E.W. 1972. Infra-red radiation to control adult stored product Coleoptera. J. Ga. Entomol. Soc. 7: 73–75.

Kirkpatrick, R.L. and Tilton, E.W. 1973. Elevated temperatures to control insect infestation in wheat. J. Ga. Entomol. Soc. 8 264–268.

Kirkpatrick, R.L., Brower, J.H. and Tilton, E.W. 1972. A comparison of microwave and infra-red radiation to control rice weevils in wheat. J. Kans. Entomol. Soc. 45: 434–438.

Kirkpatrick, R.L., Brower, J.H. and Tilton, E.W. 1973. *Gamma*, infrared and microwave radiation combinations for control of *Rhyzopertha dominica* in wheat. J. Stored Prod. Res. 9: 19–23.

Kirkpatrick, R.L., Redlinger, L.M., Zettler, J.L. and Simonaitis, R.A. 1982. CGA-20168 applied to corn for control of stored-product insects. J. Econ. Entomol. 75(2): 277–280.

Kirkpatrick, R.L., Redlinger, L.M., Simonaitis, R.A. and Zettler, J.L. 1983. Stability and effectiveness of fenitrothion on corn to control stored-product insects. J. Ga. Entomol. Soc. 18(3): 344–350.

Klotzsche, C. 1955. The toxicology of O,O-Dimethyl 2,2-Dichlorovinyl-Phosphate. Z. Angew. Zool. 1: 87–93.

Knaak, J.B. and O'Brien, R.D. 1960. Effect of EPN on the *in vivo* metabolism of malathion by the rat and dog. J. Agric. Food Chem. 8: 198.

Knipling, E.F., 1950. New developments concerning residues in dairy, meat and other food products. Proc. N. Central States Br. Amer. Assoc. Econ. Entomol. 4.

Kockum, S. 1958. Control of insects attacking maize on the cob in crib stores. East. Afr. Agric. J. 23: 275.

Kockum, S. 1965. Crib storage of maize. A trial with pyrethrin and lindane formulations. East. Afr. Agric. Forestry J. 31: 8.

Koivistoinen, P. and Aalto, H. 1970. Malathion residues and their fate in cereals. In: Nuclear techniques for studying pesticide residue problems, IAEA, Vienna. p. 11.

Kokhtyuk, F.P. and Sukhomlinova, G.K. 1977. The diagnosis of poisoning of poultry with Gardona. Veterinariya, Moscow, USSR (10): 101–102.

Koransky, W. and Portig, J. 1963. Resorption, verteilung and ausscheidwig von *alpha* und *gamma* hexachlorocyclohexane. Arch. Exp. Pathol. Pharmakol. 244: 564.

Koransky, W. and Portig, J., Vohland, H.W. and Klempau, J. 1964. Die Elimination von *alpha* und *gamma* HCH und ihr Beeinfluzsung durch Enzyme der debermicrosomen. Naunyn-Schmiedebergs Arch. Exp. Pathol. Pharmakol. 247: 49.

Kosolapova, G. Ya. 1980. Valexon and Basudin against stored-product pests. Zashch. Rast. (Mosc). 10: 40.

Kramer, K.J. and McGregor, H.E. 1978. Activity of pyridyl and phenyl ether analogues of juvenile hormone against Coleoptera and Lepidoptera in stored grain. J. Econ. Entomol. 71: 132–134.

Krause, C. and Kirchhoff, S. 1970. Gas chromatographic determination of oranophosphate residues in market samples of fruits and vegetables. Dt. Lebensmitt. Rdsch. 66: 194–199.

Kraybill, H.F. and Shapiro, R.E. 1969. Implications of fungal toxicity to human health. In: Goldblatt, L.A, Ed., Aflatoxin. Academic Press, New York. Chapter XV, pp. 401–441.

Kramer, K.J., Beeman, R.W. and Henricks, L.W. 1981. Activity of Ro/13–5223 and Ro/13–7744 against stored product insects. J. Econ. Entomol. 74: 678–680.

Krishnamurthy, K. 1975. Post harvest losses in food grains. Bull. Grain Technol. 13(1): 33–49.

Krueger, H.R. 1975. Phorate sulfoxidation by plant root extracts. Pestic. Biochem. Physiol. 5: 396–401.

Kruger, H.R. and O'Brien, R.D. 1959. Relationship between metabolism and differential toxicity of malathion in insects and mice. J. Econ. Entomol. 52: 1063.

Kuhr, R.J. and Dorough, H.W. 1976. In: Carbamate insecticides: Chemistry, biochemistry and toxicology. CRC Press, Cleveland, Ohio. pp. 143–160.

Kulkarni, A.P. and Hodgson, E. 1976. Spectral interactions of insecticide synergists with microsomal cytochrome P-450 from insecticide-resistant and susceptible houseflies. Pestic. Biochem. Physiol. 6(2): 183–191.

Kumar, A., Pandey, G.P., Doharey, R.B. and Varma, B.K. 1982. Field trials with some newer organophosphatic insecticides against insect pests of stored foodgrains. Pesticides. 16(1): 7–10, 13.

Kuper, A.W. 1978a. Residues of chlorpyrifos-methyl and 3,5,6-trichloro-2-pyridinol in tissues from calves fed chlorpyrifos-methyl. Dow Chemical, USA, R&D Report. No GH-C 1118, 5 July. 1978 24 pp. (Unpublished).

Kuper, A.W. 1978b. Residues of chlorpyrifos-methyl and 3,5,6-trichloro-2-pyridinol in tissues and eggs from chickens fed chlorpyrifos-methyl. Dow Chemical, USA, R&D Report. GH-C 1155, 15 November 1978 28 pp. (Unpublished).

Kuper, A.W. 1978c. Residues of chlorpyrifos-methyl and 3,5,6-trichloro-2-pyridinol in milk and cream from cows fed chlorpyrifos-methyl. Dow Chemical, USA, R&D Report. GH-C 1161. 19 pp. (Unpublished).

Kuwatsuka, S. 1970. Biochemical aspects of methylenedioxyphenyl compounds in relation to the synergistic action. In: O'Brien, R.D., Ed., Proc. U.S.-Jap. Coop. Sci. Program on Biochem. Toxicol. Insectic. Academic Press, New York. 5th, pp. 131–144.

Ladd, R., Smith, P.S., Jenkins, D.H., Kennedy, G.L., Kinoshita, F.K. and Keplinger, M.L. 1976. Teratogenic study with S-2539 in albino rabbits. Report from Industrial Bio-Test Laboratories to Sumitomo Chemical Co. (Unpublished).

LaHue, D.W. 1965. Evaluation of malathion, synergised pyrethrum and diatomaceous earth as wheat protectants in small bins. USDA, Marketing Research Report No. 726, 13 pp.

LaHue, D.W. 1966. Evaluation of malathion, synergised pyrethrum, and diatomaceous earth on shelled corn as protectants against insects — in small bins. USDA Marketing Research Report No. 768, 10 pp.

LaHue, D.W. 1967. Evaluation of malathion, synergised pyrethrum, and diatomaceous earth as protectants against insects in sorghum grain in small bins. USDA Marketing Research Report No. 781, 11 pp.

LaHue, D.W. 1969. Evaluation of several formulations of malathion as a protectant of grain sorghum against insects in small bins. USDA, Marketing Research Report No. 828. US Government Printing Office, Washington.

LaHue, D.W. 1970a. Evaluation of malathion, diazinon, a silica aerogel, and a diatomaceous earth as protectants on wheat against lesser grain borer attack in small bins. USDA Marketing Research Report No. 860.

LaHue, D.W. 1970b. Laboratory evaluation of dichlorvos as a short-term protectant for wheat, shelled corn and grain sorghum against stored-grain insects. USDA Research Service Report July 1970, pp. 51–37.

LaHue, D.W. 1971. Controlling the Indian-meal moth in shelled corn with dichlorvos PVC resin strips. USDA ARS Report 51–52. 12 pp.

LaHue, D.W. 1972. The effect of dichlorvos PVC resin strips on the abundance of Indian-meal moth *Plodia interpunctella* (Hubner), in shelled corn storage. Proc. N. Central Branch. Entomol. Soc. Am. 27: 68–71.

LaHue, D.W. 1973. Gardona (tetrachlorvinphos) as a protectant against insects in stored wheat. J. Econ. Entomol. 66: 485–489.

LaHue, D.W. 1974. Pesticide residues of potential protectants of grain. Proc. 1st Int. Working Conf. Stored-Product Entomology, Savannah, Georgia, 7–11 October 1974. pp. 635–638.

LaHue, D.W. 1975a. Angoumois grain moth: chemical control of infestation of shelled corn. J. Econ. Entomol. 68(6): 769–771.

LaHue, D.W. 1975b. Evaluating Gardona and malathion to protect wheat in small bins against stored grain insects. USDA Marketing Research Report No. 1037. 12 pp.

LaHue, D.W. 1975c. Pirimiphos-methyl as a short term protectant of grain against stored-product insects. J. Econ. Entomol. 68(2): 235–236.

LaHue, D.W. 1976. Grain protectants for seed corn. J. Econ. Entomol. 69(5): 652–654.

LaHue, D.W. 1977a. Pirimiphos-methyl: efffect on populations of *Tribolium confusum* and *T. castaneum* in wheat. J. Econ. Entomol. 70(1): 135–137.

LaHue, D.W. 1977b. Pirimiphos-methyl: gradient of effective doses on hard winter wheat against attack of four species of adult insects. J. Econ. Entomol. 70(3): 295–297.

LaHue, D.W. 1977c. Grain protectants for seed corn: field test. J. Econ. Entomol. 70(6): 720–722.

LaHue, D.W. 1977d. Chlorpyrifos-methyl: doses that protect hard winter wheat against attack by stored-grain insects. J. Econ. Entomol. 70: 734–736.

LaHue, D.W. 1978. Insecticidal dusts: grain protectants during high temperature — low humidity storage. J. Econ. Entomol. 71(2): 230–232.

LaHue, D.W. and Dicke, E.B. 1971. Phoxim as an insect protectant for stored grains. J. Econ. Entomol. 64: 1530.

LaHue, D.W. and Dicke, E.B. 1976a. Evaluation of selected insecticides applied to high moisture sorghum grain to prevent stored grain insect attack. USDA Marketing Research Report No. 1063, 10 pp.

LaHue, D.W. and Dicke, E.B. 1976b. Evaluating selected protectants for shelled corn against stored-grain insects. USDA Marketing Research Report No. 1058, 9 pp.

LaHue, D.W. and Dicke, E.B. 1977. Candidate protectants for wheat against stored-grain insects. USDA Marketing Research Report No. 1080, 16 pp.

LaHue, D.W. and Fifield, C.C. 1967. Evaluation of four inert dusts on wheat as protectants against insects in small bins. USDA Marketing Research Report No. 780, 24 pp.

LaHue, D.W. and Kadoum, A. 1979. Residual effectiveness of emulsion and encapsulated formulations of malathion and fenitrothion against four stored-grain beetles. J. Econ. Entomol. 72: 234–237.

Lai, F.S., Miller, B.S., Martin, C.R., Storey, C.L., Bolte, L., Shogren, M., Finney, K.F and Quinlan, J.K. 1981. Reducing grain dust with oil additives (treatment of wheat and maize includes insecticide malathion for the control of pests during storage). Transaction of the ASAE — American Society of Agricultural Engineers 24(6): 1626–1631.

Lai, S.P., Finney, K.F. and Milner, M. 1959. Treatment of wheat with ionizing radiations. IV — Oxidative, physical and biochemical changes. Cereal Chem. 36: 401–411.

Lam, J.J. and Stewart, P.A. 1969. Modified traps using blacklight lamps to capture nocturnal tobacco insects. J. Econ. Entomol. 62: 1378–1381.

Langley, P.A. and Maly, H. 1971. Control of the Mediterranean fruit fly using sterile males: effect of nitrogen and chilling during *gamma* irradiation of puparia. Entomol. Exp. Appl. 14: 137–146.

Latta, R. 1951. Proc. South. Corn Conf. on Insects and Diseases, Thomasville, Georgia. (Quoted by Freeman.)

Lawrence, J.F. 1977. Direct analysis of some carbamate pesticides in foods by high-pressure liquid chromatography. J. Agric. Food. Chem. 25: 211–212.

Leahey, J.P. 1985. Metabolism and environmental degradation of pyrethroids In: Leahey, J.P., Ed., The Pyrethroid Insecticides. Taylor and Francis, London. pp. 262–342.

Leahey, J.P. and Curl, E.A. 1982. The degradation of pirimiphos-methyl on stored grains. Pestic. Sci. 13(5): 467–474.

Leahey, J.P., Bewick, D., Gatehouse, D.M. and and Cameron, A.G. 1977a. Permethrin: metabolism and residues in goats. ICI Plant Protection Division Report No. TMJ1516B. (Unpublished).

Leahey, J.P., Gatehouse, D.M. and Parr, J.S. 1977b. Permethrin: metabolism in hens. ICI Plant Protection Division Report No. TMJ1509B. (Unpublished).

Leahey, J.P., Bewick, D., Carpenter, P.K., Parr, J.S. Carpenter, P.K. 1977c. Permethrin: absorption in chickens after dermal and oral treatments. ICI Plant

Protection Division Report No. TMJ1481B. (Unpublished).

Lee, Y.W., Westcott, N.D. and Reichle, R.A. 1978. Gas-liquid chromatographic determination of pydrin a synthetic pyrethroid, in cabbage and lettuce. J. Assoc. Off. Anal. Chem. 61: 869–871.

Lee, P.W., Stearns, S.M. and Powell, W.R. 1985. Rat metabolism of fenvalerate (Pydrin insecticide). J. Agric. Food Chem. 33: 988–993.

Lehman, A.J. 1948. The toxicology of the newer agricultural chemicals. Assoc. Food Drug Off. U.S., Q. Bull. 12: 82–89.

Lehman, A.J. 1951. Chemicals in foods. A report to the Association of Food and Drug Officials on current developments. II. Pesticides. Assoc. Food Drug Off. U.S., Q. Bull. 15: 122–133.

Lehman, A.J. 1954. A toxicological evaluation of household insecticides. Assoc. Food Drug Off. U.S., Q. Bull. 18: 3–13.

Lehman, A.J. 1965. Summaries of pesticide toxicity. Assoc. Food Drug Off. U.S. Q. Bull. 126.

Lemon, R.W. 1966. Laboratory evaluation of some organophosphorus insecticides against *Tribolium confusum* and *T. castaneum*. J. Stored Prod. Res. 1(3): 247–253.

Lemon, R.W. 1967a. Laboratory evaluation of malathion, bromophos and fenitrothion for use against beetles infesting stored products. J. Stored Prod. Res. 2: 197–210.

Lemon, R.W. 1967b. Laboratory evaluation of some additional organophosphorus insecticides against stored product beetles. J. Stored Prod. Res. 3: 283–287.

Leng, M. 1980. Government requirements for pesticide residue analysis and monitoring studies. In: Anson Moye, H., Ed., Analysis of Pesticide Residues. Wiley, New York. pp. 395–488.

Le Pelley, R. and Kockum, S. 1954. Experiments in the use of insecticides for the protection of grain in storage. Bull. Entomol. Res. 45: 295–311.

Lessard, F.F. 1985. Acaricidal activity of some specialty insecticides based on deltamethrin utilised for direct application to grains. Report, INRA, Labo. de Recherches sur les Insectes et Acariens des Denrees stockees, Domain de la Grande Ferrade — B.P. 131, 33140 Ponte de la Maye, France. 15 pp.

Leuck, D.B., Johnson, J.C. Jr., Bowman, M.C., Knox, F.E. and Berosa, M. 1971. Fenitrothion residues in corn, silage and their effects on dairy cows. J. Econ. Entomol. 64(6): 1394–1399.

Leuschner, F., Leuschner, A. and Poppe, E. 1967. Chronischer Reproduktionsversuch uber 3 Generationen an Wistar-Ratten bei fortdaiernder Verabreichung von Bromophos. Report C.H. Boehringer Sohn. (Unpublished).

Levine, B.S. and Murphy, S.D. 1977. Esterase inhibition and reactiviation in relation to piperonyl butoxide-phosphorothionate interactions. Toxicol. Appl. Pharmac. 40(3): 379–391.

Levison, H.Z. and Levison, A.R. 1979. Integrated manipulation of storage insects by pheromones and food attractants. Proc. 2nd Int. Working Conf. Stored-Product Entamol., Ibadan, Nigeria, 10–16 September 1978. pp. 66–74.

L'Hoste, J., Balloy, J. and Rauch, F. 1969. La protection des blés contre *Sitophilus granarius*. Congr. Pl. Prot. Org. by CIA and CITA, Milan No. 22, 6 pp.

Lichtenstein, E.P., Schulz, K.R. and Cowley, G.T. 1963. Inhibitors of the conversion of aldrin to dieldrin in soils with methylenedioxyphenyl synergists. J. Econ. Entomol. 56: 485 489.

Liggins, J.A. and Singh, H. 1971. Effects of fractionated doses of *gamma* irradiation on fecundity, mortality and longevity of the lesser grain borer. J. Econ. Entomol. 64: 1306–1307.

Lillie, R.J. 1972. Studies on the reproductive performance of White Leghorns fed malathion and carbaryl. Presented 61st Ame. Meeting Poultry Soc. Assoc. Columbus, Ohio.

Lillie, R.J. 1973. Studies on the reproductive performance and progeny performance of caged White Leghorns fed malathion and carbaryl. Poultry Sci. 52(1): 266–272.

Lindgren, D.L., Krohne, H.E. and Vincent, L.E. 1954. Malathion and chlorthion for control of insects infesting stored grain. J. Econ. Entomol. 47: 705.

Lindgren, D.L., Sinclair, W.B. and Vincent, L.E. 1968. Residues in raw and processed foods resulting from post-harvest insecticidal treatments. Residue Rev. 21: 1–121.

Linsley, E.G. 1944. Natural sources, habitats, and reservoirs of insects associated with stored food products. Hilgardia 16(4): 187–224.

Litchfield, N. and Norris, J. 1969. Chlorpyrifos-methyl- DOWCO 214. The Dow Chemical Company Report BC T35.12–46193–5 6 pp. (Unpublished).

Lloyd, C.J. 1961. The effect of piperonyl butoxide on the toxicity of pyrethrins to the cigarette beetle. Pyrethrum Post 6(1): 3–4.

Lloyd, C.J. 1973. The toxicity of pyrethrins and five synthetic pyrethroids to *Tribolium castaneum* and *Sitophilus granarius*. J. Stored Prod. Res. 9: 77–92.

Lloyd, T.S. 1973a. Accidental poisoning in birds. Vet. Rec. 92: 489.

Lloyd, C.J. and Field J. 1969. Tests on newer insecticides. Pest Infest. Res, 1968 p. 46.

Lloyd, C.J. and Hewlett, P.S. 1958. Relative susceptibility to pyrethrum in oil of Coleoptera and Lepidoptera infesting stored products. Bull. Entomol. Res. 49: 177–185.

Lloyd, C.J. and Hewlett, P.S. 1960. The susceptibility to pyrethrins of three species of moth infesting stored products. Pyrethrum Post 5(3): 12–13.

Lloyd, C.J. and Parkin, E.A. 1963. Further studies on a pyrethrum-resistant strain of the granary weevil, *Sitophilus granarius* (L.). J. Sci. Food Agric. 9: 655–663.

Lloyd, C.J. and Ruczkowski, G.E. 1980. The cross-resistance to pyrethrins and eight synthetic pyrethroids, of an organo-phosphorus-resistant strain of rust-red flour beetles, *Tribolium castaneum* (Herbst). Pestic. Sci. 11: 331–340.

Lloyd, J.E. and Matthysse, J.G. 1966. Polymer insecticide systems for use as livestock feed additives. J. Econ. Entomol. 59: 363.

Lloyd, J.E. and Matthysse, J.G. 1970. Polyvinyl chloride — insecticide pellets fed to cattle to control face fly larvae in manure. J. Econ. Entomol. 63: 1271.

Lloyd, J.E. and Matthysse, J.G. 1971. Residues of dichlorvos, diazinon, and dimetilan in milk of cows fed PVC-insecticide feed additives. J. Econ. Entomol. 64: 821.

Loaharanu, S., Sutantawong, M. and Ungsunantwiwat, A. 1972. Effect of *gamma* radiation on various stages of the rice moth, *Corcyra cephalonica*. Thai J. Agric. Sci. 5: 195–202.

Lockwood, L.M. 1973. Organophosphorus pesticides for use as grain protectants in India. Degradation of their residues during milling and cooking of cereal grains. Dissert. Abstr. 34: 2.

Lockwood, L.M., Majumder, S.K. and Lineback, D.R. 1974. Degradation of organophosphorus pesticides in cereal grains during milling and cooking in India. Cereal Sci. Today 19(8): 330–333, 346.

Loeffler, J.E., DeVries, D.M. Young, R. and Page, A.C. 1971. Metabolic fate of inhaled dichlorvos in pigs. Toxicol. Appl. Pharmacol. 19: 378.

Long, J.S., Lehman, R.M. and Manzelli, M.A. 1978. Altosid: an examination of its activity against the cigarette beetle. Tob. Int. 180: 126–127.

Long, J.S., Lehman, R.M., Tickle, M.H. and Manzelli, M.A. 1980. The efficacy of Kabat for control of the cigarette beetle under severe infestation conditions. Tob. Sci. 24: 119–121.

Longoni, A. and Michieli, G. 1969. Initial and residual activity in Cidial (phenthoate) compared with that of malathion and lindane on insects affecting stored products. J. Econ. Entomol. 62: 1258–1261.

Longstaff, B.C. 1981. The manifestation of the population growth of a pest species: an analytical approach. J. Appl. Ecol. 18: 727–36.

Longstaff, B.C. 1984. The effects of aeration on insect population growth. In: Champ, B.R., and Highley. E., Ed., Proc. Aust. Dev. Asst. Course on Preservation of Stored Cereals. CSIRO Division of Entomolgy, Canberra. pp. 459–465.

Longstaff, B.C. and Desmarchelier, J.M. 1983. The effects of the temperature/toxicity relationships of certain pesticides upon the population growth of *Sitophilus oryzae*. J. Stored Prod. Res. 19(1): 25–29.

Loong Fat Lim and Sudderuddin 1977. Comparative toxicity of six insecticides against *Sitophilus oryzae* and *Palembus dermestoides*. J. Stored Prod. Res. 13: 209–211.

Loschiavo, S.R. 1975. Tests of four synthetic insect growth regulators with juvenile hormone activity against seven species of stored product insects. Manit. Entomol. 9: 43–52.

Loschiavo, S.R. 1976. Effects of the synthetic insect growth regulators methoprene and hydroprene on survival, development of reproduction of six species of stored-product insects. J. Econ. Entomol. 69: 395–399.

Loschiavo, S.R. 1978a. Effect of Bay SRA-7660 on the survival and reproduction of three species of stored-product insects in the laboratory and small-bin experiments. J. Econ. Entomol. 71: 206–210.

Loschiavo, S.R. 1978b. Effect of disturbance of wheat on four species of stored-product insects. J. Econ. Entomol. 71: 888–93.

Lum, P.T.M. 1975. Effect of visible light on the behavior of stored product insects. Proc. 1st Int. Working Conf. Stored-Product Entomol., Savannah, Georgia, 7–11 October 1974. pp. 375–382.

Lum, P.T.M. and Flaherty, B.R. 1969. Effect of mating with males reared in continuous light or in light/dark cycles on fecundity in *Plodia interpunctella*. J. Stored Prod. Res. 5: 89–94.

Lum, P.T.M. and Flaherty, B.R. 1970. Effect of continuous light on the potency of *Plodia interpunctella* males. Ann. Entomol. Soc. Am. 63: 1470–1471.

Lum, P.T.M. and Phillips, R.H. 1972. Combined effects of light and carbon dioxide on egg production of Indian meal moths. J. Econ. Entomol. 65: 1316–1317.

Lum, P.T.M., Flaherty, B.R. and Phillips, R.H. 1973. Fecundity and egg viability of stressed female Indian meal moths *Plodia interpunctella*. J. Ga. Entomol. Soc. 8(4): 245.

McCallum-Deighton, J. 1975a. Pirimiphos-methyl — a new insecticide for controlling granary pests. Dokl. Soobshch. — Mezhdunar. Kongr. Zashch. Rast. 3(1): 175–7.

McCallum-Deighton, J. 1975b. The control of stored products insects with pirimiphos-methyl. Proc. 1st Int. Work. Conf. Stored-Products Entomol., Savannah, Georgia, 7–11 October 1974. pp. 567–576.

McCallum-Deighton, J. 1976. Stored grain insect control with 'Actellic' a safe new insecticide. Far East and Pacific Sri Lanka Conf., 1976, Paper 2.

McCallum-Deighton, J. 1978. Pirimiphos-methyl and other insecticides in the control of stored product insects. Outl. Agric. 9(5): 240–245.

McCallum-Deighton, J. 1981. The availability of suitable pesticides Proc. GASGA Seminar on the Appropriate Use of Pesticides for the Control of Stored Products Pests in the Developing Countries, Slough, 17–20 February 1981. pp. 160–169.

McCallum-Deighton, J. and Pascoe, R. 1976. The use of pirimiphos-methyl to control mites in grain stores. Ann. Appl. Biol. 82(1): 197–198.

McColl, J.C. 1982. Animal feeding studies — pyrethroid based products. Report from Director-General of Agriculture South Australia to Wellcome Australia Ltd (dated 22 December 1982). (Unpublished).

McDonald, L.L. 1968. Relative effectiveness of tropital and piperonyl butoxide as synergists for pyrethrins against stored-product insects. J. Econ. Entomol. 61: 1645–6.

McDonald, L.L. and Gillenwater, H.B. 1967. Relative toxicity of Bay 77488 and Dursban against stored product insects. J. Econ. Entomol. 60(5): 1195–6.

McDonald, L.L. and Gillenwater, H.B. 1976. Toxicity of pirimiphos-methyl and Bay SRA 7660 to six species of stored-product insects. J. Ga. Entomol. Soc. 11: 110–114.

McDonald, L.L, and Press, J.W. 1973. Toxicity of eight insecticides to Indian meal moth adults, *Plodia interpunctella*. J. Ga. Entomol. Soc. 8(3): 200.

McDonald, M.W., Dillon, J.F. and Stewart, D. 1964. Non-toxicity to poultry of malathion as a grain protectant. Aust. Vet. J. 40: 358–360

McDougall, K.W. 1973. The effect of moisture and temperature on the persistence of CGA 20168 in wheat. Unpublished Technical Report No. 73/9/419 from Ciba-Geigy Australia Limited.

McDougall, K.W. 1974. The influence on grain variety on the persistence of CGA 20168. Unpublished Technical Report No. 74/9/476 from Ciba-Geigy Australia Limited.

McDougall, K.W. 1975. Mini silo trials with CGA 20168 — Series 2. Unpublished Technical Report No. 75/8/519 from Ciba-Geigy Australia Limited.

McDougall, K.W. 1976a. The influence of grain variety on the persistence of CGA 20168 — Part 2. Drum trials. Unpublished Technical Report No. 76/2/554 from Ciba-Geigy Australia Limited.

McDougall, K.W. 1976b. The distribution and dissipation of CGA 20168 in white, brown and paddy rice. Unpublished Technical Report No. 76/2/553 from Ciba-Geigy Australia Limited.

McFarlane, J.A. 1963. Prospects for pyrethrum with particular reference to its use in the control of pests of stored foodstuffs. Trop. Stored Prod. Inf. 6: 202–210.

McFarlane, J.A. 1970. Insect control by airtight storage in small containers. Trop. Stored Prod. Inf. 19: 10–14.

McFarlane, J.A. 1970a. Treatment of large grain stores in Kenya with dichlorvos slow release strips for control of *Cadra cautella*. J. Econ. Entomol. 63: 288.

McFarlane, J.A. and Harris, A.H. 1964. Relative susceptibility malathion absorption of six animal-feed ingredients. J. Sci. Food Agric. 15: 612–619.

McGaughey, W.H. 1969. Malathion residues on rice. Rice J. 72: 3, Aug 1969.

McGaughey, W.H. 1970a. Evaluation of dichlorvos for insect control in stored rough rice. J. Econ. Entomol. 63(6): 1867.

McGaughey, W.H. 1970b. Malathion on milling fractions of three varieties of rough rice: duration of protection and residue degradation. J. Econ. Entomol. 63(4): 1200.

McGaughey, W.H. 1972a. Protectants for stored rough rice: Gardona, dichlorvos, and a Gardona:dichlorvos mixture. J. Econ. Entomol. 65: 1694.

McGaughey, W.H. 1972b. Malathion treatment of rough rice re-evaluated. Rice J. 75: 16.

McGaughey, W.J. 1973. Angoumois grain moth control in rough rice with a Gardona-dichlorvos mixture. J. Econ. Entomol. 66: 1353.

McGaughey, W.H. 1973a. Dichlorvos vapour for insect control in a rice mill. J. Econ. Entomol. 66: 1147.

McGaughey, W.H. 1980a. *Bacillus thuringiensis* for moth control in stored wheat. Can. Entomol. 112: 327–31.

McGaughey, W.H. 1980b. A granulosis virus for Indian meal moth control in stored meal and corn. J. Econ. Entomol. 68: 346–8.

McGaughey, W.H. and R.R. Cogburn, 1971. Current stored product insect research. Rough rice. DDVP-Gardona used in combination. Rice J. 74: 50.

McGregor, H.E. 1964. Preference of *Tribolium castaneum* for wheat containing various percentages of dockage. J. Econ. Entomol. 57(4): 511–513.

McGregor, H.E. and Kramer, K.J. 1975. Activity of insect growth regulators, hydroprene and methoprene, on wheat and corn against several stored-product insects. J. Econ. Entomol. 68: 668–670.

McGregor, H.E. and Kramer, K.J. 1976. Activity of Dimilin (TH6040) against Coleoptera in stored wheat and corn. J. Econ. Entomol. 69: 479–480.

McLean, K.A. 1979. Comparative efficiency and costs of different drying systems. Ministry of Agriculture Conference, Fuel Economy on the Arable Farm, St. Ives, Cambridgeshire, U.K., November 1979.

McLellan, R.H. 1964. Anthelmintic pyrethrum — a literature review. Pyrethrum Post 7(4): 23–26.

Maceliski, M. and Korunk, Z. 1972. A contribution to the knowledge of the mechanism of action of inert dusts on insects. Zast. Bilja. 23: 49–64.

Machin, A.F., Rogers, H., Cross, A.J., Quick, M.P., Howells, L.C. and Janes, N.F. 1975. Metabolic aspects of the toxicology of diazinon. I. Hepatic metabolism in the sheep, cow, pig, guinea-pig, rat, turkey, chicken and duck. Pestic. Sci. 6(5): 461–473.

Macklin, A.W. and Ribelin, W.E. 1971. The relation of pesticides to abortion in dairy cattle. J. Am. Vet. Med. Assoc. 159: 1743.

Macholz, R.M. and Kujawa, M. 1979. Recent state of lindane metabolism — Part II. Residue Rev. 72: 71.

Macholz, R.M. and Kujawa, M. 1985. Recent state of lindane metabolism — Part III. Residue Rev. 94: 119–149.

Maddox, J.V. 1975. Use of diseases in pest management. In: Metcalf, R.L, and Luckman, W., Ed., Introduction to Insect Pest Management. John Wiley & Sons, New York.

Madrid, F.J., White, N.D.G. and Sinha, R.N. 1983. Effects of malathion dust on Indian meal moth and almond moth (Lepidoptera: Phycitidae) infestation of stored wheat. J. Econ. Entomol. 76(6): 1401–1404.

Magallona, E.D. and Celino, L.P. 1977. Organophosphate insecticide residues in grains II Effect of cooking on malathion and pirimiphos-methyl residues in rice. Philipp. Entomol. 3: (5/6) 295–299.

Malathion Panel 1960. The determination of malathion residues in cereals and oilseeds. Analyst (London) 85: 915–925.

Mallikarjuna Rao, K., Jacob, S.A. and Mohan, M.S. 1972. Resistance of flexible packaging materials to some important pests of stored products. Indian J. Entomol. 34: 94–101.

Malone, J.C. and Brown, N.C. 1968. Toxicity of various grades of pyrethrum to laboratory animals. Pyrethrum Post 9: 3–8.

Mansur, N.A. 1969. Fumigating action of dichlorvos (DDVP), chloropicrin, and carbon tetrachloride on *Sitophilus oryzae* and *Tribolium confusum*. Primen. Pestits. Sel. Khoz., Mater. Nuach, Konf. 99 (CA 74: 110831).

Manual of Analytical Methods 1984. Analytical Methods — Sections 5–8. Manual of Analytical Methods for Pesticides Residues in Foods. Health Protection Branch, Health and Welfare, Canada, Ottawa, Canada.

Manzelli, M.A. 1982. Management of stored-tobacco pests, the cigarette beetle (Coleoptera: Anobiidae) and tobacco moth (Lepidoptera: Pyralidae), with methoprene. J. Econ. Entomol. 75: 721–723.

March, R.B., Fukuto, T.R., Metcalf, R.L. and Maxon, M.G. 1956a. Fate of P^{32}-labelled malathion in the laying hen, white mouse and American cockroach. J. Econ. Entomol. 49(2): 185–195.

March, R.B., Metcalf, R.L., Fukuto, T.R. and Gunther, F.A. 1956b. Fate of P^{32}-labelled malathion sprayed on jersey heifer calves. J. Econ. Entomol. 49(5): 679–682.

Margham, J.P. and Thomas, J.O. 1980. Weight related suceptibility to an insecticide in *Tribolum castaneum*. J. Stored Prod. Res. 16(1): 33–38.

Margot, A. and Gysin, H 1957. Diazinon, its degradation products and their properties. Helv. Chem. Acta. 40.: 1562.

Marion, W.W., Ning, J.M. and Ning S.M. 1968. Application of malathion to the laying hen. Poul. Sci. 47(6): 1956–1961.

Marzke, F.O., Street, M.W., Mullen, M.A. and McCray, T.L. 1973. Spectral responses of six species of stored-product insects to visible light. J. Ga. Entomol. Soc. 8: 195–200.

Marzke, F.O., Coffelt, J.A. and Silhacek, D.L. 1977. Impairment of reproduction of the cigarette beetle (*Lasioderma serricorne*) with the insect growth regulator, methoprene. Entomol. Exp. Appl. 22: 294–300.

Mason, W.A. 1973. The analysis of Bay 77488 on stored wheat and synthesis of a multi-element compound. Dissert. Absts. B34: 106.

Masud, S. Zafar and Zaki, M. Tahir 1976. Determination of the pesistence of malathion and fenitrothion insecticides in stored wheat by gas-liquid chromatography. Agric. Pak. 27(1): 101–6.

Mathlein, R. 1968. Artificial cooling of infested goods in Sweden. Rep. Int. Conf. Prot. Stored Prod., Lisbon-Oenras.

Matsubara, H. 1961. On the synergistic effect of natural and synthetic synergists on Barthrin. Studies on synergist for insecticides **I. Botyu-Kagaku 26(4): 125–132.**

Matsubara, H. 1963. Synergist for insecticides. **III. The synergistic effect of synthetic synergists on 1-naphthyl** *N*-methylcarbamate. Botyu-Kagaku (Sci. Pest Control) 28: 35–40.

Matthysse, J.G. 1974b. 1975 Stored grain insect control recommendations for New York State. N.Y. St. Insectic. Fungic. Herbic. Recomm. 36 Mtg., 1975, F02-F05.

Matthysse, J.G. and Lisk, D. 1968. Residues of diazinon, coumaphos, ciodrin, methoxychlor and rotenone in cow's milk from treatment similar to those used for ectoparasite and fly control on dairy cattle, with notes on safety of diazinon and ciodrin to calves. J. Econ. Entomol. 61(5): 1394–1398.

Maybury, R.B. 1980. Laboratory Manual for Pesticide Residues Analysis in Agricultural Products, Pesticide Laboratory, Food Production and Inspection Branch, Agriculture Canada, Ottawa, Ontario KIA OC5, Canada.

Mellon Institute 1958. Toxicology studies on carbaryl. Report from Mellon Institute of Industrial Research to Union Carbide Inc. (Unpublished).

Menn, J.J., Henrick, C.A. and Staal, G.B. 1981. Juvenoids: Bioactivity and prospects for insect management. In: Sehnal, F., Zabza, A., Menn, J.J., and Cymborowski, B., Ed., Regulation of Insect Development and Behaviour. Wrocklaw Techn. Univ. Press, Poland. Part II, pp. 735–748.

Mensah, G.W.K. and Watters, F.L. 1979a. Uptake of bromophos into bulk stored wheat from treated granary surfaces. J. Econ. Entomol. 72: 275–276.

Mensah, G.W.K. and Watters, F.L.1979b. Comparison of four organophosphorus insecticides on stored wheat for control of susceptible and malathion-resistant strains of the red flour beetle (*Tribolium castaneum*). J. Econ. Entomol. 72(3): 456–461.

Mensah, G.W.K. and White, N.D.G. 1984. Laboratory evaluation of malathion-treated sawdust for control of stored-product insects in empty granaries and food warehouses. J. Econ. Entomol. 77(1): 202–206.

Mensah, G.W.K., Watters, F.L. and Webster, G.R.B. 1979a. Translocation of malathion, bromophos and iodofenphos into stored grain from treated structural surfaces. J. Econ. Entomol. 72(3): 385–391.

Mensah, G.W.K., Watters, F.L. and Webster, G.R.A. 1979b. Insecticide residues in milled fractions of dry or tough wheat treated with malathion, bromophos, iodofenphos and pirimiphos-methyl. J. Econ. Entomol. 72: 728–731.

Merk, W.B. 1970. Properties of dichlorvos with regard to decontamination of grain stores from insects (German). Schweiz. Landwirt. Monatsh. 48(3): 102.

Mestres, R., Francois, C., Illes, S., Tourte, J. and Campo, M. 1976. Simultaneous quantitative determination of residues of hexachlorobenzene, other pesticides and chlorinated organic micropollutants in fatty materials. Trav. Soc. Pharm. Montpellier 36(1): 43–58.

Mestres, R., Illes, S., Campo, M. and Tourte, J. 1977. A multiple residue method for citrus fruits. Proc. Int. Soc. Citriculture 2. 426–429.

Mestres, R., Atmawijaya, S. and Chevallier, C. 1979a. Methods for the study and determination of pesticide residues in cereal products. XXXIV. Organochlorine, organophosphorus, pyrethrin, and pyrethroid pesticides. Ann. Fals. Exp. Chim. 72: 577–589.

Mestres, R., Illes, S., Campo, M. and Tourte, J. 1979b. Development of a method for multiple pesticide residue analysis in plants and foods of plant origin. Trav. Soc. Pharm. Montpellier 39: 323–328.

Metcalf, R.L. 1955. Organic insecticides; their chemistry and mode of action. Interscience Publishers Inc. New York & London.

Metcalf, R.L. 1967. Mode of action of pesticide synergists. Ann. Rev. Entomol. 12: 229–256.

Metcalf, R.L., Fukuto, T.R., Wilkinson, C., Fahmy, M.H., Abd El-Aziz, S. and Metcalf, E.R. 1966. Mode of action of carbamate synergists. J. Agric. Food Chem. 14(6): 555–62.

Methodensammlung 1982. Methodensammlung zur Rueckstandsanalytik von Pflanzenschutzmitteln, 6 Lieferung, Verlag Chemie GmbH, Weinheim/Bergstrasse, Federal Republic of Germany. (The numbers in parentheses in the text refer to the numbers of the methods in this manual.)

Metwally, M.M., Sehnal, F. and Landa, V. 1972. Reduction of fecundity and control of khapra beetle by juvenile hormone mimics. J. Econ. Entomol. 65: 1603–5.

Meyer, H. 1970. Bibliographie der Aflatoxine. Aus dem Institut fur Bakteriologie und Histologie der Bundesanstalt fur Fleischforschung, Kulmbach, West Germany (Mimeographed).

Mian, L.S. and Kawar, N.S. 1979. Persistence of dichlorvos and tetrachlorvinphos on wheat grains. J. Sci. Technol. (Peshawar, Pak.) 3(1–2): 31–2.

Mian, L.S. and Mulla, M.S. 1982a. Biological activities of IGRs against four stored-product coleopterans. J. Econ. Entomol. 75(1): 80–85.

Mian, L.S. and Mulla, M.S. 1982b. Residual activity of IGRs against stored-products beetles in grain commodities. J. Econ. Entomol. 75: 599–603.

Mian, L.S. and Mulla, M.S. 1983a. Persistence of three IGRs in stored wheat. J. Econ. Entomol. 76(3): 622–625.

Mian, L.S. and Mulla, M.S. 1983b. Effects of insect growth regulators on the germination of stored wheat. Protection Ecology 5(4): 369–373.

Mihara, K., Misaki, Y. and Miyamoto, J. 1977. Metabolism of fenitrothion in Japanese quails (female). Report of Sumitomo Chem. Co. (Unpublished).

Mihara, K., Okuno, Y., Misaki, Y. and Miyamoto, J. 1978. Metabolism of fenitrothion in goats. J. Pestic. Sci. 3: 233–242.

Miller, J.A., Chamberlain, W.F., Beadles, M.L., Pickens, M.O. and Gingrich, A.R. 1976. Methoprene for control of horn flies: application to drinking water of cattle via a tablet formulation. J. Econ. Entomol. 69: 330–332.

Miller, J.A., Eschle, J.L., Hopkins, D.E., Wright, F.C. and Matter, J.J. 1977a. Methoprene for control of horn flies: a suppression program on the island of Molokai. Hawaii. J. Econ. Entomol. 70: 417–423.

Miller, J.A., Beadles, M.L., Palmer, J.S. and Pickens, M.O. 1977b. Methoprene for control of the horn fly: a sustained-release bolus formulation for cattle. J. Econ. Entomol. 70: 589–591.

Miller, K.R. 1963. Detection and distribution of ^{32}P-labelled diazinon in dog tissues after oral administration. N.Z. Vet. J. 11: 141–148.

Miller, R.W. and Gordon, C.H. 1973. Effect of feeding Rabon to dairy cows over extended periods. J. Econ. Entomol. 66(1): 135–138.

Miller, R.W., Gordon, C.H., Bowman, M.C., Beroza, M. and Morgan, N.O. 1970. J. Econ. Entomol. 63: 1420.

Miller, R.W., Pickens, L.G. and Hunt, L.M. 1978. Methoprene: field tested as a feed additive for control of face flies. J. Econ. Entomol. 71: 274–278.

Miller, W.W. 1981. Analysis of methoprene residues in stored peanuts and peanut hulls. Tests Nos. SE-GR

340–79; SE-IR 399–79, SE-IR 300–81. Zoecon Corporation, Palo Alto, CA. (Unpublished).

Miler, W.W. 1983a. Analysis of methoprene residue in stored shelled corn. Reports nos. RM-1054, RM-0159 and RM-0167. Zoecon Corporation, Palo Alto, California. (Unpublished).

Miller, W.W. 1983b. Analysis of methoprene residues in wheat grain. Zoecon Corporation, Report no. R7–0130. (Unpublished).

Miller, W.W. 1983c. Analysis of methoprene residues in stored wheat grain: Report No. RM-0132. Zoecon Corporation, Palo Alto, CA. (Unpublished).

Miller, W.W., Wilkins, J.S. and Dunham, L.L. 1975. Determination of Altosid insect growth regulator in waters, soils, plants, and animals by gas-liquid chromatography. J. Assoc. Off. Anal. Chem. 58: 10–18.

Mills, I.H. 1976. Pirimiphos-methyl : Blood concentrations and tissue retention in the rat. ICI Central Toxicology Report No. CTL/P/247.

Mills, R.B. 1978. Potential and limitations of the use of low temperature to prevent insect damage in stored grain. Proc. 2nd Int. working Conf. Stored-Product Entomol., Ibadan, Nigeria, 10–16 September 1978. pp. 244–259.

Milner, C.K. and Butterworth, S.T.G. 1977. Toxicity of pyrethroid insecticide. Investigation of the neurotoxic potential of WL43775. Report from Shell Development Co. (Unpublished).

Minett, W. 1975. Some factors influencing the effectiveness of grain protectants. Proc. 1st Int. Working Conf. Stored-Product Entomol., Savannah, Georgia, 7–11 October 1974. pp. 297–300.

Minett, W. and Belcher, R.S. 1969. Determination of malathion and dichlorvos residues in wheat grains by gas-liquid chromatography. J. Stored Prod. Res. 5: 417–421.

Minett, W. and Belcher, R.S. 1970. Loss of dichlorvos residues in stored wheat. J. Stored Prod. Res. 6: 269.

Minett, W. and Williams, P. 1971. Influence of malathion distribution in the protection of wheat grains against insect infestation. J. Stored Prod. Res. 7: 233–242.

Minett, W. and Williams, P. 1976. Assessment of non-uniform malathion distribution for insect control in a commercial wheat silo. J. Stored Prod. Res. 12: 27–33.

Minett, W. and Williams, P. 1982. Application of insecticide concentrate to grain during transfer in commercial storage. Proc. 1st Aust. Stored Grain Pest Control Conf., Melbourne, May 1981, pp. 7/26–32.

Minett, W., Belcher, R.S., O'Brien, E.J. 1968. A critical moisture level for malathion breakdown in stored wheat. J. Stored Prod. Res. 4: 179–181.

Minett, W., Williams, P. and Amos, T.G. 1981. Gravity feed appliction of insecticide concentrate to wheat in a commercial silo. Gen. Appl. Entomol. 13: 59–64.

Minett, W., Wiliams, P., Murphy, G.D. and Stahle, P.P. 1984. Application of insecticide concentrate by gravity feed to wheat in a bucket elevator. Gen. Appl. Entomol. 16: 17–22.

Ministry of Agriculture 1969. Insects and mites in farm stored grain. Ministry Agric. Advisory Leaflet 368. UK Ministry of Agriculture, Fisheries and Food, London.

Minor, M.F., Manzelli, M.A., Lehman, R.M. and Long, J.S. 1983. A large scale evaluation of KABAT for the control of *Lasioderma serricorne* (F.). Tob. Sci. 27: 64–65.

Mitsuda, H., Kawai, F. and Vamamoto, A. 1971. Hermetic storage of cereals and legumes under the water and ground. Mem. Coll. Agric. Kyoto Univ. 100: 49–69.

Miyamoto, J. 1964. Studies on the mode of action of organophosphorus compounds. Part III. Activation and degradation of Sumithion (fenitrothion) and methyl parathion in vivo. Agric. Biol. Chem. (Tokyo) 28: 411–421.

Miyamoto, J. 1975. Terminal residues of bioresmethrin. Submission to IUPAC Terminal Residues Commission, Madrid, September 1975.

Miyamoto, J. 1977a. Proceedings of a Symposium on Fenitrothion. NRCC/CNRC No.16073. pp. 459.

Miyamoto, J. 1977b. Proceedings of a Symposium on Fenitrothion. NRCC/CNRC No.16037. pp. 307.

Miyamoto, J. 1981. The chemistry, metabolism and residue analysis of synthetic pyrethroids. IUPAC Reports on Pesticides (15). Pure Appl. Chem. 53: 1967–2022.

Miyamoto, J. and Mihara, K. 1977. The fate of fenitrothion administered to rats. Unpublished Report of Research Department, Pesticides Division, Sumitomo Chemical Co., Ltd.

Miyamoto, J. and Sato, Y. 1969. Determination of insecticide residues in animal and plant tissues VI. Determination of Sumithion residues in cattle tissues. Botyu-Kagaku (Sci. Pest Control) 34: 3–6.

Miyamoto, J., Sato, Y., Kadota, T., Fujinami, A. and Enco, M. 1963a. Studies on the mode of action of organophosphorus compounds Part 1. Metabolic fate of ^{32}P-labelled Sumithion and methyl parathion in Guinea pig and white rat. Agric. Biol. Chem. (Tokyo) 27: 381–389.

Miyamoto, J., Sato, Y., Kadota, T. and Fujinami, A. 1963b. Studies on the mode of action of organophosphorus compounds. Part II. Inhibition of mammalian cholinesterase in vivo following the administration of Sumithion and parathion-methyl. Agric. Biol. Chem. (Tokyo) 27: 669–676.

Miyamoto, J., Sato, Y. and Suzuki, S. 1967. Determination of insecticide residues in animal and plant tissues. IV: Determination of residual amount of

Sumithion and some of its metabolites in fresh milk. Botyu-Kagaku (Sci. Pest Control) 32: 95–100.

Miyamoto, J., Nishida, T. and Ueda, K. 1971. Metabolic fate of resmethrin in the rat. Pestic. Biochem. Physiol. 1: 293–306.

Miyamoto, J. Suzuki, T. and Nakae, C. 1974. Metabolism of phenothrin in mammals. Pestic. Biochem. Physiol. 4: 438–450.

Miyamoto, J., Hosokawa, S., Kadota, T., Kohda, H., Arai, M., Sugihara, S. and Hirao, K. 1976a. Studies on cholinesterase inhibition and structural changes at neuromuscular junctions in rabbits by subacute administration of Sumithion. J. Pestic. Sci. 1: 171–178.

Miyamoto, J., Mihara, K. and Hosokawa, S. 1976b. Comparative metabolism of m-methyl-carbon-14-Sumithion in several species of mammals in vivo. J. Pestic. Sci. 1: 9–21.

Miyamoto, J., Okuno, Y., Kadota, T. and Mihara, K. 1977a. Experimental hepatic lesions and drug metabolizing enzymes in rats. J. Pestic. Sci. 2: 257–269.

Miyamoto, J., Mihara, K., Kadota, T. and Okuno, Y. 1977b. Toxicity and metabolism in vivo of fenitrothion in rats with experimental hepatic lesion. J. Pestic. Sci. 2: 271–277.

Moellhoff, E. 1968. Residues and their determination in plants treated with E-605 (parathion) and Agritox. Pflanzensch. Nachr. Bayer (Engl. Ed.) 21: 327–354.

Mondal, K.A.M.S.H. 1984a. Repellent effect of pirimiphos- methyl to larval *Tribolium castaneum* Herbst. Int. Pest Control 26(4): 98–99.

Mondal, K.A.M.S.H. 1984b. Dose-mortality response of *Tribolium castaneum* (Herbst) larvae to pirimiphos-methyl. Int. Pest Control 26(5): 128–129, 137.

Monro, H.A.U. 1969. Manual of Fumigation for Insect Control. FAO Agricultural Studies No 79, FAO, Rome.

Mookherjee, P.B., Beri, Y.P., Sharma, G.C. and Dewan, R.S. 1965. Preliminary studies on the efficacy of malathion against *Sitophilus oryzae* and *Tribolium castaneum* and its persistence in stored wheat. Indian J. Entomol. 27: 476.

Moore, J.B. 1970. Terminal residues of pyrethrin-type insecticides and their synergists in foodstuffs. Residue Rev. 33: 87.

Moore, J.B. 1972. Terminal residues of pyrethrin-type insecticides and their synergists in foodstuffs. Pyrethrum Post 11(3): 106–112, 124.

Moore, J.B. 1973. Residue and tolerance considerations with pyrethrum, piperonyl butoxide and MGK-264. In: Casida, J.E., Ed., Pyrethrum, the Natural Insecticide. Academic Press, New York. pp. 293–306.

Moore, B. and McDougall, K.W. 1974a. Further studies on factors affecting the dissipation of CGA 20168 in wheat. Unpublished Technical Report No. 74/3/433 from Ciba-Geigy Australia Limited.

Moore, B. and McDougall, K.W. 1974b. Australian Wheat Board Collaborative Insecticide Studies — Determination of residue dissipation. Unpublished Technical Report No. 74/7/461 from Ciba-Geigy Australia Limited.

Moore, B. and McDougall, K.W. 1974c. The Australian Wheat Board Collaborative Insecticide Studies — Determination of residue dissipation. Part II. Unpublished Technical Report No. 74/9/477 from Ciba-Geigy Australia Limited.

Moore, B. and McDougall, K. 1974d. CGA-20168 residues in chickens. Unpublished Technical Report from Ciba-Geigy Australia Limited. 74/7/462.

Moore, B. and McDougall, K.W. 1975. Australian Wheat Board Collaborative Studies 1975: The effect of aeration on the dissipation of CGA-20168 in wheat. Unpublished Technical Report No. 75/12/538 from Ciba-Geigy Australia Limited.

Moore, S., Petty, M.B., Luckmann, W.H. and Byers, J.M. 1966. Losses caused by the Angoumois grain moth in dent corn. J. Econ. Entomol. 59(4): 880–882.

Morallo-Rejesus, B. 1973. Evaluation of five insecticides as protectants of shelled corn during storage. Ann. Rep. Corn Project Philippines.

Morallo-Rejesus, B. 1981. Insecticidal evaluations on stored grain insects in the Philippines. Proc. BIOTROP Symp. on Pests of Stored Products, Bogor, Indonesia. pp. 175–182.

Morallo-Rejesus, B. 1982a. Special problems with control of stored grain insects in the tropics. In: Rottger, U., and SEARCA Team, Ed., Proc. GASGA Seminar on Paddy Deterioration in the Humid Tropics, Baguio, Philippines, 11–18 October 1981. GTZ, Eschborn. pp. 77–80.

Morallo-Rejesus, B. 1982b. Problems with the control of pests of corn and sorghum in storage. Paper presented during the First National Symposium-Workshop on Corn and Sorghum Crop Protection, University of Southern Mindanao, Kabacan, North Cotabato, Philippines, 17–20 March, 1982.

Morallo-Rejesus, B. and Carino F.O. 1974. The residual toxicity of five insecticides in three varieties of corn. Fifth Ann. Conf. Intensified Corn Production Programme.

Morallo-Rejesus, B. and Carino, F.O. 1976a. Evaluation of five contact insecticides as protectants against insects on stored corn. Philipp. Agric. 60: 105–111.

Morallo-Rejesus, B. and Carino, F.O. 1976b. The residual toxicity of five insecticides on three varieties of corn and sorghum. Philipp. Agric. 60(3&4): 96–104.

Morallo-Rejesus, B. and Eroles, D.C. 1974. Evaluation of the combined effectiveness of grain protectant and insecticide impregnated sack for the control of storage pests of shelled corn. Fifth Ann. Conf. Intensified Corn Production Programme.

Morallo-Rejesus, B. and Eroles, D.C. 1976. Direct grain treatment and sack impregnation with insecticides to control corn storage pests. Philipp. Agric. 59(9&10): 356–363.

Morallo-Rejesus, B. and Nerona, E.H. 1973. The storage of shelled corn in synthetic sacks treated with insecticides. Fourth Ann. Conf. Intensified Corn Production Programme.

Morallo-Rejesus, B. and Santhoy, O. 1972. Toxicity of six organophosphorus insecticides to field collected DDT-resistant strains of rice weevil, *Sitophilus oryzae* and the red flour beetle *Tribolium castaneum*. Philipp. Entomol. 2(4): 283–290.

Morallo-Rejesus, B., Varca, L.M. and Nerona, L.H. 1975. Insecticide impregnation of sacks and use of plastic lining for the protection of stored corn against insect damage. Philipp. Agric. 59: 196–204.

Moreau, C. 1968. Moisissures Toxiques dans l'Alimentation. Editions P. Lechavalier, Paris 371 pp.

Morel, J.L. 1975. Determination of residues of chlorpyrifos-methyl in bread following treatment of grain in a silo with Reldan insecticide. Report from Dow Chemical, France. GHE-P 326. (Unpublished).

Morel, J.L. and Galet, G. 1975. Results of residue analysis carried out in France on grains treated with chlorpyrifos-methyl until May 1975. Communication from Dow Chemical Europe S.A.

Morley, J. 1979. Storage losses — does the farmer see them differently? NOMA 2(1): 11–12.

Morrison, F.O. 1972. A review of the use and place of lindane in the protecting of stored products from the ravages of insect pests. Residue Rev. 41: 113–180.

Mourot, D., Delepine, B., Boisseau, J. and Gayot, G. 1979. High performance liquid chromatography of deltamethrin. J. Chromatog. 173: 412–414.

Muacevic, G. 1964. Reports of C.H. Boehringer. (Unpublished).

Muacevic, G. 1967. Reports of C.H. Boehringer. (Unpublished).

Muacevic, G. and Glees, P. 1968. Bericht uber die Prufung von Bromophos an Huhnern. Report C.H. Boehringer Sohn. (Unpublished).

Muacevic, G., Dirks, E., Kinkel, H.J. and Leuschner, F. 1970. Anticholinesterase activity of bromophos in rats and dogs. Toxicol. Appl. Pharmacol. 16: 585.

Mucke, W., Alt, K.O. and Esser, H.O. 1970. Degradation of ^{14}C-labelled diazinon in the rat. J. Agric. Food Chem. 18(2): 208–212.

Muirhead-Thomson, R.L., Gordon, R.M. and Davey, T.H. 1952. A plea for the standardisation of experimental work on residual insecticides. Trans. Roy. Soc. Trop. Med. Hyg. 46: 271–274.

Mukhameshin, R.A. 1970. The toxic action of DDVP on the organism of hens. Veterinariya 47(4): 83–86 (USSR); (Rev. Appl. Entomol. B60: 1222).

Mukherjee, G., Banerjee, A., Mukherjee, A. and Mathew, T.V. 1973. Loss of pesticide residues from rice and flour during baking and cooking. Res. and Ind. India. 18: 85.

Mukherjee, A.B. and Saxena, V.S. 1970. Relative toxicity of some organophosphorus insecticides to the adults of *Tribolium castaneum* a pest of stored cereals. Indian J. Entomol. 32(3): 246.

Mullen, M.A. and Arbogast, R.T. 1984. Low temperature to control stored-product insects. In: Baur, F.J., Ed., Insect Pest Management for Food Storage and Processing. American Association of Cereal Chemists, St. Paul, Minnesota. pp. 255–263.

Muller, W. 1952. Die Giftigkeit von Hexapreparaten (Hexachlorocyclohexane) beim Geflugel. Geflugelhof 15: 299

Murakami, J., Ito, S., Okuno, Y., Arai, M. Ito, N. and Miyamoto, J. 1980. Eighteen-month chronic oral toxicity and tumourigenicity study of S-2539 in mice. Unpublished report from Sumitomo Chemical Co. of a study performed by Industrial Bio-Test Laboratories, Inc., validated by Sumitomo Chemical Co. and submitted to the World Health Organization by Sumitomo Chemical Co.

Murray, W.J. 1979. Infestation patterns in the Australian wheat industry during the past two decades — immediate short term solutions and a prospective for the future. In: Evans, D.E., Ed., Australian Contributions to the Symposium on the Protection of Grain Against Insect Damage During Storage, Moscow 1978. CSIRO Division of Entomology, Canberra. pp. 7–14.

Murray, W.J. 1981. Industry co-ordinated loss prevention programmes. In: Champ, B.R., and Highley, E., Ed., Grain Storage Research and its Application in Australia. CSIRO Division of Entomology, Canberra. pp. 101–106.

Murray, W.J. and Snelson, J.T. 1978. Fenitrothion and bioresmethrin residues in milled products from a commercial flour mill. Report to Department of Primary Industry, Canberra, Australia, July 1979.

Muthu, M.Y. and Pingale, S.V. 1955. Control of insect pests of grains stored in insecticide impregnated jute bags. J. Sci. Food Agric. 6: 637.

Nahal, E.L. and Duguet, J.S.. Evaluation of the efficacy of deltamethrin compared with malathion and with Katel Suss against *Callosobruchus chinensis*, *Callosobruchus maculatus* and *Bruchidius alfierii* in Egypt. Joint Report, Faculty of Agronomy, Cairo University and Roussel Uclaf (In French). 9 pp.

Naidu, M.B. 1965. Site of action of piperonyl butoxide and malathion in insects. Indian J. Exp. Biol. 3: 67–68.

Nakajima, E., Shingehara, E. and Shindo, S. 1974. Whole body autoradiography in rats medicated with the organophoric insecticide DOWCO 214-^{14}C. Unpublished report from the Drug Metabolism Res. Lab., Sankyo Co. Ltd., Shinagawa, Tokyo, Japan. Submitted to the WHO by the Dow Chemical Co., Midland, Michigan, USA.

Nakamoto, N., Kato, T and Miyamoto, J. 1973. Teratogenicity study of S-2539 Forte in mice. Report from Sumitomo Chemical Co. (Unpublished).

Nambu, K., Takimoto, Y. and Miyamoto, J. 1978. Degradation and fate of phenothrin in stored wheat grains (Interim Report 1978, Sumitomo Technical Report Doc. Code EM-90 Ref. No. -0003, 20 pages). J. Pestic. Sci. (Nihon Noyakugaku Kaishi) 6(2): 183–192.

Nambu, K., Takimoto, Y., and Miyamoto, J. 1981. Degradation and fate of phenothrin in stored wheat grains. J. Pestic. Sci. (Nihon Noyakugaku Kaishi) 6(2): 181–192.

Narahaski, T. 1971. Mode of action of pyrethroids. Bull. WHO 44; 337–345.

Narahashi, T., and Haas, H.G. 1968. Interaction of DDT with the components of lobster nerve membrane conductance. J. Gen. Physiol. 51:177–198.

Nasir, M.M. 1946. DDT, 666 and insect pests of stored grains. Current Sci. (India) 15, 98.

Nasir, M.M. 1954. Responses of pests to fumigation V. The toxicity of the free and sorbed vapours of BHC and DDT to some insects infesting stored products. Bull. Entomol. Res. 45, 639.

National Research Council of Canada 1975. Fenitrothion: the effects of its use on environmental quality and its chemistry. Publication No. NRCC 14104 of the Environmental Secretariat. 162 pp.

Navarro, S. 1974. Aeration of grain as a non-chemical method for the control of insects in the grain bulk. Proc. 1st Int. Working Conf. Stored-Products Entomol., Savannah, Georgia, 7–11 October 1974. pp. 341–353.

Neher, M.B., Pheil, R.W. and Watson, C.A. 1973. Thermoanalytical study of moisture in grain. Cereal Chem. 50:617–628.

Nellist, M.E. 1979. The design of heated-air grain dryers. National Institute of Agricultural Engineering Meeting for Grain Dryer Manufacturers, Silsoe, U.K., February 1979.

Nelson, S.O. 1972. Possibility for controlling stored-grain insects with RF energy. J. Microwave Power 7: 231–239.

Nelson, S.O. 1973. Insect-control studies with microwave and other radio-frequency energy. Bull. Entomol. Soc. Amer., 19: 157–163.

Nelson, S.O. 1974. Radiofrequency, infra-red, and ultraviolet radiation for control of stored-product insects: prospects and limitations. Proc. 1st Int. Working Conf. Stored-Products Entomol., Savannah, Georgia, 7–11 October 1974. pp. 325–329.

Newton, J. 1981. Pirimiphos methyl as a protectant for oatmeal rodent baits against merchant grain beetle (*Oryzaephilus mercator*). Int. Pest Control. 23(4): 96, 98.

Nir, I., Weisenberg, E., Hadani, A., and Egyed, M. 1966. Studies of the toxicity, excretion and residues of savin in poultry. Poultry Sci. 45: 720

Nishizawa, Y., Fujii, K., Kadota, T., Miyamoto, J. and Sakamoto, N. 1961. Studies on the organophosphorus insecticides. Part VII. Chemical and biological properties of new low toxic organophosphorus insecticide, fenitrothion. Agric. Biol. Chem. (Japan) 25: 605–10.

Noble, R.M., Hamilton, D.J. and Osborne, W.J. 1982. Stability of pyrethroids on wheat in storage. Pestic. Sci. 13, 246–252.

Norment, B.R. and Chambers, H.W. 1970. Temperature relationships in organophosphorus poisoning in boll weevils. J. Econ. Entomol. 63. 502–504.

O'Brien, R.D. 1957. Properties and metabolism in the cockroach and mouse of malathion and malaoxon. J. Econ. Entomol. 50: 159.

O'Brien, R.D., Dauterman, W.C. and Niedermeier, R.P. 1961. The metabolism of orally administered malathion by a lactating cow. J. Agric. Food Chem. 9(1): 39–42.

Oden, T. and Sahin. 1964. Determination of malathion residues on stored grain by bioassay. Bitki Koruma Bulteni 4: 88.

O'Donnell, M.J. 1980. The toxicities of four insecticides to *Tribolium confusum* in two sets of conditions of temperature and humidity. J. Stored Prod. Res. 16: 71–74.

OECD 1981a. Guidelines for Testing of Chemicals. Organization for Economic Co-operation and Development, Paris.

OECD 1981b. Principles of Good Laboratory Practice. Annex 2 of previous reference.

Oehler, D.D. 1979. Gas-liquid chromatographic determination of permethrin in bovine tissues. J. Assoc. Off. Anal. Chem. 62: 1309–1311.

Oehler, D.D., Eschle J.L., Miller, J.A., Claborn, H.V. and Ivey, M.C. 1969. Residues in milk resulting from ultra-low-volume sprays of malathion, methoxychlor, coumaphos or Gardona for control of horn fly. J. Econ. Entomol. 62: 1481.

O'Farrell, A.F. Jones, B.M. and Brett, G.A. 1949. The persistent toxicity under standard field conditions of pyrethrum, DDT and "Gammexane" against pests of stored food. Bull. Entomol. Res. 40: 135.

Ofosu, A. 1977. Persistence of fenitrothion and pirimiphos-methyl on shelled maize. Ghana J. Agric. Sci. 10(3): 213–216.

Ohkawa, H., Kaneko, H. and Miyamoto, J. 1977. Metabolism of permethrin in bean plants. J. Pestic. Sci., 2: 67–76.

Ohkawa, H., Kaneko, H., Tsuki, H. and Miyamoto, J., 1979. Metabolism of fenvalerate in rats. J. Pestic. Sci. 4(2): 143.

Okuno, Y., and Kadota, T. and Miyamoto, J. 1977. Neurotoxic effects of some synthetic pyrethroids and natural pyrethrins by oral administration in rats. Reports from Sumitomo Chemical Co Ltd. (Unpublished).

Olson, K.J., 1964. Results of toxicological tests on chlorpyrifos-methyl. Biochemical Research Laboratory, Dow Chemical Company USA. Report T35.12–46193–2. 26 pp. (Unpublished).

Olson, K.J., and Taylor, M.L. 1964. Results of range-finding toxicological tests on chlorpyrifos-methyl. Biochemical Research Laboratory, Dow Chemical Company USA. Report T35.12–46193–1. 6 pp. (Unpublished).

Onsager, J.A. and Day, A. 1973. Efficiency and effective radius of blacklight traps against Southern potato wireworm. J. Econ. Entomol. 66: 403–409.

Oppenoorth, F.J. 1954. Metabolism of *gamma*-benzene hexachloride in susceptible and resistant houseflies. Nature 173: 1000.

Oshima, S. Nakamura, Y. and Sugita, K. 1970. Effect of DDVP strips on stored product pests under semi-field conditions. Shokuhin Eiseigaku Zasshi 11: 129. (CA 75: 109229).

Otieno, D.A. and Pattenden, G. 1979. Degradation of natural pyrethrins. Pyrethrum Post 15(2): 30–37.

Otieno, D.A. and Pattenden, G. 1980. Degradation of natural pyrethrins. Pestic. Science 11: 270–278. Reprinted in Pyrethrum Post 15(2): 30–37.

Owen, E. 1947. "Gammexane" control of insect pests of grain. J. Inst. Brewing 53: 236.

Padget, T.J. 1968. Evaluation of dichlorvos as a grain treatment for protection against white-fringed beetle. USDA Plant Pest Control Laboratory. (Unpublished).

Page, A.C. 1970. Metabolic fate and tissue residue following oral administration of dichlorvos. Collection of unpublished reports prepared and submitted by Shell Chemical Co.

Page, R.K. and Bush, P.B. 1978. Effect of malathion in feed on lay, fertility and hatch, University of Georgia. Report in Poultry Tips. Reviewed in Poultry Digest, March 1978, p 146, and in Australian Poultry World, February 1978, p 5.

Page A.C., DeVries, D.M., Young, R. and Loeffler, J.E. 1971. Metabolic fate of ingested dichlorvos in swine. Toxicol. Appl. Pharmacol. 19: 378. (Abstract).

Pagliarini and Hrlec, G. 1982. Results of studies on the effectiveness of some pesticides for the control of stored-product mites. Zastita Bilja 33(1) 51–56.

Pallos, F.M., Menn, J.J., Letchworth, P.E. and Miaullis, J.B. 1971. Synthetic mimics of insect juvenile hormone. Nature 232: 486–7.

Panciera, R.J. 1967. Determination of teratogenic properties of orally administered carbaryl in sheep. Report from Department of Veterinary Pathology, Oklahoma State University, Stillwater. Submitted by Union Carbide Corporation. (Unpublished).

Pandey, G.P., Srivastava, J.L. and Varma, B.K. 1979. Efficacy of Baythion 50% EC against insect pests of stored foodgrains. Pesticides 13(8): 21–24.

Panel 1973. On malathion and dichlorvos residues in grain. Analyst 98: 19–24.

Panel 1977. On determination of residues of certain organophosphorus pesticides in fruits and vegetables. Analyst 102: 858–868.

Panel 1979. On determination of organophosphorus pesticides in foodstuffs of animal origin. Analyst 104: 425–453.

Panel 1980. Determination of a range of organophosphorus pesticides residues in grain. Analyst 105: 515–517.

Papworth, D.S. 1961. The protection of stored cereals by malathion admixture techniques. Agric. Vet. Chem. 2: 160.

Pardo, E. and Nordlander, G. 1979. Surface application of the juvenile hormone analogue methoprene as a protectant against a stored product insect *Tribolium confusum* (crushed wheat). Vaxtskyddsnotiser. Sveriges Lantbruksuniversitet. 43(5/6): 116–124.

Parker, B.L., Gauthier, N.L., and Kydonieus, A., 1983. Controlled release for suppression of rice storage pests. In: Roseman, T.J., and Mansdorf, S.Z., Ed., Controlled Release Delivery Systems. Marcel Dekker, New York/Basle. pp. 301–313.

Parkin, E.A. 1958. A provisional assessment of malathion for stored-product insect control. J. Sci. Food. Agric. 5: 370–375.

Parkin, E.A. 1960. The susceptibility of stored-product insects to contact insecticides. Chem. Ind. Jan 30, 1960: 108–11.

Parkin, E.A. 1961. The potentialities of pyrethrum in the bag storage of grain. Trop. Stored Prod. Inf. No. 3. p. 77.

Parkin, E.A. 1965. The onset of insecticide resistance among field populations of stored-products insects. Stored Prod. Res. 1: 3.

Parkin, E.A. 1966. The relative toxicity and persistence of insecticides applied as water-dispersible powders against stored-product beetles. Ann. Appl. Biol. 57: 1–14.

Parkin, E.A. and Bills, G.T. 1953. Resistance of some grain insects to BHC dust. Report Pest Infest. Res. Board 1952 (London). p. 21.

Parkin, E.A. and Forster, R. 1964. Tests with newer insecticides: carbaryl. Pest Infest. Res. 1963: 32.

Parkin, E.A. and Horler, D.F. 1967. Milling and baking tests of wheat treated with fenitrothion. Pest Infest. Res. 1966: 27.

Parkin, E.A., Scott, E.I.C., Bates, A.N. and Rowlands, D.G. 1962. The malathion treatment of malting barley. Pest Infest. Res. 1961: 32.

Parkin, E.A., Warman, R., and Forster, R. 1963. Tests with newer insecticides: dichlorvos vapours. Pest Infest. Res. 1962: 39.

Parkinson, G.R. 1978. Permethrin: acute toxicity to male rats. Report of ICI Central Toxicology Laboratory. (Unpublished).

Parkinson, G.R. et al. 1976. PP 557 (Permethrin): acute and sub-acute toxicity. Report of ICI Central Toxicology Laboratory No. 538/75 (Feb. 1976). (Unpublished).

Partington, G.L., Redbond, M.R., and Boase, C. 1979. Etrimfos — a new insecticide for stored grain pest control. Proc. Brit. Crop Prot. Conf. 1979 Vol. 2, pp. 525–32.

Pasarela, N.R., Brown, R.G. and Shaffer, C.B. 1962. Feeding of malathion to cattle: residue analyses of milk and tissue. J. Agric. Food Chem. 10(1): 7–9.

Passlow, T. 1958. Destruction of sorghum midge in seed grain. Queensl. J. Agric. Sci 15: 37.

Paul, B.S. and Vadlamudi, V.P. 1976. Fenitrothion injected into eggs. Bull. Environ. Contam. Toxicol. 15: 223.

Pawley, W.H. 1963. Possibilities of increasing world food production. FFHC Basic Study No. 10. FAO, Rome. 231 pp.

Paulson, G.D, and Feil, U.J. 1969. The fate of a single oral dose of carbaryl in the chicken. Poultry Sci. 48: 1593.

Peng, W.K. 1983. Relative toxicity of ten insecticides against six coleopterous stored-rice insect pests. National Science Council Monthly 11(7): 638–644.

Perry, A.S. 1960. Metabolism of insecticides by various insect species. J. Agric. Food Chem. 8: 266.

Perti, S.L., Wal R.C. and Nigam, B.S. 1965. Toxicity of synthetic contact insecticides to food pests. Labder. J. Sci. Technol. 3: 213.

Pesticide Analytical Manual 1979a. Multiresidue methods for chlorinated and organophosphorus pesticides in fatty and non-fatty foods. Volume 1, Table 201-A and Sections 211, 212, 231, 232.1 and 252. Food and Drug Administration, Washington, D.C.

Pesticide Analytical Manual 1979b. Storherr organophosphate/ carbon clean-up for non-fatty foods. Volume 1, Table 201-H and Section 232.3. Food and Drug Administration, Washington, D.C.

Pesticide Analytical Manual 1979c. Various pesticides in non-fatty foods. Volume 1, Table 201-I and Sections 232.4. Food and Drug Administration, Washington, D.C.

Pesticide Analytical Manual 1979d. Methods under compound name. Volume II. Food and Drug Administration, Washington, D.C.

Pesticide Analytical Manual 1979e. Volume 1, Table 201-J and Section 242.2. Food and Drug Administration, Washington, D.C.

Phillips, G.L. 1959. Control of insects with pyrethrum sprays in wheat stored in ships holds. J. Econ. Entomol. 52: 557–9.

Phillips, J.K. and Burkholders, W.F. 1984. Health hazards of insects and mites in food. In: Baur, F.J., Ed., Insect Pest Management for Food Storage and Processing. American Association of Cereal Chemists, St. Paul, Minnesota. pp. 279–290.

Phipers, R.F., and Wood, M.C. 1957. An investigation into reported stabilisation of pyrethrins by piperonyl butoxide. Pyrethrum Post 4(2): 11–12.

Pier, A.C., Cysewski, S.J., Richard, J.L. and Thurston, J.R. 1977. Mycotoxins as a veterinary problem. In: Rodricks, J.V. et al., Ed., Mycotoxins in Human and Animal Health. Pathotox Publishers, Park Forest South, Illinois. pp. 745–750.

Pieterse, A.H. and Schulten, G.G.M. 1972. A study on insecticide resistance in *Tribolium castaneum* in Malawi. J. Stored Prod. Res. 8(3): 183–191.

Pingale, S.V. 1964. Maintenance of buffer stocks of cereals under tropical conditions of India. Proc. 12th Int. Congr. Entomol., London 1964.

Pinniger, D.B. 1975. An assessment of residual insecticide treatments for the control of the saw-toothed grain beetle *Oryzaephilus surinamensis* in farm grain stores. Proc. 8th Brit. Insect. Fung. Conf. pp. 365–370.

Pinniger, D.B. and Halls, G., 1981. The use of pesticides at farmer and village level. Proc. GASGA Seminar on the Appropriate Use of Pesticides for the Control of Stored Products Pests in Developing Countries, Slough, 17–20 February 1981. pp. 63–66.

Pixton, S.W. 1967. Moisture content — its significance and measurement in stored products. J. Stored Prod. Res. 3: 35–47.

Pixton, S.W. and Griffiths, H.J. 1971. Diffusion of moisture through grain. J. Stored Prod. Res. 7: 133–152.

Pixton, S.W. and Warburton, S. 1971a. Moisture content/ relative humidity equilibrium of some cereal grains at different temperatures. J. Stored Prod. Res. 6: 283–293.

Pixton, S.W. and Warburton, S. 1971b. Moisture content/ relative humidity equilibrium, at different temperatures of some oilseeds of economic importance. J. Stored Prod. Res. 7: 261–269.

Pixton, S.W. and Warburton, S. 1975. The moisture content/ equilibrium humidity relationship of soya. J. Stored Prod. Res. 11: 249–251.

Plant Protection Division (1974). Actellic, containing pirimiphos-methyl, an insecticide for stored products. Tech. Data Sheet ICI Ltd, March 1974.

Pomeranz, Y. 1982. Cereal science and technology at the turn of the decade. Food Technol. Aust. 34(10): 456–469.

Porter, M.L. and Burke, J.A. 1973. An isolation and cleanup procedure for low levels of organochlorine pesticide residues in fats and oils. J. Assoc. Off. Anal. Chem. 56: 733–738.

Potter, J.C. 1976a. Tissues of rats fed 14C-SD43775 b) milk and tissues from cows fed 14C-SD43775 c) milk and tissues from cows fed 14C-SD43775 d) eggs and tissues from laying hens fed 14C-SD43775 e) cream from the milk of cows fed 14C-SD43775. Shell Development Co. Reports. (Unpublished).

Potter, J.C. and Arnold, D.L. 1977. Residues of 14C-in tissues of rats fed 14C-SD43775 for 28 days. Report of Shell Development Co. (Unpublished).

Potter, J.C. and Sauls? 1978. Residues of 14C in tissues of rats fed SD43775 for 28 days. Shell Development Co. Report. (Unpublished).

Potter, J.C., Boyer, A.C., Marxmiller, R.L., Young, R. and Loeffler 1973. Radioisotope residues and residues of dichlorvos and its metabolites in pregnant sows and their progeny dosed with dichlorvos-14C or dichlorvos-36C- formulated as PVC pellets. J. Agric. Food Chem. 21(4): 734–738.

Potter, J.C., Loeffler, J.E., Collins, R.D., Young, R. and Page, A.C. 1973a. Carbon-14 balance and residues of dichlorvos and its metabolities in pigs dosed with dichlorvos-14C. J. Agric. Food Chem. 21: 163.

Pradhan, S. 1949. Studies on the toxicity of insecticide films III. Effect of relative humidity on the toxicity of films. Bull. Entomol. Res. 40: 431–444.

Pradhan, S., Chatterji, S.M., Sethi, G.R., Bhamburkar, M.W. and Prasad, H. 1971. Feasibility of controlling stored grain pests by the sterile-male technique. In: Sterility Principle for Insect Control or Eradication, Proc. Int., Atomic Energy Agency Symp; Athens 1970. IAEA, Vienna. pp. .

Prakash, A. and Kauraw, L.P. 1983. Compatibilities between certain pesticides used for control of insect pests and seed borne fungi in stored paddy. Pesticides 17(11): 21–22.

Prakash, A. and Pasalu, I.C., 1981. Evaluation of insecticides against *Sitotroga cerealella* in stored paddy. Oryza 18: 173–194.

Prakash, A., Pasalu, I.C. and Mathur, K.C. 1983. Satisfar: a paddy seed protectant in storage. Pesticides 17(2): 24–25.

Pranata, R.I., Haines, C.P., Roesli, R. and Sunjaya 1983. A study of the use of a dust admixture treatment with permethrin for the protection of rough rice and milled rice. In: Teter, N.C., et al., Ed., Maintaining Good Grain Quality. Proc. 6th Ann. Workshop Grain Post-Harvest Technol., Bogor, Indonesia, 3–6 May 1983. Manila, Southeast Asia Co-operative Post-Harvest Research and Development Programme. pp. 132–146.

Press, A.F. and Childs, D.P. 1966. D.P. Control of the tobacco moth with dichlorvos. J. Econ. Entomol. 59: 264.

Prevett, P.F. 1959. An investigation into storage problems of rice in Sierra Leone. Colon. Research Station No. 28. Ind. Colonial Office. HMSO, London.

Prevett, P.F. 1974. Tecnologia de graos armazenados — alguns melhoramentos recentes. (Grain storage technology — some recent developments.) Bolm Inf. Soc. Bras. Cienc. e Tecnol. Alimentos, No. 30. pp. 1–21.

Proctor, D.L. and Rowley, J.Q. 1983. The thousand grain mass (TGM): a basis for better assessment of weight losses in stored grain. Trop. Stored Prod. Inf. 45, 19–23.

Purchase, I.F.H., Ed. 1974. Mycotoxins. Elsevier Scientific Publishing Company, Amsterdam. 443 pp.

Puschel, R., 1981. On the problem of "Genauigkeit" (exactness) of chemical analysis. FDA By-Lines No. 3: 157–174.

Pym, R.A.E., Gilbert, W.S. and Singh, D. 1976. The effect of dichlorvos as a contaminant in the feed on the performance of laying hens. Proc. 1st Australas. Poultry and Stock Feed Convention, Melbourne. Vol. 2, pp. 313–316.

Pym, R.A.E., Singh, G., Gilbert, W.S., Armstrong, J.P. and McCleary, B.V. 1984. Effect of dichlorvos, maldison and pirimiphos-methyl on food consumption, egg production, egg and tissue residues and plasma acetylcholinesterase inhibition in layer strain hens. Aust. J. Exp. Agric. Anim. Husb. 24: 88–92.

Qadri, S.S.H. 1971. Bioassay technique to estimate vapour toxicity of dichlorvos (DDVP) and contact toxicity of insecticides. Pesticides 5(11): 22.

Qayyum, H.A. 1978. Losses caused to stored products by insect pests in Pakistan and measures for their control. Proc. 2nd Int. Working Conf. Stored-Products Entomol., Ibadan, Nigeria, 10–16 September 1978. pp. 49–53.

Queensland Department of Primary Industries 1975. Stored Products Australia 1975. Queensland Dep. Primary Indust. Report, 1974/75. 47 pp.

Quinlan, J.K. 1977. Surface and wall sprays of malathion for controlling insect populations in stored shelled corn. J. Econ. Entomol. 70(3): 335–336.

Quinlan, J.K. 1978. Chlorpyrifos-methyl and malathion applied as protectants for high moisture stored wheat. (Abstract). Proc. Entomol. Soc. Amer. 33: 27.

Quinlan, J.K. 1979a. Malathion thermal aerosols applied to (stored) corn, soybeans, wheat and sorghum using aeration. J. Kans. Entomol. Soc. 32(3): 523.

Quinlan, J.K. 1979b. Malathion aerosols applied in conjunction with vertically placed aeration for the control of insects in stored corn. J. Kans. Entomol. Soc. 52(1): 648–652.

Quinlan, J.K. 1980. A preliminary study with malathion aerosols applied with a corn drying system for the control of insects (*Tribolium castaneum*, *Sitophilus oryzae*). J. Kans. Entomol. Soc. 15(3): 252–257.

Quinlan, J.K. and Miller, R.F. 1958. Evaluation of synergised pyrethrum for the control of Indian meal moth in stored shelled corn. US Dep. Agric. AMS Market Research Report 222. 13 pp.

Quinlan, J.K., White, G.D., Wilson, J.L., Davidson, L.J. and Hendricks, L.H. 1979. Effectiveness of chlorpyrifos-methyl and malathion as protectants for high moisture stored wheat. J. Econ. Entomol. 72: 90–93.

Quinlan, J.K., Wilson J.L. and Davidson, L.I. 1980. Pirimiphos-methyl as a protectant for high moisture stored wheat (in controlling insects). J. Kans. Entomol. Soc. 53(4): 825–832.

Quistad, G.B., Staiger, L.E. and Schooley, D.A. 1974a. Environmental degradation of the insect growth regulator methoprene (isopropyl (2E,4E)-11-methoxy-3,7.11-trimethyl 1–2,-4dodecadienoate). I. Metabolism by alfalfa and rice. J. Agric. Food Chem. 22: 582–589.

Quistad, G.B., Staiger, L.E. and Schooley, D.A. 1974b. Cholesterol and bile acids via acetate from the insect juvenile growth hormone analog, methoprene. Life Sciences 15(10): 1797–1804.

Quistad, G.B., Staiger, L.E., Bergot, B.J. and Schooley, D.A. 1975a. Environmental degradation of the insect growth regulator methoprene. VII. Bovine metabolism to cholesterol and related natural products. J. Agric. Food Chem. 23(4): 743–749.

Quistad, G.B., Staiger, L.E. and Schooley, D.A. 1975b. Environmental degradation of the insect growth regulator methoprene (isopropyl (2E,4E)-11-methoxy-3,7 11-trimethyl 1-2, 4 dodecadienoate). III. Photodecomposition. J. Agric. Food chem. 23: 299 303.

Quistad, G.B., Staiger, L.E. and Schooley, D.A. 1975c. Environmental degradation of the insect growth regulator methoprene. VIII. Bovine metabolism to natural products in milk and blood. J. Agric. Food Chem. 23(4): 750–753.

Quistad, G.B., Staiger, L.E. and Schooley, D.A. 1975d. Environmental degradation of the insect growth regulator methoprene. V. Metabolism by houseflies and mosquitoes. Pestic. Biochem. Physiol. 5: 233–241.

Quistad, G.B., Staiger, L.E. and Schooley, D.A. 1976a. Environmental degradation of the insect growth regulator methoprene. X. Chicken metabolism. J. Agric. Food Chem. 24: 644–648.

Quistad, G.B., Schooley, D.A., Staiger, L.E., Bergot, B.J., Sleight, B.H. and Macek, K.J. 1976b. Environmental degradation of the insect growth regulator methoprene. IX. Metabolism in bluegill fish. Pestic. Biochem. Physiol 6: 523–529.

Qureshi, A.H. 1967. Effects and persistence of dichlorvos vapours liberated from Vapona pest strips on *Tribolium castaneum* and other stored products pests. Rep. Nigerian Stored Prod. Res. Inst. 143.

Radeleff, R.D. 1951. Effects of various levels of lindane in the feed of beef cattle. Vet. Med. 46: 105.

Radeleff, R.D. 1958. The toxicity of insecticides and herbicides to livestock. Adv. Vet Sci. 4: 265–276.

Radeleff, R.D., Woodard, G.T., Nickerson, W.J. and Bushland, R.C. 1955. The acute toxicity of chlorinated hydrocarbon and organic phosphorus insecticides to livestock. USDA Tech. Bull. 1122. USDA, Washington, D.C.

Ragunathan, A.N., Srinath, D. and Majumder, S.K. 1974. Inhibition of storage fungi by some fungicides. J. Food Sci. Technol. 11: 19.

Rai, B.K. 1977. Control of storage pests (*Rhyzopertha dominica, Tribolium castaneum, Sitotroga cerealella*) of paddy by admixture with pirimiphos-methyl. Bull. Grain Technol. 15(3): 176–181.

Rai, B.K. and Croal, K. 1973. Chemical control of paddy moth *Sitotroga cerealella* infesting paddy bags. Agric. Res. Guyana 123–124.

Rai, L. and Roan, C.C. 1959. Report included in Geigy Chemical Co. Pesticide Petition No. 232 to the US Food and Drug Administration. (Unpublished.)

Rai, L. and Roan, C.C. 1959a. Interactions of topically applied piperonyl butoxide and malathion in producing lethal action in houseflies. J. Econ. Entomol. 52: 218–220.

Rai, L. and Roan, C.C. 1960. Fate of diazinon fed to dairy cows and steers. Report included in submission by Geigy Chemical Co. Pesticide Petition No. 232 to the US Food and Drug Administration. (Unpublished).

Rai, S.E., Afifi, E.D., Fryer, H.C. and Roan, C.C. 1956. The effect of different temperatures and piperonyl butoxide on the action of malathion on susceptible and DDT-resistant strains of houseflies. J. Econ. Entomol. 49(3): 307–10.

Rai, R.S., Panchi Lal and Srivastava, P.K. 1985. Impregnation of jute bags with insecticides for protecting stored food grains. II. Comparative efficacy of some safe organophosphorus insecticides. Pesticides April 1985: 45–48.

Rajini, P.S. and Krishnakumari, M.K. 1981. Malathion poisoning in non-target species: an acute oral study in poultry (*Gallus domesticus*). J. Food Sci. Technol. 18(6): 237–239.

Raju, P. 1984. Storage pest menace — how to combat? Role of grain protectants in its perspective. Pesticides 18(8): 16–21.

Rangawamy, J.R. 1973. Observations on the sorption of water vapour by rice and sorghum. J. Food Sci. Technol. India 10(2): 59–61.

Ratnadass, A. 1984. Les problems entomologiques lies aue stockage paysan des vivriers en Cote d' Ivoire. Note Technique No 06.841 CV IDESSA. Institu des Savanes, Idessa, Republique de Cote d' Ivoire. 47 pp.

Raucourt, M. 1945. Revue de phytopharmacie. Ann. Agron. (Paris) 15: 379.

Ray, D.E., and Cremer, J.E. 1979. The action of deltamethrin (a synthetic pyrethroid) on the rat. Pestic. Biochem. Physiol. 10: 333–340.

Reader, R.A. 1971. Survey of damage to maize stored under village conditions. Rep. Lilongwe Land Dev. Proj. Malawi 6. 45 pp.

Redlinger, L.M. and Womack, H. 1966. Evaluation of four inert dusts for the protection of shelled corn in Georgia from insect attack. US Dept. Agric. Agric. Res. Serv 51–7. 25 pp.

Rehfeld, B.M. 1971. The effect of malathion, P.C.B. and iron on growing chicks. Diss. Abs. Int. B31: 7397.

Rehfeld, B.M., Pratt, D.E. and Sunde, M.L. 1969. Effect of various levels of dietary malathion on performance of chicks. Poultry Science 48: 1718–1723.

Reichel, W.L., Kolbe, E. and Stafford, C.J. 1981. Gas-liquid chromatographic — mass spectrometric determination of fenvalerate and permethrin residues in grasshoppers and duck tissue samples. J. Assoc. Off. Anal. Chem. 64: 1196–1200.

Reichenbach, N.G. and Collins, W.J. 1984. Multiple logit analysis of the effects of temperature and humidity on the toxicity of propoxur to German cockroaches and western spruce budworm larvae. J. Econ. Entomol. 77: 31–35.

Reinhardt, R. and Esther, H. 1971. Studies of the sensitivity of some insects to organophosphorus insecticides throughout the day. Zool. Jahrb. Abst. Syst. Cekol. Georgr. Tiere. 98(3), 511.

Renfer, A. 1977. Biological evaluation of methacrifos as a grain protectant under field conditions in Switzerland — I. Basel. Unpublished Report 9.25 VS 135 from Ciba-Geigy, Basle, Switzerland.

Renfer, A., Wyniger, R., Knusli, F., Gfeller, W. and Hart, R.J. 1978. Methacrifos — A new insecticide for grain protection. Proc. 2nd Int. Working Conf, Stored-Products Entomol., Ibadan, Nigeria, 10–16 September 1978. pp. 420–431.

Rensburg, S.J. van 1974. Role of epidemiology in the elucidation of mycotoxin health risks. In: Rodricks, J.V., Hesseltine, C.W. and Mehlaman, M.A., Ed., Mycotoxins in Human and Animal Health. Pathotox Publishers, Park Forest South, Illinois. pp. 699–744.

Rheenen, H.A. van, Pere, W.M. and Magoya, J.K. 1983. Protection of stored bean seeds against the bean bruchid. FAO Plant Protection Bulletin 31(3): 121–125.

Ribelin, W.E. and Macklin, A.W. 1971. Relation of pesticides to abortion in diary cattle. Toxicol. Appl. Pharmacol. 19: 419. (Abstract).

Richards, O.W. 1947. Observations on grain weevils, *Calandra* (Col. Curculionidae) 1. General biology and oviposition. Proc. Zool. Soc. London 117: 1–43.

Ripp, B.E. 1981. Hygene in central storage systems. In: Williams, P., and Amos, T., Ed., Proc. 1st Australas. Stored Grain Pest Control Conf., Melbourne, May 1981. CSIRO, Melbourne. pp. 3/8–3/11.

Roan, C.C. 1964. Comparison of performance of various carbaryl formulations against rice weevil and confused flour beetle. Report from Entomology Department, Kansas State University to Union Carbide Corporation.

Roan, C.C. and Srivastava, B.P. (1965). Dissipation of diazinon residues in wheat. J. Econ. Entomol. 58: 996.

Robbins, W.E., Hopkins, T.L. and Eddy G.W. 1957. Metabolism and excretion of phosphorus-32-labelled diazinon in a cow. J. Agric. Food Chem. 5:509.

Robertson, G. 1974. Effect of carbon-14 on the survival and reproduction of red scale males. Entomol. Exp. Appl. 17: 31–35.

Robinson, A.S. 1973. Increase in fertility with repeated mating of *gamma* irradiated male codling moths *Laspeyresia pomonella*. Can. J. Zool. 51: 427–430.

Robinson, W. 1926. Low temperature and moisture as factors in the ecology of the rice weevil, *Sitophilus oryzae* and the granary weevil (*Sitophilus granarius*) Minn. Agric. Exp. Stn. Tech. Bull. 41.

Rodricks, J.V., Ed. 1974. Mycotoxins and other fungal related food problems: a symposium sponsored by the American Chemical Society, Atlantic City, September 1974. American Chemical Society, Washington, DC. Advances in Chemistry Series No. 149. 409 pp.

Rodricks, J.V., Hesseltine, C.W. and Mehlaman, M.A. Ed. 1977. Mycotoxins in Human and Animal Health. Pathotox Publishers, Park Forest South, Illinois. 807 pp.

Rodriguez, J.G., Potts, M.F. and Patterson, C.G. 1984. Sorptive coatings as protectants against stored-product pests. Proc. 3rd Int. Working Conf. Stored-Products Entomol., Manhattan, Kansas, October 1983. pp. 572–628.

Roger, J.-C., Upshall, D.G. and Casida, J.E. 1969. Structure/activity and metabolism studies on organophosphate teratogens and their alleviating agents in developing hen eggs with special emphasis on Bidrin. Biochem. Pharmacol. 18: 373.

Rohrlich, M., Suckow, P. and Hertel, W. 1971. Investigation of the distribution of lindane in grain and its processed products. Getreide und Mehl 21: 109.

Rosen, J.D. 1972. Metabolism of bioresmethrin in plants. In: Matsumura, F., Bosh, G.M. and Misato, T., Ed., Environmental Toxicology of Pesticides. Academic Press, New York. 438 pp.

Rosival, L., Vargova, M., Szokolayova, J., Cerey, K., Hladka, A., Batora, V., Kovacicova, J. and Truchlik, S. 1976. Contributions to the toxic action of S-methyl fenitrothion. Pest. Biochem. Physiol. 6: 280–286.

Ross, D.B and Roberts, N.L. 1974. The acute oral toxicity (LD/50) of Reldan (DOWCO 214) to the chicken. Huntington Research Centre Report. DWC 234/74822, Nov. 15, 1974. 8 pp. (Contracted by Dow Chemical, Ltd).

Ross, D.B., Burroughs, S.J. and Roberts, N.L. 1974. Egg production and hatchability in the laying hen following dietary inclusion of pirimiphos-methyl at various levels. Huntington Research Centre Report. ICI 31/74646. (Unpublished).

Ross, D.B., Christopher, D.H., Cameron, D.M., Dollery, R., Almond, R.H. and Roberts, N.L. 1976a. Egg production and hatchability following inclusion of pirimiphos-methyl at various levels in the diet of the laying hen. Report Huntington Research Centre. ICI 31A/75924, dated 18/3/76. (Unpublished).

Ross, D.B., Cameron, D.M. and Roberts N.L. 1976b. Acute and sub-acute oral toxicity tests of PP557 (permethrin). Reports of Huntington Research Centre, Huntington, Cambridgeshire. Nos. ICI 68/WL/7636–7640, 75837 and 75839 (5 March 1976). Commissioned by ICI Central Toxicology Laboratory. (Unpublished).

Ross, D.B., Prentice, D.E., Majeed, S.K., Gibson, W.A., Cameron, D.M., Cameron, M.McD. and Roberts, N.L. 1977. The incorporation of permethrin in the diet of laying hens. Huntington Research Centre Report ICI/15277387 to ICI Central Toxicology Laboratory. (Unpublished).

Ross, D.B., Roberts, N.L., Cameron, M.McD., Prentice, D.E., Cooke, L., and Gibson, W.A. 1978. Deltamethrin — LD/50 determination and assessment of neurotoxicity in the domestic hen. Report Huntington Research Centre, Huntington, England. No RSL 293-NT/7830/A. Submitted to WHO by Roussel Uclaf. (Unpublished).

Ross, E. and Sherman, M. 1960. The effect of selected insecticides on growth and egg production when administered continuously in the feed. Poultry Sci. 39: 1203–1211.

Roussel Uclaf,1976. Deltamethrin. Test to determine the toxicity in the chicken by oral route. Report RU-76.05.05/A, Roussel Uclaf. (Unpublished).

Rowe, L.D. 1978. Study of residues in milk and cream from cows fed chlorpyrifos-methyl. Dow Chemical, USA, R&D Report. No. TA-606 (26 April, 1978) 27 pp. (Unpublished).

Rowlands, D.G. 1964. The degradation of malathion on stored maize and wheat grains. J. Sci. Food Agric. 15: 824–829.

Rowlands, D.G. 1965a. The *in vitro* and *in vivo* oxidation and hydrolysis of malathion by wheat grain enzymes. J. Sci. Food Agric. 16: 325.

Rowlands, D.G. 1965b. Effect of triphenyl phosphate on the phosphatase degradation of malathion *in vitro* and the metabolism of triphenyl phosphate. Pest Infest. Res. 1964 p. 28.

Rowlands, D.G. 1966a. The metabolism of bromophos in stored wheat grains. J. Stored Prod. Res. 2: 1–12.

Rowlands, D.G. 1966b. Metabolism of insecticides on stored cereals. Pest. Infest. Res. 1965: 37–39.

Rowlands, D.G. 1966c. The activation and detoxification of three organic phosphorothionate insecticides applied to stored wheat grains. J. Stored Prod. Res. 2: 105–116.

Rowlands, D.G. 1967a. The metabolism of contact insecticides in stored grains. Residue Rev. 17: 105.

Rowlands, D.G. 1967b. Fate of insecticide residues in stored grain. Pest Infest. Res. 1967: 31–32.

Rowlands, D.G. 1969. Effect of prior fumigation with methyl bromide on the desmethylation of organophosphorus insecticides. Pest Infest. Res. 1968: 40.

Rowlands, D.G. 1970a. Comparision of rates of degradation of dichlorvos on wheat and barley. Pest Infest. Res. 1969: 37.

Rowlands, D.G. 1970b. The metabolic fate of dichlorvos on stored wheat grains. J. Stored Prod. Res. 6: 19–32.

Rowlands, D.G. 1971. The metabolism of contact insecticides in stored grains. Part II. 1966–9 Residue Rev. 34: 91.

Rowlands, D.G. 1975. The metabolism of contact inseticides in stored grains 1970–1974. Residue Rev. 58: 113–155.

Rowlands, D.G. 1976. The uptake and metabolism by stored wheat grains of an insect juvenile hormone and two insect hormone mimics. J. Stored Prod. Res. 12: 35–41.

Rowlands, D.G. 1980. The metabolism of pirimiphos-methyl by stored wheat grains under laboratory conditions. Pest Control Chemistry Department Report No. 54. Pest Infestation Laboratory, MAFF, Slough U.K.

Rowlands, D.G. 1986. Background studies on metabolism of residual insecticides in grain. In: Champ, B.R., and Highley, E., Ed., Pesticides and Humid Tropical Grain Storage Systems: Proc. Int. Seminar, Manila, 27–30 May 1985. ACIAR Proceedings No. 14. pp. 139–150.

Rowlands, D.G. and Bramhall, J.S. 1977. The uptake and translocation of malathion by the stored wheat grain. J. Stored Prod. Res. 13: 13–22.

Rowlands, D.G. and Clarke, J.H. 1968. Metabolism of insecticides in stored cereals: effect of bromophos residues on grain fungi. Pest Infest. Res. 1967: 32.

Rowlands, D.G. and Clements, J.E. 1965a. The degradation of malathion in rice brans. J. Stored Prod. Res. 1: 101–103.

Rowlands, D.G. and Clements, J.E. 1965b. Metabolism of insecticides in stored cereals. Malathion in maize and wheat. Pest Infest. Res. 1964: 32.

Rowlands, D.G. and Clements, J.E. 1966. Effect of triphenyl phosphate on malathion metabolism in wheat. Pest Infest. Res. 1965: 39.

Rowlands, D.G. and Dyte, C.E. 1979. The metabolism of two methylenedioxyphenyl compounds in susceptible and resistant strains of *Tribolium castaneum*. Proc. 10th British Insecticide and Fungicide Conf. Vol 1, pp 257–264.

Rowlands, D.G. and Horler, D.F. 1967. Penetration of malathion into wheat grains. Proc. 4th Brit. Insecticide and Fungicide Conf. Vol. 1, pp. 331–335.

Rowlands, D.G. and Wilkin, D.R. 1975. Metabolism of pesticides in stored cereals. Report of Director Pest Infestation Control Lab., Slough, U.K., 1971–3.

Rowlands, D.G., Sharma, R.K. and Dean, P.R. 1967. Metabolism of insecticides in stored cereals — Nature of oxidative activity in the seed coat. Pest Infest. Res. 1967. 29.

Rowlands, D.G. et al. 1974. Preliminary report of studies with pirimiphos-methyl as a grain protectant. Submission by United Kingdom to FAO.

Rueckert, W. and Ballschmitter, K. 1973. Abbau und Toxizitaet von Hexachloro-bicyclo (2,2,2)-hept.-5-en-Derivaten (cyclodien-Insektizide). Z. Naturforsch. Teil. C. Biochem. Bioplys. Biol. Virol. 28: 107–112.

Rusiecki, W. and Bronisz, H. 1964. Zeszyty Problemowe Postepon Nauk Roiniczych 51: 55.

Russell, D.G. 1980. Socio-economic evaluation of grains post-production, loss-reducing systems in South-East Asia. Paper presented at the E.C. Stakman Commemorative Symposium on Assessment of Losses which Constrain Production and Crop Improvement in Agriculture and Forestry. University of Minnesota, Minneapolis, Minnesota, 20–23 August, 1980. 10 pp.

Rutter, H.A. 1974. Teratogenicity study in rabbits: S-2539 Forte (phenothin). Report from Hazleton Laboratories Inc. to Sumitomo Chemical Co. (Unpublished).

Ruzo, L.O., Unai, T. and Casida, J.E. 1978. Deltamethrin metabolism in rats. J. Agric. Food Chem. 26: 918.

Ruzo, L.O., Engel, J.L. and Casida, J.E., 1979. Oxidative, hydrolytic and conjugative reactions in mice. J. Agric. Food. Chem. 27: 725–731.

Rybakova, M.N. 1966. Toxic effect of Sevin on animals (in Russian). Gig. Sanit. 31(9): 42–47.

Saba, F., and Matthaei, H.D. 1979. The use of insecticides for the protection of stored products in relation to the method of application and storage. Congress sur la lutte contre les insectes en milieu tropical. Chambre de Commerce et d'Industrie de Marseille, 13–16 Mars 1979. pp. 635–648.

Sachsse, K. and Ullmann, L. 1974b. 8-day feeding toxicity of technical GCA20168 in the Japanese quail. Unpublished report (13 July, 1974) submitted by Ciba-Geigy to WHO.

Saha, J.G. and Sumner, A.K. 1974. The fate of lindane-^{14}C in wheat flour under normal conditions of breadmaking. Can. Inst. Food Sci. Technol. J. 7: 101.

Saivaraj, K., Thirumurthy, S., Rajakannu, K., Raguraj, R., Subramanian, T.R., Dhamodhiran, L. and Gopalakrishnan, C.V. 1977. Cabaryl residues in maize fodder at feeding and at lactation in milch cows. Pesticides 11(6): 36–37.

Saldarriaga, A. 1958. Factors influencing the effectiveness of insecticides used in the production of stored grains. Agric. Trop. (Bogota) 14: 619–631. (CA 53: 22705g).

Sales, F.M. 1979. Effects of relative humidity on the toxicity of organosynthetic insecticides to the bug *Lygus hesperus*. Fitossanidade 3(1–2): 55–56.

Salunkhe, G.N. 1982. Chemical control of rust red flour beetle on stored sorghum. Journal of Maharashtra Agricultural Universities 7(3): 274–276.

Samson, P.R. 1985. Biological efficacy of residual pesticides in stored grain at high humidities and moisture contents. In: Champ, B.R., and Highley, E., Ed., Pesticides and Humid Tropical Grain Storage Systems: Proc. Int. Symp., Manila, 27–30 May 1985. ACIAR Proceedings No. 14. pp. 157–172.

San Antonio, J.P. 1959. Demonstration of lindane and a lindane metabolite in plants by paper chromatography. J. Agric. Food Chem. 7: 322.

Sandoz 1977. Feeding study with etrimfos in cattle. Sandoz Report Agro. Dok CBK 3011/77. (Unpublished).

Sandoz 1979. Residue studies with Satisfar submitted to FAO. Reports Agro Dok CBK 3630/79, 3770/79, 4121/79, 4149/79, 4118/79, 4117/79, 4148/79, 4272/80, 3630/79, 3707/79, 4209/80, Sandoz Ltd, Basle, Switzerland. (Unpublished).

Sandoz 1981. Satisfar Insecticide for Stored Crops. Technical Bulletin from Information Services, Agric. Division, Sandoz Ltd, Basle, Switzerland. 71 pp.

Sandoz 1984. Etrimfos residues in wheat grain. Reports Agro Dok CBK 10754/84, 10755/84, Sandoz Ltd. Basle, Switzerland. (Unpublished).

Sarkar D.K. Handa, S.K. and Mukerjee, S.K. 1984. Residual toxicity and persistence of fenvalerate on jute bags and its persistence on wheat grains. Pesticides 18(5): 30–31.

Sarles, M.P. 1949. Studies on the pyrethrum synergist, piperonyl butoxide, and their bearing on use of these two insecticides in the control of arthropod pests. J. Parasitol. 35(6) 56.

Sarles, M.P. and Vandergrift, W.B. 1952. Chronic oral toxicity and related studies on animals with the insecticide and pyrethrum synergist, piperonyl butoxide. Amer. J. Trop. Med. Hyg. 1952 1(5): 862–883.

Sarles, M.P., Dove, W.E. and Moore, D.H. 1949. Acute toxicity and irritation tests on animals with the new insecticide, piperonyl butoxide. Amer. J. Trop. Med. Hyg. 1952 29(1): 151–166.

Sasinovich, L.M. 1972. Hygienic evaluation of DDVP in grain and grain products. Vop. Pitan. 31(5): 60.

Sauter, E.A. and Steele, E.E. 1972. The effect of low level pesticide feeding on the fertility and hatchability of chicken eggs. Poultry Sci. 51: 71.

Saxena, S.C. 1967a. Synergized and unsynergized pyrethrum III. Comparative evaluation of piperonyl butoxide and MGK 264 as pyrethrum synergists. Appl. Entomol. Zool. 2(3): 158–162.

Saxena, S.C. 1967b. Synergized and unsynergized pyrethrum: their deposit and the persistency of the toxicity of the films. Zool. Anz. 178(5–6): 434–442.

Schaefer, C.H. and Wilder, W.H. 1972. Insect development inhibitors. A practical evaluation as mosquito control agents. J. Econ. Entomol. 65: 1066–1071.

Schafer, E.W. 1972. Acute oral toxicity of 369 pesticidal, pharmaceutical and other chemicals to wild birds. Toxicol. Appl. Pharmacol. 21(3): 315–330.

Schaffer, C.B. 1955. Malathion: summary of toxicity studies to January 1, 1985. Special Report from Industrial Toxicology Section, American Cyanamid Company, dated 8 April 1955. (Unpublished).

Schecter, M.S., Green, N. and Laforge, F.B. 1949. Cinerolene and the synthesis of related cyclopentenolones. J. Amer. Chem. Soc. 71: 3165–3173.

Schecter, M.S., Hayes, D.K. and Sullivan, W.N. 1971. Manipulation of photoperiod to control insects. Israel J. Entomol. 6: 143–146.

Schesser, J.H., Priddle, W.E. and Farrell, E.P. 1958. Insecticidal residues in milling fractions from wheat treated with methoxychlor, malathion and lindane. J. Econ. Entomol., 51: 516.

Schlinke, J.C. and Palmer, J.S. 1971. Preliminary toxicologic study of three organophosphate insecticidal compounds in turkeys. Amer. J. Vet. Res. 32(3): 495–498.

Schmidt, H.U. and Wohlgemuth, R. 1979. Dichlorvos slow emission resin strips for the control of stored product moth pests in grain (*Ephestia elutella*, *Ephestia kuehniella*, *Plodia interpunctella*) stored in bulks and stacks. Nachrichtenblatt des deutschen Pflanzenschutzdienstes 31(6): 82–89.

Schnabel, D. and Formica, G. 1979a. Methacrifos (CGA 20168). Gas chromatographic determination of total residues in wheat and its derivatives seven months after treatment with the formulation SO 050. Unpublished Report RVA 610/78 A from Ciba-Geigy Limited, Basle, Switzerland.

Schnabel, D. and Formica, G. 1979b. CGA 20168. Gas chromatographic determination in malt, beer and intermediate products after a "sandwich" treatment of malt with CGA 20168 EC 950. Unpublished Report RVA 4013/79 from Ciba-Geigy Limited, Basle, Switzerland.

Schnabel, D. and Formica, G. 1979c. Gas chromatographic determination of CGA20168 in eggs following oral application via feed. Unpublished. Report RVA 4017/79 from Ciba-Geigy Limited, Basle, Switzerland.

Schnabel, D. and Hormann, W.H. 1980. Residue determination of methacrifos and its dealkylated metabolites in wheat grain after treatment with Damfin® EC 950. Unpublished Report RVA 4017/80 A from Ciba-Geigy Limited, Basle, Switzerland.

Schoeggl, G., Scheidl, I., Pfannhauser, W. and Woidich, H. 1983. Decomposition of pyrethrins during cereal processing. Ernaehrung (Vienna) 7(10): 565–8.

Schooley, D.A., Creswell, K.M. Staiger, L.E. and Quistad, G.B. 1975. Environmental degradation of the insect growth regulator (isopropyl (2E,4E)-11-methoxy-3,7,11-trimethy -1–2,4-dodecadienoate) (Methoprene). IV. Soil metabolism. J. Agric. Food Chem. 23: 369–373.

Schrader, J. 1963. Die Entwicklung neuer insektizider Phosphorsaure-Ester. Verlag Chemie, Weinheim.

Schroeder, H.O. 1955. Some factors influencing the effectiveness of piperonyl butoxide/pyrethrins com binations for the control of insects in stored grains. J. Econ. Entomol. 48: 25.

Schulten, G.G.M. 1973. Further insecticide trials on the control of *Ephestia cautella* in stacks of bagged maize in Malawi. Int. Pest Control March/April, 18–20.

Schulten, G.G.M. 1975. Losses in stored maize in Malawi and work undertaken to prevent them. Bull. Env. Mediterr. Plant Prot. Org. 5(2): 113–120.

Schulten, G.G.M. 1981. The use of pesticides at farmer and village level. Proc. GASGA Seminar on the Appropriate Use of Pesticides for the Control of Stored Products Pests in Developing Countries, Slough, 17–20 February 1981. pp. 32–59.

Schulten, G.G.M. 1982. Post-harvest losses in tropical Africa and their prevention. Food Nutr. Bull. 4(2): 2–9.

Schwarz, H. and Dedek, W. The breakdown and excretion of trichlorphon-P32 in the pig (in German). Zentr. Veterinaermed. Reihe. B. 12: 653.. [CA 64: 11706]

Scoble, G.P.W. and Crawford, G.F. 1967. Vapona strips for continuous stored products pest control. Span 10(1): 29.

Scott, H.G. 1962. How to control insects in stored foods. Part 1. Pest Control. 30(12): 24–34.

Scott, J.G. and Georghiou, G.P. 1984. Influence of temperature on the knockdown, toxicity and resistance to pyrethriods in the housefly. Pestic. Biochem. Physiol. 21: 53–62.

SEAMEO 1985. Efficacy trial with Ekamet 50EC against post- harvest pests in rice. Report prepared by Biotrop, Bogor, Indonesia for Sandoz Ltd, Jakarta. (Unpublished).

Seenappa, M., Stoobes, L.W. and Kempton, A.G. 1979. The role of insects in the bio deterioration of Indian red peppers by fungi. Int. Biodeterior. Bull. 15(3): 96–103.

Segawa, T. 1976. Acute toxicity studies of S-2539. Forte (phenothrin) in rats and mice. Report from Hiroshima University School of Medicine. Submitted to WHO by Sumitomo Chemical Co. (Unpublished).

Sehnal, F. 1976. Action of juvenoids on different groups of insects. In: Gilbert, L.I., Ed., The Juvenile Hormones. Plenum Press, New York. pp. 301–322.

Semple, R.L. 1985. Problems related to pest control and use of pesticides in grain storage: the current situation is ASEAN and future requirements. In Champ, B.R., and Highley, E., Ed., Pesticides and Humid Tropical Grain Storage Systems: Proc. Int. Symp., Manila, 27–30 May 1985. ACIAR Proceedings No. 14. pp. 45–75.

Seth, A.K. 1973. Chemical control of stored product insect pests. Presented at FAO Seminar, Bangkok, September 1973.

Seth, A.K. 1974. Use of Actellic for stored rice insect problems in S.E. Asia. Proc. 1st Int. Working Conf. Stored-Products Entomol., Savannah, Georgia, 7–11 October 1974. pp. 656–668.

Sethi, G.R. Prasad, H. and Bhatia, P. 1979. Studies on the control of stored grain pests by *gamma* radiations under different storage conditions. J. Nuclear Agric. Biol. 8(4): 123–125.

Sharp, A.K. 1979. Disinfection of grain with carbon dioxide, in freight containers. In: Cook, J.R. and O'Reilly, M.V., Ed., Proc. Ann. Sci. Agric. Conf., Aust. Inst. Agric. Sci., Sydney, 10–12 February 1979. AIAS, North Ryde. pp. 78–84.

Sharp, A.K. and Banks, H.J. 1979. Trial use of CO_2 to control insects in exported, containerized wheat. CSIRO Food Res. Quarterly 39(1): 10–16.

Shejbal, F., Ed. 1980. Controlled Atmosphere Storage of Grains. Elsevier, Amsterdam.

Shell Chemical Company USA 1966. Fate of dichlorvos residues in cereal products subjected to processing and cooking. (Unpublished).

Shell Research Ltd 1970. Residues of Vapona in meals prepared in domestic kitchens. Report of Woodstock Laboratory. (Unpublished).

Shell Research Ltd 1971, 1972, 1973, 1974. 2,2-dichlorovinyl dimethyl phosphate bibliography (1510 references) Shell Research Ltd, Sittingbourne Laboratories, London.

Shellenberger, T.E. 1970. Toxicological evaluations of DOWCO 214 with wildlife and DOWCO 179 with mallard ducklings. Gulf Research Institute. New Iberia, Louisiana, GSRI Project NC-378. (Contracted by Dow Chemical Company) 10 pp.

Sherjugjit Singh and Chawla, R.P. 1980.Comparative persistence of residues of pirimiphos-methyl on stored wheat, maize and paddy. Bull. Grain Technol. 18(3): 181–187.

Sherman, M. and Ross, E. 1961. Acute and sub-acute toxicity of insecticides to chicks. Toxicol. Appl. Pharmacol. 3: 521.

Shikrenov, D. and Sengalevich, G. 1970. A comparative trial of preparations for wet disinfection of empty storehouses. Rastit. Zasht. 18(2): 32. [Rev. Appl. Entomol. A. 60, 4898 (1972)].

Shimkin, M.B. and Anderson, H.H. 1936. Acute toxicities of rotenone and mixed pyrethrins in mammals. Proc. Soc. Exp. Biol. 34: 135.

Shono, T., Ohsawa, K. and Casida, J.E. 1979. Metabolism of *trans*-permethrin and *cis*-permethrin *trans*-cypermethrin and *cis*-cypermethrin, and decamethrin by microsomal enzymes. J. Agric. Food Chem. 27(2): 316–25.

Shuyler, H.R., Corbett, G.G., Reusse, E. and Barreveld, W. 1976. Action versus its justification: which comes first. Proc. 15th Int. Congr. Entomol, Washington, D.C. August 1976. pp. 19–27.

Sieper, H. 1972. Residues and metabolism of lindane. In: Ulmann, E., Ed., Lindane — Monograph of an Insecticide. Verlag K. Schillinger. pp. 79–112.

Silva E. and Sousa, M.E. 1965. Losses in effectiveness found in treatments of wheat with insecticidal powders containing DDT, lindane and malathion. Agricultura Lisb. 27, 28. [Through Rev. Appl. Entomol. A56, 43 (1968).]

Simonaitis, R.A. 1983. Recovery of piperonyl butoxide residues from bread made from cornmeal and wheat flour. Pyrethrum Post 15(3): 66–70.

Simpson, B.W. 1979. Method for pyrethoid residues on wheat, sorghum, pollard, flour and bread. Submission to FAO for evaluation by JMPR (1980). (See FAO/WHO 1981, pp. 369, 382, 383).

Sinclair, E.R. 1982. Population estimates of insect pests of stored products on farms on the Darling Downs, Queensland. Aust. J. Exp. Agric. Anim. Husb. 22(114/115): 127–132.

Sinclair, E.R. and Haddrell, R.L. 1985. Flight of stored products beetles over a grain farming area in southern Queensland. J. Aust. Entomol. Soc. 24: 9–15.

Singh, H. 1973. Preirradiation temperature-induced radiosensitivity changes in *Rhyzopertha dominica* adults. J. Ga. Entomol. Soc. 8: 317–320.

Singh, H. and Liles, J.N. 1972. Temperature modifications of the post-irradiation survival of the lesser grain borer adults. Environ. Entomol. 1: 395–397.

Singh, K. 1979. Pre-harvest spray of malathion and DDVP for controlling the insect infestation of stored grains (*Sitophilus oryzae, Rhyzopertha dominica, Sitotroga cerealella*). Int. Pest Control 21(3): 66–67.

Singh, N. and Rai, B.K. 1967. Effect of methylenedioxyphenyl compounds on the respiratory rate of *Tribolium castaneum*. Indian J. entomol. 29: 339–340.

Singh, R.H. and Benazet, J. 1974. Chemical intervention on all stages and on all scales of tropical storage practice. Proc. 1st Int. Working Conf. Stored-Products Entomology, Savannah, Georgia, 7–11 October 1974. pp. 41–46.

Singh, S.R., Van Emden, H.F. and Taylor T.A. 1978. Pests of Grain Legumes: Ecology and Control. Academic Press, London. 454 pp.

Singh, V.K. 1964. Residue project from Bio/Toxicological Research Associates (in cooperation with Shell Development Co.) (Unpublished).

Singh, V.K. and Rainier, R.H. 1966. Three year chronic oral toxicity of formulated dichlorvos in swine with special reference to possible effects upon fertility and the viability of the offspring of animals fed continuously on diets containing the drug. Report from Bio/Toxicological Research Associates, Spencervile, Ohio, Prepared for Shell Chemical Co. (Unpublished).

Singh, V.K. Perkins, C.T. and Schooley, M.A. 1968. Effects of DDVP fed to gravid sows on performance of their offspring to weaning. J. Anim. Sci. 27: 1779.

Sinha, R.N. 1961. Insects and mites associated with hot spots in farm stored grain. Can. Entomol. 93: 609–621.

Sinha, R.N. and Campbell, A. 1975. Energy loss in stored grain by pest infestation. Can. Agric. Spring 1975, 3 pp.

Sinha, R.N. and Wallace, H.A.H. 1965. Ecology of fungus-induced hot spot in stored grain. Can. J. Pl. Sci. 45: 48–59.

Sissons, D.J. and Telling, G.M. 1970. Rapid procedures for the routine determination of organophosphorus insecticide residues in vegetables. I. Determination of hexane-soluble insecticides by gas-liquid chromatography and total phosphorus procedures. J. Chromatog. 47: 328–340.

Sissons, D.J., Telling, G.M. and Usher, C.D. 1968. A rapid and sensitive procedure for the routine determination of organochlorine pesticide residues in vegetables. J. Chromatog. 33: 435–449.

Slade, R. 1945. A new British insecticide the *gamma* isomer of benzene hexachloride. Chem. Trade J. 116: 279.

Slama, K., Romanuk, M. and Sorm, F. 1974. Insect Hormones and Bioanalogues. Springer-Verlag, Berlin and New York.

Smalley, H.E. 1970. Diagnosis and treatment of carbaryl poisoning in swine, J. Amer. Vet. Med. Assoc. 156: 339.

Smalley, H.E., O'Hara, P.J., Bridges, C.H. and Radeleff, R.D. 1969. Effects of chronic carbaryl administration on the neuromuscular system of swine. Toxicol. Appl. Pharmacol. 14: 409.

Smallman, B.N. 1948. Effectiveness of residual insecticides for the control of spider beetles (Ptinidae) in flour warehouses. J. Econ. Entomol. 42: 618.

Smetana, P. 1969. The use of pickled wheat in poultry feeding. J. Agric. W. Australia 10: 228.

Smith, C.S., Shaw, F., Lavigne, R., Archibald, J., Fenner, H. and Stern, D. 1960. Residues of malathion on alfalfa and in milk and meat. J. Econ. Entomol. 53(4): 495–496.

Smith, G.N., Taylor, Y.S. and Watson, B.S. 1970. An analytical method for the determination of 3,5,6-trichloro-2-pyridinol in animal tissues and the metabolism of the pyridinol in rats. Unpublished report submitted to the World Health Organization by The Dow Chemical Co., Midland, Michigan, USA.

Smith, G.N. and Williams, R.T. 1954. Studies in detoxication : the metabolism of aliphatic alcohols. The glucuronic acid conjugation of chlorinated and some unsaturated alcohols. Biochem J. 56: 618–621.

Smith, L.B. 1974. The role of low temperature to control stored food pests. Proc. Ist Int. Working Conf. Stored-Products Entomol., Savannah, Georgia, 7–11 October 1974. pp. 418–437.

Smith, L.W., Pratt, J.J., Nu, I. and Umina, A.P. (1971). Baking and taste properties of bread made from hard wheat flour infested with species of *Tribolium, Tenebrio, Trogoderma* and *Oryzacphilus*. J. Stored Prod. Res. 6: 307–316.

Smittle, B.J., Labrecque, G.C. and Carrol, E.E. 1971. Comparative effectiveness of fast neutrons and *gamma* rays in producing sterility in houseflies. J. Econ. Entomol. 64: 1030–1032.

Snelson, J.T. 1971. Sampling of bulk grain. Codex Committee on Pesticide Residues, Room Document. Copenhagen. Available as Document PB 157 from Agricultural and Veterinary Chemicals Section, Department of Primary Industry, Canberra, Australia.

Snelson, J.T. 1974. Collaborative study of analyses of malathion residues in wheat. Report to Codex Committee on Pesticide Residues, The Hague.

Snelson, J.T. 1979a. Pesticide Residue Survey. Department of Primary Industry, Canberra, July 1979.

Snelson, J.T. 1979b. Pesticide residues in stored grain : their significance and safety. In: Evans, D.E., Ed., Australian Contributions to the Symposium on the Protection of Grain Against Insect Damage During Storage, Moscow 1978. CSIRO Division of Entomology, Canberra. pp. 99–107.

Snelson, J.T. 1979c. The effect of processing on fenitrothion residues in raw bran. Report to Department of Primary Industry, Canberra, October 1979.

Snelson, J.T. 1979d. Attitudes and prospects for use of chemicals in protecting world food supplies. Proc. 2nd Int. Working Conf. Stored-Products Entomol. Ibadan, Nigeria, 10–16 September 1978. pp. 116–137.

Snelson, J.T. 1980. Chemical residues in international trade. Proc. XVI Int. Congr. Entomol. Kyoto, Japan, 1980.

Snelson, J.T. 1981a. Results of National Residue Survey — Department of Primary Industry, Canberra, Australia.

Snelson, J.T. 1981b. Regulation of chemicals and chemical residues. Proc. First Aust. Stored Grain Pest Control Conf., Melbourne, May 1981 pp. 9/17–20.

Snelson, J.T. 1983. The quantity and quality of residue data required for the establishment and enforcement of maximum residue limits. In: Myanoto, J., et al., Ed., IPUAC Pesticide Chemistry — Human Welfare and the Environment. Pergamon Press, Oxford. pp. 13–22.

Snelson, J.T. 1984a. Use of insecticides — general principles. In: Champ, B.R., and Highley, E., Ed., Proc. Aust. Dev. Asst. Course on Preservation of Stored Cereals. CSIRO Division of Entomology, Canberra. pp. 589–604.

Snelson, J.T. 1984b. Use of insecticides — properties. In: Champ, B.R., and Highley, E., Ed., Proc. Aust. Dev. Asst. Course on Preservation of Stored Cereals. CSIRO Division of Entomology, Canberra. pp. 605–636.

Snelson, J.T. 1984c. Use of insecticides — preparation and application. In: Champ, B.R., and Highley, E., Ed., Proc. Aust. Dev. Asst. Course on Preservation of Stored Cereals. CSIRO Division of Entomology, Canberra. pp. 637–657.

Snelson, J.T. 1984d. Pesticide residues and their significance. In: Champ, B.R., and Highley, E., Ed., Proc. Aust. Dev. Asst. Course on Preservation of Stored Cereals. CSIRO Division of Entomology, Canberra. pp. 658–680.

Snelson, J.T. 1986a. Safety considerations in insecticide usage in grain storage. In: Champ, B.R., and Highley, E., Ed., Pesticides and Humid Tropical Grain Storage Systems: Proc. Int. Seminar, Manila, 27–30 May 1985. ACIAR Proceedings No. 14. pp. 87–100.

Snelson, J.T. 1986b. Regulatory requirements for pesticide use. In: Champ, B.R., and Highley, E., Ed., Pesticides and Humid Tropical Grain Storage Systems: Proc. Int. Seminar, Manila, 27–30 May 1985. ACIAR Proceedings No. 14. pp. 101–120.

Snelson, J.T. and Desmarchelier, J.M. 1975. The significance of pesticide residues. Proc. 1st Int. Working Conf. Stored-Products Entomol., Savannah, Georgia, 7–11 October 1984. pp. 465–477.

Soderlund, D.M. and Casida, J.E. 1977a. Synthetic pyrethroids, ACS Symposium Series No. 42. American Chemical Society, Washington. pp. 162–172.

Soderlund, D.M. and Casida, J.E. 1977b. Effects of pyrethroid structure on rates of hydrolysis and oxidation by mouse liver microsomal enzymes. Pestic. Biochem. Physiol. 7: 391–401.

Soderstrom, E.L. 1970. Effectiveness of green electroluminescent lamps for attracting stored-product insects. J. Econ. Entomol. 63: 726–731.

Solomon, M.E. and Adamson, B.E. 1955. The powers of survival of storage and domestic pests under winter conditions in Britain. Bull. Entomol. Res. 46: 311–355.

Solomon, K.R. and Metcalf, R.L. 1974. The effect of piperonyl butoxide and triorthocresyl phosphate on the activity and metabolism of Altosid (isoprophyl 11-methoxy, 3,7,11-trimethyldodeca-2,4-dienoate) in *Tenebrio molitor* L. and *Oncopeltus fasciatus* (Dallas). Pestic. Biochem. Physiol. 4(2): 127–134.

South African Dep. Agric. 1975. Report on a trial of bromophos on stored maize and sorghum. South African Department of Agricultural Technical Services, Pretoria, South Africa — Report 39/4/9.

Sparks, T.C., Pavloff, A.M., Rose, R.L. and Clower, D.F. 1983. Temperature/toxicity relationships of pyrethroids on *Heliothis virescens* and *Anthonomas grandis grandis*. J. Econ. Entomol. 76: 243–246.

Speirs, D.R. and Zettler, J.L. 1969. Toxicity of three organophosphorus compounds and pyrethrins to malathion-resistant *Tribolium castaneum* (Herbst) (Coleoptera, Tenebrionidae). J. Stored Prod. Res. 4: 279–283.

Spratt, E.C. 1979. Some effects of a mixture of oxygen, carbon dioxide and nitrogen in the ratio 1:1:8 on the ovideposition and development of *Sitophilus zeamais*. J. Stored Prod. Res. 15: 73–80.

Sprecht, W. and Tillkes, M. 1980. Fresenius Z. Anal. Chem. 301: 300–307.

Srivastava B.P. and Dadhich, S.R. 1973. Laboratory evaluation of malathion as a protectant for the prevention of damage by pulse beetles to stored gram (*Cicer arietinum*). I. Biological effectiveness. Bull. Grain Technol. 11(1): 8–13.

Srivastava, M.K. and Gopal, K. 1984. Relative toxicity of organophosphorus insecticides against stored grain pests. J. Adv. Zool. 5(1): 23–27.

Srivastava, C.P. and Perti, S.L. 1979. Effect of temperature and humidity on the susceptibility of insects of public health importance to insecticides. Labdev J. Sci. Technol. 9(2):B 86–93.

Srivastava, A.S. and Wilson, H.F. 1947. Benzene hexachloride as a fumigant and contact insecticide. J. Econ. Entomol. 40: 569.

Srivastava, B.P., Kavadia, V.S. and Sharma, G.K. 1970. Persistence and dissipation of insecticidal residue under different temperature and humidity conditions. Indian J. Entomol. 32(1): 51–57.

Staal, G.B. 1975. Insect growth regulators with juvenile hormone activity. Ann. Rev. Entomol. 20: 417–460.

Staal, G.B. 1977. Insect control with insect growth regulators based on insect hormones. Pontif. Acad. Sci. Scr. Varia. 41: 353–383.

Stables, L.M. 1980. The effectiveness of some recently developed pesticides against stored-product mites. J. Stored Prod. Res. 16: 143–146.

Stables, L.M., Good, E.A.M. and Chamberlain, S.J. 1979. Evaluation of etrimfos as an acaricide on stored oilseed rape. Proc. 1979 Brit. Crop Prot. Conf. pp. 145–151.

Stadelman, W.B., Liska, B.J., Langlois, G.E., Mostert, G.C. and Stemp, A.R. 1964. Persistence of chlorinated hydrocarbon residues in chicken tissues and eggs. Poultry Sci. 43: 1365.

Stankovic, A., Sovljanski, R. and Stojanovic, T. 1965. Teksikoloska svojstva brasna dobwenog od psenice tretirane malationom. Savremena Poljoprevreda, No. 9 [through FAO Grain Storage Newsletter and Abstracts, 8, 171 (1966)].

Stanley, J.M. and Dominick, C.B. 1970. Funnel size and lamp wattage influence on light-trap performance. J. Econ. Entomol. 63: 1425–1426.

Starks, K.J. and Lilly, J.H. 1955. Some effects of insecticide seed treatment on dent corn. J. Econ. Entomol. 48: 549.

Staudinger, H. and Ruzicka, L. 1924. Insektotende Stoffe. Helv. Chim. Acta 7: 177–458.

Stehr, F.W. 1975. Parasitoids and predators in pest management. In: Metcalf, R.L., and Luckman, W., Ed., Introduction to Insect Pest Management. Wiley, New York.

Stein, A.A. 1977. Histophathology evaluation of animals from 3-generation reproduction study. Report from Microscopy for Biological Research Ltd, under contract to Sumitomo Chemical Co. Ltd, submitted as part of study by Beliles et al. 1978. (Unpublished).

Sternberg, J., Vinson, E.B. and Kearns, C.W. 1953. Enzymatic dehydrochlorination of DDT by resistant flies. J. Econ. Entomol. 46: 513.

Sternersen, J. 1969. Degradation of ^{32}P-bromophos by micro-organisms and seedlings. Bull. Env. Contam. Toxicol. 4: 104.

Stevenson, J.H. 1958. Chem. Ind. Lond. 26: 827.

Stiasni, M., Deckers, W., Schmidt, K. and Simon, H. 1969. Translocation, penetration, and metabolism of bromophos in tomato plants. J. Agric. Food Chem. 17(5): 1017–1020.

Stiasni, M., Rehbinder, D. and Deckers, W. 1967. Adsorption, distribution and metabolism of bromophos in the rat. J. Agric. Food Chem. 15(3): 474–478.

Stojanovic, T. 1966. The effect of the initial population density of *Sitophilus granarius* and *S. oryzae* on the loss in weight in infested wheat. Rev. Appl. Entomol. A. 58(10): 3016.

Storey, C.L. 1972. The effect of air movement on the biological effectiveness and persistence of malathion in stored wheat. Proc. North Central Branch Entomol. Soc. Amer. 27. pp. 57–62.

Storey, C.L., Sauer, D.B., Quinlan, J.K. and Ecker, O. 1982a. Incidence, concentration and effectiveness of malathion residues in wheat and maize exported from the United States. J. Stored Prod. Res. 18: 147–151.

Storey, C.L., Sauer, D.B., Quinlan, J.K. Ecker, O., and Fulk, D.W. 1982b. Infestations in wheat and corn exported from the United States. J. Econ. Entomol. 75(5): 827–832.

Storey, C.L., Sauer, D.B., and Walker, D. 1984a. Present use of pest management practices and their effectiveness in wheat, corn and oats stored on the farm. J. Econ. Entomol. 77(3): 784–788.

Storey, C.L., Sauer, D.B., and Walker, D. 1984b. Insect populations in wheat, corn and oats stored on the farm. J. Econ. Entomol. 76(6): 1323–1330.

Stratil, H., Vogel, B. and Wohlgemuth, R. 1981. Investigations on the effects of dichlorvos strips on the development stages of the dried-fruit moth (*Plodia interpunctella* Hbn.) living externally on stored products in granaries. Anzeiger fur Schadlingskunde Pflanzenschutz Umweltschutz 54(1): 1–5.

Strittmatter, J. and Gfeller, W. 1975. Sixty-three days dietary toxicity study in chicks with compound CGA 20168. Report from Ciba-Geigy Ltd, Basle, Switerland, submitted to WHO. (Unpublished).

Strong, M.B. 1985. Tolerability of calves to rations containing wheat medicated with methacrifos. Report from Ciba-Geigy Research Station, Kemps Creek, NSW, dated 26 July 1985. (Unpublished).

Strong, R.G. 1969. Relative susceptibility of five stored-product moths to some organophosphorus insecticides. J. Econ. Entomol. 62(5): 1036–1039.

Strong, R.G. 1970. Relative susceptibility of confused and red flour beetles to twelve organophosphorus insecticides with notes on adequacy of test methods. J. Econ. Entomol. 63: 258–263.

Strong, R.G. and Diekman, J. 1973. Comparative effectiveness of fifteen insect growth regulators against several pests of stored products. J. Econ. Entomol. 66: 1167–1173.

Strong, R.G. and Sbur, D.E. 1960. Influence of grain moisture content and storage temperature on the effectiveness of malathion as a grain protectant. J. Econ. Entomol. 53: 341.

Strong, R.G. and Sbur, D.E. 1961. Evaluation of insecticides as protectants against pests of stored grains and seeds. J. Econ. Entomol. 54: 235–238.

Strong, R.G. and Sbur, D.E. 1963. Protection of wheat seed with diatomaceous earth. J. Econ. Entomol. 56(3): 372–374.

Strong, R.G. and Sbur, D.E. 1964a. Influence of grain moisture and storage temperature on the effectiveness of five insecticides as grain protectants. J. Econ. Entomol. 57(1): 44–47.

Strong, R.G. and Sbur, D.E. 1964b. Protective sprays against internal infestations of grain beetle in wheat. J. Econ. Entomol. 57: 544–548.

Strong, R.G. and Sbur, D.E. 1965a. Evaluation of insecticides as protectants against pests of stored grain and seeds II. J. Econ. Entomol. 58(1): 18–22.

Strong, R.G. and Sbur, D.E. 1965b. Interrelation of moisture content, storage temperature and dosage on the effectiveness of diazinon as a grain protectant against *Sitophilus oryzae* L. J. Econ. Entomol. 58: 410.

Strong R.G. and Sbur, D.E. 1968. Evaluation of insecticides for control of stored product insects. J. Econ. Entomol. 61: 1034–1041.

Strong, R.G., Sbur, D.E., and Arndt, R.G. 1961. Influence of formulation on the effectiveness of malathion, methoxychlor and synergised pyrethrin sprays for stored wheat. J. Econ. Entomol. 54: 489.

Strong, R.G., Sbur, E.C. and Partida, G.J. 1967. The toxicity and residual effectiveness of malathion and diazinon used for protection of stored wheat. J. Econ. Entomol. 60: 500–505.

Subrahmanyan, V. 1962. The place of food technology in combatting hunger and malnutrition. Food Technol. 6(10): 24.

Subrahmanyam, B. and Cutkomp, L., 1985. Moth control in stored grain and the role of *Bacillus thuringiensis*: an overview. Residue Rev. 94: 1–47.

Sudershan, P. and Naidu, M.B. 1961. Antagonistic action of piperonyl butoxide on malathion. Sci. Cult. 27: 100–101.

Sumitomo 1971. Sumithion: its toxicity, metabolism and residues. Bull. Sumitomo Chemical Co., Osaka, Japan.

Sumitomo 1972. Six-month feeding study of Sumithion in rats. Report from the Pesticides Research Department, Sumitomo Chemical Co. Ltd., Osaka, Japan. (Unpublished).

Sumitomo 1977. Delayed neurotoxicity with Sumithion in hens. Sumitomo Chemical Co. Ltd, Technical Report No. HT-70–0200. (Unpublished).

Sumitomo 1978. Impact of Sumithion (fenitrothion) on the whole environment including humans. Bulletin (Sept. 1978) Sumitomo Chemical Co. Ltd., Osaka, Japan. 71 pp.

Summit, L.M. and Albert, J.R. 1977. a) Oral lethality of WL43775 in the rat. b) Determination of acute oral lethality of WL43775 in the male and female mouse. Report, Shell Development Co. (Unpublished).

Sun, Y.P. and Johnson, E.R. 1960. Synergistic and antagonistic action of insecticide-synergist combinations and their mode of action. J. Agric. Food Chem. 8: 261.

Sun, F., Wang, S.C. and Ku, T.Y. 1984. Effects of degradation in toxicity of insecticides on the survival and reproductive rate of maize weevils. Plant Protection Bull. Taiwan 26(1): 55–62.

Sutherland, J.W. 1984. Aeration principles. In: Champ, B.R., and Highley, E., Ed., Proc. Aust. Dev. Asst. Course on Preservation of Stored Cereals. CSIRO Division of Entomology, Canberra. pp. 426–428.

Sutherland, J.W., Pescod, D. and Griffiths, H.J. 1970. Refrigeration with bulk stored wheat. Aust. Refrig. Air. Condit. Heat. 24(8): 30–34, 43–45.

Suzuki, T. and Miyamoto, J. 1974. Metabolism of tetramethrin in houseflies and in rats in vitro. Pestic. Biochem. Physiol. 4: 86.

Suzuki, T. and Miyamoto, J. 1978. Purification and properties of pyrethroid carboxyesterase in rat liver microsome. Pestic. Biochem. Physiol. 8: 186–198.

Suzuki, T., Ohno, N. and Miyamoto, J. 1976. New metabolites of phenothrin in rats. J. Pestic. Sci. 1: 151–152.

Swaine, H., Francis, P.D., Rippington, D. and Burke, S.R. 1980a. Permethrin: residue levels of the major metabolites of the insecticide in the milk and tissues of cows fed on a treated diet. ICI Plant Protection Division Report No. RJ0124B. (Unpublished).

Swaine, H., Francis, P.D., Rippington, D., Burke, S.R., Robertson, S. and Ford, J.P. 1980b. Permethrin metabolites: eggs and tissues of hens. ICI Plant Protection Division Residue Data Report No. RD/557/27. (Unpublished).

Swart, R.W., McGregor, W.S., Boswell, C.R. and Rowe, L.D. 1976a . Chlorpyrifos-methyl chicken feeding study: Residues in eggs and tissues. Dow Chemical USA, R&D Report TA-569 8 pp. (Unpublished).

Swart, R.W., McGregor, W.S., Boswell, C.R. and Rowe, L.D. 1976b . Chlorpyrifos-methyl beef feeding study: residues in beef tissues. Dow Chemical USA, R&D Report TA-572. (Unpublished).

Sze-Peng, T. 1979. The rice weevil, *Sitophilus oryzae* on stored rice in Malaysia, M.S. Thesis, Simon Frazer University, Burnbury, B.C., Canada.

Takatsuka, M., Okuno, Y., Susuki, T., Kadota, T., Yasuda, M. and Miyamoto, J. 1980. Three-generation reproduction study of S-2539 (phenothrin) in rats. Report from Industrial Bio-Test Laboratories, validated by Sumitomo Chemical Co. (Unpublished).

Takimoto, Y. and Miyamoto, J. 1976. Residue analysis of Sumithion. Residue Rev. 60: 83–101.

Takimoto, Y., Ohshima, M. and Miyamoto, J. 1978. Degradation and fate of fenitrothion applied to harvested rice grains. J. Pestic. Sci. Japan. 3: 277–290.

Talekar, N.S. 1977. Gas-liquid chromatographic determination of *a*-cyano-3-phenoxybenzyl *a*-isopropyl-4-chlorophenylate residues in cabbage. J. Assoc. Off. Anal. Chem. 60: 908–910.

Talekar, N.S. and Mookherjee, P.B. 1969. Effect of temperature and grain moisture on the deterioration of grain protectant based on pyrethrins. Indian J. Entomol. 31(1): 64–68.

Tan, H.-C. et al. 1965. Evaluation of lindane dust as a protectant for stored grain — effectiveness and residues of various methods of mixing. Acta Phytophyl. Sin. 4, 93.

Tan, K.H. 1975. Effects of a synthetic juvenile hormone and some analogues on *Ephestia* spp. Ann. Appl. Biol. 80: 137–145.

Tan, N. and Tan, K.H. 1980. Environmental persistency of some juvenile hormone analogues — biological activity against the Mediterranean flour moth, *Ephestia kuehniella* under storage conditions. Malaysian Agric. J. 41(4) 343–350.

Tani, Y., Yoshikawa, S., Kimura, S., Chikubu, S., Tsuruta, O. and Endo, I. 1972. Underwater storage of brown rice. On the changes in qualities in stored rice. Rep. Nat. Food Res. Inst. Tokyo 27: 1–10.

Taylor, R.W. 1981. The availability of insecticides to protect farm-stored crops. Proc. GASGA Seminar on the Appropriate Use of Pesticides for the Control of Stored Products Pests in Developing Countries, Slough, 17–20 February 1981. pp. 170–177.

Taylor, R.W.D. and Evans, N.J. 1980. Laboratory evaluation of pirimiphos-methyl and permethrin dilute dusts for control of bruchid beetles attacking stored pulses. Int. Pest Control 22(5): 108–110.

Taylor, R.W.D. and Webley, D.J. 1979. Constraints on the use of pesticides to protect stored grain in rural conditions. Trop. Stored Prod. Inf. No. 38.

Taylor, T.A. 1974. Motivation and method in ensuring protection of tropical produce from grower to his market. Proc. 1st Int. working Conf. Stored Products Entomol., Savannah, Georgia, 7–11 October 1984. pp. 11–17.

Taylor, T.A. 1978. Tropical storage entomology by the end of the twentieth century. Proc. 2nd Int. Working Conf. Stored-Products Entomol., Ibadan, Nigeria, 1–16 September 1978. pp. 37–43.

Taylor, T.A., Egwuatu, R.I. and Boshoff, W.H. 1978. Significant infestation by *Araecerus fasciculatus* following the treatment of maize with pirimiphos-methyl for weevil control. J. Stored Prod. Res. 14: 159–161.

Telford, H.S., Zwick, R.W., Sikorowski, P. and Weller, M. 1964. Laboratory evaluation of diazinon as a wheat protectant. J. Econ. Entomol. 57: 272.

Telling, G.M. 1979. Good analytical practice in pesticide residue analysis. Proc. Anal. Div. Chem. Soc. 16: 37–42.

Telling, G.M., Sissons, D.J. and Brinkman, W.H. 1977. Determination of organochlorine insecticide residues in fatty foodstuffs using a clean-up technique based on a single column of activated alumina. J. Chromatog. 137: 405–423.

Tempone, M.J. 1979. Studies on the effects of insecticides on barley malting. Research Project Series No. 50, Victorian Department of Agriculture, Melbourne, Australia.

Teotia, T.P.S. and Kewal Dhari 1968. Some observations on the joint action of insecticides against pulse beetle, *Callosobruchus chinensis*. Labdev. J. Sci. Technol. 6B(3): 169.

Teotia, T.P.S. and Pandey, K.K. 1967. The influence of temperature and humidity on the contact toxicity of some insecticide deposits to *Tribolium castaneum*. Bull. Grain Technol. 5(3): 154.

Teotia, T.P.S. and Singh, R. 1969. Relative toxicity of some modern insecticides to two species of storage pests. Labdev. J. Sci. Technol. 7(B): 1, 50 [through Biol. Abstr. 51, 535(1970)].

Thomas, P.M. 1974. Public warehousing. Role of pesticides. Pesticides Annual pp. 81–84.

Thomas, P.J. and Bhatnagar-Thomas, P.L. 1968. Use of a juvenile hormone analogue as insecticide for pests of stored grains. Nature 219: 949.

Thompson, J.F. 1975. Storage, handling and preparation of pesticide reference standards. In: Environmental Quality and Safety. Georg Thieme, Stuttgart. Supp. Vol. III, pp. 47–50.

Thompson, J.F. and Mann, J.B. 1975. Analytical quality control in a pesticide residue laboratory. In: Environmental Quality and Safety. Georg Thieme, Stuttgart. Supp. Vol. III, pp. 32–39.

Thompson, R.H. and Hill, E.G. 1969. Pesticides residues in foodstuffs in Great Britain — XI. Further studies on malathion in imported cereals. J. Sci. Food. Agric. *20*(5): 293–295.

Thorpe, G.R. 1979. The effects of heat and moisture transfer on the degradation of pesticides applied to stored grain. Nat. Conf. Pub. Inst. Eng. Aust. Volume 79–8, pp. 17–20.

Thorpe, G.R. and Elder, W.B. 1980. The use of mechanical refrigeration to improve the storage of pesticide treated grain. Int. J. Refrig. 3(2): 99–106.

Thorpe, G.R. and Elder, W.B. 1982. Modelling the effects of aeration on the persistence of chemical pesticides applied to stored bulk grain. J. Stored Prod. Res. 18 103–114.

Tilton, E.W. 1975. Achievements and limitations of ionising radiation for stored-product insect control. Proc. 1st Int. Working Conf. Stored-Products Entomol., Savannah, Georgia, 7–11 October 1974. pp. 354–361.

Tilton, E.W. 1979. Current status of irradiation for use in insect control. Proc. 2nd Int. Working Conf. Stored-Products Entomol., Ibadan, Nigeria, 10–16 September 1978. pp. 218–221.

Tilton, E.W. and Schroeder, H.W. 1963. Some effects of infra-red irradiation on the mortality of immature insects in kernels of rough rice. J. Econ. Entomol. 56: 727–730.

Tilton, E.W. and Vardell, H.H. 1982a. Combination of microwaves and partial vacuum for control of four stored-product insects in stored grain. J. Ga. Entomol. Soc. 17(1): 106–112.

Tilton, E.W. and Vardell, H.H. 1982b. An evaluation of a pilot-plant microwave vacuum drying unit for stored-product insect control. J. Ga. Entomol. Soc. 17(1): 133–138.

Tilton, E.W., Burkholder, W.E. and Cogburn, R.R. 1966. Effects of *gamma* radiation on *Rhyzopertha dominica, Sitophilus orzyae, Tribolium confusum* and *Lasioderma serricorne*. J. Econ. Entomol. 59: 1363–1368.

Tilton, E.W., Brower, J.H., Brown, G.A. and Kirkpatrick, R.L. 1972. Combination of *gamma* and microwave radiation for control of the Angoumois grain moth in wheat. J. Econ. Entomol. 65: 531–533.

Tilton, E.W., Brower, J.H. and Cogburn, R.R. 1974. *Gamma* irradiation for control of insects in wheat flour. J. Econ. Entomol. 67(3): 430–432.

Tomita, M., Someya, N., Nishimura, M. and Ueda, K. 1974. Studies on the retention deposit of organophosphorus pesticides in animal and human bodies. Jap. J. Ind. Health 16: 547–556.

Tournayre, J.C. 1978. Dissipation curve of CGA 20168 in wheat grain and residues in issues. Unpublished Reports No. 13/78 A, B, C from Ciba-Geigy Agrochemicals Division, France.

Tracy, R.L. 1960. Soap Chem. Spec. 36: 74.

Tracy, R.L., Woodcock, J.G. and Chodroff, S. 1960. Toxicological aspects of 2,2-dichlorovinyl phosphate (DDVP) in cows, horses and white rats. J. Econ. Entomol. 53: 593.

Treon, J.F., Shafer, F.E., Dutra, Cappel, J., Gahegan, T. and Nedderman, N. 1951. The effects of repetitive exposure of animals to lindane vapour with some observations on distribution of lindane in their tissues. Rep. Kettering Lab., Univ. Cincinnati.

Trivelli, H.D'O. 1974. Species of insects attacking different types of stored wheat and their control. Proc. Ist Int. Working Conf. Stored-Products Entomol., Savannah, Georgia, 7–11 October 1974. pp. 219–233.

Tropical Products Institute 1981. TPI team tackles first recorded infestation by *Prostephanus truncatus* in Africa. Tropical Products Institute Newsletter No. 22.

Trottier, B.L. and Jankowska, I. 1980. In vivo study on the storage of fenitrothion in chicken tissue after long-term exposure to small doses. Bull. Environ. Contam. Toxicol. 24: 606–610.

Tsvetkov, D. 1983. Protection of stored grain against stored-product pests. Rastit. Zasht. 31(11): 13–16.

Tsvetkov, D. and Bogdanov, V. 1961. Tests on the control of the granary weevil, *Sitophilus granarius* with chemicals for fumigating and dusting wheat grain. Isv. Tsentr. Nauchnoizsled. Inst. Zasht. Rast. 1. 153.

Turtle, E.E. 1965. Pesticides, filth and the miller. Milling 145: 94.

Tyler, P.S. 1982. Misconception of food losses. Food Nutr. Bull. 4(2): 21–24.

Tyler, P.S. and Binns, T.J. 1977. The toxicity of seven organophosphorus insecticides and lindane to eighteen species of stored product beetles. J. Stored Prod. Res. 13: 39–43.

Tyler, P.S. and Binns, T.J. 1982. The influence of temperature on the susceptibility to eight organophosphorus insecticides of *Tribolium castaneum, Oryzaephilus surinamensis* and *Sitophilus granarius*. J. Stored Prod. Res. 18: 13–19.

Tyler, P.S. and Boxall, R.A. 1984. Post-harvest loss reduction programmes: a decade of activities — what consequences? Trop. Stored Prod. Inf. No. 50. pp. 4–13.

Tyler, P.S. and Green, A.A. 1968. The effectiveness of fenitrothion and malathion as grain protectants under severe practical conditions. J. Stored Prod. Res. 4: 119–126.

Tyler, P.S. and Rowlands, D.G. 1968. Protective treatment of grains. Pest Infest. Res. 1967: 25.

Tyler, P.S., Bristow, S.G. and Clements, J.E. 1965a. Field trials with grain protectant insecticides Pest Infest. Res. 1965: 26.

Tyler, P.S., Bristow, S.G., Horler, D.F. and Clements, J.E. 1965b. Persistence of insecticide residues in grain. Pest Infest. Res. 1965: 26.

Tyler, P.S., Bristow, S.G., Horler, D.F. and Clements, J.E. 1966a. Fate of malathion and fenitrothion on newly harvested grain. Pest Infest. Res. 1966: 26.

Tyler, P.S., Green, A.A., Wilkin, D.R., Bristow, S.G., Horler, D.F. and Clements, J.E. 1966b. Field trials with malathion and fenitrothion. Pest Infest. Res. 1966: 27.

Tyler, P.S., Green, A.A., Wilkin, D.R., Bristow, S.G. and Rowlands, D.G. 1967. Protective treatment of grain — field trial. Pest. Infest. Res. 1967: 25–26.

Tyler, P.S., Horler, D.R., Rowlands, D.G. and Robbins, B. 1969. The distribution of insecticides among grains. Pest Infest. Res. 1968: 33–34.

Udeaan, A.S. and Bindra, O.S. 1971a. Malathion residues in different fractions of treated food grains and their finished derivatives. In: Bindra, O.S. and Kalra, R.L., Ed. Proc. Symp. Progress and Problems in Pesticide Residue Analysis, Punjab Agric. Univ., Ludhiana Indian Comm. Agric. Res., New Delhi. pp. 61–67.

Udeaan, A.S. and Bindra, O.S. 1971b. Malathion residues in wheat grains in relation to dosages applied, storage period and receptacles used for storage. Bull. Grain Technol. 9: 13.

Udeaan, A.S. and Kalra, R.L. 1983. Joint action of piperonyl butoxide and lindane in susceptible and resistant strains of *Tribolium castaneum* Herbst. J. Entomol. Res. 7(1): 21–24.

Udeaan, A.D., Chawla, R.P. and Kalfa, R.L. 1974. Variation in insecticidal deposit on treated wheat. J. Food Sci. Technol. 11: 17.

Ueda, K., Gaughan, L.C. and Casida, J.E. 1975a. Metabolism of four resmethrin isomers by liver microsomes. Pestic. Biochem. Physiol. 5: 280–294.

Ueda, K., Gaughan, L.C. and Casida, J.E. 1975b. Metabolism of (+)-*trans* and (+)-*cis* resmethrin in rats. J. Agric. Food Chem. 23(1): 106–115.

Ulmann, E. 1972. Lindane — Monograph of an Insecticide. Verlag K. Schillinger, Freiburg in Breisgau. 384 pp.

Union Carbide. 1976. Results of tests investigating Sevin as a stored grain insecticide 1958–1962. Extract of petitions to US Department of Agriculture. Submitted to FAO by Union Carbide Corporation.

US Army 1973. Preliminary assessment of relative toxicity of candidate louse toxicant DOWCO214 — evaluation of animal exposure data and recommendations concerning a sleeve test program. Study No. 51–19–68/74. US Army Environmental Hygiene Agency. USAEHA-LAT, Aberdeen Proving Ground Maryland. 38 pp.

US Department of Agriculture 1965. Losses in agriculture. Agriculture Handbook No.291 USDA, Washington, DC. 120pp.

US Public Health Service 1958. Commonwealth Disease Centre, Atlanta, Ga., Technology Branch, Summary of Investigations No.15. (October 1958-March 1959).

Ussary, J.P. and Braithwaite, G.B. 1980a. Residues of permethrin and permethrin metabolities in milk from Ectiban treated cows (Trial No. 35NC79–001). ICI Americas Inc. Report No. TM0490/B. (Unpublished).

Ussary, J.P. and Braithwaite, G.B. 1980b. Residues of permethin and 3-phenoxybenzyl alcohol in cow tissues (Trial No. 35NC79–001). ICI Americas Inc. Report No. TMU0493/B. (Unpublished).

Ussary, J.P. and Braithwaite, G.B. 1980c. Residues of permethin and 3-phenoxybenzyl alcohol in tissues from Ectiban treated swine (Trial No. 35NC79–002). ICI Americas Inc. Report No. TNU0491/B. (Unpublished).

Ussary, J.P. and Braithwaite, G.B. 1980d. Residues of permethin and 3-phenoxybenzyl alcohol in tissues and eggs from Ectiban treated chickens (Trial No. 35NC79–003). ICI Americas Inc. Report No. TNU0492/B. (Unpublished)

Vadlamudi, V.P. and Paul, B.S. 1979. Acute oral toxicity of malathion and its effect on blood cholinesterases in Indian buffalo. Indian Vet. J. 56(8): 650–655.

Van der Pauw, C.L., Dix, K.M., Blanchard, K. and McCarthy, W.V. 1975. Toxicity of WL-43775: teratological studies in rabbits given WL-43775 orally. Report from Shell Development Co. Ltd. (Unpublished).

Vardell, H.H. and Tilton, E.W. 1981. Control of the lesser grain borer, *Rhyzopertha dominica* (F.), and the rice weevil, *Sitophilus oryzae* (L.), in rough rice with a heated fluidized bed. J. Ga. Entomol. Soc. 16(4): 521–524.

Vardell, H.H., Gillenwater, H.B., Whitten, M.E., Cagle, A., Eason, G. and Cail, R.S. 1973. Dichlorvos degradation on stored wheat and resulting milling fractions. J. Econ. Entomol. 66(3): 761.

Verma, A.H. and Ram, H. 1974. Biology and susceptibility to some safer insecticides of malathion-resistant and susceptible strains of *Tribolium castaneum*. Haryana Agric. Univ. J. Res. 1973 3(3): 112–125.

Verschoyle, R.D. and Barnes, J.M. 1972. Toxicity of natural and synthetic pyrethrins to rats. Pestic Biochem. Physiol. 2(3): 308–11.

Vigne, J.P., Chouteau, J., Tabau, R.L., Rancien, P. and Karamanian, A. 1957. Bull. Acad. Vet. Fr. 30: 84.

Vikhanskii, Y.D. 1967. Comparative toxicity of organophosphorus pesticides for grain and grain product pests (in Russian). Khim. Sel. Khoz. 5(5): 336. [CA 67: 107639]

Vikhanskii, Y.D. 1968. Stability of some organophosphorus preparations used for the wet disinfection of storehouses (in Russian). Khim. Sel. Khoz. 6(10): 748. [CA 70: 56659]

Vikhanskii, Y.D. and Sturua, L.I. 1969. Toxicity of some organophosphorus preparations for the flour beetle *Tribolium confusum* (in Russian). Khim. Sel. Khoz. 7(4): 277. [CA71: 69629]

Viljoen, J.H., Coetzer, J.J., De Beer, P.R., Prinsloo, S., Basson, A.J. and Vermaak, C. 1981b. The toxicity of contact insecticides to seed-infesting insects. Series No.6. Primiphos-methyl as a grain protectant. Technical Communication, Department of Agriculture and Fisheries, South Africa 173 11 pp.

Viljoen, J.H., De Beer, P.R., Du Toit, D.M. and Van Tonder, H.J. 1981a. Contact insecticides for the control of stored-product pests. Series No. 5. The protection against reinfestation of maize and groundnuts in bag stacks. Technical Communication, Department of Agriculture and Fisheries, South Africa 168 10 pp.

Vinopal, J.H.A. and Fukuto, T.R. 1971. Selective toxicity of phoxim. Pestic. Biochem. Physiol. 1: 44–60.

Vogel, W., Masner, P., Graf, O. and Dorn, S. 1979. Types of response of insects on treatment with juvenile hormone active insect growth regulators. Experientia 35: 1254–12566.

Vyas, H.J. 1984. Evaluation of different indigeneous materials and insecticides against *Rhyzopertha dominica* on wheat during storage. Presented at Post Harvest Technology Symposium, Junagadh, India, 8 February 1984.

Wachs, H. 1947. Synergestic insecticides. Science 105: 530.

Wachs, H. 1949. Methylenedioxyphenyl derivatives and methods for the production thereof. U.S. Patent 2,485,681. 25 October 1949.

Wachs, H. 1951. Insecticidal compositions containing pyrethrins and a synergist therefor. U.S. Patent 2,550,737, 1 May 1951.

Wachs, H., Jones, H.A. and Bass, L.W. 1950. New safe insecticides. Amer. Chem. Soc. Advances in Chem. Ser. No.1 (1950): 43–48.

Wagstaff, D.J. and Short, C.R. 1971. Induction of hepatic microsomal hydroxylating enzymes by technical piperonyl butoxide and some of its analogs. Toxicol. Appl. Pharmacol. 19: 54–61.

Wakid, A.M. 1973. Effect of nitrogen during *gamma* irradiation of purparia and adults of the Mediterranean fruit fly on emergence, sterility, longevity and competitiveness. Environ. Entomol. 2: 37–40.

Walkden, H.H. and Nelson, H.D. 1958. Evaluation of lindane for the protection of stored wheat and shelled corn from insect attack. US Dep. Agric. Market. Res. Rep. No. 234. 27 pp.

Walkden, H.H. and Nelson, H.D. 1959. Evaluation of synergised pyrethrum for the protection of stored wheat and shelled corn from insect attack. US Dep. Agric. Market. Res. Rep. No. 322. 48 pp.

Walker, B.J., Hend, R.W. and Linnett, S. 1975. Toxicity studies on the insecticide WL34775: summary of results of preliminary experiments. Report of Shell Development Co. (Unpublished).

Walker, D.W. and Locke, R. 1959. Evaluation of off-odour in malathion-treated wheat. J. Econ. Entomol. 52: 1013.

Wallbank, B.E. 1981. Gas chromatographic determination of carbaryl residues on stored wheat by on-column transmethylation. J. Chromatog. 208(2): 305–311.

Wallbank, B.E. 1983. Distribution of intermittently — or continuously applied grain protectant residues in Australian wheat bulks. In: Miyamoto, J., and Kearney, P.C., Ed., Pesticide Chemistry — Human Welfare and the Environment. Perganon Press, Oxford. pp. 217–220.

Wallwork, L., Chesher, B.C. and Malone, J.C. 1970. Toxicity of NRDC107 to laboratory animals. Report from Cooper Technical Bureau. (Unpublished).

Ware, G.W. and Naber, E.C. 1961. Lindane in eggs and chicken tissues. J. Econ. Entomol. 54: 675.

Ware, G.W. and Naber, E.C. 1962. Lindane and BHC in egg yolks following recommended uses for louse and mite control. J. Econ. Entomol. 55: 568.

Ware, G.W. and Roan, C.C. 1957. The interaction of piperonyl butoxide with malathion and five analogs applied topically to male houseflies. J. Econ. Entomol. 50: 825–827.

Warner, J.L. 1954. Protection of stored grain with pyrethrin insecticides. World Crops 6: 251–2.

Waterhouse, D.F. 1973. Insects and wheat in Australia. Farrer Memorial Oration, 1973. J. Aust. Inst. Agric. Sci. 39(4): 215–226.

Watt, M. 1962. Grain protection with malathion. Agric. Gaz. N.S.W. 73: 529.

Watters, F.L. 1956. Pyrethrins/piperonyl butoxide as a residual treatment against insects in elevator boots. Cereal Chem. 33: 145.

Watters, F.L. 1959a. Effects of grain moisture content on residual toxicity and repellency of malathion. J. Econ. Entomol. 52: 131.

Watters, F.L. 1959b. Stored grain pests and their control in the prairie provinces: Bulletin No. 9. Published by Line Elevators Farm Service, Winnipeg, Canada, December 1959.

Watters, F.L. 1961. Effectiveness of lindane, malathion, methoxychlor and pyrethrins-piperonyl butoxide against the hairy spider beetle, *Ptinus villiger*. J. Econ. Entomol. 54: 397.

Watters, F.L. 1972. Control of storage insects by physical means. Trop. Stored Prod. Inf. 23: 13–28.

Watters, F.L. 1975. Research on pesticides and future requirements for chemicals in the protection of stored products from insects. Proc. 1st Int. Working Conf. Stored-Products Entomol., Savannah, Georgia, 7–11 October 1974. pp. 272–280.

Watters, F.L. 1976a. Microwave radiation for control of *Tribolium confusum* in wheat and flour. J. Stored Prod. Res. 12: 19–25.

Watters, F.L. 1976b. Persistence and uptake in wheat of malathion and bromophos applied on granary surfaces to control the red flour beetle. J. Econ. Entomol 69(1): 353–356.

Watters, P.L. 1977. Comparison of acephate and malathion applied to stored grain for control of rusty grain beetles and red flour beetles. J. Econ. Entomol. 70(3): 377–80.

Watters, F.L. 1984. Potential of ionising radiation for insect control in the cereal food industry. In: Bauer, F.J., Ed., Insect Pest Management for Food Storage and Processing. American Association of Cereal Chemists, St Paul, Minnesota. pp. 265–274.

Watters, F.L. and Bickis, M. 1978. Comparison of mechanical handling and mechanical handling supplemented with malathion admixture to control rusty grain beetle infestations in stored wheat. J. Econ. Entomol. 71: 667–669.

Watters, F.L. and Grussendorf, O.W. 1969. Toxicity and persistence of lindane and methoxychlor on building surfaces for stored grain insect control. J. Econ. Entomol. 62: 1101.

Watters, F.L. and Mensah, G.W.K. 1979. Stability of malathion applied to stored wheat for control of rusty grain beetles. J. Econ. Entomol. 72(5): 794–797.

Watters, F.L. and MacQueen, K.F. 1967. Effectiveness of *gamma* irradiation for control of five species of stored product insects. J. Stored Prod. Res. 3: 223–234.

Watters, F.L. and Nowicki, T.W. 1982. Uptake of bromophos by stored rapeseed. J. Econ. Entomol. 75(2): 261–4.

Watters, F.L., Adem, E. and Uribe, R. 1978. Potential of accelerated electrons for insect control in stored grain. Proc. 2nd Int. Working Conf. Stored-Products Entomol., Ibadan, Nigeria, 10–16 September 1978. pp. 278–287.

Watts, C.N. and Berlin, F.D. 1950. Piperonyl butoxide and pyrethrins to control rice weevils. J. Econ. Entomol. 43: 371.

Weaving, A.J.S. 1970a. Susceptibility of some bruchid beetles of stored pulses to powders containing pyrethrins and piperonyl butoxide. J. Stored Prod. Res. 61(1): 71–7.

Weaving, A.J.S. 1970b. Susceptibility of some bruchid beetles of stored pulses (legumes) to powders containing pyrethrins and piperonyl butoxide. Pyrethrum Post 10(4): 17–21.

Weaving, A.J.S. 1975. Grain protectants for use under tribal storage conditions in Rhodesia. I. Comparative toxicities of some insecticides on maize and sorghum. J. Stored Prod. Res. 11: 65–71.

Weaving, A.J.S. 1980. Grain protectants for use under tribal storage conditions in Zimbabwe (Rhodesia). 2. Zimbabwe J. Agric. 18: 111–121.

Weaving, A.J.S. 1981. Grain protectants for use under tribal storage conditions in Zimbabwe (Rhodesia). 3. Evaluation of admixtures with maize stored in traditional grain bins. Zimbabwe J. Agric. Res. 19(2): 205–224.

Webley, D.J. 1981a. A statement on formulation of dusts. Proc. GASGA Seminar on the Appropriate Use of Pesticides for the Control of Stored Products Pests in Developing Countries, Slough, 17–20 February 1981. pp. 172–173.

Webley, D.J. 1981b. The importance of approved chemicals. Proc. GASGA Seminar on the Appropriate Use of Pesticides for the Control of Stored Products Pests in Developing Countries, Slough, 17–20 February 1981. p. 62.

Webley, D. 1984. Loss assessment — an approach to more efficient storage. In: Champ, B.R., and Highley, E., Ed., Proc. Aust. Dev. Asst. Course on Preservation of Stored Cereals. CSIRO Division of Entomology, Canberra. pp.899–926.

Webley, D.J. 1985. Results of working party trials. Report to Australian Wheat Board Working Party on Grain Protectants, May 1985. Australian Wheat Board, Melbourne. (Unpublished).

Webster, G.R.B. and Abdel-Kader, M.H.K. 1983. Analysis of malathion and fenitrothion and metabolites in stored wheat. Proc. 30th Ann. Meet. Can. Pest Manage. Soc., Truro, Nova Scotia, 11–13 July 1983.

Weeks, M.H., Leland, T.M., Boldt, R.E. and Mellick, P.W. 1972. Toxicological evaluation of pyrethroid insecticide SPP 1382. US Army Environmental Hygiene Agency Special Study No. 51–127–71/72.

Weir, R.T. 1966a. Reproduction study — rabbits. Neopynamin and pyrethrin. Report from Hazleton Laboratories, Inc. to S.C. Johnson and Son, Inc. (Unpublished).

Weir, R.T. 1966b. Acute potentiation study — oral administration. Neopynamin and pyrethrin. Report from Hazleton Laboratories, Inc. to S.C. Johnson and Son, Inc. (Unpublished).

Wellcome Foundation Ltd 1979. Residue in a cow's milk resulting from intrarumenal and later, dermal administration of ^{14}C-labelled deltamethrin. Group Research and Development Report WELL-79. 04H1BH/A. (Unpublished).

Wetters, J.H. 1980. Residues of chlorpyrifos-methyl and 3,5,6-trichloro-2-pyridinol in barley malting and brewing fractions following treatment of grain with Reldan grain protectant. Dow Chemical, USA, R&D Report GH-C 1322 (15 May 1980). 43 pp. (Unpublished).

Wetters, J.H. and McKeller, R.L. 1985. Residues of chlorpyrifos-methyl and 3,5,6-trichloro-2-pyridinol in stored grains and their process fractions following treatment of the grains with Reldan 4E insecticide. Paper presented at Fall Scientific Meeting, Amer. Chem. Soc., November 1985. 12 pp.

Wharton, D.R.A. 1971. Ultraviolet repellent and lethal action on the American cockroach. J. Econ. Entomol. 64: 252.

Whetstone, R.R., Philips, D.D., Sun, Y.P., Ward, L.R. and Shellenberger, T.E. 1966. 2-chloro-1-(2,4,5- trichlorophenyl)vinyl dimethyl phosphate, a new insecticide with low toxicity to mammals. J. Agric. Food Chem. 14: 352.

White, G.D., Berndt, W.L. and Wilson, J.L. 1975. Evaluating diatomaceous earth, silica-aerogel dusts and malathion to protect stored wheat from insects. US Dep. Agric. Market. Res. Rep. No. 1038. 18 pp.

White, L.D. and Hutt, R.B. 1972. Effects of treating adult codling moth with sterilizing and substerilizing doses of *gamma* irradiation in a low temperature environment. J. Econ. Entomol. 65. 140–143.

White, N.D.G. 1984. Residual activity of organophosphorus and pyrethroid insecticides applied to wheat stored under simulated western Canadian conditions. Can. Entomol. 116(10): 1403–1410.

White, N.D.G. and Nowicki, T.W. 1985. Effects of temperature and duration of storage on the degradation of malathion residues in dry rapeseed. J. Stored Prod. Res. 21(3): 111–114.

Whitehurst, W.E., Bishop, E.T., Critchfield, F.E., Gyrisco, G.G., Huddleston, E.W., Arnold, H. and Lisk, D.J. 1963. The metabolism of Sevin in dairy cows. J. Agric. Food Chem. 11: 167–169.

Whitney, W.K. 1974. A general survey of physical means for the control of storage pests. Proc. 1st Int. Working Conf. Stored-Products Entomol., Savannah, Georgia, 7–12 October 1974. pp. 385–411.

WHO 1978. Principles and methods of evaluating the toxity of chemicals, Part 1. Environmental Health Criteria No. 6. World Health Organization, Geneva.

WHO 1983. Guidelines on studies in environmeal epidemiology. Environment Health Criteria No.27. World Health Organization, Geneva.

Wilbur, D.A. 1952a. Effectiveness of dusts containing piperonyl butoxide and pyrethrins in protecting wheat against insects. J. Econ. Entomol. 45: 913–20.

Wilbur, D.A. 1952b. Effect of moisture content of wheat on toxicity of insecticidal dusts containing piperonyl butoxide and pyrethrins. J. Kans. Entomol. Soc. 25: 121–5.

Wilkin, D.R. 1975a. The control of stored product mites by contact acaricides. Proc. 8th Brit. Insect. Fung. Conf. 1975. pp. 355–363.

Wilkin, D.R. 1975b. The effects of mechanical handling and admixture of acaricides on mites in farm stored barley. J. Stored Prod. Res. 11(2): 87–95.

Wilkin, D.R. and Fishwick, F.B. 1981. Residues of organophosphorus pesticides in wholemeal flour and bread produced from treated wheat. Proc. Brit. Crop Prot. Conf. 1981. p. 183–187.

Wilkin, D.R. and Green, A.A. 1970. Polythene sacks for the control of insects in bagged grain. J. Stored Prod. Res. 6: 97–101.

Wilkin, D.R. and Haward, A. 1975. The effect of temperature on the action of four pesticides on three species of storage mites. J. Stored Prod. Res. 11(3/4): 235.

Wilkin, D.R. and Hope, J.A. 1973. Evaluation of pesticides against stored product mites. J. Stored Prod. Res. 8: 323–327.

Wilkin, D.R., Green, A.A., Gradidge, J.M.G., Kane, K., Aggarwal, S.L., Thomas, K, Horler, D.F. and Clarke, J.S. 1970. Insecticidal treatment of grain: admixture of iodfenphos. Pest Infest. Res. 1969 p. 31.

Wilkin, D.R., Gradidge, J.M., Kane, M.J., Hope, J.A., Aggarwal, S.L. and Clark, J.S. 1971. Admixture of iodofenphos in field and laboratory experiments. Report of Director, Pest Infestation Control Lab., Slough, 1968–70. p. 109.

Wilkin, D.R., Cruickshank and Dyte, C.E. 1983. Pesticide use on grain in commercial grain stores. Int. Pest Control May/June 1983. pp. 82–85.

Wilkinson, C.F. 1968. Detoxication of pesticides and the mechanisms of synergism. In: Hodgson, E., Ed., Enzymatic Oxidations of Toxicants. North Carolina State Univ. Press, Raleigh. pp. 113–149.

Wilkinson, C.F. 1971. Insecticide synergists and their mode of action. Proc. 2nd Int. Congr. Pestic. Chem. Vol. 2, pp. 117–159.

Williams, C.M. 1956. The juvenile hormone of insects. Nature 178: 212–213.

Williams, C.M. 1967. Third-generation pesticides. Sci. Am. 217: 13–17.

Williams, G.C. and Gladidge, J.M.G. 1965. Recent experience in the use of malathion to protect stored grain against insect attack. Proc. 12th Int. Cong. Entomol. 1964. p. 634.

Williams, I.H. 1976. Determination of permethrin with gas-liquid chromatography with electron capture detection. Pestic. Sci. 71(4): 336–338.

Williams, P. and Amos, T.G. 1974. Some effects of synthetic insect juvenile hormones and hormone analogues on *Tribolium castaneum*. Aust. J. Zool. 22: 147.

Williams, P., Minett, W. and Amos, T.G. 1975. Heterogeneous insecticide treatment of stored wheat which decreases harmful insect infestation. Dokl. Soobshch. Mezhdunar. Kongr. Zashch. Rast. 3(1): 131–7.

Williams, P., Amos, T.G. and Du Gueselin, P.B. 1978. Laboratory evaluation of malathion, chlorpyrifos and chlorpyrifos-methyl for use against beetles infesting stored wheat. J. Stored Prod. Res. 14: 163–168.

Williams, P., Semple, R.L. and Amos, T.G. 1983. Relative toxicity and persistence of three pyrethroid insecticides on concrete, wood and iron surfaces for control of grain insects. Gen. Appl. Entomol. 15: 7–10.

Willson, H.R., Singh, A., Bindra, O.S. and Everett, T.R. 1970. Rep. Rural Wheat Storage in Ludhiana District, Punjab. New Delhi, Ford Foundation. 39 pp.

Wilson, P. 1983. Insect infestation and pesticide usage in East Anglian farm grain stores. Int. Pest Control. 25(4): 116–118.

Wilson, A.D. and Desmarchelier, J.M. 1981. The use of railway wagons in obtaining representative samples of grain for chemical residues. In: Williams, P., and Amos, T.G., Ed., Proc. 1st Aust. Stored Grain Pest Control Conf., Melbourne, May 1981. CSIRO, Melbourne. pp. 1/22–1/28.

Wilton, B. 1980. Energy considerations in grain storage. SPAN 23 March: 109–111.

Winks, R.G. and Bailey, S.W. 1965. Treatment and storage of export wheat in Australia. Trop. Stored Prod. Inf. 11: 431–9.

Winks, R.G. and Bailey, S.W. 1974. Some aspects governing the use of pesticides on stored products in Australia. Proc. 1st Int. Working Conf. Stored-Products Entomol., Savannah, Georgia, 7–11 October 1974. p. 458.

Winteringham, F.P.W. and Harrison, A. 1946. Sorption of methyl bromide by wheat. J. Soc. Chem. Ind. 65: 140.

Winteringham, F.P.W. and Harrison, A. 1959. Mechanisms of resistance of adult houseflies to the insecticide dieldrin. Nature 184: 608.

Witt, P.R., Case, L. and Adamic, E. 1960. Malathion treatment of barley as related to malt quality. Proc. 1960 Ann. Meeting Amer. Soc. Brew. Chem. p. 51.

Wogan, G.N., Ed. 1965. Mycotoxins in Foodstuffs. Massachusetts Institute of Technology Press, Cambridge Mass. 291 pp.

Wohlgemuth, R. 1984. Comparative laboratory trials with insecticides under tropical conditions. Proc. 3rd Int. Working Conf. Stored-Products Entomol., Manhattan, Kansas, October 1983. pp. 286–289.

Woke, P. 1939. Inactivation of pyrethrum after ingestion by the southern army worm and during incubation of its tissues. J. Agric. Res. 58: 289.

Wolpert, V. (1967). Needless losses. Far Eastern Econ. Rev. 55: 411–412.

Womack, H. and LaHue, D.W. 1959. Tests with malathion and methoxychlor protective treatments for shelled corn stored in metal bins in the Southeast. US Dep. Agric. Market. Res. Rep. No. 357 11 pp.

Worth, H.M., Kehr, C.C. and Gibson, W.R. 1967. Effect of a single dose of bromophos. Report Ely Lilly and Co. (Unpublished).

Wrathall, A.E., Wells, D. and Anderson, P.H. 1980. Effect of feeding dichlorvos to sows in mid pregnancy. Zentralbl. Veterinaermed. Reihe A 27(8): 675–80.

Wright, F.N. 1963. Report on food storage in Malaya, Federation of Malaysia. Tropical Products Institute. London (Cited by Hall 1970).

Wyckoff, G.H. and Anderson, R.D. 1970. Fumigation or famine. J. Am. Vet. Med. Assoc. 156: 1261.

Wyniger, R., Renfer, A., Kruesli, F., Gfeller, W. and Hart, R.H. 1977. Methacrifos. A new insecticide for the control of insect pests in stored products. Proc. 9th Brit. Insect. Fungic. Conf. Vol. 3, pp. 1033–1040.

Yadav, T.D. 1980. Toxicity of DDT and lindane against thirteen species of stored product insects. Indian J. Entomol. 42(4): 671–4.

Yadav, T.D., Singh, S., Khanna, S.C. and Mookherjee, P.B. 1979. Efficacy of grain protectants against larval stages of three species of moth pests. Indian J. Plant Protect. 7(1): 15–18.

Yadav, T.D., Pawar, C.S., Khanna, S.C. and Singh, S. 1980. Toxicity of organophosphorus insecticides against stored product beetles. Indian J. Entomol. 42(1): 28–33.

Yadava, C.P. and Shaw, F.R. 1970. Residues of Rabon (tetrachlorvinphos) in tissues and egg yolks of poultry. J. Econ. Entomol. 63: 1097.

Yamamoto, I. 1973. Mode of action of synergists in enhansing the insecticidal action of pyrethrum and pyrethroids. In: Casida, J.E., Ed., Pyrethrum, the Natural Insecticide. Academic Press, New York and London. pp. 195–210.

Yamamoto, I. and Casida, J.E. 1966. O-demethyl-pyrethrin II analogs from oxidation of pyrethrin I, allethrin, dimethrin and phalthrin by a housefly enzyme system. J. Econ. Entomol. 59: 1542–1553.

Yamamoto, I., Kimmel, E.C. and Casida, J.E. 1969. Oxidative metabolism of pyrethroids in houseflies. J. Agric. Food. Chem. 17: 1227–1236.

Yamamoto, I., Elliott, M. and Casida, J.E. 1971. The metabolic fate of pyrethrin I, pyrethrin II and allethrin. Bull. WHO 44: 347–348.

Young, R. Jr., Hass, D.K. and Brown, L.J. 1979. Effect of late gestation feeding of dichlorvos in non-parasitized and parasitized sows. J. Anim. Sci. 48(1): 45–51

Zakladnoi, G.A. 1984. Protecting grain against stored-product pests. Zasht. Rast. (Moscow) 7: 40–41.

Zakladnoi, G.A. and Bokarev, E.M. 1976. Organophosphorus insecticides for protection of stored grain. Zasht. Rast. Moscow. 9: 24–5.

Zambia — Department of Agriculture, 1972. Farm storage experiments: cob and shelled maize in traditional and concrete bins. Depot storage experiments: bagged maize on hard standings. Ann. Rep. Res. Branch 1970–1971: 143–145.

Zambia — Department of Agriculture, 1972a. Pirimiphos-methyl stored product trials on maize in Zambia 1972. Extracts from Ann. Rep. Res. Branch 1971–1972.

Zeid, M., Dahm, P., Hein, R. and McFarland, R. 1953. Tissue distribution, excretion of CO_2 and degradation of radioactive pyrethrins administered to the American cockroach. J. Econ. Entomol. 46: 324.

Zemanek, J. 1979. Problems of pesticide residue in agricultural products. Prum. Potravin. 30(4): 196–200.

Zettler, J.L. and Jones, R.D. 1977. Toxicity of seven insecticides to malathion-resistant red flour beetles. J. Econ. Entomol. 70(5): 536–538.

Zhang, G.-L., Xu, B.-Z., Li, Y.-H, Xue, L.-S., Ye, Z.-X. and Shen, Z.-C. 1982. Effect of pesticides on stored seeds quality. Grain Storage (Liangshi Chucang) No.3. pp. 11–16, 55.

Zhbara, A. and Shikrenov, D. 1983. *Gamma*-irradiation against the grey grain moth. Rast. Zasht. 31(11): 37–39.

Zoecon 1984a. Methoprene distribution in Australian wheat fractionation. Report RM-0248, 17 September 1984. Zoecon Corporation, Palo Alto, California. (Unpublished).

Zoecon 1984b. Methoprene distribution in Kansas State University wheat fractionation study. Report RM-0250, 18 September 1984. Zoecon Corporation, Palo Alto, California. (Unpublished).

Zoecon 1984c. Reports on environmental fate of methoprene submitted to FAO for evaluation by the FAO Working Party on Pesticide Residues, April 1984.

Zoecon 1984d. Methoprene residues in chickens. Report submitted to FAO for consideration by JMPR. Zoecon Corporation.

Zusca, J. 1973. Longevity of *gamma*-irradiated adults of *Piophila casei*. Acta Entomol. Bohemoslov 70: 189–195.

Zutzhi, M.K. 1966. Storage of wheat by farmers in Delhi. Bull. Grain Technol., 4(3): 143–145.

Zweig, G. Ed. 1974. Analytical Methods for Pesticides, Plant Growth Regulators and Food Additives, Volume VII. Academic Press, New York.

Zweig, G., Ed. 1976. Analytical Methods for Pesticides, Plant Growth Regulators and Food Additives, Volume VIII. Academic Press, New York.

Zweig, G. Ed. 1978. Analytical Methods for Pesticides, Plant Growth Regulators and Food Additives, Volume X. Academic Press, New York.

Zweig, G., Ed. 1980. Analytical Methods for Pesticides, Plant Growth Regulators and Food Additives, Volume XI. Academic Press, New York.